5th International Conference Recent Trends in Structural Materials (COMAT 2018)

IOP Conference Series: Materials Science and Engineering Volume 461

Pilsen, Czech Republic
14 – 16 November 2018

ISBN: 978-1-5108-8007-8
ISSN: 1757-8981

Printed from e-media with permission by:

Curran Associates, Inc.
57 Morehouse Lane
Red Hook, NY 12571

Some format issues inherent in the e-media version may also appear in this print version.

Printed by Curran Associates, Inc. (2019)

For permission requests, please contact the Institute of Physics at the address below.

Institute of Physics
Dirac House, Temple Back
Bristol BS1 6BE UK

Phone: 44 1 17 929 7481
Fax: 44 1 17 920 0979

techtracking@iop.org

Additional copies of this publication are available from:

Curran Associates, Inc.
57 Morehouse Lane
Red Hook, NY 12571 USA
Phone: 845-758-0400
Fax: 845-758-2633
Email: curran@proceedings.com
Web: www.proceedings.com

TABLE OF CONTENTS

Preface

COMAT history is relatively new, we started in 2010. COMAT deals with recent trends in structural materials that cover production technologies (casting, forging, rolling…), heat and thermo-mechanical treatments, computer modelling and physical simulations and materials characterisation. The conference was launched by the research institute COMTES FHT a.s. that is very active in the fields covered by COMAT. The idea was to create a base for knowledge sharing between industry and research and academic experts. It is located in the heart of the Pilsen region that is a traditional heavy industry centre in the Czech Republic with rich history spanning over 150 years.

COMAT 2018 was held at Angelo Hotel in Plzen on November 14th – 16th. The conference was organized by company TANGER s.r.o. and COMTES FHT a.s. COMAT 2018 was the fifth of the biannually held conferences. We were really pleased to welcome almost 200 people (exactly 190) from 24 countries from Asia, America, Africa and Europe attending the conference. The participants presented almost 80 lectures and 67 posters. We were really happy to see increase of number of participants by almost 30%, confirming that the conference is attractive to the scientists. The conference was divided into seven sessions and a poster session. This year we newly had dedicated sessions to additive manufacturing that is very important topic of interest nowadays. 96 papers are selected for current proceedings. We hope the papers presented here will be interesting for readers and a good base for inspiration for future developments. We are looking forward to seeing even more participants with great contributions in 2020.

Yours sincerely

Jan DŽUGAN

November 2018

Sponsors:

BVD PECE spol. s r.o.

EDAX Smart Insight
AMETEK MATERIALS ANALYSIS DIVISION

dyna weld

CSNML

LINSEIS Thermal Analysis

JEOL

Applus IDIADA

Struers Ensuring Certainty

termovize TMV SS

SOBRIETY

Zwick / Roell

Organizers:

COMTES FHT
Complete Technological Service - Forming, Heat Treatment

Tanger

IOP Conf. Series: Materials Science and Engineering **461** (2019) 011003 doi:10.1088/1757-899X/461/1/011003

Peer review statement

All papers published in this volume of *IOP Conference Series: Materials Science and Engineering* have been peer reviewed through processes administered by the proceedings Editors. Reviews were conducted by expert referees to the professional and scientific standards expected of a proceedings journal published by IOP Publishing.

Relationship between hydrogen states present in the vicinity of the fracture surface and hydrogen embrittlement susceptibility for ferrite-martensitic dual-phase steels

D Asari[1], S Mizokami[2], M Fukahori[2] and K Takai[3]

[1]Graduate School of Science and Technology, Sophia University, 7-1 Kioi-cho, Chiyoda-ku, Tokyo 102-8554, Japan
[2]Mazda Motor Corporation, 3-1 Aki-gun, Hiroshima 730-8670, Japan
[3]Faculty of Science and Technology, Department of Engineering and Applied Sciences, Sophia University, Tokyo, Japan

Corresponding author's e-mail address: asari1228@eagle.sophia.ac.jp

Abstract. The relationship between hydrogen states present in the vicinity of the fracture surface and hydrogen embrittlement susceptibility was investigated for ferrite-martensitic dual-phase steels. Hydrogen embrittlement susceptibility was evaluated in tensile tests based on the ratio of the fracture strength of hydrogen charged and not-charged specimens. Tensile tests at strain rates of 8.33×10^{-6} s^{-1} and 1.67×10^{-6} s^{-1} were conducted on specimens containing different amounts of hydrogen. The hydrogen states present near the fracture surface were analyzed by thermal desorption analysis just after the specimens fractured. The results indicated that hydrogen embrittlement susceptibility markedly increased as the amount of hydrogen increased. Additionally, only the specimen that fractured at the highest amount of hydrogen showed not only a lower temperature peak but also a higher temperature peak. Hydrogen embrittlement susceptibility also increased at slower strain rates, and also showed the higher temperature peak. The relationship between hydrogen states present in the vicinity of the fracture surface and mechanical properties indicated that the defects corresponding to higher temperature peaks probably increased hydrogen embrittlement susceptibility. Since microvoids at the crack tips were observed by SEM only in specimens that fractured at lower strain rates, the defects corresponding to higher temperature peaks were probably voids.

1 Introduction

In recent years, automobile manufacturers have been required to reduce carbon dioxide emissions while ensuring collision safety, resulting in greater demand for high-strength thin steels that satisfy both light weight and safety requirements. If hydrogen embrittlement is hindering the further strengthening of the vehicle body frame, understanding this phenomenon is an important issue in satisfying social demands. Dual-phase (DP) steel, which has the two-phase structure of ferrite-martensite, is widely used for the body frame of vehicles because it satisfies the requirements for both strength and ductility.

However, it is reported that the strength difference between martensite and ferrite promotes hydrogen concentration at ferrite/martensite interfaces and that this concentration accelerates crack propagation [1]. Furthermore, as a promoting factor of crack propagation, it is also reported that cracking along ferrite/martensite interfaces is attributed to the decohesion and boundary sliding associated with enhanced dislocation mobility [2]. For these reasons, it is clear that investigating the relationship between hydrogen and lattice defects is one approach to clarifying the hydrogen embrittlement mechanism of DP steels. Therefore, in this study, we investigated the lattice defects involved in hydrogen embrittlement from the perspective of the relationship between the hydrogen states present in the vicinity of the fracture surface and hydrogen embrittlement susceptibility.

2 Experimental

2.1 Materials

DP steels with a ferrite-martensite microstructure were used in the present study. Table 1 gives the chemical composition of the specimens used. Fig. 2 shows the shape of the specimens with a thickness of 1 mm and a gauge length of 20 mm.

2.2 Examination of hydrogen charging condition

The time taken to reach an equilibrium hydrogen condition at both the surface and center of the specimens was measured. The specimens were charged electrochemically with hydrogen at a current density of 100 $A{\cdot}m^{-2}$ in a 0.1 N NaOH aqueous solution containing a NH_4SCN additive of 5 $g{\cdot}L^{-1}$ kept at a temperature of 30 □. The charging time was 24, 48, 72 and 96 h. Immediately after hydrogen charging until the specified time, the amounts and states of hydrogen present in the samples were analyzed by thermal desorption analysis (TDA) using a gas chromatograph at a heating rate of 100 □$\cdot h^{-1}$ in the temperature range from room temperature to 300 □. TDA profiles at different charging times are shown in Fig. 2 (a). The relationship between the charging time and amount of hydrogen present is shown in Fig. 2 (b). From the results in Fig. 2 (a) and (b), the hydrogen charging time for reaching an equilibrium condition was determined to be 72 h.

Table 1. Chemical composition of specimen material (mass %)

C	Si	Mn	P	S
0.14	0.51	2.5	0.008	0.003

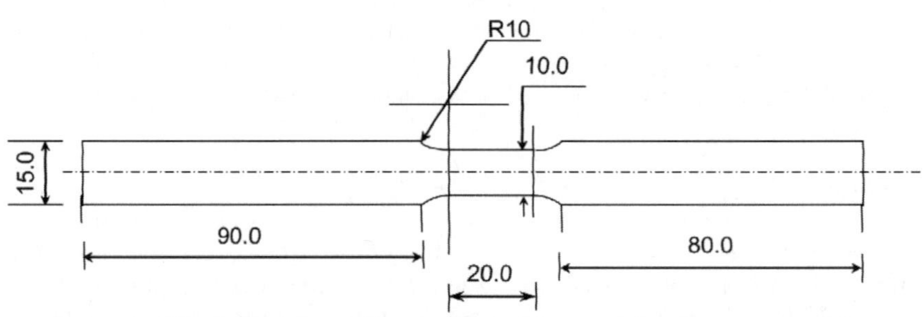

Fig 1. Specimen for tensile and constant load tests

IOP Conf. Series: Materials Science and Engineering **461** (2019) 012001 doi:10.1088/1757-899X/461/1/012001

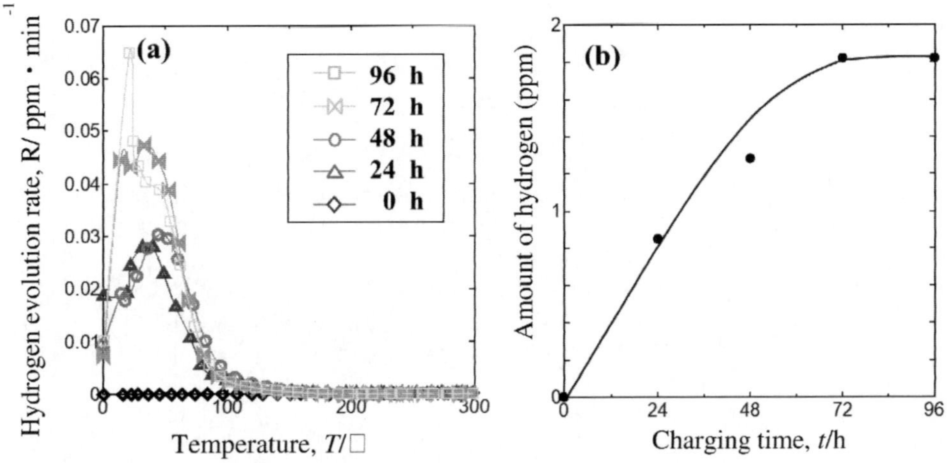

Fig 2. Effect of hydrogen charging time: (a) thermal desorption profiles and (b) amount of hydrogen present based on thermal desorption profiles

2.3 *Evaluation of hydrogen embrittlement susceptibility*

Hydrogen embrittlement susceptibility was determined as the ratio of the fracture strength of hydrogen charged and not-charged specimens obtained in tensile tests. Specimens were precharged electrochemically with hydrogen at current densities of 30 and 100 $A{\cdot}m^{-2}$ in a 0.1 N NaOH aqueous solution containing a NH_4SCN additive of 0, 0.2, 2, and 5 $g{\cdot}L^{-1}$ kept at a temperature of 30 □. After precharging, tensile tests were conducted at strain rates of 8.33×10^{-6} s^{-1} and 1.67×10^{-6} s^{-1}. Hydrogen charging was conducted concurrently during the tensile tests at 30 □ under the same charging conditions to keep the hydrogen content of the specimens constant.

2.4 *Clarifications of the vicinity of fracture surface*

Hydrogen states present near the fracture surface were analyzed by TDA just after the specimens fractured in the tensile tests under the conditions mentioned above. Before conducting tensile tests, a mark was placed at a 20-mm distance between the gauge marks, and after fracturing, the part inside the mark was cut out using a fine cutter. It was immediately cooled with liquid nitrogen and analyzed by TDA at a heating rate of 100 □$\cdot$ h^{-1} in the temperature range from room temperature to 300 □. Fracture surfaces were observed using a scanning electron microscope (FE-SEM).

3 Results and discussion

3.1 *Effect of the amounts of hydrogen*

Figure 3 (a) shows the stress-strain curves obtained in the tensile tests at a strain rate of 8.33×10^{-6} s^{-1} for different amounts of hydrogen and also shows the amount of hydrogen obtained by TDA just after fracturing. The results indicate that hydrogen embrittlement susceptibility noticeably increased as the amount of hydrogen increased. Fig. 3 (b) shows the TDA profiles of the specimens which fractured at different amounts of hydrogen. Only the specimen that fractured at the highest amount of hydrogen (4.5 mass ppm) showed not only a lower temperature peak but also a higher temperature peak. This suggests that defects corresponding to the higher temperature peak probably increased hydrogen embrittlement susceptibility.

3.2 Effect of strain rates

Figure 4 (a) shows the stress-strain curves obtained by the tensile tests at different strain rates. Hydrogen embrittlement susceptibility increased at slower strain rates. Fig. 4 (b) shows the TDA profiles of the specimens which fractured at different strain rates. At slower strain rates, a higher temperature peak around 150 □ newly appeared in the TDA profiles. Previous studies have reported that the higher temperature peaks of DP steels were presumed to be the desorption peaks of vacancy clusters introduced by deformation [3]. Therefore, the results of the present study suggest that the formation of vacancy clusters is promoted by the incursion of a certain amount of hydrogen because the higher temperature peaks appeared only for high amounts of hydrogen or at lower strain rates, which allowed sufficient time for hydrogen to diffuse. It is also inferred that susceptibility to hydrogen embrittlement increases due to the formation of vacancy clusters.

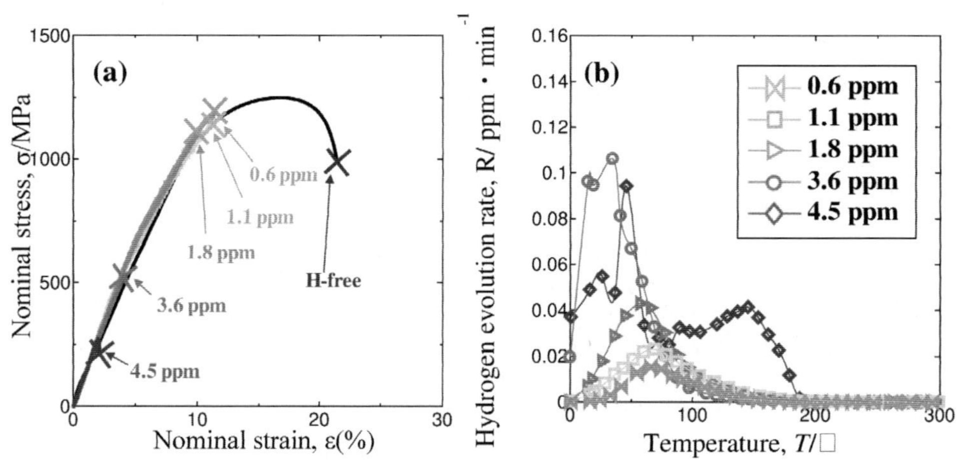

Fig 3. (a) Stress-strain curves for specimens at different amounts of hydrogen and (b) thermal desorption profiles at different amounts of hydrogen

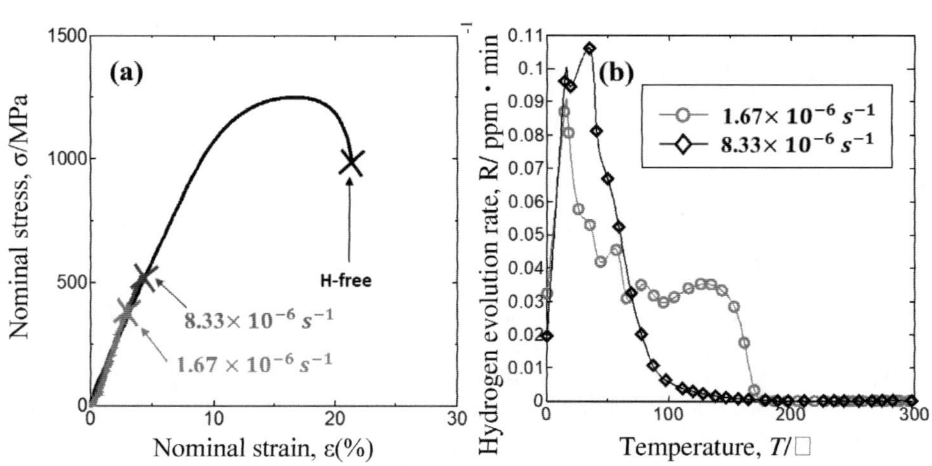

Fig 4. (a) Stress-strain curves for specimens at different strain rates and (b) thermal desorption profiles at different strain rates

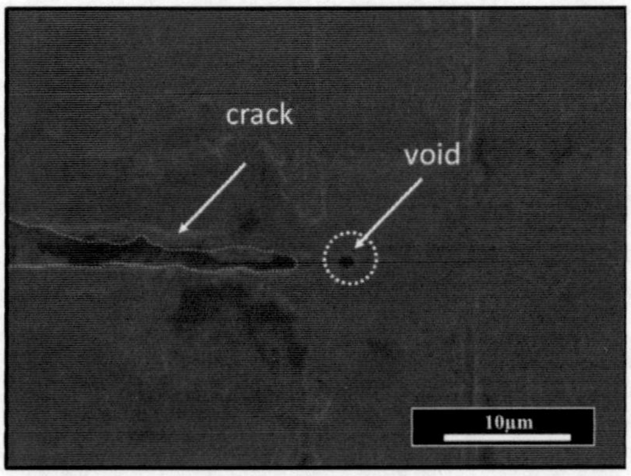

Fig 5. Microvoids at the crack tips in H-charged specimen fractured by lower strain rate SSRT

Figure 5 shows the vicinity of the fracture surface obtained in a tensile test at a strain rate of 1.67×10^{-6} s^{-1}. Microvoids with a diameter of approximately 0.5 μm are observed at the crack tips. Since microvoids were not observed at a higher strain rate of 8.33×10^{-6} s^{-1}, this is also evidence that defects corresponding to higher temperature peaks were probably microvoids.

4 Conclusions

The relationship between hydrogen states present in the vicinity of the fracture surface and hydrogen embrittlement susceptibility was examined for ferrite-martensitic dual-phase steels in relation to different amounts of hydrogen and strain rates. The results obtained in the present study can be summarized as follows.

(1) Only the specimen that fractured at the highest amount of hydrogen (4.5 mass ppm) showed not only a lower temperature peak, but also a higher temperature peak; the other specimens that fractured at lower amounts of hydrogen (3.5 ppm or less) showed only lower temperature peaks.
(2) When the strain rates were lowered, hydrogen embrittlement susceptibility increased even under the same hydrogen charging condition. The higher temperature peak appeared only in specimens that fractured at lower strain rates in the tensile tests.
(3) Microvoids with a diameter of approximately 0.5 μm at the crack tips were observed in the vicinity of the fracture surface in specimens that fractured only at lower strain rates in tensile tests.
(4) Defects corresponding to higher temperature peaks were probably microvoids. In addition, the formation of microvoids is probably a factor that increases hydrogen embrittlement susceptibility.

References

[1] Sun J, Jiang T, Sun Y, Wang Y, and Liu Y, A lamellar structured ultrafine gain ferrite-martensite dual-phase steel and its resistance to hydrogen embrittlement, *Journal of Alloys and Compounds,* **698,** p395, (2017)
[2] Koyama M, Tasan C C, Akiyama E, Tsuzaki K and Raabe D, Hydrogen-assisted decohesion and localized plasticity in dual-phase steel, *Acta Materilia* **70** p186, (2014)
[3] Okano H and Takagi S, Identification of hydrogen trapping sites in a strained ferritic-martensitic dual phase steel *Tetsu-to-Hagané* **104** No.1 p35, (2018)

IOP Conf. Series: Materials Science and Engineering **461** (2019) 012002 doi:10.1088/1757-899X/461/1/012002

Modeling of a combined process for the production of contact wires for high-speed railways

R N Asfandiyarov[1,2], G I Raab[1] and D A Aksenov[1,2]

[1]Ufa State Aviation Technical University, 12 K. Marx str., Ufa 450008 Russia

[2]Institute of Molecule and Crystal Physics - Subdivision of the Ufa Federal Research Centre of the Russian Academy of Sciences, 151 pr. Oktyabrya, Ufa 450054 Russia

E-mail: a.r.n@list.ru

Abstract. The development of high-speed railways (HSR) is an important task aimed at improving the transport system of today's global community. The most heavily-loaded element of HSR, governing their reliability and lifetime, is a contact wire. In this connection, the present study was focused on designing a new set of physico-mechanical properties of a wire made from the heat-treatable Cu-0.65Cr alloy with the use of a new procedure of continuous plastic processing, based on the principles of severe plastic deformation. Thus, an important difference from the common technical solutions used in the production of wires was the use of a combination of SPD processes – radial forging and ECAP-Conform, joined with the shape-forming of shaped sections of wire. In our study, we performed finite-element computer and physical modeling of the processes of plastic and heat treatment. Using computer modeling, we demonstrated that as a result of the implementation of the new procedure of continuous processing, a rather homogeneous strained state is formed in the workpiece, and the accumulated strain is in a range of e=6-7. At all stages of plastic processing, compressive stresses prevail in the deformation site. As a result of physical modeling, we produced laboratory samples of contact wire from the heat-treatable Cu-0.65Cr alloy with a cross-section area of 150 mm^2. Metallographic studies reveal that a banded structure of a grain-subgrain type with a fragment size below 1 micron is formed in the laboratory samples of contact wire. The ultimate tensile strength of these samples after heat treatment is 550-560 MPa, the electrical conductility is 72-75% IACS, the ductility is 16-20%.

1 Introduction

Contact wire is one of the most heavily-loaded and important parts of HSR. In today's world with the growing speeds and volume of transportation, the requirements to contact wires increase correspondingly. There emerges a need to increase a set of performance properties of this product. The current technologies for the manufacture of contact wires for railway networks, incorporating the processes of continuous casting-rolling, as well as drawing and heat treatment, have restrictions in terms of variation of processing parameters and therefore cannot influence actively the properties of final products [1]. An alternative could be the methods of plastic structure formation, such as severe plastic deformation, resulting in an active influence on the physical and mechanical properties of commony used metallic materials [2,3]. Due to an extremely non-monotonic strain, SPD enables producing ultrafine-grained states, also in copper materials, thereby providing enhancement of their properties [4-6]. In this connection, a new technological approach for the production of contact wires, using SPD methods, was developed as an alternative for the conventional approach. To meet the conditions of continuous production, the most technologically advanced methods were selected, namely radial forging with an alteration of the axe of force application by 90 degrees and ECAP-Conform [7-10]. The ECAP-Conform operation was joined with the shape-forming of shaped sections of wire. This approach provides a range of advantages related to the reduction of power consumption, number of operations, expenditures between operations and expenditures for a die-set, with a reduction of the required production floors. In this study, we performed computer and verification analyses of the processing parameters and physico-mechanical properties of the samples of contact wire from a Cu alloy produced using a combination of SPD methods, and evaluated the use of the combined method for industrial production.

2 Finite-element computer modeling

When analyzing the deformation processes, our research was focused on the study of the stress-strain state (SSS) of the workpieces during the processing and temperature analysis. As it is known, these factors are responsible for the adaptability of a process to manufacture, the durability and reliability of a deforming tool, the physical and mechanical properties of materials subjected to deformation.

2.1. Boundary conditions in the modeling.

The material under study is a heat-treatable Cu-Cr alloy in the initial state, quenched from 1000°C. For the modeling, we used the Deform 3D database. The processing parameter of the rate of workpiece feed was selected on the basis of the conditions provided by foundry equipment and amounted to 4, 5 and 6 meters per minute. Radial forging was implemented in two processing operations.

The dimensions of the square cross section of the initial workpiece were 37.2x74 mm, the length was 300 mm. The finite-element grid of the workpiece was generated from tetrahedra in the amount of 32 000. The rate of the anvil movement was selected as constant for all of the modeling options and amounted to 200 mm/s. The angular rate of rotation of the working wheel was 1 rad/s, which corresponds to 10 rev/min. The cross section of the workpiece after the first forging unit was 33x65 mm, and after the second forging unit – 26x26 mm.

The friction coefficient according to Siebel was accepted as: $f = 0.2$ for radial forging; for ECAP-C, $f=0.4$ between the wheel and the workpiece, $f=0.3$ between the workpiece, the clamp and the die.

Modeling was performed for the processing at room temperature, taking into account the increment in metal temperature due to the thermal effect of deformation. The coefficient of heat exchange between the anvils and the workpiece was accepted as equal to 5 N/s/mm/°C (5000 W/m2×°C).

2.2. Research results. Stress-strain state.

It follows from the analysis of the patterns of accumulated strain distribution that in the longitudinal section of the workpiece, the accumulated strain after the first forging unit is on average $e=1\text{-}1.3$, with the maximum values prevailing in the workpiece center, which corresponds to the procedure of simple upsetting with the formation of deformation cross (shear deformation). After the second forging unit, the accumulated strain grows to 2.5-3.5, and there is observed a high non-uniformity of strain distribution. After ECAP-C, in the steady-state flow, at the channel outlet the accumulated strain in the shaped wire workpiece is $e = 6\text{-}7$ (figure 1), i.e. the produced rather homogeneous deformed state implies also the homogeneity of the produced properties. Thus, the use of an extremely non-monotonic deformation and the revealed high level of accumulated strain are a significant prerequisite for UFG structure formation. The average stresses emerging in the deformation site are predominantly compressive.

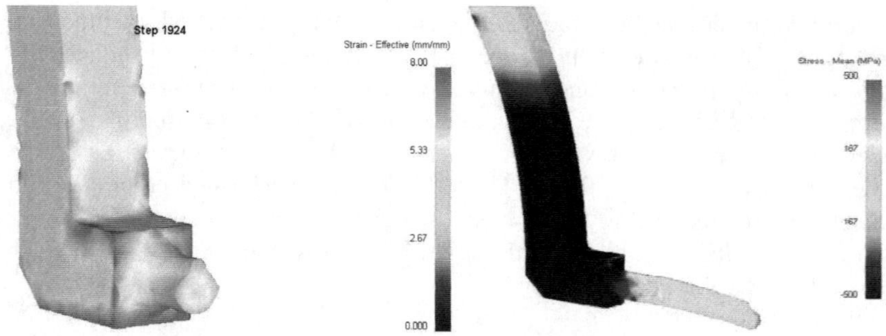

a b

Figure 1. Stress-strain state of the workpiece after ECAP-Conform with the shape-forming of a shaped section.

2.3. Analysis of the thermal fields of the workpiece during the processing.

The temperature range of deformation is another highly important factor, especially for the processing of heat-treatable materials under large severe strains. An increase in the temperature reduces flow stresses and, correspondingly, the deformation forces, accelerates the diffusion processes, solid solution breakdown and the precipitation of strengthening particles, therefore it is required to analyze these processes immediately after the procesing. In addition, the aging temperature of the chromium bronze lies in a range of 450-500 °C (figure 2), which is a limiting factor for its plastic treatment in the solid-solution state (after quenching) and should be monitored under the conditions of intensive deformational heating. In this connection, study of the temperature fields was also an important task in our research.

The modeling shows that during high-speed swaging, an intensive defromational heating of the material takes place. After deformation in the first forging unit, the heating of the workpiece is 100-150 °C. After the second forging unit, the temperature reaches 250-290 °C. Processing by ECAP-Conform with pressing takes place with an intensive heat generation, and in the steady state in the peripheral zone of pressing the temperature reaches 550 °C. This established fact calls for measures to cool the workpiece, for instance, by air flow, prior to the ECAP-Conform operation.

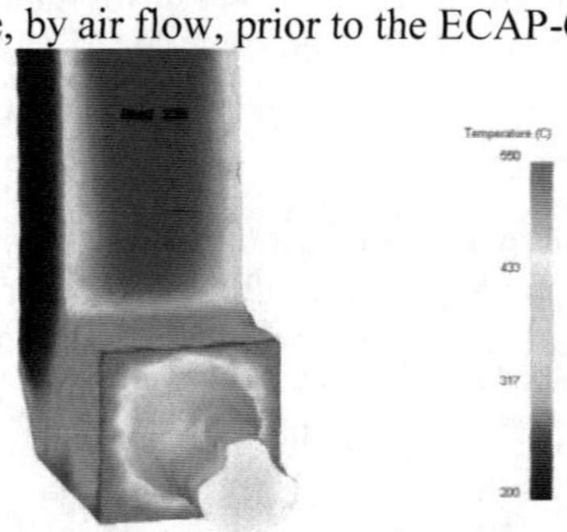

Figure 2. Thermal fields of the workpiece after ECAP-Conform with shape-forming.

3 Material and experimental procedure

3.1. Contact wire material.

The Cu-Cr alloy refers to a special class of low-alloyed conductive chromium bronzes [11]. Unlike other systems of bronzes with a high electrical conductivity, such as Cu-Ag, Cu-Sn, Cu-Cd, Cu-Mg, or heat-resistant systems Cu-Zr, Cu-Hf, the Cu-Cr alloy has simultaneously a high electrical conductivity and a high heat resistance. Besides, it is cheaper than the ternary system Cu-Cr-Zr.

It is also known that chromium affects electrical conductivity less than other alloying elements. The material in the initial state was produced by heat treatment, including holding in a Nabertherm furnace at a temperature 1000°C for 1 hour, followed by water quenching. As a result, an equiaxed grain structure was obtained. The average grain size was 120±3 μm. In the initial state, the Cu-0.65Cr alloy had the following physico-mechanical characteristics: ultimate tensile strength ~240 MPa, electrical conductivity 45% IACS, ductility 40%.

3.2. Physical modeling.

Physical modeling was performed following the principle of similarity to the processing. The processing is similar in terms of the following criteria – the pattern of forces applications, the speed parameters of processing, the temperature range of processing, as well as the geometrical similarity with consideration of true strain. The operation of radial swaging was conducted using a pneumatic hammer, turning the workpieces during the processing. The frequesncy of strokes of this hammer is 210 min^{-1} (the frequency of strokes of the anvils on the radial-forging machine is about 300 min^{-1}). The temperature of the workpiece in the beginning of the process was room temperature, 20°C. The heating of the workpiece during the processing was 300°C. The direction of drawing was maintained during the processing. The proposed procedure of plastic processing is displayed in figure 3.

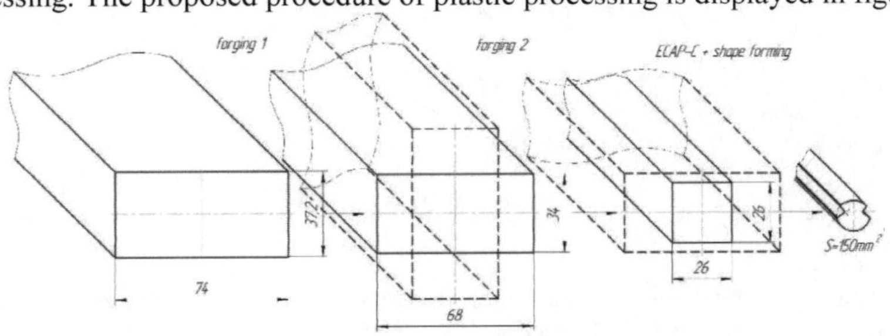

Figure 3. The propsed procedure for producing contact wire using SPD methods – the initial workpiece, after the first radial swaging, after the second radial swaging, after ECAP-C and shape-forming of shaped sections.

The next operation of the proposed technology is ECAP-Conform joined with the extrusion of shaped sections of contact wire. In order to observe the principle of similarity, the die-set and the workpiece during the processing were heated to a temperature of 450°C. The laboratory sample produced as a result of pilot studies are shown in figure 4. The wire samples were subjected to heat treatment (aging) at a temperature of 450°C for 1 hour.

Experiment results

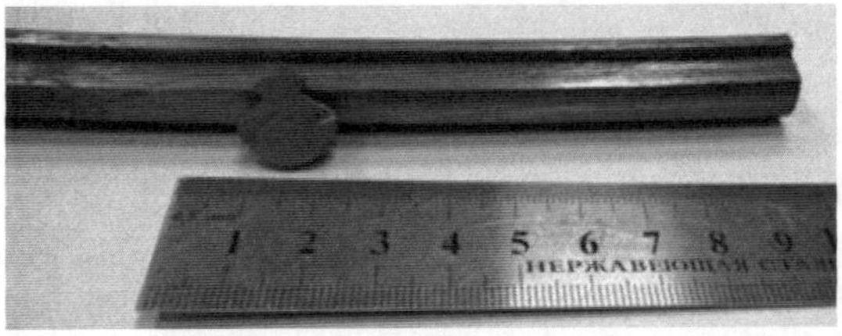

Figure 4. General view of the workpiece after ECAP.

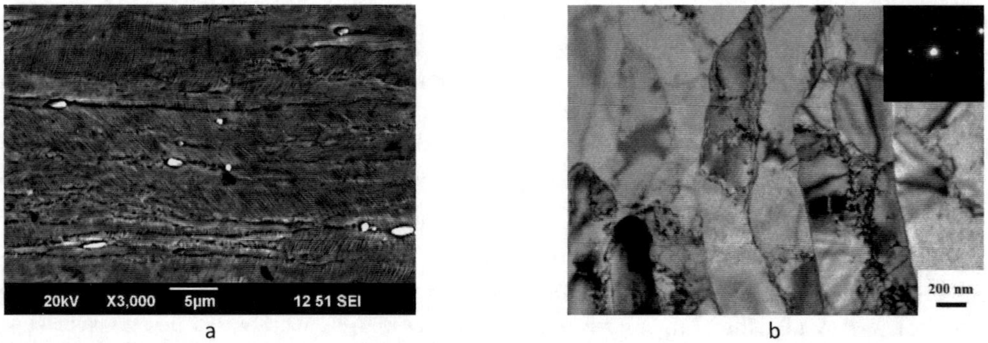

Figure 5. Structure of the sample after ECAP-Conform with shape-forming, longitudial section: a – SEM, b – TEM.

Structural studies demonstrate that as a result of plastic processing via the proposed procedure, a homogeneous banded structure is formed (figure 5), with particle precipitation predominantly at boundaries. Finer studies (TEM) (figure 5) show that the size of fragments inside the bands is below 500 nm.

Studies of the physical and mechanical properties provided the following results:

- electrical conductivity: 72-75% IACS;
- ultimate tensile strength: 550-560 MPa;
- ductility 16-20%;
- number of twists: 6;
- number of bends: 4.

In comparison with the conventional processing (quenching – drawing e=70% – aging) providing an ultimate tensile strength of 520 MPa, a ductility of 5%, an electrical conductivity of 72-75% IACS, we have attained a higher set of properties in contact wire [11].

Conclusions

- Using mathematical modeling, we have investigated the stress-strain state, thermal fields and processing parameters for the processes of radial swaging and ECAP-Conform with the extrusion of shaped sections of contact wire.

- It has been established that the intensity of strain in the process of radial forging after the second unit is distributed rather uniformly and reaches 3.5. After ECAP-C, the accumulated strain reaches e = 6-7, while in the central region of the wire's cross section slightly smaller values of e=5-6 prevail.

- It has been shown that in the process of ECAP-Conform deformational heating to 550 °C occurs in the peripheral region of the zone of shape-forming of the wire sections. It has been found that at all of the three stages of plastic processing, compressive stresses prevail (a «mild» deformation procedure), which has a beneficial effect on the quality of a produced item.

- As a result of physical modeling, we have produced samples of contact wire for HSR with a new set of physico-mechanical and performance properties, having an electrical conductivity of 72-75%, an ultimate tensile strength of 550-560 MPa, a ductility of 16-20%. For the bronze of the investigated chemical composition, this is a new enhanced set of physico-mechanical properties.

- As a result of plastic processing via the proposed procedure, there forms a homogeneous banded structure with particle precipitation predominantly at boundaries, with the size of fragments inside the bands equal to 300-500 nm.

Acknowledgements

The authors acknowledge the financial support by the Ministry of Science and Higher Education of the Russian Federation under Grant agreement No. 14.586.21.0025 (unique project identifier RFMEFI58616X0025).

References

[1] Gershman I S and Mironos N V 2011 Requirements for contact wires for high-speed rail transport *Vestnik of All-Russia R&D Institute of Railway Transport* **3** pp 13-17 (in Russian)

[2] Valiev R Z and Aleksandrov I V 2000 Nanostructured materials produced by severe plastic deformation (Moscow: Pegas) p 272 (in Russian)

[3] Gunderov D V, Nizamutdinova A M, Valeev K A and Stolyarov V V 1998 RFeB alloys in the initial state and subjected to severe plastic deformation by torsion *Bulletin of the Bashkir University* **2** pp 28-32 (in Russian)

[4] Kaybyshev O L and Utyashev F Z 2002 Superplasticity, refinement of structure and processing of hard-to-deform alloys (Moscow: Science) p 438 (in Russian)

[5] Utyashev F Z 2008 Modern methods of severe plastic deformation (Ufa: USATU) p 313 (in Russian)

[6] Zherebtsov S V 2013 Structural changes in the course of large plastic deformation and the development of methods for producing an ultrafine-grained structure in semi-finished products of titanium-based alloys: Abstract of. dis. Dr. of Tech. sciences (Ekaterinburg: Ural Federal University named after the first President of Russia B.N. Yeltsin) p 43 (in Russian)

[7] Dedyulina O K and Salishchev G A 2013 Formation of ultrafine-grained structure in medium-carbon steel 40Cr-Mn-Ni-Mo by rotary forging and its influence on mechanical properties *Fundamental research* **1** pp 701-706 (in Russian)

[8] Tyurin V A 1990 Forging on radial swaging machines (Moscow: Mechanical engineering) p 256 (in Russian)

[9] Asfandiyarov R, Raab G and Aksenov D 2018 Analysis of the stress-strained state of billets processed by rotary forging with special shape of the tool *Solid State Phenomena*, Accepted for publication

[10] Sosenushkin E N 2013 Technologies of constructional nanomaterials from the position of synergetic approach *News of National Technical University "Kharkov polytechnical institute"* **42** pp 156-166 (in Russian)

[11] Osintsev O and Fedorov V 2004 Copper and copper alloys. Domestic and foreign brands (Moscow: Mechanical engineering) p 336 (in Russian)

IOP Conf. Series: Materials Science and Engineering **461** (2019) 012003 doi:10.1088/1757-899X/461/1/012003

Research on the use of zeolite for smoothing surfaces in tumbler machining

Bańkowski D and Spadło S

Kielce University of Technology, Kielce, Poland, EU

E-mail: dbankowski@tu.kielce.pl

Abstract. The article presents the proposal of using grounded zeolite rocks for container processing. The hardness of 5-6 according to the Mohs scale classifies this material as medium hard. Which is applicable for roughing of elements and deburring and polishing processes. The aim of the research was to assess the quality of zeolite application as a cutting material. Effectiveness of removing corrosion and geometric structure of the surface changes of objects made of carbon steel were investigated. Specified mass losses depending on the length of the vibratory machining process and changes in the geometric structure of the surface. The results were compared to analogous processes carried out with the use of ceramic abrasive media.

1 Introduction

Finishing works are a significant challenge in the manufacturing of small objects and with extended geometry, what in many traditional methods of finishing treatment often presents many problems [1-2]. The machining of sharp edges and hard-to-reach places is particularly problematic. Opposite to the requirements of machining of objects with complex spatial structure, they enable their use by loose abrasive machining. Most of all, we should understand here all treatments using high pressure abrasive particles - sandblasting, shot peening, soda treatment, etc. but also container machining with using loose abrasive media [4-6]. Literature data gives [3] a significant dominance of vibratory machining over rotary due to significantly lower machining times required to achieve analogous results as with rotary machine tools only – figure 1.

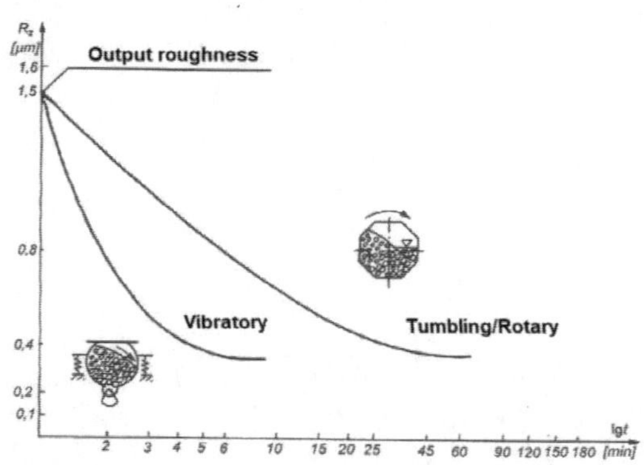

Figure 1. Change of surface roughness of St10 steel samples depending on the time of finishing in the container smoothing machines [3]

Vibratory machining is a finishing process of the object surface to the most optimal form, removing the material allowance and burrs, rounding the edges, smoothing the surface and cleaning it thoroughly [7-9]. The process takes place in containers containing workpieces, the working medium in the form of appropriately selected abrasive or polishing media and the working fluid [10-12]. The process consists in moving the workpieces to the abrasive media [13-15].

Ceramic shapes used in the tests consist of irregularly shaped grains of ceramic cullet connected by a binder [16]. In ceramic shapes there are also pores, ie empty space. It should be noted that the most important role of the binder is the deposition of abrasive grains and keeping them for a suitable time on the active surface of the grinding wheel [17-18]. The pores present in the volume are important, because allow to supply the protective and lubricating agent to the microcutting zone and they hold in their volume and remove products of processing, i.e. chips, from this zone [19]. In grinding wheels with ceramic binders, the pores are formed in an intrinsic way in the process of firing the grinding wheel body [19].

2 Zeolite

Zeolites are a large group of natural hydrated aluminosilicate minerals with different chemical compositions and properties. These are sodium and calcium aluminosilicates, to a lesser extent barium, strontium, potassium, magnesium and manganese. Zeolites were formed as a result of volcanic eruptions in millions of years under the influence of high temperature and pressure, when the volcano lava came into contact with the saltwater. The reaction of volcanic ash with the salts present in the lakes has transformed the solidified lava into various groups of aluminosilicates with unique physicochemical properties. These properties result from their specific crystalline structure, characterized by the existence in the spatial aluminosilicate skeleton of canals and chambers with strictly defined molecular dimensions. The total volume of empty spaces is 24 - 32%. The hardness of 5-6 according to the Mohs scale classifies this material as medium hard. Which is applicable for roughing of elements and deburring and polishing processes. The aim of the research was to assess the quality of zeolite application as a cutting material.

Figure 2. Zeolitic shapes with gradation 4-8 mm and ceramic media CB 20T

3. RESEARCH OBJECT

In order to test the cutting properties of zeolite, i.e. a natural mineral, it was decided to compare the results of the treatment with loose shapes in the form of crushed zeolite rock in a vibrating machine. A fraction of fine particles of zeolite rocks with gradations from 4 to 8 mm was used for the tests. The results of the obtained tests were referred to the results of vibro-abrasive treatment using the shapes with similar abrasive abilities - triangular ceramic shapes CB 20 T (with a size of approx. 20 mm). It should be emphasized that ceramic shapes are a typical composite consisting of ceramics and a binder.

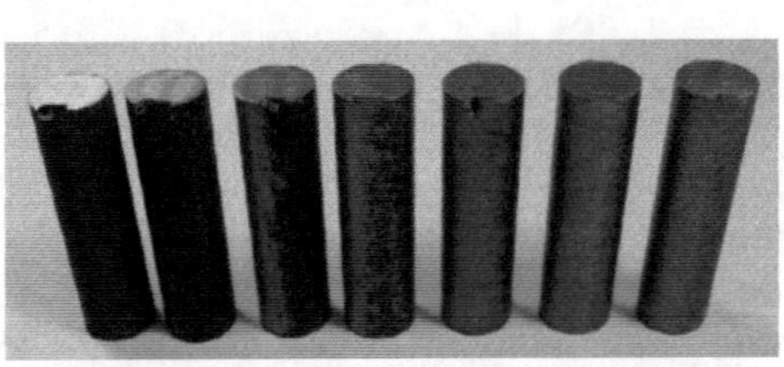

Figure 3. Steel samples a) without machining, b) after 60 min of treatment, c) 120min, d) 180 min, e) 240min, f) 300min and g) 1000 min

In the research, attention was paid to the possibility of using zeolite for finishing steel (corrosion removal of steel). The samples (figure 3) were made of carbon steel in the form of rods with a diameter of 12 mm, length approx. 45 mm, which as a result of long-term storage were affected by the environment – corrosion.

The tests were carried out on a machine dedicated for vibro-abrasive machining of Rollwasch SMD-25-R with a working tumbler capacity of 25 dm^3. Before the tests, the process parameters were determined. The vibration frequency of the tank (tumbler) was rated at 2400Hz. In order to have a reference to the zeolite's cutting ability, it was decided to examine how the mass loss and the basic parameters of the geometric structure of the surface change along with the duration of the treatment time. The tests were carried out for 0, 60, 120, 180, 240, 300 and 1000 minutes of treatment duration. A charge in the form of crushed zeolite rock (4-8mm) in an amount of approx. 13 kg was filled into the machine tank. The treatment was carried out wet with the addition of about 200 ml ME L100 A22 / NF processing aid and about 400ml of demineralised water. As a result of the zeolite's properties, i.e. the tubular structure of the mineral, all of the liquid part of the charge was absorbed almost immediately by a fixed part of the charge. Too much liquid addition leads to complete soaking of the zeolite, which in turn leads to a solidified mass and the inability to process. Therefore, based on the experience, the authors decided to use a limited amount of supplementary support.

4. RESULTS AND DISCUSSION

The samples were marked and weighed before being placed in the working tumbler. Then, after an appropriate duration of treatment the samples were pulled from the tumbler. They were thoroughly cleaned in an ultrasonic cleaner. Then the mass was re-measured. As a result of the obtained results, it was possible to determine mass losses depending on the length of time of the vibratory machining. Data of mass loss are presented in table 1. In addition, mass losses for analogous processing times using ceramic shapes were compared. A mass loss with the use of zeolites media was applied to a mass loss using ceramic shapes, which was expressed as a percentage value.

Table 1. Comparison of mass losses and the basic parameters of the geometric structure of the surface for the treatment with fittings made of zeolites and ceramics

Tumbling time, min	Weightloss, milligrams		Weightloss Zeolites/ ceramic %	Sa, µm		Sz, µm	
	Zeolites media	Ceramic media		Zeolites media	Ceramic media	Zeolites media	Ceramic media
0	0	0	0	2.024	2.024	39.037	39.037
60	47.2	80.7	58.5%	5.803	2.039	40.522	32.613
120	70.9	89.5	79.2%	3.838	1.881	37.739	22.893
180	88.2	107.9	81.7%	2.362	1.852	30.759	21.315
240	89.6	114.4	78.3%	1.802	1.547	22.52	19.537
300	96.1	115.5	83.2%	1.841	1.282	23.028	17.943
1000	97.4	164.3	59.3%	2.236	0.768	22.258	15.728

On the basis of measurements of mass losses of corroded steel samples subjected to vibratory machining using crushed zeolite and ceramic media it can be clearly stated that processing media in the form of zeolite molds causes mass losses however smaller than those considered with the use of ceramic media. Based on the measurements. it can be concluded that vibratory machining with finely divided zeolite rocks leads to approx. 80% of the mass losses of analogous processing times as in the case of ceramic media. The exception is the shortest - 60 min and the longest processing time - 1000 min where the discrepancies are different from the other machining times and amount to approx. 60%.

Long-term studies of smoothing for 1000 minutes with zeolitic media resulted in a much smaller loss of sample mass equal to 97.4 mg. i.e. about 60% weight loss of samples as in the case of ceramic media. The reason for this situation may be the fact that in the case of long-term processing the sharp edges of the zeolite media become rounded and lose their cutting properties. The weight loss in the case of processing for 300 minutes amounted to approx. 96.1 mg. whereas the treatment for 1000 min approx. 97.4 mg. Therefore, it can be concluded that longer processing times than 5 hours do not bring significant changes in the loss of mass of processed details. It is not advisable to use longer treatment times using the zeolitic media. It should be emphasized that the discrepancies in the results of mass losses can be determined by the fact that ceramic media are composite shapes consisting of a ceramic cullet and a binder - a binding material that keeps individual abrasive shapes in a uniform form. The zeolite moldings are finely crushed zeolite rock which is not a typical example of abrasive shapes dedicated to container processing.

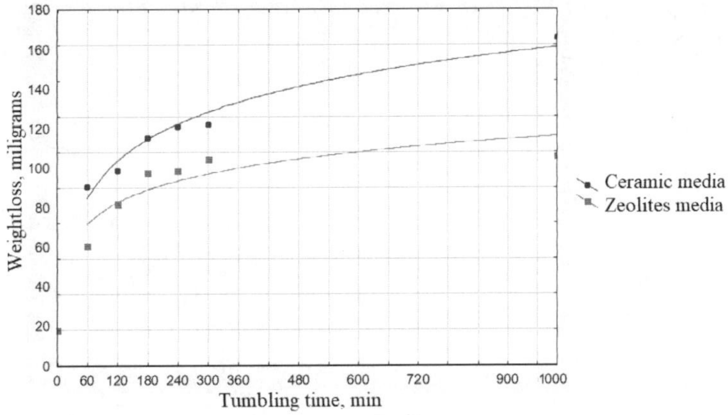

Figure 4. Diagram of the mass loss dependence on the treatment time for ceramics and zeolites

In order to obtain knowledge on which way the mass of the processed samples is reduced measurements of the geometric structure of the surface were carried out using the Taylor Hobson Talysurf CCI Lite 3D optical profilometer. As a result of the conducted analyzes 3D maps of the examined surfaces were obtained along with the SGP parameters. The results of measurements for different tumbling times of the basic surface roughness parameters are included in Table 1. According to Janecki [20] the 3D area surface texture parameters were defined. Sa is the arithmetic mean height of the surface. Parameter Sz is the maximum height of the surface. Parameter Sv is the maximum height of valleys. And parameter Sp is the maximum height of peaks (difference between Sz and Sv) [20].

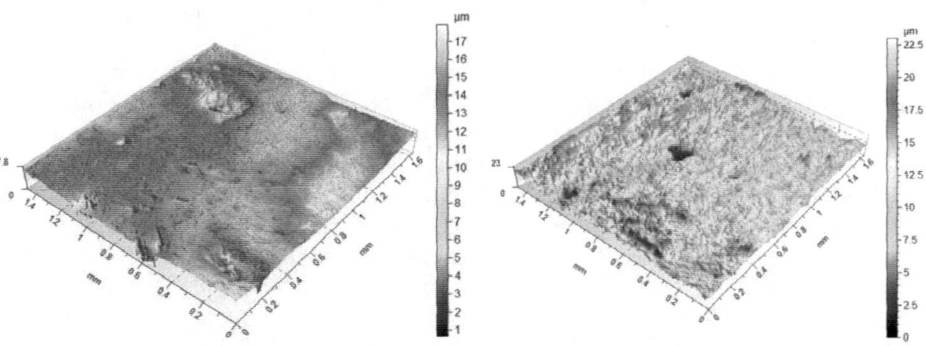

Figure 5. 3D surface texture of steel surface a) processed with ceramic media for 300 min. Sa = 1.841 µm b) processed with zeolit media for 300 min. Sa = 1.282 µm

The undetermined geometry of the media made of zeolite rock translates into a large number of sharp edges that draw surfaces causing the formation of many micro-dimples. In comparison to triangular ceramic shapes having a homogeneous construction surfaces for vibratory machining are characterized by lower surface roughness. This is confirmed by photographs of the obtained surfaces from the Talysurf CCI optical microscope. The obtained data allows to propose the use of fine fractions of zeolite fittings for surface preparation processes for coating application and removal of significant surface irregularities. In order to obtain very smooth surfaces with a low surface roughness should be carried out a two-step treatment. In the first stage of deburring with the use of e.g. zeolite and in the second stage smoothing and brightening with the use of polishing media. On the basis of graph 6, it can be concluded that the treatment with ceramic media allows to achieve smaller average arithmetic surface roughness – *Sa* than the media made of zeolite. Longer working times with ceramic media reduce Sa. The situation is similar with the zeolite media but only up to 300 minutes of treatment. However, machining for 1000 min does not reduce the surface roughness of Sa compared to machining for 300 minutes.

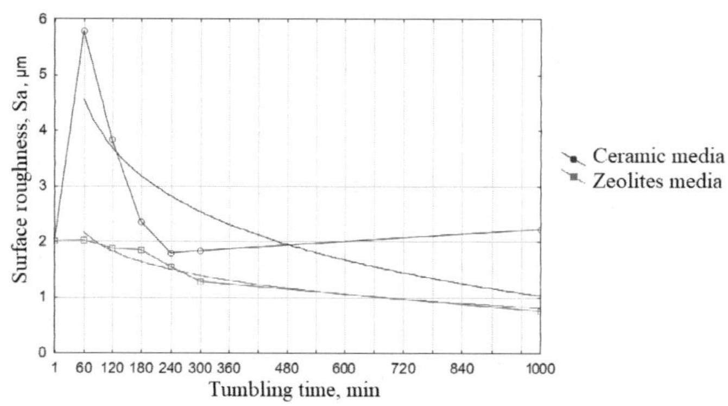

Figure 6. Graph of the dependence of changes in the arithmetic mean surface roughness *Sa* from the duration of the treatment time for ceramic and zeolite media.

Conclusions

On the basis of the conducted research it can be concluded that the shredded zeolite rock has cutting properties.

Visual observations already allow to conclude that the signs of corrosion are removed after approx. 4-5 hours of vibratory machining.

Mass losses of processed corroded parts made of steel with the use of zeolite account for approx. 80% of the mass losses for processing with ceramic media.

Extending the vibratory machining time over 5 hours is pointless due to slight changes in the weight loss of the processed workpieces.

Microscopic observations and measurements of the geometrical structure of the surface allow to state that vibro-abrasive machining with the use of zeolite rock media results in obtaining a surface with a higher average arithmetic surface roughness than analogue vibratory machining with ceramic media. The reason for this state of affairs may be a much smaller size of zeolitic media than ceramic media. Ceramic media are a combination of a binder and a ceramic pestle that plays the role of abrasive grains while the zeolite is pure mineral rock. This translates into a smaller mass of a single piece and the number and type of sharp edges responsible for the effects of machining.

Acknowledgement

The research was carried out in the framework of the project "The use of zeolite as an abrasive material for the construction of innovative abrasive tools" in the Program "INNOVATION`S+ INCUBATOR"

References

[1] Młynarczyk P and Spadło S 2016 The Analysis of the Effects Formation Iron - Tungsten Carbide Layer on Aluminum Alloy by Electrical Discharge Alloying Process. In METAL 2016: 25th Anniversary International Conference on Metallurgy and Materials, Ostrava: TANGER, , pp. 1109-14

[2] Spadło S and Młynarczyk P 2016 Analysis of the mechanical interactions of the filament brush electrode on the formation of the surface roughness. InMETAL 2016: 25th Anniversary International Conference on Metallurgy and Materials, Ostrava: TANGER, pp. 1169-74

[3] Filipowski R and Marciniak M 2000 Techniki obróbki mechanicznej i erozyjnej, Oficyna Wydawnicza Politechniki Warszawskiej, Warszawa, pp: 304-308

[4] Bańkowski D and Spadło S 2016 Influence of the smoothing conditions in vibro-abrasive for technically dry friction the parts made of steel X160CRMOV121, in: Proc. 25th Int. Conf. Metall. Mater. Metal 2016. pp 1019-24

[5] Bańkowski D and Spadło S 2017 Investigations of influence of vibration smoothing conditions of geometrical structure on machined surfaces, IOP Conference Series-Materials Science and Engineering. Volume: 179, Article Number: UNSP 012002

[6] Starosta R 2008 Surface treatment, Gdynia, Wydawnictwo Akademii Morskiej w Gdyni. (in Polish)

[7] Davidson D A 2002 Mass Finishing Processes, *Metal Finishing Guidebook and Directory*, New York, TAM surfaces are Elsevier Science

[8] Davidson D A 2016 Finishing Processes and Mechanical Surface Preparation https://www.linkedin.com/pulse/how-can-you-produce-high-quality-edge-surface-large-lots-davidson Mass

[9] Woźniak K 2017 Surface treatment in container smoothers, Warszawa, WNT (in Polish)

[10] http://www.rollwasch.it/en. Vibro Dry Experience PL. Rollwasch Italiana S.p.a. online [2017-12-14]

[11] Gillespie LK 1999 Deburring and Edge Finishing Handbook, Society of Manufacturing Engineers

[12] Brinksmeier E and Giwerzew A 2005 Hard gear finishing viewed as a process of abrasive wear, *Wear*, Vol. 258, pp 62-69

[13] Rao Suren B 2009 Repair of aircraft transmission gears via isotropic superfinishing, Gear Technology

[14] Eric C Ames 2012 Repair of high-value high-demand spiral bevel gears by superfinishing, *Gear Technology*, pp 50-59

[15] Bańkowski D and Spadlo S 2017 Vibratory machining effect on the properties of the aaluminum alloys surface. *Archives of Foundry Engineering*. Volume: 17. Issue: 4. pp 19-24

[16] Bańkowski D and Spadło S 2017 Vibratory tumbling of elements made of Hardox400 steel, Proceedings of 26th International Conference on Metallurgy and Materials METAL 2017. pp: 725-30

[17] Oryński F and Synajewski R 2010 Surface roughness testing of conventionally treated surfaces and vibration on a surface grinder, *Mechanik* 3/2010, pp 190-192 (in Polish)

[18] Krzos J 2011 Nowoczesne spoiwo ściernic ceramicznych, *Obróbka Metalu* 4/2011: pp 26-30

[19] Kacalak W, Lipiński D, Bałasz B, Rypina Ł, Tandecka K and Szafraniec F 2017. Performance evaluation of the grinding wheel with aggregates of grains in grinding of Ti-6Al-4V titanium alloy. *The International Journal of Advanced Manufacturing Technology*. 10.1007/s00170-017-0905-x 2018

[20] Janecki D, Stepien K and Adamczak S 2010 Problems of measurement of barrel- and saddle-shaped elements using the radial method, 19th IMEKO World Congress Location, Volume: 43, Issue: 5, Pages: 659–63

Microstructure and microhardness of the Ti15Mo alloy subjected to severe plastic deformation in α+β condition

K Bartha[1], A Terynková[1], J Stráský[1], I Semenova[2], S Gatina[2], M Janeček[1]

[1]Department of Physics of Materials, Charles University, Ke Karlovu 5, 121 16 Prague, Czech Republic

[2]Institute of Physics of Advanced Materials, Ufa State Aviation Technical University, K. Marx street 12, 450 000 Ufa, Russian Federation

E-mail: kristina.bartha@met.mff.cuni.cz

Abstract. Metastable β-titanium alloys are perspective candidates for use in biomedicine. For that purpose a high strength of these materials is required. The enhanced strength of these alloys can be achieved by subjecting the material to severe plastic deformation. However, due to the limited ductility of the metastable β-titanium alloys, deformation of the alloy in β solution treated condition at room temperature is sometimes inaccessible. In this study a Ti15Mo alloy in a two phase α+β condition was subjected to equal channel angular pressing at elevated temperature of 400°C. The resulting microstructure after N=4 and 8 ECAP passes consists of deformed β-phase with elongated α-phase particles. Despite the microstructure is not ultra-fine grained the microhardness of the deformed alloy exhibits higher values compared to the non-deformed counterpart.

1 Introduction

Metastable β-Ti alloys are nowadays industrially used in aircraft manufacturing industry, namely for construction of landing gears [1]. Binary Ti15Mo alloy investigated in this study is considered for biomedical applications as it contains only biocompatible elements [2]. Restriction on biocompatible elements results in lower strengthening by advanced alloying typical in high-strength alloys for aerospace industry [3]. Therefore, improvement of the strength can be achieved by significant grain refinement. For this purpose, severe plastic deformation (SPD) methods can be used which proved to be an effective approach to prepare ultra-fine grained (UFG) materials with high strength [4].

Equal channel angular pressing (ECAP) is a principal SPD method that uses simple shear deformation applied by pressing specimen through the die with the given angle [5]. For successful deformation, it is critical that material remains compact without cracking. Moreover, pressing force must be sufficient for pressing the material through the die. ECAP can be done at ambient temperature [6] or, if necessary at elevated temperatures [7]. Titanium alloys are very strong, extremely tough, with limited ductility and therefore very hard to deform by ECAP [8]. Critical is using ECAP die with the angle of 120° (i.e. not the most common 90°) [9, 10]. Despite some β-Ti alloys can be processed at room temperature [11], elevated temperatures are usually required [9, 10]. This brings another complication in case of metastable β-Ti alloys, as the material may undergo phase transformations during ECAP deformation. In this study, Ti15Mo alloy in a fully aged condition (α + β) was used to avoid phase transformations during processing, namely formation of ω-phase which causes material embrittlement [3].

2 Experimental

The Ti15Mo alloy was supplied by ATI Corp. in a form of rod with a diameter of 10 mm. The material was thermally treated in order to get a duplex α+β microstructure. A 2-step thermal treatment (TT) consisted of annealing of the material above β-transus temperature for 1 hour followed by slow cooling below 552°C. The second step of the TT was a stress-relief annealing at 816°C for 10 minutes and consequent slow cooling below 552°C.

The deformation by ECAP was accomplished at Ufa State Aviation Technical University (USATU) in Ufa, Russia. The deformation was carried out at elevated temperature of 400°C in a die with channels intersecting at angle of $\Phi = 120°$.

Samples after N = 4 and 8 ECAP passes were prepared for this study. The total achieved equivalent strain [12] was $\varepsilon_{equiv} = 2.66$ and 5.32, respectively. The principle of the ECAP process is schematically displayed in figure 1 and is described elsewhere [13].

Samples for scanning electron microscope (SEM) observations and microhardness measurement were prepared by mechanical grinding and polishing followed by a three-step vibratory polishing.

The scanning electron microscope Zeiss Auriga CrossBeam operated at 10 kV was used for microstructural observations. The microstructure was studied in two planes – parallel to the processing (rolling direction) and perpendicular to the processing (normal direction) as schematically displayed in figure 1. Microhardness measurements were carried out using the automatic microhardness tester Qness Q10a by Vickers method; 0.5 kgf load, 20 indents per sample were applied.

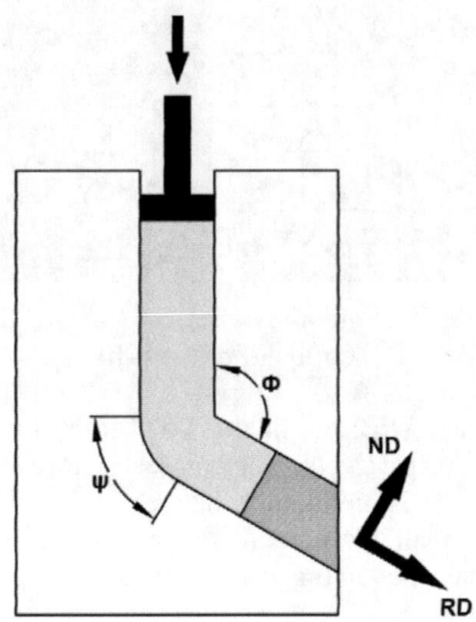

Figure 1. Schematic representation of the ECAP process and orientation of the sample (rolling-RD and normal-ND direction).

3 Results

Figure 2 shows the initial microstructure of the Ti15Mo alloy in α+β condition. The α-phase particles appear darker in the image due to the chemical contrast (Mo depletion). α-particles are elongated with the length of few micrometers and submicrometer width.

IOP Conf. Series: Materials Science and Engineering **461** (2019) 012004 doi:10.1088/1757-899X/461/1/012004

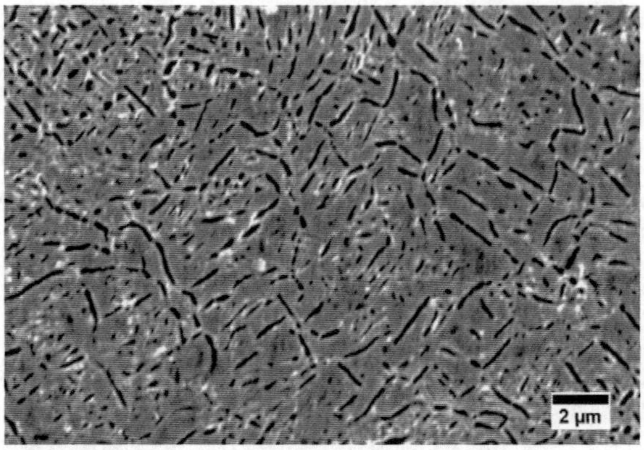

Figure 2. BSE micrograph of the as-received Ti15Mo alloy in α+β condition.

In Figure 3 a and b the microstructure of Ti15Mo alloy after N=4 ECAP passes in both rolling and normal directions is shown. Elongated α-phase particles are only partly refined. Individual grains of β-phase cannot be resolved. α-phase particles appear bigger and more equiaxed in the normal direction when compared to the rolling direction.

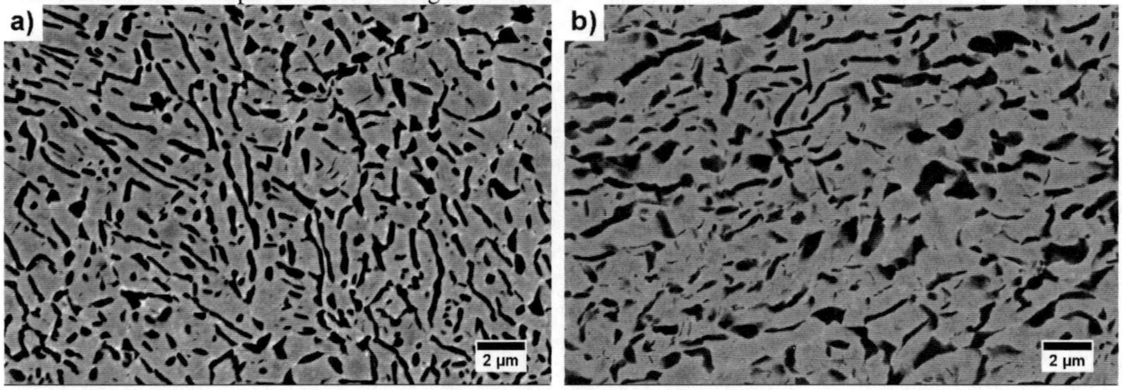

Figure 3. BSE micrograph of the Ti15Mo alloy after N=4 ECAP passes in rolling direction (a) and normal direction (b)

Figure 4 a and b shows the microstructure of the Ti15Mo alloy after N = 8 ECAP passes in both rolling and normal directions, respectively. The microstructure is highly deformed and β-phase is significantly refined as it is well visible due to changes of the grey contrast. The microstructure is not homogeneous, areas with smaller α precipitates appear. Nevertheless, in normal direction less amount of α precipitates, which are even bigger, are present.

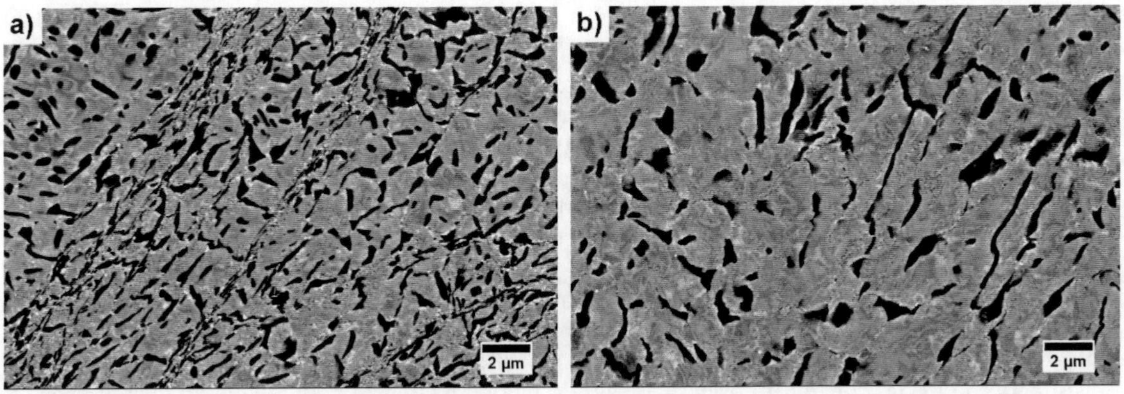

Figure 4. BSE micrograph of the Ti15Mo alloy after N=8 ECAP passes in rolling direction (a) and normal direction (b)

Evolution of the microhardness with the ECAP processing is shown in figure 5. The microhardness was measured both in rolling and normal directions. The microhardness of the Ti15Mo alloy in $\alpha+\beta$ condition reaches 360 HV [14] and is displayed by a dashed line in figure 5. The microhardness after N = 4 ECAP passes increases up to 383 HV and 374 HV in rolling and normal direction, respectively. The microhardness slightly increases with increasing number of ECAP passes, however, the increase is not significant. Microhardness of N = 8 condition is not homogeneous, which is consistent with the observation of the non-homogeneous microstructure in SEM.

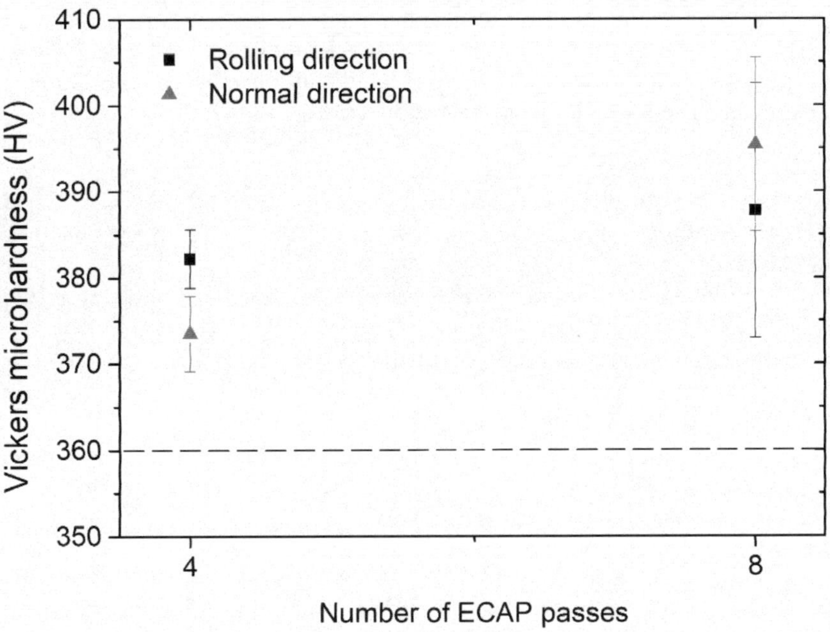

Figure 5. Evolution of the microhardness with the number of ECAP passes in both normal and rolling directions (microhardness of the non-deformed Ti15Mo alloy is displayed by a dashed line)

4 Discussion

Deformation by 8 passes by ECAP was possible thanks to using the die with the outer angle of 120°, processing temperature of 400°C and initial material with thermodynamically stable α + β phase composition. Processing of β solution treated material is comparatively more complicated, probably due to the formation of the ω-phase during ECAP processing, which causes material embrittlement [15]. After 8 ECAP passes, α-phase particles were deformed, but internal structure suggesting grain refinement was not observed. This is consistent with observation of Ti6Al7Nb alloy in α + β condition [16]. On the other hand, β matrix is refined which can be inferred from SEM observations, especially in figure 4(b).

The microhardness was measured using indentation according to Vickers with the load of 500g, which results in indents with the approximate diagonal length of diagonal of 50 μm. One indent therefore spans over several α-phase particles. Microhardness of the Ti15Mo alloy in α + β condition is superior to the microhardness of the Ti15Mo in solution treated β condition due to the presence of stronger α-phase and mainly due to strengthening by phase interfaces [3, 16]. The microhardness increases with increasing number of ECAP passes which can be attributed to the grain refinement of the β matrix. Microhardness of Ti15Mo alloy in α + β condition processed by ECAP is lower than the microhardness of β solution treated Ti15Mo alloy processed by high pressure torsion (HPT) – up to 510 HV [17]. HPT processed material is characterized by more refined microstructure with grain size of few hundreds nm and most importantly by the presence of the ω-phase. On the other hand, α-phase is not present in this condition. It can be concluded that presence of the ω-phase in β solution treated material after extreme refinement by HPT is decisive for achieving the microhardness which is significantly higher than in α + β aged structure processed by ECAP. Similar results were obtained in metastable β aerospace Ti-6.8Mo-4.5Fe-1.5Al alloy [18]. In this alloy, the presence of the ω-phase results in maximum microhardness of 550 HV, while fine α + β structure is characterized by microhardness of 500 HV, which is caused both by solid solution strengthening and very fine α particles [18]. Comparable microhardness was not achieved by ECAP processing of Ti15Mo alloy. Note that despite the specimen is rotated successively between individual ECAP passes around its longitudinal axis, significant differences both in microstructure and microhardness are achieved in rolling and normal directions.

5 Conclusion

Biomedical metastable β Ti15Mo alloy was successfully prepared by 8 passes of equal channel angular pressing (ECAP) at 400°C using the die outer angle of 120°. The following conclusion can be drawn from this investigation:

- Fully aged two phase thermodynamically stable α + β condition of Ti15Mo alloy can be processed by ECAP easier than β solution treated material.
- α-phase particles are significantly deformed after N = 8 passes, but there is no sign of grain refinement within these particles.
- β-phase matrix was significantly refined by ECAP processing.
- Microhardness increases with the strain imposed by ECAP processing, but it does not reach the microhardness of Ti-6.8Mo-4.5Fe-1.5Al aerospace alloy in aged condition.

Acknowledgement

This work was financially supported by the Czech Science Foundation under the project 17-04871S. Partial financial support by the Ministry of Education Youth and Sports under the project LTARF18010 is also gratefully acknowledged. The student K.B acknowledges financial support by Grant Agency of Charles University project no. 1106216.

References

[1] Boyer R R 1994 Aerospace applications of beta titanium alloys *JOM* **46** pp 20-23
[2] Steinemann S G 1998 Titanium — the material of choice? *Periodontology 2000* **17** pp 7–21

[3] Welsch G, Lütjering G, Gazioglu K and Bunk W 1977 Deformation characteristics of age hardened Ti-6Al-4V *Metall. Trans.* **A8** pp 169–77
[4] Kawabe Y and Muneki S 1991 Strengthening and Toughening of Titanium Alloys *ISIJ Int.* **31** pp 785–91
[5] Segal V M, Reznikov V I, Drobyshevsky A D and Kopylov V I. 1981 Plastic working of metals by simple shear *Russian Metallurgy* **1** p 99
[6] Estrin Y, Isaev N V, Lubenets S V, Malykhin S V, Pugachov A T, Pustovalov V V, Reshetnyak E N, Fomenko V S, Fomenko L S, Shumilin S E, Janecek M, Hellmig R J 2006 Effect of microstructure on plastic deformation of Cu at low homologous temperatures *Acta Mater.* **54** pp 5581-90
[7] Minárik P, Veselý J, Král R, Bohlen J, Kubásek J, Janeček M, Stráská J 2017 Exceptional mechanical properties of ultra-fine grain Mg-4Y-3RE alloy processed by ECAP *Mater. Sci. Eng.* A. **708** pp 193-8
[8] DeLo D P and Semiatin S L 1999 Hot Working of Ti-6Al-4V via Equal Channel Angular Extrusion *Metall. Mater. Trans.* A. **30** pp 2473-81
[9] Semenova I P, Raab G I, Saitova L R, Valiev R Z 2004 The effect of equal-channel angular pressing on the structure and mechanical behavior of Ti–6Al–4V alloy *Mater. Sci. Eng.* A. **387–389** pp 805-8
[10] Saitova L R, Höppel H W, Göken M, Semenova I P and Valiev R Z 2009 Cyclic deformation behavior and fatigue lives of ultrafine-grained Ti-6AL-4V ELI alloy for medical use *Int. J. Fatigue* **31** pp 322-31
[11] Li Z, Zheng B, Wang Y, Topping T, Zhou Y, Valiev R Z, Shan A, Lavernia E J 2014 Ultrafine-grained Ti–Nb–Ta–Zr alloy produced by ECAP at room temperature *J. Mater. Sci.* **49** pp 6656-66
[12] Iwahashi Y, Wang J, Horita Z, Nemoto M and Langdon T G 1996 Principle of equal channel angular pressing for processing of ultra-fine grained materials *Scripta Mater* **35** pp 143-6
[13] Valiev R Z and Langdon T G 2006 Principles of equal-channel angular pressing as a processing tool for grain refinement *Prog. Mater. Sci.* **51** pp 881–981
[14] ATI Ti15Mo Technical Data Sheet 2014 https://www.atimetals.com/Products/Documents/datasheets/titanium/alloyed/ati_15Mo_Titanium_Alloy_en_v4%20final.pdf
[15] Bartha K, Terynková A, Stráský J, Minárik P, Veselý J, Polyakova V, Semenova I, Janeček M 2018 Inhomogeneous precipitation of the α-phase in Ti15Mo alloy deformed by ECAP *Mater Sci Forum* in press
[16] Janeček M et al 2014 Mechanical Properties and Dislocation Structure Evolution in Ti6Al7Nb Alloy Processed by High Pressure Torsion *Metall. Mater. Trans.* **45A** pp 7–15
[17] Václavová K, Stráský J, Polyakova V, Stráská J, Nejezchlebová J, Seiner H, Semenova I and Janeček M 2017 Microhardness and microstructure evolution of ultra-fine grained Ti-15Mo and TIMETAL LCB alloys prepared by high pressure torsion *Mat. Sci. Eng.* A. **682** pp 220-8
[18] Šmilauerová J, Janeček M, Harcuba P, Stráský J, Veselý J, Kužel R, Rack H J 2017 Ageing response of sub-transus heat treated Ti–6.8Mo–4.5Fe–1.5Al alloy *J. Alloys Compd.* **724** pp 373-80

IOP Conf. Series: Materials Science and Engineering **461** (2019) 012005 doi:10.1088/1757-899X/461/1/012005

Fabrication and characteristics of sintered copper-silica composites

M Kargul[1], J Borowiecka- Jamrozek[1], M Konieczny[1]

[1]Department of Metal Science and Materials Technologies, Faculty of Mechatronics and Mechanical Engineering, Kielce University of Technology, Al. Tysiąclecia P.P. 7, 25-314 Kielce, Poland

E-mails: mkargul@tu.kielce.pl, jamrozek@tu.kielce.pl, mkon@tu.kielce.pl

The paper presents the results of research on the production of a sintered Cu-SiO_2 composites fabricated by powder metallurgy. The starting materials for obtaining the composites were commercial powders: Cu and SiO_2. SiO_2 particles were introduced into a copper matrix in the amount of 2.5 %, 5 %, 7.5 % and 10 % by weight. Finished powders mixtures were subjected to single pressing with a hydraulic press at a compaction pressure of 620 MPa and then subjected to sintering process at 900 °C in an atmosphere of dissociated ammonia. The sintering time was 60 minutes. The sintered compacts were subjected to the following tests: hardness, measurement of density, electrical conductivity and abrasion resistance. The highest density was obtained for a composite containing 2.5 % SiO_2 which was 7.94 g/cm^3. The highest hardness was obtained for a composite containing 10 % SiO_2 which amounted to 50 HB. For the composite containing 2.5 % SiO_2 the highest electrical conductivity was obtained which was 47 MS/m. The highest abrasive wear resistance was obtained for a composite containing 10 % SiO_2. Both density and electrical conductivity decreased along with the increase of SiO_2 content in composites.

1. Introduction

Currently, research is being carried out on the production of modern materials in order to achieve the desired properties. A single material generally does not have satisfactory properties, therefore the need for composites with specific properties increases every day. In recent years, sintered copper-matrix composites are finding an increasing interest due to its very high electrical and thermal conductivity [1-6]. It is well known that the inclusion of even a small amount of copper alloying elements results in a significant decrease in electrical conductivity. In order to increase the strength of copper without a significant decrease in electrical conductivity, ceramic particles such as Al_2O_3, SiO_2, ZrO_2, Cr_2O_3, BeO, MgO as well as carbides (SiC, TiC, Cr_7C_3, Cr_3C_2) and borides and nitrides (TiB_2, ZrB_2, CrB_2, BN) are used [7-13]. By using of ceramic particles as a strengthening phase, diffusion phenomenon does not occur between the metal matrix and reinforcing particles. As a result, composites with electrical conductivity similar to pure copper are obtained. Research is also being carried out on the preparation of composites containing intermetallic phases, the introduction of which results in the increase of strength properties similar to those made with ceramic particles [14-16]. The composites reinforced with ceramic particles combine the properties of the metal matrix with the reinforcing properties of ceramics. Ceramic particles have high hardness and resistance to high temperatures, therefore, the composites reinforced with ceramic particles are also used for working at elevated temperatures. The introduction of ceramic particles into the metal matrix increases the hardness and abrasion resistance of the composite. Therefore, the copper matrix composites reinforced with ceramic particles are mainly used in electronics and electrical engineering for relays, switches, elements of electric motors and for the tip of electrodes for resistance welding. Metal matrix composites reinforced with ceramic particles are usually made using powder metallurgy. The production of composites by the powder metallurgy method consists in preparing a powder mixture, powder compaction and then sintering. Research is also being carried out on the production of metal matrix composites using the hot pressing method and also using the SPS (spark plasma sintering) technique [17-21]. In this work, an attempt was made to produce a sintered composite with a copper matrix in which SiO_2 powder was used as strengthening particles. The obtained composites were subjected to microstructural investigations and the density, hardness, electrical conductivity and abrasion resistance were examined.

2. Materials and methods

The materials used in the experiment to fabricate the composites were commercial powders. Copper powder (99.5 % Cu) with an average particle size of less than 63 μm was used as the matrix while silica (96.5 % SiO_2), with an average particle size of less than 45 μm was used as the reinforcement. The shapes and arrangements of the powder particles used in the experiments are shown in Figure 1.

a) b)

Figure 1. Images of the tested powders: a) electrolytic copper powder, b) silica powder.

Before the fabrication of composites, the powders were observed using a JEOL JSM-7100F field emission scanning electron microscope. Powder mixtures with the following content of SiO_2 - 2.5 %, 5 %, 7.5 % and 10 % by weight were prepared for the tests. Finished powdered mixtures were subjected to single-track pressing on a hydraulic press at a compaction pressure of 620 MPa. The specimens with a dimension of Φ 20x10 mm were sintered in a sillit tubular furnace at 900 °C for 60 minutes in the dissociated ammonia atmosphere. Finally, the material was cooled in the furnace. After the sintering process, the samples were subjected to compaction process at a pressure of 620 MPa and re-sintered at 900 °C for 60 minutes. The sintered compacts were measured for hardness, electrical conductivity, density and abrasion resistance tests. The hardness of the composites was measured using the Brinell method (with a steel ball 5 mm in diameter at a load of 250 kG) in accordance with the PN EN ISO 6506-1:2014 standard. The electrical conductivity tests were performed using the GE Phasec 3D device using the eddy current method. The density was determined by weighing the specimens in air and water using WPA120 hydrostatic scales in line with the PN EN ISO 2738:2001 standard. The abrasion resistance tests of the fabricated composites was conducted using device with a roll on block configuration. The roller was made of hardened 100Cr6 steel with a hardness of

62 HRC. The blocks with a dimension of 6.35x10x15 mm were made of Cu/SiO_2 composites. The test was performed with a rotational speed of 300 rpm with a load of 5 N, while the friction path was set as 3000 m. The blocks was weighed before and after the tests to determine the loss of weight. Microstructure analisys on the metallographic specimens were conducted using the JEOL JSM-7100F field emission scanning electron microscope fitted with OXFORD INSTRUMENTS EDS X-Max Aztec software for elemental analisys.

3. Results and discussion

3.1. Microstructural characterization

The introduction of powdered silica causes a distinct change in microstructure of obtained composites. Silica particles of various sizes are very clear in a copper matrix. Differences in a particle size result from the range of the size of the introduced silica powder (<45 μm). Mixing the powders of copper and silica led to an even distribution of the particles in the matrix. In certain areas of composites , the silica particles are bonded into longitudinal agglomerates. Large clusters of silica particles were not observed. Introduced silica particles have not dissolved in the matrix in the sintering process, which is associated with high thermal resistance of silica. The microstructure of fabricated materials is shown in Figure 2.

a) b) c) d)

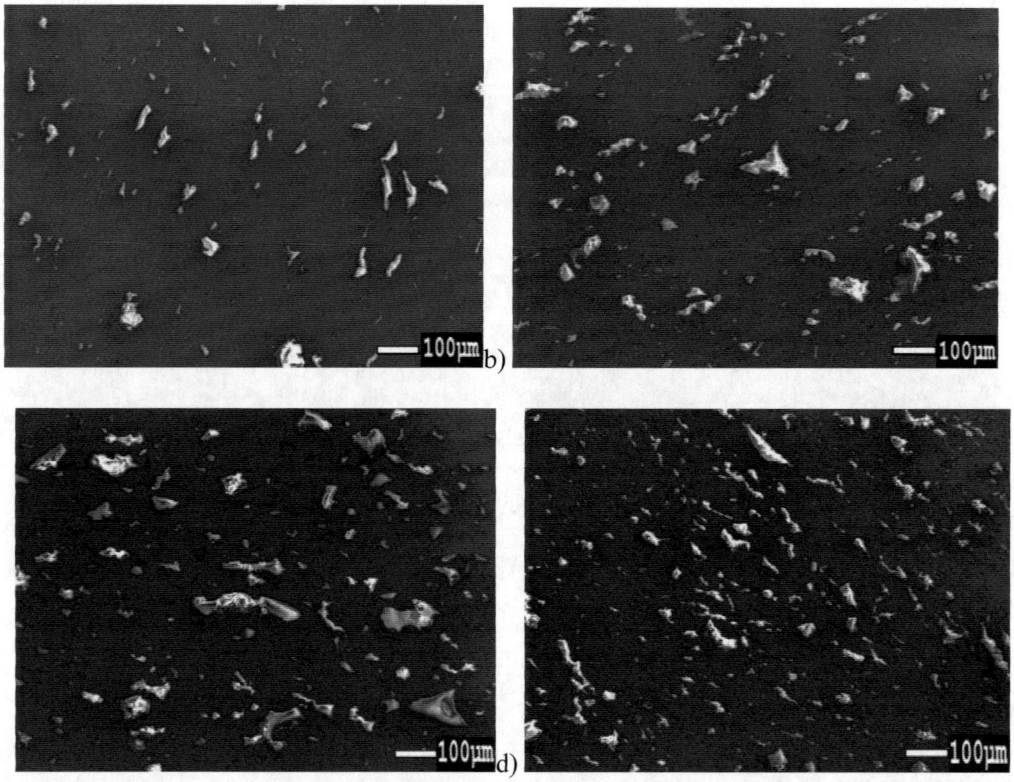

Figure 2. Microstructures of the sintered compacts observed with a SEM obtained for a) Cu+2.5 % silica, b) Cu+ 5 % silica, c) Cu+7.5 % silica d) Cu+10 % silica

An example of the distribution of elements in a composite containing 10 % SiO_2 is shown in Figure 3. From the mapping it is clear that the composite contains only copper, silicon and oxygen.

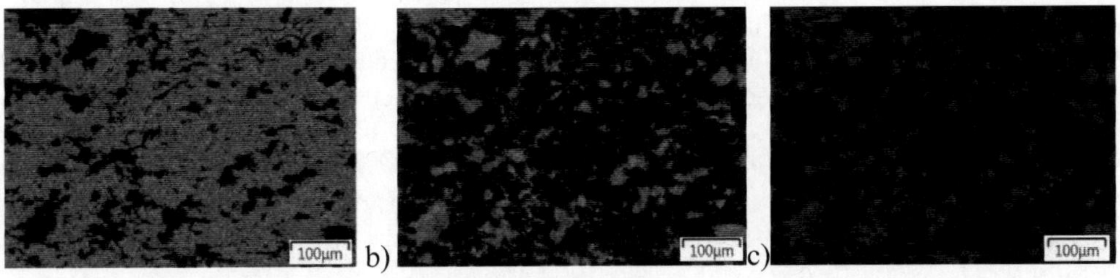

a) b) c)

Figure 3. Distribution of elements in the micro-area of the composite containing 10% SiO_2, a) copper, b) silicon, c) oxygen

3.2. *Density and hardness measurements*

The results of density and hardness measurements are presented in Table 1.

Table 1. Results of the density and hardness measurements

Material	Density (g/cm^3)	Relative density (%)	HB
Cu	8.19± 0.02	92.01	36.65± 1.5
Cu + 2.5 % of SiO_2	7.95± 0.04	90.88	44.61± 1.8
Cu + 5 % of SiO_2	7.29± 0.05	85.78	45.56± 1.3
Cu + 7.5 % of SiO_2	7.23± 0.03	84.93	48.59± 1.7
Cu + 10 % of SiO_2	6.73± 0.07	81.31	49.66± 1.8

The examination showed that, the introduction of silica particles into the copper matrix increases hardness of the composites, while reducing the density. This is due to the high hardness and low density of silica particles. The highest hardness was obtained for the composite containing 10 % of silica which amounted 50 HB (36% higher than composite made of pure copper). The same phenomenon was also noticed by other researchers [2, 4, 7, 9, 12]. The highest density was obtained for a composite made from pure copper which amounted 8.19 g/cm^3. Addition 10 % of silica particles caused decrease in density to 6.73 g/cm^3.

3.3. *Electrical conductivity test*

The results of electrical conductivity test are presented in Figure 4.

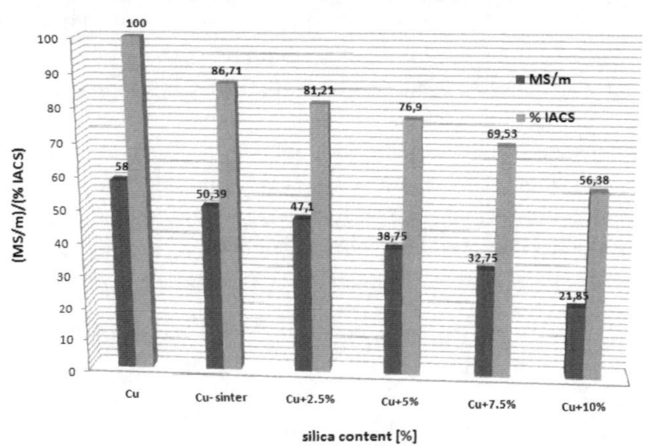

Figure 4. The results of electrical conductivity measurements of copper-silica composites

The examination showed that the introduction of silica particles into copper matrix caused decrease in electrical conductivity. This phenomenon is caused by the fact that even a small amount of alloy additives in copper causes a reduction of their electrical conductivity. The highest electrical conductivity was obtained for a composite made from copper powder which amount 50.39 MS/m (14 % less than pure copper in a solid state). Similar results were received by other researchers [2, 4, 5]. The addition of 2.5 % of silica particles caused decrease in electrical conductivity to 47.1 MS/m but addition of 10 % of silica particles caused decrease in electrical conductivity to 21.85 MS/m.

3.4. Wear characterization

Weight loss of composites after tribological test are shown in Figure 5.

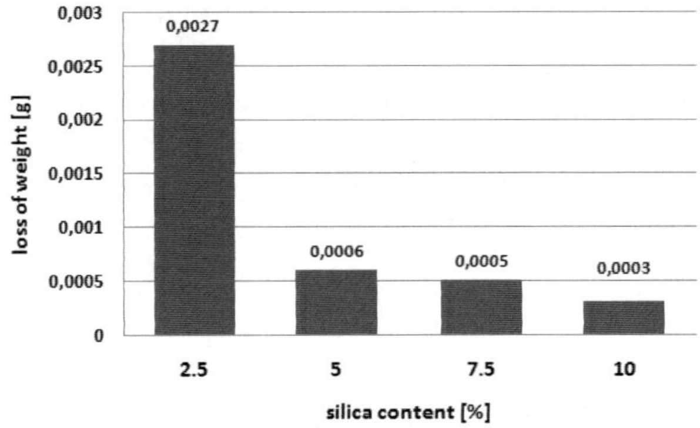

Figure 5. Weight loss of composites depending on the content of silica

Tribological tests has shown that the addition of silica particles increasing wear resistance of composites. The smallest loss of weight was observed for the composite containing 10% of silica.

4. Conclusion

The analysis of the microstructure of the composites showed that the silica powder particles combine into agglomerates. Before sintering, powders should be mixed thoroughly to break up the agglomerates and obtain a homogeneous mixture. The sintering parameters were chosen correctly based on previous research and compared with the results available in the literature. The microstructural examinations showed that there were no discontinuities at the interface between the matrix and the ceramic particles. A very good bonding of silica particles with the copper matrix was obtained, without voids, only the pores occurring in sintered metals were visible on the micrographs. Silica particles were clearly visible in the form of irregular precipitates. The introduction of silica particles resulted in an increase in the hardness of the composites and a decrease in the density and electrical conductivity. Tribological tests have shown that the introduction of silica particles increases abrasion resistance.

5. References

[1] Gao F, Fu R, Song B Y, Xu Z Y and Berthier Y, 2009, Effects of speed seguence on friction properties of sintered Cu-SiO_2, *Tribology- Materials, Surfaces & Interfaces,* **vol. 3:1**, pp 37-40.

[2] Mikuła J, Łach M and Kowalski J S, 2015, Copper matrix composites reinforced with volcanic tuff, *Metalurgija,* **vol. 54**, pp 143-146.

[3] Xuehui Z, Chenguang L, Shun C, Zengde L, 2016, Characteristics of nano-alumina particles dispersion strengthened copper fabricated by reaction synthesis, *Rare Metal Materials and Engineering,* **vol.45:4**, pp 893-896.

[4] Lee D W and Kim B K, 2004, Nanostructured Cu-Al_2O_3 composite produced by thermochemical process for electrode application, *Materials Letters,* **vol.58**, pp 378-383.

[5] Shi Z and Yan M, 1998, The preparation of Al_2O_3-Cu composite by internal oxidation, *Applied surface science*, **vol.134**, pp 103-106.

[6] Kruger C and Mortensen A, 2013, In situ copper–alumina composites, *Materials Science and Engineering A,* **vol.585**, pp 396-407.

[7] Borowiecka- Jamrozek J, Kargul M, Borowiecki S, Engel Z, 2018, Wpływ dodatku naturalnego zeolitu na mikrostrukturę i własności mechaniczne spiekanego kompozytu o osnowie żelaza, *Rudy i metale nieżelazne*, vol. 63:7, pp 12-17.

[8] Li J F, Chen B, Tang H, Zhang S and Li C, 2017, Tribological behavior of Cu- based composites with $NbSe_2$ and TiB_2, *Digest Journal of Nanomaterials and Biostructures,* **vol.43**, pp 1205-14.

[9] Taha M A and Zawrah M F, 2017, Effect of nano ZrO_2 on strengthening and electrical properties of Cu-matrix nanocomposites prepared by mechanical alloying, *Ceramics International,* **vol.43**, pp 12698-704.

[10] Samal C P, Parihar J S and Chaira D, 2013, The effect of milling and sintering techniques on mechanical properties of Cu-graphite metal matrix composite prepared by powder metallurgy route, *Journal of Alloys and Compounds*, **vol. 569**, pp 95-101.

[11] Li J F, Chen B, Tang H, Zhang S and Li C, 2017, Tribological behavior of Cu- based composites with $NbSe_2$ and TiB_2, *Digest Journal of Nanomaterials and Biostructures,* **vol.43**, pp 1205-14.

[12] Taha M A and Zawrah M F, 2017, Effect of nano ZrO_2 on strengthening and electrical properties of Cu-matrix nanocomposites prepared by mechanical alloying, *Ceramics International,* **vol.43**, pp 12698-704.

[13] Samal C P, Parihar J S and Chaira D, 2013, The effect of milling and sintering techniques on mechanical properties of Cu-graphite metal matrix composite prepared by powder metallurgy route, *Journal of Alloys and Compounds*, **vol. 569**, pp 95-101.

[14] Konieczny M and Mola R, 2007, Sintered copper matrix composites containing aluminium-ferric intermetallic phases, *Composites Theory and Practice*, **vol.7:2**, pp. 109-113.

[15] Konieczny M, 2013, Mechanical properties of Ti-(Al_3Ti+Al) and Ti-Al_3Ti laminated composites, *Composites Theory and Practice*, **vol.13**, pp 102-106.

[16] Karakulak E, 2017, Characterization of Cu-Ti powder metallurgical materials, *International Journal of Minerals, Metallurgy and Materials,* **vol.24:1**, pp 83-90.

[17] Martinez C, Briones F, Rojas P, Aguyilar C, Guzman D and Ordonez S, 2017, Microstructural and mechanical characterization of copper, nickel and Cu-based alloys obtained by mechanical alloying and hot pressing, *Materials Letters*, **vol.209**, pp 509-512.

[18] Zhou D, Wang X, Muransky O, Wang X, Xie Y, Yang C, Zhang D, 2018, Heterogeneous microstructure of an Al_2O_3 dispersion strengthened Cu by spark plasma sintering and extrusion and its effect on tensile properties and electrical conductivity, *Materials Science and Engineering A*, **vol.730**, pp 328-335.

[19] Oanh N T H, Viet N H, Kim J S, and Junior A M J, 2017, Characterization of in-situ Cu-TiH_2 and Cu-Ti-C nanocomposites produced by mechanical milling and spark plasma sintering, *Metals*, **vol.7:4**, pp 117.

[20] Borowiecka-Jamrozek J, 2013, Engineering structure and properties of materials used as a matrix in diamond impregnated tools, *Archives of Metallurgy and Materials*, **vol. 58:1**, pp 5-8.

[21] Borowiecka-Jamrozek J and Konstanty J, 2014, Microstructure and Mechanical Properties a New Iron-base Material Used for the Fabrication of Sintered Diamond Tools, *Advanced Materials Research*, **vol. 1052**, pp 520-523, doi:10.4028/www.scientific.net/AMR.1052.520.

Characterization of the Polylactic acid stretched uniaxial and annealed by Raman spectrometry and Differential scanning calorimetry

A Bouamer[1, 2], N Benrekaa[2], A Younes[1] and H Amar[1]

[1]Research Center in Industrial Technologies (CRTI), Algiers, Algeria

[2]Laboratories of Physics of Materials, Faculty of Physics, Faculty of Physics, USTHB, Algiers, Algeria

amirouchebouamer@gmail.com; nbenrekaa@gmail.com; younesabdo11@gmail.com; amarhichem16@yahoo.fr

Abstract. In this work, we have been interested in the characterization of the effect of heat treatment and mechanical treatment on the crystallinity of a polylactic acid (PLA) film by two techniques, DSC and Raman spectroscopy. The results obtained by the DSC for the stretched film shows the appearance of a broad peak of crystallization around 120 °C, a rise in melting peak in a significant way, which shows that the uniaxial stretching has increased the crystallinity of the PLA, whereas for the annealed film appearance of a double melting peak. The result obtained by Raman spectroscopy show a new peaks appears at 922 cm^{-1} and 540 cm^{-1} after stretching and annealed process, indicating crystallization process.

1 Introduction

The characterization of materials by the Raman technique is based on the measurement of the vibrational states of the molecules or crystals. The use of this technique for the characterization of polymers is of great interest, given the amount of information can be collected such as the chemical composition of the macromolecules of the polymer including the identification of basic unit, the chemical structure, the interactions intra- or intermolecular, chain conformation, crystallinity and orientation of macromolecules. This technique is non-destructive, fast and will require minimal preparation of the sample [1, 2]. On the other hand the Differential scanning calorimeters (DSC) becomes a standard tool in polymer science. This technique takes a particular place among other methods. Additionally to its universal and simple process, the advantage of DSC compared to other calorimetric techniques lies in the broad dynamic range regarding heating and cooling rates, it is an effective analytical tool for characterizing the physical properties of a polymer. It is used to determinate crystallization, melting, enthalpy, entropy changes, and characterization of the glass transition [3].

Poly lactic acid (PLA) is analiphatic polyester, which is biodegradability and biocompatibility; it is derived mainly from agricultural biopolymer resources such as corn and sugarcane [4]. The process technology such as annealing or stretching increases the rate of crystallinity of the PLA [5]. This treatment has an improvement of the mechanical properties. The aim of this work is to study the effect of thermal and mechanical treatments on the structure of PLA and crystallinity using optical and thermal techniques.

2 Experimental

2.1 Material

The polylactic acid (PLA) used in the experimental was supplied by NatureWorks for our laboratory in film form. Figure 1 presents the chemical composition of polylactic acid (PLA) with its different chemical bonds. Chemical composition between two hooks presents the repeated unit of PLA.

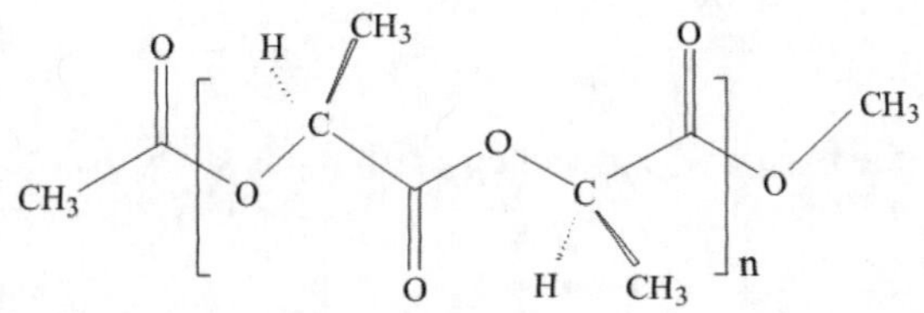

Figure 1. Chemical composition of polylactic acid (PLA).

2.2 Preparation of the annealed sample

The PLA film is cut into pieces and encapsulated in aluminium cup and saddle with a press Afterward, the sample is put in the oven a 95 °C during 4 hours.

2.3 Instrumentation

The Raman spectroscopy was made using a LabRAM HR Evolution micro-spectrometer (HORIBA). For recording spectra, we used an exciting wavelength ($\lambda = 633$ nm) that is provided by a helium-neon laser. All peaks were obtained using an x50 lens.

Differential Scanning Calorimetry (DSC) analysis gives for each sample a thermogram reflecting the variation of heat flux as a function of temperature. The device used for this analysis is a micro calorimeter DSC-7, the system is initially calibrated in temperature and heat output using high-purity Indium and Zinc sample. The baseline is determined with the capsule without sample at the same heating rate as for the analyzed samples. The weight of each sample is 8±0.5 mg. The thermal experiments were performed by heating the samples from 25 to 200°C. The value of the specific fusion enthalpy ΔHf was determined from the area under the peak of the melting curves and the reference enthalpy ΔHf° at 100% crystalline (ΔHf° = 93.1 j/g for the PLA). The degree of crystallinity is given by the following equation:

$$Cristallinity\ [\%] = \frac{\Delta H_f - \Delta H_c}{\Delta H_f^0} 100$$

2.4 Mechanical test

The PLA specimens was prepared according to the ASTM D-882, this norm is used to test thinner plastic sheets. It was carried out on specimen 0.39 mm thick and room temperature of 35°C (summer period), with relative humidity of about 50%. The tensile test was performed on an MLS machine with a load capacity of 100KN. The speed of this tensile was 10 mm/min and five specimens were tested, in order to a good reproductivity of results.

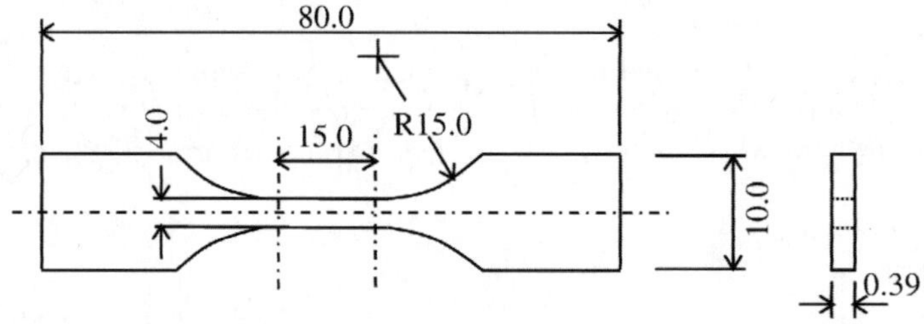

Figure 2. Geometry of the tensile test specimen used in this study according to ASTM D882.

3 Results and discussions

3.1 Raman analysis

The Raman spectra of three PLA films with different states; unstretched film, film undergoes uniaxial stretching and annealed film are shown in figure 3. The Raman active modes obtained for the three spectra are typical of those associated with PLA [6], the observed wavenumbers are given in Table 1, and assignments of vibrational wavenumbers are proposed by comparison with work on other compounds with a helical structure such as polypropylene and poly(α-L-alanine) [8;6]. The Raman bands observed at 3002 cm^{-1}, 2949 cm^{-1} were assigned to stretching asymmetric and symmetric modes of CH_3, stretching mode of CH bonds was observed at 2882 cm^{-1} [7]. The symmetric and asymmetric deformation modes of the CH_3 group which were observed at 1386 cm^{-1} and 1454cm^{-1} respectively [10]. The typical carbonyl stretching (C = O) was centred at 1766cm^{-1} [8]. In addition, the 1091 cm^{-1} and 740 cm^{-1} bands can be assigned to the COC and C = O stretching modes respectively [9-11].

Two other bands appeared around 1130 cm^{-1} and 1045 cm^{-1} are assigned to the r_s(CH3) symmetric rocking and ν C-CH_3 stretching, respectively.

The C—COO stretching of the repeated unit is responsible a very strong Raman band at 873 cm^{-1} and has been assigned to both semi-crystalline and amorphous states [10]. Another to 713 cm^{-1} which shows the out-of-plane deformation of the C = O double bond [12] .The weak bands located at 922 cm^{-1} and 540 cm^{-1} was observed only in the Raman spectra of stretched film and annealed film are attributed to the swing mode of the CH3 group and the deformation mode of C-C, and to the δ_1 C-CH_3 +δCCO respectively. The Raman bands 922 cm^{-1} and 540 cm^{-1} were associated with crystalline phase corresponding to a helical conformation (α form) by analogy with poly(α-L-alanine)[6].

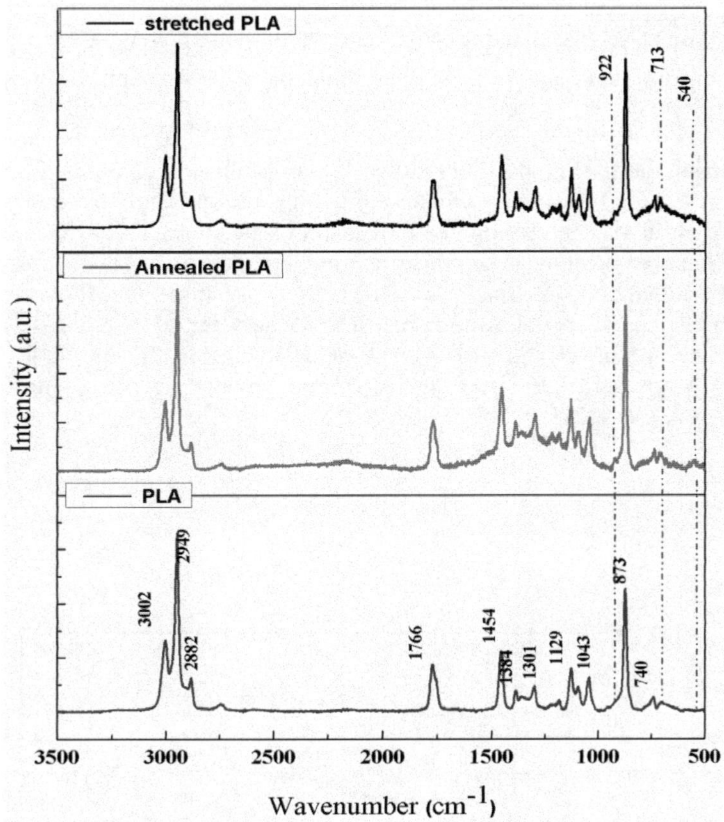

Figure 3. Raman spectra of PLA, stretched PLA and annealed PLA.

Table 1. Wavenumbers (cm^{-1}) and Vibrational Assignments of Raman Spectra of the PLA, PLA stretched and annealed PLAin the 4000–400 cm^{-1} Region

Wavenumber (cm^{-1})	Assignments
3002	ν_{as} CH_3
2949	ν_s CH_3
2882	ν CH
1766	ν (C=O)
1454	δ_{as} CH_3
1392	δ_s CH_3
1301	δ_s CH
1129	r_{as} CH_3
1043	ν C-CH_3
922	$rCH_3 + \nu CC$
873	ν C- COO
740	δ C=O
713	γ C=O
540	δ_1 C-CH_3 +δ CCO

3.2 DSC analysis

Figure 4 shows the DSC thermograms superposition of three PLA films with different states; unstretched film, film undergoes uniaxial stretching and annealed film. From the results obtained by DSC for unstretched PLA, a clear glass transition at a temperature around 64 ° C, a very low melting peak is shown at 154 °C. This result shows the PLA has an amorphous state. After stretching uniaxial, the film exhibits a broad peak of crystallization around 120 °C, and an increase melting peak in a significant way, which shows that the uniaxial stretching has increased the crystallinity. After uniaxial stretching, the movement of the chains is restricted, which can increase the glass transition temperature, which also reflects the rearrangement of macromolecular chains. Knowing that, endothermic melting peak T_f gives information on the morphology as well as information on the type of crystallization present in the polymer.

The appearance of lower and higher melting peaks at 145 °C and 154 °C respectively for annealed PLA are shown in Figure 4; that can be explained by the existence of two different crystal morphologies [13].

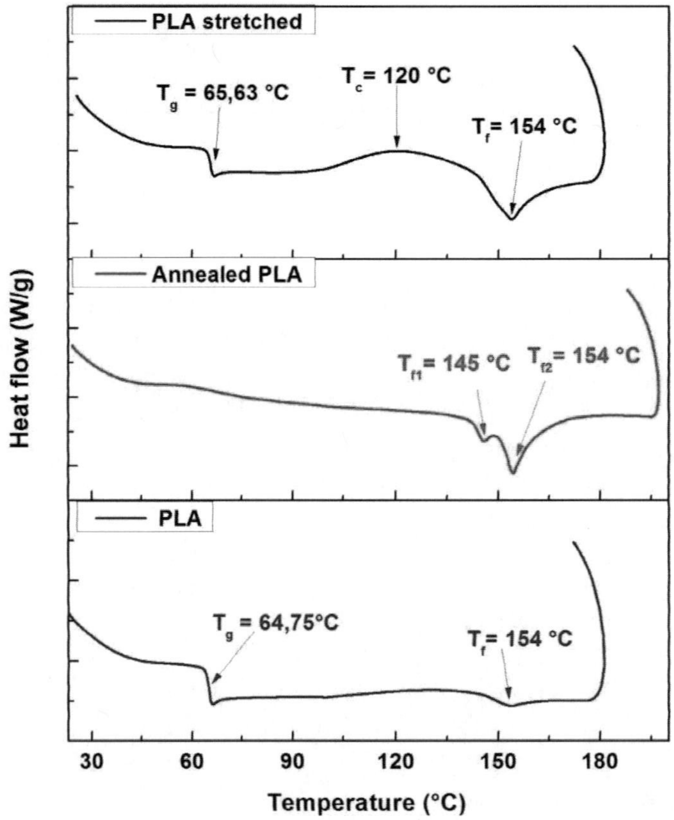

Figure 4. Superposition DSC thermograms of PLA, stretched PLA and annealed PLA.

IOP Conf. Series: Materials Science and Engineering **461** (2019) 012006 doi:10.1088/1757-899X/461/1/012006

4 Conclusion

The effect of uni-axial stretching and rubber annealed on the microstructure of PLA was studied by Raman spectroscopy, which allowed us to identify the chemical bonds of the characteristic bands, and to see the influence of uni-axial stretching and rubber annealed on the chemical bonds. Functional groups through the variation of their frequency and intensity and the appearance of new frequencies in the stretched and annealed polymer, this result is verified by scanning differential calorimetry (DSC)indicating the influence of mechanical and thermal treatment on the microstructure of PLA, change of the latter from an amorphous state to semi-crystalline state.

References

[1] Urban Marek W 1996 Attenuated Total Reflectance Spectroscopy of Polymers Theory and Practice. Polymer Surfaces and Interfaces Series Journal of the American Chemical Society**119**(45) p 11136

[2] Fontaine N H, Furtak T E 1998 Variable-angle internal-reflection Raman spectroscopy for depth-resolved vibrational characterization of polymer thin films Phys. Rev. B**57**pp 3807-3810

[3] SchickC 2009 Differential al scanning calorimetry (DSC)ofsemicrystalline polymersAnal BioanalChem**395**pp1589-1611

[4] Armentanoun I, Bitinis N, Fortunati E, Mattioli S, Rescignano N, Verdejo R, Lopez-Manchado M A, Kenny J M 2013 PLA multifunctional nanostructure materials for packaging and tissue engineering Progress in Polymer Science **38** pp1720-1747

[5] Ruijie Xu, JiayiXie and Caihong Lei 2017 Influence of melt-draw ratio on the crystalline behaviour of a polylactic acid cast film with a chi structureThe Royal Society of Chemistry **7**pp39914–39921

[6] Kister G, Cassanas G, Vert M, Pauvert B,Terol A 1995 Vibrational analysis of poly(l-lactic acid) J Raman Spectroscopy**11** pp26-307

[7] Wu Jyh-Horng, Ming-Shien Yen, Wu-Chien-Pang, Chia-Hao Li, Kuo C 2013 Effect of Biaxial Stretching on Thermal Properties, Shrinkage and mechanical properties of Poly (Lactic Acid) Films J. of Polymers and the Environment **21** pp303-311

[8] KisterG, CassanasG, GreenM1998 Effects of morphology, conformation, and configuration on IR and Raman spectra of various poly (lactic) acids Polymer**39**pp 267-273

[9] AurasR, HarteB, SelkeS 2004 An overview of polylactides as packaging materialsMacromol. Biosci. **4** pp 835-864

[10] HerreraKV, MisiunA, VogtC 2015 Preparation, and characterization of poly (lactic acid) / poly (methyl methacrylate) tablets compressed for application in quantitative analysis by Raman microspectroscopy J. Raman Spectrosc. **46** pp273-279

[11] ArrietaM P, LópezJ, LópezD, KennyJM, PeponiL 2015 Development of flexible materials based on plasticized electrostatic mixtures PLA-PHB: structural, thermal, mechanical and disintegration properties Euro. Polym. J. **73** pp 433-446

[12] SudhaMuttavarapu R 2015 Characterization of Cold Drawn Poly L Lactic Acid by Raman Spectroscopy and Raman Hyperspectral Imaging Master of Science in Chemistry for the Department of Chemistry, CLEVELAND STATE UNIVERSITY, May

[13] Radjabian M, Kish M H, Mohammadi N 2010 Characterization of Poly(lactic acid) Multifilament Yarns. I. The Structure and Thermal BehaviorJournal of Applied Polymer Science**117**pp 1516- 1525

IOP Conf. Series: Materials Science and Engineering **461** (2019) 012007 doi:10.1088/1757-899X/461/1/012007

Influence of strain rate on stress changes during Lüders bands formation and propagation

T Brlić, S Rešković, I Jandrlić, F Skender

University of Zagreb Faculty of Metallurgy, Aleja narodnih heroja 3, 44000 Sisak, Croatia

E-mail: tbrlic@simet.hr

Abstract. Characterization of the Lüders bands that occur in niobium microalloyed steel is carried out in different points of their formation and propagation. During static tensile testing infrared thermography is simultaneously used to record temperature changes and digital image correlation to determine the strain amounts in specific points of Lüders bands formation and propagation. Temperature changes which present stress changes were taken at the different strain amounts 0.005 mm/mm, 0.013 mm/mm, 0.026 mm/mm and 0.032 mm/mm. Performed static tensile tests showed the influence of different strain rates $1.8{\cdot}10^{-3}$ s^{-1}, $7{\cdot}10^{-3}$ s^{-1} and $1.85{\cdot}10^{-2}$ s^{-1} on the stress changes during Lüders bands formation and propagation. In different specific points such as Lüders band front, behind the Lüders band front (deformed area) and in front of the Lüders band front (undeformed area) tests were carried out. The studies revealed maximum stress changes behind Lüders band front and their increase with increasing strain rate. At the highest strain rate in the point of maximum stress behind the Lüders band front the greatest temperature changes was measured.

1 Introduction

Niobium microalloyed steels are low carbon steels that show the appearance of inhomogeneous deformations, i.e. Lüders bands [1,2].

It is well known that during the plastic deformation the release of heat energy is manifested as a temperature changes on the sample surface. Therefore, infrared cameras are often used to determine the temperature changes during deformation process [3,4]. In this way, it is possible to detect temperature changes and/or stress changes when Lüders band appear during cold deformation.

Thermography and digital image correlation has shown as suitable methods to detect Lüders bands in different type of steels [5] and aluminium alloys [6]. Combined these methods allowed a precise characterization of the Lüders bands and provide qualitative and quantitative analysis of Lüders bands formation and propagation at any point in the deformation zone [7].

Lüders band are clearly observed in studies on low carbon steels and niobium microalloyed steels during cold deformation at different strain rates. It was noticed that the velocity of the Lüders band propagation increase with the strain rate during cold deformation [2,8].

The contribution of this paper is determination of strain rates effect on stress changes during Lüders bands formation and propagation. Temperature changes will show stress changes in different points of Lüders bands formation and propagation at different strain rates.

2 Experimental

Samples were taken from hot rolled microalloyed strip with addition 0.6% of niobium. Tensile test is carried out on the static tensile machine EU 40mod with a nominal force value 400kN at three different strain rates $1.8\cdot10^{-3}$ s^{-1}, $7\cdot10^{-3}$ s^{-1} and $1.85\cdot10^{-2}$ s^{-1} at room temperature. The samples used had the original gauge length of 45 mm, width of 20 mm and thickness of 3 mm. Chemical composition is given in table 1.

Table 1. Chemical composition of niobium microalloyed steel [wt.%].

	C	Mn	Si	P	S	Al	Nb	N
Nb-steel	0.11	0.50	0.10	0.017	0.020	0.018	0.060	0.0083

The temperature changes on the samples were observed with an infrared camera. Infrared camera (VarioCAM M82910) with temperature sensitivity 80 mK is focused on one side of the sample surface. Uniform emissivity factor of black matte coating used for thermography was 0.95.
Strain amounts were determined with digital image correlation using digital camera Panasonic HDC-SD9 over entire deformation zone.

3 Results and discussion

Thermographic tests of niobium microalloyed steels simultaneously with static tensile tests were performed to determine temperature changes in the deformation zone during Lüders band appearance. At the same deformation amounts, the temperature changes can be observed as a stress changes in the deformation zone.

During static tensile tests proportionality limit (P) and points A, B, C, D are determined, figure 1. Up to proportionality limit (P) only elastic deformations take place in niobium microalloyed steel. By reaching proportionality limit to point D, elastic and plastic deformation occur. After point D only plastic deformation occurs in the material.

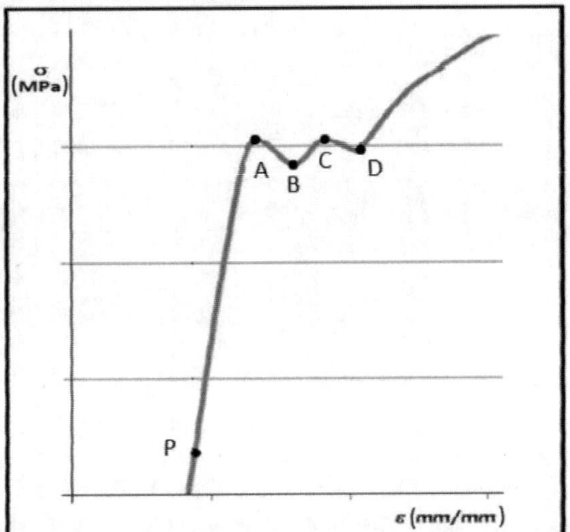

Figure 1. The area of inhomogeneous deformations during static tensile test.

Achieved strains were measured with digital image correlation in the points A, B, C, D. At the point A there was a strain of 0.005 mm/mm and at the point D strain was 0.032 mm/mm which means that the propagation of the Lüders bands through the deformation zone from point A to point D increase strain amounts, figure 1.

Temperature changes are determined with thermographic tests of niobium microalloyed steels during static tensile tests in the deformation zone. Temperature changes indicate a stress changes during deformation. Thermographic measurements have established that the Lüders band forms in the area from proportionality limit (P) to point A, figure 2. From point A to point D, Lüders band propagates through the deformation zone. Figure 1 shows the occurrence of inhomogeneous deformations in this area.

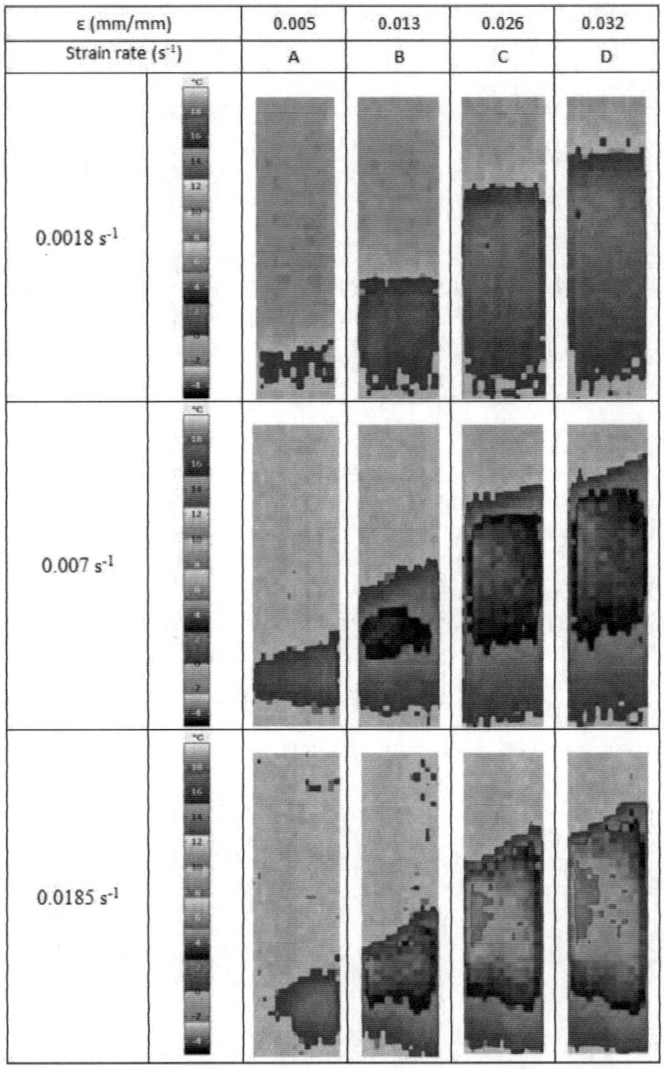

Figure 2. Lüders band propagation at various strain rates and strain amounts.

Tests were carried out at three different stretching rates in which strain rates of $1.8 \cdot 10^{-3}$ s^{-1}, $7 \cdot 10^{-3}$ s^{-1} and $1.85 \cdot 10^{-2}$ s^{-1} were achieved. The appearance of the Lüders band was observed at each strain rate. At all strain rates Lüders band is formed and propagates through the deformation zone.

Quantitative thermographic analysis during Lüders band propagation through the deformation zone showed stress changes at points A, B, C and D. The increase of the stress with Lüders band propagation from one end of the sample to the other end of the sample through the deformation zone is observed by increasing strain amounts at the same strain rate. It is clear visible higher stress changes behind Lüders front at higher strain rates, figure 2.

In order to determine the influence of strain rate on the stress changes during Lüders band propagation through the deformation zone, the stress changes in the characteristic points of the deformation zone were determined. The measurements were performed in three characteristic points: in front of the Lüders band front (R_1), on the Lüders band front (R_2) and in the area of maximum temperature changes or stress changes behind the Lüders band front (R_3), figure 3.

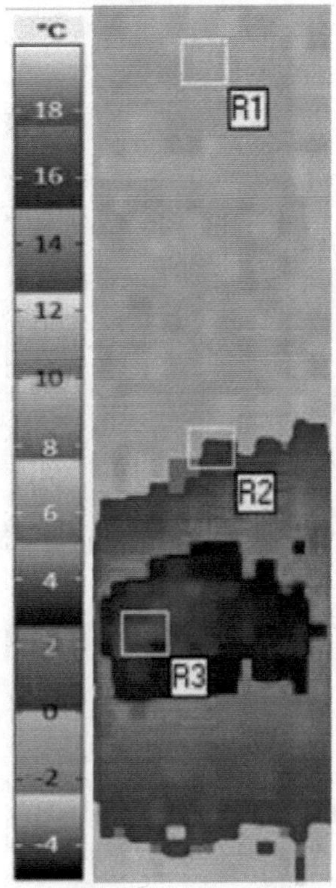

Figure 3. Thermographic measurement in characteristic points during Lüders band propagation.

In the area in front of the Lüders band front, there are no stress changes (R_1). There are also no stress changes in the area in front of the Lüders band front (R_1) by increasing strain rate. On the Lüders band front (R_2) was not observed stress changes during Lüders band propagation. There are certain changes with the change of strain rate. The greatest stress changes were found behind the Lüders band front (R_3). These changes are more pronounced at higher strain rates. At the lowest strain rate (0.0018 s^{-1}), behind the Lüders band front (R_3), no stress changes were observed during Lüders band propagation. The increase of the strain rates (0.007 s^{-1} and 0.0185 s^{-1}) showed an increase of stress changes during Lüders band propagation, figure 2. The greatest stress changes behind the Lüders band front (R_3) were obtained at the highest strain rate (0.0185 s^{-1}). The obtained maximum stress changes with the strain rate changes behind the Lüders band front are shown in figure 4.

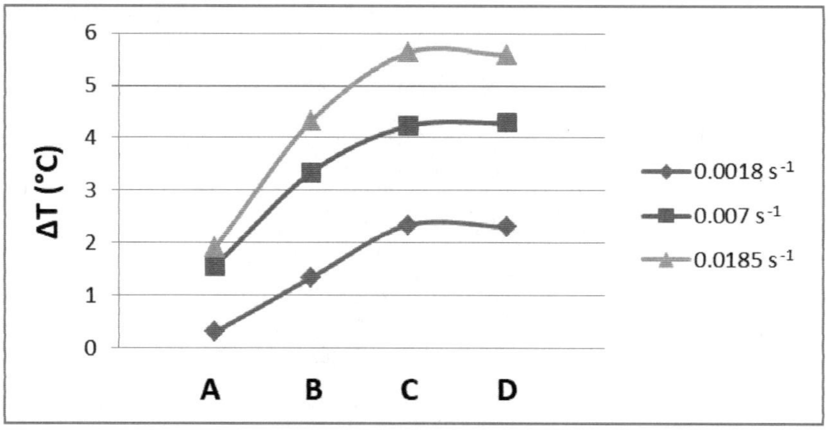

Figure 4. Maximum stress changes behind the Lüders band front with change of strain rate.

Figure 4 clearly shows that the temperature changes, i.e. stress changes behind the Lüders band front are the greatest at the highest strain rate. At the same strain rate, stress increase during Lüders band propagation through the deformation zone. This phenomenon was not previously found in niobium microalloyed steel.

Futher microstructural studies will research what is the reason of stress increase during Lüders band propagation.

4 Conclusion

The conducted tests have shown that the strain rate affects stress changes at the start of plastic flow during cold deformation in niobium microalloyed steel.

Inhomogeneous deformations occur at the start of plastic flow of material. Lüders band appearance and propagation are related with inhomogeneous deformations. It was found that the Lüders bands appear at all strain rates.

At the same strain rate, certain stress increase during Lüders band propagation through the deformation zone was determined. These stresses increase with increasing strain rate at the same strain amount.

Further research on the effect of microstructure will show the influence of microstructure homogeneity on stress changes at the start of plastic flow.

Acknowledgment

This work has been fully supported by Croatian Science Foundation under the project IP-2016-06-1270.

References

[1] Brlić T, Rešković S, Vodopivec F and Jandrlić I 2018 Lüders bands at the beginning of the plastic flow of materials *Metalurgija* **57** 357-359

[2] Rešković S, Jandrlić I and Vodopivec F 2016 Influence of testing rate on Lüders band propagation in niobium microalloyed steel *Metalurgija* **55** 157-160

[3] San Juan M, Martín O, Santos F J, De Tiedra P, Daroca F and López R 2013 Application of thermography to analyse the influence of the deformation speed in the forming process *Procedia Engineer.* **63** 821-828

[4] Kutin M M, Ristić S S, Prvulović M R, Prokolab M M, Marković N M and Radosavljević M R 2011 Application of thermography during tensile testing of butt welded joints *FME Trans.* **39** 133-138

[5] Nagarajan S, Raghu N and Venkatraman B 2013 An insight into Lüders deformation using advanced imaging techniques *J. Mater. Eng. Perform.* **22** 3085-92

[6] Nogueira De Codes R, Hopperstad O S, Engler O, Lademo O G, Embury J D and Benallal A 2011 Spatial and temporal characteristics of propagating deformation bands in AA5182 alloy at room temperature *Metall. Mater. Trans. A* **42A** 3358-69

[7] Nagarajan S, Raghu N and Venkatraman B 2013 Advanced imaging for early prediction and characterization of zone of Lüders band nucleation associated with pre-yield microstrain *Mat. Sci. Eng. A-Struct.* **561** 203-211

[8] Sun H B, Yoshida F, Ohmori M and Ma X 2003 Effect of strain rate on Lüders bands propagating velocity and Lüders strain for annealed mild steel under uniaxial tension *Mater. Lett.* **57** 4535-39

IOP Conf. Series: Materials Science and Engineering **461** (2019) 012008 doi:10.1088/1757-899X/461/1/012008

Metallurgical aspects involved in Selective Laser Melting

B Buchmayr and G.Panzl

Montanuniversitaet Leoben, Chair of Metal Forming, A-8700 Leoben, Austria

EMAIL: bruno.buchmayr@unileoben.ac.at

Abstract. Additive manufacturing using Selective Laser Melting (SLM) offers several advantages for industrial applications and has motivated many scientist and industrial R&D colleagues to add more and more innovations. Although the principles are well known, only a small number of materials have found a significant usage, others have shown serious problems during fabrication. In the paper, the influencing process parameters, the interactions of important mechanisms and the microstructural based assessment of laser weldability as well as the appearance of defect formations are considered in detail in order to understand the physical limits and to highlight the peculiarities of the SLM-process. Many problems can be solved using the lessons learned in welding technology. On one side, the high cooling rate and strong undercooling of SLM-process is quite positive to fabricate dissimilar joints and on the other side, residual stresses and distortion may be a challenge.

1 Introduction

Additive manufacturing has found a broad interest in the manufacturing community as a disruptive technology. Pros and Cons are now well understood in terms of economy and applicability. However, the fundamental understanding of the most influencing parameters and the microstructural phenomena involved are not so well-known. Based on an assessment of laser weldability, problems and typical failures using different alloying systems are mentioned, which provides good criteria for material selection. Residual stresses and distortion are critical issues for further development of the SLM technology. New alloys and joining of hybrids are seen as a big challenge for the future.

The advantages of the SLM-process comprise: freedom of design, enhanced complexity, direct/ tool-free manufacturing, lightweight potential via topology optimization, reduction of manufacturing steps, increase of functionality and the cons are: low production speed, high manufacturing costs, time-intensive trials for appropriate designs and processing parameters, limited part dimensions and high residual stresses and/or distortion. Further information can be found in the literature [1-7].

2 Influencing process parameters

The influencing process parameters can be divided into four groups [8]: a) laser related parameters (power, spot size, pulse duration and pulse frequency), b) powder related parameters (particle size distribution, shape, powder bed density, layer thickness, material properties), c) scan related parameters (scan speed, scan spacing, scan pattern) and d) temperature related parameters (powder bed temperature, powder feed temperature, temperature uniformity). All these parameters have influence on weld thermal cycle (weld pool size, cooling rate, as-welded microstructure), dendrite arm spacing, mechanical properties, residual stresses and distortion as well as on other service properties like corrosion resistance.

In principle, the laser welding process can be analyzed by the consideration of the weld thermal cycle, the microstructural changes in the weld pool area and in the heat affected zone as well as the post-weld heat treatment and their consequences on the mechanical properties.
A characteristic parameter is the heat input, which is defined as the quotient of laser power divided by scan speed multiplied by layer thickness. This parameter is typical for the weld thermal cycle with cooling rates between 10.000 to 100.000 °C/s or a cooling time $t_{8/5}$ of some microseconds. This leads to a strong undercooling (very high temperature gradient and very high solidification rate) which results in a very fine microstructure in the as-welded state. There is no time for any substantial diffusion process, i.e. there is no segregation effect and a very homogeneous structure distribution with excellent mechanical properties.
An excellent overview of the main characteristic phenomena involved and their influencing parameters, interactions and resulting effects is shown in figure 1 taken from ref [9].

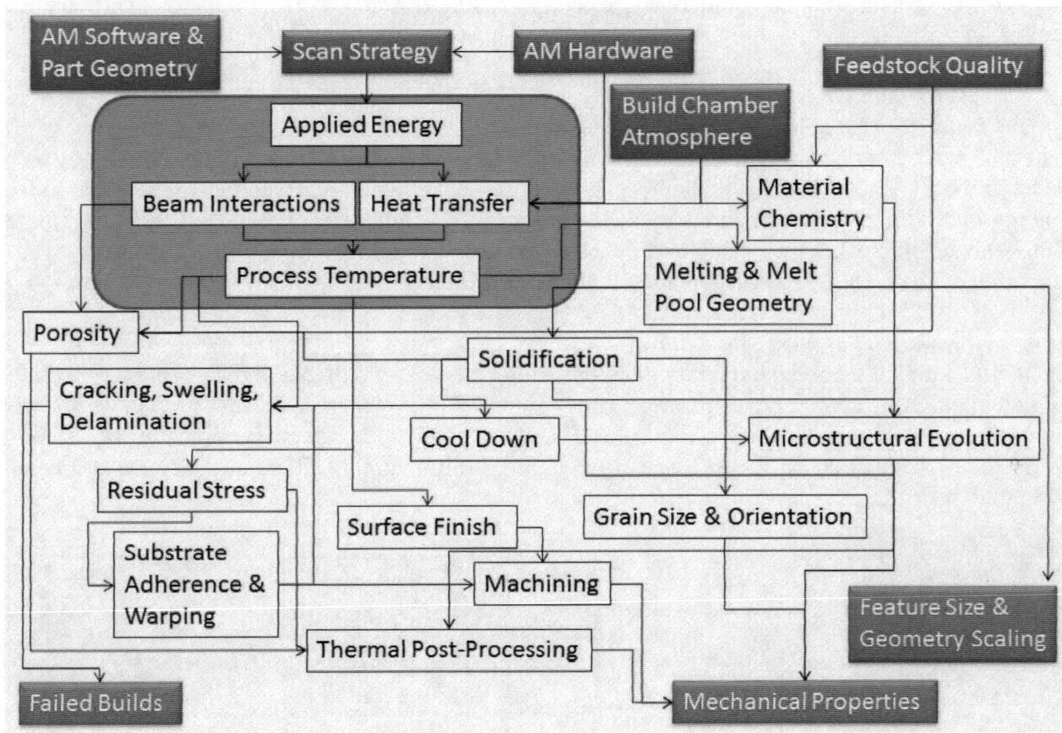

Figure 1. Processing map for selective laser melting [9].

Optimized SLM-parameters are found by trial and error by variation of laser power and scan speed. In cases of very high heat input, there will be large distortion. Low laser power and high speed will result in higher porosity or even insufficient fusion.

3 SLM printability or SLM-weldability

At the moment only a few metal powder types are commercially of interest. They cover the most important application areas. Typical examples are: Maraging steels, stainless steels incl. PH-steels, nickel-base alloys, Ti-alloys, CoCr-alloys and AlSi-alloys. The main question arises, why some alloys show a better processing performance than others?
From a metallurgical point of view the following main problems or defects can be found:

- Crack formation (i.e. cold cracking) for steels with high C-content

- Hot cracking
- Lack of fusion
- Distortion and residual stresses
- Metastable formation of brittle phases
- Evaporation effects
- and other more.

3.1 Main factors for weldability of low alloy steels

The main problem for this kind of materials is based on the rapid cooling of the weld pool, which leads to 100% martensitic microstructure when the carbon content is higher than about 0,1%. The most influencing parameters in this case are the carbon equivalent, the cooling time $t_{8/5}$, the preheating temperature, component thickness and the mechanical constraints. Therefore, only steel grades with a very low C-content can be welded successfully. Maraging steels and stainless steels are good examples for this fact.

3.2 Weldability of Al-alloys

A main factor for the weldability of Al-alloys is the high hydrogen solubility in the melt with a big change of solubility during solidification in the solid state, which leads to very high porosity. A second aspect is the likelihood for hot-cracking mainly caused by a large solidification range. Near eutectic systems like AlSi10Mg are favored because of the low melting temperature and the very fine and homogeneous microstructure with excellent mechanical properties. In addition, using PWHT by tempering at about 180°C can also increase the strength value.

3.3 *Weldability of nickel-base alloys*

The main factor for a good weldability of Ni-base alloys is the sum of Al- and Ti-content, which form the precipitation hardening phase gamma prime. Standard alloy for many cases is Inconel 718, which shows excellent laser weldability. Figure 2 shows a typical microstructure of an IN718 as welded SLM-microstructure and on the left side a processing limit diagram (alloys in the shadowed area are difficult to weld).

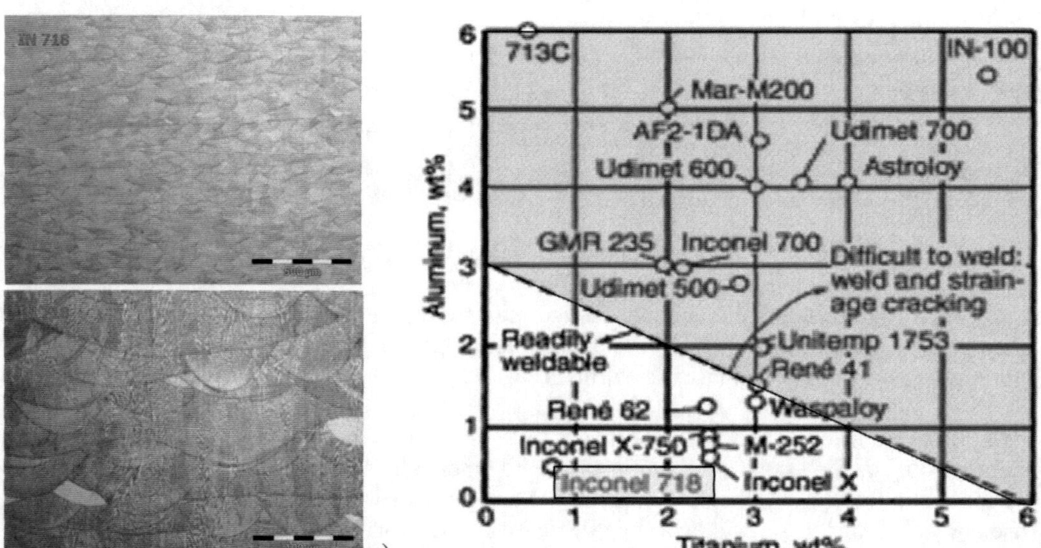

a) b)

Figure 2. a) As-welded microstructure of IN718 and b) limits of weldability of Ni-base alloys defined by their Al and Ti-content.

3.4 Post weld treatments
There are many types of PWTs applied after welding. Quite common are heat treatments for stress relieving or to reorganize the microstructure by a full heat treatment. Al-alloys may be tempered directly after the SLM-process to get additional strengthening effects by fine precipitates.

HIP (hot isostatic pressing) treatment is done for high performance components to close the remaining pores and to improve impact and fatigue properties. Surface treatment (mainly by peening or also by electro-chemical processing) is performed to reduce the surface roughness.

3.5 Quality control of SLM-processing
Applications in the areas of medicine or aerospace dictate very tough QC requirements. A direct process control is quite difficult to execute because of the closed working platform with vacuum or shield gas and the thousands of layers and the immense volume of process data. Nevertheless, laser signals and optical sensor data or acoustic emission data are recorded and analyzed by fast numerical methods.

After SLM-fabrication, the components can also be tested by conventional NDT-methods or by computer tomography. Further information on this important issue are given in [10-12].

3.6 Residual stresses and distortion
A very severe problem for SLM-use is given by the formation of high residual stresses in case of rigid construction or high distortion in case of flexible or thin designs. Heat transfer and the development of high thermal strains are responsible for high stresses during the building process, which can also lead to cracking during the welding process. These delamination cracks are visible to the naked eye and may also lead to interruptions of the process, because of a collision with the doctor blade.

The main influencing factors for the stress development are the temperature gradient, the volumetric change, the heat input per unit length, the Fourier number, the rigidity of the component and the deposition time. An analytical expression for the thermal strain was defined by Mukherjee et.al. [13].

The complexity of this issue requires very sophisticated numerical FE-models. The authors use two software systems, i.e. Amphyon (Additive Works, Bremen) and simufact additive (simufact, Hamburg). An example of the stress development during the SLM-process and removal from the building platform is shown in figure 3.

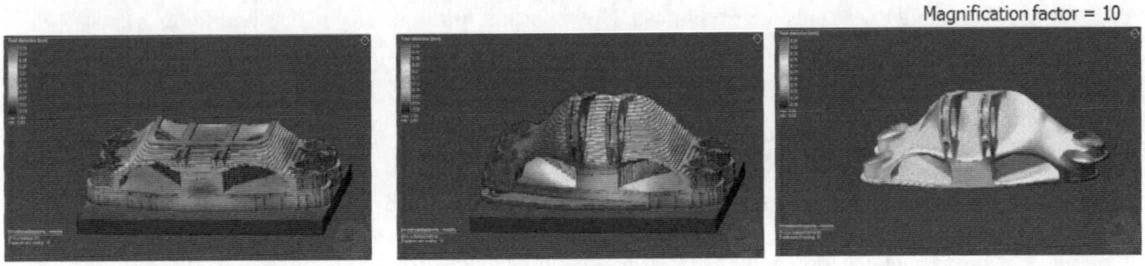

Figure 3. FEM prediction of residual stresses and distortion during building of a bracket (software used: simufact additive).

4 About the role of metal powders on the quality of SLM-products
Metal powders for the SLM-process are usually gas-atomized to achieve spherical particles with a good flow-performance. Beside the aforementioned process- and materials-related aspects, the powder quality and proper size distribution are quite essential for the properties of the final product. Based on a recent study, the authors have found quite significant differences in the powder characteristics and

the consequent component behavior. Some of the important parameters have been considered in the standard VDI 3405, which provides hints for the characterization and testing of metallic powders for additive manufacturing.

Beside chemical composition, particle size distribution, flowability testing, density, humidity and absorption rate are considered in detail. Using laser diffraction analysis, a complete size distribution curve as shown in figure 4 can be measured in a short time.

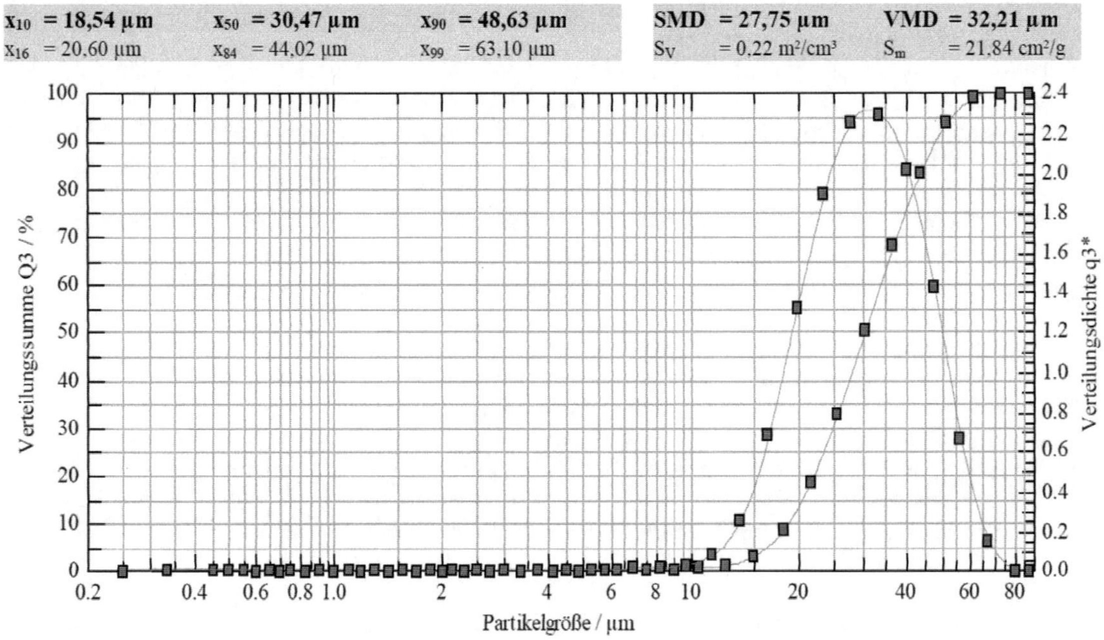

Figure 4. Measured particle size distribution of HX-alloy using a HELOS-system.

In general, small size distributions show a better flowability, a higher accuracy of geometrical reproduction and a less surface roughness. However, the productivity is less and the oxygen absorption is higher compared with powders showing a broader distribution. Coarse particles lead to higher porosity and as a consequence to lower strength values [14].

5 Conclusions

The final properties of products made by selective laser melting depend very strongly on the processing conditions/parameters, on the weldability of the considered alloys and on the properties/ characteristics of the metallic powders. Many fundamental aspects can be taken from former findings of weldability of various alloys, which need only be modified according to the special conditions of the SLM-process. The guideline VDI 3405 Bl.2.3 provides a first step for a better and unified characterization of metallic powders. Same alloys produced on different machines or same alloys with powders from different suppliers may result in significant processing performance and final product properties.

Acknowledgements
The work was done within the national Austrian leader project "addmanu". We acknowledge the financial support provided by the Austrian research fonds (FFG) for this work.

References

[1] Gebhartdt A 2013 *Generative Fertigungsverfahren* 4th ed. (Munich: Hanser)
[2] Berger U, Hartmann A and Schmid D (2013) *Additive Fertigungsverfahren* (Haan-Gruiten: Europa Lehrmittel-Verlag)
[3] Wimpenny, D I et.al. (2017) *Advances in 3D Printing & Additive Manufacturing Technologies* (Singapore: Springer Verlag)
[4] EPMA (2015) *Introduction to Additive Manufacturing Technology* (Shrewbury UK: European Powder Metallurgy Association)
[5] Frazier W E (2014) *Metal Additive Manufacturing: A Review* J. Materials Engineering and Performance **23** pp 1917-1928
[6] Yap C Y et.al. (2015) *Review of selective laser melting: Materials and applications* Applied Physics Reviews 2 041101 2015 pp 1-21
[7] Berger R (2016) *Additive Manufacturing – next generation AMnx* Study by Roland Berger GmbH April 2016 84 pages
[8] Aboulkhair N T, Everitt N M, Ashcroft I and Tuck C (2014) *Reducing porosity in AlSi10Mg parts processed by selective laser melting* Additive Manufacturing 1 pp77-86
[9] Sames W J List F A Pannala S Dehoff R R and Babu S S (2016) *The metallurgy and processing science of metal additive manufacturing* Int. Materials Reviews **61** (Taylor&Francis) pp315-360
[10] Berumen S Bechmann F Lindner S Kruth J-P Craeghs T (2010) *Quality control of laser- and powder bed-based Additive Manufacturing technologies* Physics Procedia **5** (Elsevier) pp 617-622
[11] Zenzinger G Bamberg J Henkel B Hess T Ladewig A (2014) *Online-Prozesskontrolle bei der additiven Fertigung mittels Laserstrahlschmelzen* ZfP-Zeitung **140** pp51-54
[12] Krauss H (2016) *Qualitätssicherung beim Laserstrahlschmelzen durch schichtweise thermografische In-Process-Überwachung* PhD-Thesis TU Munich
[13] Mukherjee T Zuback J S De A DebRoy T (2017) *Printability of alloys for additive manufacturing* www.nature.com/scientificreports 6: 19717
[14] Spierings A B Herres N and Levy G (2011) Influence of the particle size distribution on surface quality and mechanical properties in additive manufactured stainless steel parts, The international journal for research on additive manufacturing technologies and rapid product development, 2011, USA, ISSN1355-2546, DOI 10.1108/13552541111124770, Volume 17, Issue 3, Pages 195--202

IOP Conf. Series: Materials Science and Engineering **461** (2019) 012009 doi:10.1088/1757-899X/461/1/012009

Analysis of stresses in FeSi sheets induced by laser cutting

T Bulin[1], P Roupcova[2], C Ondrusek[1] and O. Schneeweiss[2,3]

[1] Brno University of Technology, Faculty of Electrical Engineering and Communication, Technická 3058/10, 616 00 Brno, Czech Republic

[2] Institute of Physics of Materials ASCR, Žižkova 22, 616 62 Brno, Czech Republic

[3] CEITEC IPM, Institute of Physics of Materials ASCR, Žižkova 22, 616 62 Brno, Czech Republic

E-mail: xbulin01@stud.feec.vutbr.cz

Abstract. The stresses induced by laser cutting in the non-oriented fully processed electrical FeSi sheets were investigated using X-ray diffraction, Mössbauer spectroscopy, and magnetic measurements. The magnetic measurements were used for studies of changes in magnetic properties due to changes in stresses. The original laser cut samples were compared with those after heat treatment. The magnetic measurements appeared as the most sensitive method among those selected for the present study. The influence of stresses has an impact on the shape of the hysteresis loops and the permeability in dependence on the excitation magnetic field. The selected heat treatment caused decrease of stresses and improvement of magnetic properties of the sheets.

1 Introduction

The knowledge of the stresses in material is very important. Currently, the design of conventional types of electric machines was improved so much that the greatest potential for increasing efficiency is the use of better material or improve processing of existing material [1]. During the processing of silicon steel sheets for electrical machines, the stresses are transmitted into these sheets which affect the resulting properties of the material [2]. The stress may be removed by heat treatment [3-5]. It is appropriate to know the number of stresses resulting from processing. This method can then help determine given the size of stresses whether it is worth to anneal this material.

2 Experimental

2.1 Material and samples

The tested material was non-oriented fully processed electrical steel M330-35A with thickness of sheets 0.35 mm and insulation coating on the surface. It is according [6] an alloy of Fe - 2.884 wt.% Si. Tested samples were cut using laser into shape of toroid. This method of production causes stresses, which are induced to the sheets because of the high-temperature gradient. The samples after laser cutting have almost immediately room temperature. Besides the original state of the sheets, a part of the cut samples was annealed at 650°C for 60 minutes in vacuum. Even though the sample was annealed in vacuum, the insulating layer on the surface of sheets changed color from silver to brown. Insulation was partially damaged but still worked according to the measurements with multimeter Agilent U1252B.

Geometrical dimensions of toroidal samples are shown in **Table 1.** Sample 1 is the original sample prepared by laser cutting. 23 ring sheets (**Figure 1.**) were stacked together by PVC insulation tape and then wound by Cu wire making primary and secondary winding **(Figure 2.)**

Table 1. Parameters of toroidal samples for magnetic measurements

Sample	1	2a	
Outer diameter	100	100	mm
Inner diameter	84	84	mm
Thickness	8.1	8.1	mm
Mass	141.2	141.3	g
Primary windings	279	275	
Secondary windings	50	50	

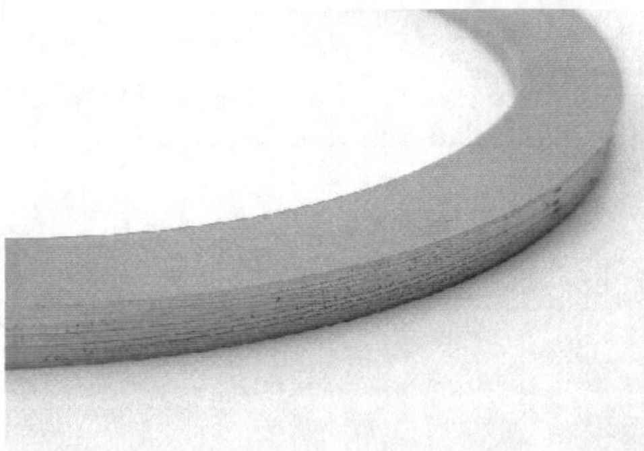

Figure 1. The shape of sample for magnetic measurement

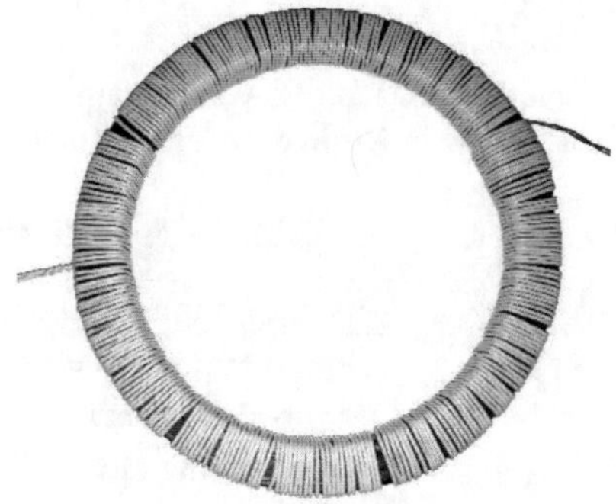

Figure 2. Toroid sample for magnetic measurement with winding

For the XRD and Mössabuer spectroscopy, the ring was cut into eight parts and stacked as shown in **Figure 3.** There is also possible to see the color of the annealed sample.

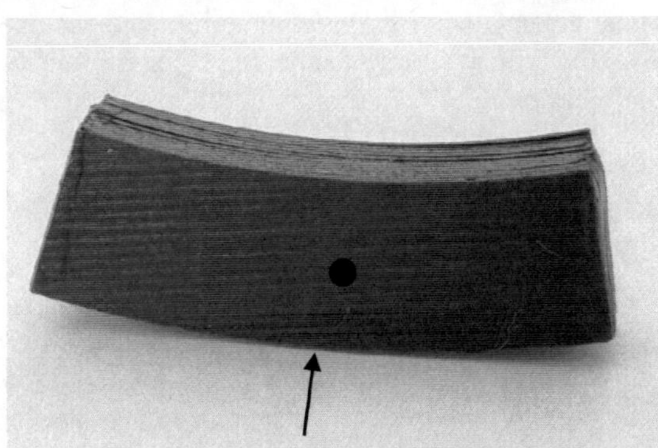

Figure 3. The shape of sample for XRD and MS

2.2 Experimental methods.

The stress and texture measurement were obtained by mean X-ray powder diffractometer *Empyrean* (PanAnalytical, Nederland) and *Stress* program using $sin^2\psi$ method. The value Youngs modulus $E = 30GPa$ and Poissons ratio $v = 0.29$ was used for calculation. Equipment was set up in parallel beam

mode with 0.5 mm capillary and 2-D detector works in 0-D mode for stress measurement and 2-D mode for texture measurement. The side inclination stress measurements were performed in two directions φ and positive and pseudo-negative tilt up to $sin^2\psi = 0.9$. Penetration dept is influenced by tilt is approximately 15μm.
^{57}Fe Mössbauer spectroscopy was applied in scattering geometry with detection of 14.4 keV gamma. The spectra were measured at room temperature and calibrated against α-Fe. The computer processing of the spectra was done using CONFIT program package [7]. This yielded the relative values of the spectrum components and values of the hyperfine parameters [8].
Magnetic measurements in this article are based on Faraday's law of electromagnetic induction [9,10] and the hysteresis graph is used for the measurements. It is called RemaGRAPH – RemaCOMP C-710 and it is made by company Magnet-Physik Dr. Steingroever. It is possible to make measurements based on quasistatic or dynamic hysteresis curves (different saturation fields and different frequencies). Hysteresis graph records the magnitude of the magnetic flux density B, magnetic field strength H, relative permeability, total magnetic loss Ps, shape of the hysteresis curve and other parameters.

3 Results and discussion

The stress and texture results were obtained from the plane surface (see the dot in **Figure 3)** and outer ring surface (arrow in **Figure 3**). The values of compressive stresses are shown in **Table 2**.

Table 2. Residual stress measured on (1 1 2) plane.

Place of measurement	Laser cut	Annealed	
Plane surface φ = 0°	45±11	85±9	MPa
Plane surface φ = 90°	50±10	66±9	MPa
Outer ring φ = 0°	37±19	18±15	MPa
Outer ring φ = 90°	24±18	17±15	MPa

The values of stress are low even in laser cut sample which confirmed sensibility of laser cutting for low-stress application. Unfortunately the preparation of the silicon-steel sheets causing preferred orientation of crystals in plates and increased the uncertainty of stress measurement as well. The stress was slightly increasing in the plane surface and decreasing in the outer ring surface by annealing sample. We are expecting more significant changes in magnetic behavior. The poles figures from plane surface, see example in **Figure 4**, shows the typical texture in sheets [11] and indicate that the annealing do not changes character of texture in the sample.

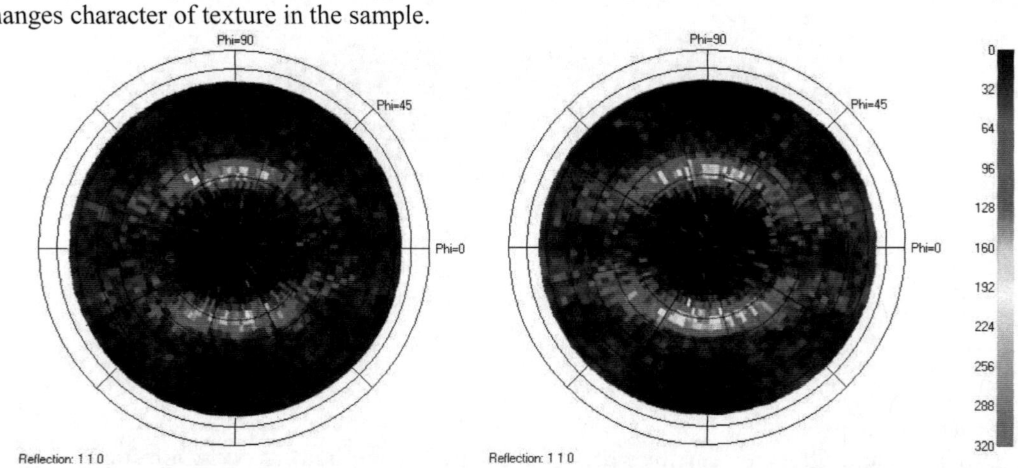

Figure 4. Pole figures of reflection (1 1 0) samples after laser cutting and after annealing at 650°C for 60 minutes in vacuum.

Mössbauer spectra are drawn in **Figure 5.** Based on the fitting parameters two main results can be derived. (i) The partial sextets (drawn by doted lines in **Figure 5.**) exhibit very similar intensities and hyperfine parameters for both samples. It indicates that the annealing does not change in chemical composition in the surface layer scanned by Mössbauer spectra (approximately 30 micrometers in depths). (ii) There are slight changes in the intensity ratio of the first and second lines (R_{12}) in the sextets (in **Figure 5.** the lines filled by forward slash and crosshatch, respective). For the sample after laser cutting $R_{12} = 0.86 \pm 0.01$ and after the annealing $R_{12} = 0.74 \pm 0.01$. It reflects fine change in the mutual directions of gamma radiation from the Mössbauer source and orientation of magnetic moments in the sample surface [8]. The magnetic moments in the sample after laser cutting are oriented closer to the sample plane.

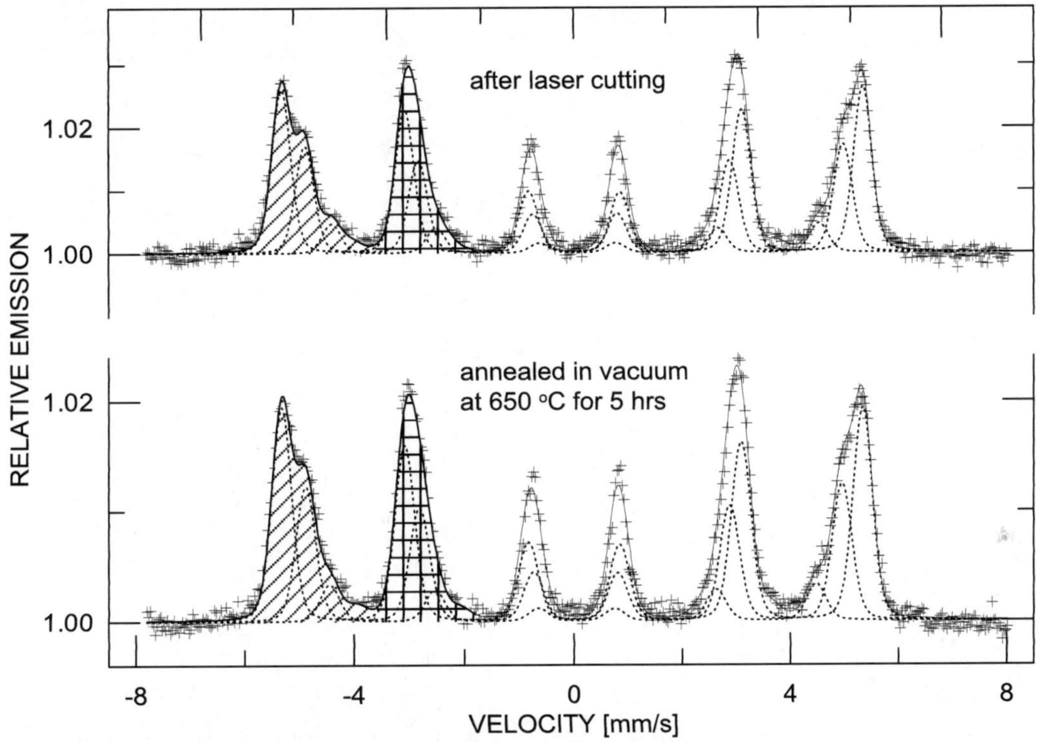

Figure 5. Mössbauer spectra of the samples after laser cutting and after annealing at 650°C for 60 minutes in vacuum. The experimental points are labeled by crosses, the solid line corresponds to the sum of the fitted components, the doted lines to partial sextets. The filled parts of the spectra represent first (forward slash) and second (crosshatch) lines in the sextets.

Results from the magnetic measurements proved that magnetic properties are significantly deteriorated after laser cutting. After annealing, all parameters are improved and returned to the expected state. Example of measured parameters is shown in **Table 3**. Other magnetic results can be seen in **Figures 6. – 9**. Sample 2-a shows better desired magnetic properties. The number of magnetic losses after annealing is decreased even if the surface insulation is damaged in some places. The biggest difference in the graphs is in the shapes of hysteresis curves **(Figure 8.)** and values of relative permeabilities (**Figure 9.**). The hysteresis curves are deformed (Sample 1) after laser cutting and after annealing (Sample 2-a) it returns to the expected shape. The same situation occurred with the permeability, which was increased again after the stress was removed.

Table 3. Magnetic properties at 50Hz

Sample	$1^{0.8T}$	$1^{1.5T}$	$2a^{0.8T}$	$2a^{1.5T}$	
B_{max}	0.801	1.498	0.803	1.498	T
H_{max}	330	2376	129	1907	A/m
μ_r	1929	502	5155	626	
Ps	1.62	4.22	1.06	3.43	W/kg
B_r	0.347	0.407	0.569	0.935	T
H_c	98.7	121	59.3	76.1	A/m

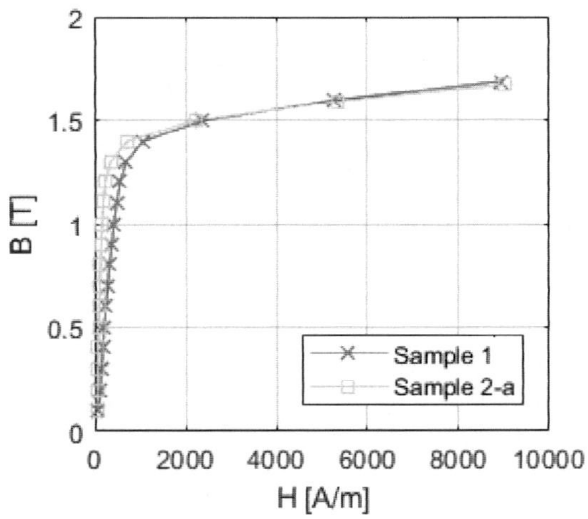

Figure 6. Commutation curves at 50Hz

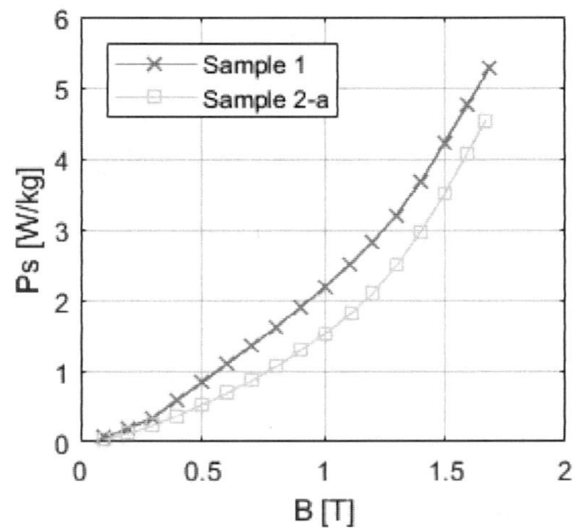

Figure 7. Total magnetic losses at 50 Hz

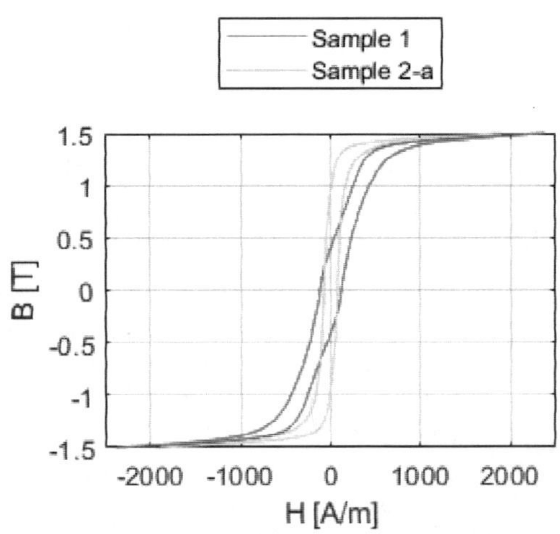

Figure 8. Hysteresis curves at 50Hz

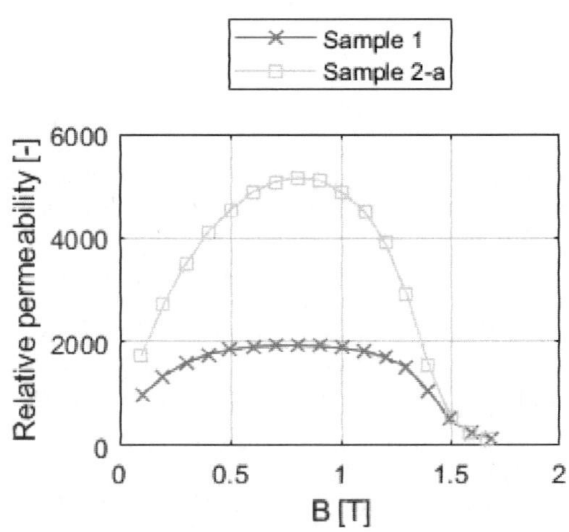

Figure 9. Relative permeability at 50Hz

4 Conclusions

Although the laser cutting method has the advantage of cheaper prototype production, it has also the worst effect on the processed material. Magnetic measurements proved that induced stresses from heat gradient significantly affect hysteresis curves and other magnetic parameters. All tested parameters have returned to the expected values after annealing. The biggest difference was measured at magnetic flux density 0.8 T and it was relative permeability, which has value 1929 for the original sample and 5155 for the annealed sample. According XRD the laser cutting method is quite sensible for this type of application. The stress changes caused by annealing do not change magnetic properties negatively. Mössbauer spectra showed that the annealing did not change the chemical composition in the surface layer and the best results are from magnetic measurements.

ACKNOWLEDGEMENTS

This research work has been carried out in the Centre for Research and Utilization of Renewable Energy (CVVOZE). Authors gratefully acknowledge financial support from the Ministry of Education, Youth and Sports of the Czech Republic under NPU I programme (project No. LO1210). They also acknowledge partner collaboration in frame of project of Technology Agency of the Czech Republic No. TE02000232.

References

[1] D. Gerada, A. Mebarki, N. L. Brown, C. Gerada, A. Cavagnino, and A. Boglietti 2014 High-speed electrical machines: Technologies, trends, and developments *IEEE Trans. Ind. Electron.* vol 61 no. 6 pp 2946–2959

[2] T. Bulín, E. Švábenská, M. Hapla, C. Ondrůšek, and O. Schneeweiss 2017 Magnetic properties and structure of non-oriented electrical steel sheets after different shape processing *Acta Phys. Pol. A* vol 131 no. 4

[3] G. Paltanea, V. Manescu, H. Gavrila, A. Nicolaide, and B. Dumitrescu 2016 Comparison between magnetic industrial frequency properties of non-oriented FeSi alloys, cut by mechanical and water jet technologies *Rev. Roum. des Sci. Tech. Ser. Electrotech. Energ.* vol 61 no. 1 pp 26–31

[4] W. Shi, J. Liu, and C. Li 2014 Effect of cutting techniques on the structure and magnetic properties of a high-grade non-oriented electrical steel *J. Wuhan Univ. Technol. Sci. Ed.* vol 29 no. 6 pp 1246–1251

[5] Pyrhönen J, Jokinen T and Hrabovcova V 2008 Design of rotating electrical machines John Wiley & Sons ltd. pp 183–218

[6] C. Schayes, J. B. Vogt, J. Bouquerel, F. Palleschi, and S. Zaefferer 2016 Cyclic plasticity mechanism of the M330-35A steel *Int. J. Fatigue* vol 82 pp 530–539

[7] Zak T and Jiraskova Y 2006 Surface and Interface Analysis vol 38 no. 4 pp 710–14

[8] Gonser U, Ed. 1975 Mössbauer Spectroscopy vol 1 (Berlin: Springer)

[9] Fiorillo F 2010 Measurements of magnetic materials *Metrologia* vol 47 no. 2 pp S114–S142

[10] IEC 2003 Methods of measurement of the magnetic properties of magnetically soft metallic and powder materials at frequencies in the range 20 Hz to 200 kHz by use of ring specimens *IEC Standard Publication* 60404-6 (Geneva: IEC Central Office)

[11] S. Umehara, H. Inoue and J. Hamadaiorillo 2018 *IOP Conf. Ser.: Mater. Sci. Eng.* 375 012015

IOP Conf. Series: Materials Science and Engineering **461** (2019) 012010 doi:10.1088/1757-899X/461/1/012010

Evolution of deformation texture in cold drawing of steel tubes using EBSD analysis and FEM simulation

Ľ Parilák[1,2a], P Burik[1b], P Bella[1c], M Mojžiš[1,3d] and P Kejzlar[4e]

[1]ŽP Research and Development Centre, Kolkáreň 35, 976 81 Podbrezová, Slovak Republic, EU, E-mail: [a]parilak@zelpo.sk, [b]burik@zelpo.sk, [c]bella@zelpo.sk, [d]mojzis.milan@zelpo.sk

[2]Technical University of Košice, Faculty of Manufacturing Technologies with a Seat in Prešov, Bayerova 1, 080 01 Prešov, Slovak Republic, EU

[3]Slovak University of Technology in Bratislava, Faculty of Materials Science and Technology in Trnava, Paulínska 16, 917 24 Trnava, Slovak Republic, EU

[4]Technical University of Liberec, Faculty of Mechanical Engineering, Department of Material Science, Studentská 2, 461 17 Liberec, Czech Republic, EU, E-mail: [e]pavel.kejzlar@tul.cz

Abstract. Cold drawing of steel tubes is the manufacturing process characterized by anisotropic material flow during drawing. In this paper, the microstructural of the E235 steel tube after recrystallization annealing and cold drawing were characterized by optical microscopy and EBSD (Electron BackScatter Diffraction) analysis performed on a scanning electron microscope. The input feedstock after recrystallization annealing with dimensions of Ø 25.0 mm × 1.5 mm (O.D × W.T.) was cold drawn in two drawing passes with fixed plug to the final dimensions of Ø 18.0 mm × 0.9 mm (O.D × W.T.). The development of crystallographic texture (both qualitative and quantitative) was analysed on a longitudinal section of the tube. The stress-strain state in the tube material during drawing was calculated by FEM (Final Element Method) software and the development of the texture depending on the stress-strain state in three different directions using cylindrical coordinates with respect to the drawing direction was evaluated. We have investigated the ratio of individual crystallographic planes depending on stress state during drawn tube. The main objective was to define the development of crystallographic grain orientation from stress- strain state during cold drawn tube.

1. Introduction

Cold tube drawing is a metalworking process used to produce high-quality seamless tubes with precise dimensions, good surface, finish and high mechanical properties. Seamless tubes are used for applications where homogeneity of strength, microstructure, and extended product life are important design parameters. Cold drawing is widely applied in the industrial production of seamless tubes, employed for various mechanical applications. Cold drawing of steel tubes is the manufacturing process characterized by anisotropic material flow during drawing. In order to achieve the final diameter and wall thickness, the extruded pre-tubes are reduced successively in several cold drawing steps. In the first passes, where the outer diameter and wall thickness is reduced, the drawing with the fixed plug is used. Hollow sinking is a final process in which the wall thickness stays the same and the outer dimension is reduced [1, 2].

The low temperature during the cold drawing process doesn't enable dynamic recrystallization and thus occurs the material hardening due to the accumulation of dislocations. During the forming process, individual grains are elongated in the direction of the main deformation. Too high level of deformation and hardening is connected with a depletion of material plasticity and the risk of defects, and therefore recrystallization annealing is required. During the annealing, the new deformation-free equiaxial grains are formed and the material is capable of being further deformed [2, 3, 4].

The aim of metal forming industry is perpetually improve productivity and product quality. In order to reach this purpose, a better understanding of the processes is necessary. On the one hand, a large series of experimental tests can be done. However, this type of approach may be time and money consuming. On the other hand, lots of tests can be performed virtually thanks to finite element (FE) modelling. It also gives access to physical values, such as stresses and strains, which are not measurable during the process. Thus, FE modelling seems to be a helpful tool [5].

2. Materials and methods

The material used in this study is E235 steel grade with ferrite and pearlite (8.9 %) microstructure. The the volume fraction of the pearlite in steel was determined by calculation based on the carbon content. The chemical composition of the used steels is in Table 1. The samples for metallographic characterization were hot fixed into a polymeric resin and subsequently grinded and polished. Mechanical-chemical polishing using colloidal silica (OP-S Struers) took place as a final step. The input tube with dimensions of Ø 25.0 mm × 1.5 mm (O.D × W.T.) was the recrystallizing anneal at temperature 580 °C / 34 min. The input feedstock was cold drawn in two drawing passes with fixed plug to the final dimensions of Ø 18.0 mm × 0.9 mm at the drawing speed 625 mm / s. The first drawing passes were dimensional from Ø 25.0 mm × 1,5 mm to Ø 22.0 × 1.25 mm (reduction 26.42 %). The second drawing passes were dimensional from Ø 22.0 × 1.25 mm to Ø 18.0 mm × 0.9 mm (reduction 40.67 %). Reduction of tube cross section at drawing was calculated according to the equation (1):

$$R = \frac{(D_0 - d_0) - (D_1 - d_1)}{(D_0 - d_0)} 100 \tag{1}$$

R – reduction of cross section (%), D_0 – outer diameter of tube before tube deformation (mm), d_0 – inner diameter of tube before tube deformation (mm), D_1 – outer diameter of tube after tube deformation (mm), d_1 – inner diameter of tube after tube deformation (mm). Mechanical properties of the material before and after drawing were measured by a tensile test.

The stress state in the material tube during drawing was calculated by finite element software – DEFORM 3D. With respect to the numerical simulation setting, the input tube was defined by flow stress which were achieved of tensile tests. The main objective was to determine the value of the stress in which the crystal lattice rotation occurs. DEFORM 3D expresses the stress state in a cylindrical coordinate system [*R*, *θ*, *Z*].

The crystallographic orientation of grains with individual orientations was examined by means of EBSD method. ESBD maps were acquired on an area ~ 200 x 150 µm using a step size of 1 µm and applying a sample tilt angle of 70° at accelerating voltages 20 kV. The obtained EBSD data were processed by HKL Channel 5 software.

Table 1 Chemical composition of E355 steel grade (acc. to STN 411353) in wt.%

Element	**C**	**Ni**	**Mn**	**Mo**	**Cr**	**Fe**
Concentration [wt. %]	≤ 0.09	≤ 0.06	≤ 0.42	≤ 0.02	≤ 0.06	Bal.

IOP Conf. Series: Materials Science and Engineering **461** (2019) 012010 doi:10.1088/1757-899X/461/1/012010

3. Result and discussion

EBSD analysis was realized with object to understand the development of crystallographic texture during the drawing of precision tubes. The development of textures is shown using inverse pole figures (IPF) in radial, tangential and axial directions of sample. The intensity of IPF is shown by the mean uniform density (MUD). MUD value above unity indicates that there are more data points of a particular orientation than would be expected from a sample that is totally random.

Figures 1 – 6 show inverse pole figures (IPF) maps of the samples after individual technological steps (initial state; after the 1st and 2nd step of cold drawing).

The input tube has a certain inheritance of the weak deformation texture, Figures 1 – 3 (IPF - on the left). There is a strong dominance of (101) to the axial direction, of (111) to the radial direction, of (001) to tangential direction as a result of the previous deformation. The material was completely recrystallized after the previous deformation.

Tube after recrystallization was draw while the components of the strain tensor in the tube during the drawing process act in mutually perpendicular directions. Considering that at the second drawing passes is a larger reduction than at first drawing passes therefore, at the second drawing passes is a higher strain than at first drawing passes, Table 2.

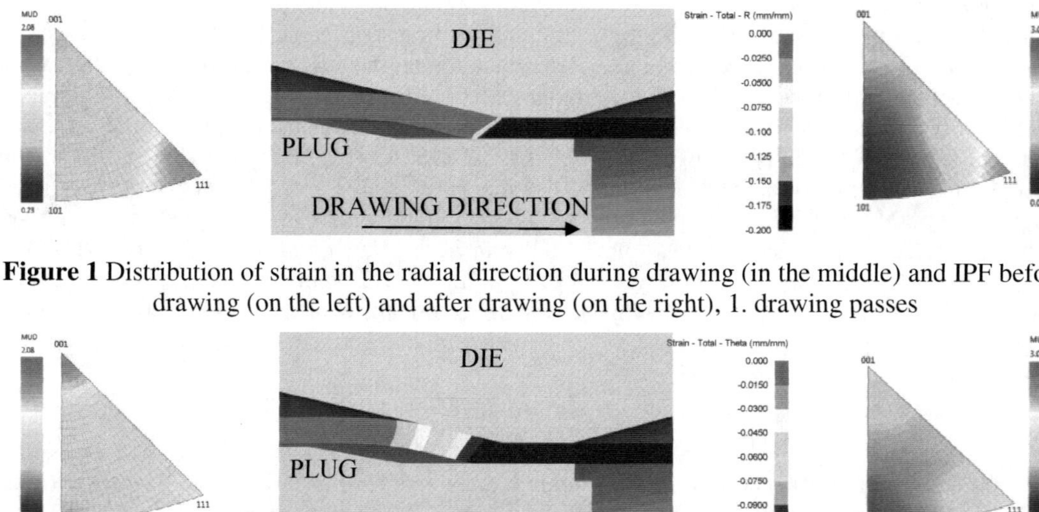

Figure 1 Distribution of strain in the radial direction during drawing (in the middle) and IPF before drawing (on the left) and after drawing (on the right), 1. drawing passes

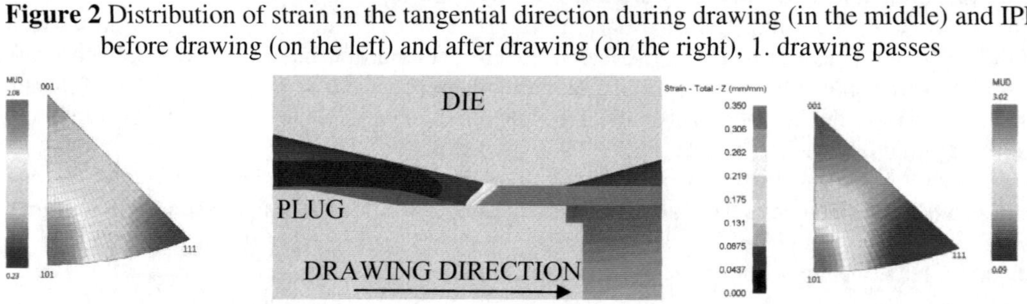

Figure 2 Distribution of strain in the tangential direction during drawing (in the middle) and IPF before drawing (on the left) and after drawing (on the right), 1. drawing passes

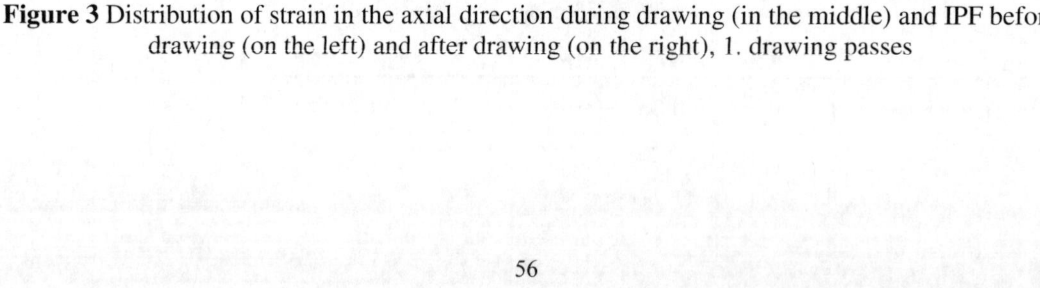

Figure 3 Distribution of strain in the axial direction during drawing (in the middle) and IPF before drawing (on the left) and after drawing (on the right), 1. drawing passes

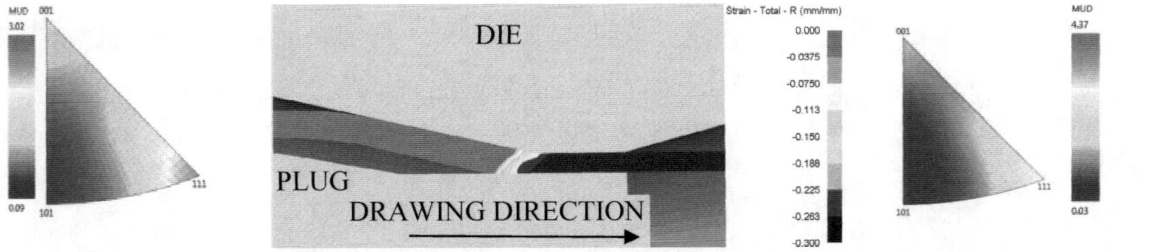

Figure 4 Distribution of strain in the radial direction during drawing (in the middle) and IPF before drawing (on the left) and after drawing (on the right), 2. drawing passes

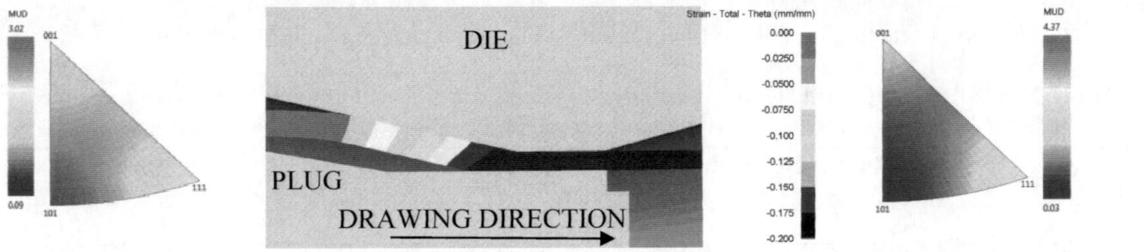

Figure 5 Distribution of strain in the tangential direction during drawing (in the middle) and IPF before drawing (on the left) and after drawing (on the right), 2. drawing passes

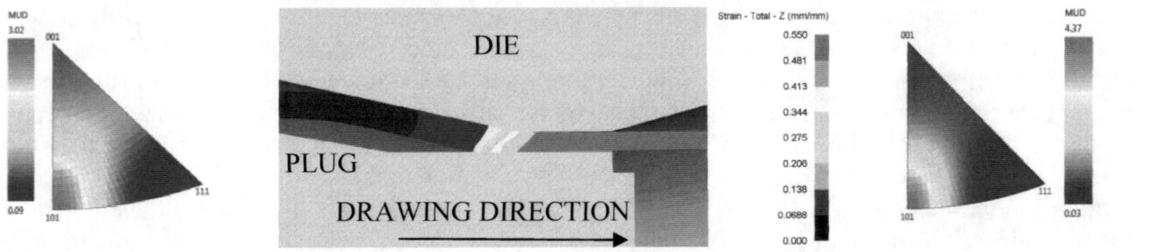

Figure 6 Distribution of strain in the axial direction during drawing (in the middle) and IPF before drawing (on the left) and after drawing (on the right), 2. drawing passes

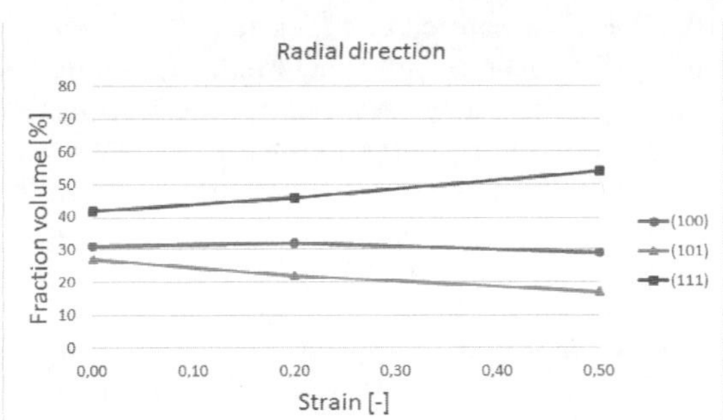

Figure 7 The diagram of the fiber texture evolution during drawing in radial direction

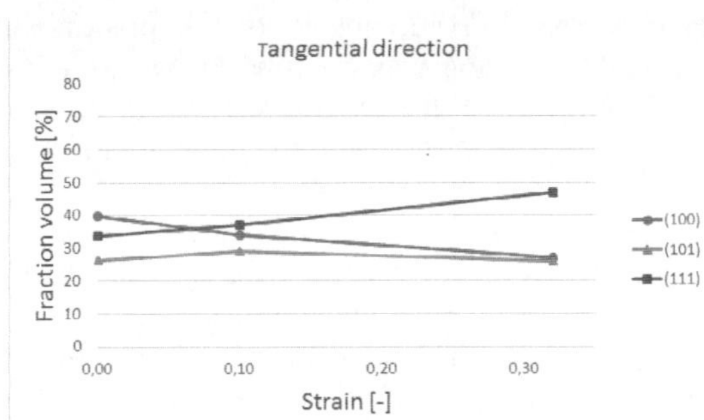

Figure 8 The diagram of the fiber texture evolution during drawing in tangential direction

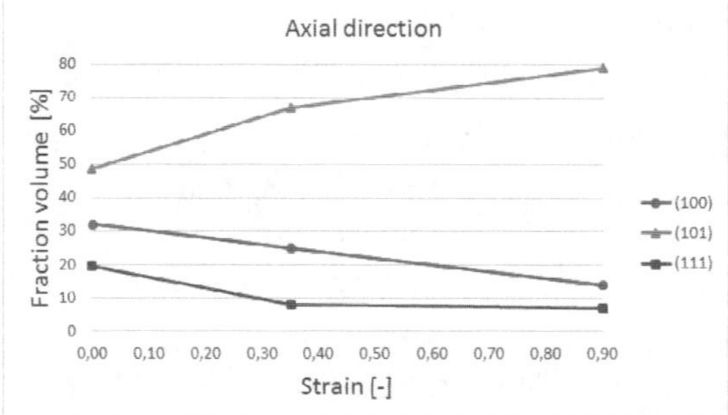

Figure 9 The diagram of the fiber texture evolution during drawing in axial direction

Table 2 Calculated approximate maximum stresses and strains at drawing in the deformation zone by finite element method (FEM)

	Radial direction		Tangential direction		Axial direction	
	1. drawing passes	2.drawing passes	1. drawing passes	2.drawing passes	1. drawing passes	2.drawing passes
Strain [-]	0.2	0.3	0.12	0.2	0.35	0.55
Stress [MPa]	450	675	375	602	351	575

In radial direction was the tube loaded by dominant pressure stress component, as illustrated, Figure 1, Figure 4. The dominant effect of pressure stress caused reduction the wall thickness of the tube after drawing. We can declare, according IPF that pressure stress has influence on increasing of texture (111) from random texture (state after recrystallization). Share of others crystallographic texture (001), (101) have been realigned into texture (111) (Figure 7). The development of the texture is steady from an increasing deformation.

In tangential direction was the tube loaded by dominant pressure stress component, as illustrated Figure 2, Figure 5. The dominant effect of pressure stress caused reduction the outer diameter of the tube after drawing. And also, pressure stress caused increasing density of poles in texture (111). The texture (001) (state after recrystallization) changed significantly and unlike texture (101) which density of poles was reduced (Figure 8). This fact is substantiated of high critical slip stress required to rotate the plane (001) compared to other planes in the BCC (Body-Centered Cubic) grid [6].

In axial direction was the tube loaded by dominant tensile stress component, Figure 3, Figure 6. The tensile stress in the axial direction caused a prolongation of the tube. We can declare, according IPF that pressure stress has influence on increasing of texture (111) from random texture (state after recrystallization). Share of others planes were decreased from dominant tensile strain/stress, Figure 9. The crystal plane (101) in the body-centered cubic (BCC) lattice of ferrite is the most densely occupied by atoms and therefore the deformation occurs by sliding in this plane.

The axial stresses/strains are higher than in radial and tangential direction, it caused that texture development in the axial direction is more distinctive than radial and tangential direction. When the material is stretched the angle between the drawing direction and the slip direction is decrease and when the material is compressed the opposite is occur [7]. For that reason, the pressure stress during drawing increases the Schmid factor. Therefore, higher pressure stress is needed to rotate the crystal in material. The rotation of the crystal was less pronounced in second drawing pass than in first drawing pass, because of deformation hardening during first drawing pass prevented it. Ultimate tensile strength during first drawing passes has increased from *UTS = 388 MPa* to *602 MPa* and during second drawing passes has increased from *UTS = 602 MPa* to *710 MPa,* (Table 3).

Table 3 Mechanical properties (R_p0.2 – yield stress, R_m – tensile strength, A_5 – elongation)

State	R_p 0.2 [MPa]	R_m [MPa]	A_5 [%]
Recrystallizing	307	388	46
1. drawing passes	596	602	8
2. drawing passes	685	710	6

Conclusion

The texture characteristic of a E235 steel grade tube is investigated based on electron backscattered diffraction results, this analysis enabled us to reach the following conclusions:

- Crystals in radial and tangential direction rotated by effect pressure the stress into a stable texture (111) that is at pressure stress in BCC grid.
- Crystals in the axial direction rotated by effect tensile stress into a stable texture (101) that is at tensile stress in BCC grid.
- Obtained data served as a valuable basis for the optimization of technological processes of production. The main aim of the study is to obtain the optimal crystallographic orientation of the drawing of seamless steel tube with properties close to deep drawing parameters.

Acknowledgements

This paper shows intermediate results of the Research Project No. 3/2018/ŽPVVC „TUMIFORM – Optimization of precision tube drawing technology considering dislocation theory, microstructure, and plasticity limit ". This project is being solved at ŽP Research and Development Centre in Podbrezová, Slovakia.

References

[1] Brokmeier H G and et. al. 2015 Texture gradient in a copper tube at maximum and minimum wall thickness *IOP conf. series: Materials science and engineering* 82

[2] Loharkar P K and Pradhan M K 2017 Modeling and analysis of cold drawing process *Handbook of research on manufacturing process modeling and optimization strategies* ed R Das, M Pradhan pp. 40-53

[3] Raji N A and Oluwole O O 2011 Influence of degree of cold-drawing on the mechanical properties of low carbon steel *Materials sciences and applications* 02 (11) pp. 1556-1563

[4] Belyakov, A, et. al. 2004 Microstructure evolution in ferritic stainless steels during large strain deformation *Materials transactions* 45 (9) pp. 2812-2821

[5] Boutenela F, Delhommeb M and Velaya V 2018 Bomanc R. Finite element modelling of cold drawing for high precision tubes *Comptes rendus mécanique* 348 (8) pp.665-677

[6] Pariľák Ľ et. all. 2017 EBSD analysis on cold drawn steel tube grades E235 *Internal report 14/2017/ŽP VVC, in Slovak*

[7] Gullberg D et. all. 2010 Influence of composition, grain size and manufacture process on the anisotropy of tube materials *Dissertation thesis (*Uppsala*)* Uppsala universitet pp. 62

The hot rolled duplex laminated steel presenting both high strength and high toughness

S L Chen[1], Z X Cao[2], M D Zhang[1], H F Xu[1], J Dzugan[3], and W Q Cao[1]

1. Central Iron and Steel Research Institute, Beijing, 100081, China, 2. Science Faculty, University of Sydney, NSW2006, Australia, 3.COMTES FHT a.s., Prumyslova 995, 33441Dobrany,Czech Republic

E-mail:caowenquan@nercast.com

Abstract: In this study, a new steel with the chemical composition of Fe-0.15C-5Mn-3Al (wt-%) was designed. Its microstructure and mechanical properties were examined by optical microscopy, scanning electron microscopy and mechanical tests. It was found that hot rolling in the dual phase region of ferrite and austenite at high temperature resulted in micro-laminated dual phase microstructure. An excellent combination of high strength (~1200MPa) and ultrahigh impact toughness (~400J) of the micro-laminated dual phase steel was obtained. Based on the measurement of the phase volume fraction and the analysis of the fractography, it was shown that the high strength was mainly dependent on the high martensite volume fraction, while the ultrahigh impact toughness was attributed to the micro-laminated microstructure.

1 Introduction

Traditional dual phase (DP) steel is a typical composite material with a soft granular ferrite (F) matrix and hard granular martensite or austenite (MA) islands, which has good mechanical properties, such as low yield strength, continuous yielding, high ultimate tensile strength and high ductility, but relative low impact toughness[1,2]. Another kind of composite material is the pack rolling (PR) steel with an obvious layered structure, which results in high impact toughness in the perpendicular direction of the rolling plane[3]. It would be a promising material with both the excellent mechanical properties of the DP steel and the ultrahigh impact toughness of the PR steel.It is well known that the laminated materials were usually fabricated by stacking and repeated rolling different steels and/or other materials[3-6].The process is very complicated and difficult to be achieved in industry, which makes the metallurgical and material researchers spend a long time to find a simple and feasible way to fabricate the laminated materials in the steel industry.

In this research, a new kind of dual phase steel was designed with the chemical composition of 0.15C5Mn3Al. The microstructure and the mechanical properties of the steel were examined after hot rolling in the α+γ region. Our aim is to explore the possibility to fabricate the micro-laminated microstructure in our new designed steel and its effects on the strength and impact toughness.

2 Materials and methods

The steel with the chemical composition of Fe-0.15C-5Mn-3Al (wt-%) was designed in the present study. The steel was first smelt in a 50 kg vacuum induction furnace and casted into an ingot. Then the ingot was forged into a square slab with thickness of 30 mm. The slab was homogenized at 1200 °C for 2 h and subsequently hot rolled into a thin slab with 11 mm thickness at the temperature between 1200 °C and 1000 °C followed by air cooling to room temperature. 5Mn-C-Fe phase diagram and 5Mn-3Al-C-Fe phase diagram were calculated by using the Thermo-Calc software with TCFE 7 database provided by the CISRI-TCS Joint Open Laboratory.

Samples for microstructure studies were polished and etched with 4% nital solution. The microstructure was analyzed by optical microscopy (OM) and scanning electron microscopy (SEM). The volume fractions and the shape characteristics of the phases were measured and described according to the gray difference in metallograph. The micro-hardness of different phases was measured in the transverse plane. Two groups (L-S and L-T) of standard specimens for Charpy V-notch impact measurement were cut from the slab to examine the anisotropy of the impact toughness. L–T (resp. L–S) configuration is investigated in which L corresponds to the axial direction and T (resp. S) to the notch direction. More details of the sampling information are shown in Fig.1. The impact toughness was measured at both -40 °C and 25 °C. The dog-bone shaped tensile specimens with gauge length of 25 mm and diameter of 5 mm were cut with their axial directions paralleling to the longitudinal direction (rolling direction). Tensile tests were conducted in a universal testing machine with a strain rate of 10^{-3} s^{-1} at room temperature.

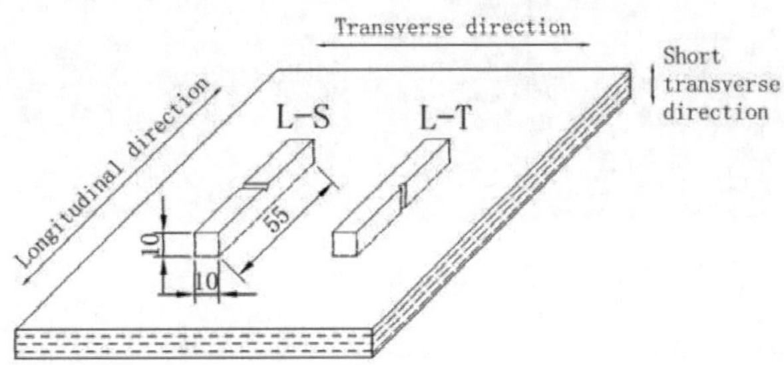

Figure 1 Charpy V-notch impact test specimens in both the L–S and the L–T directions

3 Results and discussions

5Mn-C-Fe phase diagram and 5Mn-3Al-C-Fe phase diagram were calculated by using the Thermo-Calc software with TCFE 7 database as shown in Fig.2. The dual phase region of ferrite and austenite only exist below 750°C (α+γ) or around 1450 °C (δ+γ) in 5Mn-C-Fe phase diagram as shown in Fig.2 a. Aluminium has a great effect on promoting the formation of ferrite. It does not only increase the stabilization of α-ferrite but also expands the δ-ferrite phase region. The α+γ and δ+γ dual phase regions are expanded and connected by adding 3 wt.% aluminium in medium manganese steel as shown in Fig.2 b.

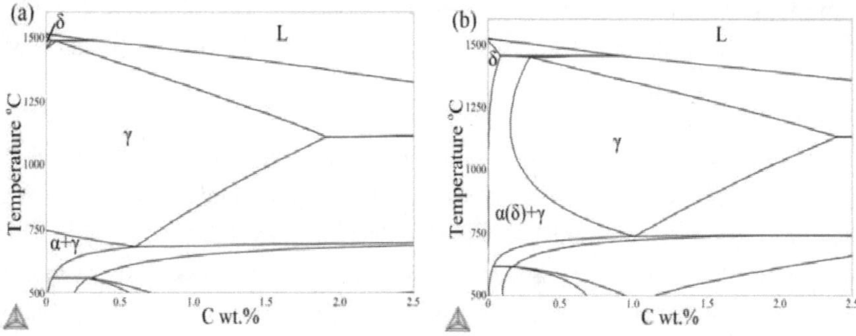

Figure 2 The calculated (a) 5Mn-Fe-C and (b) 5Mn-3Al-Fe-C phase diagrams

Fig.3 a and b show the hot rolled microstructure examined by OM. It can be seen that the steel is composed of two different phases arranged alternatively along the short transverse direction. The micro-hardness of the ferrite phase is about 191±13.8 HV(50g), while the micro-hardness of the martensite phase is about 413±14.4 HV(50g). The ferrite phase existed stably during the hot rolling due to the addition of aluminium[7], which is consistent with the calculated phase diagrams as shown in Fig.2. However, the martensite phase should be the phase transformed from the austenite during the air cooling process[8]. The ferrite is elongated seriously along the longitudinal direction and the transverse direction. Fig.3 c and d reveal the microstructure examined by SEM that shows the thin ferrite lamellae and the martensite matrix clearly. .

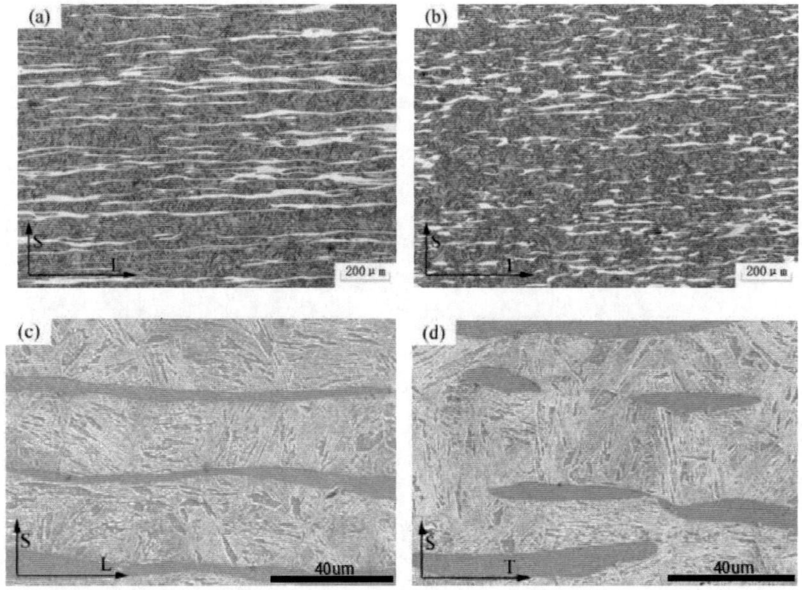

Figure 3 Laminated dual phase microstructures OM (a) L×S and (b) T×S, SEM (c) L×S and (d) T×S

Different from the traditional DP steel and the PR steel[3-6,9-10],the microstructure of the steel exhibits a large number of thin ferrite lamellae with thickness 1-10 μm paralleling to the rolling plane which locate evenly in the matrix with martensite lamellae about 10-50μm. The ferrite volume fraction of the steel is 11±1.2%. The shapes of the ferrite and the martensite in this steel are completely different from the granular shape ferrite and MA islands in the traditional DP steel[9]. The thickness of the ferrite and the martensite lamellae is much thinner than that of the millimetre-laminated steels[3,5]. In addition, the process only includes hot rolling and air cooling, which is feasible in the steel industry comparing with the traditional DP steel and the PR steel.

The mechanical properties were given in Table 1 and Table 2. It can be seen that the new dual phase steel achieves a high yield strength and a high ultimate tensile strength but a low ratio of yield strength to ultimate tensile strength. The big increase from yield strength to tensile strength implies a high work hardening and good plasticity capacity. The uniform elongation of ~6% and total elongation of ~15% indicate a good ductility of the steel. In addition, ultrahigh impact toughness (~400 J) was obtained in the L-S direction samples. However, the impact toughness in the L-T direction (~80 J) is much lower than that in the L-S direction, indicating obvious anisotropy of the impact toughness.

Table 1. Mechanical properties of the steel and traditional DP steels @25 °C

	Yield strength /MPa	Ultimate tensile strength /MPa	Yield ratio	Uniform elongation /%	Total elongation /%
This study	785	1199	0.655	6.0	14.8
DP590	340	590	-	-	21
DP780	420	780	-	-	14
DP980	550	980	-	-	8

Table 2 Impact toughness of the steel @ -40 °C and 25 °C

Impact toughness@-40 °C /J		Impact toughness @25 °C /J	
L-S	L-T	L-S	L-T
372±43	65±31	405±59	83±15

As it was well known, the strength of the traditional DP steel could be related to the strength contribution and deformation interaction of the ferrite phase and the martensite phase[11]. Under the assumption that the plastic strains of two phases are equal, the flow stress of the laminated dual phase steel can be simplified to the following equation (eqn.(1)),

$$\sigma = V_F \sigma_F + V_M \sigma_M \tag{1}$$

As it was measured, the micro-hardness is about 191 HV for the ferrite phase and 413 HV for the martensite phase. According to the relationship between hardness and tensile strength, σ_F and σ_M are equal to 623 MPa and 1344 MPa, thus the calculated ultimate tensile strength of the steel is about 623*0.11+1344*0.89=1265 MPa, which is very close to the measured ultimate tensile strength of 1199 MPa as given in Table 1. Another feature of this steel is its good ductility with the high volume fraction of martensite phase (~90%). The volume fraction of martensite phase should be limited from 10% to 30% in traditional DP steel to obtain high ductility[11]. However, the regular lamellar microstructure with about 90% martensitic phase could reduce the stress concentration and retard the formation of micro voids, thus improve the non-uniform elongation.

The impact toughness in the L-S direction of the steel is about 400J, which is about 5-6 times of the traditional DP steels[12-13]. However, the impact toughness in the L-T direction is about 80 J, which is in the similar level with other steels[14-15]. In order to understand the ultrahigh impact toughness in the L-S direction and the anisotropy of the impact toughness in different directions, the macrographs of fractured Charpy V-notch impact samples were examined as shown in Fig.4. It can be seen that the extrahigh impact toughness could be related to the cracking along the rolling plane, which absorbed most of the impact energy when the V-notch is opened in the L-S direction as shown in Fig.4a and b. Only a slightly crack along the rolling plane when the notch is opened in L-T direction as shown in Fig.4c and d, which results in a much lower impact toughness.

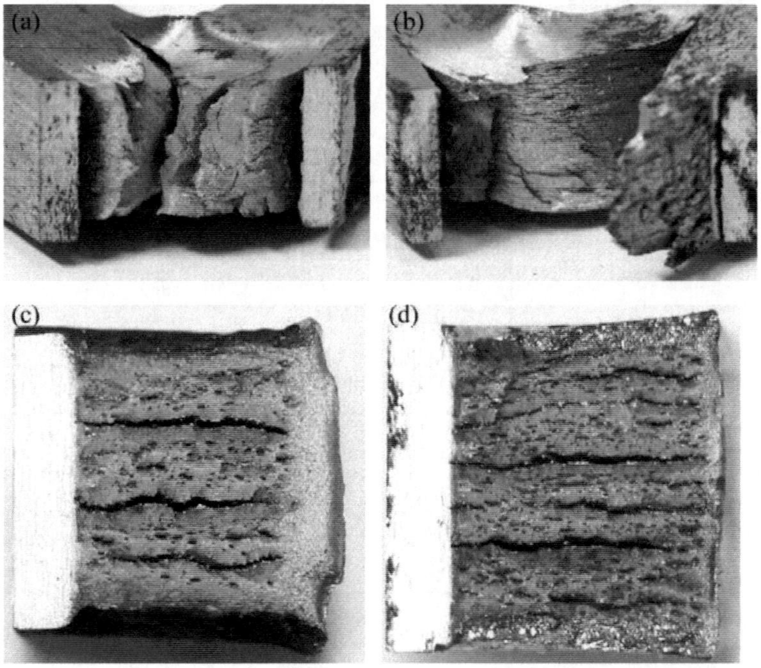

Figure 4 Fracture surfaces of specimens.(a) 25°C and (b) -40°C in L-S, (c) 25°C and (d) -40°C in L-T

4 Conclusions

A micro-laminated dual phase steel was produced by simple hot rolling and air cooling processes. And this microstructure gives a very good combination of mechanical properties and ultrahigh impact toughness. The main conclusions can be drawn as follows,

(1) A micro-laminated dual phase steel with ferrite lamellae and martensite lamellae arranged alternatively in the short transverse direction was successfully developed by the rational design of the alloy composition and the hot rolling process in the dual phase region.

(2) An ultimate tensile strength of ~1200 MPa and a total elongation of ~15% of the steel indicate a good combination of strength and ductility.

(3) The ultrahigh impact toughness of ~400 J was obtained in the L-S direction of the micro-laminated high strength dual phase steel, due to the cracking along the interfaces of the micro-laminated steel.

Acknowledgements
This research was supported by both National Key R&D Program of China (No. 2016YFB0300101) and National Natural Science Foundation of China (No.51871062).

References

[1] Tavares S S M, Pedroza P D, Teodósio J R, and Gurova T 1999 *Scripta Mater.* **40** 887.
[2] Chao Y J, Ward Jr. J D and Sands R G 2007 *Mater. Des.* **28** 551
[3] Pozuelo M, Carreño F and Ruano O A 2006 *Compos. Sci. Technol.* **66** 2671
[4] Kavarana F H, Ravichandran K S and Sahay S S 2000 *Scripta Mater.* **42** 947
[5] Inoue J, Nambu S, Ishimoto Y and Koseki T 2008 *Scripta Mater.* **59** 1055
[6] Lhuissier P, Inoue J and Koseki T 2011 *Scripta Mater.* **64** 970
[7] Yi H L, Ryu J H, Bhadeshia H K D H, Yen H W and Yang J R 2011 *Scripta Mater.* **65** 604
[8] Fang H S, Feng C, Zheng Y K, Yang Z G and Bai M Z 2008 *J. Iron. Steel Res. Int.* **15** 1
[9] Calcagnotto M, Adachi Y, Ponge D and Raabe D 2011 *Acta Mater.* **59** 658
[10] Akbarpour M R and Ekrami A 2008 *Mater. Sci. Eng. A* **477** 306
[11] Movahed P, Kolahgar S, Marashi S P H, Pouranvari M and Parvin N2009 *Mater. Sci. Eng. A* **518** 1
[12] Fereiduni E, Banadkouki S S G 2014 *Mater. Des.* **56** 232
[13] Deng X T, Wang Z D, Tian Y, Fu T L and Wang G D 2013 *Mater. Des.* **49** 220
[14] Zhou M, Du L X and Liu X H 2011 *J. Iron. Steel Res. Int.* **18** 59
[15] Maropoulos S, Ridley N, Kechagias J and Karagiannis S 2004 *Eng. Fract. Mech.* **71**, 1695.

Multistage cryogenic treatment of X153CrMoV12 cold work steel

A. Ciski[1], P. Nawrocki[1], T. Babul[1], D. Hradil[2]

[1] Institute of Precision Mechanics, Duchnicka 3, 01-796 Warsaw, Poland
[2] COMTES FHT a.s., Průmyslová 995, 334 41 Dobřany, Czech Republic

aleksander.ciski@imp.edu.pl

Abstract. The article presents the influence of multistage cryogenic treatment (MCT) on the microstructure and properties of X153CrMoV12 cold work steel. Basic mechanical properties of steel, such as hardness and wear resistance (3 rolls-cone system) were tested. The microstructure of steel was observed by means of scanning microscopy. Test results were compared with the properties of steel subjected to conventional heat treatment and deep cryogenic treatment (DCT). The cryogenic processes of X153CrMoV12 steel influenced on the percentage of strengthening phases in the material matrix microstructure. In the case of MCT and DCT, similar changes in the microstructure were observed, consisting in increasing of the amount, more uniform distribution and refinement of carbide phases. Changes in the microstructure of steel affected the mechanical properties of the material.

1 Introduction

Multistage cryogenic treatment (MCT) processes, consisting of cyclic cooling down of the load to cryogenic temperature (e.g. -180°C) and heating it up to a certain temperature (e.g. temperature of -100°C or 20°C), are one of the most promising directions of cryogenic treatment technology development.

There is no data in the literature on the description of the phenomena occurring during this modified variety of cryogenic treatment. In a few articles, mainly of a commercial nature, MCT cryogenic treatment is presented as a favorable sub-zero treatment giving an effect similar to the result obtained during DCT treatment, i.e. mainly increasing the durability of tool steels [1, 2], as well as obtaining favorable changes in the sound of musical instruments [2]. In these marginal literature reports, the authors fail to take economic aspects, as according to the authors of this work, the lower cost of such treatment is very important in comparison with DCT, while maintaining similar effects on elevated mechanical properties of tool steels. There is also no exact data on how to perform this treatment, i.e. the number of repetitions in the cooling / heating cycle and used temperatures. An example treatment scheme is available in [2], but the author does not disclose the details of the heat treatment process (Figure 1).

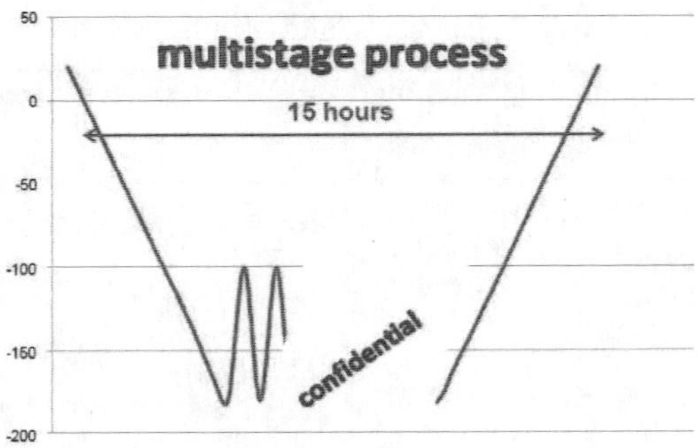

Figure 1. Exemplary MCT process [2].

Notes on MCT treatment can also be found on the websites of commercial contractors for this heat treatment [3, 4]. Due to the commercial nature of the information, this treatment is described quite enthusiastically and in comparison with conventional heat treatment or DCT is referred to as "the best" treatment [4]. The authors predict that the multistage cryogenic treatment may intensify precipitation processes occurring in tool steels. It can be assumed that repeating the process of holding steel at cryogenic temperature and heating it, e.g. to ambient temperature, i.e. severalfold repeating the carbide nucleation stage by forming clusters of carbon atoms occurring in the sequence of strengthening phases, will positively affect the mechanical properties of steel. The assessment of the influence of heat treatment on durability of tools requires a review of the relationship between the microstructure and mechanical properties of materials, such as hardness or wear resistance. The parameters of individual heat treatment operations should be analysed, as well as, the obtained microstructure, with particular emphasis on the content of residual austenite, the morphology of the martensitic matrix and the amount and distribution of carbides of alloying elements.

2 Materials and methods

The X153CrMoV12 cold work tool specimens were subjected to tests. The actual chemical composition of steel is shown in Table 1.

Table 1. Chemical composition of X153CrMoV12 steel (wt. %).

C	Si	Mn	P	S	Cr	Mo	Ni	Al	Co	Cu	V	W
1.56	0.31	0.51	<0.0019	<0.0020	11.47	0.65	0.36	0.03	0.11	0.10	>0.9624	0.25

The steel samples were austenitized in a vacuum and quenched in nitrogen at a pressure of 4 bar. The tempering processes were carried out under a nitrogen atmosphere. The sub-zero processes was carried out in an industrial cryogenic chamber. In tests a DCT process in temperature at -180 °C and a 24 hour holding time. In the case of MCT processes, the treatment consisted in cooling the charge to -180 ° C and subsequent heating and cooling of the steel several times in fixed temperature range (-180°C, -100°C or 20°C). The cryogenic processes were carried out immediately after the hardening operation.

All samples were tested in three modes of heat treatment, presented in Table 2.

Table 2. Parameters of heat treatment.

Mode	Austenitizing	Quenching	Cryogenic treatment	Tempering
Q+T	1060°C	Nitrogen, 4 bar	-	520°C
Q+DCT+T			**DCT** (-180°C, 24h)	
Q+MCT-100+T			**MCT-100** (-180°C/-100°C, 6 cycles)	
Q+MCT20+T			**MCT 20** (-180°C/20°C, 6 cycles)	

Observations of metallographic specimens etched with nital were performed using the JEOL JSM-IT100 scanning electron microscope at magnifications in the range of x2000 ÷ 10000. Residual austenite content tests were carried out using a PSF-3M Rigaku diffractometer. The sin2ψ method and the Cr lamp were used. The HRC surface hardness tests were performed using a Struers DuraScan70 hardness tester, performing 5 readings on each of the samples. Test of resistance to wear was carried out in a 3 rollers-cone system with a continuous lubrication of samples, at constant unit pressure of 100, 200 and 400 MPa. Conical counter-specimen was made of 100Cr6 steel quenched and hardened to the hardness of 64 HRC and ground to a roughness Ra = 0.32 micrometers.

3 Results and discussion

3.1 Microstructure

Microstructure images of X153CrMoV12 steel specimens subjected to heat treatment are shown in Figure 2. Structural imaging of steel subjected to nital etching allowed to determine the size of the original austenite grain and to observe the precipitation of primary carbides and larger precipitations of secondary carbides in the material matrix. The morphology of the matrix visible in the photographs can be described as lath martensite.

The average size of primary grains of austenite for all variants was approx. 10 μm, which corresponds to No. 10 in the ASTM scale. The largest amount of fine-disperse carbides in the form of plaques was observed in the case of a variant including deep cryogenic treatment DCT and multistage cryogenic treatment MCT20 (Figure 2c, 2d). This result is consistent with the observations of the authors of the work [5], in which the content of the carbide phase increased from 18 to 26%. An important difference between cryogenic treated steel and without cryogenic treatment is even distribution of carbides in the case of steel subjected to sub-zero processing, which is consistent with the results described in a number of publications [5, 6, 7, 8, 9].

a) Q + T

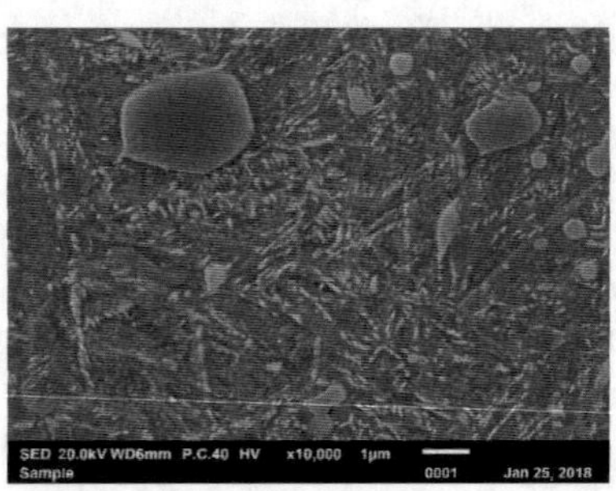

b) Q + MCT-100 + T

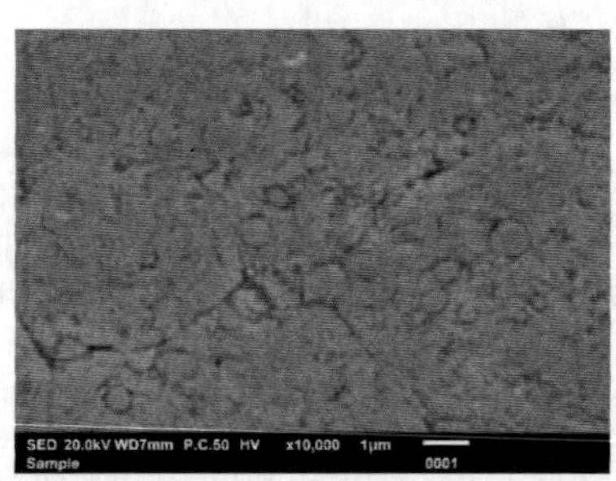

c) Q + MCT 20 + T d) Q + DCT + T

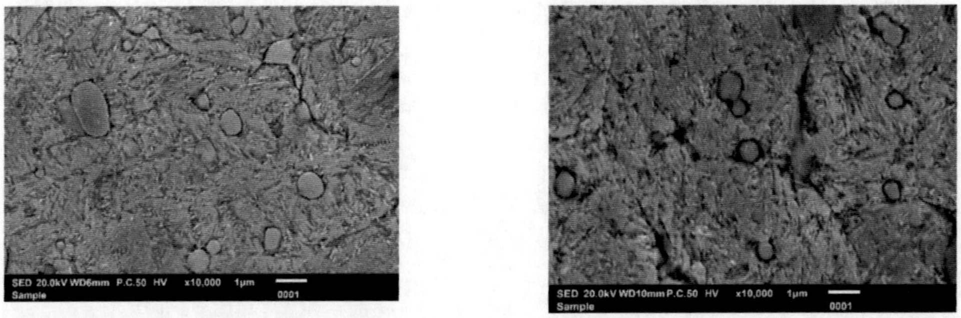

Figure 2. Microstructure of X153CrMoV12 steel after various modes of heat treatment, SEM, SE.

3.2 *Retained austenite content*

The X153CrMoV12 steel was austenitized at 1060°C to ensure high residual austenite content after the hardening operation.

X153CrMoV12 steel samples subjected to hardening contained approximately 26.2% residual austenite, the tempering process reduced this value to 7% (Figure 3). The X153CrMoV12 steel in the standard hardened condition contains up to 35% residual austenite and M7C3 carbides (primary and secondary), Mo (C, N) and MC. In the tempered condition at maximum hardness, strengthening is the result of the presence of M2(C,N) and Mo2C nanocarbides. After conventional heat treatment, more than 10% of residual austenite remains. In the article [5], it was found that the 48-hour process of cryogenic heat treatment process of D2 steel (equivalent of X153CrMoV12 steel) can promote the almost complete conversion of residual austenite into martensite. In this work, D2 steel hardened at 950°C contained 8% of residual austenite, and the cryogenic treatment allowed to lower this content below the detection capacity (> 1% volume).

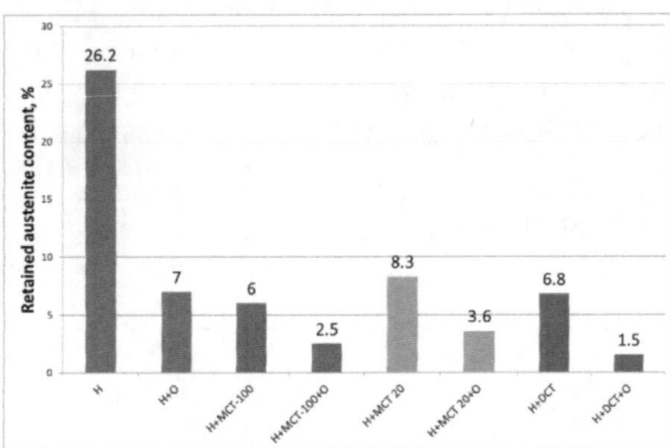

Figure 3. Retained austenite content of X153CrMoV12 steel after various modes of heat treatment, XRD.

3.3 *Hardness*

The HRC surface hardness tests (Figure 4) showed that the X153CrMoV12 tool steel in the hardened condition had a hardness of 64.2 HRC. The process of cryogenic treatment of tempered samples made it possible to obtain a much higher hardness (66.7 to 66.9 HRC), which is certainly associated with the continuation of martensitic transformation at lower temperatures and the formation of a structure with a much lower content of the soft phase component which is residual austenite (Figure 3).

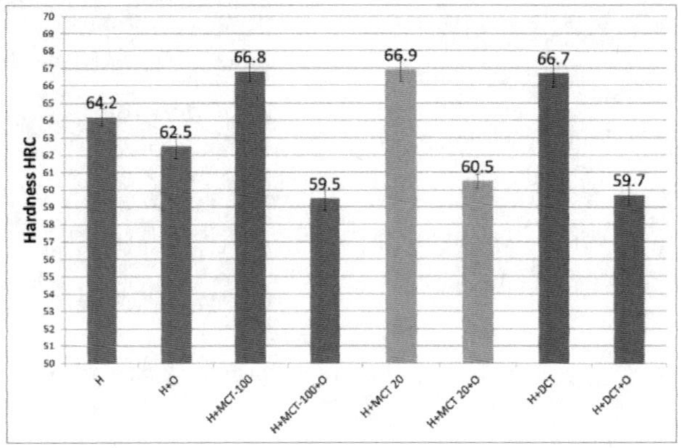

Figure 4. Hardness of X153CrMoV12 steel after various modes of heat treatment, HRC.

The tempering process of hardened samples (without cryogenic treatment) caused a hardness reduction to 62.5 HRC. For cryogenic treated samples subjected to tempering, the hardness was reduced by even lower values (in the range of 59.5 ÷ 60.5). The shift of the highest hardness peak to lower temperatures and a lower hardness after tempering at a given temperature was observed for the cold working tool steel Vanadis 6 subjected to cryogenic treatment [10]. This fact was related to the faster occurring precipitation phenomena after the cryogenic treatment. The shift to higher temperatures of thermal effects associated with the precipitation processes of cementite [11] or secondary carbides [12] was observed by the authors of the work during calorimetric studies.

3.4 Wear test

The results of the steel wear test, in the summarised form of maximum linear wear after 100 minutes of friction test, are presented in Table 3. In all heat treatment variants and at all unit pressures of 100, 200 and 400 MPa, the samples were characterized by the wear with a regular course, manifesting the occurrence of steady friction state in the entire course of wear, except running-in period and final seizure (load of 400 MPa).

Table 3. Decrease in maximum wear value, %.

Mode	Decrease in maximum wear value, %		
	100 MPa	200 MPa	400 MPa
Q+T	-	-	*Seizure after 70 min*
Q+MCT-100+T	*11.4*	*14.6*	*Seizure after 70 min*
Q+MCT20+T	*22.1*	*30.9*	*Seizure after 90 min*
Q+DCT+T	*29.0*	*25.7*	*Seizure after 100 min*

The most advantageous variant of cryogenic treatment seems to be the DCT process, where the decrease in consumption, depending on the load, ranged from 25.7 to 29.0%. Tests conducted under all unit loads showed that the cryogenic processes caused a decrease in the value of steel wear resistant compared to steel subjected to conventional heat treatment.

4 Summary

The work allowed to state that the processes of cryogenic treatment of the X153CrMoV12 cold work tool steel affect the contribution of strengthening phases in the material matrix microstructure. In the case of multi stage cryogenic treatment and deep cryogenic treatment, similar changes in the microstructure were observed, consisting in increasing the amount, more uniform distribution and refinement of carbide phases. The above changes result from the refinement of the martensite substructure and the precipitation processes occurring in a different manner in the cryogenic treated material. The microstructural changes also included the obvious influence of sub-zero treatment on the content of residual austenite.

Acknowledgements

This research was supported by CORNET Programme ("COOLBEND" Project).

References

[1] Alava L.A. 2008 Multistage cryogenic treatment of materials: process fundamentals and examples of application, Conference: Cryogenics 2008, Prague (Czech Republic), April 2008.

[2] Lausecker W. 2016 A review of DCT applications of industrial and non industrial components, Proceedings of 3rd Mediterranean Conference of Heat Treatment and Surface Engineering, Portoroz, Slovenia, 26-28 September 2016.

[3] http://www.percryo.com/lead_page.htm

[4] http://thermalprocessing.com/article/detail/6271/deep-cryogenic-treatment

[5] Amini K, Akhbarizadeh A, Javadpour S. 2012 Effect of deep cryogenic treatment on the formation of nano-sized carbides and the wear behavior of D2 tool steel, *International Journal of Minerals, Metallurgy and Materials*, Volume 19, Number 9 (2012) p 795.

[6] Meng F. et al. 1994 Role of eta-carbide precipitations in the wear resistance improvements of Fe 12Cr Mo V 1.4C tool steel by cryogenic treatment, *ISIJ International*, vol. 34 (1994) no 2, pp 205-210.

[7] Yun D. et al. 1998 Deep Cryogenic Treatment of High-speed Steel and its Mechanism, *Heat Treatment of Metals* (1998).3, pp 55-59

[8] D.Mohan Lal, S. Renganarayanan, A.Kalanidhi, 2001 Cryogenic treatment to augment wear resistance of tool and die steels, *Cryogenics 41* (2001) pp 149-155.

[9] Firouzdor V. Nejati E, Khomamizadeh F. 2008 Effect of deep cryogenic treatment on wear resistance and tool life of M2 HSS drill, *Journal of materials processing technology* 206 (2008) pp 467–472.

[10] Jurči P. et al. 2013 Tempering response of sub-zero processed cr-v ledeburitic steel Vanadis 6, Preceedings of Metal 2013 Conference, 15. - 17. 5. 2013, Brno, Czech Republic, EU.

[11] Ciski, A., Jeleńkowski, J.and Babul, T. 2015 Microstructural, thermal stability and wear resistance properties of X153CrMoV12 tool steel subjected to deep cryogenic treatment *Int. J. Microstructure and Materials Properties*, Vol. 10, No. 2, (2015) pp 140–148.

[12] Pellizzari M, Molinari A. 2002 Deep cryogenic treatment of cold work tool steel, 6th International Tooling Conference, (2002) pp 657-669.

IOP Conf. Series: Materials Science and Engineering **461** (2019) 012013 doi:10.1088/1757-899X/461/1/012013

Fabrication of Fe / Al_2O_3 composite foam by sintering techniques

W Depczyński

Faculty of Mechatronics and Mechanical Engineering, Kielce University of Technology, 7 Tysiaclecia Panstwa Polskiego Ave., 25-314 Kielce, Poland, EU

E-mail: wdep@tu.kielce.pl,

Abstract. Composite foams of iron and aluminium oxide were prepared by sintering techniques. Al_2O_3 microsphere powders added during Fe foam formation were used as starting materials. Samples containing various compositions were prepared (Al_2O_3 microspheres bonding in Fe foam). The metallic porous structure was obtained by sintering on the basis of iron-based powder mixtures with the addition of a copper catalyst and Fe (III) oxide as a space carrier. Sintering was carried out at 1130°C and 50 minutes long, in a dissociated protective gas with ammonia. Observation of the microstructure and phase identification of the synthesized samples was carried out using optical microscopy (OM), scanning electron microscopy (SEM) and energy dispersion spectroscopy (EDS). The porosity of the Fe / Al_2O_3 composite products for the three different compositions was measured by means of the Cavalleri principle on transverse samples prepared from sintered materials.

1 Introduction

Metallic porous materials are a new class of engineering materials that are dynamically researched due to their innovative and useful properties [1-6]. The increase of interest in the porosity of materials was influenced by understanding of porous structures can be found in living organisms, for instance human bones or wood morphology, where there is unusual combination of properties: high rigidity with minimal weight. The development of metallic foams (as high porosity materials, ranging from 40% to 98% by volume) began more than 20 years ago [7]. These unique materials combine properties such as energy absorption, fluid permeability, wavelength, low thermal conductivity, and good insulating properties. They can be used as: kinetic energy absorbers, fluid filters and impurities, porous welding electrodes, high temperature sealers, heat exchangers [8 - 10]. As the field of metallic foams application dynamically grows, there are many methods of obtaining them [11-13]. To obtain new physical and chemical properties or improve the integrative properties of metallic foams, ceramic particles were added into metallic foams to form composite foams [14 - 16]. In addition because of low densities and low-pressure drops, metal foams offer excellent flow characteristics as catalyst structural supports. The Fe/Al_2O_3 system has recently received attention because of its potential magneto-transport properties and uses as surface coatings by granular films [17]. In addition, several researches have been carried out to investigate catalytic performance of Fe–Al_2O_3 system. [18,19] in work [20] Iron–alumina composite foam was prepared through combination of combustion synthesis (CS) and spark plasma sintering (SPS) techniques. Al, Fe_2O_3 and Fe powders were used as starting materials. Samples containing 5–20 wt % (Al + Fe_2O_3) powders and balanced content Fe powder were sintered by SPS apparatus. Metal foam was formed by a typical base metal such as Al or Cu [21-24]. The subject of the experiment was the use test of producing techniques of the porous metal foams by reduction of metal oxides during sintering. The mixture was sintered in a dissociated ammonia atmosphere. Fe foam was prepared according to the method described in the Polish patent No. 199720 [25]. This allows for composing of irregular cellular structures with pores open or closed. The range of the porosity depends strongly on used materials, particle size and type of particulate material. However crucial influence on the porosity has the ratio between quantity of metal oxide powder and amount of matrix metal powder, which is a basic structure of produced sinter [26]. A new and promising method to produce open- cell metallic foams on the base of iron powder, low and high alloyed steel powders was the research objective The foam material can be stacked and co-sintered with top layers to sandwich structures. The experiment used Ferrum powder ASC 100.29, Fe_20_3 powder and Al_2O_3 microspheres to produce Fe-Al_2O_3 composite foam for potential magnetotransport and catalyst structural support applications. The object of the research was to examine new and promising method of fabrication an open-cell metallic Fe /Al_2O_3 composite foam.

2 Equipment and material

The initial stage for the production of porous sinter is to weigh and mix the powders together in appropriate proportions. There are four types of mixtures, labelled no. I, no. II and no. III. Table no. 1 shows types and proportions of powders that are components of each blend. Sample no. II disintegrated by too large share of microspheres Al_2O_3 in the mixture.

Table 1. Types and proportions of mixing powders.

	Sample designation		
Composition	**No. I**	**No. II**	**No. III**
ASC 100.29 [g]	140	140	140
Cu [g]	8	8	8
Fe_20_3 [g]	16,8	16,8	16,8
Al_2O_3[g]	3	12	6

Fe foam was prepared from mixture of iron powder ASC 100.29 (source Höganäs company) and iron oxide Fe_2O_3 + microspheres Al_2O_3. The sintering of prepared samples was done in a laboratory tube furnace at 1180°C, in a reducing gas shield of dissociated ammonia. Ammonia dissociates to nitrogen and hydrogen at 850°C in the presence of an iron catalyst. Such residual gases were directed to the sintering chamber into which a sample was placed. Hydrogen was a reducing oxide agent, iron oxide was reduced, thus providing protective atmosphere for the sintering process. Excess ammonia as well as the process derived products: the nitrogen and water vapour were subject to combustion. After sintering the samples situated in furnace were moved to an area where it was cooled in protective atmosphere. Cooling took place in with an average speed of about 25 degrees per minute. To determine pore size and shape from the foam specimens, the image analyse software was used.

3 Micro and macro structure investigation

To illustrate fusion structures, investigators used the optical microscopy and SEM methods. For tests of metallographic specimens, Nikon MA 200 Eclipse microscope with the image analysis system NIS 4.20 was used. All samples were mounted in a vacuum, using Buehler EpoThin resin during preparation process for the protection of the porous structure. SEM examination was performed using the JEOL JSM 7100F microscope (with Schottky field emission) with EDS OXFORD X-MAX microprobe. The results showed that there is a possibility of obtaining the binding between the Fe foam and Al_2O_3 microsphere.

4 The observation with optical microscope and image analysis.

The porosity was studied using an NIS 4.20 image analyser. Figure 1 shows an image of the porous structures of formed Fe sample and Al_2O_3 microsphere binding.

Sample No. I

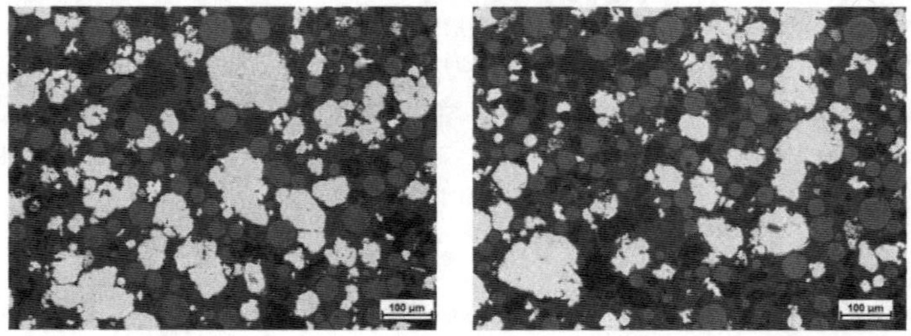

Sample No. III

Figure 1. The porous samples OM microstructures (mag. 100x). Not etched.

Considerable expansion of the surface inside the structure was achieved. It should be noted that non-uniform distribution of pores was due to diversity of the composition powder particle size. There are bridges connecting porous layer with a steel substrate surface. The relative volume of particles was determined in accordance with the Cavalieri-Hacquert principal. The result of the measure of the porous metal foam showed the porosity in the range of 51-59% with bulk density. The chemical composition and shape of the sintered material particles affect the creation of the diffusion bridges. A quantitative analysis of the sample microstructure was conducted to determine its porosity. It was necessary to determine the number of pores per unit considering polished surface for different cross-sections. 5 areas so called *Region of Interests* (ROI) of the structure were analysed. The average porosity was: 51% for specimen no. I and was 59% for specimen no. III. Fig. 2. shows location of ROI for analysed specimen No. I.

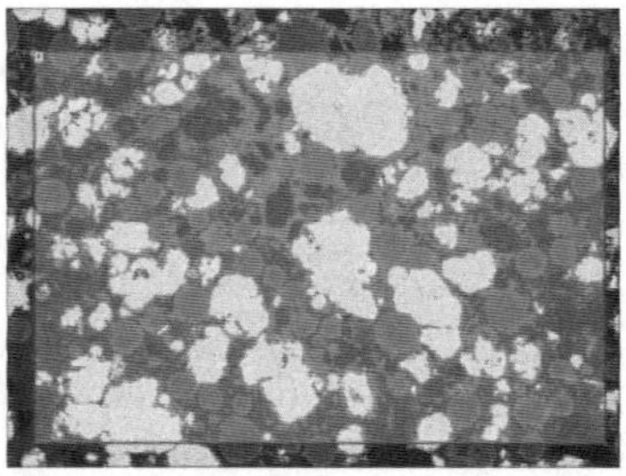

Figure 2. The analysis of ROI of sample No. I.

5 The observation with SEM microscope and EDS analysis.

Figure 3 shows the SEM micrographs of sintered foam surfaces of composition no. I (Figure 3), imaged using the JEOL JSM 7100F electron microscope. The depth of the field obtained by SEM allows us to assess the complex structure of the foam with visible parent material, voids forming porosity and Al_2O_3 microspheres.

Figure 4 shows the microstructures of sample No. III of metallic foam (500x and 1000x magnification) with invisible parent material, voids forming porosity and Al_2O_3 microspheres.

Figure 3. SEM photography of sample No. I (mag. 500x and 1000x).

Figure 4. SEM photography of sample No. III (mag. 500x and 1000x).

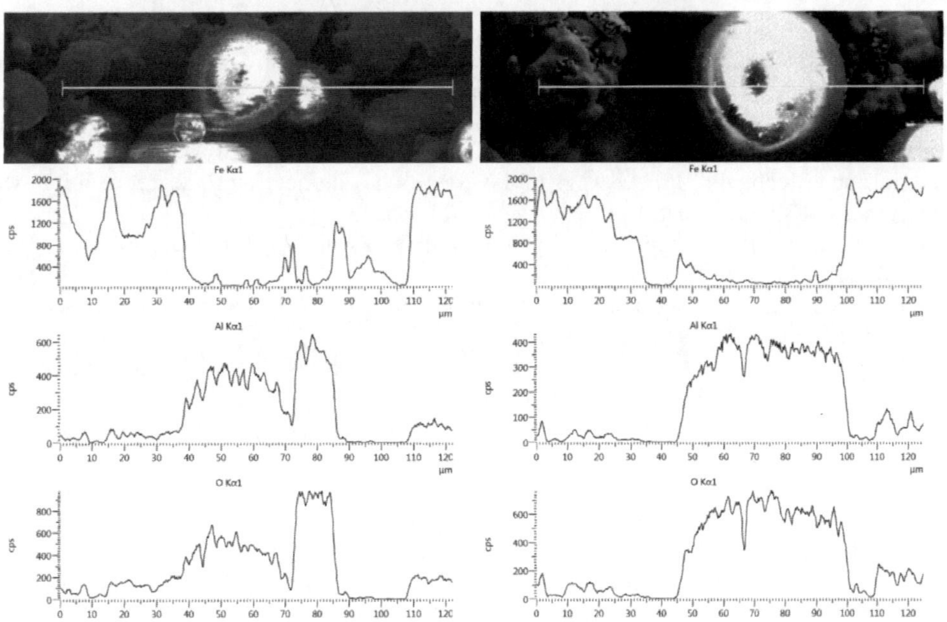

Sample No. I Sample No. III

Figure 5. The microstructure of the elements SEM and EDS distribution for Fe, Al and O, (samples No. I and III).

During the SEM observation and the EDS microlens tests, there was no diffusion of the surface finish from the Fe / Fe_2O_3 mixture to the Al_2O_3 microsphere material. Lack of diffusion is evident in case of Fe / Al_2O_3 association. Fig. 5 shows the microstructure of the elements SEM and EDS distribution for Fe, Al and O. Connections inside the Fe foam obtained by sintering in dissociated ammonia have shown morphological characteristics of the foam connections with open cells [17, 18]. The pores remained open to the outside. Similarly to works [17-20], a combination with acceptable parameters was obtained. When using techniques such as "joining metals" [21], the metal used for binding the foam pores, causes it's closing. However, any comparison with literature data from papers [17-21] is difficult to perform due to significant differences in the materials and techniques used. Available publications consider bindings such as foam-foam and foam-ceramics and foam-base metal configurations [27-30]. Metallographic examination and EDS analysis have demonstrated the possibility of using sintering techniques for materials difficult to bind, such as metallic foams with a non-metallic component of Al_2O_3.

6 Conclusions

There is a wide range of possible applications for presented materials, e.g. as heat exchangers, filters or catalysts. New and promising method of producing open-cell metallic foams on the base of iron powder, low and high alloyed steel powders as well as nickel alloy powder was used. Macroscopic and microscopic observations were provided insight into the morphology of the structure. In comparison with literature reports presenting other methods of producing porous composites Fe/Al_20_3, relatively simple and available manufacturing technology was used. The obtained effects are encouraging and allow assuming that the compositional composites can be made on the basis of other sintered metal powders using the technology used.

References

[1] Ashby, M.F.; Evans, A.G.; Fleck, N.A.; Gibson, L.J.; Hutchinson, J.W.; Wadley, H.N.G. Metal Foams: A Design Guide; Buterworth-Heineman: Boston, MA, USA, 2000.

[2] Arwade, S.R.; Hajjar, J.F.; Schafer, B.W.; Moradi, M.; Smith, B.H.; Szyniszewski, S. Steel foam material processing, properties and potential structural applications. In Structural Materials and Mechanics, Proceedings of the 2011 NSF Engineering Research and Innovation Conference, Atlanta, GA, USA, 4–7 January 2011; pp. 1–7.

[3] Banhart, J. Manufacture, characterisation and application of cellular metals and metal foams. Prog. Mater. Sci. 2001, 46, pp. 559–632.

[4] Davies, G.J.; Zhen, S. Metallic foams: Their production, properties and applications. J. Mater. Sci. 1983, 18, pp. 1899–1911.

[5] Nakajima, H. Fabrication, properties and application of porous metals with directional pores. Prog. Mater. Sci. 2007, 52, pp. 1091–1173.

[6] Qin, J.; Chen, Q.; Yang, C.; Huang, Y. Research process on property and application of metal porous materials. J. Alloys Compd. 2016, 654, pp. 39–44.

[7] Lefebvre, L.P.; Banhart, J.; Dunand, D.C. Porous metals and metallic foams: Current status and recent developments. Adv. Eng. Mater. 2008, 10, pp. 775–787.

[8] Piasecka M., Laser texturing, spark erosion and sanding of the surfaces and their practical applications in heat exchange devices, 2014, Advanced Material Research, vol. 874, pp. 95-100

[9] Piasecka M., Heat transfer mechanism, pressure drop and flow patterns during FC-72 flow boiling in horizontal and vertical minichannels with enhanced walls, 2013, Int. Jour. of Heat and Mass Transfer, vol. 66, pp. 472-488.

[10] Depczyński W., Piasecki A.,Piasecka M., Strąk K.," Impact of Fe powder sintering and soldering in production of porous heating surface on flow boiling heat transfer in minichannels", E3S Web of Conferences, Vol.: 19, p.: 1-6

[11] Depczynski, W.; Kazala, R.; Ludwinek, K.; Jedynak, K.: Modelling and Microstructural Characterization of Sintered Metallic Porous Materials, MATERIALS, Volume: 9 Issue: 7 Art:

567, DOI: 10.3390/ma9070567, 2016.

[12] Depczyński W., Nowakowski L., Hepner P., Miko E. The influence of porosity on machinability of sintered Fe foam elements, Metalurgija 56 (2017) 3-4, pp. 364-366

[13] Depczyński W., The Selected Properties Of Fusion Of Fe Foam And Sheet Metal With Use Of Sintering In Dissociated Ammonia, Proceedings Of 25th Anniversary International Conference On Metallurgy And Materials Metal 2016, pp. 682-687 Published: 2014

[14] Banhart J. "Manufacture, characterisation and application of cellular metals and metal foams," Progress in Materials Science 46 (2001) 559–632

[15] Y. Luo, S. Yu, W. Li, J. Liu, M. Wei, J." Compressive behavior of SiCp/AlSi9Mg composite foams" Alloys Compd. 460 (2008) 294–298

[16] A. Essakhi, A. Löfberg, S. Paul, B. Mutel, P. Supiot, V. Le Courtois, P.Rodriguez, V. Meille, E. Bordes-Richard, „Materials chemistry for catalysis: Coating of catalytic oxides on metallic foams" Micropor. Mesopor. Mater. 140 (2011) 81–88.

[17] A. Paesano Jr., C.K. Matsuda, J.B.M. da Cunha, M.A.Z. Vasconcellos, B. Hallouche, S.L. Silva, J., "Synthesis and characterization of Fe-Al_2O_3 composites", Magn. Magn. Mater. 264 (2003) 264–274.

[18] W. Li, H. Wang, Z. Ren, G. Wang, J. Bai, "Co-production of hydrogen and multi-wall carbon nanotubes from ethanol decomposition over Fe/Al2O3 catalysts", Appl. Catal. B 84 (2008) 433–439.

[19] T.V. Reshetenko, L.B. Avdeeva, V.A. Ushakov, E.M. Moroz, A.N. Shmakov, V.V. Kriventsov, D.I. Kochubey, Y.T. Pavlyukhin, A.L. Chuvilin, Z.R. Ismagilov, "Coprecipitated iron-containing catalysts (Fe-Al2O3, Fe-Co-Al2O3,Fe-Ni-Al2O3) for methane decomposition at moderate temperatures Part II. Evolution of the catalysts in reaction" Appl. Catal. A. Gen. 270 (2004) 87–99.

[20] A. Babakhani, E. Zahabi, H. Yavari Mehrabani, Fabrication of Fe/Al2O3 composite foam via combination of combustion synthesis and spark plasma sintering techniques", Journal of Alloys and Compounds 514 (2012) 20– 24

[21] Parvanian A.M., Sadatfar M., Panjepour M. Kingston A., Sheppard A.P. The effects of manufacturing parameters on geometrical and mechanical properties of copper foams produced by space holder technique. Materials and Design 53 (2014) p. 681-690

[22] Parvanian A.M., Panjepour M., Mechanical behavior improvement of open pore copper foams synthesized trough space holder technique., Materials and Design 49 (2013) p. 834-841

[23] Hangai Y., Zushida K., Fuji H., Ueji R., Kuwazuru O., Yosikawa N. Friction powder compaction process for fabrication open celled Cu foam by sintering – dissolution process route using NaCl space holder, Material Science & Engineering A 585 (2013) p. 468-474

[24] Hangai Y., Koyama S., Hasegawa M., Utsonomiya T., Fabrication of Aluminum Foam/Dense Steel Composite by Friction Stir Welding, Metalurgical and Materials Transactions A volume 41A, September 2010 p. 2184 - 2186

[25] Chatys R., Depczyński W., Żórawski W.: „A method of producing porous structures (in Polish), Patent of the Republic of Poland No. 199720

[26] Depczyński W., Sintering of copper layers with a controlled porous structure, Metal 2014: 23rd International Conference On Metallurgy And Materials Pages: 1219-1224 Published: 2014

[27] Murakami T., Nakata K., Ikeda T.,Nakajima H., Ushio M. Weld fusion property of lotus-type copper by laser beam irradiation, Material Science and Engineering A357 (2003) p. 134-140

[28] Campana G., Ascari A., Fortunato A., Laser forming for joining aluminum foam cores inside a hollow profile, Optics & Laser Technology 48 (2013) p. 331-336

[29] Cambronero L.E.G., Cañdas I., Ruiz-Román J.M. ,Cisneros M., Corpas Iglesias F.A., Weld structure of joined aluminium foams with concentrated solar energy., Journal of Materials Processing Technology 214 (2014) p. 2637 - 2643

[30] Shirzadi A.A., Zhu Y., Bhadesshia H.K.D.H., Joining ceramics to metals using metallic foam, Materials Science and Engineering A, 496 (2008) p. 501 - 506

IOP Conf. Series: Materials Science and Engineering **461** (2019) 012014 doi:10.1088/1757-899X/461/1/012014

Ferritic-martensitic structure in steel hardenable by copper precipitation

J Dlouhy, T Studecky and P Podany

COMTES FHT a.s., Průmyslová 995, 334 41, Plzeň, Czech Republic

E-mail: jaromir.dlouhy@comtesfht.cz

Abstract. The copper precipitation offers possibility of effective and easily achievable ferrite strengthening. There is reported yield point rise by 150 MPa thanks to formation of nanometric copper precipitates in ferrite for Cu content 1.5 wt.%. Such rise of yield strength is most significant for soft and ductile materials, containing free ferrite. Ferritic-martensitic dual steels are example of such material. They are widely used for their favourable combination of strength and ductility. Their yield strength is defined by the soft phase, the ferrite. Ferrite strengthening in these steels is achieved mainly by grain refinement and by solution strengthening caused by substitutional alloying elements. This article describes preparation of dual phase structure in 0.2% C steel alloyed by Cu, Mn, Ti and B. Dual phase structure was prepared by controlled rolling with rolling finishing temperatures in the intercritical region. Samples were analysed by light and scanning electron microscopy. Hardness was measured and tensile tests were performed.

1 Introduction

Copper precipitation is widely studied mean of strengthening ferrite [1 – 4]. Solubility of copper in ferrite decreases with temperature. Copper can precipitate from supersaturated ferrite solution when its concentration rises above 0.5 wt. %. Nanometer-size copper precipitates can strengthen ferrite and increase tensile strength and yield point of steel. Micro alloying by boron and titanium was reportedly successful approach in experiments seeking the most effective way for copper precipitation and structural refinement during thermomechanical processing [5, 6]. Boron increases hardenability and copper suppresses formation of pearlite during continuous cooling. These effects facilitates formation of ferritic-martensitic structure. Titanium is added as binding agent for nitrogen, because nitrogen can form boron nitride and remove in such way boron from solid solution.

This article is devoted to the preparation and annealing of dual ferritic-martensitic structure of copper alloyed steel. Dual F/M steels are widely used as structural steels combining high-strength with sufficient formability and ductility. Their yield strength is limited by yield strength of their weaker constituent – ferrite. Copper precipitation can offer strengthening of the ferrite and thus enhancement of the yield stress of F/M microstructure.

2 Experiment

2.1 Material preparation

Two experimental melts were casted in vacuum induction furnace – lower ("L" melt) and higher ("H" melt) alloyed by copper and manganese. Their chemical composition is in the Table 1. The melts do not differ in content of other elements.

Casted ingots weighing 500 kg were hot forged by hydraulic press into slabs 300 x 80 mm in crossection. The slabs were hot rolled into sheets 30 mm thick. Rolling mill in duo configuration was used with 550 mm rolls diameter and rolling speed 1 m/s. Forging and rolling temperature was 1050°C.
30 mm thick sheets were cut into samples 150 mm wide and 250 mm long. They were milled into thickness 25 mm to remove decarburized layer.
Temperatures A1 and A3 were determined by dilatometric measurement. Samples were 20 mm long and 4 mm in diameter. Both heating and cooling rates were 3 °C/min. There was significant difference between temperatures $Ac_{1,3}$ during heating and temperatures $Ar_{1,3}$ during cooling, as can be expected due to the boron content. Temperature Ar_1 for the melt H could not even be determined, because proeutectoid ferrite formation was followed by bainitic transformation. Thus temperatures $Ac_{1,3}$ will be close to the equilibrium temperatures $A_{1,3}$ and temperatures $Ar_{1,3}$ illustrates stability of austenite. Transformation temperatures are in the Table 1.

Table 1. Chemical composition and temperatures Ac_1, Ac_3, Ar_1, Ar_3 of the experimental materials. Concentrations in wt. %.

Melt	C	Mn	Cu	Si	Ti	B	N	Ac3 [°C]	Ac1 [°C]	Ar3 [°C]	Ar1 [°C]
L	0.21	0.98	1.08	0.08	0.025	0.0013	0.0063	812	719	779	629
H	0.21	2.00	1.49	0.08	0.023	0.0015	0.0057	779	695	691	-

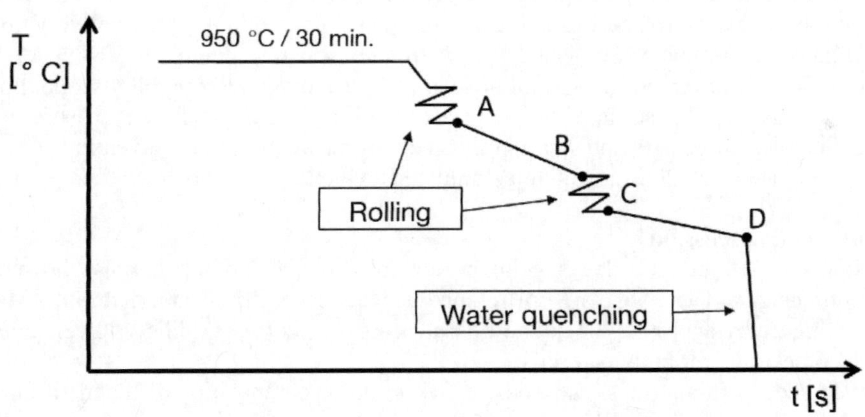

Figure 1. Schematic description of the cotrolled rolling regime.

Table 2. Regimes of controlled rolling.

Regime	A	B	C	D
L 1	830	800	730	680
L 2	830	760	690	650
H 1	830	800	730	700
H 2	830	740	680	650

2.2 *Controlled rolling and thermal treatment*

The controlled rolling was performed by rolling mill described in the section 2.1. Samples were covered by protective coating against decarburization and heated in atmospheric furnace. The scheme of the experimental rolling is in the Figure 1. Heating temperature was 950°C, dwell at the temperature was 30 minutes.

First part of the controlled rolling simulated conventional hot rolling. The rolling operation above point A consisted of two passes through the rolling mill wit summary reduction of 50%. Thickness of the sample in point A was 12.5 mm, temperature 830°C. Sample was cooled in still air until it reached the temperature for the second intercritical rolling operation (see in Table 2). This second rolling consisted in three passes with summary reduction of 55% into final thickness 5.6 mm. Sample was transferred into cooling tunnel and rapidly cooled by water spraying. Delay between intercritical rolling and water spraying was 10 seconds.

Two regimes were performed for each material. Regime 1 included rolling above A_1 temperature, but mainly or entirely under temperature A_3. Regime 2 included rolling around A_1 temperature. Specimens for tensile testing was machined from the samples. Three specimens were tested in as-rolled conditions, three samples were annealed at 400 °C for 1 hour.

2.3 *Sample characterization*

Flat specimens for tensile testing were machined from samples in as-rolled state and annealed state. The specimens had gauge length 45 mm, width 8 mm and thickness of the sample – from 5,3 to 5,6 mm. Ultimate tensile strength (UTS), yield stress (YS), homogeneous plastic deformation (Ag), ductility (A5) and reduction of area (Z) were measured. YS was determined as proof stress at 0.2 % offset. In case of distinctive yield point the upper yield point was determined as YS.

Metallographic analysis was performed on the sections prepared in longitudinal direction. Section were mechanically ground and polished. Microstructure was revealed by 3 % Nital solution. Microstructure was observed by light microscope and scanning electron microscope. Phase composition was roughly estimated on the base of the image analysis. Exact measurement was not possible due to heterogeneity and small grain size in some areas. Values given in following chapter are orientation values with estimated uncertainty ±10 vol.%.

3 Results and Discussion

Tensile test results of the as-rolled specimens are in Table 3. Both materials showed significant deformation hardening. There was large difference between strength of both materials. Material 2 had UTS and YS higher by approx. 500MPa and its ductility was correspondingly lower. Interestingly, Z values were comparable for both materials.

Large difference in strength was caused by different phase composition of L and H samples (see Figure 2). Structure of material L was composed of ferrite, martensite, bainite and pearlite (sorted from the most abundant to the rarest). Structure was heterogeneous. Ferrite content was 50 vol.% in the center, but increased towards the surfaces of the rolled sheets up to 75 vol.%. Pearlite was rare in the middle, but its content also rose towards the sheet surfaces. Difference between regimes 1 and 2 was observable only in pearlite content. Pearlite grains in regime 1 was up to 5 vol.% nearby the surface and only about 1 vol.% in case of regime 2. This can be caused by fast cooling of the surface area by contact with the rolls. Temperature could drop so low that pearlite transformation was not possible. Cooling was slower in the middle of the sheet and therefore pearlite transformation had time to start.

H samples were formed mostly by mixture of martensite, acicular ferrite and bainite with minor content of free polygonal ferritic grains. Structure was highly heterogeneous. Bands of pure martensite were present in the structure as well as narrow bands of ferritic grains. There was clear difference between regimes H 1 and H 2. Regime H1 resulted in mostly bainitic microstructure in whole thickness of the sheet. There was only small amount of free polygonal ferrite, most of the ferrite was acicular or bainitic. Ferrite content was ca 25 vol.%. Regime H2 resulted in significant higher content of polygonal ferrite in vicinity of the sheet surface. Centre of the sheet was similar to the sample H1. Overall ferrite content was also ca 25 vol.%.

Table 3. Results of mechanical testing in as-rolled state. UTS – ultimate tensile strength, YS – yield stress at 0.2% offset, Ag – homogeneous plastic deformation, A5 – ductility, Z – reduction of area.

Regime	**UTS [MPa]**	**YS [MPa]**	**Ag [%]**	**A5 [%]**	**Z [%]**
L 1	1059	516	11.8	16.8	25
L 2	1084	533	10.0	14.2	20
H 1	1604	1158	2.9	8.6	23
H 2	1552	1048	4.0	7.0	16

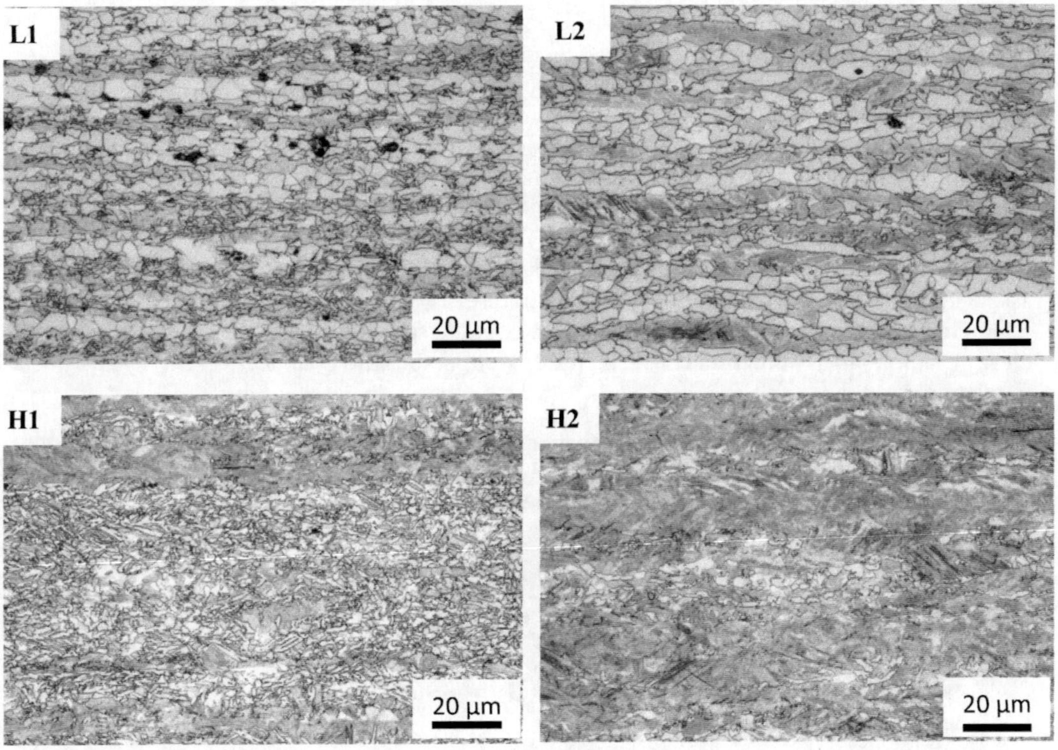

Figure 2. Micrographs of specimens in as-rolled state. Ferrite is white, martensite gray and pearlite forms dark areas (in L1 and L2)

Difference between materials L and H was not only in phase composition, but also in ferrite grain size (see Figure 3). Mostly polygonal ferrite grains in material L changed their diameter gradually from 8 µm in the center to 3 µm nearby the surface of the sheet. Polygonal ferrite grains in material H had size max. 3 µm, but most of the ferrite was acicular. Width of the ferrite plates was usually up to 2 µm, their length up to 10 µm. Acicular ferrite occurred in mixture with M/A islands. Large individual ferrite plates 1 µm wide could be categorized as free acicular ferrite, but fine ferrite sheaves mixed with M/A plates fulfills definition of granular bainite.

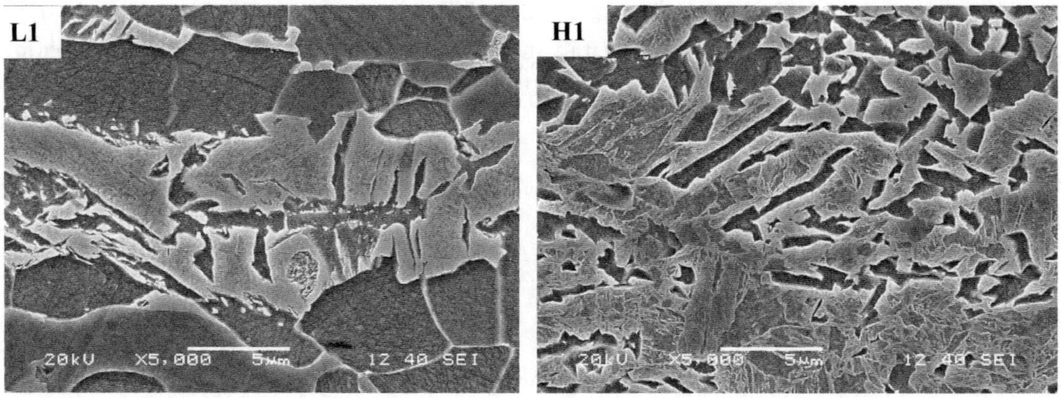

Figure 3. Micrographs of specimens in as-rolled state from SEM.

Annealing at 400°C for 60 minutes caused tempering of martensite and also provide favorable conditions for copper precipitation, as can be seen in Figure 4. Tempered martensite in materials L and H had different spatial density of precipitated carbides. It can be attributed to the higher carbon content of martensite in material L. There was less martensite in the material L compared to material H, thus all carbon remained concentrated in smaller volume. Carbon content in martensite was circa 0.4 wt.% for material L and 0.27 wt.% for material H based on estimated ferrite contents for these materials.

Copper precipitates cannot be identified by means of SEM analysis. Their observation can be only assumed on the basis of their appearance in literature. There were observed small particles less than 50 nm in diameter in ferrite in all samples. However, these particles were not well defined and clear in samples in as-rolled state. Much clearer were in the case of annealed samples, particularly in material L.

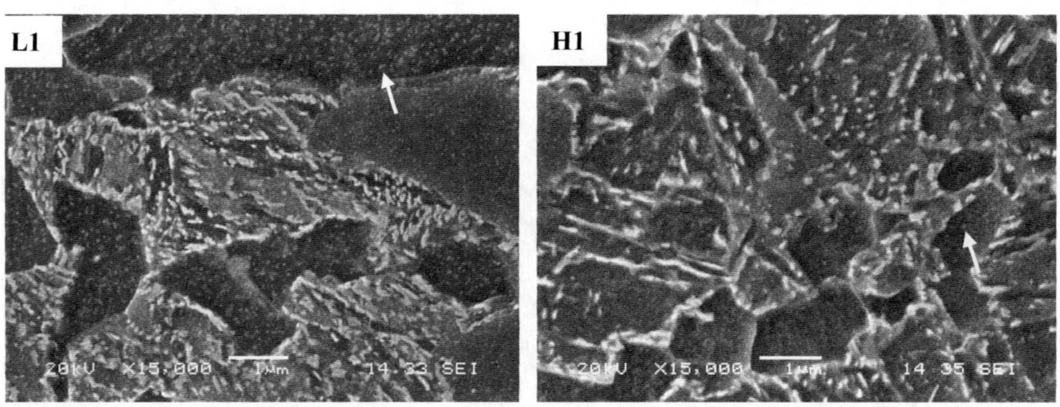

Figure 4. Micrographs of specimens in annealed state taken at higher magification. Aroows points to one of the particles in the ferrite, assumed to be copper precipitates.

Mechanical properties after annealing showed the same trend for all samples (see Table 4). YS increased and UTS decreased. In fact, they were almost identical for the material H. Material L retained more of the deformation hardening capability. Homogeneous plastic deformation slightly decreased in all cases, but ductility A5 increased as well as reduction of area. Also, there were much less difference between regimes 1 and 2 in comparison with as rolled state for both materials.

Table 4. Results of mechanical testing after annealing. Values of YS marked by asterisk (*) mean upper yield point. The specimens exhibited distinctive yield point in that case.

Regime	**UTS [MPa]**	**YS [MPa]**	**Ag [%]**	**A5 [%]**	**Z [%]**
L 1	839	785*	7.9	18.2	52
L 2	842	776*	7.8	18.6	48
H 1	1246	1231	2.6	11.8	40
H 2	1206	1186	2.6	10.8	47

Annealing also lead to the appearance of distinctive yield point in case of material L and to the sharp bend of material H stress-strain curve.

4 Conclusion

Complex-phase structure was prepared by controlled rolling. Prevailing ferritic-martensitic structure contained also small amount of bainite and pearlite. Alloying had strong effect on the resulting microstructure and properties. On the other hand, there was no significant difference upon changing final rolling temperature (FRT) from intercritical region between A3 and A1 to the subcritical temperature below A1. Both materials are favourable from the technological point of view because their properties does not depend strongly on FRT in examined range. There was significant heterogeneity observed between the centre and nearby the surface of sheets. Final thickness 5.6 mm was apparently too large and homogeneous structure could be achieved by rolling to thinner sheets.

Annealing at 400°C almost erased differences in sheet properties gained from rolling with different finish rolling temperatures. Copper precipitation might contribute to the significant rise of yield stress of ferritic-martensitic microstructure. It was accompanied by rapid decrease of deformation strengthening capability. Structure with approximate ferrite : martensite ratio 1:1 retained some of it, but structure with prevailing martensite exhibited almost none deformation strengthening after annealing.

Acknowledgements

The results presented in this paper arose under the project “Synergy of precipitation, deformation and transformation hardening in steels with higher copper content” No. 17-19002S funded by Czech Science Foundation.

These results were also created under the project entitled Development of West-Bohemian Centre of Materials and Metallurgy No.: LO1412, financed by the Ministry of Education, Youth and Sports of the Czech Republic.

References

[1] Takahashi, J., *et al.*, “Consideration of particle-strengthening mechanism of copper-precipitation-strengthened steels by atom probe tomography analysis,“ *Materials Science and Engineering A*, Vol. 532, (2012), pp. 144-152.

[2] Rana, R., *et al.*, “Development of high strength interstitial free steel by copper precipitation hardening,“ *Materials letters*, Vol. 61, (2007), pp. 2919-2922.

[3] Fine, M.E., *et al.*, “Origin of copper precipitation strengthening in steel revisited,“ *Scripta Materialia*, Vol. 53, (2005), pp. 115-118.

[4] Isheim, D., *et al.*, “An atom-probe tomographic study of the temporal evolution of the nanostructure of Fe–Cu based high-strength low-carbon steels,“ *Scripta Materialia*, Vol. 55, (2006), pp. 35-40.

[5] Podany, P. and Martinek, P. “Thermomechanical processing of micro-alloyed steel,“ *Materiali in Tehnologije*, Vol. 48, No. 6 (2014) pp. 855-859.

[6] Ghosh, S.K., *et al.*, “Effects of Cu addition on the synergistic effects of Ti–B in thermomechanically processed low carbon steels,“ *Materials Science and Engineering A*, Vol. 527, (2010), pp. 1082-1088.

Comparative research on abrasive blasting of 145Cr6 steel for various working media

D Dudek

Kielce University of Technology, Kielce, Poland, EU

E-mail: dominik_dudek@op.pl

Abstract. This paper deals with the influence of testing elements made of 145Cr6 steel after heat treatment. The elements have been machined with various abrasive materials using abrasive blasting. The abrasive materials were: quartz sand, two corundum sizes and two sizes of Brown Fused Alumina. The treatment of all types of abrasives took place at the same working pressure of 0.55 MPa. The 2D and 3D roughness parameters obtained after the abrasive blasting were analyzed. Roughness measurements were made using the Talysurf CCI Lite optical profilometer. Microscopic observations were performed on a Nicon Eclipse MA 200 inverted microscope.

1 Introduction

Modern technology development confronts many engineers with problems that they have to solve. The development of modern technologies is associated with the emergence of modern construction materials. There are also modifications and improvements of materials known for a long time [1,2,3]. Advances in materials engineering have resulted that currently produced materials are characterized by: increasingly higher strength properties, increased hardness, abrasion resistance, etc. For the processing of modern materials, modifications of well-known methods of their processing are often used, such as microwelding [4,5,6,7]. One of the types of known machining methods are surface treatments. Surface treatments are widely used in modern production processes. Due to their diversity and versatility, it is possible to use complicated technologies such as laser technology [8] or relatively simple thermal methods to implement them. Their diversity means that surface treatments are used in many cases, for example, changes in the functional properties of the native material [9,10] (eg surfacing) [11], applying various coatings (protective, decorative, technical, etc.) or as finishing treatment [12].

Abrasive blasting is used in the processing of metal elements and it can be also used as a treatment aimed at preparing the surface before the next stages of the technological process or to remove oxides, lacquer coatings, prior to gluing processes to develop the surface in relation to the surface nominal as well as finishing treatment of the surface of workpieces [13,14].

2 Blasting - abrasive - characteristics

In the blasting process, the prepared abrasive grains suspension in the working medium (which may be liquid or gas) is sprayed onto the workpiece surface of the object. If water is used as the carrier medium, it is advisable to add corrosion inhibitors. The reinforced abrasive grit, due to its properties, allows machining of complex shapes. It also allows one to process large surfaces. The technological parameters of processing (such as process efficiency, roughness) are influenced by many factors. These are: the type and size of grains of the working medium, the distance of the spraying device from the work surface, working pressure, treatment time, an angle of the abrasive stream relative to the surface to be machined and the type of workpiece. Abrasive blasting is used mainly in cases where there are no excessive requirements as to the dimensional accuracy of the workpieces.

In the abrasive blasting section, the working fluid pressure which is usually applied is in the range 0.2 - 1.2 MPa. The abrasive jet sprayed by the nozzle at the given pressure range reaches the speed of 200 - 800 m / s. During the process, the material layer being removed is usually 0.01 - 0.1 mm. Typically, the angle of the nozzle axis of the spray device relative to the work surface (κ) is in the range of 15 - 90°. At κ angles above 30°, we obtain an additional effect of strengthening the surface layer to a depth of a few micrometres. In addition, for the described values of the angle κ, a favorable distribution of compressive stresses is obtained.

3 Experimental research

The tests were carried out on cylindrical samples made of 145Cr6 steel after heat treatment. This steel is a tool steel for a cold work. It is characterized, inter alia, by a high abrasion resistance.

The chemical composition of steel is shown in Table 1.

Table 1. Chemical composition (%) 145Cr6 steel

C	Si	Mn	P	S	Cr	Ni	Mo	Al	Cu	V	W	Ti	Co
1.33	0.25	0.49	0.018	0.005	1.42	0.02	0.032	0.14	0.14	0.11	0.01	0.004	0.01

The samples before the experimental tests were heat treated (hardening, low tempering - the obtained hardness of the samples was 62 HRC). The front faces of the samples have been sanded. The blasting process was carried out in the PK-700 sandblasting machine.

Prior to the actual tests, preliminary tests were performed in which machining parameters used in further experimental work were determined as follows:

• air stream pressure p = 0.55 MPa, the distance of the nozzle of the element gushing from the surface of the processed sample equals 80 mm;
• the angle of the nozzle axis of the gushing element relative to the work surface κ ≈ 45°;
• the nozzle diameter of the spout element equals 6 and 7 mm;
• processing time up to 40 s.

There were five different working materials used in the study. Their summary is presented in Table 2.

Table 2. A summary of the working media used in the experiment and the nozzles of the gushing device used accordingly

	Quartz sand	Brown Fused Alumina A-20B	Brown Fused Alumina A-20B	Corundum	Corundum
the size of the abrasive grains	0.2-0.8 mm	F10	F24	0.2-0.5	0.5-1.4
diameter of the nozzle	6 mm	7 mm	6 mm	6 mm	6 mm

4 Results and discussion

After the abrasive blasting process, all samples were cleaned and subjected to observations, using light microscopy techniques and roughness measurements. The observations were made on a Nicon Eclipse MA 200 micrograph of an inverted type microscope. In Figure 1, there is shown an example of a photo of the surface of a C45 steel element after blasting.

Figure 1. Surface of the sample of 145Cr6 steel after abrasive blasting (working medium - Corundum 0.2 - 0.5 mm, diameter of the nozzle 6 mm)

The table 3 presents examples of 2D and 3D roughness parameters.

Table 3. A summary of exemplary measured parameters of 2D and 3D roughness

	Quartz sand	Brown Fused Alumina A-20B	Brown Fused Alumina A-20B	Corundum	Corundum
Ra, μm	1.095	3.703	3.850	0.884	0.851
Rz, μm	8.529	23.622	22.753	8.177	6.943
Sa, μm	1.173	5.085	5.324	0.899	0.891
Sz, μm	13.102	66.967	39.846	16.105	9.235

The surface roughness and a *3D* surface profile was measured for all the samples. Roughness measurement was carried out using a Talysurf CCI Lite *3D* non-contact surface profiler.

The measured and analyzed *2D* roughness parameters were: *Rp* (μm) – Maximum peak height of the roughness profile, *Rv* (μm) –- Maximum valley depth of the roughness profile, *Rz* (μm) – Maximum height of the roughness profile, *Rc* (μm)– Mean height of the roughness profile elements, *Rt* (μm) – Total height of roughness profile, *Ra* (μm) – Arithmetic mean deviation of the roughness profile, *Rq* (μm) –Root-Mean-Square (*RMS*) Deviation of the roughness profile, *Rsk* – Skewness of the roughness profile, *Rku* – Kurtosis of the roughness profile, *Rmr* (%) – Relative Material Ratio of the roughness profile, *Rdc* (μm) – Roughness profile Selection Height difference,

RSm (mm) – Mean Width of the roughness profile elements, *Rdq* (deg) – Root – Mean – Square slope of the roughness profile.
The measured and analyzed *3D* roughness parameters were: *Sq* (μm) – Root mean square height, *Ssk* (μm) – Skewness, *Sku* (μm) – Kurtosis, *Sp* (μm) – Maximum peak height, *Sv* (μm) – Maximum pit height, *Sz* (μm) – Maximum height, *Sa* (μm) – Arithmetic mean height [15]
Figure 2 presents an example of a graph of the polar roughness distribution for sample No. 3, (a variant of an abrasive blasting using A-20 B ordinary corundum, size F24). One can see no roughness directionality - randomness.

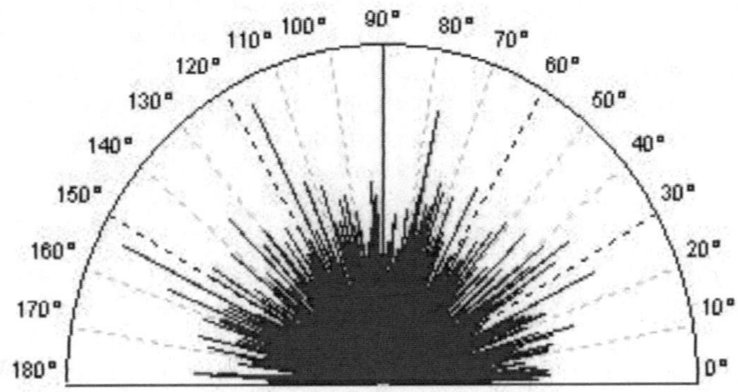

Figure 2. The graph of the polar roughness distribution for sample No. 3

Figure 3 presents exemplary roughness results for sample no. 3 (a variant of abrasive blasting using A-20 B ordinary corundum, size F24).

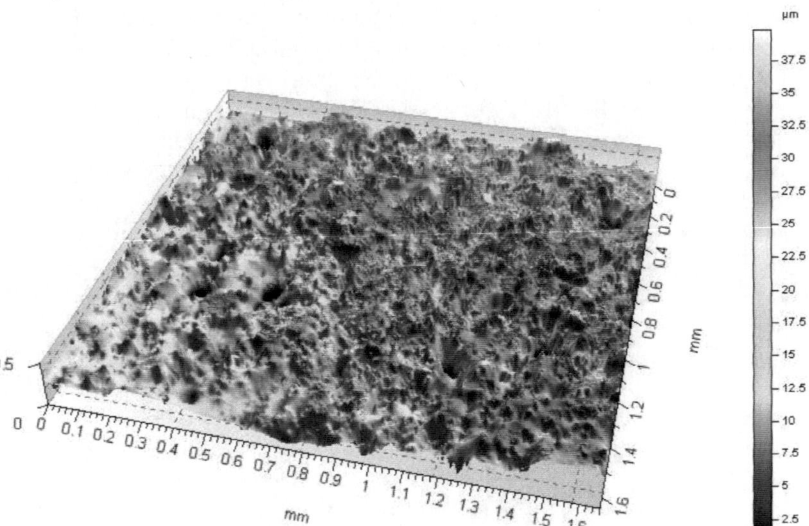

Figure 3. The 3D roughness results for sample no. 3

The tests carried out and the analysis of the obtained results show that depending on the working medium used, abrasive grains obtain different energy at the same working pressure. It results directly from the dependence on the kinetic energy of the particle, which depends on the mass of the particle and its velocity. It was found that it is necessary to select the machining parameters, such as: type and size of abrasive grains, air working pressure, the angle of the sprinkler nozzle to the work surface and the distance of the device gushing from the work surface depending on the type of material and the assumed technological effects of the process. By an appropriate selection of the size of abrasive grains, one can regulate both the intensity of the layer remove from the surface of the workpiece and the degree of surface development of the workpiece.

The analysis of 3D surface topography and the polar graph of roughness distribution for all samples shows the lack of directivity of the geometric structure of the surface.

The use of abrasive grains with low granulation results in a reduction in the intensity of interactions with the treated surface. As a result of the treatment with low granulation grains, an even distribution of roughness on the work surface is obtained. In the case of using larger sizes of abrasive grains, due to their relatively high kinetic energy, they may be formed as a result of their interactions with the surface treated with deep craters. Abrasive blasting can also be used for relatively hard materials, such as heat-treated cast iron, etc., but this is associated with a much greater wear of the working medium in the form of abrasive grains.

Conclusions

The size of abrasive grains regulate both the intensity of the layer remove from the surface of the workpiece and the degree of surface development of the workpiece. The surface roughness after the abrasive blasting process is undetermined - is random. As a result of the treatment with low granulation grains, an even distribution of roughness on the work surface is obtained. The using larger sizes of abrasive grains on the surface may be make deep craters. The surface after blasting is more susceptible to corrosion, hence the abrasive slurry should contain corrosion inhibitors.

References

[1] Szwed B and Konieczny M 2016 Influence of bonding pressure on the dissimilar joints of titanium and stainless steel with an aluminium interlayer In METAL 2016: 25rd International Conference on Metallurgy and Materials. Brno Czech Republic May 25-27: TANGER pp 1552-57

[2] Konieczny M and Szwed B 2016 Effect of cooling rate on titanium and stainless steel joints performed with copper interlayer In METAL 2016: 25rd International Conference on Metallurgy and Materials. Brno Czech Republic May 25-27: TANGER pp 1456-61.

[3] Szwed B and Konieczny M 2016 Microstructure and Mechanical Properties of Joints of Titanium with Stainless Steel Performed Using Nickel Filler Archives of Metallurgy and Materials 61(2) pp 997-1001

[4] Depczyński W, Młynarczyk P, Spadło S, Ziach E and Hepner P 2015 The selected properties of porous layers formed by pulse microwelding technique. In METAL 2015: 24rd Anniversary International Conference on Metallurgy and Materials, Ostrava: TANGER, pp. 1087-92.

[5] Młynarczyk P, Spadło S, Depczyński W, Śliwa E and Strzębski P 2015 The selected properties of the connection superalloy Haynes h 230 (r) using microwelding title of paper. In METAL 2015: 24rd Anniversary International Conference on Metallurgy and Materials, Ostrava: TANGER, 2015, pp. 792-7.

[6] Młynarczyk P, Spadło S and Depczyński W 2017 Investigations into the effects of spot welding on thin sheet of superalloys Hastelloy x and Haynes 230 (r). In METAL 2017: 26th International Conference on Metallurgy and Materials. Ostrava: TANGER, 2017, pp. 1881-1886.

[7] Spadło S, Młynarczyk P, Depczyński W, Skowron E and Mijas R 2017 Research on the influence of electron beam welding of titanium alloy microstructure connection. In METAL 2017: 26th International Conference on Metallurgy and Materials. Ostrava: TANGER, 2017, pp. 1939-1944.

[8] MUCHA Z, WIDLASZEWSKI J, KURP P and Mulczyk K 2016 Mechanically-assisted laser forming of thin beams. LASER TECHNOLOGY 2016: PROGRESS AND APPLICATIONS

OF LASERS. Book Series: Proceedings of SPIE Volume: 10159. Article Number: 101590U. Published: 2016.

[9] Młynarczyk P and Spadło S 2016 The Analysis of the Effects Formation Iron - Tungsten Carbide Layer on Aluminum Alloy by Electrical Discharge Alloying Process. In METAL 2016: 25th Anniversary International Conference on Metallurgy and Materials, Ostrava: TANGER, 2016, pp. 1109-1114

[10] Spadło S, Depczyński W and Młynarczyk P 2017 Selected properties of high velocity oxy liquid fuel (hvolf) - sprayed nanocrystalline wc-co infralloy(tm) s7412 coatings modified by high energy electric pulse, METALURGIJA, vol: 56 Issue: 3-4 pp: 412-414, Published: JUL 2017

[11] Dudek D, Stefańczyk K and Skowron E 2017 An analysis of the properties of chromium - based coatings deposited by TIG welding. METAL 2017: 26th International Conference on Metallurgy and Materials. 2017. pp: 1246-51.

[12] Dudek D, Spadło S and Czernikiewicz K 2017 The use of vibratory tumbling to produce a smooth surface finish on sprocket wheels. METAL 2017: 26th International Conference on Metallurgy and Materials. 2017. pp: 1240-45.

[13] Achtsnick M, Geelhoed P.F, Hoogstrate A.M and Karpuschewski B 2005 Modelling and evaluation of the micro abrasive blasting process- WEAR Volume 259, Issues 1–6, July–August 2005, Pages 84-94

[14] Celotta D.W, Qureshi U.A, Stepanov E.V, Goulet D.P, Hunter J, Buckberry C.H, Hill R, Sherikar S.V, Moshrefi-Torbati M and Wood R.J.K. 2007 Sand erosion testing of novel compositions of hard ceramics. WEAR Volume 263, Issues 1–6, 10 September 2007, Pages 278-28

[15] Adamczak S, Miko E and Cus Franc 2009 Strojniski Vestnik, A model of surface roughness constitution in the metal cutting process applying tools with defined stereometry; -Journal of Mechanical Engineering; No 55, pp. 45-54.

Improved corrosion resistance of WE43 magnesium alloy with continuous network of MgF_2 prepared by powder metallurgy

Drahomir Dvorsky*[1], Jiri Kubasek[1], Eva Kristianová[1], Dalibor Vojtech[1]

[1]Department of Metals and Corrosion Engineering, Faculty of Chemical Technology, University of Chemistry and Technology Prague, Technická 5 166 28 Praha 6 – Dejvice, Czech Republic.

E-mail: dvorskyd@vscht.cz* (corresponding author), Jiri.Kubasek@vscht.cz, Dalibor.Vojtech@vscht.cz, Eva.Kristianova@vscht.cz

Abstract. This paper deals with the preparation of the composite materials with a continuous network of MgF_2 in magnesium matrix. Powder of the WE43 magnesium alloy was boiled in NaOH to obtain a thick layer of $Mg(OH)_2$ and it was subsequently immersed in HF. $Mg(OH)_2$ was conversed into a thick layer of MgF_2. The chemically treated powders were processed by spark plasma sintering. As a result, composite materials with a continuous network of MgO and MgF_2 respectively were formed. The corrosion resistance of the sample containing the network of MgF_2 was superior compared to the products from untreated powder.

Keywords: Composites; Corrosion; Sintering; Powder metallurgy.

1 Introduction

Magnesium alloys are considered as suitable biodegradable materials as they have similar mechanical properties as a bone tissue [1]. The main disadvantage of magnesium lies in the high corrosion rate which is associated with hydrogen release. The corrosion rate may be reduced by alloying, preparation method or by coatings. Alloying is associated with including another element which may result in lowered biocompatibility, but it can significantly improve corrosion resistance. WE43 alloy (2.4-4.4 wt.% of Nd, 3.7-4.3 wt.% of Y, >0.4 wt.% of Zr, >0.5 wt.% of Gd and Dy) is example of magnesium alloy with improved corrosion resistance due to the positive effect of neodymium and yttrium dissolved in the matrix [1], [2], [3].

The preparation method is associated with the structure of the material. Finer and more homogeneous structures are usually characterized by higher corrosion resistance [4]. The preparation method has also influence on the formation of intermetallic phases. Intermetallic phases should be small and homogeneously dispersed in the matrix and have potential close to the magnesium matrix. Powder metallurgy method resulted in many cases in the improved mechanical and corrosion properties [4], [5]. The atomized powder is the input material used for the sample preparation. It is characterized by a very fine structure which can be preserved if the compacting method is fast enough. Spark plasma sintering (SPS) is a progressive method of sintering by high current [6]. The advantage of SPS is a short time of exposure to the high temperature (several minutes) and therefore the material preserves its fine structure after atomization. The compact sample is usually characterized by low porosity.

Coatings may further reduce the corrosion rate of substrate materials. Magnesium fluoride conversion coating was reported to have a very positive effect on the corrosion resistance even in low thicknesses of the coating [7], [8], [9], [10], [11]. The fluoride coating can be easily prepared by immersion in hydrofluoric acid for several hours. As a result, a very thin film (0.1 to 4 μm) of MgF_2 is created on the surface of the material [7], [9], [10], [11]. The thickness of the coating can be adjusted by immersion time and by the concentration of HF as well as by temperature. Another advantage of fluoride coating is in the good adhesive properties (33 to 43 MPa) [7], [12], [8], [9]. The corrosion rate is reduced by the barrier effect of the hardly soluble MgF_2 layer. The positive effect of fluoride coating was reported for the majority of magnesium alloys [7], [8], [9]. On the other hand, Trinidad et al. [13] and Höhlinger et al. [14] showed some difficulties with the coating of the WE43 alloy. Both authors observed inhomogeneous coating. The fluorine was concentrated around intermetallic phases. Nevertheless, the coating was effective even though it was inhomogeneous. The good biocompatibility of MgF_2 coatings was proven by cytotoxicity tests as well as by animal tests [10], [12].

The disadvantage of every coating is its localized disruption by possible mechanical loading. The disturbed coating no longer protects the coated material and the corrosion is localized on the one spot with increased intensity. Uneven corrosion may lead to premature implant failure due to the reduced loading area. The solution to this danger can be a composite material.

We have introduced the innovative way of preparation of the composite material with the continuous network of MgF_2 via powder metallurgy [15]. It was proven that this structure has a strong impact on the corrosion resistance and mechanical properties. This paper applies this method to the WE43 alloy. To prevent the inhomogeneity of the coating, the powder was first boiled in the NaOH as was proposed by Weber et al. [16] for MgNd2 alloy. The pretreatment converted the surface into $Mg(OH)_2$ which could be easily converted into MgF_2.

2 Materials and methods

2.1 Fluoride treatment of the powder

Gas atomized powder of WE43 alloy (4.0 wt.% of Y, 2.2 wt.% of Nd, 0.4 wt.% of Gd, 0.4 wt.% of Dy, 0.5 wt.% of Zr) was characterized by laser diffraction to determine the particle size, which ranged between 10 and 200 μm. Fifty grams of powder was boiled in NaOH solution with a concentration of 200 g/l for 2 hours. The boiled powder was then rinsed with distilled water and ethanol. The powder was dried at 50 °C in air. Part of the boiled powder was immersed and stirred in the 300 ml of 40% HF for 24 h. The HF was then poured out and the powder was rinsed with distilled water multiple times, filtered through filter paper and rinsed with ethanol multiple times. The powder was then dried at 50°C in air. In the following text, the uncoated material is referred as WE, the boiled powder as WE-BT and boiled with fluoride treatment as WE-BT+FT.

2.2 Compacting of samples

Both coated and uncoated WE43 powders were processed by the SPS method at 500 °C with a heating rate of 100 °C/min and 7 kN pressure level and with the operation time of 10 minutes. The SPS machine HP D 10 FCT system GmbH was used. The final cylindrical samples had a diameter of 20 mm and a height of 10 mm.

2.3 Microstructure characterization

The input powders and microstructure of the compact materials were characterized by scanning electron microscope SEM TescanVEGA3 LMU with energy dispersion spectrometry (EDS, AZtec). Samples were mounted in an epoxy resin. Subsequently, they were ground on SiC grinding papers (P80-P2500) and polished on diamond paste D3, D2, D0.7. The final polishing was performed on Etosil F. Porosity was determined by image analysis of 10 sections of each sample. Phase analyses were performed using X-ray diffraction (XRD, X'Pert Philips, 30 mA, 40 kV, CuKα X-ray radiation).

2.4 Mechanical properties
Compressive tests were measured using LabTest 5.250SP1-VM at room temperature on rectangular samples with dimensions of 5×5×7 mm. Constant deformation speed of 0.001 s^{-1} was used. Compressive yield strength (CYS), ultimate compressive strength (UCS) and total deformation to fracture were determined.

2.5 Immersion test
Immersion tests were performed in simulated body fluid at 37 °C for 14 days. The ratio of solution volume to the surface area was 100 ml/cm^2. Samples were removed from the immersion solution and were rinsed in distilled water and dried. The corrosion products were removed by the solution of 200 g/l CrO_3, 10 g/l $AgNO_3$, 20 g/l $Ba(NO_3)_2$ according to ASTM-G31-72. Samples were then dried and weighted. The corrosion rate was calculated from weight changes after immersion and removing of corrosion products.

3 Results and discussion

3.1 Microstructure characterization
The atomized powder was characterized by round shape particles (**Figure 1A**). The very fine dendritic microstructure was presented in each particle. XRD analysis revealed the presence of α-Mg solid solution and metastable β^1-phase, with a composition close to β-phase ($Mg_{14}Nd_2Y$). Boiling in NaOH resulted in a thick layer of $Mg(OH)_2$ which was detected by XRD and is also displayed in **Figure 1B**. Subsequent immersion in HF led to the conversion of $Mg(OH)_2$ into the thick and homogeneous coating of MgF_2. There were residues of $Mg(OH)_2$ which were not detected by XRD analysis. However, they were noticeable by EDS analysis especially around small particles (**Figure 1C**).

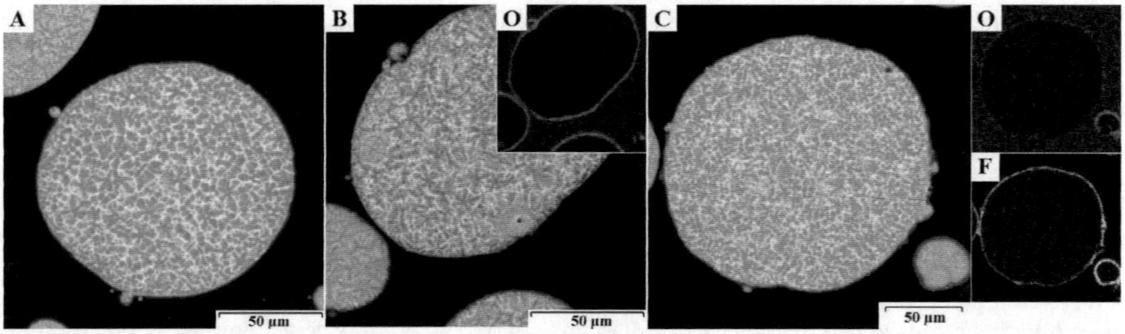

Figure 1. The microstructure of the powder particles in cross-sections: A) WE, B) WE-BT with EDS distribution map of O, C) WE-BT+FT with EDS distribution map of O and F.

All powders were processed by spark plasma sintering in order to obtain compact samples. The final structures are displayed in **Figure 2**. Individual particles are distinguishable in all samples. The WE sample is characterized by a small porosity of 0.1 % and by shells rich on Y and partially O around each particle **(Figure 2A**). This behavior is typical for Mg-Y alloys as a higher content of Y improves resistance to oxidation at higher temperatures by the formation of Y_2O_3 on the surface [17]. However, as the Y in the solid solution is consumed for the formation of this layer, the area beneath it (about 4 µm) is depleted on Y. Nevertheless, the Y_2O_3 layer should have a positive effect on the corrosion resistance [17]. On the other hand, the network of depleted areas has a negative effect on the corrosion behavior, because the micro-galvanic corrosion is promoted by the inhomogeneous concentrations of Y. The insides of the particles are the same for all samples and they consist of the α-Mg solid solution and very fine intermetallic phases. These phases were identified as $Mg_{24}Y_5$ and β-phase ($Mg_{14}Nd_2Y$).

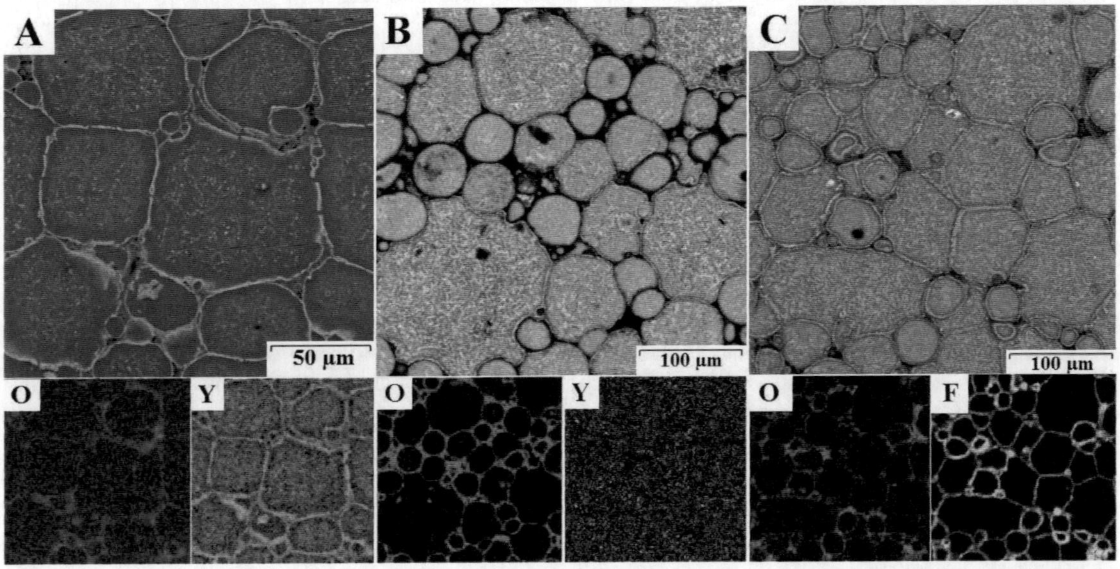

Figure 2. The microstructure of A) WE, B) WE-BT, C) WE-BT+FT. The colour figures represent the occurrence of specific elements in the microstructure.

Material prepared by compacting of the powder boiled in NaOH (WE-BT) resulted in structure displayed in **Figure 2B**. The porosity of this sample was about 4.6 %. This can be explained by the decomposition of $Mg(OH)_2$ into MgO and H_2O at an elevated temperature during sintering, which disturbed the sintering process. Nevertheless, the structure with a continuous network of thick MgO and without any yttrium rich shells is created. The total amount of oxygen in the surface layer of the sample was measured by the EDS analysis and it was determined to be about 14.0 ± 2.3 %. However, part of it comes from the partially oxidized surface after polishing.
WE-BT+FT sample was characterized by a porosity of 2.8 % due to the presence of residues of $Mg(OH)_2$, which is decomposed. Otherwise, the continuous network of thick MgF_2 was successfully formed without any yttrium enrichment of the shells (**Figure 2C**). The enrichment of the shells was prevented by the thick layer of magnesium oxide and magnesium fluoride respectively. The thickness of the MgF_2 interface was measured by line scan analysis by EDS and it was determined to be 4-6 µm. The total amount of oxygen and fluorine in the compacted WE-BT+FT sample, measured by EDS analysis, was about 3.2 ± 0.3 wt.% and 5.8 ± 0.5 wt.% respectively. The amount of F is almost by 3 wt.% higher value than that measured for the Mg-MgF_2 composite material prepared by the simple immersion of the powder in HF and processing by SPS [15]. The increased value of F can be attributed to the pretreatment by boiling of the powder.

3.2 Mechanical properties

Compressive properties are summarized in Tab.1. Properties of human bone are displayed for comparison. The properties of prepared fluoride composite meet the properties of bone tissue and may be therefore used in medicine as materials for biodegradable implants. The mechanical properties of oxide composite are very low due to the high porosity and worse adhesive properties of the oxide layer. The compressive yield strength of the WE-BT+FT composite material is comparable with the uncoated sample. However, the ultimate compressive strength and plasticity are severely reduced. This reduction may be associated primarily with porosity and the residues of MgO in the interfaces as well as with the adhesion strength of the fluoride coating. Also there might be slower diffusion between magnesium fluoride layers due to its higher melting point. Contrary, improvement of mechanical properties after application of fluoride network was observed in previous study with pure Mg [15].

Table 1. Mechanical properties of the prepared materials compared with the bone [18], [19].

	CYS [MPa]	UCS [MPa]	D [%]
WE	225 ± 5	421 ± 10	16.0 ± 0.9
WE-BT	-	125 ± 23	1.2 ± 0.6
WE-BT+FT	210 ± 8	297 ± 18	7.2 ± 1.4
Bone [18], [19]	130 - 180	-	-

3.3 Corrosion properties

The corrosion tests were performed in simulated body fluid at 37°C for 14 days. The corrosion rate of bare WE was 3.15 ± 0.27 mm/y. This is a relatively high corrosion rate as was expected due to the existence of depleted areas on Y. Nevertheless, the assumption that the Y_2O_3 layer improves corrosion resistance was confirmed in **Figure 3A**, as the corrosion front is slowed down on the yttrium rich interface. This interface works as a hardly soluble barrier, however, after disturbing the barrier the corrosion quickly progress through Y depleted areas as they are prone to corrosion.

The corrosion rate of WE-BT composite was 6.5 ± 0.44 mm/y, which is twice higher than the bare WE material. This may be due to the high porosity. The corrosion progresses through the material then the corrosion products may peel the whole layer of uncorroded material. Otherwise, the corrosion front is also slowed down on the MgO interface (**Figure 3B**).

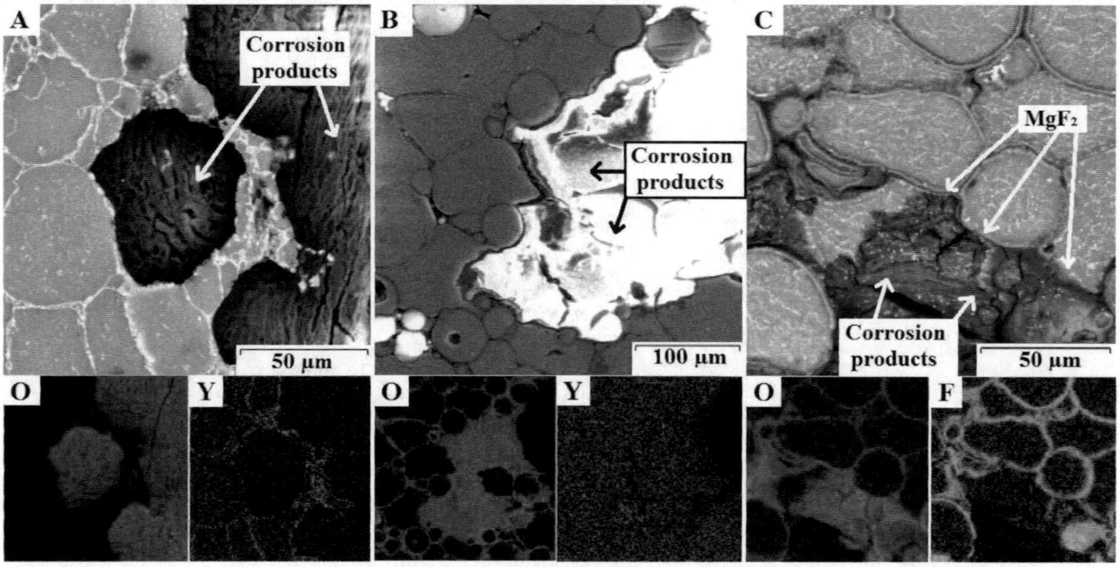

Figure 3. Cross-section of the material after corrosion A) WE, B) WE-BT, C) WE-BT+FT. The colour figures represent the occurrence of specific elements in the microstructure.

The WE-BT+FT composite material was characterized by the lowest corrosion rate of 1.69 ± 0.05 mm/y. The reduced corrosion rate is associated with the fluoride interfaces which effectively slowed down the corrosion (**Figure 3C**). The effectiveness can be demonstrated by the amount of fluorine on the surface after removing of the corrosion products. The amount of F raised from 5.8 to 19.5 wt.%. The corrosion of the material is expected to be generally more uniform as the corrosion is slowed down on each barrier. Such material should not suffer from the deep localized attacks caused by impurities like Fe or Ni [20], as the effect of these impurities should be localized to just one particle. The corrosion rate was reduced on the half-value compared with the bare WE. In the previous study with Mg, the corrosion rate was reduced on one third [15]. This might be connected with the porosity of the material which promotes corrosion.

4 Conclusion

Powder of the WE43 alloy was boiled in NaOH and subsequently chemically treated in hydrofluoric acid. As a consequence, the coating of MgF_2 was formed on the surface. Treated powders were compacted by spark plasma sintering. Specimen prepared from non-treated atomized powder was characterized by the oxide shells rich in yttrium and areas depleted in alloying elements beneath them, which degrade corrosion properties of the material. Materials prepared from powders processed by boiling in NaOH were characterized by increased porosity due to the decomposition of $Mg(OH)_2$ which highly deteriorate mechanical properties. However, materials containing MgF_2 (WE-BT+FT) showed best corrosion resistance as the fluoride barriers successfully slowed down the corrosion progress and promising CYS slightly exceeding the corresponding value for human bone.

Acknowledgment

Authors wish to thank the Czech Science Foundation (project no. GA16-08963S) and specific university research (MSMT No 21-SVV/2018) for the financial support of this research.

References

[1] Avedesian M M and Baker H, Magnesium and magnesium alloys. In *ASM International, Materials Park, OH*, 1999.
[2] Qian M, Hildebrand Z C G and StJohn D H 2009 *Metall and Mat Trans A* 40, 2470-2479.
[3] T. Rzychoń A K 2007 *J. Achiev. Mater. Manuf. Eng.* 21 (1).
[4] Liao J, Hotta M and Mori Y 2012 *Mat. Sci. Eng. A* 544, 10-20.
[5] Kubásek J, Dvorský D, Čavojský M, Vojtěch D, Beronská N and Fousová M 2017 *J. Mater. Sci. Technol.* 33 (7), 652-660.
[6] Průša F, Vojtěch D, Bláhová M, Michalcová A, Kubatík T F and Čížek J 2015 *Mater. Des.* 75, 65-75.
[7] Ren M, Cai S, Liu T, Huang K, Wang X, Zhao H, Niu S, Zhang R and Wu X 2014 *J. Alloys Compd.* 591, 34-40.
[8] Bakhsheshi-Rad H R, Idris M H and Abdul-Kadir M R 2013 *Surface and Coatings Technology* 222, 79-89.
[9] Jo J-H, Kang B-G, Shin K-S, Kim H-E, Hahn B-D, Park D-S and Koh Y-H 2011 *Journal of Materials Science: Materials in Medicine* 22 (11), 2437-2447.
[10] Sun J e, Wang J, Jiang H, Chen M, Bi Y and Liu D 2013 *Materials Science and Engineering: C* 33 (6), 3263-3272.
[11] Dvorsky D, Kubasek J and Vojtech D 2017 *Manuf. Technol.* 17 (4), 440-446.
[12] Yan T, Tan L, Zhang B and Yang K 2014 *J. Mater. Sci. Technol.* 30 (7), 666-674.
[13] Trinidad J, Arruebarrena G, Marco I, Hurtado I and Argandoña E S d 2013 *Proceedings of the Institution of Mechanical Engineers, Part H: Journal of Engineering in Medicine* 227 (12), 1301-1311.
[14] Höhlinger M, Heise S, Wagener V, Boccaccini A R and Virtanen S 2017 *Applied Surface Science* 405, 441-448.
[15] Dvorsky D, Kubasek J and Vojtech D 2018 *Materials Letters* 227, 78-81.
[16] Weber C M, Eifler R, Seitz J-M, Maier H J, Reifenrath J, Lenarz T and Durisin M 2015 *Acta Biomaterialia* 18, 249-261.
[17] Yu X, Jiang B, He J, Liu B and Pan F 2018 *J. Alloys Compd.* 749, 1054-1062.
[18] Staiger M P, Pietak A M, Huadmai J and Dias G 2006 *Biomaterials* 27 (9), 1728-1734.
[19] Yusop A H, Bakir A A, Shaharom N A, Abdul Kadir M R and Hermawan H 2012 *International Journal of Biomaterials* 2012, 10.
[20] Song G L and Atrens A 1999 *Adv. Eng. Mater.* 1 (1), 11-33.

Force-to-Stress Conversion Methods in Small Punch Testing Exemplified by Creep Results of Fe-Al Alloy with Chromium and Cerium Additions

F Dobeš[1], P Dymáček[1*] and M Friák[1]

[1]Institute of Physics of Materials, Academy of Sciences of the Czech Republic, Žižkova 22, 616 62 Brno, Czech Republic

[*]Corresponding author's e-mail address: pdymacek@ipm.cz

Abstract. Application of the small punch test in technical practice is critically dependent on the existence of reliable methods for conversion of data measured by means of such a biaxial test to equivalent quantities of uniaxial tests. In the paper, various methods for recalculation of force in small punch test to stress in conventional creep test are investigated. The methods include: (i) classical "CEN Workshop Agreement" model, (ii) empirical "force to stress conversion" method, (iii) modified Chakrabarty method and (iv) "constant deflection rate" method. The methods are examined using the data obtained by testing Fe-Al based alloy with additions of chromium and cerium. The tested alloy enables verification of conversion methods for different crystallographic lattices, namely $D0_3$ at lower temperatures and B2 at elevated temperatures.

1 Introduction

An attractive possibility for estimation of the residual lifetime of facilities operating at elevated temperatures includes testing of miniaturized discs detracted non-invasively from the surface of construction parts, i.e., small punch (SP) testing [1-5]. An important prerequisite for practical application of such testing is establishing reliable correlative means for comparison of force in SP test and stress in conventional creep test. A range of force-to-stress conversion methodologies were recently compared in connection with the preparation of new European standard for small punch testing [6]. In the present contribution we would like to verify the methodologies using data obtained in research of Fe-Al-based alloy. Furthermore, another method is proposed that starts from results of SP tests at constant rate of deflection.

2 Material and experimental technique

The alloy containing 28.4 % Al, 2.6 % Cr, 0.02 % Ce and 0.16 % C (at. %, Fe balanced) was prepared by vacuum melting, casting in an argon atmosphere and rolling of the original sheet (thickness 40 mm) to the final one (13 mm) at 1100 °C. Details of its preparation and final microstructure can be found elsewhere [7]. SP creep tests were performed at temperatures ranging from 500 °C to 800 °C and applied forces from 40 N to 500 N. Specimens of 8 mm in diameter and thickness of 0.500 ± 0.002 mm were used, the geometry of SP setup was as follows: punch radius 1.25 mm, lower die radius 2 mm. The central deflection was measured through the relative motion of the punch. This is in contrast to measurements in ref. [6] where it was done through the movement of a ceramic rod attached to the specimen from the opposite side. Results of SP tests are described in the previous paper [8]. For the sake of extending the conversion methodology, additional tests at constant deflection rate of 0.0055 mm/s were performed at both room temperature and elevated temperatures. For a direct comparison with uniaxial tests, the results of conventional creep tests in tension - performed on the same sheet are used [9]. In a similar way, results of constant strain rate tests published in conference contribution [10] are also used.

3 SPC conversion methodologies

3.1 Classical "CEN Workshop Agreement" model

In the preliminary European Code of Practice [11], a rough estimate of the relationship between force F in small punch test and stress σ in conventional test is made using the Chakrabarty membrane stretch model [12]

$$F/\sigma = 3.332 k_{SP} r^{-0.2} R^{1.2} h_0 \tag{1}$$

where r is radius of lower die, R is punch radius and h_0 is initial thickness of the disc. k_{SP} is a correlation factor introduced to adjust the theoretical model to the experimental results. The original idea of k_{SP} equal to unity is indefensible of view of available experimental results [13-15]. Comparing force F in our SP tests and stress in uniaxial tests [9] of the same duration, the values of k_{SP} for the present geometry can be found. They are presented in figure 1 as a function of time to rupture. k_{SP} is clearly a function of temperature and time to rupture (i.e., force). A striking difference is evident between values at temperature of 500 °C and at temperatures in the range from 600 to 800 °C. This can be ascribed to difference in crystal lattices: at temperatures above 540 °C, the less-ordered FeAl phase with B2 lattice exists, whereas at lower temperatures the stable compound is ordered Fe_3Al with $D0_3$ lattice [16].

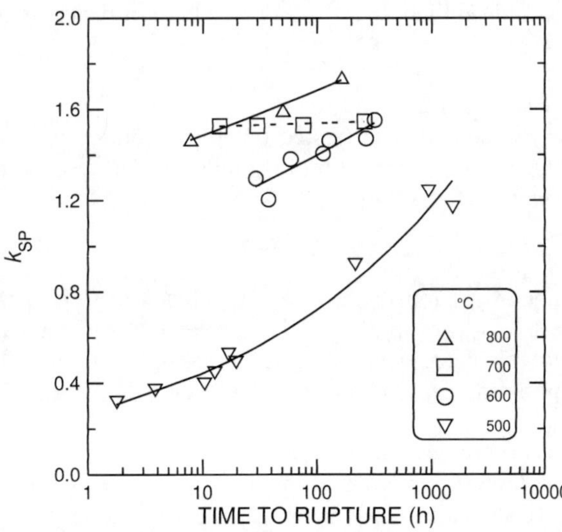

Figure 1. Dependence of ductility parameter k_{SP} on time to rupture at different temperatures.

3.2 Empirical "force to stress conversion" method (EFS)

For the prepared European small punch standard a fully empirical force-to-stress conversion model was chosen as the preferred methodology [6]. It was shown that the force-to-stress ratio is dependent on the value of deflection u_m measured at the location of the minimum deflection rate. The optimized formula is then [6, 15]

$$F/\sigma = 1.9162 u_m^{0.6579}. \tag{2}$$

Comparison of the stress dependence of time to rupture measured by uniaxial creep tests (UAC) [9] and the results of small punch creep tests (SPC) dependent on equivalent stress calculated by means of equation (2) is given in figure 2a. The agreement of uniaxial data and recalculated small punch data in the B2 range of temperatures is excellent. On the other hand, the results in $D0_3$ range (500 °C) do not match well. This is not surprising because the empirical formula (2) is based entirely on the results of ductile steels while the present alloy is evidently brittle in $D0_3$ state.

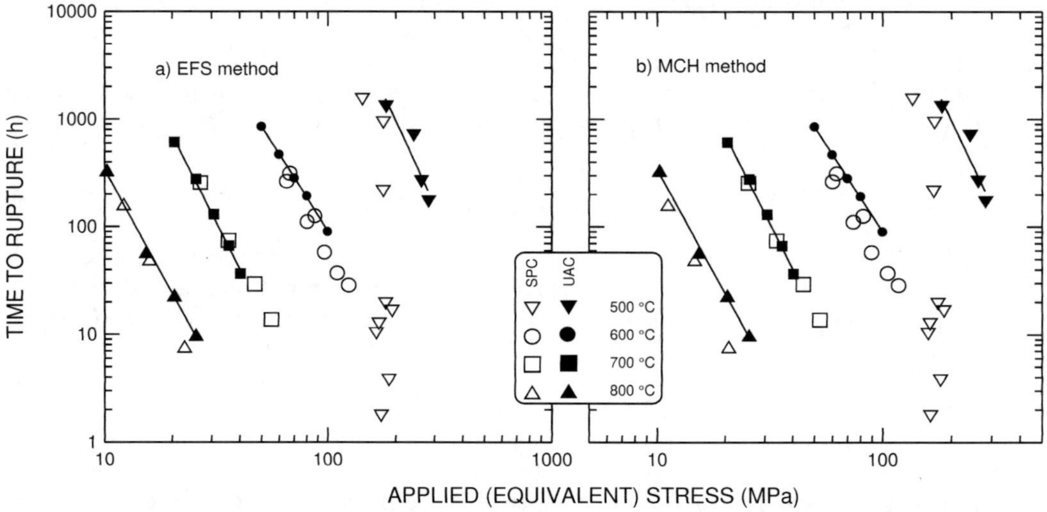

Figure 2. Dependence of time to rupture in small punch creep tests and uniaxial creep tests on calculated equivalent stress and applied stress, respectively. Equivalent stress calculated using: a) equation (2) and b) equation (3).

3.3 *Modified Chakrabarty method (MCH)*

A linearized version of EFS formula was suggested by Holmström et al. [6] in order to simplify the approach and to open the possibility to account for different types of test set-ups. The MCH optimized force to stress ratio for the standard test set-up is

$$F / \sigma = 0.6143 + 1.2954 u_{\mathrm{m}} \quad (3)$$

Comparison of the small punch data, recalculated to equivalent stress using the equation (3), with the results of uniaxial creep tests is given in figure 2b. Similar conclusions can be deduced about the conformity of measured and calculated data as in the foregoing paragraph: The agreement in the B2 range of temperatures is slightly inferior, the results in $D0_3$ range do not match well.

3.4 *Methods based on constant deflection rate tests*

In principle, three main different testing modes of small punch can be operated:

- First, Constant Force (CF) is a test in which the punch penetrates under a constant force F and the time dependence of the deflection is recorded. This test is similar to conventional constant load creep tests.
- Second, the punch penetrates through the disc at a given constant rate of motion and the force F is measured [17-19]; this test is termed Constant Deflection Rate (CDR) in what follows. It has certain analogy with the conventional tensile test under a constant strain rate. The tests can be used for estimation of yield stress σ_{Y} and ultimate tensile strength σ_{UTS} [20-22].
- Third, the constant deflection mode is the analogy to stress relaxation testing. The deflection of the disc (preferably in conditions of the membrane stretching regime) is held constant and the force relaxes as the elastic strain is replaced by inelastic creep strain [23].

Komazaki [24] recently suggested that the ratio of maximum force in CDR test to ultimate tensile strength can be used as a conversion factor of force in small punch CF tests to stress in uniaxial creep test:

$$\frac{F}{\sigma} = \frac{F_{max}}{\sigma_{UTS}} \tag{4}$$

This proposal can be of enormous importance, especially if its validity is demonstrated for room temperature values of F_{max} and σ_{UTS} to conversion of elevated temperature creep data. The application of room temperature data is certainly problematic in the present intermetallic alloy due to existence of different crystal lattices at different test temperatures [16]. That is why the data at specific temperature should be preferred in our investigation: Temperature dependence of σ_{UTS} was published by Hakl et al. [10]; dependence of force on deflection in CDR tests is illustrated in figure 3.

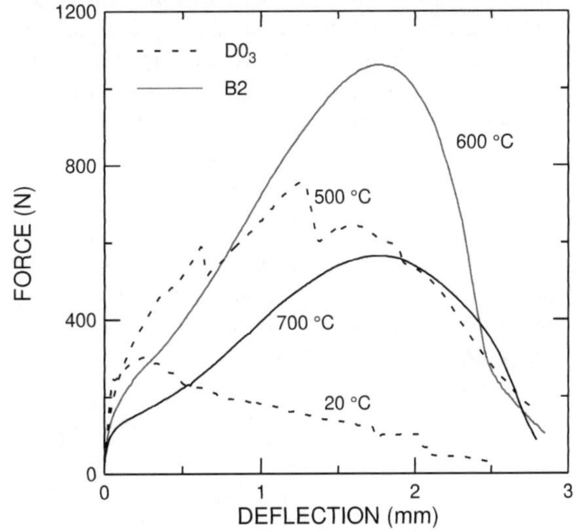

Figure 3. Force vs. deflection curves at various temperatures.

Comparison of small punch data recalculated using the equation (4) and uniaxial creep data is given in figure 4a. The SPC points at 800 °C are not included in the figure because the ultimate tensile strength is not available at this temperature. The agreement of the SPC and UAC points is the worst of all the methods tested in the present investigation. Moreover, the application of the method requires a value of the ultimate tensile strength at a given temperature that is usually not known.

The idea of conversion SPC into UAC data using the results of CDR tests can be modified in the following way. It was shown previously [20-22] that the ultimate tensile strength correlates with the results of CDR tests by means of a simple linear relationship

$$\sigma_{UTS} = \beta \frac{F_{max}}{h_0 u_{max}}, \tag{5}$$

where h_0 is the initial thickness of the disc specimen, u_{max} is the deflection corresponding to the maximum force and β is a constant. The conversion factor which follows the idea of equation (4) is then

$$\frac{F}{\sigma} = \frac{h_0 u_{max}}{\beta}. \tag{6}$$

To estimate the conversion factor for a series of SPC tests at given temperature it is necessary to perform only a single CDR test and to determine the value of deflection u_{max} at the maximum force. The values of the constant β found in available studies are reviewed in our previous paper [21]. The value of $\beta = 0.32$ was applied in the present investigation following the recommendation of Komazaki [24]. Comparison of uniaxial and small punch creep data using this value of β is given in figure 4b. Because the CDR test at 800 °C is not available, the same value of conversion factor is used as for the temperature of 700 °C. It can be seen that the agreement of UAC and SPC data is very good. Contrary to the above version of CDR-based method (equation 4), the equation (6) may offer a possibility of conversion factor F/σ dependent on the force applied in the small punch creep test. This can be realized either through a series of CDR tests at different deflection rates or by refining the equation (5). The tests to confirm the method for heat-resistant steels are in progress.

As can be seen in Fig. 3, several force drops are observed at the temperature of 500 °C. The deflection at the maximum observed force was used as the input in eq. 6. It remains a matter of discussion whether the deflection at the first force drop is not more suitable for the conversion. This would increase the equivalent stress by a factor of two and may improve the fit of very short-term tests that probably cracked at loading. Apparently, the conversion of SPC data of brittle D03 phase requires additional attention and it is beyond the scope of the present contribution. The fractographs of ruptured SP specimens are illustrated in our previous paper [8].

4 Conclusions

Various methods for conversion of force in small punch test to equivalent stress in conventional uniaxial creep tests are compared. The problems emerge in the temperature range of the existence of $D0_3$ lattice. New method based on estimation of deflection corresponding to maximum force in constant deflection rate test is suggested as a potential route for considering the ductility of tested material.

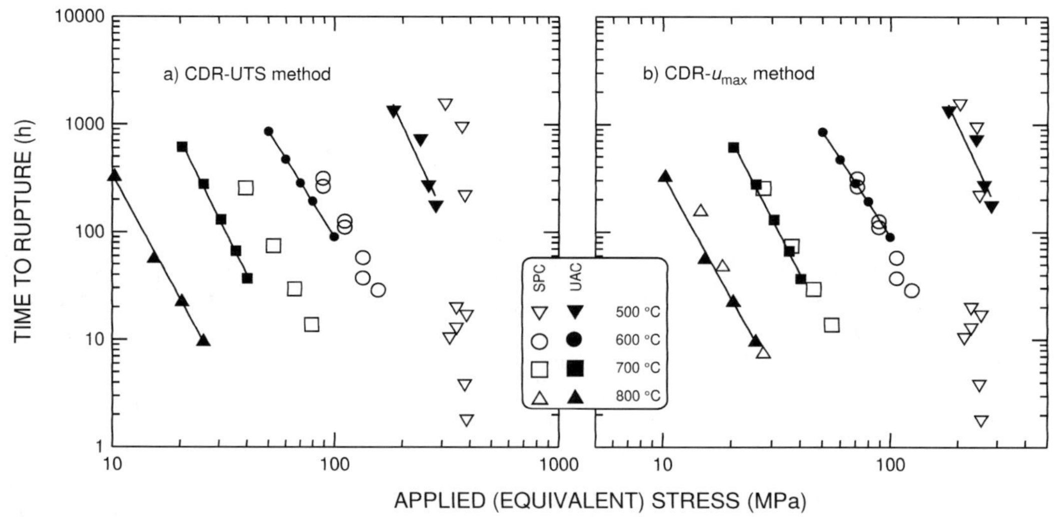

Figure 4. Dependence of time to rupture in small punch creep tests and uniaxial creep tests on calculated equivalent stress and applied stress, respectively. Equivalent stress calculated using: a) equation (4) and b) equation (6).

Acknowledgements
Financial support from the Czech Science Foundation (project No. 17-22139S) is gratefully acknowledged. Professor P. Kratochvíl (Charles University, Prague) kindly supplied the experimental alloy.

References

[1] Dyson C C, Sun W, Hyde C J, Brett S J and Hyde T H 2016 *Mater. Sci. Technol.* **32** 1567
[2] Ma Y W, Shim S and Yoon K B 2009 *Fatigue Fracture Engng. Mater. Struct.* **33** 951
[3] Komazaki S-I and Kato T 2010 *Mater. High Temper.* **27** 205
[4] Wen C, Xu T and Guan K 2016 *Materials* **9** 796
[5] Andrés D, Lacalle R and Álvarez J A 2016 *Mater. Design* **96** 122
[6] Holmström S *et al.* 2018 *Mater. Sci. Engng.* A **731** 161
[7] Hotař A and Kratochvíl P 2002 *Kovove Mater.* **40** 45
[8] Dobeš F, Dymáček P and Friák M 2018 *Kovove Mater.* **56** 205
[9] Kratochvíl P, Šedivá I, Hakl J and Vlasák T 2002 *Kovove Mater.* **40** 124
[10] Hakl J, Vlasák T and Kratochvíl P 2001 *Proc. Conf. Metal 2001*, ed. T Prnka (Ostrava: Tanger) paper No. 79
[11] CEN 2006 *Small Punch Test Method for Metallic Materials*. Part A: A Code of Practice for Small Punch Creep Testing, CEN Workshop 21 document (Brussels: Comité Européen de Normalisation) p 23
[12] Chakrabarty J 1970 *Int. J. Mech. Sci.* **12** 315
[13] Blagoeva D 2009 *Development of a Residual Lifetime Prediction Methodology for Creep and Fracture Behaviour of Ferritic-Martensitic Steels using Small Punch Testing Technique*, Ph.D. thesis (Pisa, Universita di Pisa) p 88
[14] Komazaki S-I, Ohkawa Y and Yonemura M 2016 Effect of ductility on load/stress conversion coefficient of small punch creep test for Ni-base alloys *Tetsu-To-Hagane/Journal of the Iron and Steel Institute of Japan* **102** 646-652
[15] Dymáček P, Li Y, Dobeš F and Stevens P 2018 New approach to determine uniaxial creep properties from small punch creep curves *Mater. High Temp.* in press
[16] Stein F and Palm M 2007 *Int. J. Mater.* Res. **98** 580
[17] Mao X and Takahashi H 1987 *J. Nucl. Mater.* **150** 42
[18] Bruchhausen M, Holmström S, Simonovski I, Austin T, Lapetite J-M, Ripplinger S and de Haan F 2016 *Theor. Appl. Fract. Mech.* **86** 2
[19] Altstadt E, Ge H E, Kuksenko V, Serrano M, Houska M, Lasan M, Bruchhausen M, Lapetite J-M and Dai Y 2016 *J. Nucl. Mater.* **472** 186
[20] García T E, Rodríguez C, Belzunce F J and Suárez C 2014 *J. Alloys Compd.* **582** 708
[21] Dobeš F, Dymáček P and Besterci M 2015 *Mater. Sci. Engng.* A **626** 313
[22] Altstadt E, Houska M, Simonovski I, Bruchhausen M, Holmström S and Lacalle R 2018 *Int. J. Mech. Sci.* **136** 85
[23] Dymáček P 2016 *Theor. Appl. Fract. Mech.* **86** 25
[24] Obata K, Komazaki S, Tomobe M, Yaguchi M and Kumada A 2017 Creep Property Assessment of Gr.91 Boiler Pipings by Small Punch Test, *Proceedings of the 55th Symposium on Strength of Materials at High Temperature*, The Committee on High Temperature Strength of Materials, The Society of Materials Science, Japan pp 3-7

Calibration of the constitutive equations for materials with different levels of strength and plasticity characteristic based on the uniaxial tensile test data

Ihor Dzioba[1] and Sebastian Lipiec[1]

[1]Kielce University of Technology, Al. 1000-lecia PP 7, 25-314 Kielce, Poland

E-mail: slipiec@tu.kielce.pl

Abstract. Today finite elements method (FEM) modeling and calculation are used for analysis current state and forecasting reliability lifetime period of components and constructions. Essential issues for correctness of obtained results in FEM analysis are: proper interpretation of boundary conditions and right constitutive equation of material in terms of high stresses and strains. Standard procedure obtaining of material constitutive relation directly from experimental tensile test data does not proper result for every type of material. Method of constitutive equation creation for elastic-plastic material for a high level of plasticity based on data obtained during tensile test is presented. The proposed calibration method was verified on materials with different levels of strength and plasticity characteristics: S355JR and 42CrMo4 steels and ADI cast iron. Plastic deformation level, triaxiality stress state parameter and Lode factor were taken into account during created of the constitutive equation.

1 The idea of the constitutive equations calibration

When starting the analysis based on numerical modeling and calculation, especially if components are made from high-plasticity material, it is necessary to implement the calibration of the stress-strain constitutive equation, primary received from uniaxial loading test data. Some of the researchers propose fitted of the experimental and numerical test results by iteration selection of correlation parameters for the true stress-strain relationships obtained from uniaxial loading test data [1,2,3]. This method needs a few repetitions of the numerical calculations, until then numerical and experimental loading curves coming closer together (would be convergent).

Another method was proposed by Bai and Wierzbicki [4,5,6] and next developed by Neimitz and co-workers [7]. The yield stress were defined according to the next formula:

$$\sigma_{yld} = \overline{\sigma}(\bar{\varepsilon}_p)[1 - c_\eta(\eta - \eta_0)]\left[c_\theta^s + (c_\theta^{ax} - c_\theta^s)\left(\gamma - \frac{\gamma^{m+1}}{m+1}\right)\right] \tag{1}$$

In the presented formula, the function $\bar{\sigma}(\overline{\varepsilon_p})$ describes relation between effective real stress and plastic strain. Symbol $\eta = \frac{\sigma_m}{\sigma_e}$ is the triaxiality factor, where $\sigma_m = \frac{1}{3}(\sigma_{xx} + \sigma_{zz} + \sigma_{yy})$ and σ_e is a effective stress; η_0 is a reference value of the triaxiality coefficient (for the uniaxial tensile test cylindrical specimen, η_0=1/3).

Bai and Wierzbicki postulated that the γ function presents in the form:

$$\gamma = 6.46[sec(\theta - \pi/6) - 1] \tag{2}$$

where θ is the Lode angle, which must satisfy the inequality $0 \leq \theta \leq \pi/3$. The Lode angle could be calculated using the below relationships (3)-(5):

$$\bar{\theta} = 1 - \frac{6\theta}{\pi} = 1 - \frac{2}{\pi} arccos\xi \tag{3}$$

$$\xi = L\frac{9 - L^2}{\sqrt{(L^2 + 3)^3}} \tag{4}$$

$$L = \frac{2\sigma_{II} - \sigma_I - \sigma_{III}}{\sigma_I - \sigma_{III}} \tag{5}$$

The σ_I, σ_{II}, σ_{III} are principal stresses and σ_I, and σ_{III} – maximum and minimum values of principal stresses, respectively. Quantity L called as the Lode parameter.
The parameters c_η, c_θ^s, c_θ^{ax} and m in Eq. 1 must be determined experimentally.
The equation describes the effect of material softening immediately before damage caused by growth and connection of voids was proposed by Neimitz [7].

$$c_\eta{}' = c_\eta[1 + H(\varepsilon_{pl\_0})(\varepsilon_{pl\_i} - \epsilon_{pl\_0})]^\alpha \tag{6}$$

In this equation, ε_{pl_0} is a value of plastic strain at which material softening beginning occurs (see Fig. 1) and ε_{pl_i} are further plastic strain values. $H(\varepsilon_{pl_0})$ is the H'eaviside function.

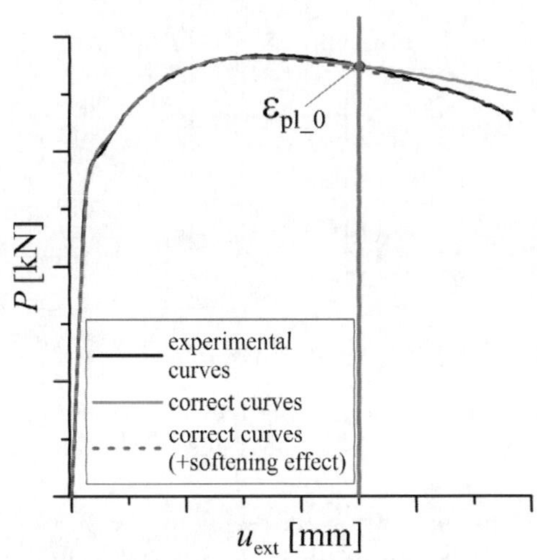

Figure 1. Scheme for determining the ε_{pl_0} value.

2 Materials used in experimental tests and numerical analysis

Experimental tests and following numerical calculations were carried out on specimens made of steel types – S355JR, 42CrMo4 and cast iron ADI EN-GJS-1059-6. The steel S355JR after normalization has ferrite-pearlite microstructure type. Microstructure of 42CrMo4 steel after quenching (850 °C, oil) and tempering (500 °C) was tempered martensite type. Selection of those materials was justified of different levels of strength and plasticity characteristics. Uniaxial tension tests were carried out used cylindrical specimens with a 5 mm in diameter and base length of 25 mm (Fig. 2) at temperature 20 °C. The characteristics of strength and plasticity are presented in Tab. 1. Relationship between true stress and true strain was determined on the basis of the nominal stress and strain from uniaxial tensile tests results [8]. The diagrams of stress-strain relationships for nominal and true values are shown in figure 3. The nominal and true stress and strain calculated from recorded signals of load F and extensometer elongation ΔL as (Eq. 7):

$$\varepsilon_{nom} = \frac{L-L_0}{L_0};\ \sigma_{nom} = \frac{F}{S_0};\ \varepsilon_{true} = \ln(1+\varepsilon_{nom});\ \sigma_{true} = \sigma_{nom}(1+\varepsilon_{nom}) \tag{7}$$

Table 1. The strength and plasticity properties of analyzed materials.

Materials	σ_{LYS}/ σ_{HYS} [MPa]		σ_{UTS}[MPa]		E [GPa]	n		A_5 [%]
	Nominal	True	Nominal	True	True	Nominal	True	Nominal
S355JR	366/375	377/380	495	613	201	7.90	4.78	35.43
42CrMo4	1126	1135	1211	1287	202	25.64	15.24	14.94
ADI	775	781	1125	1232	159	7.89	6.11	9.55

σ_{LYS} - lower yield strength, σ_{HYS} - higher yield strength, σ_{UTS} - ultimate strength, E - Young's modulus

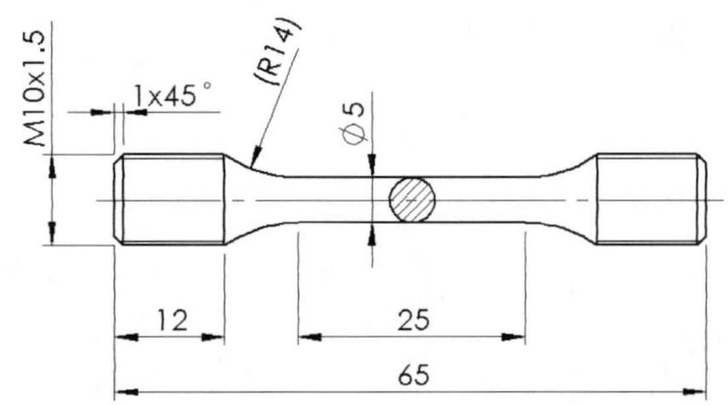

Figure 2. The specimen for uniaxial loading test.

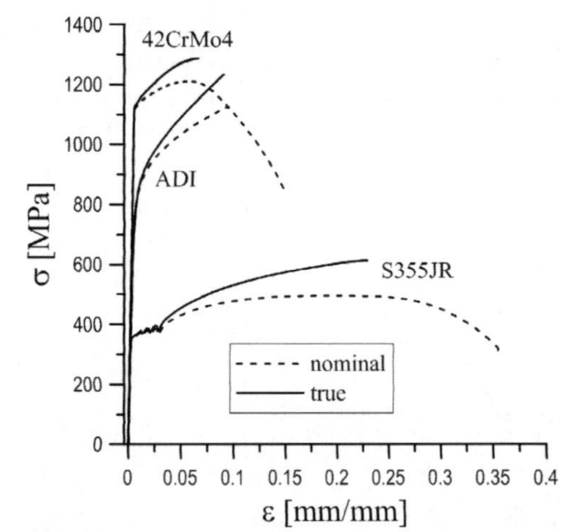

Figure 3. The nominal and true stress-strain diagrams of tested materials.

The lowest strength characteristic and the highest plasticity were obtained for S355JR steel. The highest strength characteristics showed 42CrMo4 steel. The pronounced necking was presented in both steel types specimens before damage during uniaxial test, while the necking was not observed before cracking for specimens of cast iron ADI.

3 Calibration process of material constitutive equations

The true stress-strain curves obtained on the basis data of uniaxial tensile test. Taking into account the data set which are limited of maximum stress value, that is associated with necking process formation during tension-loaded specimen process. The stress-strain relationship in range from start yielding to necking initiation is written by power function and known as Ramberg-Osgood or Ludvik laws [9].

IOP Conf. Series: Materials Science and Engineering **461** (2019) 012018 doi:10.1088/1757-899X/461/1/012018

Further strain process in necking area takes place in 3D stress state with participation of voids nucleation, growth and coalescence, and required more complex mathematical expression to written these physical processes.

Based on the true stress-strain relationship was conducted the procedure of material constitutive equations creation in domain of large strain and stress values.

In the first stage the numerical simulation of specimens same as uniaxial loading were carried out and true stress-strain relationships obtained from tensile tests were used with data approximation according to liner dependence. The program ABAQUS [10] was used for numerical modeling and calculation. It was modeled of ¼ of the tensile specimens because its axially symmetrical [Fig. 4].

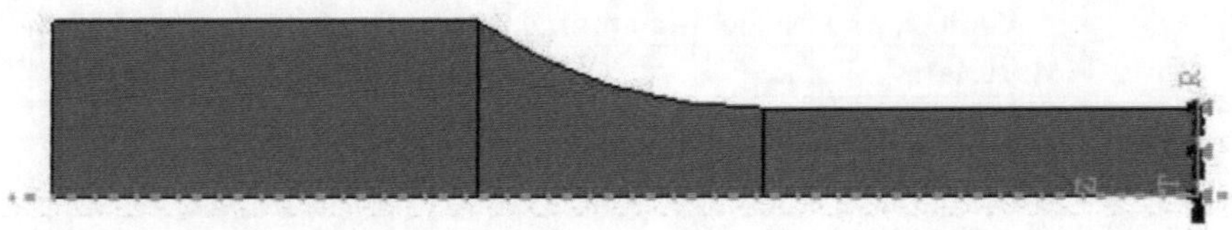

Figure 4. Numerical model in Abaqus program for uniaxial tensile specimen.

The load and elongation data obtained of numerical calculation were compared to the experimental results. Only for specimens made of cast iron ADI numerical and experimental data are completely overlaps. For specimens made of S355JR and 42CrMo4 steels were observed significant differences between calculated and experimental data, namely for data which corresponding with necking process.

For the specimens in which did not occur necking before damage the constitutive equation of material can be used in form of true stress-strain relationship without any corrections. The data obtained of numerical calculation and experimental testing are similar (Fig. 5).

Other situation has place for specimens with necking (S355JR and 42CrMo4 steels). Takes into account a fact, that strain level during necking process reaches more 2.0, then true strain obtained from uniaxial tensile test are limited to 0.2, it is necessary extrapolated the true stress-strain relationship data to higher values. Initially the extrapolation was carried out in form of power function of *true stress – plastic strain* dependence to 2.5 and more of plastic strain level. As we can see after this type extrapolation are significant differences in numerical calculated and experimental data in ranges graphs corresponding to necking process.

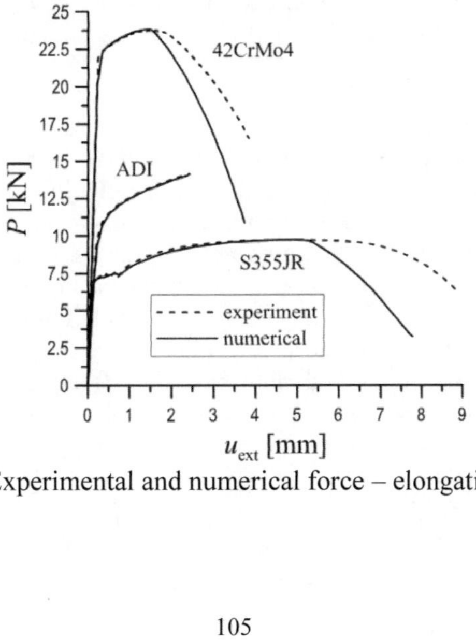

Figure 5. Experimental and numerical force – elongation curves.

The second type of the extrapolation was done in liner function form based on fitting of ~200 true stress-strain data points directly before initial necking. As a result of implementing to numerical calculation the extrapolation in liner form, the graphs obtained of numerical data are located slightly higher than one of experimental data in necking process range (Fig. 5).

So for specimens in which necking are observed before damage, during creation the constitutive equations, should be introduced correction, which takes into account processes leading to material softening. For this purpose the equation 6 was used [7]. The appropriate coefficients values for equation 1, which calculated by iteration method, are presented in table 2. As a result introduced this correction the experimental and numerical simulation data are similar (Fig. 6).

Table 2. The use coefficients values for equation 1.

Material:	c_η	c_θ^s	c_θ^{ax}	**m**	ε_{pl_0}	α
S355JR steel	0.20	1.03	1.00	1.00	0.06	2.35
42CrMo4 steel	0.20	1.01	1.00	1.00	0.06	1.40

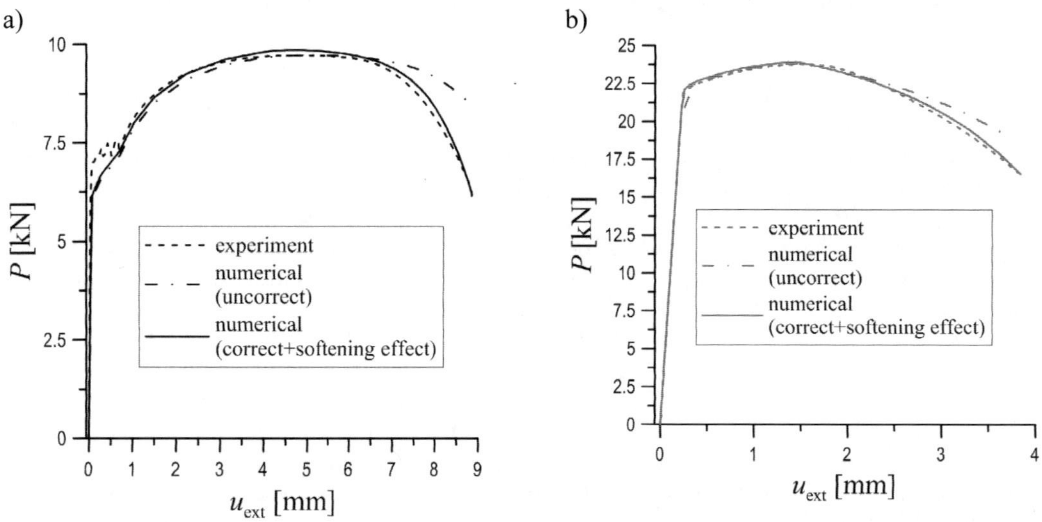

Figure 6. Experimental and numerical corrected curves of force – elongation: a) S355JR steel, and b) 42CrMo4 steel.

4 Summary

The tests carried out in presents paper clearly indicate on the necessity to perform the right calibration procedure of stress-strain constitutive relation for obtained properly result during numerical calculation. For materials, which do not shown necking and softening effect, calibration procedure can be limited to interpolation and extrapolation of the true stress – logarithmic strain relationship by linear function. While for materials, which shown necking during tensile loading [11,12], calibration procedure should additionally take into account members writing softening effect. The softening process of metallic materials by damage increase are consideration also in GTN-model [13,14]. Taking into account of softening effect in steels allows obtained correct stress and strain distributions directly in front of the crack or defects and right described fracture process construction members [7,13].

Acknowledgements
The work was performed as part of research projects: 01.0.08.00/2.01.01.01.0008 MNSP.MKPK.16.001 and 01.0.08.00/2.01.01.02.0021 MNSC.MKPK.16.002.

References

[1] Zhao K and Wang L and Chang Y and Yan J 2016 Identification of post-necking stress–strain curve for sheet metals by inverse method *Mechanics of Materials* 92 pp 107-118
[2] Joun MS and Eom JG and Lee M Ch 2008 A new method for acquiring true stress–strain curves over a large range of strains using a tensile test and finite element method *Mechanics of Materials* 40 pp 586-593
[3] Depreński Ł and Seweryn A 2011 Experimental Research into Fracture of EN-AW 2024 and EW-AW 2007 Aluminum Alloy Specimens with Notches Subjected to Tension *Experimental Mechanics* 51 pp 1075-1094
[4] Bao Y and Wierzbicki T 2004 On fracture locus in the equivalent strain and stress triaxiality space *International Journal of Mechanical Science* 46 pp 81-98
[5] Bai Y and Wierzbicki T 2010 Application of extended Mohr–Coulomb criterion to ductile fracture *International Journal of Fracture* 161 pp 1-20
[6] Bai Y and Wierzbicki T 2008 A new model plasticity and fracture with pressure and Lode dependence *International Journal of Plasticity* 24 pp 1071-1096
[7] Neimitz A and Gałkiewicz J and Dzioba I 2018 Calibration of constitutive equations under conditions of large strains and stress triaxiality *Archives of Civil and Mechanical Engineering* 18 pp 1123-1135
[8] ASTM E8 2003 Standard test method for tension testing of metallic materials International West Conshohocken
[9] Anderson TL 2008 Fracture Mechanics New York
[10] Abaqus 6.12 Getting Started with Abaqus Interactive Edition.
[11] Dzioba I and Lipiec S and Furmańczyk P and Pała R 2018 Investigation of fracture process of S355JR steel in transition region using metallographic, fractographic tests and numerical analysis *Acta Mechanica et Automatica* 12 pp 145-150
[12] Dzioba I 2010 Properties of the 13KHMF steel after operation and degradation under the laboratory conditions *Materials Science* 12 46 (3) pp 357-364
[13] Defaisse C and Maziere M and Marcin L and Besson J 2018 Ductile fracture of an ultra-high strength steel under low to moderate stress triaxiality *Engineering Fracture Mechanics* 194 pp 301-318
[14] Zhang Y and Lorentz E and Besson J 2017 Ductile damage modelling with locking-free regularized GTN model *International Journal for Numerical Methods in Engineering* 113 (3) pp 1871-1903

IOP Conf. Series: Materials Science and Engineering **461** (2019) 012019 doi:10.1088/1757-899X/461/1/012019

Determination of forming limits of TWIP steel sheet using linear and nonlinear strain paths

P Farahnak, M Urbanek, M Rund and J Dzugan

COMTES FHT a.s., Prumyslova 995, 334 41 Dobrany, Czech Republic

Corresponding author's E-mail: pedram.farahnak@comtesfht.cz

Abstract. The main goal of the present paper is investigation of the mechanical behaviour of high manganese content steel sheet which is well known in the literature as Twinning-Induced Plasticity (TWIP) under linear and nonlinear loading path. Contrary to Nakajima experiment, cruciform specimen under biaxial loading can provide linear and nonlinear strain path. This work aims to determine the numerical Forming Limit Curve (FLC) for TWIP sheet using cruciform specimen under biaxial loading. For this purpose, uniaxial tensile tests are performed on standard dog-bone samples which are extracted from rolling, transversal and diagonal directions to identify the flow curve as well as Lankford ratios. Data are used as input for numerical simulation of cruciform specimen. Results show the applicability of cruciform specimen for linear and nonlinear strain paths.

1 Introduction

The Forming Limit Diagram (FLD) introduced by Keeler [1], has been used extensively in sheet metal forming. Forming Limit Curve (FLC) is a plot of the major and minor principal strains in the plane of deformed sheet where necking takes place. The Nakajima experiment is often adopted as the conventional experiment for FLC determination. Thus, different samples with a variety of widths have to be tested in order to cover whole FLC from the uniaxial to equi-biaxial strain range. However, in this experiment contact between the punch and sheet influences the real formability of the material. As reported in the literature [2], the main challenge is to experience a pure biaxial mode and predictions are difficult as a consequence. Contact condition can be in the form of either pressure or friction [3, 4]. Friction can cause variation and the test has to be repeated while punch pressure on sheet leads to overestimation of the material formability [5].

The biaxially loaded cruciform specimen was designed as an alternative for the Nakajima test as a friction-free test. The original sheet thickness of the investigated TWIP steel is 1.5 mm in the present work. Based on a special design of cruciform specimen, necking is induced by groove construction in the central region. The influence of thickness on material formability and fracture were investigated in [6-8]. In the central region, the geometry has thickness variation from minimum 0.3 mm to maximum 1.5 mm. Afterwards, the displacement ratio of the actuators can control the strain path in the region.

TWIP steel is known for its combination of very high strength with extreme elongation capacity. This steel was selected for shock tower apron reinforcement. Chemical composition is summarized in table 1. In this work, uniaxial samples were extracted from sheet in Rolling, Transverse and Diagonal

Directions and are denoted by (RD), (TD) and (DD), respectively. Afterwards, uniaxial test were done on samples. Flow curve and Lankford ratios were characterized by using Digital Image Correlation (DIC) system. These data are used as input for numerical simulation of cruciform specimen under linear and nonlinear loading conditions. The experimental setup and details will be explained in section 2.

Table 1. Chemical composition of TWIP steel

C [%]	Mn [%]	P [%]	S [%]	Cr [%]
0.068	4.96	0.009	0.004	0.049

2 Material characterization

Figure 1 depicts the geometry of standard dog-bone specimen. For characterization of Lankford ratios, samples are extracted from the experimental sheet in the Rolling, Diagonal and Transverse Directions and values are denoted by r_0, r_{45} and r_{90}, respectively. Electro-mechanical testing system, Mayes, is used to conduct the uniaxial tests and the GOM Aramis™ system with a 12 megapixel camera is utilized to monitor $2D$ strain fields with high resolution. The data acquisition rate is set at 100 Hz in the testing system for accurate measuring of the tensile behaviour. It is noticed that for the standard dog-bone geometry, initial length of extensometers is chosen as 80 mm. Figure 2(a) represents the engineering stress-strain curves of tensile tests for rolling, transverse and diagonal directions. The Abaqus FEM base software is utilized for numerical simulation with conventional plasticity [9] based on von Mises material. Flow curve in figure 2(b), is extrapolated using linear combination of Voce-Swift (equation (2)) and parameter γ in equation (3), is calibrated through inverse analysis by comparing numerical and experimental force-displacement curves (figure 2(c)) and parameters are summarized in table 2. Mesh sizes in the centre of specimen was 0.8 mm and in surrounding area was 1.2 mm.with 4 C3D8R elements (according to Abaqus library) along the thickness.

Gauge sections are defined in transversal direction as well. Three experiments are performed on the specimens in rolling, diagonal and transversal directions. GOM Aramis™ captures 2D displacement fields during deformation, e.g., axial and transversal extensometers with facet size of 16 $pixels$ provide engineering surface strains and these are then transformed into true values. Finally, thickness strain is measured based on the incompressibility condition. The Lankford ratio can be calculated from the average slope of true plastic width (ε_w^p) and thickness (ε_{th}^p) strain in equation (1) in a specific range of equivalent plastic strain ($0.1 \leq \bar{\varepsilon}_p \leq 0.2$) and values are depicted in Figure 2 (d).

$$r = \frac{d\varepsilon_w^p}{d\varepsilon_{th}^p} \tag{1}$$

$$K_{Voce} = k_0 + Q\big(1 - \exp(-\beta\bar{\varepsilon}_p)\big) \tag{2}$$

$$K_{Swift} = A(\varepsilon_0 + \bar{\varepsilon}_p)^n$$

$$\bar{\sigma}_p = \gamma K_{Swift} + (1-\gamma)K_{Voce} \tag{3}$$

3 Cruciform specimen

One of the ways how to achieve experimentally a linear and nonlinear strain path is the biaxial test. Biaxial testing machines can depict the real behavior of materials without effect of the friction [10]. Moreover, for linear and nonlinear strain paths measurement, cruciform specimen can be utilized without friction. The main difficulty of the cruciform specimen is necking imposition in the central region of the specimen due to thickness reduction or thinner gauge. If necking is induced to the central region, then the loading ratios of the machine can control the strain paths in the center of gauge. In previous work s[11], the cruciform geometry was optimized for biaxial test, using FEM simulations in order to achieve maximum deformation in the central region and not in the shoulders. The optimized

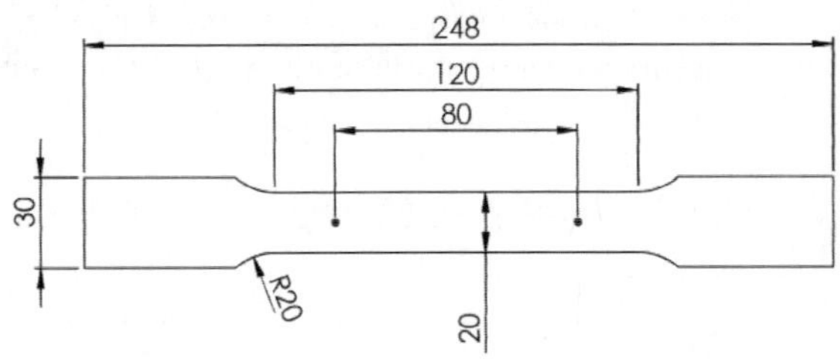

Figure 1. Geometry of dog-bone specimen

Table 2. Voce- Swift flow curve parameters

Voce			Swift			Linear combination
k_0 [MPa]	Q [MPa]	β	A [MPa]	ε_0	n	γ
410	900	6	2400	0.2	0.9	0.08

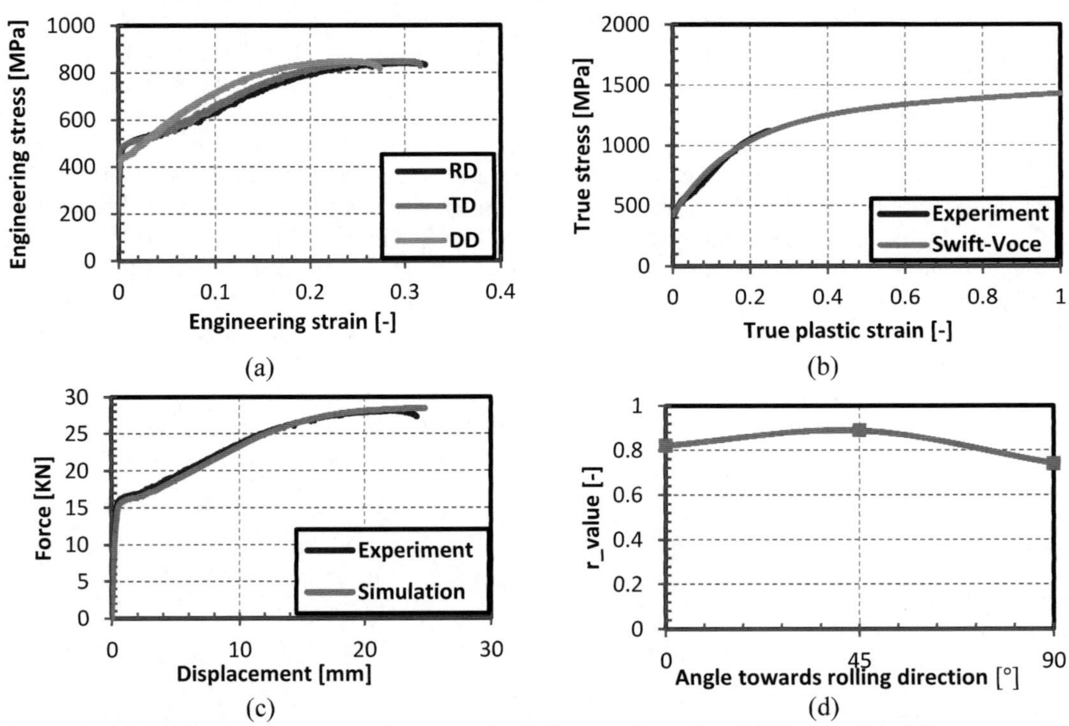

Figure 2. (a) Engineering stress-strain curves in different orientation (b) Extrapolated flow curve (c) Experimental and numerical force-displacement curves (d) Lankford ratios

gauge is similar to cruciform specimen [12], but the thickness ratios are different. The thickness of the thinner gauge is 0.3 *mm*. Geometry of the cruciform with thickness variations is depicted figure 3.

3.1 Numerical simulation of the cruciform specimen

Plasticity parameters from Section 2 are used as input for FEM simulation of cruciform specimen with the thickness variation to clarify how material characterization based on medium and small specimens can influence the strain path. The aim of numerical simulation is to represent the real-world experiment as accurately as possible. Theory of Hill 48 and the influence of anisotropy parameters on strain path was investigated by Farahnak et al [13].

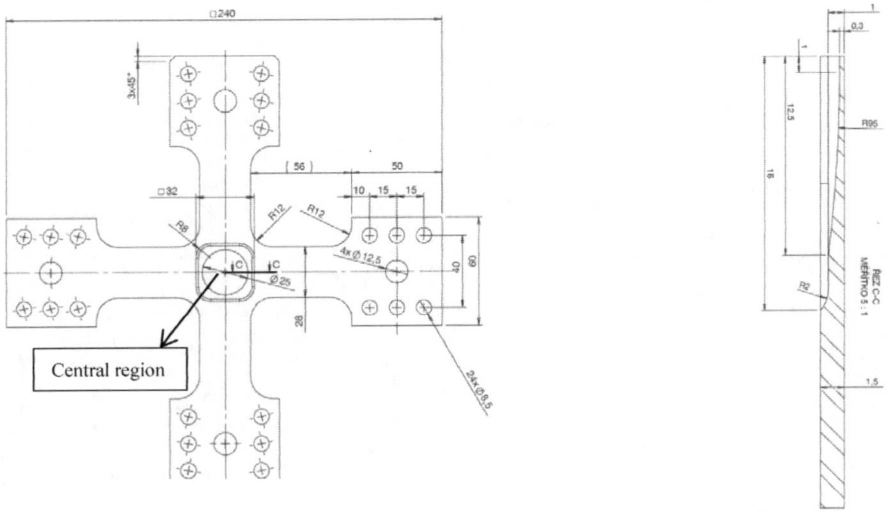

Figure 3. Geometry of cruciform specimen with longitudinal thickness variations

Mesh size plays a prominent role in numerical simulation of cruciform specimen, e.g. with a coarse mesh necking can take place in higher level of strains and subsequently it can affect the strain path's slope and necking. Therefore, for all simulations, one quarter of the cruciform geometry is modelled using 0.8 mm mesh size in the central region , 1 mm for surrounding areas and 6 elements along the thickness with $3D$ elements with 8 nodes (C3D8 according to Abaqus). A mesh size smaller than these values leads to time-consuming calculation and does not have significant influence on the results. In all simulations, strain paths are plotted for central elements where the deformation is maximum.

For linear strain paths, major strains are plotted in terms of minor strains for uniaxial, plane strain and equi-biaxial loading conditions and are indicated in figure 4(a). Beginning time of instability is detected using the first and second derivation method in the thinnest element [14] and as an example for equi-biaxial loading condition is depicted in figure 4(b). As can be seen in the figure, peaks are rigorously localized in terms of simulation time.

For nonlinear strain paths, combinations of equi-biaxial tension-plane strain tension and uniaxial tension-equi-biaxial tension are taken into account in the present paper and strain paths are shown in figure 4(c) and (d), respectively. As can be seen in figure 4(c) and (d), different numerical steps are highlighted with different colors. It is noticeable that for both nonlinear strain paths, force imposed on RD. Take equi-biaxial tension-plane strain tension experiment as an example, the experiment is started under equi-biaxial tension with force imposition in RD and TD and then for plane strain tension, cruciform specimen in TD is clamped for numerical simulation. Therefore, RD was subjected to main deformation during numerical simulations.

Besides, stress triaxiality (λ) and equivalent plastic strain ($\bar{\varepsilon}_p$) as ductile damage parameters are plotted in the center of cruciform specimen and shown in figure 5. Stress triaxiality can be defined as the ratio of hydrostatic pressure to the equivalent stress and its values corresponding to uniaxial, plane strain and equi-biaxial tensions are summarized in table 3 as follow:

Table 3. Stress triaxiality values in different loading conditions

Loading condition	Uniaxial	Plane strain	Equi-biaxial
λ	0.33	0.57	0.66

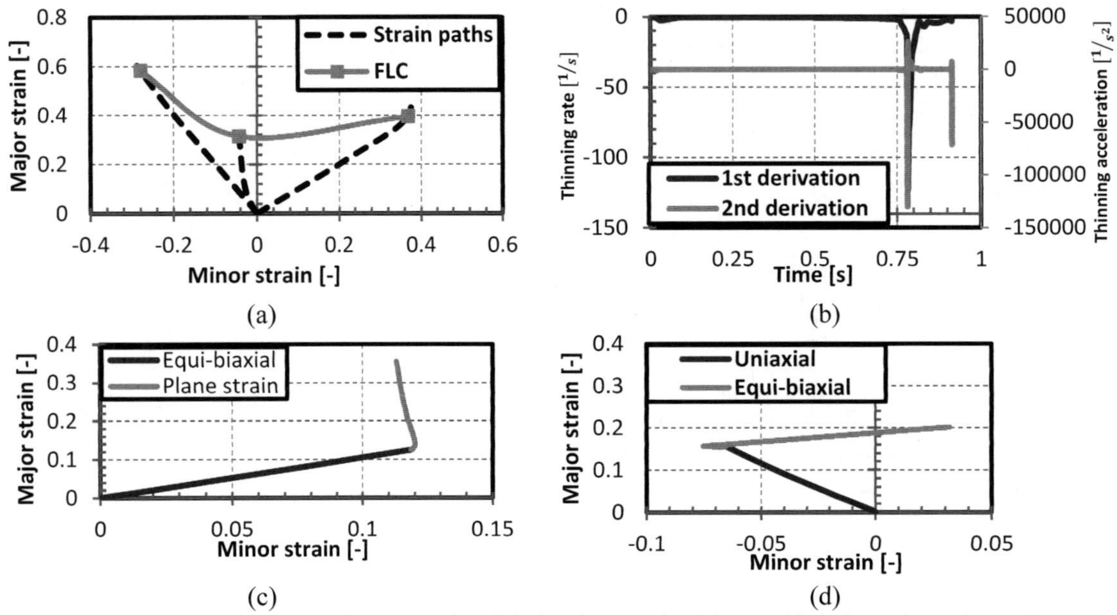

Figure 4. (a) Numerical FLC (b) Example of derivation method for necking detection (c) Nonlinear strain path, equi-biaxial tension- plane strain tension (d) Nonlinear strain path, uniaxial tension - equi-biaxial tension

It is obvious in figure 5 how the stress triaxiality changes by changing the loading condition. In figure 5(a), after changing the loading condition from equi-biaxial tension to plane strain tension, there is an evolution for stress triaxiality to attain the ideal value in table 3. However, there is a discrepancy in figure 5(b) between numerical and ideals values for stress triaxiality, which is most common for cruciform specimens during uniaxial tensile tests. As can be seen in the figure, after changing the loading condition from uniaxial tension to equi-biaxial tension, stress triaxiality almost attains 0.66 corresponding to equi-biaxial loading condition.

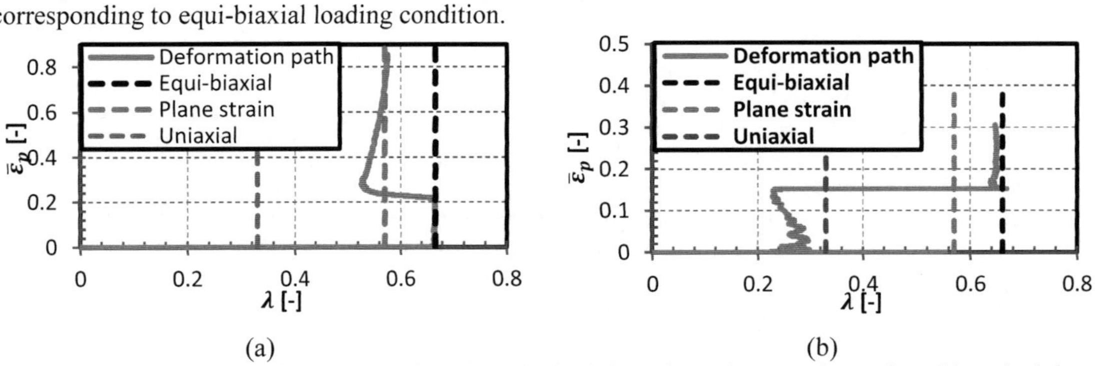

Figure 5. Nonlinear deformation paths (a) Equi-biaxial tension- plane strain tension (b) Uniaxial tension- equi-biaxial tension

4 Conclusion

This work introduce cruciform specimen as an alternative for Nakajima experiment since it can provide frictionless FLC from uniaxial tension to equi-biaxial tension. In addition, it can provide nonlinear and complex loading paths. The cruciform geometry was optimized in previous works and the idea was as follows: If necking is induced to the central region, then the loading ratios of the machine can control the strain paths in the center of gauge. Therefore, uniaxial tensile test in different material directions were performed. Flow curve and Lankford ratios were characterized for TWIP steel sheet. Those parameters were used as input for numerical simulation of the cruciform specimen with using Hill 48 anisotropic yield criterion. First and second derivation methods were utilized to detect the beginning time of instability. Peaks in first and second derivations are rigorously localized in terms of time. Therefore, a numerical FLC was presented for the material under investigation. Cruciform specimen was simulated under nonlinear loading paths as well. In the present work, equi-biaxial tension- plane strain tension and uniaxial tension- equi-biaxial tension were taken into account. Deformation paths in terms of strain and stress triaxiality were plotted. It was shown how stress triaxiality changes by changing the loading condition. This work only investigates the plastic behavior of a TWIP steel sheet considering anisotropic effects. Extensions to fracture as well as experiments are deferred to future research.

Acknowledgements

This paper is a result of the project "Methods development for formability assessment of thin sheets considering anisotropy and non-linear loading path" No.: TF02000072, financed by Technology Agency of the Czech Republic.

References

[1] Keeler S and Backofen W 1963 *J. Transaction of American society for metals,* Plastic instability and fracture in sheets stretched over rigid punches, **56** 25-48.

[2] Lester S, Nikhare C 2014 *International Deep Drawing Research Group (IDDRG),* Investigation of strain development and hole expansion ratio in mild steel and AA5083, 210-15

[3] Kuwabara T, Hashimoto K, Iizuka E and Yoon J 2014 *J. Mater. Process. Technol* Effect of anisotropic yield functions on the accuracy of forming simulations of hole expansion **81** 2433-38.

[4] Ram S and Kang H 2010 *ASME International Mechanical Engineering Congress and Exposition,* Investigation of hole expansion characteristics of DP 600 with testing and modelling **3** 47-51.

[5] Nikhare C, Vorisek E, Nolan J and Roth J 2017 *Journal of Engineering Materials and Technolgy* Forming limit differences in hemispherical dome and biaxial test during equibiaxial tension on cruciform **139** 0410111-9.

[6] Farahnak P, Urbanek M, Konopik P and Dzugan J 2018, *Int J Mater Form*, Determination of formability limits of mild steel DC01 using Nakajima and cruciform experiments with consideration of thickness influence on strain path (under submission).

[7] Urbanek M, Farahnak P, Rund M and Dzugan J 2018, *Journal of physics*, Sheet thickness reduction influence on fracture strain determination, **1063**.

[8] Konopik P, Farahnak P, Rund M, Prochazka R and Dzugan J 2018,*Ubiquity Proceedings*, Application of micro-tensile test for material characterization of mild steel DC01, DOI: https://doi.org/10.5334/uproc.33

[9] Farahnak P, Azinpour E, Belinha J and sa J 2016 The numerical non-linear analysis of elasto-plastic materials using the radial point interpolation method *International book of Material Modelling: Applications, Challenges and Research* 215-248 ISBN: 978-1-53612-161-2.

[10] AbuFarha F, Hector L and Kharisheh M 2009 *Met. Mater, Soc. (TMS),* Cruciform-shaped specimen for elevated temperature biaxial testing for lightweight materials, **61**(8) pp.48-56.

[11] Prantl A, Urbanek M, Prochazka R and Dzugan J 2016 *International Deep Drawing Research Group (IDDRG),* Evaluation of influence of strain path on FLD diagram.

[12] Zidane I, Guines D, Leotoing L and Ragneau E 2010 *Measurement Science and Technology* Development of an in-plane biaxial test for forming limit curve (FLC) characterization of metallic sheets, **21,** 1-11.

[13] Farahnak P, Dzugan J and Prantl A 2016 *International conference of COMAT*, **179** 012023. Sheet necking prediction in forming limit diagrams with the anisotropy influence incorporation

[14] Farahnak P, Dzugan J and Urbanek M 2017 *International conference of Deep Drawing Research Group* **896** 012082. Investigation study on determination of fracture strain and fracture forming limit curve using different experimental and numerical methods

Determination of edge strain in DP steel sheet for predicting edge cracking

M. Fedorko 1,a, M. Urbánek 1, J. Machálek 2

[1] COMTES FHT a.s, Dobřany 334 41, Czech Republic

[2] PWO Czech Republic a.s, Valašské Meziříčí 757 01, Czech Republic

E-mail: mikulas.fedorko@comtesfht.cz, miroslav.urbanek@comtesfht.cz, Jakub.Machalek@pwo.cz

Abstract. This contribution is concerned with determining the edge strain in edges with different finish qualities associated with crack formation in the free edge of a deep drawing blank. The sheared edge quality is of major importance, particularly in high-strength steels which are used for safety components. The purpose of this study was to determine for various parting processes for sheet metal the limit strain which leads to cracking in the free edge. The parameter which was varied in this investigation was the parting process that produced the edge: shearing and machining. A special fixture was employed for measuring the strain value. The values were verified using FE modelling with AutoForm software. The surface finish qualities of the edges were compared using a special test procedure, the Diabolo test, which had been previously described in publications by Held [1] and Liewald and Gall [2]. In this test, the sheared edge is subjected to loading under plane strain conditions. Another test of susceptibility to edge cracking is the hole expansion test [3], in which crack is induced in a circular hole in the centre of a part. The limit edge strain can be identified using a DIC optical measuring system which monitors strain increments in the course of the test. Thanks to these results, the factors which affect the susceptibility of the sheared edge to edge cracking material have been identified and validated in the DP1000 material.

Keywords: edge cracks, shear cutting, Diabolo test, finite element simulation.

1 Introduction

In recent years, several approaches have been developed to predict the formation of a free-edge curve by numerical simulation [1]. They share the use of FLC as the basis for calculating reduced forming ability. These models require experimental data obtained from tests most commonly used for this purpose and the hole expansion test according to ISO 16630. In this test, a hole is extended by a conical piston until it breaks in its circumference. The evaluation criterion is defined as the ratio of the extension of the opening between the original diameter and the fracture size. However, this test does not provide the maximum strain value, which is achieved before crack formation. Therefore, several alternative tests [5] have been developed to measure this value. In the present article, the so-called Diabolo test described by Held [1] was used and DP1000 steel was analyzed. The susceptibility of the sheared edge was tested on two test specimens. Each was produced by a different cutting method: shearing and chip cutting. Each specimen had a size of 40×200 mm. The Diabolo test set-up involved optical measurement. Thanks to a special punch geometry, the strain was localized into a small region adjacent to the sheared edge. This strain can be measured by the GOM Aramis strain measurement system. In this study, the maximum major strain φ1 before cracking was evaluated to characterize the edge cracking sensitivity. The moment of necking was detected by a time-dependent approach similar to the approach suggested by Volk and Hora for the Nakazima test [6]. The configuration of the punch and sample before the forming process is shown in Fig. 1a. The maximum edge strain values were then fed into a numerical model of the Diabolo test implemented in AutoForm software.

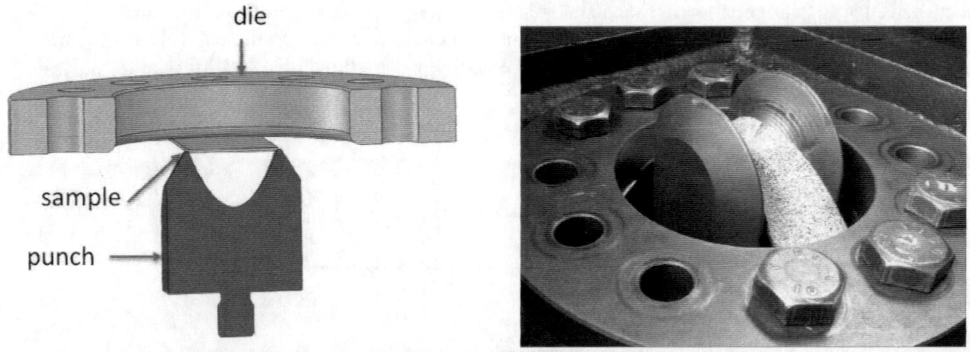

Fig. 1: a) Configuration of tools and sample before forming, b) Sample after measurement

2 Experimental material and method

Diabolo test was performed using our Nakazima test machine [7], in which the classical spherical punch was replaced by a diabolo punch. The measurement and evaluation methodology was based on ISO12004, which describes the determination of the FLC curve from measurements. The ARAMIS DIC system and the time dependent method described by Prof. Hora were used for online measurement and evaluation. The measurement procedure consisted of assembling and calibrating of the setup, measurement, evaluation and processing of the measurement data, and finally their application in an FEM software. Edges created with two types of processes have been tested. One specimen had been sheared and the other machined. The outcome of the measurement was the dependence of force on the piston displacement.

The shear cutting procedure can be defined as two edges working towards each other to separate material into smaller pieces. The material between the edges is deformed so that fractures appear and eventually the material will separate. Material properties such as strength, Young's modulus and ductility have great impact on the cutting procedure. When the two edges engage in the material, see Fig 3, a fracture will appear on each side and will continue to grow as the edges continue to move through the material. A basic condition that must be met to perform a shear cutting operation is that the material of the cutting tool needs to have larger hardness and strength than the material to be separated.
After a shear cut operation, an edge with several different zones appear, see Fig. 2, and depending on the adjustment of the cutting tool the lengths of these zones can vary. With optimal adjustments of the cut tool, the shear zone represents 25–35% of the edge surface. An edge with a large burr zone is a sign that the adjustment of the shearing tool was suboptimal.

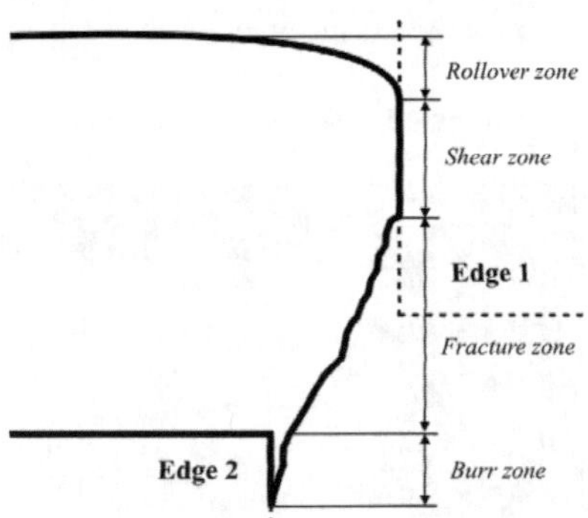

Fig. 2 Zones in a shear cut edge

With the shearing knives appropriately adjusted, the cracks running from both sides meet, resulting in material separation, as in Fig. 3. With increasing clearance between or wear in the knife edges, the cracks are less likely to meet. This leads to higher forces needed for separation and poorer quality of the sheared surface. The resultant sheared surface quality can be altered by making changes related to the gap (clearance) between the knife edges, the wear in the knives, the shearing edge geometry and the amount of lubricant.

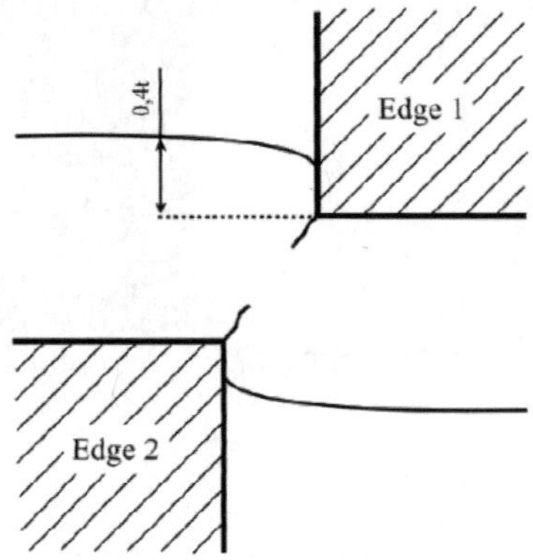

Fig. 3 With optimal adjustments, the fractures will meet and give a fine cut. When the edge advances approx. 40% (0.4t) of the thickness of the material, a fracture will appear.

2.1 *Sheared edges of specimens*

Both specimens were examined under an optical microscope to evaluate the effect of the surface condition on the risk of edge cracking. The edge qualities were analyzed under a microscope and micrographs were taken. Fig. 4 below shows the quality of the sheared surface in the sheared specimen

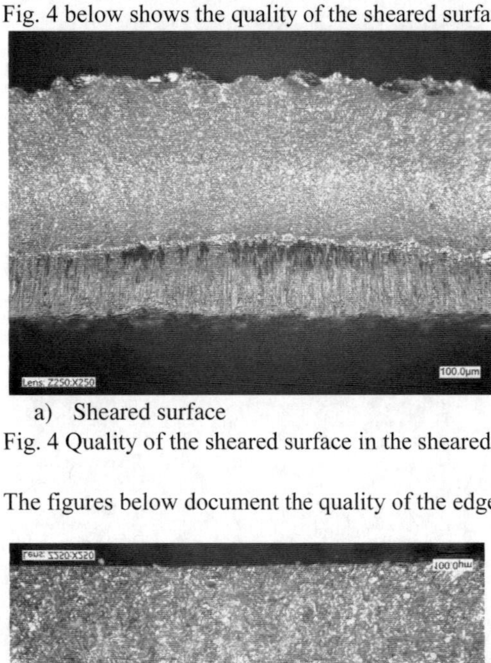

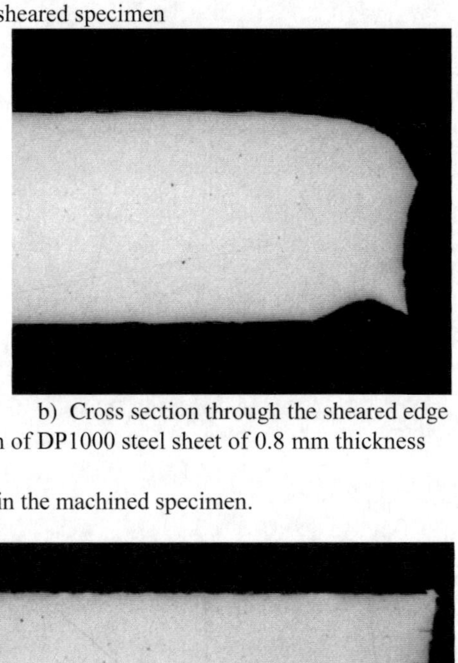

a) Sheared surface b) Cross section through the sheared edge

Fig. 4 Quality of the sheared surface in the sheared specimen of DP1000 steel sheet of 0.8 mm thickness

The figures below document the quality of the edge surface in the machined specimen.

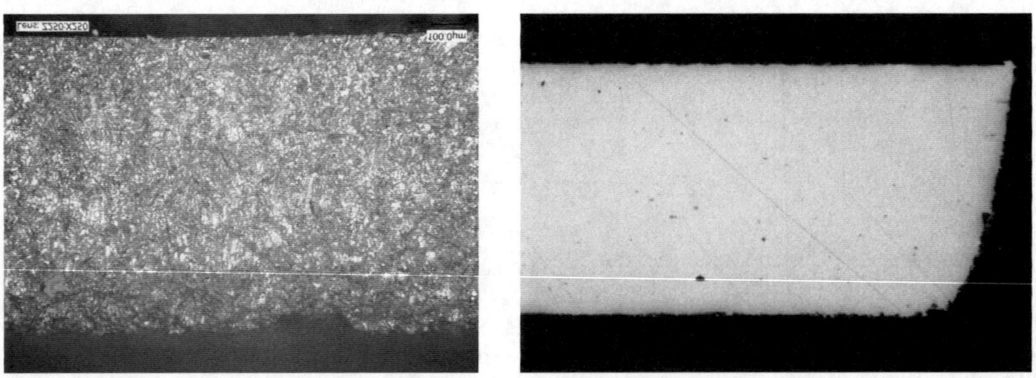

a) Machined surface b) Cross section through the edge

Fig. 5 Quality of the machined surface in the machined specimen of DP1000 steel sheet of 0.8 mm thickness

2.2 Measurement results

The measurement revealed considerable differences resulting from the processes that had been used for creating the edge surfaces. The maximum force prior to cracking depended strongly on how the specimen edge had been created. For the specimen with a machined edge, the force was almost 2.5 times higher than for the sheared specimen. The first response recorded early in the graph was the specimen bending. Then, once the entire shaping punch surface had come into contact with the specimen, the force increased steeply, after which a crack initiated and began spreading from the free edge (Fig. 4).

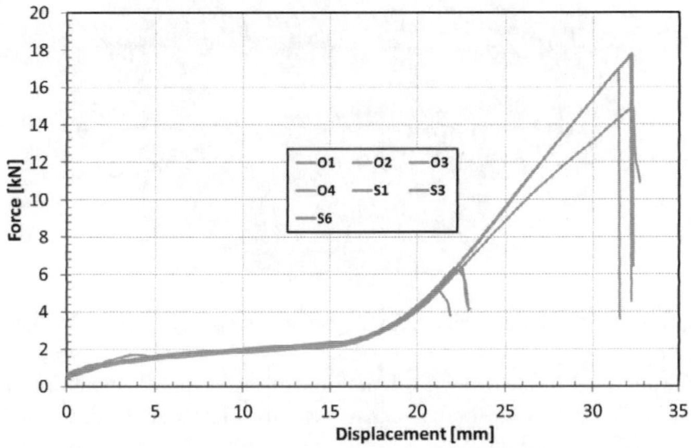

O - machined specimen S – sheared specimen

Fig. 6 Force vs. piston displacement

ARAMIS is a Digital Image Correlation (DIC) system developed to measure displacements, surface strain, velocity and accelerations of a test object. The system creates a 3D measurement and uses digital images to measure changes of the material specimen of just a few millimeters to several meters in size. The measured data is used to determine material properties of the test specimen such as Young's modulus (elastic modulus) and Forming Limit Curves (FLC).

An example of what major strain distribution prior to cracking may look like in the ARAMIS system is shown in Fig. 7.

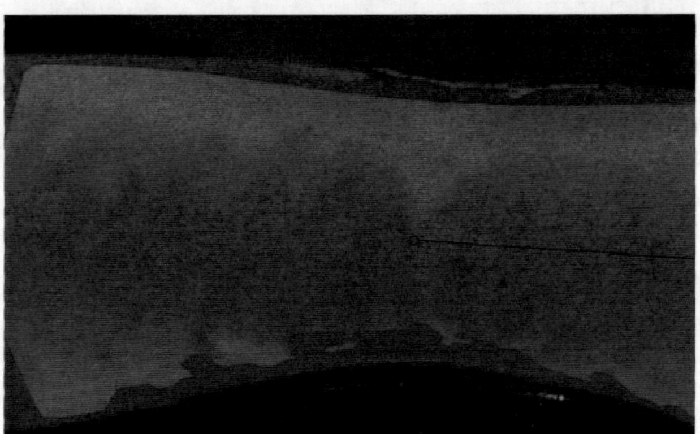

Fig. 7 The major strain from ARAMIS before a large fracture

3 Numerical simulation

Numerical modelling of the Diabolo test was performed using the AutoForm software. The longitudinal axis of the specimen was parallel to the rolling direction. The "roll angle" in the software was therefore set at 90°. Only a single-action draw was simulated, where the positions of the punch, the die and the blank holder are precisely defined. In this case, the specimen is held firmly by two high-strength bolts, which is why the blank holder was affixed securely. The friction value in AutoForm was set to different values in order to test which value gives results similar to the actual tests. This was done by exporting force and punch depth data and comparing it with results from the real tests in a diagram.

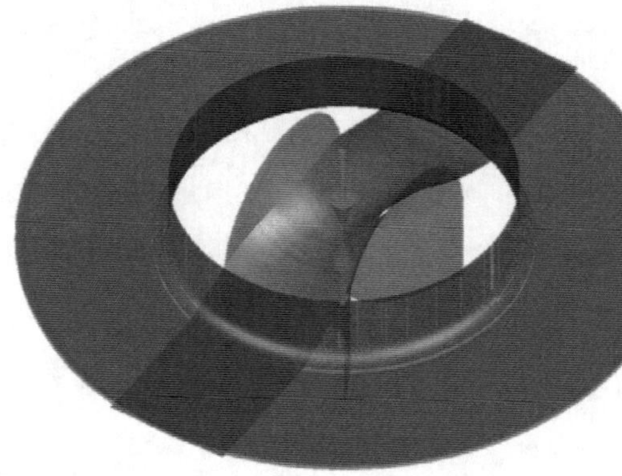

Fig. 8 An overview of the die, blank holder, punch and specimen used for the simulation

The material model embodied in the work-hardening curve has a major impact on the force magnitude computed in numerical simulation. The simulated and actually measured forces were in relatively good agreement. The damage parameters in the AutoForm software were only considered for visual evaluation of the threshold. They had no effect on the force calculation. The main goal was to adjust the damage parameters in the AutoForm software to ensure that the simulated crack forms at the same time and piston position as in the real-world test. Input values were based on the measurement and had to be defined for two manufacturing processes. The threshold strain was determined directly from the DIC ARAMIS data and then calibrated with the AutoForm software. When the threshold value is correct, red regions in the simulation identify the same critical locations for cracking at the same piston positions as in the real-world measurement. The critical region is usually associated with values near unity. This threshold value also depends on the left side of the FLC diagram which depicts the uniaxial stress state generated in the free edge and in its vicinity.

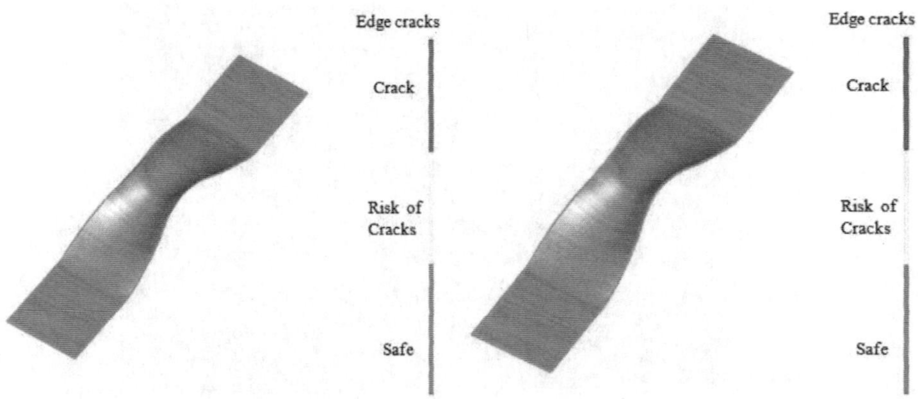

a) Sheared specimen b) Machined specimen

Fig. 9 Evaluation of the risk of cracking by simulation

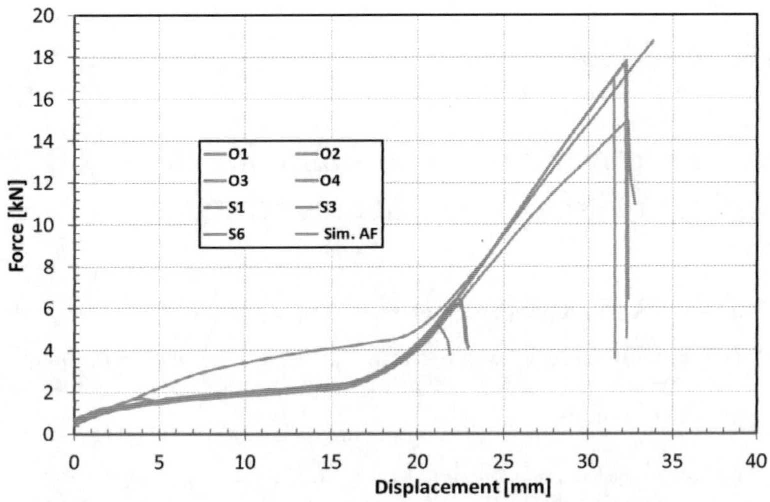

O - machined specimen S – sheared specimen

Fig. 10 Comparison between values from simulation in AutoForm and the Diabolo test

4 Conclusion

This paper evaluates the effects of various cutting processes on crack formation in a blank edge using the Diabolo test. The test specimens of 0.8 mm thickness were made of DP1000 steel and had two types of edges. Those were produced by shearing and chip cutting. The microstructures in their material were examined, which revealed new aspects related to this issue. The measured strain values were fed into numerical simulation of the Diabolo test carried out using AutoForm software which offers the capability to use such values for modelling deep drawing processes. Measured values can be used for any computational software that allows calculation of material damage.

Acknowledgement
This study was created under the project "Development of West-Bohemian Centre of Materials and Metallurgy" No.: LO1412, financed by the Ministry of Education of the Czech Republic.

References

[1] Held, C., Schleich, R., Sindel, M., Liewald, M.: Investigation on advanced forming limit prediction techniques for high strength steels, in: Proceedings of the IDDRG 2009 Conference, 461–470.

[2] Liewald, M., Gall, M.: Experimental investigation of the influence of shear cutting parameters on the edge crack sensitivity of dual phase steels, in: Proceedings of the IDDRG 2013 Conference, Zurich, Switzerland 219–224.

[3] ISO 16630, Metallic materials - Sheet and strip - Hole expanding test, Berlin, Beuth, 2009.

[4] Falk, J.: Fracture prediction of stretched shear cut edges in sheets made of Dual-Phase steel, Master's degree thesis, Blekinge Institute of Technology, Karlskrona, Sweden 2017.

[5] Bouaziz, O., Douchamps, S., Durrenberger, L., Bui-Van, A.: The double bending test: a promising new way for an optimal characterization of cut-edges ductility, in: Proceedings of the IDDRG 2010 Conference. Graz, 575–582.

[6] Volk, W., Hora, P.: New algorithm for a robust user independent evaluation of beginning instability for the experimental FLC determination. International Journal of Material Forming, 4, 339–346, 2011.

[7] Fedorko, M., Urbánek, M., Rund, M., Prediction of thinning of the sheet metal in the program AutoForm and its experimental verification, in: IOP Conference Series: Materials Science and Engineering, ISSN: 17578981, 2017

Microstructure and tribological behavior of HVOF sprayed and laser treated CoCrTaAlCSiY coatings

A Glanc[1], Š Houdková[1], M Vostřák[1]

[1]University of West Bohemia, Plzeň, Univerzitní 8, Czech Republic, EU

glancale@kmm.zcu.cz

Abstract Co-based alloys are used over a wide range of application because of their excellent corrosion and wear properties. Especially the Stellites (Co-Cr-W) are well known and quite well researched. The CoCrTaAlCSiY is a new material on the field of thermal spray coatings and due to its composition was chosen for studding of tribological behavior. In this paper the HVOF sprayed and post heat treated coatings of CoCrTaAlCSiY are presented. Specimens were treated by a laser with two different laser parameters in order to reach different thickness of laser treated layers. Erosion test and linearly oscillating ball-on-flat test were applied to evaluate the wear behavior of CoCrTaAlCSiY coating. Microstructure of as-sprayed and heat treated coatings were also studied. The evaluation revealed a positive effect on wear coefficient for both laser treated coatings. A significant improvement of volume loss during erosion test was also observed for both laser treated coatings.

1. Introduction

Today's industry applications claim still higher demands on material properties and low purchase prices therefore materials with high wear and corrosion resistance are highly desirable. Their usage is in wide range of applications such as power generation, aerospace, chemical industry or medicine. There are the materials exposed to high mechanical load or corrosion environment or a combination of both. Only materials with excellent combination of mechanical and oxidation properties can fulfill such requirements but these special materials are quite expensive. That is why they are often applied on the surface of cheaper components as a coating material improving the surface properties.

Thermal spray coatings represent really good possibility how to combine low manufacturing costs and excellent material properties. The HVOF (High Velocity Oxygen Fuel spraying) is widely used method of thermal spraying. The main advantages of HVOF is preparation of coatings with low oxidation during the spraying process, high adhesion to substrate and low coating porosity [1].

Additional heat treatment of thermal spray coatings could lead to improvement of material properties. One of possible processes is laser heat treatment which is a technology that modifies the surface properties of materials. Usage of laser heat treatment in combination with thermal sprayed coatings results in elimination of oxides and porosity at the intersplat boundaries. Thus the internal coating cohesion is higher and subsequently the corrosion and wear resistance can be improved [2].

There exists a wide variety of wear resistant materials which differs in chemical composition and purpose of usage. The most commonly used wear resistant materials are composites of hard tungsten carbides WC or chromium carbides Cr_3C_2 (also a combination of both is possible) and Ni, Cr or Co-based metallic matrix (or a combination of these metals) where are the carbides distributed [1]. The Cr-based wear resistant materials with high content of chromium carbides

Cr_3C_2 are characterized by resistance to high temperatures. These Cr-based materials are represented by Cr_3C_2–25NiCr and Cr_3C_2–25CoNiCrAlY [3] [4]. The Ni-based alloys are generally based on the Ni-Cr solid solution. Typical representatives of such alloys for thermal spraying are NiCrBSi and Hastelloy C-276 [5].

And last but not least there is a family of Co-based alloys which are called Stellites, typically represented by Stellite 6 which is a Co-Cr-WC alloy [6] [7] [8]. One of the Co-based materials is CoCrTaAlCSiY which is a new material on the field of thermal spray coatings with the maximal operating temperature 1050 °C [9]. It is characterized by Co-based matrix with high content of Cr which provides oxidation resistance. On the other hand there is a considerable content of Al (7.5 wt.%) and Ta (8.2 wt.%), especially the effect of tantalum on the tribological behavior of HVOF sprayed coatings is not researched sufficiently. Addition of yttrium to the limit of 1 wt.% within Co-based alloys can enhance mechanical properties of upper oxidation layer which can cause improvement of wear resistance [10]. Tantalum is very reactive and carbide-forming element [11]. Tantalum is forming the tantalum carbide TaC which is harder than carbides Cr_3C_2 [10].

Target of this paper is to evaluate and describe results of laser heat treatment process which was applied on the HVOF sprayed coatings of CoCrTaAlCSiY material with the focus on mechanical properties and tribological behavior of the coatings. Two different laser parameters settings were selected in order to carry out coatings with different thickness of laser treated layers.

2. Experimental procedures

2.1. Coatings preparation

The CoCrTaAlCSiY coatings were sprayed by HP/HVOF TAFA JP5000 spraying equipment in VZÚ Plzeň s.r.o. The grit blasted steel 11 523 plate with the dimensions 200 mm x 100 mm x 10 mm was used as a substrate. These dimensions were chosen to ensure sufficient cooling rates during laser treatment process. The powder F22 with a grain size 0.8-1 mm was used for grit blasting of substrates. For HVOF spraying was used commercially available gas atomized powder FST 469.001 with nominal chemical composition of 25 wt.% Cr, 8.2 wt.% Ta, 7.5 wt.% Al, 0.75 wt.% C, 0.7 wt.% Si, 0.75 wt.% Y and balanced Co. The laser treatment was applied using two different laser parameters (named LR1 and LR2). For the parameters LR1 was chosen the laser power 1070 W, specific energy 5.4 J/mm^2, traverse speed 16.7 mm/s and spot size 12 mm x 1 mm. The parameters LR2 are specified with laser power 1070 W, specific energy 17.8 J/mm^2, traverse speed 5 mm/s and spot size 12 mm x 1 mm. For laser treatment was used the HPDD 4 kW laser Coherent HighLight ISL-4000L with wave length 808 ± 10 nm. Preheating of the substrates up to 350 °C was performed in order to prevent coatings from high thermal gradient which could cause cracks in the coated layer. After laser treatment were the samples cooled in air atmosphere

2.2. Coatings analyses

The same experimental procedure was applied to as-sprayed coating and also to both laser treated specimens (with parameters LR1 and LR2). Automated LECO grinding and polishing equipment was used for grounding and polishing of the cross sections specimens. Quality of specimens polishing was continuously checked in optical microscope Nikon Epiphot 200 which was also used for optical micrographs of specimens microstructure. The microhardness was evaluated from cross sections by digital optical 3D microscope Hirox KH7700. Corresponding SEM images was prepared using the microscope EVO MA25 from Zeiss with LaB6 thermal filament and equipped by EDX detector SDD X-Max 20 Oxford Instruments.

The indents for microhardness measurement were made in 50 μm equidistant steps and started 50 μm below the coatings surface which resulted in 6 x 6 indentation matrix. The microhardness was evaluated by HV0.1 measurement and mean values are plotted in the graph in Figure 3.

Sliding wear resistance was evaluated using the linearly oscillating ball on flat wear test (ASTM G-133). Subsequently the wear coefficient (K) and coefficient of friction (COF) were determined at room temperature. The parameters of ASTM G-133 wear test were as follows: steel 100Cr6 counterpart (6 mm diameter ball), 25 N normal force, 5 Hz oscillating frequency, 10 mm stroke length, 1000 s testing time and 100 m total sliding distance. Three different measurements were applied for each specimen then the depth profile of each wear track was measured by profilometer KLA-Tencor P-6 Profiler at three different places and finally the total volume loss was calculated. Surfaces of specimens for ASTM G-133 wear test were at first ground and polished to ca. 0.04 ± 0.02 Ra value. Created wear tracks were observed in SEM in order to identify wear mechanism.

Abrasive wear resistance was evaluated using the dry sand rubber wheel test (ASTM G-65) and subsequently the wear coefficient (K) was determined at room temperature. The parameters of ASTM G-65 wear test were as follows: wheel diameter 231.89 mm, wheel width 12 mm, wheel hardness 65 Shore A, abrasive media Al_2O_3, size of abrasive particles 200 μm, load 22 N, feeding rate 500 g/min and 718 m total sliding distance. Surfaces of specimens for ASTM G-65 wear test were left without any treatment.

The erosion test (Figure 1) was carried out at VZÚ Plzeň s.r.o. with their own testing equipment. Principle of the erosion test is as follows: the reservoir at the top of the device provides a given constant amount of abrasive media (white corundum F70), the abrasive particles enter in between two rigidly connected rotating discs containing four accelerating channel, there are the abrasive particles accelerated by centrifugal forces and at the outlet of channels the particles stream encounter the specimens. The weight of specimens is several times measured and the weight loss is evaluated and because the density of tested material is known the volume loss can be determined.

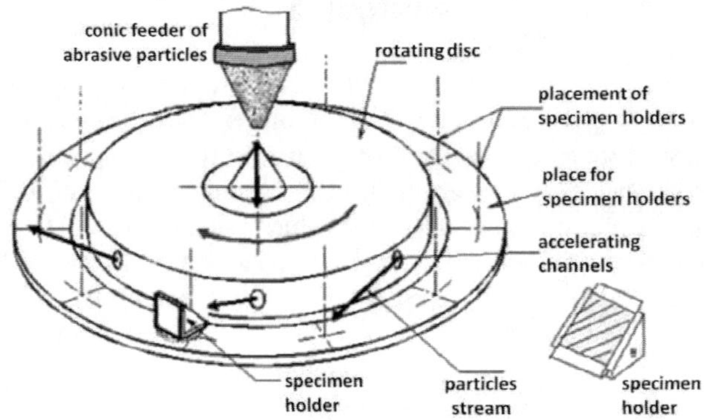

Figure 1. Schematic drawing of erosion test at VZÚ Plzeň s.r.o.

3. Results and discussion

3.1. Microstructure and microhardness

Microstructure of HVOF as-sprayed CoCrTaAlCSiY coating (Figure 2a) is characterized by dense web of splats and containing intersplat boundaries potentially lowering the coatings cohesion. For HVOF coatings is typical small amount of porosity in whole coating. The Al_2O_3 particles used for grit blasting can be occasionally observed on the substrate-coating interface.

Laser parameters LR1 (specific energy 5.4 J/mm^2) led to heat treatment of ca. third of coating thickness whereas LR2 (specific energy 17.8 J/mm^2) resulted in heat treatment of complete coating thickness. Both laser treatments (Figure 2b, 2c) led to decrease of internal coating porosity and final coatings are denser and more compact. Another consequence of laser treatment is that the coating surface is much smoother than as-sprayed coatings. The laser treated coating layer is characterized by recrystalized dendritic structure inside individual splats. In specimen with LR2 parameters (Figure 2c) is noticeable the upper coating layer where the microstructure was completely remelted. Additional effect of higher specific energy is that the porosity from substrate-coating interface propagated towards the coating surface.

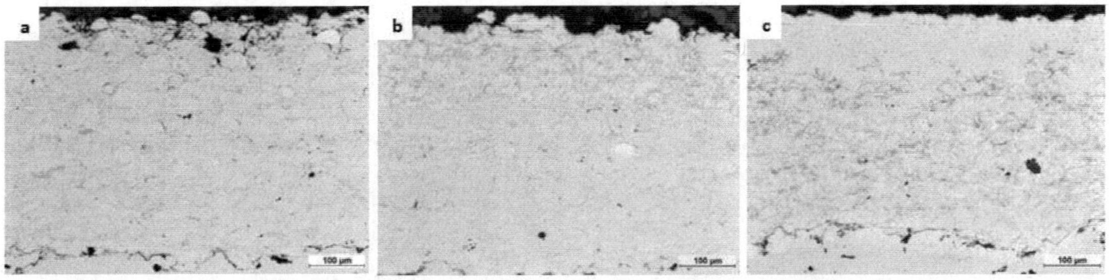

Figure 2. Optical micrographs of microstructure for HVOF CoCrTaAlCSiY coatings: (a) as-sprayed, (b) laser parameters LR1, (c) laser parameters LR2.

Results of microhardness measurement are shown in the graph in Figure 3. For both laser treated coatings (parameters LR1 and LR2) there is an increase of microhardness which is more evident for the LR1 with only partly laser treated coating. One of the reasons for microhardness improvement is perhaps the more compact coating with lower porosity than in the as-sprayed coating. Although scattering of values is quite high, there are two facts resulting from the microhardness measurement. First fact is the confirmation of laser treatment of whole coating thickens for parameters LR2 because in the layer near the substrate-coating interface are values lower than in as-sprayed specimen which is an effect of dilution of Fe from substrate material into coating. Second fact is that the microhardness of completely laser remelted coating (upper layer in LR2 specimen) is lower than the only laser treated coating which is typical for deeper layers.

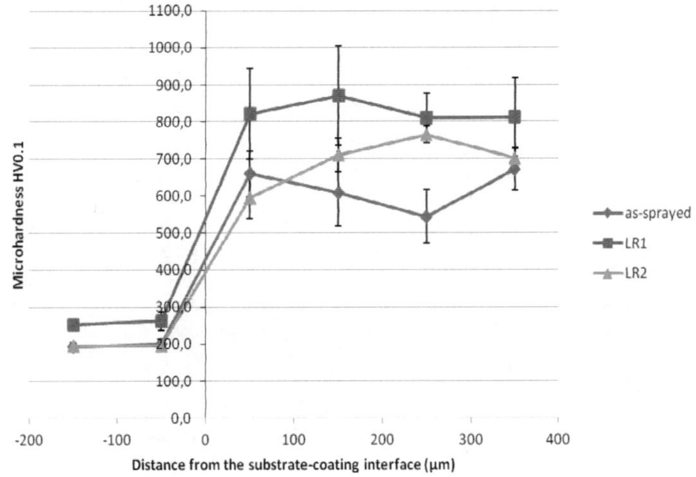

Figure 3. Dependence of microhardness HV0.1 on distance from the substrate-coating interface.

3.2. Wear and abrasion resistance

Graphs in Figure 4 revealed results of the sliding wear resistance measurement. Coefficient of friction (COF) which was evaluated according the ASTM G-133 remains the same for parameters LR2 as for the as-sprayed coating whereas the COF for LR1 parameters shows slight reduction. In wear coefficient (K) can be observed evident trend of decreasing K with increasing amount of laser treated coating layer. The completely laser remelted coating layer (Figure 2c) is beneficial for wear resistance of CoCrTaAlCSiY coating. It could be explained by vanishing of intersplat boundaries in laser remelted layer which avoided delamination of entire splats during wear test.

SEM images of wear tracks after ASTM G-133 are in Figure 5. The main wear mechanism of all specimens is abrasion and for the as-sprayed coating (Figure 5a) was additionally revealed partial delamination of individual splats as a consequence of intersplat decohesion (Figure 2a). Both wear mechanisms contributed to higher K for as-sprayed coating despite its lower microhardness.

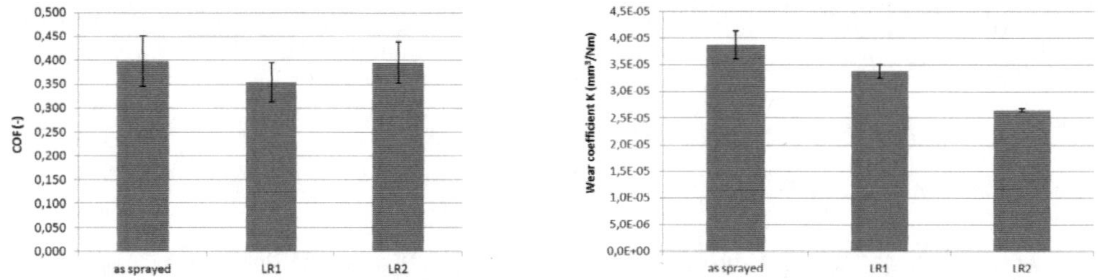

Figure 4. Coefficient of friction (COF) and wear coefficient K evaluated from ASTM G-133.

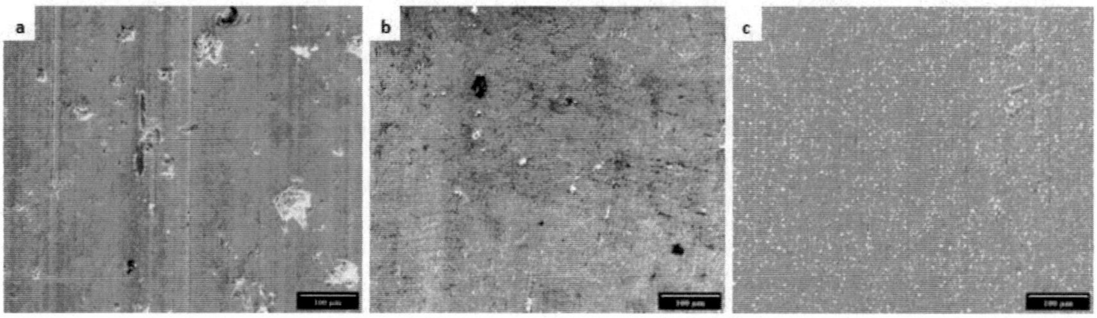

Figure 5. SEM of wear tracks for HVOF CoCrTaAlCSiY coatings: (a) as-sprayed, (b) laser parameters LR1, (c) laser parameters LR2.

Abrasive wear resistance was evaluated according the ASTM G-65 and the wear coefficient is plotted in Figure 6a. The beneficial effect of completely laser remelted coating layer was confirmed for LR2 parameters whereas parameters LR1 show slightly increased K which correspond to microhardness measurement.

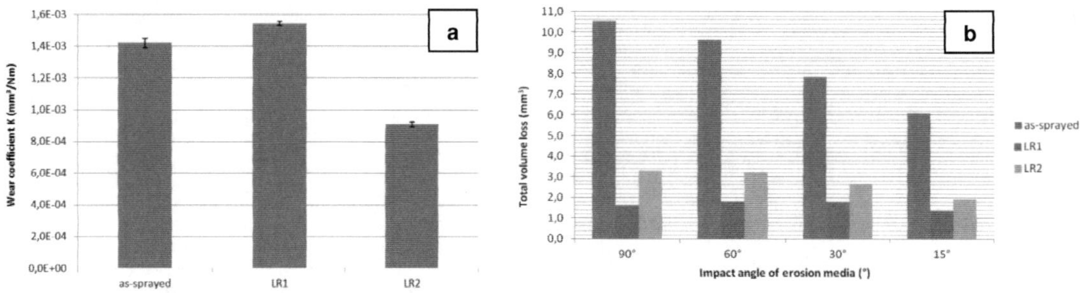

Figure 6. (a) Wear coefficient K evaluated from ASTM G-65 and (b) volume losses dependent on the impact angle of erosion media.

3.3. *Erosion resistance*

Total volume losses with dependency to impact angle of erosion media are reported in graph in Figure 6b. There is a significant improvement of erosion behavior for both laser treatment coatings. It could be explained as a consequence of enhanced internal coating cohesion and elimination of porosity after laser treatment. The reduction of volume losses is more pronounced for LR1 parameters which could be a consequence of higher microhardness, it means that more erosion particles reflect from the coating surface without any interaction with it. The tendency of decreasing volume losses with decreasing impact angle of erosion media is more or less evident for all specimens.

4. Conclusion

The aim of this study was to reveal tribological behavior of HVOF sprayed and laser treated CoCrTaAlCSiY coatings. An improvement of sliding wear resistance was proved for both laser parameters compared to as-sprayed coating. The specimen with parameters LR2 (higher specific energy) is characterized by significant reduction of wear coefficient which was also proved by ASTM G-65 wear test. The erosion test revealed beneficial influence of enhanced internal coating cohesion and elimination of porosity inside laser treated coatings. The consequence is markable reduction of volume losses for both laser parameters and all impact angles of erosion media. Enhanced erosion and wear behavior of both laser treated parameters is accompanied by higher microhardness. Further tests and especially the XRD analyses should be applied to reveal the phase composition of CoCrTaAlCSiY coatings.

Acknowledgement

The result was developed within the CENTEM project, reg. no. CZ.1.05/2.1.00/03.0088, co-funded by the ERDF as part of the Ministry of Education, Youth and Sports OP RDI programme and, in the follow-up sustainability stage, supported through CENTEM PLUS (LO1402) by financial means from the Ministry of Education, Youth and Sports under the "National Sustainability Programme I." and project no. SGS 2016-005.

References

[1] Oksa M, Turunen E, Suhonen T, Varis T and Nannula S P 2011 Optimization and characterization of high velocity oxy-fuel sprayed coatings: techniques, materials and applications *Coat.* **1** no 1 pp 17–52

[2] Poza P, Munez C J, Maneiro M A G, Vezzu S, Rech S and Trentin A 2014 Mechanical properties of Inconel 625 cold-sprayed coatings after laser remelting depth sensing indentation analysis. *Surf. Coat. Technol.* **243** pp 51–7

[3] Hong S, Wu Y, Wang Y, Ying G, Li G, Gao W, Wang B and Guo W 2013 Microstructure and cavitation-silt erosion behavior of high-velocity oxygen-fuel (HVOF) sprayed Cr_3C_2–NiCr coating *Surf. Coat. Technol.* **225** pp 85–91

[4] Picas J A, Punset M, Manergues E, Martin E and Baile M T 2015 Microstructural and tribological studies of as-sprayed and heat-treated HVOF Cr_3C_2–CoNiCrAlY coatings with a CoNiCrAlY bond coat *Surf. Coat. Technol.* **268** pp 317–24

[5] Glanc A, Houdková Š, Vostřák M and Bystriansky M 2018 Microstructure and tribological behavior of HVOF sprayed and laser remelted Hastelloy C-276 coating Proc. Int. Conf. METAL 2018 (Brno) 2018 (Ostrava: Tanger)

[6] Razavi R S 2016 Laser surface treatment of Stellite 6 coating deposited by HVOF on 316L alloy *J. of Mat. Eng. and Performance* **25** pp 2583–95

[7] Houdková Š, Pala Z, Smazalová E, Vostřák M and Česánek Z 2016 Microstructure and sliding wear properties of HVOF sprayed, laser remelted and laser clad Stellite 6 coatings *Surf. Coat. Technol.* **318** pp 129–41

[8] Sassatelli P, Bolelli G, Gualtieri M L, Heinonen E, Honkanen M, Lusvarghi L, Manfredini T, Rigon R and Vippola M 2018 Properties of HVOF-sprayed Stellite-6 coatings *Surf. Coat. Technol.* **338** pp 45–62

[9] H C Starck Product Information Amperit 469 CoCrTaAlCSiY [cit. 18.08.2018] Available on-line https://www.hcstarck.com/amperit_469

[10] Farnia A. Ghaini F M, Rao J C, Ocelík V and Hosson J T M 2012 Effect of Ta on the microstructure and hardness of Stellite 6 coating deposited by low power pulse laser treatments *Surf. Coat. Technol.* **213** pp 278–84

[11] Steenkiste T V and Gorkiewicz D V 2003 Analysis of tantalum coatings produced by the kinetic Spray process *J. of Thermal Spray Technol.* **13** pp 265–73

INVESTIGATION OF THE EFFECT OF ALUMINUM ALLOY POSITION ON RESIDUAL STRESSES IN DISSIMILAR FSW WELD BY USING THE ULTRASONIC METHOD

I Hadji[1,2], R Badji[1], M Gaceb[2], N Kherrouba[1], L Rabahi[1]

[1] Research Center in Industrial Technologies CRTI. P.O. Box 64, Cheraga 16014 Algiers, Algeria
[2] University M'hamed Bougara, Boumerdes, Faculty of Hydrocarbons and Chemistry, Independence Avenue, 35000, Algeria

E-mail: [1] I.hadji@crti.dz, [2] gaceb_m@yahoo.fr

Abstract. The main goal of this study is to show the effect of the advancing side (AD) and the retreating side (RT) position on the Residual stresses and local mechanical behaviour of dissimilar friction stir welds of aluminum alloys AA2024-T3 and AA7075-T6. Different samples were produced by varying the rotational speed of the tool (1200 and 1400 rpm) and the alloy position regarding the advancing side of the tool. Ultrasonic Method has been used to evaluate Residual Stresses. This method is based on the acoustoelastic effect, which measures the velocity variation of the elastic waves according to the stress state of the material. This can be achieved through a calibration test, which permits the determination of the acoustoelastic coefficient (K). The results show a tensile stress in the Nugget (N), the heat affected zone (HAZ) and a compression stress in the base metal (BM). Increasing the rotational speed reduces the amplitude of the longitudinal residual stresses with a high reduction in the case where AA7075 - T6 is in the advancing side whit 1400 rpm. This has been directly associated to the increase in the heat input and the reduction of the thermal mismatch between different areas of the weld. The microstructure effect of aluminum alloy position acts on the acoustoelastic constant K. The choice of t_0 corrects the overestimated residual stresses in the (HAZ) and (N).

1 Introduction

The Aluminum alloys are renowned for their ratio weight / high strength and their high corrosion resistance with structural applications in aeronautical, automotive and marine industries. The assembly of aluminum alloys by the Fusion welding is known to be problematic with the formation of secondary brittle phases and cracking during solidification. In addition, the thermal cycle of the weld process causes a development of the residual stresses in the weld. The presence of the residual stresses can impact mechanical and corrosion properties of the weldment, they play an important role in the strength and service life of structures [1]. FSW is in joining low weldability materials, such as AA2024 and AA7075 aluminums alloys [2], it takes place in the solid-phase, at temperatures below the melting point of the material. The reduced welding temperature during this process improves the mechanical properties and affects considerably the residual stresses [3]. Therefore, the development of several methods to measure the residual stress such as X-ray diffraction, incremental hole drilling, and more recently ultrasonic methods. The latter is based on the acoustoelastic effect, which measures the velocity variation of the elastic waves according to the stress state of the material. However, few studies on residual stresses of aluminum alloys friction stir welding through ultrasonic method have been done. Javadi et al. [4] presented in their works different experimental configuration needed to measure the longitudinal residual stresses of plates. In his work, H Jamshidi [1] used the X-ray diffraction to measure residual stress in dissimilar FSW of AA6082-T6 and AA7075-T6, it was found that the residual stress distribution at different zones is a function of the local mechanical properties and the temperature distributions. Similar studies were developed by J Zapata et al [5]. Recently, welding dissimilar aluminum alloy has been widely studied. Guo et al. [6] investigated the effect of exchanging materials location of dissimilar aluminum alloy on material flow, microstructure and tensile properties. Zhixia et al. [7] found that the location of materials has an important effect on mechanical properties and the microstructure of joints in FSW dissimilar joint of AA6013/AA7003 aluminum alloys. The motion of the tool in the dissimilar aluminum alloy, lead to local strains zones an asymmetric residual stresses distribution and mechanical properties between the advancing side (AD) and the retreating side (RT) [8]. The fracture in the dissimilar weld AA2024-T3/AA7075-T6 is located in the HAZ on the side of the aluminum alloy AA2024-T3 where the hardness is low. Maximum tensile strength of the joint was achieved when AA2024 was located on the advancing side [2]. Furthermore, by using a digital image correlation, Niu et al. study the response of the local mechanical properties to determine the global one on the aluminum alloys welded with FSW process [9]. Researchers studied the effect of the material position on the mechanical properties [10], but they did not study it on the distribution of residual stresses. Accordingly, ultrasonic method for residual stress analysis and digital image correlation (DIC) for analyses of heterogeneity deformation are performed to evaluate the effects of FSW process parameters and material position on distribution of residual stress in welded joint.

2 Experimental Procedures

2.1 Sample description

Two aluminum alloys, AA7075 and AA2024 have been tested. Different configurations of dissimilar welds obtained by friction stir welding of 5 mm thick sheets. A sheet of 200 mm length 100mm width were welded parallel to the rolling direction by employing a machine three-axis. The tool material used was AISI H11 tempered steel with 6 mm diameter threaded cylindrical pin and 20 mm shoulder and is positioned in weld line centre. In the present study, AA7075-T6 and AA2024-T3 each one positioned on the advancing side of the tool for different welding.

2.2 Measurement Device

The measurement device, shown in Figure 1, includes a wave train generator synchronized with a digital oscilloscope having a sampling frequency of 5 GHz. Measurements were performed with a piezoelectric transducer, which is composed of one emitter (E) and two receivers (R1 and R2). The nominal frequency and diameter of the piezoelectric elements are 5 MHz and 10 mm, respectively.

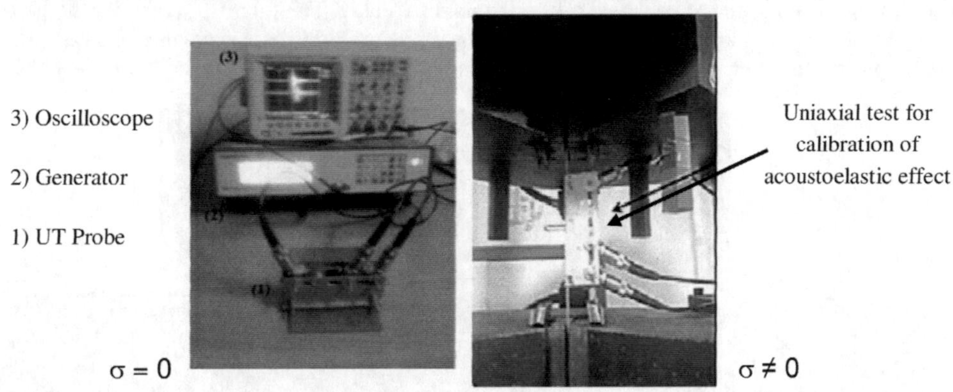

Figure 1. Schematic representation of the Lcr wave measurement setup for $\sigma = 0$ and calibration of acoustoelastic effect for $\sigma \neq 0$.

2.3 Evaluation of the Calibration Constants

The evaluation of residual stresses is based on the acoustoelastic effect [11 - 14]. The acoustoelastic constant K is deduced experimentally from a uniaxial tensile test (Figure 1). In this case, the calibration samples for tensile test were taken in the direction parallel to the weld figure 2. In this study, to show the effect of the advancing side (AD) and the retreating side (RT) position on the residual stresses, we need to evaluate the calibration constants corresponding to the BM and N, equation (1):

$$K = -1/\sigma[(t - t_0)/t_0] \qquad (1)$$

Where, t and t_0 are the time-of-flight measured between the two receivers for stressed and unstressed samples, respectively. σ is the applied stress.

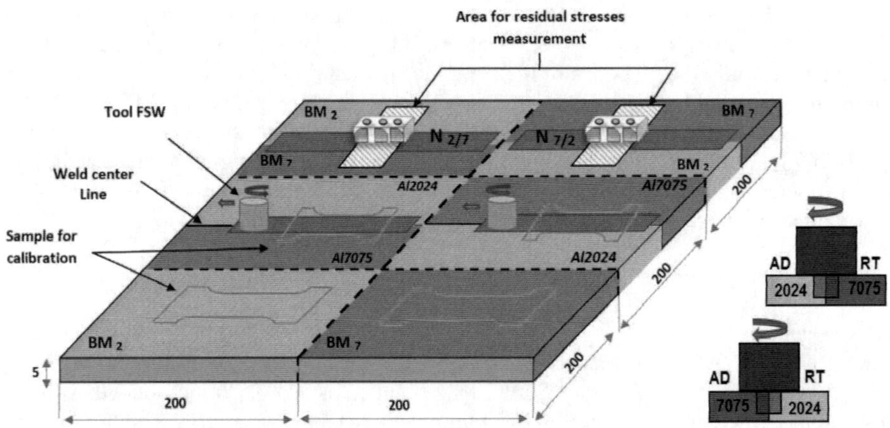

Figure 2 Positions of the calibration samples and area for residual stresses measurement on the original plate.

3. Results and Discussion

3.1 Macro and microstructures of joints

Figure 3 shows the cross section of a weld obtained with 1200 rpm and 40 mm / min. The FSW seal is characterized by four different zones: Nugget (N), affected thermomechanical zone (TMAZ) and heat affected zone (HAZ). In addition, the size of the ring flux pattern in the nugget region increases with the speed of rotation tool, due to increase in heat inputs. obviously. The microstructures of the base metals of the advancing side and retreating side are shown in Figure 3. it's consisted an elongated grain along the rolling direction with a random distribution of small black particles.

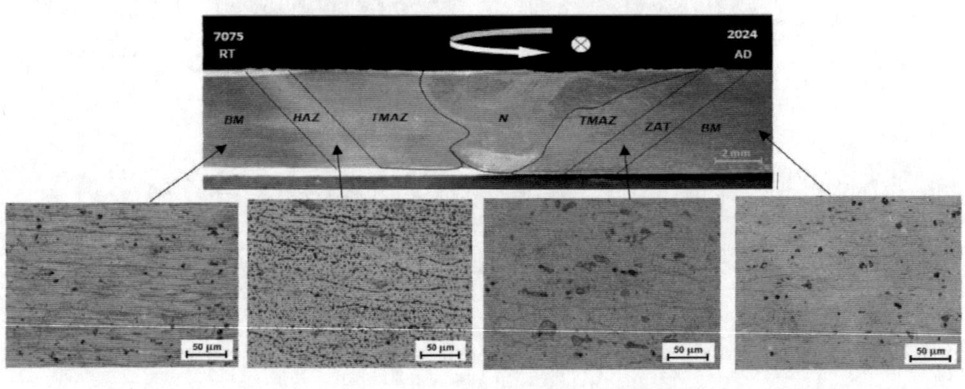

Figure 3 Macro and microstructures of the various weld zones microstructures.

3.2 Microstructure Effect on the Acoustoelastic Constant

The results of tensile test performed an BM 2024, BM 7075, N 7AD/2RT and N 7RT/2AD samples to measure acoustoelastic constant are presented in Figure 4. Linear curves were obtained during the loading and unloading phases, where the slope of the lines represents K acoustoelastic constants. Figure 4 shows that the acoustoelastic behaviour of the material is very affected by the microstructure. That is, acoustoelastic constants K (N 7AD/2RT) and K (N 7RT/2AD) are respectively different by 10%, 12% from the K_{BM2} constant and 17, 20% for K_{BM7}. Furthermore, 4% of the difference between the two-nugget constant, show the weak effect of the alloy position in the nugget zone. However, is difficult to measure the constant of the HAZ because is less than 3mm.

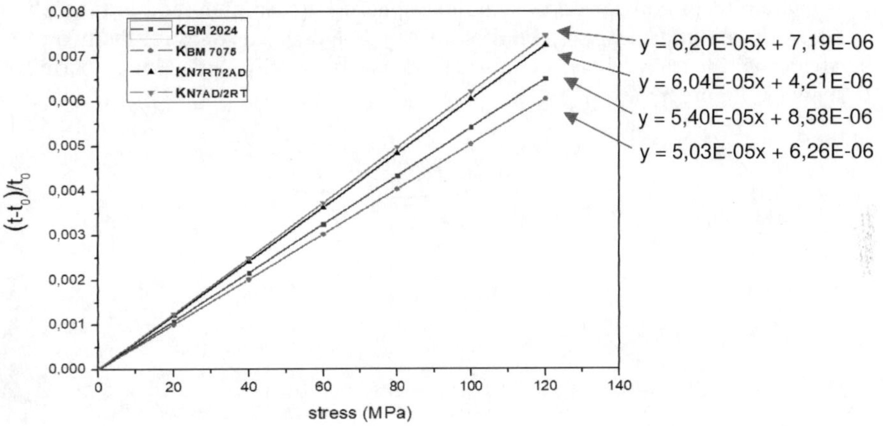

Figure 4. Influence of the microstructure on the acoustoelastic calibration constant.

3.3 Residual stresses characterization

Regarding the Figure 5, it is seen that the stress profiles displayed the characteristic "M" shape for the longitudinal residual stresses, the asymmetry between advancing and retreating stress peaks was clearly visible as already fond by H. Jamshidi et al [1]. As can be seen when AA 2024 is in advancing side, the profile shows a lower magnitude of stresses on the advancing side compare to the retreating side. in the advancing side, the direction of tangential component of tool rotation and the velocity vector associated with the welding speed coincide this causes more friction. These factors create higher temperature on AD as compared to RT.

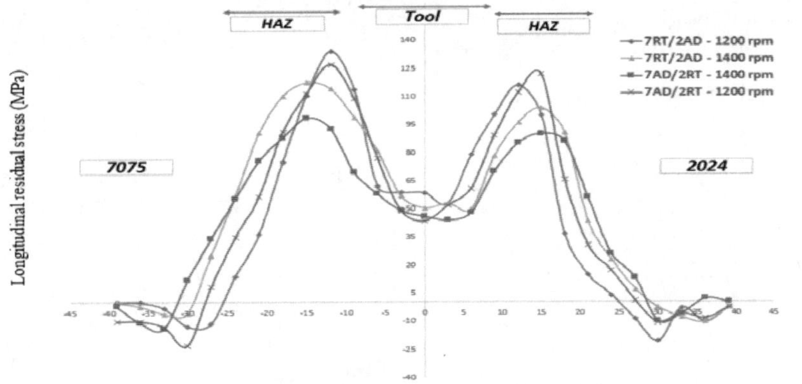

Figure 5. Longitudinal residual stress distributions.

It can be observed that the maximum value of the residual stress is located on the HAZ for both materials, whit significant variation between the different profile. However, there is very small variation in nugget zone. For the case when AA2024-T3 is in advancing side the maximum tensile residual stress occurs in the AA7075 retreating side to 135 MPa. the magnitude of the residual stresses decreases (for both side AD and RT) upon a change in velocity from 1200 to 1400 rpm whit a "M" shape increases in width. However, when AA2024-T3 located in retreating side, the residual stress peaks on the advancing and retreating sides were nearly equal. Furthermore, the magnitude of the residual stresses in AA2024 RT side is higher than AA2024 AD side. Of observations to note is the considerable reduction the magnitude of the residual stress for AA7075 advancing side upon a change in velocity from 1200 to 1400 rpm, this is related with the mechanical resistance of AA7075 which is higher compared to the AA2024. thus, causing higher heat input.
The strain maps correspond to strain distribution at maximum load is shown in Figure 7 a), and the related local mechanical properties of different welding subzones for the two case when AA2024 is in advancing side and when is in retreating side is achieved by Digital image correlation (DIC) method are shown in Figure 6 b).

a)

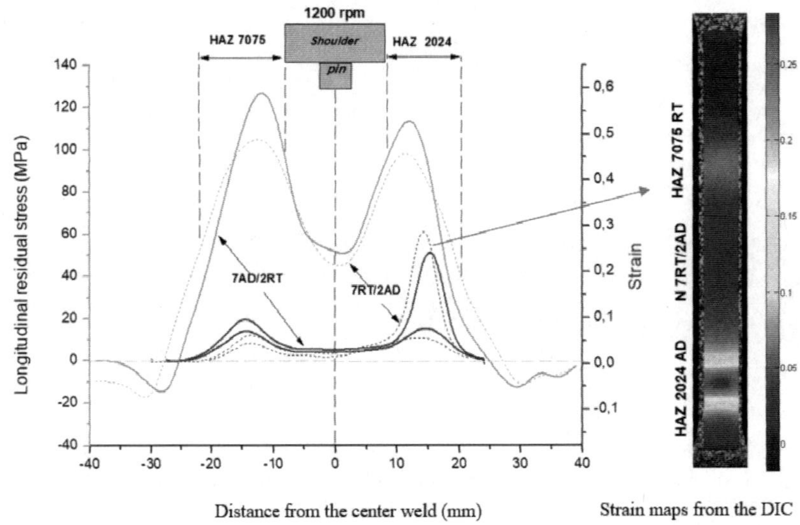

b)

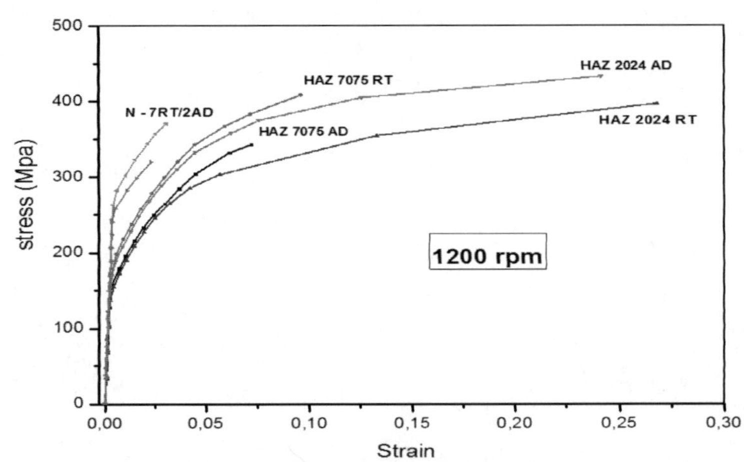

Figure 6 for the two case: AA2024 located in advancing side and in retreating side
a) Strain maps resulting from the DIC measurement according to the residual stress.
b) the local mechanical properties of different weld sub-zones.

According to the strain maps, the tensile sample fracture in the HAZ of the AA2024 side where a minimum tensile residual stress in the weld zone are observed is illustrated in Figure 6a). It can be seen that the residual stress distribution at the region HAZ 2AD and HAZ 2RT is a function of the local mechanical properties and the inhomogeneous temperature distributions during and after welding. The HAZ 2AD with lower residual stress profile show higher peak yield stress than the related region in the HAZ 2RT side (Figure 6 a).and figure 6b). furthermore, N 7RT/2AD has a rather higher yield stress than N 7AD/2RT.

4. CONCLUSION

The main goal of this study is the effect of the material position and rotational speeds on residual stresses wish is evaluated by using the ultrasonic method. The following conclusions could be drawn:

- The microstructure effect acts on both the acoustoelastic constant K.
- The increase in rotation speed results in more reduction in the amplitude of the residual stress when AA7075 is in advancing side compared to AA7075 is in retreating side.
- The residual stresses become compressive as the distance from the HAZ zone increases thus become tension stresses.
- The HAZ 2AD with lower residual stress profile show higher peak yield stress than the related region in the HAZ 2RT side.
- The maximum value of the residual stress is located on the HAZ for both materials.

5. REFERENCES

[1] Jamshidi H A 2015 Microstructure and residual stress distributions in friction stir welding of dissimilar aluminium alloys, *Mater. Des.* **87** 405–413

[2] Khodir S A, Shibayanagi T 2008 Friction stir welding of dissimilar AA2024 and AA7075 aluminum alloys, *Mater. Sci. Eng. B Solid-State Mater. Adv. Technol.* **148** 82–87

[3] Zadpoor A A, Sinke J, Benedictus R, Pieters R 2008 Mechanical properties and microstructure of friction stir welded tailor-made blanks, *Mater. Sci. Eng. A.* **494** 281–290

[4] Javadi Y, Sadeghi S, Najafabadi M A 2014 Taguchi optimization and ultrasonic measurement of residual stresses in the friction stir welding, *Mater. Des.* **55** 27–34

[5] Zapata J, Toro M, Lopez D 2016 Residual stresses in friction stir dissimilar welding of aluminum alloys, *J. Mater. Process. Technol.* **229** 121–127

[6] Guo J F, Chen H C, Sun C N, Bi G, Sun Z, Wei J 2014 Friction stir welding of dissimilar materials between AA6061 and AA7075 Al alloys effects of process parameters, *Mater. Des.* **56** 185–192

[7] Zhao Z, Liang H, Zhao Y, Yan K 2018 Effect of Exchanging Advancing and Retreating Side Materials on Mechanical Properties and Electrochemical Corrosion Resistance of Dissimilar 6013-T4 and 7003 Aluminum Alloys FSW Joints, *J. Mater. Eng. Perform.* **27**, 1777–1783

[8] Niu P, Li W, Yang X, Vairis A 2018 Effects of microstructural asymmetries across friction stir welded AA2024 joints on mechanical properties, *Sci. Technol. Weld. Join.* **23** 58–62.

[9] Niu P, Li W, Zhang Z, Yang X 2017 Global and local constitutive behaviors of friction stir welded AA2024 joints, *J. Mater. Sci. Technol.* 2–6

[10] Rodriguez R I, Jordon J B, Allison P G, Rushing T, Garcia L 2015 Microstructure and mechanical properties of dissimilar friction stir welding of 6061-to-7050 aluminum alloys, *Mater. Des.* **83** 60–65

[11] Hughes D, Kelly J L 1953 Second-Order Elastic Deformations of Solids, *Phys. Rev.* **92** 1145–1150

[12] Francis M D 1951 Finite Deformation of an elastic solid, Wiley, New York

[13] Belahcene F, Lu J 2000 Determination of Residual Stress in Z8CDWV12 Steel Using Critically Refracted Longitudinal Waves, *JSME Int. J. Ser. A.* **43** 367–373

[14] Basatskaya E I, LV, 1981 Theoretical study of ultrasonic longitudinal subsurface waves in solid media, Defektoskopiya

The states of hydrogen and hydrogen embrittlement susceptibility of precipitation hardened SUS630 stainless steel

Y Hayashi[1, a)], T Ito[2] and Y Nishimura[2], K Takai[3, b)]

[1]Graduate Student, Sophia University, 7-1 Kioi-cho, Chiyoda-ku, Tokyo 102-8554
[2]Materials Engineering R&D Division, DENSO CORP.
[3]Department of Engineering and Applied Sciences, Faculty of Science and Technology, Sophia University

a) dog.tennis.y1109uc@gmail.com

b) takai-k@sophia.ac.jp

Abstract. To elucidate the influence of precipitates on the states of hydrogen and hydrogen embrittlement susceptibility of SUS630, solution treatment, sub-aging (410 °C, 1 h), peak-aging (470 °C, 1 h), and over-aging (550 °C, 1 h) methods were used to prepare specimens. States of hydrogen were analyzed by thermal desorption analysis (TDA). Hydrogen embrittlement susceptibility was evaluated using tensile tests. Specimens were electrochemically precharged with hydrogen and then tensile tests were carried out simultaneously with hydrogen charging under the same conditions as those after precharging. Susceptibility was evaluated as the elongation ratio with/without hydrogen. The solution-treated and sub-aged specimens displayed a single peak near 100 °C, whereas peak- and over-aged specimens displayed two peaks, one near 100 °C and the other near 200 °C. The peak near 100 °C corresponded to hydrogen desorbed from the matrix and the peak near 200 °C corresponded to hydrogen desorbed from Cu precipitates. Ductility of peak- and over-aged specimens with hydrogen did not decrease, however, that of solution-treated and sub-aged specimens with hydrogen did. These findings indicate that specimens containing hydrogen with a higher temperature peak associated with Cu precipitates showed less hydrogen embrittlement susceptibility than those containing hydrogen with only a lower temperature peak.

1 Introduction

It has been reported that nanosize precipitates were observed in steel specimens by adding Cu and performing aging treatment [1, 2]. In recent years, precipitation strengthening by Cu precipitates has attracted attention in that it can improve the balance between strength and ductility of steel materials. Though hydrogen embrittlement can occur in high strength materials, Cu precipitates may improve resistance to it by trapping hydrogen in the same way as Ti and V carbides [3-5]. However, the hydrogen trapping capacity and hydrogen embrittlement susceptibility of Cu precipitates have been poorly understood to date. In addition, it is also expected that the hydrogen trapping capacity changes by increasing the aging temperature because the coherence of the precipitate-matrix interface changes with the crystal structure of the transforming precipitates. In this study, we investigated the influence of precipitates on the states of hydrogen and hydrogen embrittlement susceptibility of SUS630 under various aging conditions.

2 Experimetal.

2.1 Material

The chemical composition of the specimens is listed in Table 1. Four kinds of specimens with different aging treatments were prepared (solution-treated, sub-, peak-, and over-aged specimens). The aging conditions and the tensile strengths are shown in Table 2. Two kinds of test pieces were prepared with different shapes. Cylindrical test pieces (10 mm diameter, 0.8 mm thick) were prepared for the analysis of the states of hydrogen. Specimens were prepared as shown in Fig. 1 for use in evaluating hydrogen embrittlement susceptibility.

2.2 Analysis of States of Hydrogen

The specimen surface was mechanically polished using emery paper (#800, 1000 and 2000). TDA was applied to the cylindrical specimens (0.8 mm thickness and 10 mm diameter). Hydrogen charging into the specimens was carried out by immersion in a 10 mass % NH_4SCN solution maintained at 30 °C for 48 h until hydrogen reached equilibrium at the center of the specimens. Immediately after hydrogen charging, the hydrogen content was analyzed by thermal desorption analysis (TDA) using a gas chromatograph calibrated with a standard gas containing 50 vol ppm of hydrogen in Ar used as a detection system. The temperature was raised from room temperature to 400 °C at 100 °C · h^{-1}.

2.3 Evaluation of Hydrogen Embrittlement Susceptibility

Hydrogen embrittlement susceptibility was evaluated in tensile testing. Two methods, tensile testing and TDA, were applied to the specimens as shown in Fig.1. They were electrochemically precharged with hydrogen in a 0.1N NaOH solution maintained at various current potentials (-0.9 - -2.0 V vs Ag/AgCl) and then tensile tests were carried out at a crosshead speed of 0.09 mm · min^{-1} and a temperature of 30 °C with/without hydrogen precharging. To keep the hydrogen concentration inside the specimen constant during tensile testing, the tests were conducted simultaneously with hydrogen charging under the same conditions as those after hydrogen precharging. Susceptibility was evaluated as the elongation ratio with/without hydrogen. The hydrogen content was analyzed by TDA. In this study, tensile tests were carried out under two conditions, which were a low hydrogen content (hydrogen charged at -0.9 - -1.0 V vs Ag/AgCl) and a high hydrogen content (hydrogen charged at -1.2 - -1.5 V vs Ag/AgCl), so as to evaluate the influence between the states of hydrogen and hydrogen embrittlement susceptibility. The fracture surfaces were observed using a field emission scanning electron microscope (FE-SEM).

Table 1. Chemical composition of SUS630 specimen (mass %).

C	Si	Mn	P	S	Cu	Ni	Cr	Nb+Ta
0.05	0.24	0.86	0.034	0.004	3.32	4.25	15.66	0.24

Table 2. Heat treatment conditions and tensile strength of SUS630 specimens (MPa).

	Solution	Sub-aged	Peak-aged	Over-aged
ST	1050 °C	1050 °C	1050 °C	1050 °C
AT	—	410 °C	470 °C	550 °C
TS	1160 MPa	1265 MPa	1421 MPa	1168 MPa

3 Results and Discussion.

3.1 *States of hydrogen in SUS630*

Figure 2 shows the TDA profiles of hydrogen added by immersion in a 10 mass % NH_4SCN solution maintained at 30 °C for 48 h. The results show that the solution-treated and sub-aged specimens had a single peak near 100 °C, whereas the peak- and over-aged specimens had two peaks, one near 100 °C and the other one near 200 °C. It has been reported that hydrogen desorbed near 100 °C is trapped at atomic vacancies, the elastic stress field around the dislocation core, the dislocation core, and grain boundaries in the matrix of medium-carbon tempered martensitic steel [6, 7]. In the case of the tested martensitic stainless steel SUS630, the peak near 100 °C presumably corresponded to hydrogen desorbed from the same trapping sites. The peak near 200 °C corresponded to hydrogen desorbed from Cu precipitates because it appeared with increasing aging temperature. In addition, the peak near 200 °C did not appear in the case of sub-aged specimens. These findings indicate that the states of Cu precipitates affected the hydrogen trapping capacity.

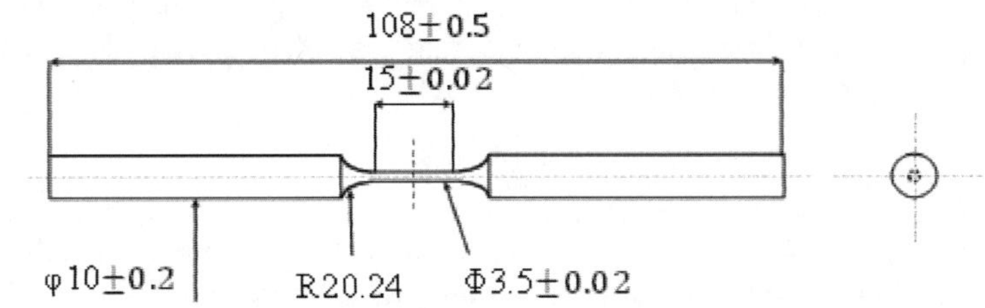

Fig. 1 Geometry and dimensions of specimen for the tensile tests.

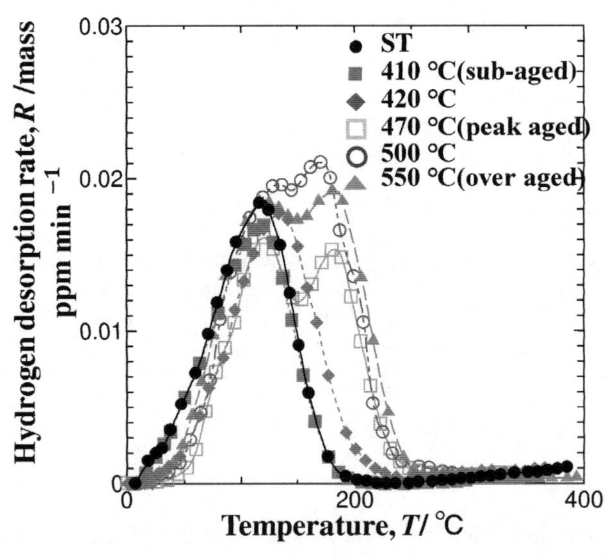

Fig. 2 Hydrogen desorption profiles of specimens immersed for 48 h in a 10 mass% NH_4SCN solution at a temperature of 30 °C for various aging temperatures.

3.2 Hydrogen Embrittlement Susceptibility in SUS630

3.2.1 Change in Ductility with/without Hydrogen

Figure 3 and Figure 4 show nominal stress-strain curves at a crosshead speed of 0.09 mm · min^{-1} under low and high hydrogen content conditions, respectively, for (a) solution-treated, (b) sub-aged, (c) peak-aged, and (d) over-aged specimens. The tensile test results show that the ductility of peak- and over-aged specimens did not decrease, whereas the ductility of solution-treated and sub-aged specimens decreased under a low hydrogen content. At a high hydrogen content, the ductility of all specimens clearly decreased.

3.2.2 TDA profiles of Tensile Testing

TDA profiles under low and high hydrogen content conditions are shown in Fig. 5 and Fig. 6, respectively. The results indicate that hydrogen was trapped in the matrix of solution-treated and sub-aged specimens under a low hydrogen content, whereas it was mainly trapped in the Cu precipitates of peak- and over-aged specimens. Peak separations were attempted using the Gaussian function to ascertain the states of hydrogen under a high hydrogen content for further investigation. As a result, peaks were separated into two peaks consisting of one at 137 °C and the other one at 233 °C. Hydrogen was trapped not only in the Cu precipitates but also in the matrix of peak- and over-aged specimens. For these reasons, when hydrogen is trapped only in the precipitates, there is probably an effect of suppressing hydrogen embrittlement. However, that effect disappears when hydrogen is trapped not only in the precipitates but also in the matrix.

3.2.3 Fracture Surface Observation

Figure 7 and Figure 8 show photographs of fracture surfaces observed by the FE-SEM under the low and high hydrogen content conditions, respectively. Under a low hydrogen content, a quasi-cleavage fracture surface was observed in (a) solution-treated and (b) sub-aged specimens that showed hydrogen embrittlement, whereas (c) peak-aged and (d) over-aged specimens displayed a fracture surface with microvoid coalescence. Under a high hydrogen content, a quasi-cleavage fracture surface was observed for all aged specimens. An intergranular fractured surface was observed for over-aged specimens.

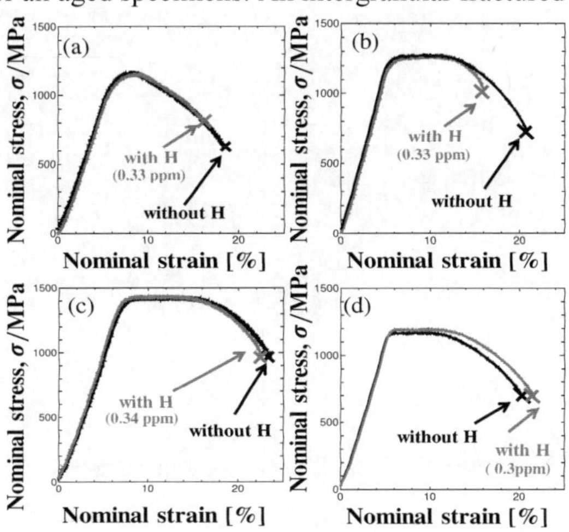

Fig. 3 Nominal stress-strain curves with/without hydrogen charging at -0.9 - -1.0 V vs Ag/AgCl in a 0.1N NaOH solution at a temperature of 30 °C. (a) Solution-treated, (b) sub-aged, (c) peak-aged and (d) over-aged specimens.

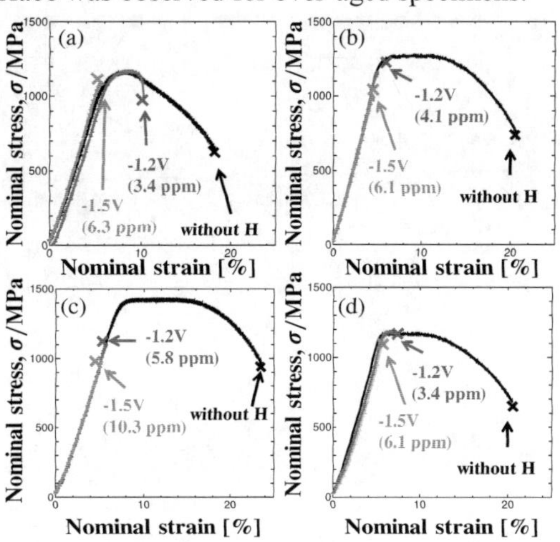

Fig. 4 Nominal stress-strain curves with/without hydrogen charging at -1.2 - -1.5 V vs Ag/AgCl in a 0.1N NaOH solution at a temperature of 30 °C. (a) Solution-treated, (b) sub-aged, (c) peak-aged and (d) over-aged specimens.

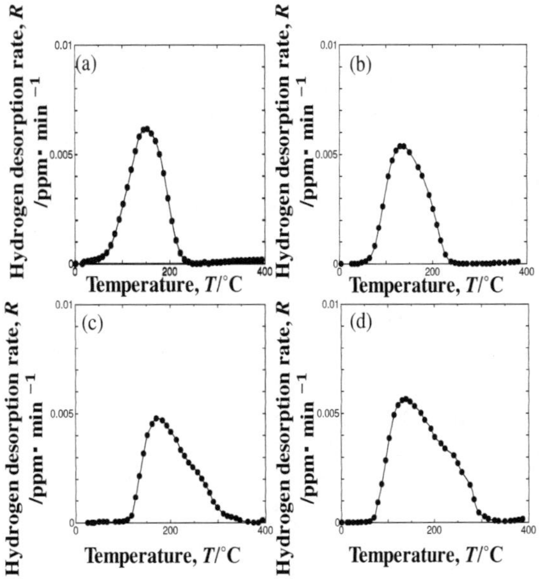

Fig. 5 Hydrogen desorption profiles of hydrogen-charged at -0.9 - -1.0 V vs Ag/AgCl in 0.1N NaOH solution at a temperature of 30 °C. (a) Solution-treated, (b) sub-aged, (c) peak-aged and (d) over-aged specimens.

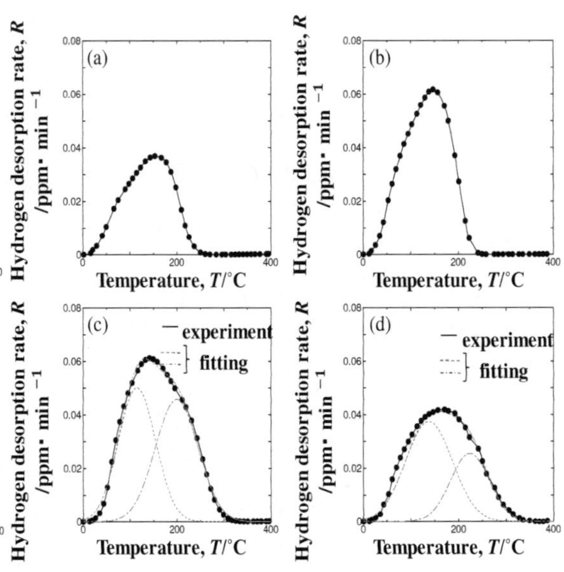

Fig. 6 Comparison of thermal desorption profiles between experiment and fitting using the Gaussian function of hydrogen - charged at -1.2 V vs Ag/AgCl in 0.1N NaOH solution at a temperature of 30 °C. (a) Solution-treated, (b) sub-aged, (c) peak-aged and (d) over-aged specimens.

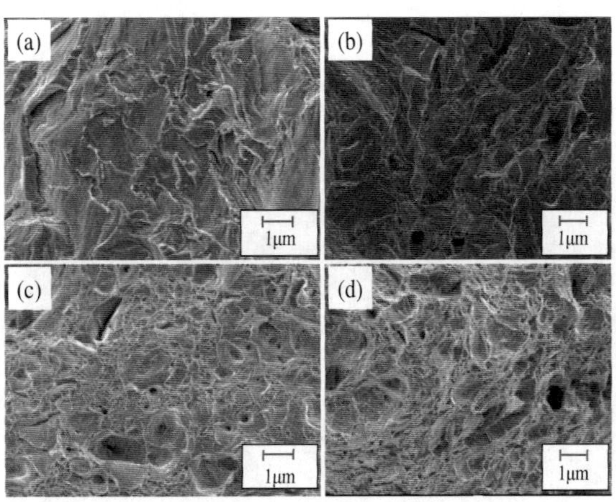

Fig. 7 Microscopic fracture surfaces with/without hydrogen charging at -0.9 - -1.0 V vs Ag/AgCl in a 0.1N NaOH solution at a temperature of 30°C. (a) Solution-treated, (b) sub-aged, (c) peak-aged and (d) over-aged specimens.

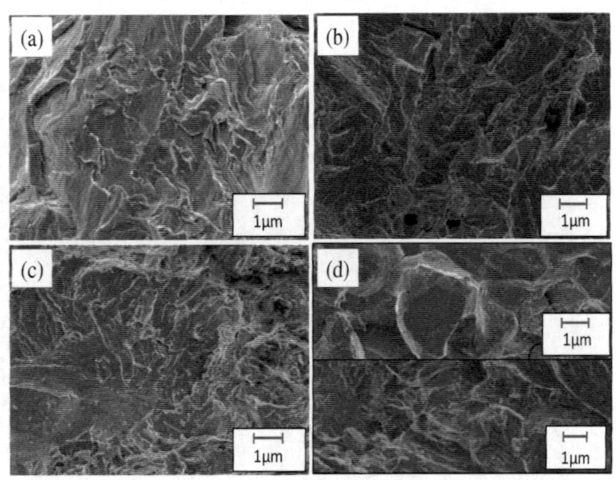

Fig. 8 Microscopic fracture surfaces with/without hydrogen charging at -1.2 V vs Ag/AgCl in a 0.1N NaOH solution at a temperature of 30 °C. (a) Solution-treated, (b) sub-aged, (c) peak-aged and (d) over-aged specimens.

4 Conclusion

(1) Solution-treated and sub-aged specimens displayed a single peak near 100 °C, whereas peak- and over-aged specimens displayed two peaks, one near 100 °C and the other one near 200 °C. The peak near 100 °C corresponded to hydrogen desorbed from the matrix and the peak near 200 °C corresponded to hydrogen desorbed from Cu precipitates.

(2) Ductility of peak-aged and over-aged specimens was not decreased, however, ductility of solution-treated and sub-aged specimens was decreased under low hydrogen content. With high hydrogen content, ductility of all specimens were decreased.

(3) When hydrogen was trapped only in precipitates, there was the effect of suppressing hydrogen embrittlement. However, that effect was disappeared when hydrogen was trapped both in precipitates and matrix.

5 References

[1] Goodman R S, Brenner S S and Low J R Jr 1973 Metallurgical and Materials Transactions **4** 2363.
[2] Osamura K, Okuda H, Takashima M, Asano K and Furusaka M 1993 *Materials Transactions* **34** 305.
[3] Hirth P J 1980 *Materials Transactions A* **11A** 861.
[4] Tsuchida T, Hara T and Tsuzaki K 2002 *Tetsu to Hagane* **88** 771.
[5] Yokota T and Shiraga T 2003 *The Iron and Steel Institute of Japan* **43** 534.
[6] Takai K 2011 *Zairyo-to-Kankyo* **60** 230.
[7] Koyama M, Rohwerder M, Tasan Cem C, Bashir A, Akiyama E, Takai K, Raabe D & Tsuzaki K 2017 *Material Science and Technology* **33** 1483.

IOP Conf. Series: Materials Science and Engineering **461** (2019) 012024 doi:10.1088/1757-899X/461/1/012024

Comparison of hydrogen behavior trapped at precipitated and undissolved vanadium carbide in vanadium-bearing high strength steels

K. HOKAZONO[1),a)], M. KAWAMORI[2)], Y. MATSUMOTO[3)], K. TAKAI[4),b)]

[1]Graduate Student, Sophia University, 7-1 Kioi-cho Chiyoda-ku Tokyo 102-8554
[2]Materials Research Lab., Kobe Steel, Ltd.
[3]Research and Development Lab., Kobe Steel, Ltd.
[4]Department of Engineering and Applied Sciences, Faculty of Science and Technology, Sophia University.
Email:[a)]hokakoda.7.15@gmail.com, [b)]takai-k@sophia.ac.jp

Abstract. Vanadium carbide (VC) in high strength steels is divided into two types, i.e., finely formed precipitated VC and coarsely formed undissolved VC. The behavior of hydrogen trapped at the two types has been investigated using thermal desorption analysis (TDA). Specimens containing different amounts of V (0, 0.10 and 0.15 %) were prepared. Specimens with 0 % V (QT) and 0.10 % V (QT) were quenched at 900 °C and tempered at 600 °C. Some specimens with 0.15 % V were quenched at 900 °C and are denoted here as 0.15 % V (Q). Hydrogen was charged into specimens by immersion in a NH_4SCN aqueous solution and analyzed by TDA. The end temperatures of hydrogen release for 0 % V (QT) and 0.15 % V (Q) specimens were in a range from 76 to 79 °C, and that of the 0.10 % V (QT) specimen was 104 °C. The amounts of hydrogen were 0.10 ppm in the 0 % V (QT) specimen, 0.28 ppm in the 0.15 % V (Q) specimen and 0.54 ppm in the 0.10 % V (QT) specimen. The specimens containing precipitated VC had a higher end temperature of hydrogen release and larger amounts of hydrogen than those containing undissolved VC and matrix. These findings suggest that precipitated VC in high strength steels has a higher hydrogen trapping capacity than undissolved VC.

1 Introduction

Bolts and other structural members made of high strength steels have been widely used in recent years. Hydrogen embrittlement, whereby materials suddenly break down due to the incursion of hydrogen, has thus become a serious problem. In this regard, the hydrogen trapping effects of alloy carbides (V, Ti and Nb) have been noted for developing high strength steels with excellent resistance to hydrogen embrittlement [1]. The behavior of hydrogen trapped at vanadium carbide (VC) has been actively investigated in recent years in connection with the use of the hydrogen trapping effects of VC as a technique for suppressing hydrogen embrittlement. For example, Asahi *et al*. reported that the activation energy of hydrogen desorption from trapping sites in 0.3 and 0.10 % V-bearing steels under various heat treatment conditions was calculated to be 35.3 and 32.6 $kJ \cdot mol^{-1}$, These values indicated there was no appreciable change in activation energy in those steels due to the amounts of V and the tempering temperature [2]. Tsuchida *et al*. reported that V-bearing high strength steels tempered at 600 °C showed TDA profiles having a hydrogen desorption temperature peak at 200 °C due to hydrogen trapping sites at precipitated (V, Mo) C. In addition, V-bearing high strength steels tempered at 700 °C showed a TDA profile having two hydrogen desorption temperature peaks at 100 °C and 260 °C. The temperature peak at 260 °C was due to hydrogen trapping sites at undissolved (V, Mo) C. These TDA profiles suggest that undissolved (V, Mo) C has a higher hydrogen trapping capacity than precipitated (V, Mo) C [3]. However, these findings have still not provided a fundamental understanding. Accordingly, the behavior of hydrogen trapped at precipitated and undissolved VC was investigated by TDA for high strength steels containing various amounts of V.

2 Experimental

2.1 Materials

The chemical composition of the steels used is listed in Table 1. Three types of high strength steels specimens with various amounts of V were prepared. The amounts of V associated with precipitated and undissolved VC in the γ phase as a function of the various amounts of V in specimens quenched at 900 °C was calculated by Thermo-Calc and the results are shown in Table 2. 0.10 mass % V associated with precipitated VC was present in the 0.10 % V specimen because 0.10 % V completely dissolved when quenched at 900 °C. On the other hand, 0.15 % V did not completely dissolve when quenched at 900 °C. Therefore, 0.122 mass % V associated with precipitated VC and 0.028 mass % V associated with undissolved VC were present in the specimens.

Table 1 Chemical composition of specimens (mass %).

	C	Si	Mn	P	S	Cr	Mo	V
0 % V	0.39	0.15	0.70	0.011	0.010	0.97	0.21	0.001
0.10 % V	0.39	0.15	0.71	0.009	0.013	0.96	0.20	0.10
0.15 % V	0.40	0.15	0.71	0.010	0.014	0.97	0.20	0.153

Table 2 Amounts of vanadium associated with precipitated and undissolved vanadium carbide in the γ phase as a function of the various amounts of vanadium in specimens quenched at 900 °C as calculated by Thermo-Calc.

Steel types	Amounts of V (mass %)	Amounts of V associated with precipitated VC (mass %)	Amounts of V associated with undissolved VC (mass %)
0 % V	0.000	0.000	0.000
0.10 % V	0.100	0.100	0.000
0.15 % V	0.150	0.122	0.028

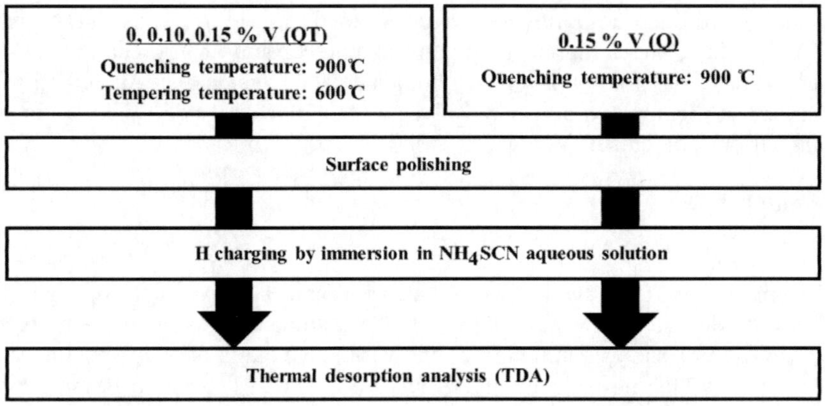

Figure 1 Procedure for hydrogen trapping state analysis.

2.2 *Material characterization*
The microstructure of the V-bearing high strength steels was examined by means of transmission electron microscopy (TEM). The chemical composition of precipitates observed by TEM was determined using energy dispersive X-ray analysis (EDX)

2.3 *Hydrogen trapping state analysis*
Figure 1 shows the procedure for the hydrogen trapping state analysis. Specimens containing 0 % V, 0.10 % V and 0.15 % V were quenched at 900 °C and tempered at 600 °C and are denoted here as 0 % V (QT), 0.10 % V (QT) and 0.15 % V (QT). Some specimens containing 0.15 % V were quenched at 900 °C and are denoted here as 0.15 % V (Q). Specimens were machined to 10 mm in diameter and 0.3, 0.5 and 1.0 mm in thickness. Specimens were polished with emery paper (# 2000) and hydrogen was charged into the specimens by immersion in 1 or 5 mass % NH_4SCN aqueous solution at 30 °C for 24 h until hydrogen reached equilibrium at the center of the specimens. After that, the specimens were analyzed by TDA at a heating rate of 100 $°C \cdot h^{-1}$.

3 Results and discussions

3.1 *Material characterization*
TEM and EDX images of the 0.15 % V (Q) specimen quenched at 900 °C are shown in Fig 2. Coarse spherical precipitates of approximately 50 nm in diameter were observed by TEM. Element analysis by EDX was performed on coarse spherical precipitates indicated by an arrow in Fig 2(a). Coarse spherical precipitates were VC because they were found to consist of much larger quantities of V and C than other elements. As a result, the coarse spherical precipitates were regarded as undissolved VC in the 0.15 % V (Q) specimen. A TEM observation was attempted for precipitated VC obtained by quenching and tempering. Precipitated VC consisted of very fine precipitates that could not be observed.

3.2 *Thickness dependence of TDA profiles*
Figure 3 shows the TDA profiles of 0.15 % V (QT) specimens having various thicknesses in case of charging hydrogen into those specimens by immersion in 1 mass % NH_4SCN aqueous solution. The TDA profiles indicate that the end temperatures of hydrogen release decreased with decreasing specimen thickness. It is assumed that the TDA profiles indicate a transition from hydrogen diffusion-controlled to hydrogen desorption-controlled due to decreasing specimen thickness. According to a previous study, the specimen thickness that determined the hydrogen desorption peak temperature due to hydrogen-desorption control at hydrogen trapping sites was 0.3 mm [4]. Thus, we decided to compare the hydrogen trapping behavior in V-bearing high strength steels for specimens of 0.3 mm in thickness.

IOP Conf. Series: Materials Science and Engineering **461** (2019) 012024 doi:10.1088/1757-899X/461/1/012024

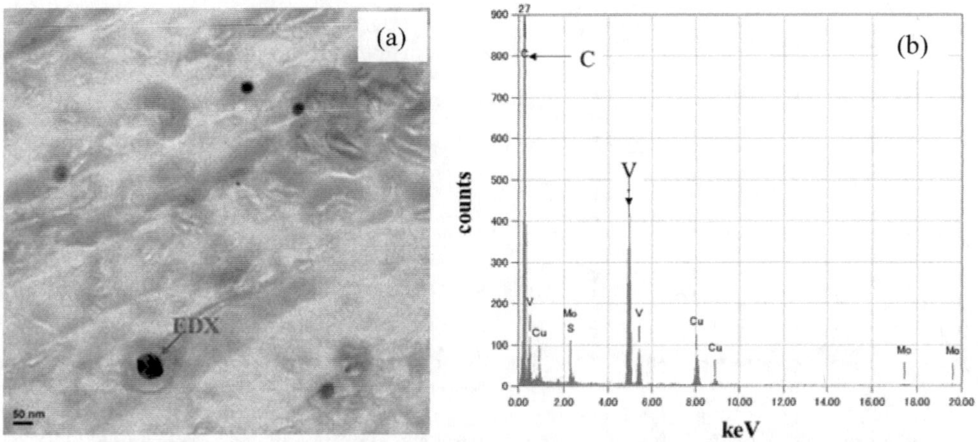

Figure 2 Undissolved vanadium carbide of the specimen containing 0.15 % vanadium quenched at 900 °C as observed by (a) TEM and (b) EDX.

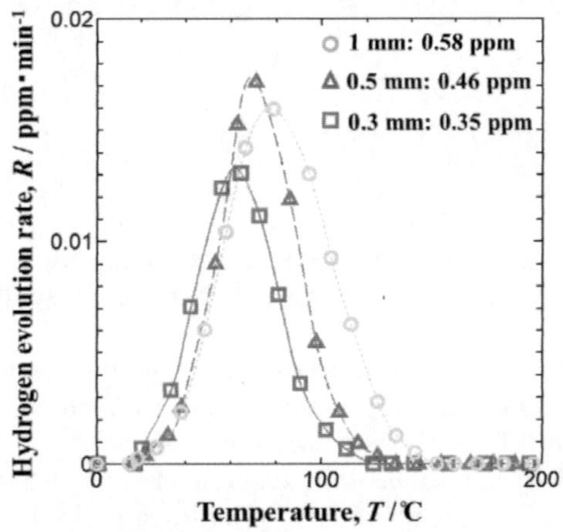

Figure 3 TDA profiles and hydrogen contents of 0.15 % vanadium specimens of various thicknesses quenched at 900 °C and tempered at 600 °C.

3.3 Comparison of behavior of hydrogen trapped at precipitated and undissolved VC in TDA profiles

Figure 4 shows the TDA profile of 0 % V (QT) specimen containing no VC, fig 5 shows the TDA profile of the 0.15 % V (Q) specimen containing undissolved VC and fig 6 shows the TDA profile of the 0.10 % V (QT) specimen containing precipitated VC. The TDA profiles in figs 3-5 are the results for specimens of 0.3 mm thickness charged with hydrogen by immersion in a 5 mass % NH_4SCN aqueous solution. The results in these figs showed that the end temperatures of hydrogen release of the 0 % V (QT) and 0.15 % V(Q) specimens were in a range from 76 to 79 °C, and that the end temperature of hydrogen release of the 0.10 % V (QT) specimen was 104 °C. The amounts of hydrogen were 0.10 ppm in the 0 % V (QT) specimen, 0.28 ppm in the 0.15 % V (Q) specimen and 0.54 ppm in the 0.10 % V (QT) specimen. The specimens containing precipitated VC had higher end temperatures of hydrogen release and larger amounts of hydrogen than the specimens containing undissolved VC and matrix. These findings suggest that precipitated VC in high strength steels has a higher hydrogen trapping capacity than undissolved VC.

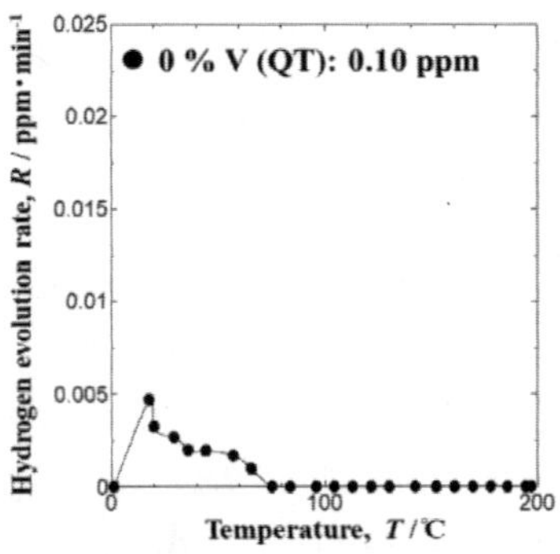

Figure 4 TDA profile and amounts of hydrogen for 0.3 mm thick specimen containing 0 % vanadium quenched at 900 °C and tempered at 600 °C.

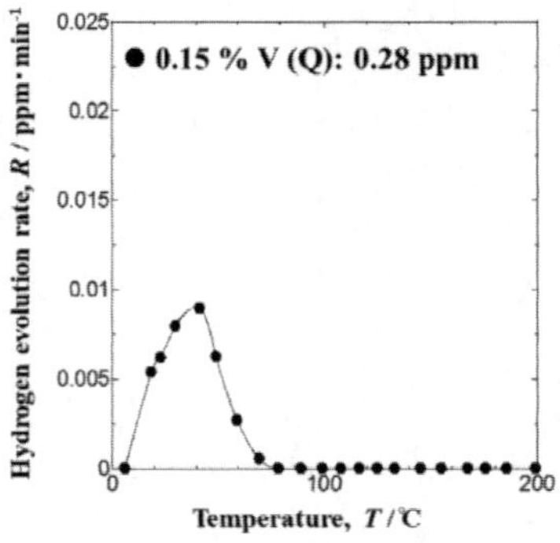

Figure 5 TDA profile and amounts of hydrogen for 0.3 mm thick specimen containing 0.15 % vanadium quenched at 900 °C.

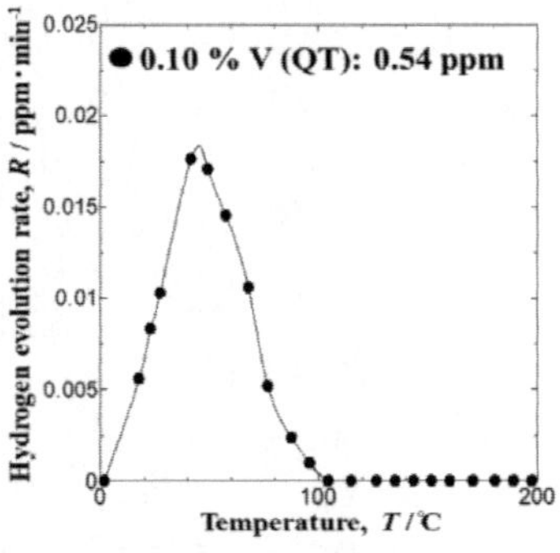

Figure 6 TDA profile and amounts of hydrogen of 0.3 mm thick specimen containing 0.10 % vanadium quenched at 900 °C and tempered at 600 °C.

4 Conclusions

The behavior of hydrogen trapped at precipitated and undissolved VC was compared for V-bearing high strength steels in the present study. The results obtained are summarized as follows.

(1) In the TDA profiles, the end temperatures of hydrogen release decreased with decreasing specimen thickness. Thus, this result suggested a transition from hydrogen diffusion-controlled to hydrogen desorption-controlled due to decreasing specimen thickness.
(2) When hydrogen was charged into specimens by immersion in a 5 mass % NH_4SCN aqueous solution, TDA profiles showed the end temperatures of hydrogen release of 0 % V (QT) and 0.15 % V (Q) specimens were in a range from 76 to 79 °C, and that the end temperature of hydrogen release for 0.10 % V (QT) specimen was 104 °C. The amounts of hydrogen were 0.10 ppm in the 0 % V (QT) specimen, 0.28 ppm in 0.15 % V(Q) specimen and 0.54 ppm in 0.10 % V (QT) specimen. In comparison of the behavior of hydrogen trapped at finely formed precipitated VC and coarsely formed undissolved VC in V-bearing high strength steels, the results suggested that precipitated VC in high strength steels had the higher trapping capacity than undissolved VC.

References

[1] T. Omura, T. Kushida, K. Miyata, and Y. Komizo, "Hydrogen absorption of high Nb bearing Steel", *Tetsu-to-Hagane*, **90**, No. 2 (2004), 106-112.
[2] H. Asahi, D. Hirakami and S. Yamasaki "Hydrogen trapping behavior in vanadium-added Steel", *ISIJ International* **43**, No. 4 (2003), 527-533.
[3] T. Tsuchida, T. Hara and K. Tsuzaki, "Relationship between microstructure and hydrogen absorption behavior in a V-bearing high strength steel", *Tetsu-to-Hagane*, **88**, No. 11 (2002), 771-778.
[4] K. Saito, H. Suzuki and K. Takai: *CAMP-ISIJ*, **27** (2014), 955, CD-ROM.

IOP Conf. Series: Materials Science and Engineering **461** (2019) 012025 doi:10.1088/1757-899X/461/1/012025

The nature of hydrogen-related fracture in X80 pipeline steel with stress concentration

T Homma[1], S Onuki[2], H Suzuki[3] and K Takai[3]

[1]Graduate School of Science and Technology, Sophia University, 7-1 Kioi-cho, Chiyoda-ku, Tokyo 102-8554, Japan
[2]Tokyo Gas Co., Ltd., 1-7-7, Suehiro-cho, Tsurumi-ku, Yokohama, Kanagawa 230-0045, Japan
[3]Faculty of Science and Technology, Department of Engineering and Applied Sciences, Sophia University, Tokyo, Japan

E-mail: t-homma-fc4@eagle.sophia.ac.jp

Abstract. The nature of hydrogen-related fracture in X80 pipeline steel with stress concentration was investigated through observation of fracture modes at hydrogen-related crack initiation sites and crystallographic analysis of fracture surfaces. The fracture modes and crack initiation sites were observed using a field emission scanning electron microscope (FE-SEM). The hydrogen-related crack initiation sites of X80 specimens were quasi-cleavage (QC) fracture at the notch tip. In addition, distributions of plastic strain and principal stress near the notch tip were calculated by finite element method (FEM) analysis. Since maximum plastic strain occurred at the notch tip where the initiation of QC fracture was observed, the main factor causing QC fracture was probably plastic strain. In order to clarify the crystallographic features of hydrogen-related fracture surfaces of X80, a trace analysis of the QC fracture surface was carried out using electron backscatter diffraction patterns (EBSD). Since not all points on the fracture surface were parallel to {011} slip planes, hydrogen-related fracture consisted of not only {011} slip planes but also various crystal planes due to cross slip in the body-centered cubic lattice. These findings indicate that the nature of hydrogen-related fracture in X80 is probably associated with plastic strain on various slip planes.

1 Introduction

For the development of the hydrogen economy, transport of hydrogen using gas pipelines would enable it to be transported in large quantities at low cost. However, hydrogen embrittlement is a concern when large earthquakes occur that can induce large strain in pipelines. Hattori et al. evaluated hydrogen embrittlement susceptibility using smooth specimens of X80 pipeline steel and reported that hydrogen-enhanced strain-induced lattice defects are involved in the reduction of ductility [1]. However, corrosion pits and scratches may exist on the pipeline surface in an actual environment, so it is more important to investigate hydrogen embrittlement susceptibility and the fracture mechanism of X80 under stress concentration. Therefore, in the present study, the effect of the hydrogen content and tensile rate on hydrogen embrittlement susceptibility was investigated along with examining the nature of hydrogen-related fracture in X80 under stress concentration.

2 Experimental

2.1 Materials

X80 pipeline steel with a ferrite-bainite microstructure was used in the present study. The detailed chemical composition of this steel is shown in Table 1. Specimens with gauge dimensions of 20 × 10 × 2.0 mm were machined from actual pipelines. Notches (K_t i.e., the stress concentration factor: 3.6) were introduced in the center of the specimen gauge region as locations of stress concentration.

2.2 Hydrogen embrittlement susceptibility

The first set of tensile tests was carried out to investigate the influence of hydrogen content. Specimens were electrochemically precharged with hydrogen in an aqueous solution of 0.1 N NaOH and from 0 to 10 $g \cdot l^{-1}$ NH_4SCN with a current density of 5 $A \cdot m^{-2}$ for 72 h. Hydrogen content was measured by thermal desorption analysis (TDA) at a heating rate of 100 $°C \cdot h^{-1}$. Tensile tests were carried out at a crosshead speed of 0.05 $mm \cdot min^{-1}$ at 30 °C. The second set of tensile tests was carried out by changing the crosshead speed. Specimens were electrochemically precharged with hydrogen in an aqueous solution of 0.1 N NaOH and 1 $g \cdot l^{-1}$ NH_4SCN with a current density of 5 $A \cdot m^{-2}$ for 72 h. Tensile tests were carried out at a crosshead speed of 0.005-1 $mm \cdot min^{-1}$ at 30 °C with/without hydrogen precharging. Hydrogen charging was conducted concurrently with the tensile tests after precharging in order to keep the hydrogen content constant. For the uncharged specimens, the tensile tests were conducted in air at 30 °C. Hydrogen embrittlement susceptibility was evaluated based on the maximum fracture strength.

2.3 Controlling factor of crack initiation induced by hydrogen

The crack initiation site of hydrogen-related fracture in X80 was investigated based on the method reported by Kurokawa et al. [2]. The load was applied with hydrogen charging and unloaded just before fracture. After unloading, the samples were kept in a thermostatic chamber in air at 30 °C for 144 h for degassing and tensile testing was conducted at a crosshead speed of 0.05 $mm \cdot min^{-1}$ in air. Hydrogen-related crack initiation sites of fracture surfaces were observed with a field emission scanning electron microscope (FE-SEM). Furthermore, the ABAQUS modeling software was used to conduct a simulation of the distribution of principal stress and plastic strain in the vicinity of the notch root.

2.4 Crystallographic analysis of fracture surface at hydrogen-related crack initiation site

A trace analysis was carried out using electron backscatter diffraction patterns (EBSD) to clarify the crystallographic features of hydrogen-related fracture surfaces of X80. A Ni layer was electrodeposited onto the tensile-tested hydrogen-charged specimen with an aqueous solution of 50 $g \cdot L^{-1}$ Ni_2SO_4 + 15 $g \cdot L^{-1}$ NH_4Cl + 15 $g \cdot L^{-1}$ H_3BO_4 to preserve the fracture surface. One side of the electrodeposited specimen was mechanically polished and then polished with colloidal silica. The EBSD measurement and analysis were performed with TSL OIM Data Collection and TSL OIM Analysis, respectively.

Table 1. Chemical composition of X80 steel.

Element	C	Si	Mn	P	S	Nb	Mo	Ti
wt. %	0.08	0.22	1.85	0.007	0.002	0.05	0.12	0.02

3 Results and discussion

3.1 Hydrogen embrittlement susceptibility

Figure 1 shows the relationship between the maximum fracture strength and hydrogen content. As the hydrogen content increased, the maximum fracture strength decreased and then became constant. Figure 2 shows the relationship between the maximum fracture strength and the crosshead speed. As the crosshead speed decreased, the maximum fracture strength decreased only for hydrogen-charged specimens, resulting in increased hydrogen embrittlement susceptibility. Possible causes of these results are the hydrogen concentration in a stress concentrated region and the interaction between dislocations and hydrogen. During the tensile test, hydrogen accumulated at the notch tip due to stress-induced diffusion. At a low crosshead speed, hydrogen accumulated sufficiently because there was ample time for it to diffuse [3]. In addition, at a low crosshead speed, hydrogen can move with moving dislocations and the interaction between dislocations and hydrogen is expected to have a large effect [4].

3.2 Controlling factor of crack initiation induced by hydrogen

Figure 3 shows a photograph of crack initiation sites of a fracture surface at the notch tip of a hydrogen-charged specimen to which the load was applied until just before fracture; the specimen was then dehydrogenated and fractured in the air. Figure 3 (a) shows QC fracture at the notched tip region and microvoid coalescence (MVC) inside specimen. Furthermore, Figure 3 (b) shows QC fracture surface in high magnification. Figure 4 shows the distribution of principal stress and plastic strain at the notch tip obtained by FEM analysis. The plastic strain was maximum at the notch root and the principal stress was maximum at about 0.13 mm away from the notch root. QC fracture is observed in the region from the notch root to about 75 μm away in Figure 3, and this region does not include the maximum principal stress point. Therefore, the main factor causing QC fracture was probably plastic strain.

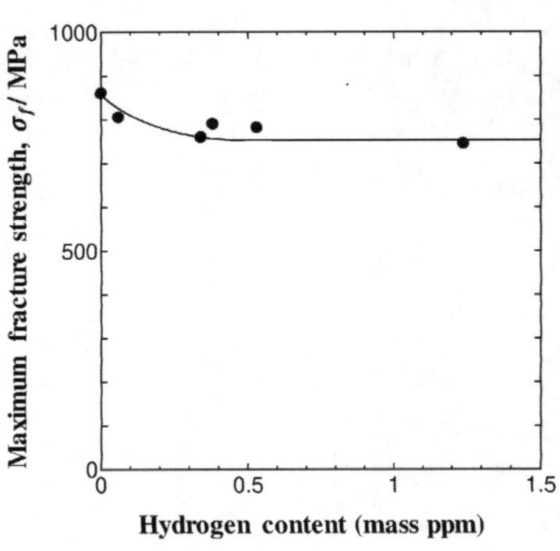

Figure 1. Relationship between maximum fracture strength and hydrogen content.

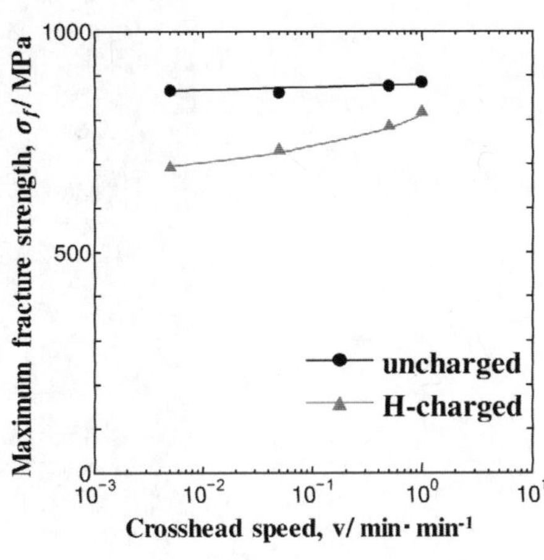

Figure 2. Relationship between maximum fracture strength and crosshead speed.

3.3 Crystallographic analysis of fracture surface at hydrogen-related crack initiation site

Figure 5 shows (a) an SEM image and (b) an EBSD orientation map of the QC fracture surface caused by hydrogen. The light gray area in the SEM image and the black area in the EBSD orientation map correspond to the Ni layers. The orange line in the SEM image shows the boundary on the first surface and the blue line shows the boundary on the second surface. The lengths of the white bars show the calculated translation distance values. The traces of {011} planes and cubic lattices showing crystal orientation near the fracture surface are indicated by the red lines in the EBSD orientation map. The measurement point numbers parallel to the {011} planes are indicated in red, and the measurement point numbers not parallel to the {011} planes are indicated in white. The measurement points parallel to the {011} planes are 3/16 points. Since not all points on the fracture surface are parallel to {011} slip planes, hydrogen-related fracture probably consisted of not only {011} slip planes but also various crystal planes due to cross slip in the body-centered cubic lattice.

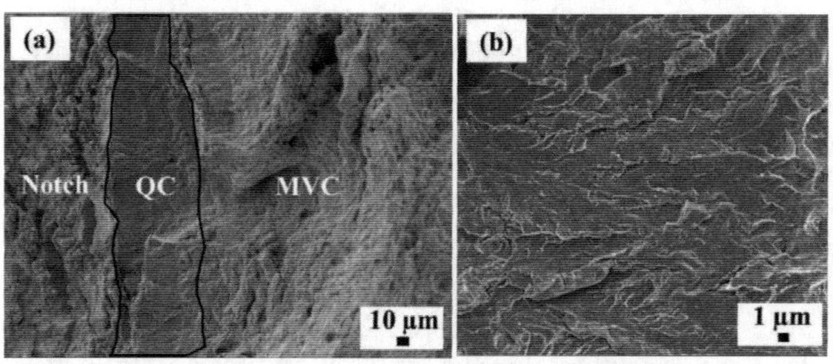

Figure 3. Crack initiation sites of fracture surface at the notch tip in (a)low magnification and (b)high magnification of QC fracture.

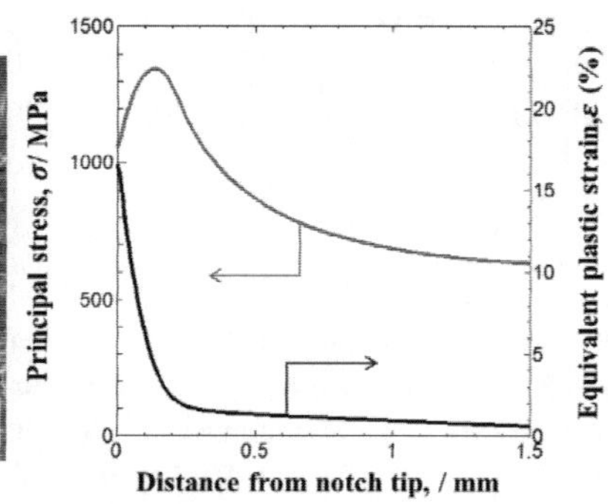

Figure 4. Distribution of principal stress and equivalent plastic strain at the notch tip.

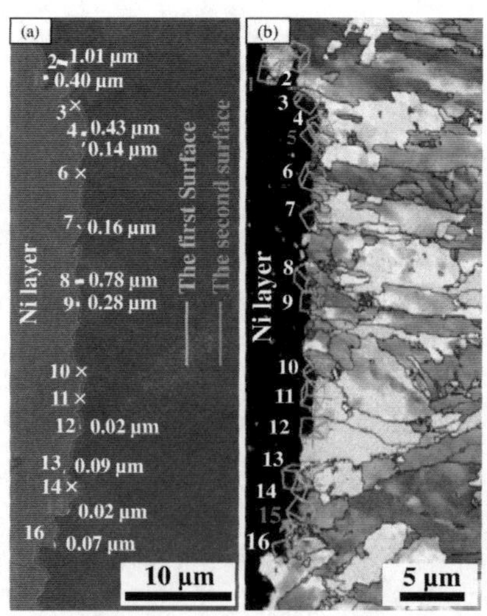

Figure 5. (a) SEM image and (b) EBSD orientation map of quasi-cleavage fracture surface caused by hydrogen.

4 Conclusion

The nature of hydrogen-related fracture in X80 pipeline steel with stress concentration was investigated through observation of fracture modes at the hydrogen-related crack initiation sites and crystallographic analysis of fracture surfaces. The results can be summarized as follows.

(1) Hydrogen embrittlement susceptibility increased with increasing hydrogen content and decreasing crosshead speed.

(2) The hydrogen-related crack initiation sites of X80 specimens were QC fracture at the notch tip. Plastic strain was maximum at the notch root and the principal stress was maximum at about 0.13 mm away from the notch root. Therefore, the main factor causing QC fracture was probably plastic strain.

(3) QC fracture in X80 consisted of not only {011} slip planes but also various crystal planes.

References

[1] Hattori M, Suzuki H, Seko Y and Takai K 2017 The role of hydrogen-enhanced strain-induced lattice defects on hydrogen embrittlement susceptibility of X80 pipeline steel *The journal of the Minerals, Metals & Materials Society* **69** 1375-1380.

[2] Kurokawa A, Doshida T, Hagihara Y, Suzuki H and Takai K 2017 Main factors causing intergranular and quasi-cleavage fractures at hydrogen-induced cracking in tempered martensitic steels Proc. Int. Conf.on the Science and Engineering of Materials (Kuala Lumpur:Malaysia) **1958** 020027-1–020027-9

[3] Wang M, Akiyama E and Tsuzaki K, 2005 Crosshead speed dependence of the notch tensile strength of a high strength steel in the presence of hydrogen *Scripta Materialia* **53** 713–718.

[4] Shoda H, Suzuki H, Takai K and Hagihara Y 2010 Hydrogen Desorption Behavior of Pure Iron and Inconel 625 during Elastic and Plastic Deformation *ISIJ International* **50** 115-123

The effect of heat treatment on mechanical properties and microstructure of additively manufactured components

T Hrbáčková[1], T Simson[2], J Koch[2], G Wolf[3]

[1]COMTES FHT a.s., Průmyslová 995, 334 41 Dobřany

[2]Ostbayerische Technische Hochschule Amberg-Weiden;

[3]Fraunhofer-Institut für Umwelt-, Sicherheits- und Energietechnik UMSICHT

Corresponding author: thrbackova@comtesfht.cz

Abstract. Components from maraging steel MS1 (1.2709) and aluminium alloy AlSi10Mg were fabricated by additive manufacturing (AM). As-built and heat-treated components were analysed in terms of mechanical properties and microstructure. The heat treatment of the material MS1 was performed by a solution annealing at 850°C for 1 hour, ageing temperature at 480°C for 3 hours and air cooling. The heat treatment of the AlSi10Mg was performed by stress relieve at 300°C for 2 hours. Ostbayerische Technische Hochschule Amberg-Weiden carried out tensile tests of as-built and heat-treated samples. Samples were afterwards analysed in COMTES FHT a.s. for further tests of porosity and microstructure using optical and scanning electron microscopy. The aim of the present work is to show the change of mechanical properties and the microstructure with a heat treatment after AM.

1 Introduction

Additive manufacturing (AM) is technology, which builds up 3D objects for example of plastic, metal or concrete. Using AM technology, three dimensional parts are fabricated directly from CAD models and built in a layer-by-layer manner. AM technology allows freeform fabrication of geometrically complex parts without special fixtures as required in material removal processes. The term AM includes many technologies which cover: 3D Printing, Rapid Prototyping (RP), Direct Digital Manufacturing (DDM), layered manufacturing and additive fabrication. AM is being used to fabricate end-use products in aircraft, dental restorations, medical implants, automobiles, and even fashion products [1]. Currently, the most used alloys for additive manufacturing are stainless steels (e.g. GP1, 316L, 17-4PH), cobalt chrome (e.g. MP1, SP2), Inconel 625 and 718, titanium Ti6Al4V, maraging steel MS1 and aluminium AlSi10Mg [2].

This work presents the effect of heat treatment on mechanical properties and microstructure of maraging steel MS1 and aluminium AlSi10Mg parts additively manufactured from powders produced by EOS. The change of yield strength (YS), ultimate tensile strength (UTS), elongation (El) and hardness will be shown in this work. Light microscope and the scanning electron microscopy will be used for documenting fractography, microstructure and porosity change after heat treatment.

Maraging steel MS1 (1.2709) studied in this work is martensitic hardened steel, with excellent strength combined with high toughness, very good mechanical properties, and easily heat treatability by means of simple age hardening process to obtain excellent hardness and strength. The nominal composition is given in **Table 1** [2].

Table 1. Chemical composition of maraging steel MS1

Element (wt.%)	Ni	Co	Mo	Ti	Al	Cr, Cu	C	Mn, Si	P, S
Maraging steel MS1	17-19	8.5-9.5	4.5-5.2	0.6-0.8	0.05-0.15	≤0.05	≤0.03	≤0.1	≤0.01

Aluminium alloy AlSi10Mg is a typical casting alloy with good casting properties and is typically used for cast parts with thin walls and complex geometry. In the frame of aluminium alloys family, AlSi10Mg offers good strength, hardness and dynamic properties and is therefore also used for parts subjected to high loads. The nominal composition is shown in **Table 2** [2].

Table 2. The chemical composition of aluminium alloy AlSi10Mg

Element (wt.%)	Si	Fe	Cu	Mn	Mg	Ni	Zn	Pb, Sn	Ti
AlSi10Mg	9-11	≤ 0.55	≤ 0.05	≤0.45	0.2-0.45	≤0.05	≤0.10	≤0.05	≤0.15

2 Experimental Procedure

The maraging steel MS1 and AlSi10Mg had different processing history. Samples of tested materials were annealed (HT) or they stayed untreated (As-built). **Table 3** summarized the heat treatment conditions. As-built and heat-treated components were analysed in terms of mechanical properties and microstructure.

Table 3. Heat treatment condition of the analyzed samples

Sample	Heat treatment condition
MS1-AB (as-built)	No heat treatment
MS1-HT	Solution annealing 850°C for 1 hour, ageing temperature 480°C for 3 hours, air cooling
AlSi10Mg-AB (as-built)	No heat treatment
AlSi10Mg –HT	Stress relieve 300°C for 2 hours

2.1 Mechanical properties

Ostbayerische Technische Hochschule Amberg-Weiden carried out tensile tests of as-built (AB) and heat-treated (HT) samples. They tested 3 AB samples and 3 HT samples of each material extracted in a horizontal direction. Average mechanical properties such as yield stress (YS), ultimate tensile strength (UTS), elongation (El) and hardness with HV5 load are summarized in **Table 4** and shown in **Figure 1**. The results of maraging steel MS1 shows, that the heat treated samples have higher hardness, yield stress and ultimate tensile strength, which increased about 61%. Elongation was reduced by about 4.1%. Heat treated samples of the aluminium alloy AlSi10Mg have lower values of hardness, ultimate tensile strength and yield stress. Elongation increased by about 0.9% and ultimate tensile strength was reduced about 18% after heat treatment.

Table 4. Mechanical properties of analyzed samples

Sample	YS (MPA)	UTS (MPa)	El (%)	Hardness HV5
MS1 AB	1061,3 ± 6,9	1178,8 ± 5,4	9,3 ± 0,1	361
MS1 HT	1826,0 ± 17,3	1896,6 ± 4,9	5,2 ± 0,1	542
AlSi10Mg AB	202,8 ± 8,1	282,6 ± 41,6	2,5 ± 1,5	95
AlSi10Mg (HT)	148,5 ± 1,6	231,0 ± 16,0	3,4 ± 1,6	75

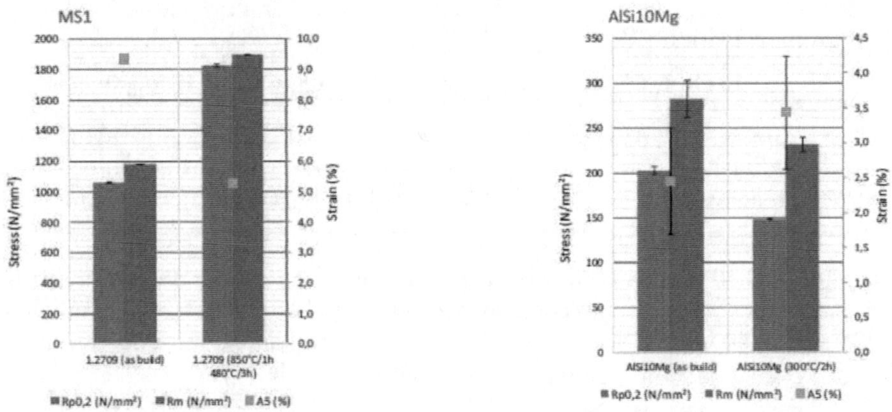

Figure 1. Mechanical properties of the AB samples and HT samples materials MS1 and AlSi10Mg

2.2 *Fractography*

After the tensile test, a fractographic analysis was performed on AB and HT samples using the scanning electron microscopy (SEM). Although yield stress and ultimate tensile strength of the maraging steel MS1 increased after heat treatment, ductile fracture with dimple morphology was proved in both conditions. Also, AlSi10Mg as-built and heat treated samples have a similar ductile fracture in both conditions with high porosity content (see red arrows in **Figure 2**)

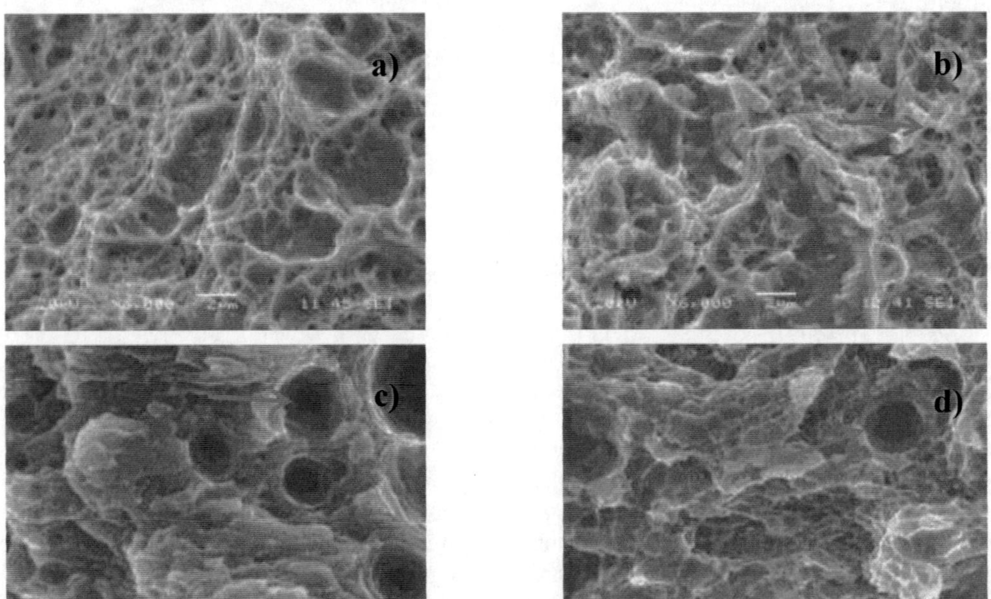

Figure 2. Fractured samples with a dimpled ductile fracture (arrows indicate pores) a) MS1-AB, b) MS1-HT, c) AlSi10Mg-AB, d)AlSi10Mg-HT

.2.3 *Porosity and microstructure*

An important parameter for a material fabricated by additive manufacturing is porosity. The porosity greatly affects the mechanical properties. The pores could be the crack initiators. Samples from maraging steel MS1 exhibit very low porosity. Samples from AlSi10Mg exhibit porosity close to 5% for untreated samples and about 3.5% for samples after the heat treatment. **Table 5** shows the values of porosity of all tested samples.

Table 5 Comparison of the porosity

	MS1 AB	MS1 HT	AlSi10Mg AB	AlSi10Mg HT
Porosity (%)	0,26	0,23	4,94	3,54

For the microstructure analysis, the samples were cut by the scheme shown in Figure 3 in three directions. Planes XZ and YZ were taken along the building axis (axis Z) whereas XY has a normal parallel to the building axis. Samples from each plane underwent standard metallographic preparation consisted of grinding followed by polishing. The microstructure of the maraging steel was revealed by etching in the Vilella-Bain etchant and Picral 4%. Aluminium alloy AlSi10Mg was etched with Keller´s reagent. The metallographic observation was performed by means of a light microscope and scanning electron microscope.

Figure 3. Scheme of sampling

3D microscopic images in **Figure 4** and **Figure 5** document the microstructure of the maraging steel MS1 and aluminium alloy AlSi10Mg. Images show the structures of untreated materials with semicircularly shaped melt pools along the Z axis. XY-plane shows the tracks of the laser scan. After the heat treatment, the maraging steel MS1 melt pool borders disappeared, the melt pool borders of the AlSi10Mg are less visible. Higher magnification images in the XZ plane are shown in **Figure 6**, **Figure 7** and **Figure 8**. **Figure 6** a) demonstrates the maraging steel MS1 in as-built state with melt pools visible after etching in Picral 4%. Microstructure revealed by etching in Vilella-Bain reagent is martensitic with no visible retained austenite (**Figure 6** b). **Figure 6** c) shows the maraging steel MS1 cellular dendritic structure with different grain shapes, which is due to the solidification rate of a local melted region. **Figure 7** demonstrates the change of microstructure after the heat treatment when the melt pools boundaries disappeared and the carbides are precipitated along the grain boundaries as was proved by the electron microscopy [3, 4, 5].

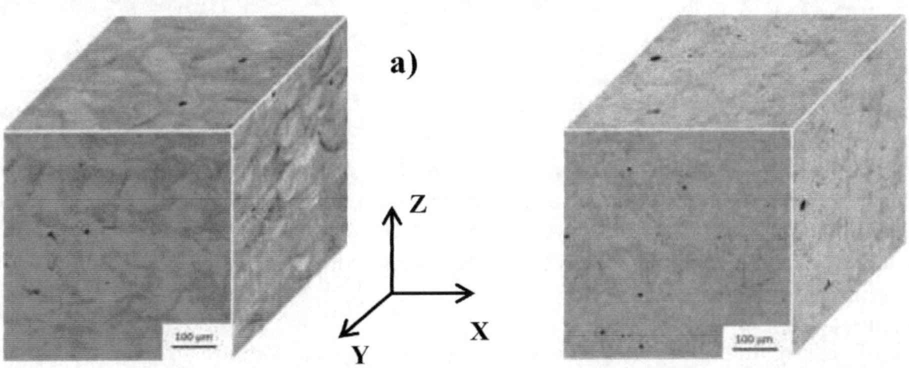

Figure 4. 3D images a) untreated maraging steel MS1 (AB) b) heat treated maraging steel MS1

IOP Conf. Series: Materials Science and Engineering **461** (2019) 012026 doi:10.1088/1757-899X/461/1/012026

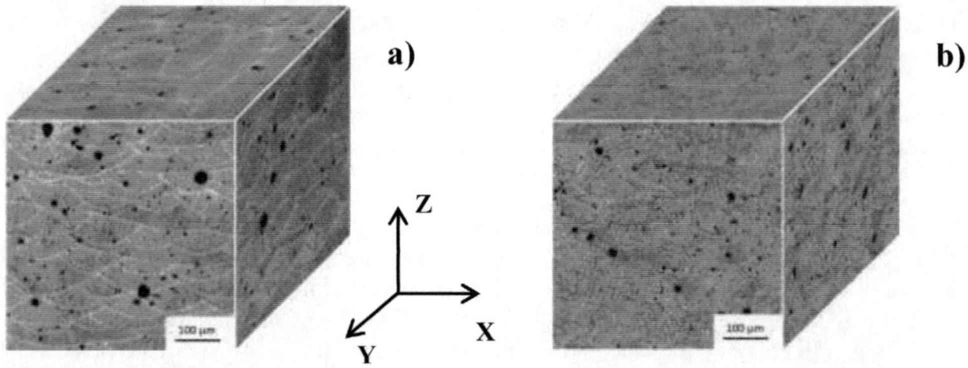

Figure 5. 3D imagines a) untreated AlSi10Mg (AB) b) AlSi10Mg after heat treatment (HT)

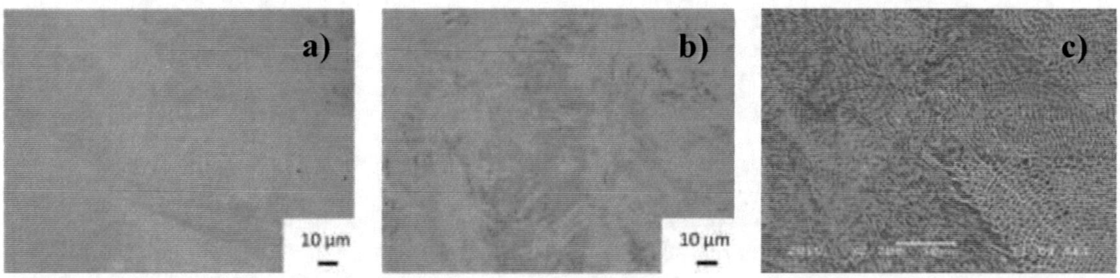

Figure 6. The maraging steel MS1 microstructure in as-built state (XZ plane) a) etched in Picral 4%, b) etched in a Vilella-Bain reagent, c) SEM image of the cellular structure with visible melt pools boundaries etched in Picral 4%

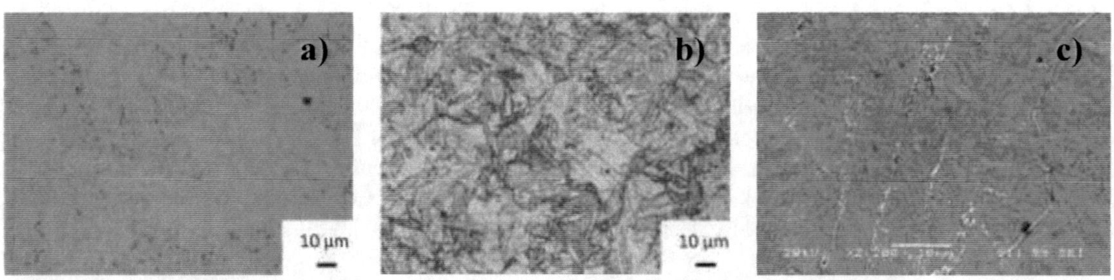

Figure 7. The maraging steel MS1 microstructure in heat treated state (XZ plane) a) etched in Picral 4%, b) etched in a Vilella-Bain reagent, c) SEM image of the structure with carbides precipitated along the grain boundaries

Aluminium alloy AlSi10Mg microstructure is composed of fine cellular dendritic structure inside the melt pools with a larger dendritic cell on the melt pools boundaries due to the thermal gradient. The cells are mainly composed of aluminium with fine silicon network (**Figure 8** a). **Figure 8** b) shows similar structure with less visible melt pools boundaries and with small Si based particles [6, 7].

3 Conclusion

The effects of heat treatment on the hardness, tensile properties and microstructure of maraging steel MS1 and aluminium alloy AlSi10M have been investigated in this research. Tensile test and hardness

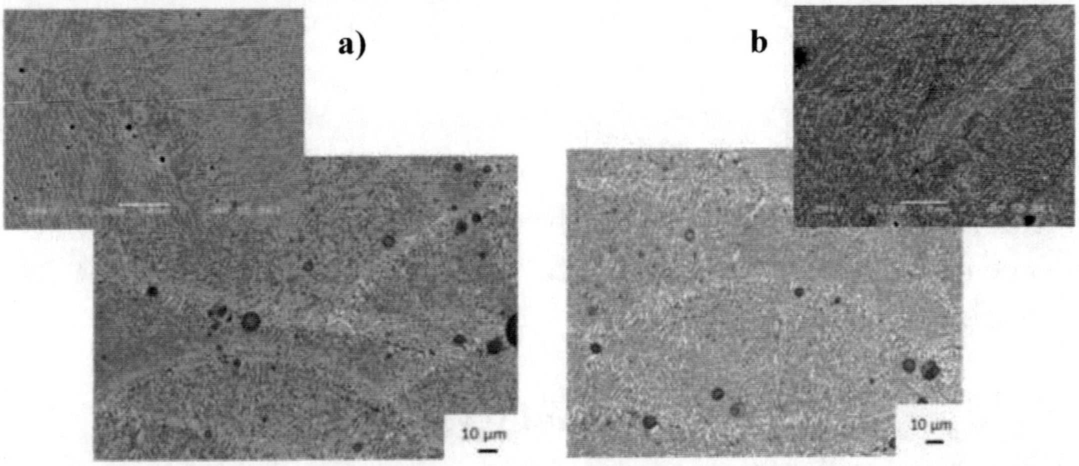

Figure 8. AlSi10Mg microstructure of XZ plane a) light and electron microscopy of the aluminium alloy in as-built state b) light and electron microscopy of the aluminium alloy in annealed state

results revealed that heat treated samples of maraging steel have higher yield stress about 58%, ultimate tensile strength increased about 61% and elongation reduced about 4,1%. Hardness increased by 181 HV5. Stress relived aluminium alloy AlSi10Mg has lower values of yield stress of about 13.8%, ultimate tensile strength decreased 18%, elongation increased about 0,9% and hardness decreased from 95 HV to 75 HV. This result of aluminium alloy might be mainly due to the relief of internal stress. Fractography shows that all analysed fractures have a ductile fracture with dimple morphology. The porosity of the maraging steel is similar in both conditions. The porosity of aluminium alloy AlSi10Mg is close to 5% for untreated samples and about 3.5% for heat treated samples. Martensitic maraging steel MS1 changed typical microstructure of melt pools composed of cellular dendritic structure to the microstructure with carbides precipitated along the grain boundaries after heat treatment. Aluminium alloy AlSi10Mg has similar microstructure in both conditions.

Acknowledgement
The results were created under the project entitled 3D COVER – Metallic materials in the process chain of additive manufacturing, Project No.: 185, financed by the ERDF.

References

[1] Nannan Guo, Ming C. Leu. Additive manufacturing: technology, applications and research needs, Frontiers of Mechanical Engineering 2013, 215-243
[2] Material data sheet [https://www.eos.info/material-m]
[3] TAN, Chaolin, Kesong ZHOU, Wenyou MA, Panpan ZHANG, Min LIU a Tongchun KUANG. Microstructural evolution, nanoprecipitation behavior and mechanical properties of selective laser melted high-performance grade 300 maraging steel. Materials & Design. 2017, 23–34.
[4] BAI, Yuchao, Yongqiang YANG, Di WANG a Mingkang ZHANG. Influence mechanism of parameters process and mechanical properties evolution mechanism of maraging steel 300 by selective laser melting. Materials Science and Engineering. 2017, 703, 116–123.
[5] Mutua James, Nakato Shinya, Onda Tetsuhiko, Chen Zhong-Chun.Optimization of selective laser melting parameters and influence of post heat treatment on microstructure and mechanical properties of maraging steel. Materials & Design. 139 (2018), 486–497.
[6] K.G. Prashanth, S. Scudino, H.J. Klauss, K.B. Surreddi, L. Löber, Z. Wang, et al., Microstructure and mechanical properties of Al-12Si produced by selective laser melting: Effect of heat treatment, Mater. Sci. Eng. A. 590 (2014) 153–160.
[7] Brandl Erhard, Heckenberger Ulrike, Holzinger Vitus, Buchbinder Damien. Additive manufactured AlSi10Mg samples using Selective Laser Melting (SLM): Microstructure, high cycle fatique, and fracture behavior. Materials & Design. 34 (2012), 159-169

IOP Conf. Series: Materials Science and Engineering **461** (2019) 012027 doi:10.1088/1757-899X/461/1/012027

The influence of oxide layer pattern size created on the surface of C45 steel on laser light absorption during laser quenching procedure

Hruška M, Vostřák M, Tesař J

University of West Bohemia, Univerzitní 8, 306 14 Pilsen, Czech Republic, EU

maslej@ntc.zcu.cz

Abstract. An increase in surface absorptivity can be helpful with a number of laser heat treatment processes. This study examines the relationship between the surface condition and its influence on interaction with a laser beam. The aim of this paper is to examine the effect of the oxide layer pattern size on the absorption of IR radiation and the influence of the ambient atmosphere. The process was carried with inert gas shielding. The oxide layers are created by the controlled heating method. The controlled heating method uses a scanning laser system to create oxide layers of different thickness on steel EN 10083-2: C45. Subsequently, the absorbency of the formed oxide layer was measured, depending on the irradiation energy emitted into the surface. According to recent experiments has been found that the higher the thickness of the oxide layer increases absorption. These results have shown that the use of surface oxidation increases the coupling of laser energy to material surface in the order of tens of percent. The thickness and hardness of the laser hardened layer were considered. That paper describes possibilities of precise laser quenching of the oxidized C45 material surface.

1 Introduction

The formation of the oxide layer on the surface of the material results in a change of its properties as a whole. Of course, the oxide layer has different electrical, optical and mechanical properties. There are three sets of optical properties (three different wavelengths) for the Kirchhoff's law [1]. The three wavelengths are the wavelength of engraving laser 1064 nm, the wavelength of quenching laser 808 nm, and the measuring range of infrared camera 1 μm.

A sequential method of surface absorption determination in the laser quenching process is described in [2]. This paper presents an efficient algorithm for determining the surface absorptivity in the process of laser quenching. A regression model for the reflectivity coefficient evaluation in laser surface quenching is presented in [3]. The model is numerically calculated by comparing the actual surface temperature to the theoretical prediction obtained by process simulation. Paper [4] describes the numerical-experimental analysis of the surface oxidation effect on the laser quenching. The model is able to offer the temporal course of the oxide thickness and the calculation of the increment of the absorption as a consequence of the development of the oxide layer. The relative error of this method for determining the maximum temperature is less than 5%. Two different mechanisms of radiative absorption for the surface oxidation material were described in [5]. The thickness of oxide layer influences the absorptive mechanism. The oxide layer of low thickness (in the order of 100 nm) has high transmission and absorption on the surface of the

underlying material dominates. The absorptivity of the oxide layer increases with it's growing thickness. Absorption through the oxide layer is predominant for thicker layers (in the order of 500 nm or more) of oxide.
The phenomenon called total reflection [6] can occur for a certain thickness of the oxide layer and the given value of the wavelength of the incident light.
An increase in the surface absorption of the material can be used, for example, in the laser quenching process, especially for precise heat treatment [7].
The absorptivity can be obtained by the method of measurement of spectral normal hemispherical reflectivity at room temperature (SNHRRT)[8] or for high-temperature assessment the method of measurement of the spectral normal emissivity at high temperature (SNEHT) [9] can be used.
The aim of the paper is the description of the behavior of different oxide layer pattern sizes created by the method of controlled heating. The influence of the width of the oxide layer and the gap between oxide layers on the absorption of laser radiation during the heat treatment process using argon as shielding gas is investigated. The aim is to enable very precise choice of the area which should be hardened. It will be done using the selective surface oxidation procedure. The method of processing can change optical properties of the material surface (absorptivity).

2 Experimental procedure

Tests were performed on a rectangular block from rolled steel EN 10083-2: C45 of 200x100x20 mm^3. The oxide layer was created by marking laser SPI-G3-SP-20P with MOPA arrangement by the method of controlled heating [10]. Heating parameters of the creation of the oxide layer were chosen based on the previous experiment the oxide layer with the highest thickness (fluence - 50 J/mm^2) was selected. Oxide layer patterns were formed on the sample surface according to Figure 1. After that the surface layer of the sample was modified by controlled oxidation procedure, the modified sample surface was subjected to a laser quenching. The sample was processed by the circular laser beam with a diameter of 8 mm. Four separated hardened tracks were created on the sample. The tracks were put perpendicularly to oxide pattern direction.

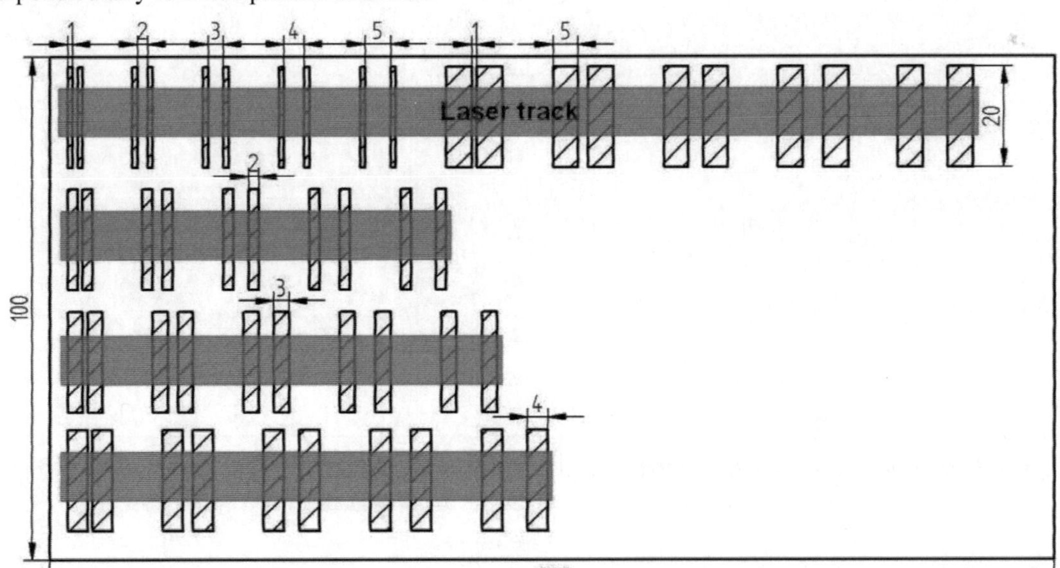

Figure 1. Sample with different oxide layer patterns, created with different width and gap distances between them, laser tracks are marked in red, the dimensions are in mm

Laser quenching procedure was implemented using argon as a shielding gas. A nozzle was used which feeds the shielding gas argon directly to the interaction area of the laser beam and the surface of the treated material. The argon as shielding gas prevents the surface oxidation of the processed material.
The thickness of the oxide layer was measured by scanning electron microscope Analytical SEM Hitachi SU-70 [10].
The surface temperature was measured during the quenching process. The measurement was made by the thermal camera. The thermal camera measures on 1 μm wavelength, so that it is not very sensitive to selected emissivity value. Because metals have an emissivity around 0.9 for temperatures around 1000°C and the measured temperature changes by 50°C for emissivity in range 0.7-1.
The chemical composition of an oxide layer on the surface of the sample was inspected by grazing-incidence x-ray diffraction. The composition was measured for basic material and for oxidated surface (fluence – 50 J/mm^2).
The absorptivity value was obtained by the method of measurement of spectral normal hemispherical reflectivity at room temperature (SNHRRT).

3 Results and discussion

The absorptivity was measured in ultraviolet-visible – near-infrared light spectral region (Figure 2.). It was measured on surface modified by controlled oxidation in comparison to the surface of the basic unmodified material. The oxide layer was created by laser fluence of 50 J/mm^2. The absorptivity at 808 nm (wavelength of quenching laser) measured for the modified surface was around 20 percent higher in comparison to absorptivity of the unmodified material surface.

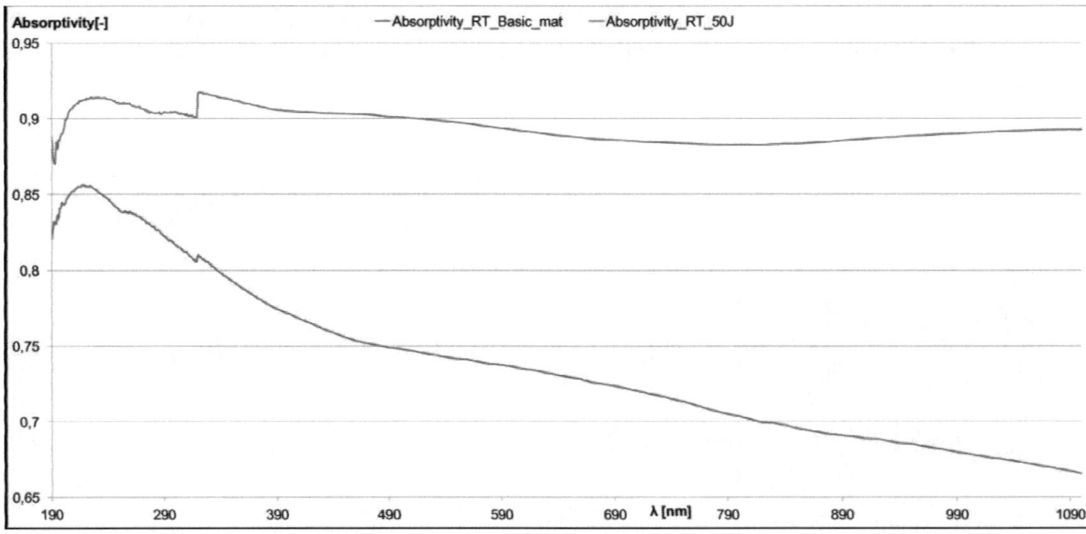

Figure 2. Measured absorptivities in the UV-VIS-NIR light spectral region for surface modified by controlled oxidation(blue color) in comparison to the basic unmodified material surface (red color)

Table 1. Surface hardness and depth of hardened layer – for surfaces modified by controlled oxidation

The width of the oxide layer pattern	1 mm pattern	2 mm pattern	3 mm pattern	4 mm pattern	5 mm pattern
Hardness [HV1]	515	512	526	541	563
Depth [μm]	300	350	390	410	420

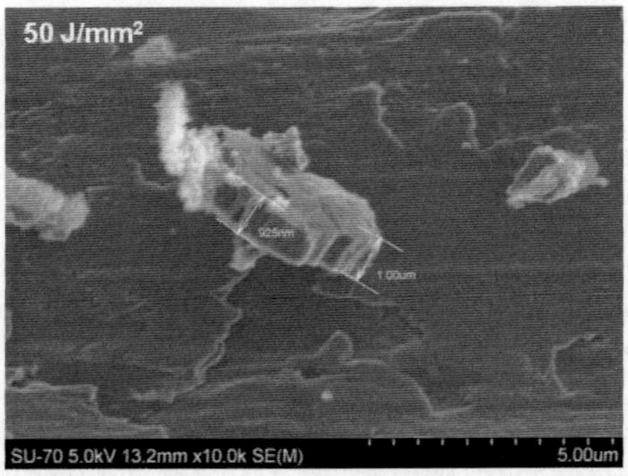

Figure 3. ASEM picture of oxide layer thickness for 50 J/mm^2 fluence

Surface hardness and depth of the hardened layer were measured in cross-section of the sample (Table 1.). The depth of the hardened layer was determined as the distance from the surface where 80% of the surface hardness was achieved. The hardness parameters were measured for modifications (oxide) widths from 1 mm to 5 mm and gaps of basic material between them from 1 mm to 5 mm (Figure 1.). The laser quenching process respects the lowest gap area for all widths of oxide patterns, 1 mm area of unhardened material has been preserved.

The measured thickness of the oxide layer formed by 50 J/mm^2 laser fluence was around 1 μm (Figure 3.). A scratch was made into the oxide layer and the thickness of the fragments was measured.

The surface temperature was measured during the quenching process with shielding gas. The temperature measured on the modified surface was around 1100°C (austenitization occurs) and on the unmodified basic material was measured temperature around 600°C (no structural changes occurs) [10].

The composition of the created oxide layer is around 45% FeO and 55% Fe_3O_4 before laser quenching process. The composition changes to 85% FeO and 15% Fe_3O_4 after laser quenching process.

4 Conclusion

The paper deals with the effect of oxide layer pattern size created on the surface of the material by the method of controlled heating on the surface absorptivity of the laser beam during the process with argon shielding.

The using of shielding gas allows us higher selectivity of laser quenching procedure. The work focuses on the determination of the modification size or size of the gap between them that will be respected by laser quenching process.

In the paper were analyzed thickness of the oxide layer and the composition of the layer created by the method of controlled heating. The surface hardness and depth of hardened layer were measured after laser quenching process. The influence of created oxide layer pattern size on these material properties was inspected.

Acknowledgments

The result was developed within the CENTEM project, reg. no. CZ.1.05/2.1.00/03.0088, cofunded by the ERDF as part of the Ministry of Education, Youth and Sports OP RDI programme and, in the follow-up sustainability stage, supported through CENTEM PLUS (LO1402) by financial means from the Ministry of Education, Youth and Sports under the "National Sustainability Programme I." and project no. SGS-2016-005.

References

[1] RINALDI R, VAINER G B, KUNC P, KNIZKOVA I, GOTTSCHALK K, CARLOMAGNO M G, ALLOUIS CH, PAGLIARA R, ROTOLANTE R A, GRINZATO E, Infrared Thermography (Recent Advances and Future Trends). *Infrared Thermography*. 2012. Vol. 241. Carosena Meola.

[2] NGUYEN Q and Y, A sequential method to determine the surface absorptivity in the process of laser surface hardening. *International Journal of Heat and Mass Transfer*. 2016. Vol. 95, p. 224–229. Ching-yu. DOI 10.1016/j.ijheatmasstransfer.2015.11.087.

[3] FORTUNATO A, ASCARI A, ORAZI L, CAMPANA G , and CUCCOLINI G, Numerical evaluation of the reflectivity coefficient in laser surface hardening simulation. Surface & Coatings Technology. 2012. Vol. 206, p. 3179–3185. DOI 10.1016/j.surfcoat.2011.12.043.

[4] CORDOVILLA F, GARCÍA-BELTRÁN Á, DOMINGUEZ J, and SANCHO P, Numerical-experimental analysis of the effect of surface oxidation on the laser transformation hardening of Cr – Mo steels. Applied Surface Science. 2015. Vol. 357, p. 1236–1243.

[5] PANTSAR H and KUJANP V, Effect of oxide layer growth on diode laser beam transformation hardening of steels. Surface & Coatings Technology. 2006. Vol. 200, p. 2627–2633. DOI 10.1016/j.surfcoat.2004.09.001.

[6] GOOS F and HÄNCHEN H, Über das Eindringen des totalreflektierten Lichtes in das dünnere. *Medium Ann. Physik*. 1947. Vol. 435, no. 5, p. 1947.

[7] HRUŠKA M, VOSTŘÁK M, SMAZALOVÁ E and ŠVANTNER M, 3D scanning laser hardening. *METAL 2014 - 23rd International Conference on Metallurgy and Materials, Conference Proceedings*. 2014. P. 921–926.

[8] ZHANG K, YU K, LIU Y, and ZHAO Y. An improved algorithm for spectral emissivity measurements at low temperatures based on the multi-temperature calibration method. *International Journal of Heat and Mass Transfer* [online]. 2017. Vol. 114, p. 1037–1044. DOI 10.1016/j.ijheatmasstransfer.2017.06.133.

[9] HONNER M, HONNEROVÁ P, KUČERA M and MARTAN J. Laser scanning heating method for high-temperature spectral emissivity analyses. *Applied Thermal Engineering*. 2016. Vol. 94, p. 76–81. DOI 10.1016/j.applthermaleng.2015.10.121.

[10] HRUŠKA M and VOSTŘÁK M, The impact of controlled en 10083-2: C45 steel surface oxidation on NIR light absorption. *METAL 2017 - 26th International Conference on Metallurgy and Materials, Conference Proceedings*. 2017. P. 1153-1158.

IOP Conf. Series: Materials Science and Engineering **461** (2019) 012028 doi:10.1088/1757-899X/461/1/012028

Coatings on Steels T91 and 316L in Lead-Bismuth Eutectic Environment

M Chocholousek, L Rozumova, Z Spirit, F Di Gabriele

Centrum Vyzkumu Rez, CVR, Hlavni 130, Husinec-Rez, Czech Republic

E-mail: michal.chocholousek@cvrez.cz

Abstract. The work is aimed on the coatings used as a barrier against Heavy Liquid Metal (HLM) influence and coated material behaviour under mechanical loading. The comparison of two kinds of coatings deposited on two different steels is examined: ferritic-martensitic steel T91 coated with AlTiN and austenitic steel 316L coated with Al_2O_3. Both samples underwent tensile tests with constant extension rate in the HLM autoclave-cell for material testing. The testing environment in the cell was liquid PbBi eutectic at the temperature of 550 °C. The examination by Scanning electron microscopy and Energy-dispersive X-ray spectroscopy followed after exposure. The yield strength of the steel was exceeded in both cases, cracks and adhesion were studied in depth to characterize their resistance to the simultaneous effect of load and environment. Both coatings offered protection against corrosion; however, localized cracking gave rise to the damage of the substrate. These coatings provided a corrosion barrier, which once damaged is not able of self-healing.

1 Introduction

Heavy liquid metals (HLM) are being studied as candidate coolants for Generation IV fast reactors. The nuclear industry remains generally conservative in terms of using new materials and qualification under irradiation. Thus the use of coatings as protective barriers is being investigated [1-3] as a response to the issue of material dissolution in HLM environment [3-5]. The main adopted principle is the use of Al_2O_3 forming materials, as the oxide is fairly stable and resistant to high temperature exposure [1, 6 and 7].

The main candidates among structural materials proposed for HLM applications include ferric-martensitic steel T91 and austenitic steel 316L, mainly because of their qualified properties under irradiation and the existence of a large database for their corrosion behaviour in environment [4].

AlTiN coatings as a perspective solution for HLM application are proposed due to the suitable mechanical properties and chemical stability. Pure Al_2O_3 has a potential due to substantially insolubility in lead and compatibility from the chemical point of view [8, 9].

Experiments in the frame of the FP7-MATISSE project were aimed to compare the performance of two coatings under loading condition in lead-bismuth eutectic (LBE). Moreover, the characterisation of the corrosion resistance of coatings was carried out, due to insufficiency of literature data on the stability and overall performance of coatings under loading conditions.

2 Experiments

2.1 Materials

IOP Conf. Series: Materials Science and Engineering **461** (2019) 012028 doi:10.1088/1757-899X/461/1/012028

The coating-substrate combinations were chosen according to the agreement of the project (FP7 MATISSE) members. The selection was limited on a small amount due to the project schedule.

2.1.1 Ferritic-martensitic steel T91. (Grade 91 Class 2/S50460) of nominal composition (wt. %) Fe-8.9Cr-0.9Mo-0.4Mn-0.2Si-0.2V was produced by Industeel, Arcelor Mittal group. The material as 15 mm thick plate was normalized at 1050°C with a holding time of 1 min/mm followed by water quenching to room temperature and then annealed at 770°C with a holding time of 3 min/mm followed by air cooling. T91 steel was used as a substrate for application of AlTiN coating.

2.1.2 Austenitic steel 316L. (ASTM A240-Ed02) was produced by Industeel, Alcelor Mittal group. The austenitic stainless steel was received as hot rolled and heat treated plates with a thickness of 15 mm. The solution annealing was done at 1050-1100 °C. The microstructure obtained was fully austenitic with a small fraction of delta ferrite. 316L stainless steel was used as a substrate for application of Al_2O_3 coating.

2.2 Coatings

2.2.1 AlTiN. (more than 50% of Al) coating normally used for abrasive and high temperature applications (>800 °C) was deposited on the steel T91 tensile specimens by using combination of reactive HiPIMS (High Power Impulse Magnetron Sputtering) and pulsed-DCMS (Direct Current Magnetron Sputtering) technologies in the CNR (Centro Nazionale Ricerca) in Italy. The thickness of the layer was about 7 μm.

2.2.2 Al_2O_3. Coating for wide range of applications naturally provides a passivation layer, which prevents the direct interaction between the steel and the environment. The coating was deposited on the austenitic 316L steel by detonation gun (a high-velocity thermal spray process with an extremely good adhesive strength, usually low porosity and coating surface with compressive residual stresses).

2.3 Specimen

Rod tensile specimens (figure 1) with 4 mm diameter were fabricated in longitudinal direction of the plates. The shape of the specimen was designed to fit in alignment system in HLM.

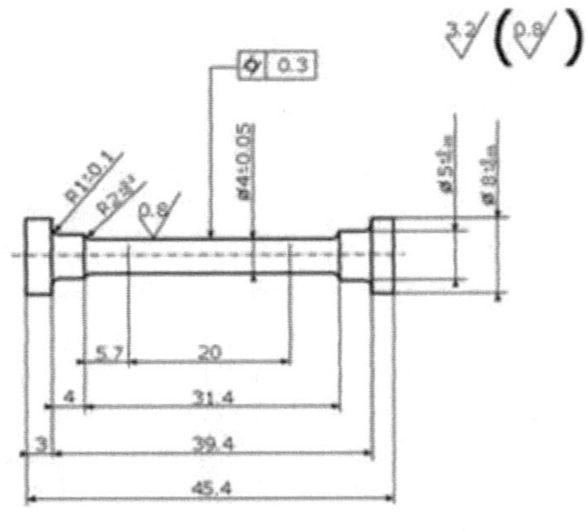

Figure 1. Rod tensile specimen with coating for HLM environment.

2.4 Experimental procedure

The tensile tests were performed in LBE medium at 550 °C in order to characterize the thermo-mechanical behavior of materials' coatings. As a testing equipment, the autoclave cell built on the Kappa 50DS Electromechanical Creep Testing machine was used. The testing system is based on 2-

tank concept, where the first tank is for medium preliminary preparation and the medium is subsequently transferred to the second tank with the specimen and holders, where the oxygen dosing is regulated to obtain required environmental conditions. The test is then initialized after reaching the precise conditions (table 1) and performed at the constant extension rate. One of the specimens with Al_2O_3 coating was firstly tested in air environment to quickly verify the adhesive properties of the coating. The test was performed due to previous insufficient preparation of another Al_2O_3 coating.

Table 1. Conditions of performed tensile tests.

Sample design.	Environment	Temperature [°C]	Extension rate [s^{-1}]	C_{O2} wt.%
AlTiN 1	LBE	550	10^{-4}	10^{-8}
AlTiN 2	LBE	550	10^{-4}	10^{-8}
Al_2O_3 1	Air	550	10^{-6}	-
Al_2O_3 2	LBE	550	10^{-6}	10^{-13}

2.5 *Post-test evaluation*

After exposure, the samples were cleaned in a solution H_2O_2, CH_3COOH and CH_3CH_2OH, in the ratio 1:1:1, in order to remove residuals of LBE from the surface. Samples were subjected to the surface analysis using Scanning electron microscopy (SEM) LYRA 3, TESCAN. Subsequently, samples were embedded in the acrylic resin and their cross-section polished for SEM analysis, followed by Energy-dispersive X-ray (EDX) spectrometer. The integrity of the coatings and their adhesions were examined.

3 Results

Figure 2 summarizes the stress-deformation results from both materials at 550 °C. Due to very thin layer thickness the effect of the coatings to material stress-dependence is being considered as negligible. Thus the maximum stress thresholds can be evaluated as approximate Ultimate tensile strength (UTS) of steels T91 and 316L at the given temperature and extension rate.

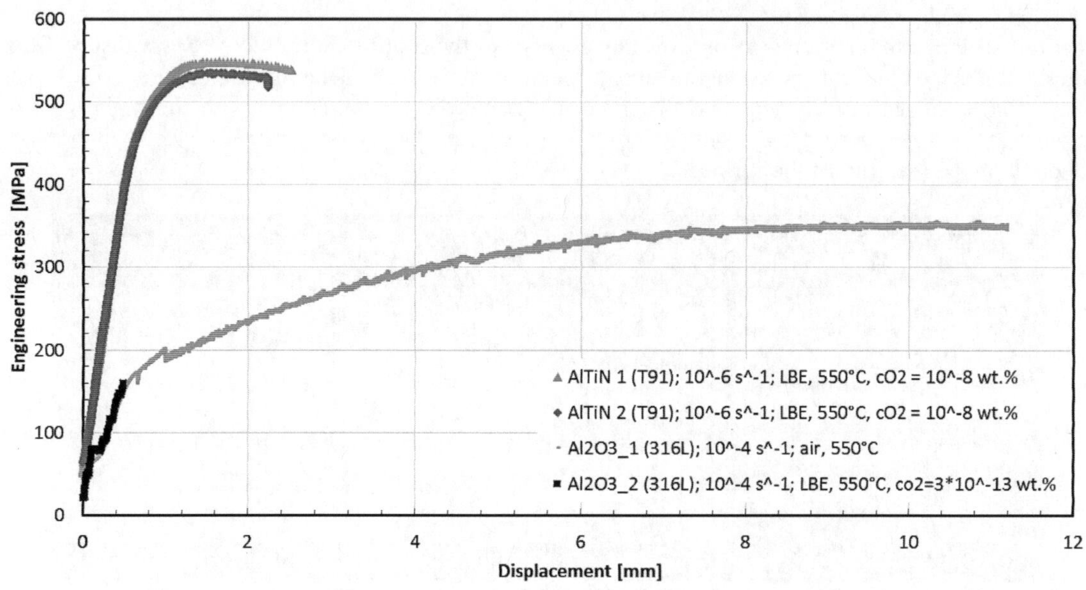

Figure 2. Tensile test with constant extension rate at 550°C on the steels with coatings.

The 316L specimen with Al_2O_3 coating was loaded in air up to UTS, as a reference test. However, 316L steel has significant plastic deformation up to the UTS (approx. 11 mm extension), therefore it was decided to load the specimen in LBE up to Yield stress (YS) of the material at 550 °C.

3.1 AlTiN

AlTiN surface layer on the sample showed cracks of regular rhomboidal shapes (Figure 3a). The cracks are distributed over the coating surface, though without signs of significant delamination. Cross-section examination indicates that the process of corrosion immediately started due to the contact of LBE with substrate through the cracking (figure 3b, c). The direct contact between the T91 and LBE at 550°C, with 10^{-8} wt. % of oxygen amount, induce a fast reaction which is driving the Fe and Cr out into the liquid metal and make space for the LBE to penetrate quickly into the steel.

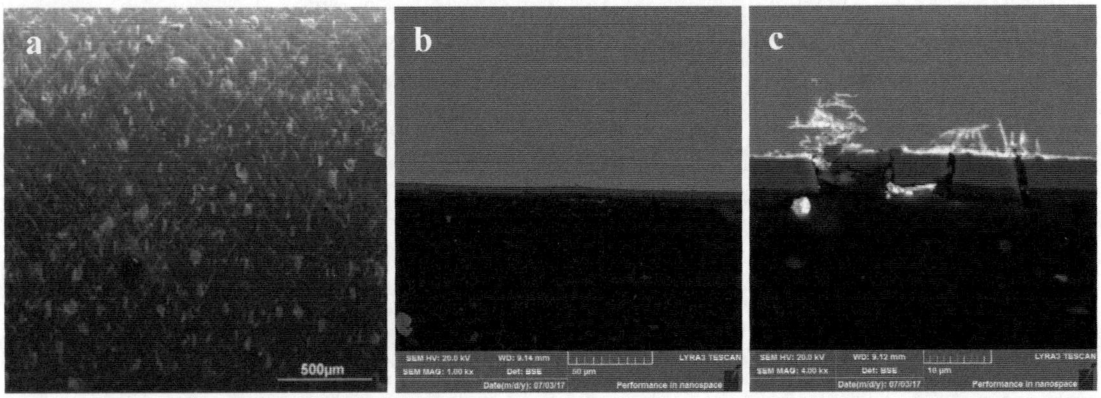

Figure 3. a) Surface of the T91 steel + AlTiN coating with homogeneously distributed cracks; b) detail of cross-section area; c) detail of cracks in the coating and the underneath damage.

3.2 Al_2O_3 in Air environment

Post-test surface examination did not show any changes of the coating morphology. Several superficial cracks of the coating can be observed and the coating is locally delaminated (figure 4b, c). The concentration of the coating loss was observed in the middle of the sample with the highest stress concentration. Considering the extension of the specimen, the coating remain mostly intact and adherent on the majority of the surface.

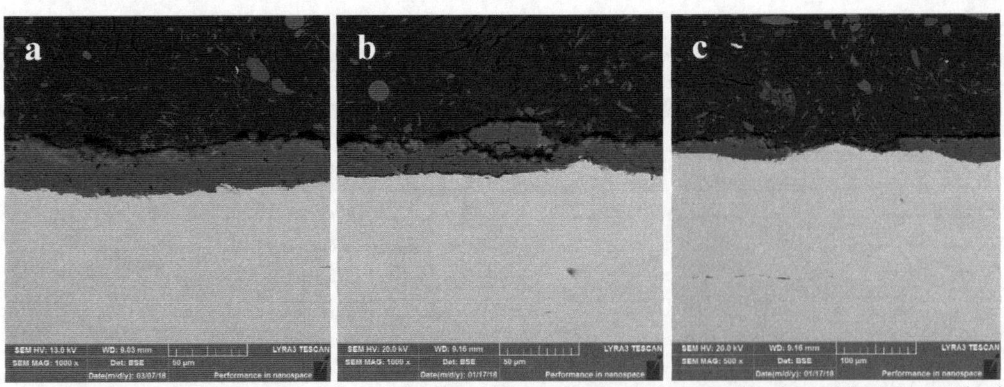

Figure 4. Cross-section of steel 316L + Al_2O_3 coating after test in air at 550 °C; a) The intact Al_2O_3 coating layer, b) detailed cracked layer of Al_2O_3; c) detail of delaminated coating layer.

3.3 Al_2O_3 in LBE environment

The cracking of Al_2O_3 coating after LBE exposition differs from AlTiN coating. Al_2O_3 coating is partly delaminated and LBE came to contact with the substrate 316L (figure 5c). The process of corrosion in the steel 316L started as well, though with the minor effect compared to the T91 steel.

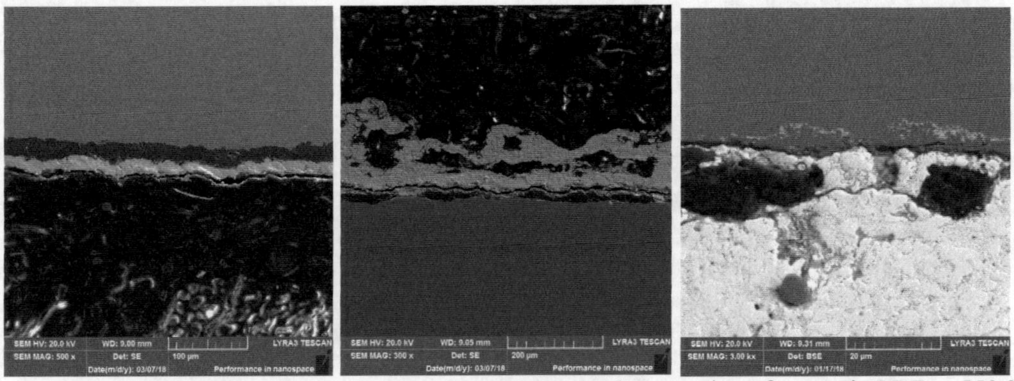

Figure 5. Cross-section of steel 316L + Al_2O_3 coating after test in LBE at 550 °C; a) the coating layer with the LBE layer on it; b) cracked and partly delaminated coating layer; c) detail of cracks in the coating and the LBE-substrate contact

4 Discussion

The difference in depths of penetration in the steels is possibly due to the different steel-LBE contact time. This may indicate earlier failure of the AlTiN coating and the LBE had more time to interact with the steel. In fact, the strain rates are very different: the AlTiN was tested at faster rate, therefore it can be assumed, that the coating failed very early in the loading process. On the other hand, the Al_2O_3 coated specimens were tested at lower strain rates, but at much lower YS. Therefore it could be assumed that this coating failed at a much later stage, since the test was also very slow and the steel-LBE contact was only shortly before stopping the test.

The T91 and 316L steels are different with different mechanical properties: the T91 has significantly higher YS than the 316L; moreover the extension necessary to reach the UTS is significantly different too, very short for T91 (0.5 mm) and very long for the 316L (10 mm). However, both reached the YS after about 0.5 mm of deformation. These parameters have a different effect on the coating and in this case it is difficult to compare. However, the main observation was that both coatings failed in the loading conditions and the LBE was in direct contact with the substrate and consequently, interacted with the steel.

5 Conclusion

- We may say that the methods used for application of the AlTiN coating and for Al_2O_3 was chosen correctly because of the great adhesion of both coatings to the substrates.
- The coatings in combination with the structural materials (AlTiN with ferritic-martensitic steel T91; Al_2O_3 with stainless steel 316L) had not only good adhesions but also high hardness, corrosion resistances in contact with liquid metals.
- In both cases (AlTiN and Al_2O_3), the coatings in atmosphere and also in LBE were very stable at testing temperature and no chemical reactions occurred.
- These kinds of coatings provide a corrosion barrier, which once damaged is not able of self-healing.
- Coatings are a valuable solutions for high temperature HLM, however, application limits have to be established and quantified by a wider experimental campaign.

Acknowledgments

The presented work was financially supported by the Ministry of Education, Youth and Sport Czech Republic - project LQ1603 Research for SUSEN. This work has been realized within the SUSEN Project (established in the framework of the European Regional Development Fund (ERDF) in project CZ.1.05/2.1.00/03.0108 and of the European Strategy Forum on Research Infrastructures (ESFRI) in the project CZ.02.1.01/0.0/0.0/15_008/0000293, which is financially supported by the Ministry of Education, Youth and Sports - project LM2015093 Infrastructure SUSEN. This work was also supported by the project FP7 MATISSE, g.a. n. 604862.

References

[1] Weisenburger A, Heinzel A, Müller G, Muscher H and Rousanov A 2008 T91 cladding tubes with and without modified FeCrAlY coatings exposed in LBE at different flow, stress and temperature conditions, *Journal of Nuclear Materials* 376 (2008) pp 274-81

[2] Rivai A K and Takahashi M 2010 Corrosion investigations of Al–Fe-coated steels, high Cr steels, refractory metals and ceramics in lead alloys at 700 °C, *Journal of Nuclear Materials* 398 (2010) pp 146-52

[3] Hojna A, Di Gabriele F and Klecka J 2016 Characteristics and Liquid Metal Embrittlement of the steel T91 in contact with Lead–Bismuth Eutectic, *Journal of Nuclear Materials* 472 (2016) pp 163-70

[4] OECD/NEA 2015 Handbook on Lead-bismuth Eutectic Alloy and Lead Properties, materials compatibility, thermal-hydraulics and technologies, OECD (2015), NEA No. 7268

[5] Gromov B F, Belomitcev Y S, Yefimov E I, Leonchuk M P, Martinov P N, Orlov Y I, Pankratov D V, Pashkin Y G, Toshinsky G I, Chekunov V V, Shmatko B A and Stepanov V S 1997 Use of lead-bismuth coolant in nuclear reactors and accelerator-driven systems, *Nuclear Engineering and Design* 173 (1997) pp 207-17

[6] Engelko V, Mueller G, Rusanov A, Markov V, Tkachenko K, Weisenburger A, Kashtanov A, Chikiryaka A and Jianu A 2011 Surface modification/alloying using intense pulsed electron beam as a tool for improving the corrosion resistance of steels exposed to heavy liquid metals, *Journal of Nuclear Materials* 415 (2011) pp 270–5

[7] Weisenburger A, Schroer C, Jianu A, Heinzel A, Konys J, Steiner H, Müller G, Fazio C, Gessi A, Babayan S, Kobzov A, Martinelli L, Ginestar K, Balbaud-Célerier F, Martín-Muñoz F J, Soler Crespo L, Long term corrosion on T91 and AISI1 316L steel in flowing lead alloy and corrosion protection barrier development: Experiments and models, *Journal of Nuclear Materials* 415 (2011) pp 260-9

[8] García Ferré F, Ormellese M, Di Fonzo F and Beghi M G 2013 Advanced Al_2O_3 coatings for high temperature operation of steels in heavy liquid metals: a preliminary study, *Corrosion Science* 77 (2013) pp 375–8

[9] García Ferré F, Mairov A, Iadicicco D, Vanazzi M, Bassini S, Utili M, Tarantino M, Bragaglia M, Lamastra F R, Nanni F, Geseracciu L, Serruys Y, Trocellier P, Beck L, Sridharan K, Beghi M G and Di Fonzo F 2017 Corrosion and radiation resistant nanoceramic coatings for lead fast reactors, *Corrosion Science* 124 (2017), pp 80-92

Modelling and parametric analysis of cohesive behaviour of interface delamination of graphene monolayer nanocomposites

E Kirilova[1], T Petrova[1,3], W Becker[2], J Ivanova[3]

[1]Institute of Chemical Engineering, Bulgarian Academy of Sciences, Acad. G. Bonchev str., Bl.103, 1113 Sofia, Bulgaria

[2] Institute of Structural Mechanics, TU Darmstadt, Franziska-Braun-Str. 7, L5|01 347a, 64287 Darmstadt, Germany

[3] European Polytechnical University, 2300 Pernik, Bulgaria

Corresponding author e-mail: ivanovajordanka@yahoo.com

Abstract. The engineering importance of the problem of interface delamination for graphene monolayer nanocomposites is also of high level interest and it is directly connected with the optimal choice of the structure geometry and the choice of the matrix material. In this work modelling and parametric analysis of cohesive interface delamination of graphene monolayer nanocomposites are presented, using shear lag method. The external excitation of the matrix is an axial displacement, controlling the interface delamination. The statement of the considered problem is given in displacements, the interface behaviour is assumed to be a bi-linear cohesive one. Three zones along the interface are defined and for them the solutions of the system of two ordinary differential equations about the displacements of the monolayer of graphene and one for the matrix are fitted by appropriate boundary, symmetry and continuity conditions.

All analytical results about the interface shear stress, the lengths of the elastic, cohesive and delaminated zones are calculated for a given structure and are illustrated in figures. The parametric analysis and the influence of the interface length, applied load and the thickness of the host layer are performed. After detailed discussion some recommendations are given.

1 Introduction

During the last decades the new materials, used for the creation of the new type of nanocomposites containing monolayer of graphene is of great interest in engineering society. The use of such materials is very important for sensors, solar cells, etc. and they have different possible industrial applications in automotive, building and aerospace industries, etc.

Due to mismatch of the thermo-mechanical properties the damage and failure (Modes I and II) in layered structures or coatings on substrate constitute to be one of the most important problems for this type of composite materials. In most cases the delamination at interface and layer cracking are the main modes of degradation of structure performance.

An extensive investigation has been conducted in the last decades with application of different approaches, generally based on the fracture mechanics models or on Cohesive Zone Model (CZM) approach. Numerous studies have also been devoted to the fatigue failure of coated layers and correlation between the capability of cracks to traverse an interface [1] and the authors successfully predicted the crack growth toward the interface in ductile bimaterial systems. Based on the linear

IOP Conf. Series: Materials Science and Engineering **461** (2019) 012029 doi:10.1088/1757-899X/461/1/012029

elasticity and fracture mechanics, the solutions for interface cracks in fact exhibit singular oscillatory behavior of stress field in the crack tip zone, generating some difficulties in crack growth modelling. A useful alternative to these solutions is the shear lag or the cohesive zone model. The shear lag approach has been frequently applied in the analysis of stress transfer through the interfacial shear stress in composite or bonded materials [2]. By Volkersen [3] and later by Cox [4] in a simple one dimensional model of analysis of stress transfer between a matrix and a fiber assuming the linear elastic relation between shear stress and tangential slip were proposed. The subsequent treatments by Hedgepeth [5] and Kelly [6] were devoted to composite analysis, assuming elastic or elastic–plastic interaction models. The cohesive zone models first proposed by Dugdale [7] and Barrenblatt [8] were addressed to analyze the localized damage or plastic flow in the front of crack tip.

Based on [9, 10] the authors implemented in this work 1D shear lag method to investigate the cohesive behavior of the interface. The external excitation of the matrix is an axial displacement, controlling the interface delamination. As a consequence, the statement of the considered problem is given in displacements. The interface behavior is assumed to be a bi-linear cohesive one. In the three defined zones along the interface, the solutions of the system of two ordinary differential equations about the displacements of the monolayer of graphene and one for the matrix are fitted by appropriate boundary, symmetry and continuity conditions. The lengths of elastic, cohesive and delaminated zones are calculated in a new manner by using the system of non-linear equations.

All analytical results about the interface shear stress, the lengths of the elastic, cohesive and delaminated zones are calculated for a given structure and are illustrated in figures. After detailed discussion some recommendations are given.

2 Problem formulation in displacements

Let us consider the mechanical model of the structure, shown in Figure 1(a). The layered structure has a length $2l$, thicknesses of the layer A and B are $2h_A, 2h_B$ respectively and the thickness of the interface is h_I.

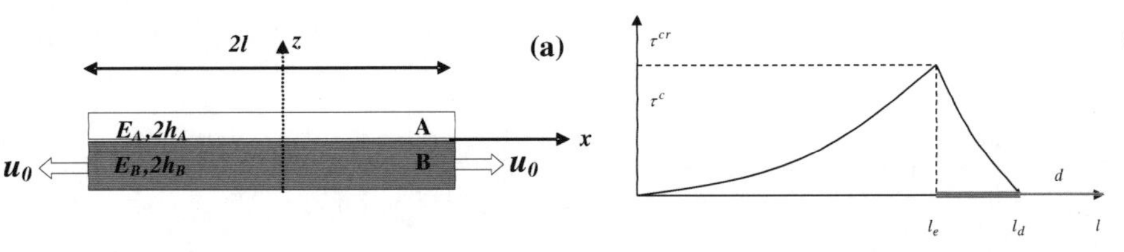

Figure 1. Scheme of the considered bi-layer structure (a) and interface shear stress behaviour (b)

The structure is loaded by monotonically increasing load as follows:

$$u'_B(l)=u'_B(-l)=0; \quad u_B(l)=u_B(-l)=u_0 \tag{1}$$

Under shear lag hypothesis, the interface works only in shear, while layers A and B work on extension. The layers A and B remain linear elastic during the loading process satisfying the Hooke's law $\sigma_A = 2h_A E_A \varepsilon_A, \sigma_B = 2h_B E_B \varepsilon_B$, where E_A, E_B are Young's modulus of layers A and B, respectively. The interface mechanical behaviour is assumed to follow the bilinear elastic-cohesive law, such as:

$$\tau_I = \begin{cases} K_t u_I^e & for \quad u_I \le u_I^c = \dfrac{\tau_I^c}{K_t} \\ \tau_I^c - K_s\left(u_I^s - u_I^c\right) & for \quad u_I > u_I^c, \quad u_I^f = \tau_I^c\left(\dfrac{1}{K_t} + \dfrac{1}{K_s}\right) \\ 0 \quad for \quad u_I^d \ge u_I^f & \end{cases}; \tag{2}$$

Since the problem is assumed to be symmetric, only the half of the length of the structure will be considered. The mechanical behaviour of the structure remains linear elastic until the uniaxial loading reaches some value u_0, at which the interfacial shear stress takes the value $\tau_I^c(l)=\left|\tau^c\right|$. Then the formation of the softening zone begins. Under this load u_0 and equation (2), the interface can be divided into 3 zones: elastic, softening and delaminated. The formation of the damage zone (softening+delaminated) begins at $x=l$ and finishes at $x=l_d$.

Increasing the load u_0, the picture of the behaviour of interface remains the same, only the lengths l_e, l_d and $d=l-l_d$ change until the possible full degradation of the interface (Figure 1(b)). Again, the defined regions for all three zones are: elastic zone - $0\le x\le l_e$; softening zone - $l_e\le x\le l_d$; delaminated zone - $l_d\le x\le l$. Solving consequently the problem given schematically in Figure 1(b), the final result consists in determination of the lengths l_e, l_d and d.

The behaviour of the interface is presented for the loading u_0 with symmetric, continuity and boundary conditions in three different zones in the following equations (3):

Elastic zone - $0\le x\le l_e$	**Softening zone** - $l_e\le x\le l_d$	**Delamination zone** - $l_d\le x\le l$	
$u_A^e(0)=0$	$\tau_I^e(l_e)=\tau_I^s(l_e)$	$u_A^s(l_d)=u_A^d(l_d)$	
$\tau_I^e(l_e)=-\tau_I^c$	$u_A^e(l_e)=u_A^s(l_e)$	$u_B^s(l_d)=u_B^d(l_d)$	(3)
$u_B^e(0)=0$	$u_A'^e(l_e)=u_A'^s(l_e)$	$2h_AE_Au_A'^s(l_d)+2h_BE_Bu_B'^s(l_d)=u_B'^d(l_d)$	
	$u_B'^e(l_e)=u_B'^s(l_e)$	$u_B^d(l)=u_0$	
	$\tau_I^s(l_d)=0$		

For displacements and interface shear stress in the elastic zone, we have:

$$u_I^e(x)=-u_I^c\frac{\sinh(\lambda_e x)}{\sinh(\lambda_e l_e)} \qquad \tau_I^e(x)=-K_tu_I^c\frac{\sinh(\lambda_e x)}{\sinh(\lambda_e l_e)}$$

$$u_A^e(x)=-\frac{u_I^c}{(1+\rho)}\frac{\sinh(\lambda_e x)}{\sinh(\lambda_e l_e)}+\frac{\left(u_0+u_I^f\right)(1+\rho)+u_I^c\left[1+\left(\frac{\lambda_e}{\lambda_s}\right)^2\right]}{(1+\rho)\left[l(1+\rho)-l_d\rho\right]}.x$$

$$u_B^e(x)=\frac{\rho u_I^c}{(1+\rho)}\frac{\sinh(\lambda_e x)}{\sinh(\lambda_e l_e)}+\frac{\left(u_0+u_I^f\right)(1+\rho)+u_I^c\left[1+\left(\frac{\lambda_e}{\lambda_s}\right)^2\right]}{(1+\rho)\left[l(1+\rho)-l_d\rho\right]}.x \tag{4}$$

Respectively, in the softening zone, solutions are:

$$u_I^s(x)=\frac{u_I^c\lambda_e^2}{\lambda_s^2}\frac{\sin(\lambda_s(l_d-x))}{\sin(\lambda_s(l_d-l_e))}+u_I^f\ ;\quad \tau_I^s(x)=-K_s\frac{u_I^c\lambda_e^2}{\lambda_s^2}\frac{\sin(\lambda_s(l_d-x))}{\sin(\lambda_s(l_d-l_e))}$$

$$u_A^s(x)=\frac{1}{(1+\rho)}\frac{u_I^c\lambda_e^2}{\lambda_s^2}\frac{\sin(\lambda_s(l_d-x))}{\sin(\lambda_s(l_d-l_e))}+\frac{\left(u_0+u_I^f\right)(1+\rho)+u_I^c\left[1+\left(\frac{\lambda_e}{\lambda_s}\right)^2\right]}{(1+\rho)\left[l(1+\rho)-l_d\rho\right]}.x-u_I^c\left[1+\left(\frac{\lambda_e}{\lambda_s}\right)^2\right] \tag{5}$$

$$u_B^s(x) = -\frac{\rho}{(1+\rho)}\frac{u_I^c\lambda_e^2}{\lambda_s^2}\frac{\sin(\lambda_s(l_d - x))}{\sin(\lambda_s(l_d - l_e))} + \frac{\left(u_0 + u_I^f\right)(1+\rho) + u_I^c\left[1+\left(\frac{\lambda_e}{\lambda_s}\right)^2\right]}{(1+\rho)[l(1+\rho) - l_d\rho]}.x - u_I^c\left[1+\left(\frac{\lambda_e}{\lambda_s}\right)^2\right] - u_I^f$$

And in the delamination zone, solutions are:

$$u_I^d = \tau_I^d = 0\,; \qquad u_A^d = \frac{\left(u_0 + u_I^f\right)l_d - u_I^c(l - l_d)\left[1+\left(\frac{\lambda_e}{\lambda_s}\right)^2\right]}{[l(1+\rho) - l_d\rho]}$$

$$u_B^d = \frac{\left(u_0 + u_I^f\right)(1+\rho) + u_I^c\left[1+\left(\frac{\lambda_e}{\lambda_s}\right)^2\right]}{[l(1+\rho) - l_d\rho]}.x - \frac{u_0 l_d\rho + l\left\{u_I^f(1+\rho) + u_I^c\left[1+\left(\frac{\lambda_e}{\lambda_s}\right)^2\right]\right\}}{[l(1+\rho) - l_d\rho]} \tag{6}$$

Equations for determining the lengths l_e, l_d and d of the three zones, where $d = l - l_d$, are:

$$\lambda_e u_I^c \coth(\lambda_e l_e) = \lambda_s \cot g[\lambda_s d]\,; \qquad \frac{\sin(\lambda_s d)}{\lambda_s[l(1+\rho) - l_d\rho]} = \left(\frac{\lambda_e}{\lambda_s}\right)^2 \frac{u_I^c}{\left(u_0 + u_I^f\right)(1+\rho) + u_I^c\left[1+\left(\frac{\lambda_e}{\lambda_s}\right)^2\right]} \tag{7}$$

For simplicity, in the abovementioned analytical solutions the following notations are taken into account:

$$\lambda_e^2 = \left(\frac{1}{2h_A E_A} + \frac{1}{2h_B E_B}\right)K_t\,; \quad \rho = \frac{2h_A E_A}{2h_B E_B}\,; \quad \lambda_s^2 = \left(\frac{1}{2h_A E_A} + \frac{1}{2h_B E_B}\right)K_s\,; \tag{8}$$

3 Illustrative numerical examples

The structure consists of two layers, A and B. The data for model solutions validation are taken from [11] for graphene monolayer/PET and they are listed in Table 1. The values of u_I^c, u_I^f correspond to the values of l_e, l_d, at which τ_I reaches the τ_I^c and 0, or u_I reaches the u_I^c, u_I^f, respectively.

Table 1. Material properties and geometry of considered structure.

E_A, TPa	E_B, GPa	K_t TPa/m	K_s, TPa/m	u_I^c m	u_I^f MPa	τ_I^c MPa	$2h_A$ nm	$2h_B$ m	$2l$ μm
1	3	74	1.596	6.75e-09	0.32e-06	0.5	0.35e-09	0.35e-04	21.8; 60

In Figure 2(a) the interface shear stress at three different values of u_0 is shown, for graphene length of 60 μm. The embedded figure here shows the behavior of interface shear stress near l_e. The additional axis at the top is related to the interface stress at the middle value of u_0. The obtained values for l_e, l_d are given in Table 2. The values of elastic zone length for three different u_0 are at first, too small compared to the full graphene length, and second, too close to each other. It is seen, that increasing the value of u_0, the length of cohesive zone decreases, so the interfacial debonding length d increases.

In Figure 2(b), the behavior of interfacial shear stress at constant load and for three different values of the graphene length $2l$ is shown. For better visibility here only essential part of graphene length is presented. The parametric analysis of the influence of the length of the structure on the length of cohesive zone l_d and delamination length d shows that increasing the length of the structure, the cohesive length increases too, and respectively, the delamination length d decreases.

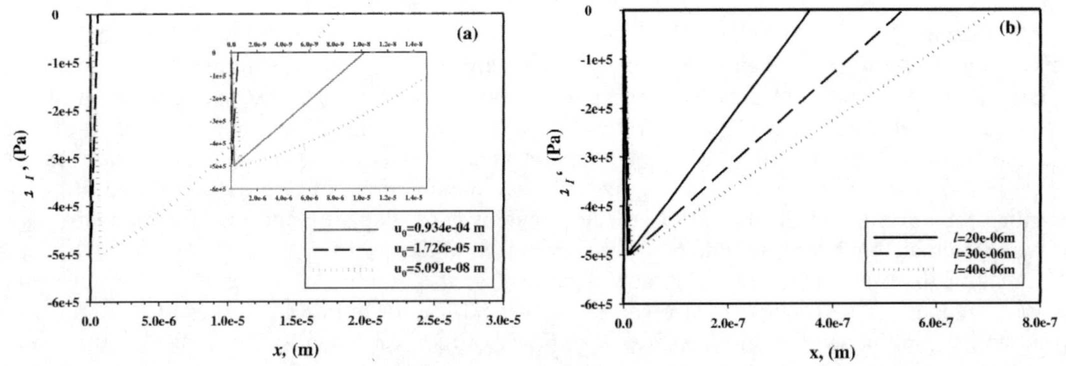

Figure 2. Interface shear stress along the graphene length at three different u_0 (a), and influence of graphene length on the interface shear stress at u_0 =1.726e-05m, (b)

Table 2. Values of l_e , l_d for graphene length 60 μm at different displacements

u_0, m	l_e, m	l_d, m
0.934E-04	2.166E-10	1.026E-08
1.73E-05	1.132E-08	5.358E-07
5.09E-08	7.757E-07	1.801E-05

In Figure 3(a), the behaviour of the displacements u_A,u_B at constant value of the external displacement is illustrated, for graphene length of 60 μm.

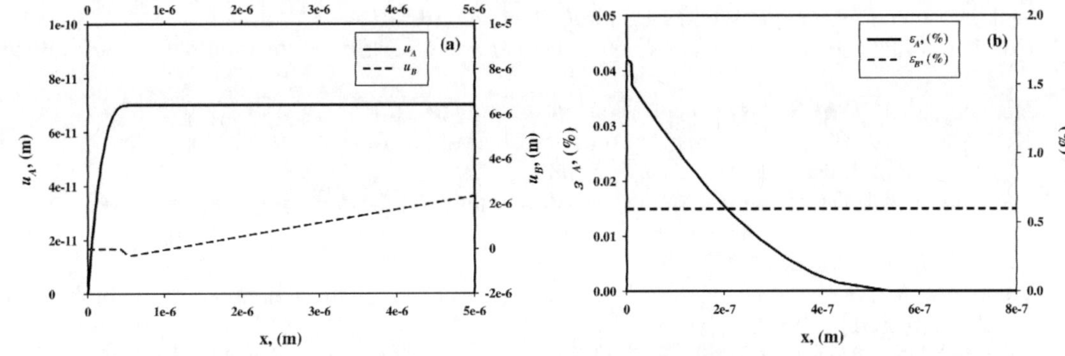

Figure 3. Displacements in graphene u_A and PET u_B along the structure length (a), and strains in graphene and PET along the structure length (b), both at u_0 =1.726e-05m and $2l$=60 μm

The results in Figure 3(b), are related to the strains in layers as functions of x (along the essential part of structure). It is seen, that ε_B practically doesn´t depend on x.

In addition, the influence of thickness of host layer is investigated. It is proved that the influence of this parameter is negligible. The total relative error between values of the interface shear stresses, calculated at constant u_0=1.726e-05m and $2l$=60 μm, for both $2h_B = 0.35e-04m$ and $2h_B = 0.35e-08m$ does not exceed 0.02%. When respective deformations in graphene layer are compared, the total relative error is less than 4%.

4 Conclusions

In this study modelling and parametric analysis of cohesive interface delamination of graphene monolayer nanocomposites are presented, using 1D shear lag method. The external excitation of the matrix is an axial displacement, controlling the interface delamination. The statement of the considered problem is given in displacements, the interface behaviour is assumed to satisfy a bi-linear cohesive law. In three defined zones along the interface the solutions of the system of two ordinary differential equations about displacements of the monolayer of graphene and one for the matrix are fitted by appropriate boundary, symmetry and continuity conditions.

All analytical results about the interface shear stress, the lengths of the elastic, cohesive and delaminated zones are calculated for a given structure and are illustrated in figures. The parametric analysis and the influences of the interface length, magnitude of given external displacement and substrate thickness are performed. It is shown, that increasing the values of u_0, the length of cohesive zone decreases, so the interfacial debonding length d increases. When increasing the length of the structure, the cohesive zone length increases too, and respectively, the delamination length d decreases. The influence of host layer thickness on the zone lengths is negligible.

Acknowledgements: This study is carried out by the financial support of DFG project "Shear lag analysis by a cohesive bilinear model for the interface stress transfer in a graphene monolayer nanocomposite", BE 1090/46-1, 2018.

References

[1] Mroz K and Mroz Z 2015 Analysis of delamination and damage growth in joined bi-linear systems *Geomechanics for energy and environment* **4** 4
[2] Suresh S, Sugimara Y, Ogawa T 1993 Fatigue crack in coated materials with brittle surface coatings *Scr. Metall. Mater* **29** 237
[3] Volkerson O 1938 Die Nietkraftverteilung von zugbeanspruchten Nietverbindungen mit konstanten Laschenquerschnitten *Luftfahrtforschung* **15** 41
[4] Cox L H 1952 The elasticity and strength of paper and other fibrous materials *Br. J. Appl. Phys* **3** 72
[5] Hedgepeth J M 1961 Stress concentrations in filamentary structures NASA TN **D882** 1
[6] Kelly A, Tyson W R 1965 Tensile properties of fibre reinforced metals: Copper/tungsten and copper/ molybdenum *J. Mech. Phys. Solids* **13** 329
[7] Dugdale D S 1960 Yielding of steel sheets containing slits *J. Mech. Phys. Solids* **8** 100
[8] Barenblatt G I 1962 The mathematical theory of equilibrium cracks in brittle fracture *Advances in Applied Mechanics* **7** 55
[9] Ivanova J, Mroz Z, Valeva V 2006 Mechanical modeling of the delamination of bi-material plate structure, *J. Theor. Appl. Mechanics* **36** 4
[10] Ivanova J, Nikolova G, Gambin B 2011 Interface delamination of bi-material structure under dynamic time harmonic loading *ZAMM* **91** 2
[11] Guo G and Zhu Y 2015 Cohesive shear lag modeling of interfacial stress transfer between a monolayer graphene and polymer substrate *Journal of Applied Mechanics* **82** 031005-1

IOP Conf. Series: Materials Science and Engineering **461** (2019) 012030 doi:10.1088/1757-899X/461/1/012030

Effect of deformation rate on low carbon steels mechanical properties

I Jandrlić, S Rešković, T Brlić and V Furlan

University of Zagreb Faculty of Metallurgy, Aleja narodnih heroja 3 Sisak, Croatia

E-mail: ijandrli@simet.hr

Abstract. Study shows the influence of stretching rate on the tensile properties for two low carbon steels during cold plastic deformation. Studies were carried out at laboratory conditions at room temperature with a static tensile test, at stretching rates ranging from 5 to 30 mm / min. Influence of the stretching rate on the elongation (A), tensile strength (R_m), yield stress (R_e) and the ratio R_e/R_m were analysed by MATLAB software. For both steels, yield stress and tensile strength increase with the increase in stretching rate, while the elongation decreases for both steels. The ratio R_e/R_m for niobium microalloyed steel remains constant. It can be concluded that both steels are strain rate sensitive, and that only R_e/R_m ratio cannot be used to determine the strain rate sensitivity of the steel.

1 Introduction

Depending on the need to achieve a better surface quality, narrow dimensional tolerance and/or increasing the strength, subsequently after the hot deformation the cold plastic processing of steel is carried out [1-2].

The final mechanical properties of steel products depend on the parameters during the plastic processing of steel. Considering there are no recrystallization mechanisms during cold plastic deformation, the strengthening is mainly due to the increase in dislocation density [3]. Strain rate, or rate at which the deformation is carried out during forming the steel products, is one of the most important parameters during plastic deformation [2].

Therefore, it is essential to know whether the steel to be shaped is sensitive to strain rate or not. It has been shown that low-carbon steels have a high sensitivity to strain rate. Depending on the chosen strain rate and the degree of reduction, the final mechanical properties of the product can be influenced [4]. Studies show that increasing the strain rate in the case of low-carbon steels increases the yield stress and tensile strength, while elongation is reduced [5-7].

The strain rate sensitivity of the steels depends primarily on the homogeneity of the steel and the presence of obstacles for dislocations movement [2,6]. Accordingly, the steels with a lower content of the alloying elements and the larger grain size show a greater sensitivity to the increase in the strain rate. Studies on microalloyed steels show that these steels exhibit less or no sensitivity to changes in strain rate. They claim that if there is less obstacles to the dislocation movement the strain rate sensitivity is higher [2].

As the decrease in plasticity is possible due to the inhomogeneities of steel, most research shows the sensitivity of the steel to increase of strain rate through the R_e/R_m ratio, which is completely dependent only on the material itself [5]. It is shown that increase of strain rate for low-carbon steels increases the R_e/R_m ratio [6,7]. Some studies show that this increase is more pronounced for steel with R_e <300 MPa and R_e/R_m > 0.82, in which case it is recorded a significant drop in the elongations [6]. It is well known that with microalloying, due to the presence of precipitates as obstacles to dislocation movement, the significantly higher values in the yield stress and tensile strength are achieved [8-10]. Due to the assertion that strain rate sensitivity is largely dependent on steel homogeneity, more specifically on the obstacles to dislocation motion [2], it is expected that niobium microalloyed steel is less susceptible to changes in strain rate.

This study aims to examine the strain rate sensitivity of tensile properties of two low carbon steels, one with and another without the presence of niobium precipitate. The influence of stretching rate (i.e. strain rate) on the basic mechanical properties of the studied steels will be analyzed using the MATLAB software package.

2 Experimental material and methods

The influences of the increase in stretching rate on the tensile properties were carried out on two steels with the same base chemical composition, one of which is additionally microalloyed with niobium.

The chemical compositions of the tested steels are given in table 1.

Table 1. Chemical composition of the tested steels.

Steel / Element (mas %)	C	Mn	Si	P	S	Al	Nb	N
Niobium microalloyed steel	0.12	0.78	0.18	0.011	0.018	0.02	0.048	0.008
Low carbon steel	0.13	0.77	0.18	0.010	0.019	0.02	-	-

The tests were carried out by static tensile test at room temperature, on the mechanical universal tensile testing machine Zwick 50kN. Tests were carried out at 4 different stretching rates ranging from 5 to 30 mm/min. Used stretching rates correspond to the strain rates in the range from $1.8 \cdot 10^{-3}$ s^{-1} to $1.1 \cdot 10^{-2}$ s^{-1}. Samples of rectangular cross-section were taken in the direction of rolling from the 3 mm thick hot-rolled steel strips. Samples having original gage length of 45 mm, width 20 mm and thickness of 3 mm were prepared on CNC milling and turning machine. From previous research on the same steels [7], it is known that they have a fine grained ferrite-pearlite microstructure, figure 1.

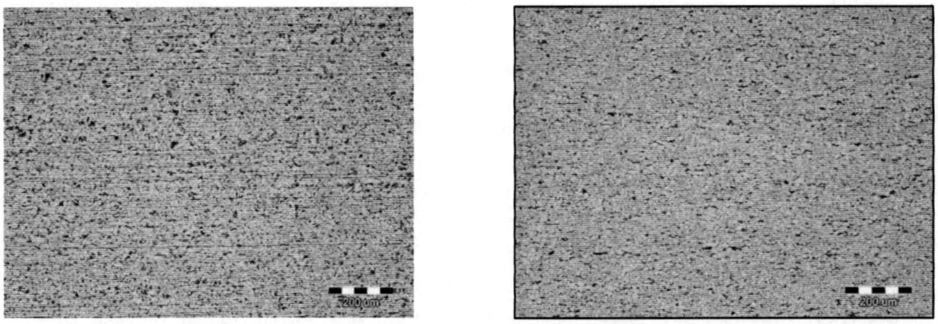

Low carbon steel Niobium microalloyed steel

Figure 1. Microstructure of the initial materials for low-carbon and niobium microalloyed steel

Tensile testing machine records force-elongation diagrams from which stresses and strains can be obtained. DIC method was used for determination of onset of plastic deformation during tensile testing. Using DIC the point of yielding was determined, and from force – elongation diagrams stress was obtained.

The analysis of the results was performed using the MATLAB software package, and the dependence of the mechanical properties on stretching rate was determined.

3 Results and Discussion

From the recorded force-extension graphs and from measured extension, values of yield stress, tensile strength and elongation were calculated, table 2.

Table 2. Average values of measure R_e, R_m an A at different stretching rates.

Steel / Element (mas %)	Stretching rate mm/min	R_e (MPa)	R_m (MPa)	A (%)
Niobium microalloyed steel	5	510	582	26.33
	10	537	589	26.31
	15	531	593	25.35
	30	540	615	24.30
Low carbon steel	5	394	538	31.18
	10	395	541	28.66
	15	413	546	27.00
	30	434	656	26.33

The stretching rate dependence of the yield stress is shown for the tested steels by figure 2.

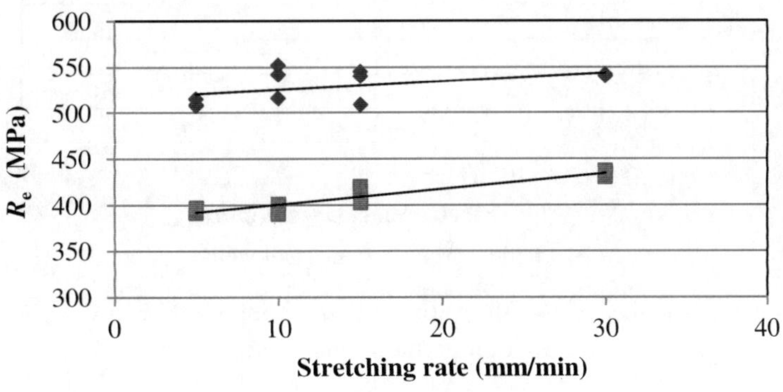

Figure 2. Dependence of the yield stress on stretching rate for low-carbon steel and niobium microalloyed steel

Previous investigations have shown the presence of niobium precipitates in the examined microalloyed steel [7,10]. By comparing the values of the measured yield stress, the higher amounts are observed in the case of microalloyed steel. This was expected as it is precipitation hardened steel. It can be noticed that in both steel with the increase of the stretching rate there is increase in the measured values of the yield stress. In the case of microalloy steel, this increase is somewhat lover. Similar observations are also measured on the values of tensile strength, figure 3.

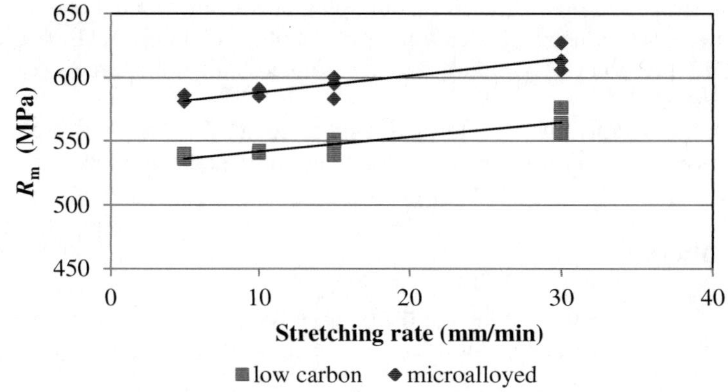

Figure 3. Dependence of the tensile strength on stretching rate for low-carbon steel and niobium microalloyed steel

Higher values of tensile strengths are also measured in the case of microalloyed steel. This was expected due to the presence of niobium precipitates. By increasing the stretching rate, the increase in tensile strength for the both tested steels was measured. It is clearly seen that increase of strength is of the same intensity for both steels. On the other hand, the elongation for both steels shows a drop by increasing the stretching rate, figure 4.

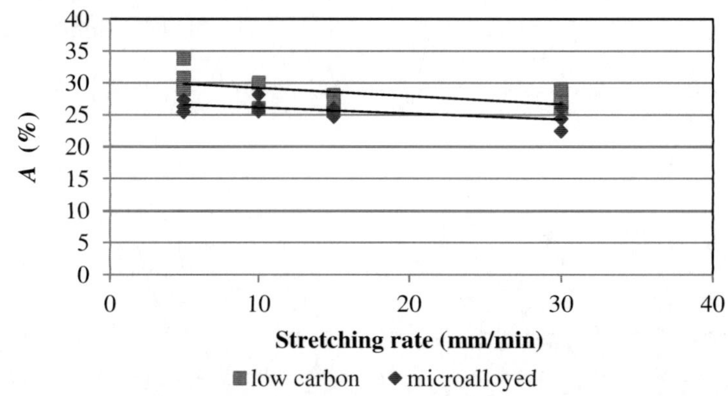

Figure 4. Dependence of the elongation on stretching rate for low-carbon steel and niobium microalloyed steel

From the obtained dependences, it can be concluded that both steel have a similar dependence on the change of stretching rate. Studies on other microalloyed steels have shown the opposite, i.e. that those steels are less sensitive to strain rate [2]. This is explained by the fact that due to the existence of obstacles to the dislocation movement, the steels become less susceptible to changes in strain rate [2]. On the other hand, some inhomogeneity in the material may also affect the drop in deformability as indicated in [5]. Most of the research on hardening due to change of strain rate is expressed by the R_e/R_m ratio [2,5-7]. The resulting dependences of the measure R_e/R_m ratio with the increase in the stretching rate from our research are given in figure 5.

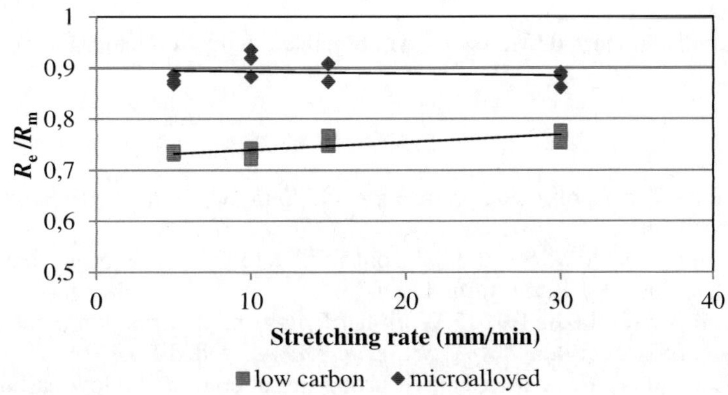

Figure 5. Dependence of the R_e/R_m ratio on stretching rate for low-carbon steel and niobium microalloyed steel

It is noted that the R_e/R_m ratio in the case of ordinary low-carbon steel shows a significant increase with increase of stretching rate. From this it can be concluded that the increase in the strain rate results strong strengthening of this steel. The yield stress increases more significantly and at higher strain rates there is a need for more significantly forces to be applied for the onset of plastic deformation. In the case of microalloyed steel, it can be seen that the R_e/R_m ratio remains constant. According to earlier claims [2,5] it could be concluded that the tested microalloyed steel is insensitive to changes in strain rate. On the other hand, measured changes in the amount of all mechanical properties show strong dependency on the stretching rate.

Previous research on this niobium microalloyed steel has shown that there are no inhomogeneities that could cause a decline in plasticity [7,10]. It can be concluded that in this case the R_e/R_m ratio is not a good indicator of the steels strain rate sensitivity. It is very important to consider other values of mechanical properties as well. The constant R_e/R_m ratio is explained by the fact that with the increase of the stretching rate there was an equal increase in values of yield stress and the tensile strength, and the ratio remained unchanged.

4 Conclusion

Both tested steel shows a clear increase in the values of yield stress and tensile strength with the increase in the stretching rate.

On the other hand measured drop in the elongation is a clear proof of strengthening during cold plastic deformation. In the case of low-carbon steel without the presence of a precipitate, increase in strain rate promotes elastic deformation. This is indicated by decrease of the R_e/R_m ratio. This should be taken into account in the subsequent cold processing of this steel.

In the case of microalloyed steel, where the presence of small precipitates of niobium has been shown by earlier studies, the dependence on the stretching rate is equally expressed. It is shown slightly smaller increase in the amount of yield stress. On the other hand, the tensile strength increases and the elongation decrease with the increase of the stretching rate. In this case the R_e/R_m ratio cannot be taken as a true indicator of strain rate sensitivity, since in the case of microalloyed steel it has a constant amount for all used stretching rates. In case of microalloyed steel other tensile values have an expected trend with the increase in the stretching rate. This is characterized by an equal amount of increase in the yield stress and tensile strength.

When considering strain rate sensitivity we believe that all the values of the mechanical properties must be taken into account.

Acknowledgements
This work has been fully supported by the Croatian Science Foundation under the project number IP-2016-06-1270.

References

[1] Čaušević M 1979 Teorija plastične prerade metala (Sarajevo, Sarajevo Svjetlost OOUR) chapter 1 pp 7–40
[2] Buršák M, Mamuzić I and Michel J 2008 Contribution to evaluation of mechanical properties during impact loading, *Metalurgija* **47** 19-23
[3] Su-Fen W, Yan P and Zhi-Jie L 2013 Work-hardening and deformation mechanism of cold rolled low carbon steel *Res. J. Apll. Sci. Eng. Technol.* **5** 823-8
[4] Çetinarslan C S and Güzey A 2013 Tensile properties of cold-drawn low-carbon steel wires under different process parameters *Mater. Technol.* **47** 245–52
[5] Mihalikova M 2007 Analysis of the influence of the loading rate on the mechanical properties of microalloyed steel *Journal of Metals, Materials and Minerals* **17** 25–28
[6] Buršák M and Michel J 2010 Influence of the strain rate on the mechanical and technological properties of steel sheets *Metalurgija* **49** 317–320
[7] Jandrlić I 2015 The stress distribution in the deformation zone of niobium microalloyed steel doctoral thesis (Sisak: University of Zagreb Faculty of metallurgy)
[8] Rešković S and Prelošćan A 1990 Međuutjecaji mehanizama ojačanja, omekšanja i izlučivanja u završnoj termomehaničkoj obradi čelika mikrolegiranog niobijem *Strojarstvo* **32** 137–142
[9] Cuddy L J 1984 Grain refinement of Nb steel by control of recrystallization during hot rolling Metall. Mater. Trans. A **15** 87–98
[10] Rešković S and Jandrlić I 2013 Influence of niobium on the beginning of the plastic flow of material during cold deformation, *The Scien. Wor. J.* **2013** (Article ID 723725)

IOP Conf. Series: Materials Science and Engineering **461** (2019) 012031 doi:10.1088/1757-899X/461/1/012031

Instrumentation for SCC testing in low pressure superheated hydrogen steam environments

J Janoušek[1, 2], F Scenini[3], L Volpe[3], A Hojná[1] and T Trojan [4]

[1]Research Centre Rez, Hlavní 130, 250 68 Husinec-Řež, Czech Republic

[2]University of West Bohemia, Faculty of Mechanical Engineering, Department of Material Science and Technology, Univerzitní 22, 306 14 Plzeň, Czech Republic

[3]The University of Manchester, Material Performance Centre, Manchester, UK

[4]ŠKODA JS a.s., Orlík 266/15, Bolevec, 316 00 Plzeň, Czech Republic

E-mail: jas@cvrez.cz

Abstract. This contribution is focused on the issue of Environmentally Assisted Cracking (EAC). For EAC testing in low pressure superheated hydrogen steam environments a special corrosion apparatus is being built in cooperation with University of Manchester and Škoda JS. The whole concept is being prepared for the MEACTOS project. One section concerns the problems with determining an environment with a mixture of hydrogen, argon and steam. The presented equations are necessary for controlling the environment and for the settings of the area of test interest. The test instrumentation is described.

1 Introduction

Environmentally assisted cracking (EAC) is the dominant issue in determining the reliability of most commercial equipment and applications which are controlled by the interactions between structural materials, cooling environments and operating stresses [1, 2]. The goal of the MEACTOS project (Mitigating Environmentally-Assisted Cracking Through Optimisation of Surface Conditions) [3] is to improve the safety and reliability of Generation II and III nuclear power plants (NPPs) by increasing the resistance of critical locations, including welds, to EAC through the application of optimised surface machining and improved surface treatments. This project will quantify the effects of various surface machining and treatment techniques on the EAC behaviour of specific structural materials in the primary circuit of NPPs. The gained knowledge will be summarized in practical guidelines, which can be incorporated into key nuclear design and manufacturing codes.

Pressurized Water Reactor (PWR) nuclear power plants operate with primary cooling water in the proximity of the Ni/NiO electrochemical transition. This is also where Ni based alloys are most susceptible to EAC. Hydrogen is often added to high temperature water to maintain low levels of dissolved oxygen, thereby minimizing corrosion of the structural metals. Primary water stress corrosion cracking (SCC) of Alloy 600 and similar materials is a form of EAC that can occur in essentially pure hydrogenated water at high temperatures [4]. Hydrogen gas produces rapid growth of EAC in high strength steels, and this EAC initiates on surfaces which are absolutely smooth and

requires no previous defects such as pits, intergranular penetrations or mechanical defects [2]. These facts and a lack of clarity has led to careful, detailed investigation.
The area of interest is focused on significant acceleration of the EAC process in a low pressure H_2-steam environment. In fact, Economy et al. [5] showed a monotonic dependence of crack growth rate (CGR) and EAC initiation time in both water and steam, suggesting that the EAC initiation mechanism is similar for both environments. A special corrosion apparatus built in cooperation with University of Manchester will be used within the MEACTOS project. The final implementation is performed by Škoda JS.

2 Definition of the Environment

It is necessary to perform accelerated testing to determine the PWR structural material resistance to EAC initiation. That is why this device, where tests can be carried out in a hydrogen atmosphere simulating the primary circuit of PWR at temperatures up to 480 °C, was built. Material specimens will be especially tested at 480°C in a steam-hydrogen mixture where the oxygen partial pressure (p_{O_2}) will be controlled. This control will be obtained by manipulating the steam to hydrogen ratio (R_{steam/H_2}).
Equilibrium oxygen partial pressure is achieved by the following equation

$$p_{O_2\ Ni/NiO} = \left[exp\left(\frac{\Delta G^0_{NiO}}{RT}\right)\right]^2 \tag{1}$$

where R = 8.314 J.mol^{-1}. K^{-1} is the gas constant (also known as the molar, universal, or ideal gas constant), T corresponds with testing temperature, and ΔG^0_{NiO} is the Gibbs free energy which can be calculated as

$$\Delta G^0_{NiO} = A + BT\ log\ T + CT \tag{2}$$

where constants A, B and C are shown in Table 1 for NiO.

Table 1. Constants for computation of the Gibbs free energy for NiO and H_2O.

	NiO	$H_2$0
A	-232 450	-239 648
B	0	18.75
C	83.435	-9.25

The area of interest is the ratio between the equilibrium oxygen partial pressure and the oxygen partial pressure in the system in a range of 0.078 up to 50. The partial pressure of oxygen in the system is calculated similarly to equation (1) except for the following details

$$p_{O_2} = R_{steam/H_2}\, e^{\left(\frac{\Delta G^0_{H_2O}}{RT}\right)} \tag{3}$$

where the hydrogen ratio (R_{steam/H_2}) can be calculated as

$$R_{steam/H_2} = \frac{F_{H_2O} \cdot \rho_{H_2O} \cdot T_{room}}{F_{H_2} \cdot \rho_{steam} \cdot T_{test}}. \tag{4}$$

In equation (4) there are variables which depend on the system and which can be selected. The flow of water F_{H_2O} is provided by a pump in millimetres per minute. The flow of hydrogen F_{H_2} depends on the mixture (for example if the mixture is 4% H_2 + 96% Ar, the flow of hydrogen will be 0.04*total flow F_{H_2+Ar}). The density of steam can be determined by the following relationship

$$\rho_{steam} = AT^0 + BT^1 + CT^2 + DT^3 + ET^4 + FT^5 + GT^6 \quad (5)$$

where constants *A*, *B*, *C*, *D*, *E*, *F*, *G*, are presented in Table 2.

Table 2. Constants for density of steam determination.

A	*B*	*C*	*D*	*E*	*F*	*G*
8.26774 E-04	-2.83640 E-06	7.75890 E-09	-1.48420 E-11	1.79870 E-14	-1.22340 E-17	3.50000 E-21

The Gibbs free energy for H_2O is established in the same way as in equation (2) but different numbers in the H_2O column in Table 1 have to be used as constants.

3 Configuration of the Experimental Device

The environment defined above requires building a new corrosion cell. This technological system consists of a test chamber, a fluid filling system, an organized leaks system and a system invoking the loading of the tested samples.

The new cell for low the pressure superheated hydrogen steam environment is currently installed on an electromechanical creep testing machine Kappa SS-CF (see Figure 1) with a load capacity of up to 100 kN and with a speed range from 1 µm/h to 100 mm/min. The creep machine consists of four columns to which the test chamber is attached. The upper crosshead is fixed and the bottom one is sliding. The creep machine invokes the necessary load for carrying out the experiments by moving the bottom crosshead.

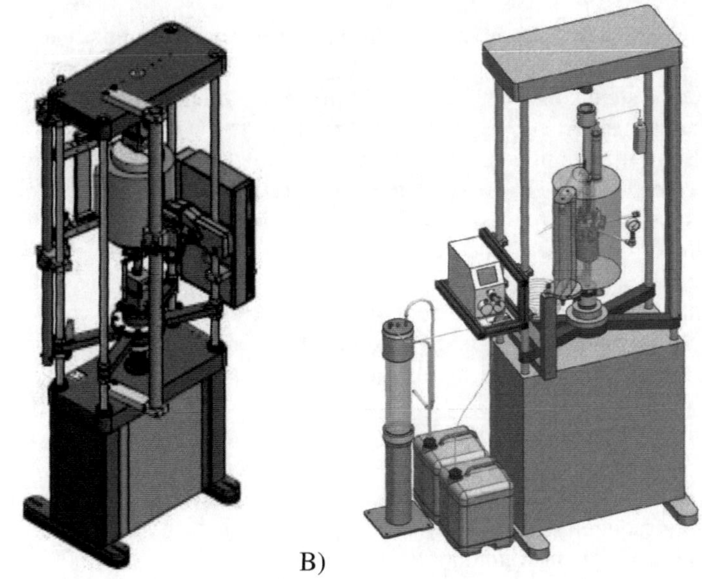

A) B)

Figure 1. A) Electromechanical creep testing machine Kappa SS-CF and B) its modification with corrosion cell for low pressure superheated hydrogen steam environment.

The experiments are performed in the test chamber. The test chamber is shown in Figure 2. It can be seen from the figure that the test chamber consists of two main parts – a cover and a vessel. The test chamber vessel primarily serves as a pressure envelope. The test chamber cover is tightened to the vessel using strength bolts and packed with a torus seal. The tested sample is placed in the test chamber cover with prismatic reductions. A spring-bellows provides the sealing of the test chamber during linear movement. The prismatic attachment of the sample reduction allows testing of different types of samples - cylindrical samples for micro-tensile test such as in [6] or flat samples. The strain gauge, thermocouples and the filling mixture system are located in the test chamber cover.

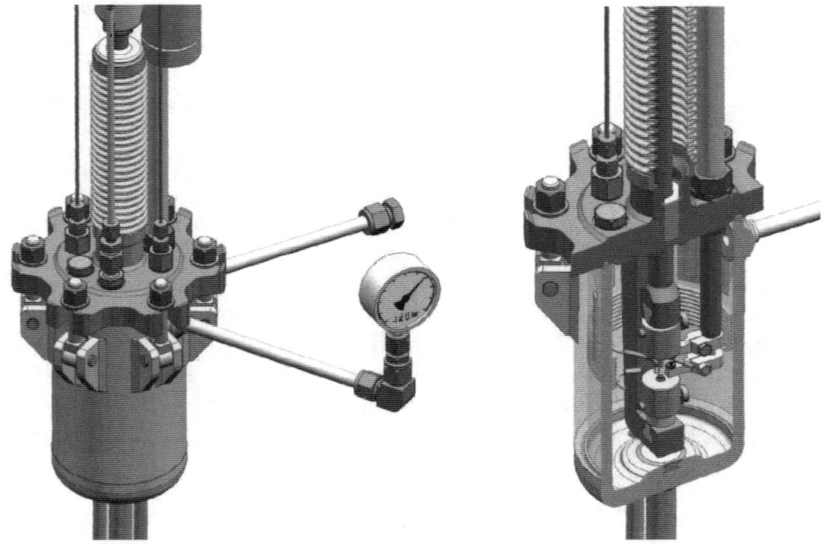

Figure 2. External and internal assembly of test chamber.

The filling mixture system consists of a storage tank, a dosing pump, a steam generator, $Ar+H_2$ gas dosing system and a blender. A special chromatographic pump sucks demi water from the storage tank and doses it into the steam generator. Water evaporates in the steam generator and it passes to the blender as steam. The $Ar+H_2$ gas is heated in a parallel pipe and proceeds to the blender where it is mixed with steam. The mixture is mixed in the blender and heated to 200°C at which it enters into the test chamber. Using a coiled heat exchanger the mixture can be heated up to 480°C and aimed at the tested sample.

Since it is a continuous fluid dosing system, the steam with the $Ar+H_2$ gas will further be drained from the chamber through the drainage system. The drainage system consists of a cooler, a condenser, an air-leak chamber and a waste demi water storage tank. The cooler and condenser form a piping system where the gaseous mixture is cooled down and as the temperature drops below boiling point it liquidizes and proceeds to the air-leak chamber. The air-leak chamber has a free level at a height of approximately 1 metre by which it keeps hydrostatically internal overpressure of approximately 0.1 bar. Once bubbled through the $Ar+H_2$ gas, the mixture separates from the water. Subsequently, the $Ar+H_2$ gas flows to the air condition exhaust and water through the overflow to the waste demi water storage tank.

The entire system is controlled autonomously using a control system controlled by thermocouples and it also controls the equipment using feedback. The system also controls the $Ar+H_2$ gas and demi water dosing. All the analysed data will be completely backed-up.

4 Conclusions

A low pressure superheated H_2-steam system is used to accelerate the oxidation kinetics while keeping the conditions similar to PWR primary water. One of the most important requirements for this environment is that it needs to replicate the Ni/NiO transition. Nevertheless, despite several studies being carried out by different research groups in H_2-steam environments, there is still a considerable level of uncertainty over the thermodynamics of the oxidation process. The new corrosion cell is being built for this purpose. Two materials will be investigated in the MEACTOS project: Alloy 182 weld metal and type 316L stainless steel. It is planned to test the EAC with two types of specimen – tapered and single-edge notch tension (SENT) with U-notch [7]. Each tapered specimen will have two flat surfaces treated with different final treatments. One flat surface will be a reference and the other one will be treated with one of the investigated surface techniques. A Single Edge Notch Tensile specimen will be used for active constant load testing to study EAC crack initiation on a limited area of the surface - only on the notch surface. Testing will be performed at temperatures of 360, 400 and 440°C.

Acknowledgements

This work was financially supported by the Ministry of Education, Youth and Sport Czech Republic - project LQ1603 (Research for SUSEN), project LM2015093 (SUSEN) and within the MEACTOS H2020 project (GA 755151).

References

[1] Hojná A, Burda J and Ernestová M 2007 CITLIVOST AUSTENITICKÉ OCELI 08CH18N10T K PRASKÁNÍ VLIVEM PROSTREDÍ PRIMÁRNÍHO OKRUHU JE, *KOROZE*

[2] Raja V S and Shoji T 2011 Stress corrosion cracking, Theory and practice, Woodhead Publishing Limited

[3] https://meactos.eu/

[4] Scenini F 2006 THE EFFECT OF SURFACE PREPARATION ON THE OXIDATION AND SCC BEHAVIOUR OF ALLOY 600 AND 690 IN HYDROGENATED STEAM, PhD thesis, The University of Manchester

[5] Economy G, Jacko R J, Pement F W 1987 IGSCC Behavior of Alloy 600 Steam-Generator Tubing in Water or Steam Tests above 360 C, *Corrosion*, **vol. 43**, no. 12, pp. 727–73

[6] Jirková H, Rubešová K, Konopík P and Opatová K 2018 Effect of the parameters of semi-solid processing on the elimination of sharp-edged primary chromium carbides from tool steel, *Metals*, **Vol. 8**, Issue 9

[7] BS 8571:2014, Method of test for determination of fracture toughness in metallic materials using single edge notched tension (SENT) specimens, BSI Standards Publication

IOP Conf. Series: Materials Science and Engineering **461** (2019) 012032 doi:10.1088/1757-899X/461/1/012032

Analysis of the initial cavitation erosion period of selected nickel alloys

R Jasionowski[1] and W Depczyński[2] and D Zasada[3]

[1] Faculty of Marine Engineering, Institute of Basic Technical Sciences, Maritime University of Szczecin, 2-4 Willowa Str, 71-650 Szczecin, Poland, EU
[2] Faculty of Mechatronics and Mechanical Engineering, Kielce University of Technology, 7 Tysiaclecia Panstwa Polskiego Ave., 25-314 Kielce, Poland, EU
[3] Faculty of Advanced Technology and Chemistry, Military University of Technology, 00-908 Warsaw, 2 Gen. W. Urbanowicza Str., Poland, EU

r.jasionowski@am.szczecin.pl, wdep@tu.kielce.pl, dariusz.zasada@wat.edu.pl

Abstract. Cavitation wear resistance, of four different nickel-based alloy during the incubation period was investigated with the use of stream-impact device and results are presented in this paper. Four nickel alloys with excellent resistance corrosion very high corrosion resistance and very good mechanical properties were selected for the tests: Monel 400, Inconel 600, Incoloy 800H and Hasteloy C22. Nickel alloys examined for the period of 600 minutes. Mechanism of formation and destruction of the surface layer under working liquid influence is described. It was noticed, that first microcracks are formed on the grain boundaries, especially in the triple contact points. Effects of plastic strain were also noted on the surface of samples as a uplifting and collapsing in boundary area. It was stated, that among nickel alloys examined, a better resistance to cavitational erosion in initial period of cavitation damage has the alloys: Hasteloy C22 and Incoloy 800H.

1 Introduction

Nickel-based alloys are construction materials characterized by very high strength and high corrosion resistance [1-7]. Due to their properties, nickel-based alloys are used in the construction of aircraft gas turbines, steam turbine power plants, nuclear power systems, chemical and petrochemical industries. Nickel-based alloys are known by their trademark names such as: Monel®, Hastelloy®, Inconel®, Incoloy® and Nimonic®.

Monel® alloys are a nickel-copper alloy with high strength and excellent corrosion resistance in a range of media including sea water, hydrofluoric acid, sulfuric acid, and alkalis. It is used by marine engineering, chemical and hydrocarbon processing equipment, valves, pumps, shafts, fittings, fasteners, and heat exchangers. There are three Monel® alloys are: Monel 400, Monel K500 and Monel R-405

Hastelloy® - the name of a group of corrosion-resistant nickel alloys, namely the Ni-Mo and the Ni-Cr-Mo alloys. Depending on the desired combination of properties, Hastelloy® alloys contain varying amounts of Mo (up to 30%), Cr (up to 23%), Fe (up to 29%), Hastelloy® alloys are characterized by high resistance to hydrochloric, sulfuric, phosphoric, acetic, and formic acids, to media containing ions of chlorine and fluorine, and to many organic media. The most popular among Hastelloy® alloys are: Hastelloy B2/B3, Hastelloy C22 and Hastelloy C-276.

Inconel® alloys are a group of austenitic nickel-chromium-based superalloys. Inconel® alloys characterized by excellent mechanical properties at both extremely low and extremely high temperatures, outstanding resistance to pitting, crevice corrosion and intercrystalline corrosion high resistance to oxidation at elevated temperatures and good resistance to acids, such as sulfuric, phosphoric, nitric, and hydrochloric. Inconel 600, Inconel 625, Inconel 690, Inconel 718 and Inconel 751 are alloys used in gas turbine blades, seals, and combustors, as well as turbocharger rotors and seals, high temperature fasteners, chemical processing and pressure vessels, heat exchanger tubing, steam generators and core components in nuclear pressurized water reactors.

Incoloy® and Nimonic® is registered trademark name of Special Metals Corporation. Incoloy® alloys are a nickel-chromium-iron alloys, which used in chemical and petrochemical processing, nuclear fuel reprocessing, acid production, in power plants for super-heater and reheater tubing, oil and gas well piping, in industrial furnaces and for heat-treating equipment. The most popular among Incoloy® alloys are: Incoloy 800H/800HT/800AT, Incoloy 825, Incoloy 925 and Incoloy 945.

Nimonic® alloys typically are a nickel-chromium or nickel-chromium-cobalt alloys. Nimonic® alloys consist of more than 50% nickel chromium (up to 30%) and cobalt (up to 32%) with additives such as titanium (up to 6%) and aluminum (up to 6%) and molybdenum (up to 5%). Nimonic® alloys use is in gas turbine components such as turbine blades and exhaust nozzles on jet engines and extremely high performance reciprocating internal combustion engines, where the pressure and heat are extreme. Nimonic® alloys characterized by very good corrosion and oxidation resistance at high temperatures and easy machined. In Nimonic® alloys are: Nimonic 75, Nimonic 80, Nimonic 86, Nimonic 90, Nimonic 105, Nimonic 115.

Machine parts made of nickel alloys very often operate under extreme conditions and are exposed to corrosive and erosion phenomena caused by aggressive environments. Nickel alloys used in marine engineering, chemical and petrochemical processing, pulp and paper industry, food processing, nuclear engineering are exposed to cavitational erosion caused by cavitation.

Cavitation is the process of formation, growth and implosion of bubbles containing steam, gas or steam-gas mixture due to cyclic pressure changes in the flowing liquid. The cavitation implosion is an effect of pressure change from the area of its low value to a region of elevated pressure, causing condensation of steam filing the cavitation bubble. Implosion phenomenon occurs at very high velocity (exceeding 100 m/s), and in such case time of growth and decay of cavitation bubble is
in milliseconds. The dynamics of the formation and sealing of bubble is dependent on the physicochemical properties of liquid and the distance between the wall and the interaction of bubbles. In the vicinity of the wall, a microstructure implosion bubble forms, which can reach speeds of 300 to 500 m/s. Microstreams formed during the implosion of cavitational bubbles transmit on the wall pressure impulses in the order of 1÷4 GPa. Multiple repetitive cavitation implosion causes material destruction [8-13].

The aim of the paper to analyze of the initial cavitation erosion period of selected nickel alloys.

2 Investigated materials

The detailed examinations of resistance to cavitational wear was carried out on four deliberately selected nickel-based alloys:

- Monel 400 – alloy is a solid solution alloy that can only be hardened by cold working. This nickel alloy exhibits characteristics like good corrosion resistance, good weldability and high strength. A low corrosion rate in rapidly flowing brackish or seawater combined with excellent resistance to stress-corrosion cracking in most freshwaters, and its resistance to a variety of corrosive conditions led to its wide use in marine applications.
- Inconel 600 - alloy used for applications that require corrosion and high temperature resistance. It is non-magnetic, has excellent mechanical properties, and presents the desirable combination

of high strength and good weldability under a wide range of temperatures. The high nickel content in Inconel 600 provides excellent resistance corrosion.

- Incoloy 800H - alloy with good strength and excellent resistance to oxidation and carburization in high-temperature exposure. Incoloy 800H is intend for high temperature structural applications. The nickel content makes the alloys highly resistant to both chloride stress-corrosion cracking and to embrittlement from precipitation of sigma phase.
- Hasteloy C22 - alloy with enhanced resistance to pitting, crevice corrosion and stress corrosion cracking. The high chromium content provides good resistance to oxidizing media while the molybdenum and tungsten content give good resistance to reducing media. This nickel alloy also has excellent resistance to oxidizing aqueous media including wet chlorine and mixtures containing nitric acid or oxidizing acids with chlorine ions.

Chemical composition and mechanical properties of tested nickel alloys are shown in Table 1.

Table 1. Chemical composition and mechanical properties of tested materials

Chemical element [% Wt.]	Nickel alloys			
	Monel 400	**Inconel 600**	**Incoloy 800H**	**Hasteloy C22**
Ni	bal.	bal.	30.67	bal.
Fe	2.07	8.95	bal.	4.09
Cu	29.94	-	0.32	-
Si	0.25	0.28	0.46	0.048
Mn	1.18	0.20	0.87	0.31
C	0.14	0.06	0.071	0.009
Cr	-	18.96	19.57	21.48
Co	-	0.021	0.10	0.03
Mo	-	-	0.26	13.78
Ti	-	0.020	0.60	-
Al	-	0.22	0.42	-
V	-	-	-	0.018
W	-	-	-	3.20
Mechanical properties	**Monel 400**	**Inconel 600**	**Incoloy 800H**	**Hasteloy C22**
Rm [MPa]	616	713	554	780
Re [MPa]	335	361	219	426
A [%]	41	44	44	50

3 Methods of investigation

The examination of cavitational erosion was carried out on stream-impact device [14]. This device consists of rotating arm, equipped with a sample holder. The samples rotate with high speed and are subjected to the impact of a liquid, flowing through a nozzle. The impact of flowing liquid simulates cavitational impulses. The intensity of destruction primarily depends on rotational speed of the samples, the distance between nozzle and samples, the flow rate and the physio-chemical properties of the liquid. Examinated samples had cylindrical shape with 20 mm diameter and 6 ±0.5 mm height. Surface roughness of samples before the experiment, measured by PGM-1C profilometer, was in range

of 0.010÷0.015 μm. The samples were vertically mounted in rotor arms, parallel to the axis of water stream, pumped continuously at 0.06 MPa through a 10 mm diameter nozzle located 1.6 mm from the edge of the sample. The rotating samples were hitting by the water stream. The samples were tested for the period of 30 minutes, took out from the fixtures, de- greased in an ultrasonic washer for 10 minutes at 30°C, dried in a laboratory drier for 15 minutes at 120°C and weighed, than mounted again in the rotor arms, maintaining the initial position in relation to the water stream. The analyses included 4 samples of each alloy, examined for the period of 600 minutes.

4 Study results and their analysis

Microscopic observations of cavitation process effects have shown that the main reason of alloys destruction was mechanical interaction between water stream and surface of specimen. First minutes of the test show, that water stream action on the surface of samples induces the strengthening of surface layer and the increase of micro-hardness. The plastic strain and single pits was observed on the surface of all nickel alloys samples. The plastic deformation on Monel 400 and Inconel 600 alloys surface are showed in figure 1.

a) b)

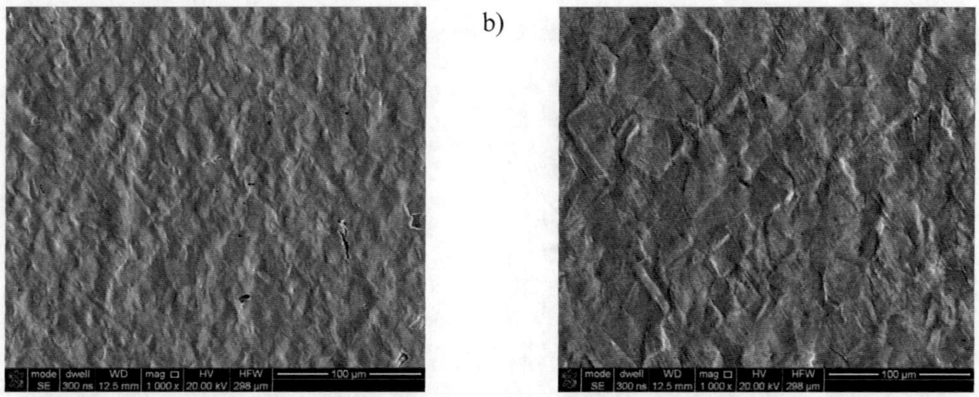

Figure 1. SEM microphotographs present the surface changes after 120 min, a) Monel 400 alloy, b) Inconel 600 alloy.

a) b) c) d)

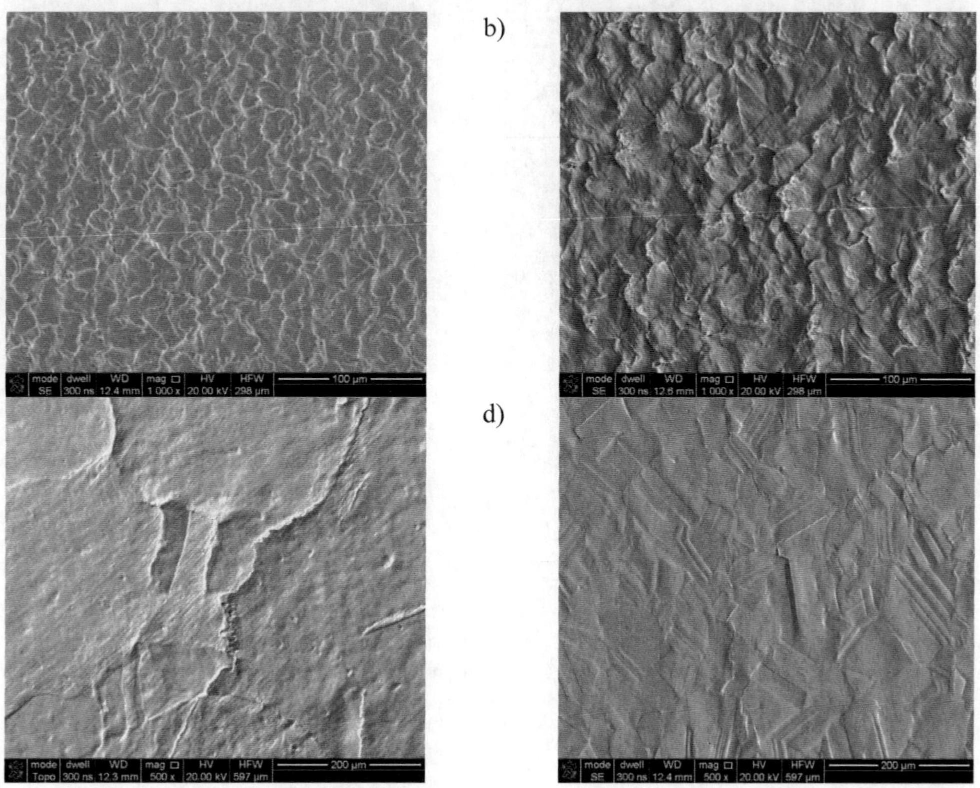

Figure 2. SEM microphotographs present the surface changes after 300 min, a) Monel 400 alloy, b) Inconel 600 alloy, c) Incoloy 800H alloy, d) Hasteloy C22 alloy.

Further exposition of the alloy surface on water stream has caused:

- higher plastic deformation of the surface of Monel 400 and Inconel 600 alloys led to roughness increase. Higher surface roughness of the tested samples changed the angle of the water stream interaction and accelerated the process of cavitation destruction;
- first microcracks are formed on the grain boundaries, especially in the triple contact points of the surface of Incoloy 800H alloys;
- effect of material uplifting and collapsing in a boundary area of the surface of Hasteloy C22 alloys.

These changes on surface condition of nickel alloys samples are showed in figure 2.

a) b) c) d)

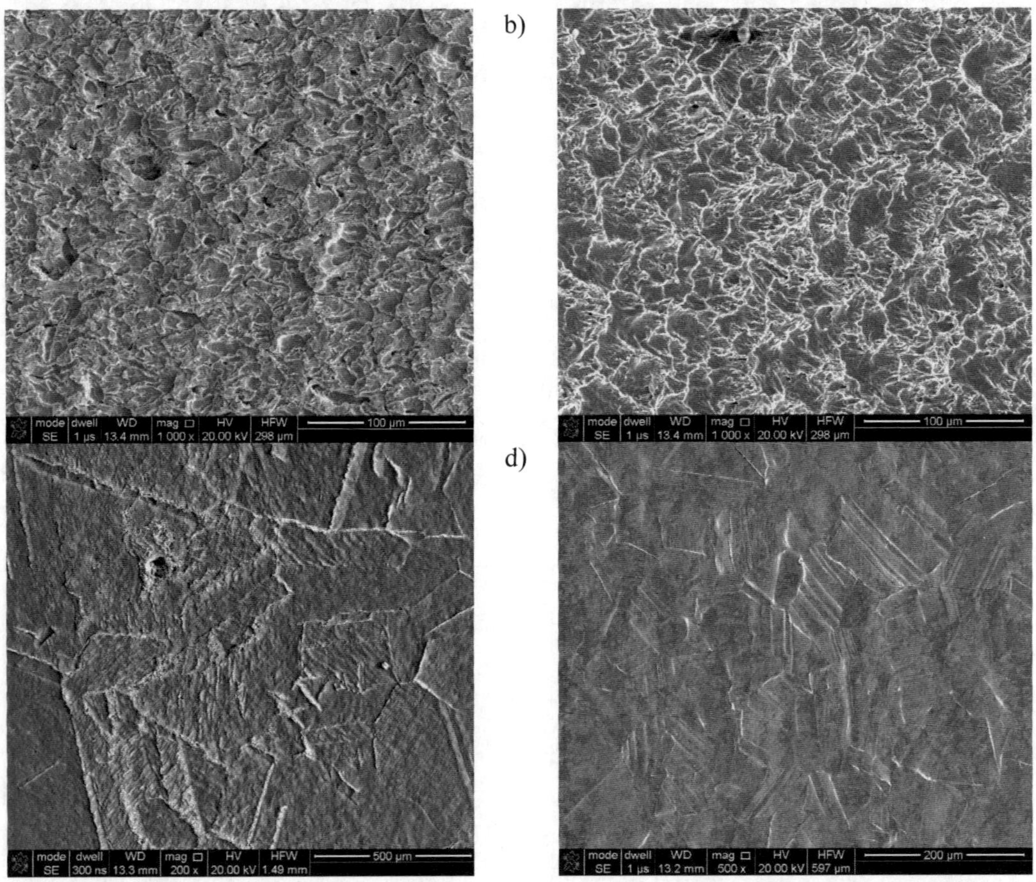

Figure 3. SEM microphotographs present the surface changes after 600 min, a) Monel 400 alloy, b) Inconel 600 alloy, c) Incoloy 800H alloy, d) Hasteloy C22 alloy.

In further stage of tested nickel alloys there were observed below mentioned processes:

- slow erosion of the Monel 400 and Inconel 600 alloys surface;
- destruction of uplifted grain boundaries on the Incoloy 800H alloys surface;
- first microcracks and first single losses were formed on twins grain boundaries of the Hasteloy C22 alloys surface

These changes on surface condition of nickel alloys samples are showed in figure 3. After 600 minutes of examination the average mass loss of four samples of each nickel alloys was: 0.5, 0.4, 0.2 and 0.1 mg for Monel 400, Inconel 600, Incoloy 800H and Hasteloy C22 respectively.

5 Summary

First period of cavitation destruction - incubation period – was characterized by no mass decrement (mass decrement rate was equal to zero), fatigue processes and plastic strain effects occurred and micro-cracks developed on the surface. All these phenomena were observed on tested samples of nickel alloys. Tests of resistance to cavitation erosion proved, that materials with a long incubation period have a very good resistance to this type of destruction. Determination of the resistance to cavitation erosion of the nickel alloys tested would be possible in case of the studies continuation. The analysis of initial period of tested nickel alloys cavitation erosion process has proved, that two alloys have the highest resistance to cavitation erosion: Incoloy 800H and Hasteloy C22.

Acknowledgements
Scientific work funded by the Polish Ministry of Science and Higher Education for statutory activities No. 1/S/IPNT/16.

References

[1] Everhart J 1971 Engineering Properties of Nickel and Nickel Alloys (Springer)
[2] Davis J 2000 Nickel, Cobalt, and Their Alloys (ASM International)
[3] Nunes R 1990 Properties and Selection: Nonferrous Alloys and Special-Purpose Materials Vol. 2 (ASM International) pp 1362-1404
[4] DuPont J and Lippold J and Kiser S 2009 Welding Metallurgy and Weldability of Nickel-Base (John Wiley&Sons)
[5] Boyce M 2002 Gas Turbine Engineering Handbook (Gulf Professional Publishing) p 422
[6] Lai G 1990 High-Temperature Corrosion and Materials Applications (ASM International) chapter 9 pp 249–258
[7] Pollock T 2006 Nickel-Based Superalloys for Advanced Turbine Engines: Chemistry, Microstructure, and Properties *J. of Propulsion and Power* **22**(2006)361
[8] Brennen C 1995 Cavitation and Buble Dynamics (Oxford University Press)
[9] Trevena D 1987 Cavitation and tension in liquids (IOP Publishing Ltd)
[10] Plesset M and Chapman R 1971 Collapse of an Initially Spherical Vapour Cavity in the Neighbourhood of a Solid Boundary, *J. Fluid Mech.* **47**(1971)283
[11] Burka E, 2002 Cavitation in hydraulic machinery *Trans. Inst. Fluid-Flow Machinery* **110**(2002)7
[12] Young F 1989 Cavitation (McGraw-Hill Book Company)
[13] Jones O and Zuber N 1978 Bubble growth in variable pressure fields *ASME J. Heat Transfer* **100**(1978)453
[14] Jasionowski R and Polkowski W and Zasada D 2018 Effect of Crystalographic Texture on Cavitation Wear Resistance of as-Cast CuZn10 Alloy *Arch. Metall. Mater.* **63**(2018)935

IOP Conf. Series: Materials Science and Engineering **461** (2019) 012033 doi:10.1088/1757-899X/461/1/012033

Study of lamination in beta-phase of Ti-5553 titanium alloy

M Kalienko[1,2], A Volkov[1], M Leder[1], A Zhelnina[1,2], P Panfilov[2]

[1] PSC VSMPO-AVISMA Corporation, Parkovaya st. 1, Verkhnaya Salda 624760, Russia
[2] Ural Federal University, 19 Mira street, Ekaterinburg 620002, Russia

E-mail: kamak@yandex.ru

Abstract. The present study primarily focuses on lamination in beta-phase of Ti-5553 titanium alloy. Microstructure and chemical composition of Ti-5553 alloy after heat-treatment was investigated by electron microscopy (SEM, TEM), x-ray diffraction (XRD) and energy dispersive x-ray analysis (EDX). It was revealed that after annealing and slow cooling from β phase field the β phase separated into β-enriched and β-lean regions. This effect does not effect on precipitation of secondary α phase.

1 Introduction

High-strength titanium alloy Ti-5553 is widely used in the aircraft industry for parts such as the chassis and structural elements. This alloy belongs to the group of metastable β alloys [1]. The microstructure of this type of alloys is very sensitive to thermal treatment. Ti-5553 alloy is used after deformation above beta transus temperature and in α+β fields and the mechanical properties resulting from such treatments are strongly depend of the morphology, volume fraction, size and distribution of the α phase in the β matrix [2-5].

Decomposition and lamination of the β phase solid solution to β lean and β rich phases was observed by transmission electron microscope in form lamellae with light and dark contrast [6-8]. There has been only a limited number of previous investigations addressing the decomposition and lamination of β phase in titanium alloys. To identify the effect of lamination requires a precision approach, methods such as electron transmission microscopy (HR-TEM, HAADF-STEM), atomic tomography and microanalysis (3DAP Tomography, APT), x-ray analysis. Illarionov A. G. et. al. [6] studied structural and phase transformations during cold plastic deformation of VT22 titanium alloy, which was taken as the basis for development of the alloy Ti-5553 [1]. They found the formation of fine lamellae within the layers of the β matrix by TEM. The observed structure in the VT22 alloy according to the electron diffraction and XRD was ascribed to the precipitation of a body-centered-tetragonal τ phase via a shear transformation. Zherebtsov S. V. et al. [7] reported that a TEM diffraction pattern from the β phase did not show any other phases, so the presence of "bands" between reflections of the β phase may imply the presence of areas of short-range order with a structure varying somewhat from the matrix. In the work of P. Barriobero-Vila et al. [8], the presence of "bands" in the structure of β phase of the investigated Ti-5553 alloy is associated with the fluctuation of the chemical composition, but does not provide experimental evidence.

The phenomena of solid solution lamination, spinodal decomposition observed in the form of separation in β phase, were found in titanium alloys with high concentration of β stabilizers [9-14]. The theoretical description of the observed phenomena is faced with great difficulties, so new experimental facts describing the variety of possible thermodynamically unstable states are needed to clarify the theory guidelines. In our work [15] we observed lamination in Ti-5553 alloy with different fraction of the primary α phase, so the objective of this work is to study the effect of lamination using XRD and TEM analysis

2 Materials and methods

The work is carried out on the forged material Ti-5553 alloy, the chemical composition of the alloy is presented in table 1. The β transus temperature of the alloy was confirmed as equal to 823 °C. In order to formation lamellar structure of the alloy, sample was first solution treated in the β field at 870 °C for 1 hour and then furnace cooled to a temperature at 700 °C, kept at this temperature and then cooled in air (the first stage of heat treatment). Then aging was carried out at a temperature of 530 °C for 8 hours (the second stage of heat treatment).

Table 1. Chemical composition of Ti-5553 alloy (at. %).

Ti	Al	Mo	V	Cr	Fe	Zr	O ppm	Mo eq.
Bal.	8.55	2.53	4.83	2.64	0.36	<0.0010	142	9.5

The microstructure and composition of the alloy were characterized using SEM Quanta 3D FEG with EBSD Pegasus XM4 detector and TEM Tecnai G2 20F S-TWIN. The TEM was equipped with both a high angle annular dark field HAADF-STEM detector and energy-dispersive spectroscopy (EDS) capability. EBSD map was taken at a step size of 0.1 µm. The EBSD data was analyzed using the Orientation Imaging Microscopy (OIM 5.31) software. The boundaries with misorientations between 1.5° and 15° were defined as low-angle grain boundaries, and those of misorientations larger than 15° as high-angle grain boundaries.

Specimens for SEM and XRD analysis were prepared by applying the standard mechanical polishing techniques. Thin foils for TEM study were cut by electrical discharge machine, grinding to thickness up to 100 microns, then prepared in TenuPol-5 at a voltage of 30 V and a temperature of -40 °C.

X-ray diffraction measurements were performed on the Bruker D8 Advance diffractometer with a goniometer radius 217.5 mm in Bragg–Brentano diffraction geometry. XRD patterns were recorded at room temperature using Cu-Kα radiation with following measurement conditions: tube voltage of 40 kV, tube current of 40 mA, stepscan mode with a step size of 0.02° 2θ and counting time of 1 s/step by LynxEye-detector. XRD analysis was carried out using the commercial software TOPAS by "Pawley method" for whole pattern modelling.

3 Results and discussion

Figure 1a shows backscattered electron micrograph of the Ti-5553 alloy after slow cooling from β phase field. The microstructure of the alloy consisted of lamellar α colonies within prior β grains. The β grains are equiaxed and the grain size in this condition has been estimated to be 260 µm. The volume fraction of primary α phase was 33 %, the average thickness of α lamellae in colonies was 0.4 ±0.02 µm. After air cooling according to the EBSD method, only the primary α phase and matrix β phase are present in the alloy structure. Within β grain, the orientation of the crystallographic lattice of the β phase in the study by EBSD remains unchanged, the plates of the primary α phase have different crystallographic orientation and are oriented in accordance with the crystallographic ratio of Burgers, figure 1b. It should be noted that the calculated disorientation of the angle for the analysis of the subgrain structure of the β phase was taken to be equal to 1.5 degrees. I.e., of the immutability of the orientation of the crystal lattice within a single β grain it is possible to speak with accuracy of 1.5 degrees.

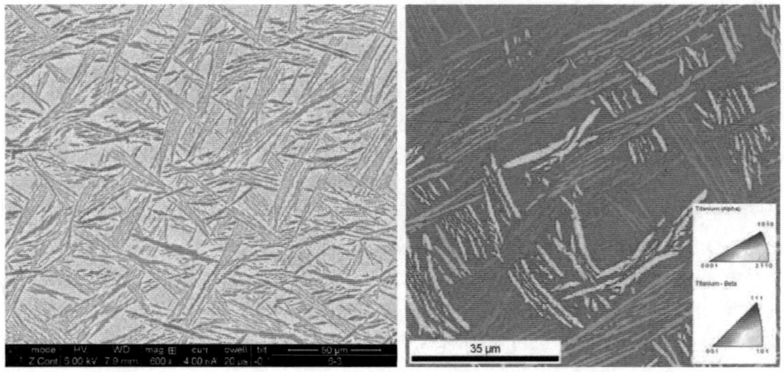

a) b)

Figure 1. The microstructure of Ti–5553 alloy after first stage of heat treatment: (a) SEM image, (b) EBSD orientation map.

The XRD pattern of the alloy after the first stage of heat treatment is shown in figure 2. The observed asymmetry of peaks of the β phase on the XRD pattern is caused by two adjacent diffraction lines resulting in overlapping peaks appearing as a single asymmetric peak. The two lines caused by one type of crystallographic β phase but with two different lattice parameters. Using the method of full-profile analysis, the diffraction lines of the β phase are well described using a model that includes the presence of several phases of cubic symmetry (BCC, Im-3m), figure 2b. Two different lattice parameters are caused by different chemical composition known as β phase decomposition and known to result in different lattice parameters and different angular position. When the concentration of β stabilizers increase in β phase then its unit volume decreases and parameter of the crystal lattice and the position of the diffraction lines of pattern is modified.

For the Ti-5553 alloy after air cooling from 700 °C, the best description of the profile of β phase diffraction lines is used two phases of cubic symmetry in the model. Thus, with a full-profile description of the XRD pattern the model well describes the line β(310) by two phases with parameter of lattice 3.225 and 3.228 Å, figure 2b. Probably, lamination of β phase occurred during cooling from β field and holding at α+β field.

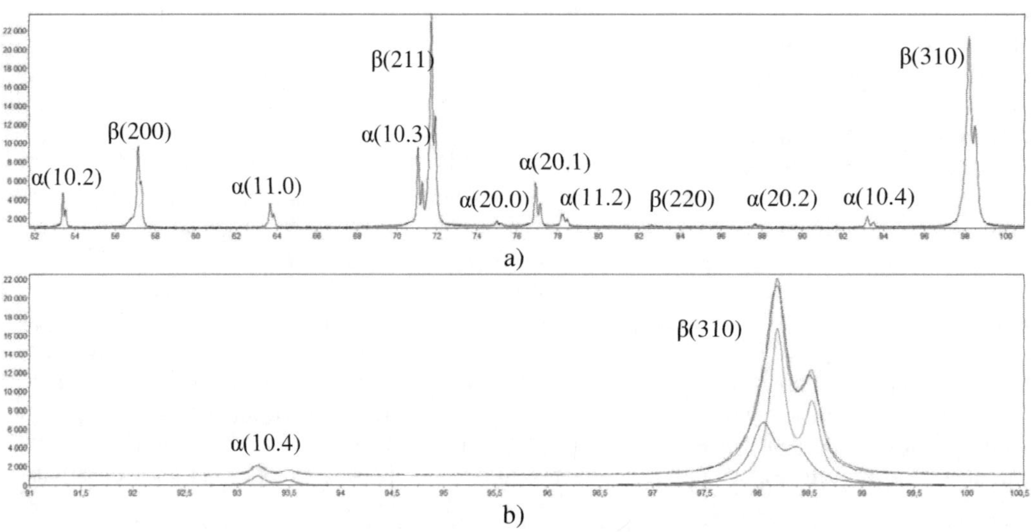

Figure 2. XRD pattern of Ti-5553 alloy after first stage of heat treatment (a) and deconvolution of the (310) diffraction peaks of the β phase (b).

The study of microstructure by transmission electron microscope in the structure of the β phase observed bands that differ in contrast and thickness, figure 3a (HAADF-STEM images). By electron diffraction pattern was compared to the orientation of the crystal lattice of two nearby "bands" of different contrast, the received results is identical for the nearby "bands" and presented figure 3b, c. The obtained result confirms the data obtained in [7]. It is well known that the density of the β phase is greater than the density of the α phase, since the β phase contains more β stabilizers and less aluminum, therefore, in the images in the reflected electrons, the β phase has a lighter contrast than the α phase. According to these arguments, the light contrast bands in the β phase should contain more heavy chemical elements. Indeed, according to the results of EDX analysis of the chemical composition of the β-phase regions, a difference in their composition was revealed, table 2. In all studied areas of the β phase with a lighter contrast, the molybdenum equivalent of the chemical composition is higher than in a nearby area with a darker contrast. Molybdenum equivalent was calculated taking into account aluminum [1]. According to the obtained data, the chemical composition of investigated areas of β phase has the maximum difference in the content of aluminum and vanadium, the difference in the content of molybdenum and chromium is less pronounced. The iron content in the studied areas was not estimated due to its small content in the alloy (0.36 at. %) and as a consequence of the large error of its analysis.

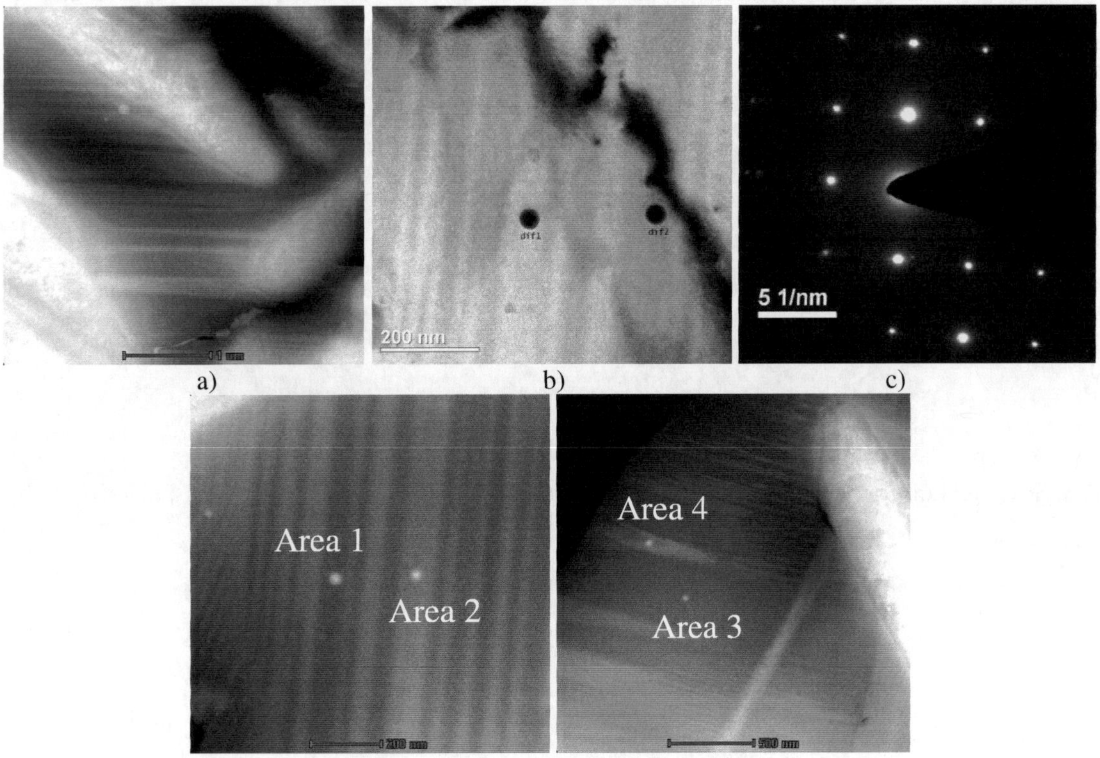

Figure 3. HAADF-STEM image showing the lamination of β phase in Ti-5553 alloy after first stage of heat treatment. (a). Bright-field TEM images showing the lamination region (b), electron diffraction pattern of the region (c) and areas on HAADF-STEM images for EDX analysis (d-e).

In the structure of the Ti-5553 alloy after the second stage of heat treatment, aging at 530 °C, there is observed of fine secondary α phase, figure 4a. Precipitation of secondary α phase is resulted in dispersion hardening of the alloy. The dispersed particles have a large number of interfacial boundaries, which serve as barriers to the movement of dislocations. In the β phase between the plates of the primary α phase, the secondary phase is allocated in limited volumes, mainly due to the size limitations of the areas (first factor) and the lower defect density near the interfacial boundaries. During the aging process secondary α phase first precipitated in the free areas between the colonies, so the concentration of β stabilizers in unit volume of the β phase increases, which is the second limiting factor for the formation of secondary α phase between the plates in the colonies. The lamination which was observed after annealing at 700 °C was observed also after subsequent aging at 530 °C. Secondary α phase and contrast "bands" (lamination) in the β phase can be observed between the primary α phase (large bright particles) in Figure 4b. The density of secondary α phase formation in the free regions of the matrix β phase is typical. There is no relation between a place of precipitate of the secondary α phase and the lamination of the β solid solution, figure 4a, b. Analysis of chemical composition of β phase after aging revealed increase content of vanadium, chromium, while the content of molybdenum decreased slightly. It should be noted that the EDX analysis is a semi-quantitative method of calculating the chemical composition, so the results which can be influenced by various factors and the calculation model itself is not ideal. Thus, for example, the calculated molybdenum content depends on the energy line which can be selected for analysis (K or L-line). In this study for analyzed elements were used K-lines of spectrum. The main aim of present work was analyzed the chemical composition of two closely spaced regions of the β phase, so for this analysis the influence of the geometry of experiment conditions is minimal and the obtained data revealed the limitations of the β phase on β-enriched and β-lean regions.

a) b)

Figure 4. SEM (a) and HAADF- STEM (b) micrographs of solution treated and aged specimen. In (b) showing areas for EDX chemical analysis.

Table 2. Compositions of β lean and β rich phases regions (at. %).

Area	Al (at. %)	V (at. %)	Cr (at. %)	Mo (at. %)	Mo eq.
№ 1	6.60	5.95	2.14	3.72	11.5
№ 2	**5.10**	**6.30**	**3.35**	**3.68**	**14.5**
№ 3	7.52	4.97	3.58	3.40	12.2
№ 4	**6.64**	**6.78**	**3.60**	**3.45**	**14.1**
№ 5	6,81	7.77	3.88	2.72	13.8
№ 6	**7.30**	**10.22**	**3.75**	**2.91**	**15.4**

4 Conclusions

The decomposition of β phase and subsequent α precipitation in the Ti-5553 alloy have been investigated by TEM and XRD technics. A new effect for lamination of β phase is revealed. It was found that the β phase separates into periodic "bands" with variation of alloying elements and represent lean and rich β phase regions. The lamination of β phase does not have impact to homogeneity precipitation of the secondary α phase.

Acknowledgements
This article was prepared with the financial support of Ural Federal University Competitiveness Enhancement Program – CEP 3.1.1.1-18.

References

[1] J. D. Cotton, R. D. Briggs, R. R. Boyer, et al., State of the art in beta titanium alloys for airframe applications, JOM 67 (6) (2015) 1281–1303.

[2] J.D. Cotton, R.R. Boyer, R.D. Briggs, R.G. Baggerly, C.A. Meyer, M.D. Carter, W. Wood, G. Tewksbury, V. Li, and X. Yao: Ti-2007: Science and Technology, M. Ninomi, S. Akiyama, M. Ikeda, M. Hagiwara, and K. Maruyama, eds., The Japan Institute of Metals, Kyoto, 2007, pp. 471–74.

[3] M. Harper, R. Williams, G.B. Viswanathan, J. Tiley, R. Banerjee, D.J. Evans, and H.L. Fraser: Titanium 2003: Science and Technology, G. Lutjering and J. Albrecht, eds., Wiley-VCH, Hamburg, 2003, vol. 3, pp. 1559–66.

[4] S. Shekhar, R. Sarkar, S. Kumar Kar, A. Bhattacharjee, Effect of solution treatment and aging on microstructure and tensile properties of high strength β titanium alloy, Ti–5Al–5V–5Mo–3Cr, Materials and Design 66 (2015) 596–610.

[5] J.C. Fanning and R.R. Boyer: Titanium 2003: Science and Technology, G. Luetjering and J. Albrecht, eds., Wiley-VCH, Hamburg, Germany, 2003, vol. IV, pp. 2635–42.

[6] A. G. Illarionov, I. V. Narygina, M. S. Karabanalov, et al., Structural and Phase Transformations in a Titanium Alloy of the Transition Class under the Effect of Deformation, The Physics of Metals and Metallography, 110 (2010) 279–288.

[7] S. V. Zherebtsov, M. A. Murzinova, M. V. Klimova, et al., Microstructure evolution during warm working of Ti–5Al–5Mo–5V–1Cr–1Fe at 600 and 800 °C, Materials Science & Engineering A 563 (2013) 168–176.

[8] P. Barriobero-Vila, G. Requena, S. Schwarz, et al., Influence of phase transformation kinetics on the formation of α in a β-quenched Ti–5Al–5Mo–5V–3Cr–1Zr alloy, Acta Materialia 95 (2015) 90–101.

[9] V. P. Skripov, A. V. Skripov, Spinodal decomposition, Advances in physical Sciences, Vol. 128, № 2, 1979, pp. 193–231.

[10] Z. Fan, A. P. Miodownik, TEM study of metastable β-phase decomposition in rapidly solidified Ti-6AI-4V alloy, Journal of materials science 29 (1994) 6403–6412.

[11] C. R. M. Afonso, P. L. Ferrandini, A. J. Ramirez, R. Caram, High resolution transmission electron microscopy study of the hardening mechanism through phase separation in β-Ti–35Nb–7Zr–5Ta alloy for implant applications, Acta Biomaterialia 6 (2010) 1625–1629.

[12] A. Devaraj, S. Nagand R. Banerjee, Alpha phase precipitation from phase-separated beta phase in a model Ti–Mo–Al alloy studied by direct coupling of transmission electron microscopy and atom probe tomography, Scripta Materialia 69 (2013) 513–516.

[13] C. Ghosh, J. Basu, D. Ramachandran, E. Mohandas, Phase separation and ω transformation in binary V-Ti and ternary V-Ti-Cr alloys, Acta Materialia 121 (2016) 310–324

[14] Y. M. Zhu, S. M. Zhu, M. S. Dargusch, J. F. Nie, HAADF-STEM study of phase separation and the subsequent α phase precipitation in a β-Ti alloy, Scripta Materialia 112 (2016) 46–49.

[15] M.S. Kalienko, A.V. Volkov, V.A. Kropotov, et al., Primary α-Phase VST5553 Alloy with Lamellar Structure Properties Effect // Titanium'2011: Science and technology. Proc. 12-th Int. Conf. of Titanium. Beijing, China, 2011. – pp. 1303–1311.

Change in dislocation mobility of tempered martensitic steel through charged hydrogen

H Kamei[1], Y Matsumoto[1], H Suzuki[2] and K Takai[2]

[1] Graduate School of Science and Technology, Sophia University, 7-1 Kioi-cho, Chiyoda-ku, Tokyo 102-8554, Japan

[2] Faculty of Science and Technology, Department of Engineering and Applied Science, Sophia University, Tokyo, Japan

E-mail address:k17.makh.6@gmail.com

Abstract. To clarify the relationship between dislocation slip and hydrogen for elucidation of hydrogen embrittlement mechanism, the effect of hydrogen on dislocation slip was examined by comparing changes in the stress-strain curve and stress relaxation behavior with/without hydrogen charging. Tempered martensitic steel with the tensile stress of 1480 MPa was used. Hydrogen charging was conducted by the electrolysis charging. Tensile tests were conducted for hydrogen-charged specimens containing hydrogen from 0 to 5.6 mass ppm at strain rates from 1.83×10^{-5} to 1.83×10^{-2} s^{-1}. In the stress-strain curve, the elastic modulus did not change, but the upper and lower yield points increased through charged hydrogen. The tensile strength increased and the work hardening rate did not change through charged hydrogen. The yield points and tensile strengths increased with increasing the amount of hydrogen. It suggests that hydrogen suppressed dislocation slip, that is, hydrogen caused hardening. Stress relaxation tests were conducted for hydrogen-charged specimens containing hydrogen from 0 to 4.7 mass ppm. Even though the amount of hydrogen increased, the amount of stress relaxation remained constant. Therefore, hydrogen probably did not promote the increase in dislocation density. These results reveal that hydrogen causes not softening but hardening.

1 Introduction

Mechanical properties of steel include various indexes such as yield strength and work hardening rate. It is important to examine the influence of hydrogen on these indexes in detail in considering the hydrogen embrittlement mechanism from the viewpoint of the relationship between dislocation slip and hydrogen. It has been reported that dislocation slip is suppressed [1] or promoted [2] by hydrogen charging. Since these results were influenced by the concentration of impurities, hydrogen content, temperature, and strain rate, among other factors, a unified view has not been obtained. Several methods, for example, tensile tests and stress relaxation tests, are utilized for evaluating the effect of hydrogen on mechanical properties [1]. The relationship between dislocation slip and hydrogen can be evaluated in more detail by comparing the results obtained with these methods. Therefore, in this study, the influence of hydrogen on dislocation slip was investigated by several experimental methods for tempered martensitic steel, which is a practical steel containing a lot of impurities.

2 Experimental

Induction quenched and tempered martensitic steel with a tensile stress of 1480 MPa was used as the specimens. Table 1 gives the chemical composition of the specimens and figure 1 shows their shape. The specimens had a parallel part with a diameter of 3 mm and a length of 20 mm. The specimen shows quasi-cleavage fracture in hydrogen environment [3]. The specimen surface was polished using emery paper (800, 1000, and 2000 grit). They were electrochemically charged with hydrogen in an aqueous solution of 0.1 N NaOH and 5.0 $g{\cdot}L^{-1}$ NH_4SCN with a current density from 0 to 30 $A{\cdot}m^{-2}$ at 30 °C for 72 h. Tensile tests were carried out at strain rates from 1.83×10^{-5} to 1.83×10^{-2} s^{-1}. The stress relaxation test was carried out in the following procedure. Specimens were electrochemically charged with hydrogen in an aqueous solution of 0.1 N NaOH and 5.0 $g{\cdot}L^{-1}$ NH_4SCN with a current density from 0 to 15 $A{\cdot}m^{-2}$ at 30 °C for 72 h. They were subjected to stress of 0.92 σ_B (elastic region) or 0.98 σ_B (plastic region) at a strain rate of 1.83×10^{-5} s^{-1} and locked at the gripper interval for 1800 s. σ_B is a tensile stress. The specimen was deformed up to 0.92 σ_B or 0.98 σ_B then the machine was stopped and the specimen allowed relaxing. The stress decrease with time was recorded. The relaxation stress was defined as the difference between the initial subjected stress and the stress after the test. The hydrogen content was measured by thermal desorption analysis (TDA) using a gas chromatograph at a heating rate of 100 $°C{\cdot}h^{-1}$. The hydrogen contents of specimens charged with hydrogen at current densities of 5, 10, 15 and 30 $A{\cdot}m^{-2}$ were 1.1, 3.1, 4.7 and 5.6 mass ppm, respectively.

Table 1. Chemical composition of specimen material (mass %)

C	Si	Mn	P	S
0.32	1.60	0.75	0.014	0.005

3 Results

3.1 Tensile test

Figure 2 shows the stress-strain curves of tempered martensitic steel specimens with various hydrogen contents at a strain rate of 1.83×10^{-5} s^{-1}. Elongation decreased with increasing hydrogen content, whereas the upper and lower yield points and the tensile strength increased. The elastic modulus did not change with increasing hydrogen content.

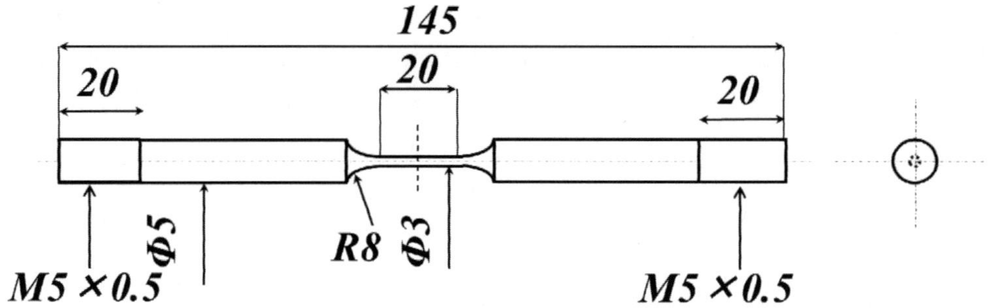

Figure 1. Shape of specimen

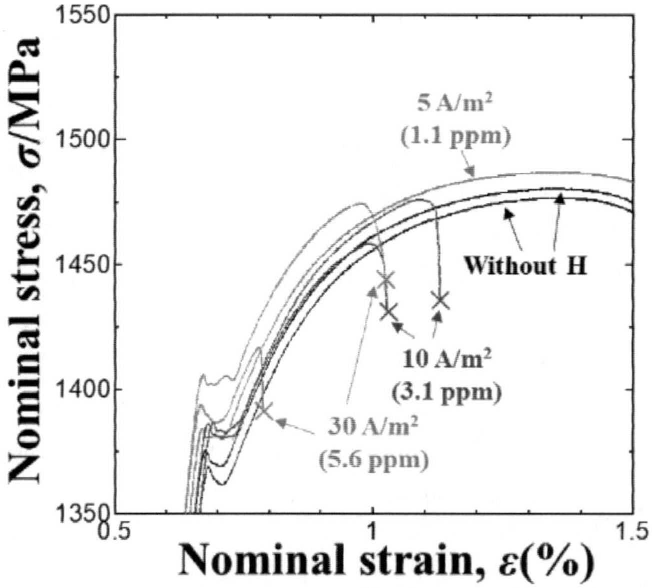

Figure 2. Stress-strain curves of tempered martensitic steel specimens with various hydrogen contents at a strain rate of 1.83×10^{-5} s^{-1}.

Figure 3 shows the relationship between the modulus of elasticity and strain rate. The modulus of elasticity was defined as the slope of straight line connecting the point at 0 MPa and 1000 MPa in the stress-strain curve. The modulus of elasticity did not change with increasing strain rate at any hydrogen content, nor did it change with increasing hydrogen content at any strain rate. Figure 4 shows the relationship between the upper yield point and strain rate. The upper yield point increased with increasing strain rate at any hydrogen content and it increased with increasing hydrogen content at any strain rate. Figure 5 shows the relationship between the lower yield point and strain rate. The lower yield point increased with increasing strain rate at any hydrogen content and it increased with increasing hydrogen content at any strain rate. Figure 6 shows the relationship between the tensile strength and strain rate. The tensile strength increased with increasing strain rate at any hydrogen content and it increased with increasing hydrogen content at any strain rate. Therefore, the upper and lower yield points and the tensile strength increased with increasing hydrogen content at any strain rate. The results suggest that hydrogen suppressed dislocation slip,

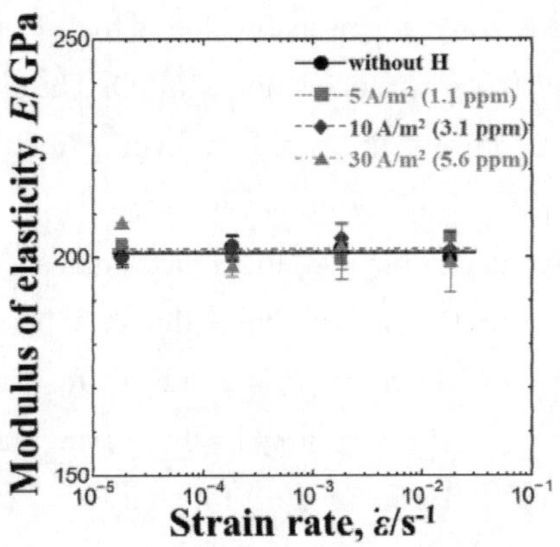

Figure 3. Relationship between modulus of elasticity and strain rate.

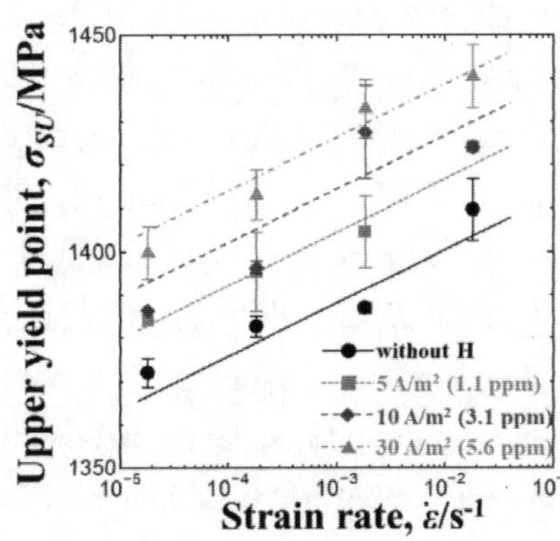

Figure 4. Relationship between upper yield point and strain rate.

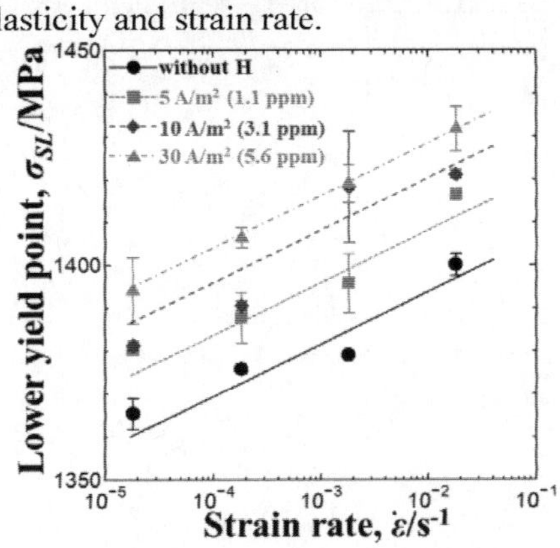

Figure 5. Relationship between lower yield point and strain rate.

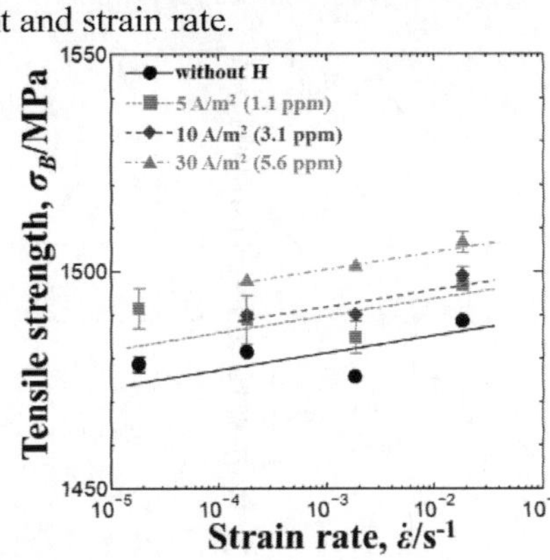

Figure 6. Relationship between tensile strength and strain rate.

i.e., it caused hardening.

3.2 Stress relaxation test

Figure 7 shows the relationship between relaxation stress and hydrogen content. Even though the amount of hydrogen increased, the amount of stress relaxation remained constant under all stress conditions. Hence, hydrogen probably did not promote any increase in dislocation density.

4 Discussion

A previous study reported that the formation of lattice defects is accelerated by plastic strain in the presence of hydrogen in pure iron [4], Inconel 625 [4], SUS304 [5], and SUS316L [5]. It has been reported that vacancy defects such as vacancies or vacancy clusters form in cold-drawn pearlitic steel at the rupture location in the plastic region in the presence of hydrogen [6]. It has been reported that resistance to deformation increases through the introduction of vacancies by neutron irradiation in SUS304, SUS316 and SUS347 [7]. Therefore, the cause of the increase in the lower yield stress and tensile strength with increasing hydrogen content in the present study was probably due to hydrogen or vacancy-type defects. Presumably, such defects became resistance to dislocation slip. On the other hand, it has been reported that strain-induced lattice defects are not increased by hydrogen in the elastic region in cold-drawn pearlitic steel. Several models have been proposed for the relationship between the concentration of solute atom and solid solution strengthening [8, 9]. Hence, the cause of the increase in the upper yield stress with increasing hydrogen content was probably due to hydrogen. It implies that hydrogen promoted resistance to dislocation slip. However, it is a subject for future study whether strain-induced lattice defects are promoted or not by straining in the elastic/plastic region in the presence of hydrogen in tempered martensitic steel. A further subject for study is whether these defects affect dislocation slip or not.

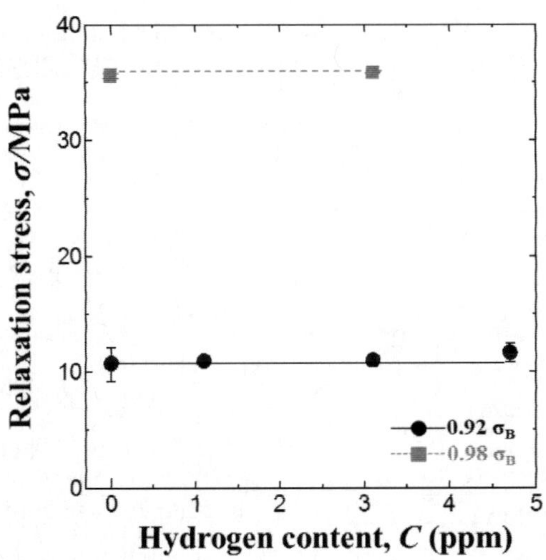

Figure 7. Relationship between relaxation stress and hydrogen content.

In high strength steels which are frequently used in practice, few studies have been reported about hardening or softening by hydrogen. However, in the present study, it was revealed that in the tempered martensitic steel, softening phenomenon by hydrogen charging was not observed in a wide range from the elastic region to the plastic region, and hardening phenomenon was specifically observed in the plastic region. From these results, it was suggested that hydrogen did not promote the movement of dislocation but obstruct it. These are important results that can lead to elucidation of the mechanism of hydrogen embrittlement in the real environment.

5 Conclusion

The following findings were obtained concerning the influence of hydrogen on dislocation slip in tempered martensitic steel.

(1) Tensile tests were conducted for hydrogen-charged specimens containing hydrogen from 0 to 5.6 mass ppm at strain rates from 1.83×10^{-5} to 1.83×10^{-2} s^{-1}. The modulus of elasticity did not change with increasing strain rate at any hydrogen content. The upper and lower yield points and the tensile strength increased with increasing strain rate at any hydrogen content. The modulus of elasticity did not change with increasing hydrogen content at any strain rate. The upper and lower yield points and the tensile strength increased with increasing hydrogen content at any strain rate.

(2) Stress relaxation tests were conducted for hydrogen-charged specimens containing hydrogen from 0 to 4.7 mass ppm. Even though the amount of hydrogen increased, the amount of stress relaxation remained constant at both stress conditions of 0.92 σ_B and 0.98 σ_B.

(3) Hydrogen caused not softening but hardening. It was suggested that hydrogen did not promote the movement of dislocation but obstruct it.

References

[1] Asano S, Nishino Y and Otsuka R 1979 *J. Japan Inst. Metals* **43** 241-8
[2] Hagi H, Hayashi Y and Ohtani N 1979 *J. Japan Inst. Metals.* **43** 320-5
[3] Matsumoto Y and Takai K 2017 *Metall. Mater. Trans. A* **48** 666-77
[4] Takai K, Shoda H, Suzuki H and Nagumo M 2008 *Acta Mater.* **56** 5158-67
[5] Hatano M, Fujinami M, Arai K, Fujii and Nagumo M 2014 *Acta Mater.* **67** 342–53
[6] Konno R, Manabe T, Hirakami D and Takai K 2018 *Tetsu-to-Hagane* **104** 36-45
[7] Higgy R H and Hammad H F 1975 *J. Nucl.Mater.* **55** 177-186
[8] Friedel J 1964 *Dislocations* (Oxford: Pergamon Press) p 351-66
[9] Labusch R 1970 *Phys. Status. Solid. B* **41** 659-69

The effect of reinforcement particle size on the properties of Cu-Al_2O_3 composites

M Kargul[1], J Borowiecka[1]- Jamrozek, M Konieczny[1]

[1]Department of Metal Science and Materials Technology, Faculty of Mechatronics and Mechanical Engineering, Kielce University of Technology, Al. Tysiąclecia P.P. 7, 25-314 Kielce, Poland

e-mails: mkargul@tu.kielce.pl, jamrozek@tu.kielce.pl, mkon@tu.kielce.pl

Abstract. The main aim of this paper was to investigate the effect of reinforcement particle size on the properties of Cu-Al_2O_3 composites. The starting materials for obtaining the composites were commercial powders: Cu and Al_2O_3. Experiments were performed using specimens containing 2.5, 5, 7.5 and 10 % of alumina. The average particle sizes of alumina were 3.5, 22.5, and 65 μm. Before the sintering process, single pressing was performed with a hydraulic press at a compaction pressure of 620 MPa. The obtained sinters were subjected to the compaction process at a pressure of 620 MPa, and then again sintered at 900 °C for 60 minutes. The sintering process was carried out for 60 minutes at a temperature of 900 °C. The sintered compacts were subjected to the following tests: measurement of density, hardness and electrical conductivity. Observations of the microstructure on metallographic specimens made from the sintered samples were also performed using a scanning electron microscope (SEM). The density of all the composites increased when the particle size of alumina was reduced. Higher hardness were observed in samples containing finer Al_2O_3 particles. The electrical conductivity of the fabricated composites increased with the reduction of the alumina particle size.

1. Introduction

In recent years, sintered copper matrix composites are increasing interest due to its very high electrical and thermal conductivity. It is well known that the introduction of even a small amount of alloy additives into copper results in a significant reduction in electrical conductivity. In order to increase the strength of copper without a significant decrease in electrical conductivity, among others particles such as metal oxides (Al_2O_3, SiO_2, ZrO_2, Cr_2O_3, BeO, MgO), carbides (SiC, TiC, Cr_7C_3, Cr_3C_2), nitrides and borides (TiB_2, ZrB_2, CrB_2, BN) are used like a reinforcement phases [1-3].

Research is also being carried out on the preparation of composites containing intermetallic phases, the introduction of which results in the increase of strength properties similar to those made with ceramic particles [4]. The composites reinforced with particles combine the properties of the metal matrix with the reinforcing properties of ceramics. Ceramic particles have high hardness and resistance to high temperatures, therefore, the composites reinforced with ceramic particles are also used for working at elevated temperatures. The introduction of ceramic particles into the matrix increases the hardness and abrasion resistance of the composite without a significant decrease in electrical conductivity. For this reason, copper-based composites reinforced with ceramic particles are mainly used in electronics and electrotechnics for relays, switches, elements of electric motors and for the tip of resistance welding electrodes [5-7]. Metal matrix composites reinforced with ceramic particles are usually made using powder metallurgy techniques [8]. The main advantage of powder metallurgy compared to casting is better control of the distribution of the reinforcement in powder metallurgy compacts. Solid state diffusion, dependent on time and sintering temperature, plays a major role in powder metallurgy in the formation and growth of interparticle bonding. Currently, apart from the basic technique of making composites in powder metallurgy technology consisting in the preparation of mixtures, pressing and sintering, modern technologies such as hot pressing (HIP) and SPS (spark plasma sintering) are also used [9, 10]. The strength properties of composites with a metal matrix reinforced with particles depend, inter alia, on the retention properties of the matrix as well as on the size of the reinforcing particles. For composite materials, the strengthening factor depends on the diameter of the reinforcing particles [11, 12]. The best properties are obtained for dispersion-strengthened composites in which the size of strengthening particles is in the range from 0.01 μm to 0.1 μm [13-15]. In the case of particles larger than 1μm, composites are reinforced with large particles. The main difference between dispersion and large particles is that in the case of composites reinforced with large particles, the loads are transferred both by the particles and the metal matrix. The aim of this work was to fabricate copper matrix composites reinforced with Al_2O_3 particles using powder metallurgy technology and to analyze the influence of strengthening particle size on their properties.

2. Experimental procedure

Alumina powder with the particle size of 3.5, 22.5 and 65 μm and 99.3 % purity from 0 to 10 wt. %, along with the copper powder with average 63 μm particle size and 99.5 % purity have been used. The density of copper powder was 8.9 g/cm^3 and the density of alumina powders was 3.95 g/cm^3.
Electron- microscopic images (SEM) of the starting powders are shown in Figure 1.

Figure 1. Images of the tested powders: a) electrolityc copper powder, b) alumina (3.5 μm particle size), c) alumina (22.5 μm particle size), d) alumina (65 μm particle size)

Before the fabrication of composites, the powders were observed using a JEOL JSM-7100F field emission scanning electron microscope. Powder mixtures with the following content of Al_2O_3 - 2.5 %, 5 %, 7.5 % and 10 % by weight were prepared for the tests. Finished powdered mixtures were subjected to single-track pressing on a hydraulic press at a compaction pressure of 620 MPa. The specimens with a dimension of Φ20x10 mm were sintered in a sillit tubular furnace at 900 °C for 60 minutes in the dissociated ammonia atmosphere. Finally, the material was cooled in the furnace. After the sintering process, the samples were subjected to compaction process at a pressure of 620 MPa and re-sintered at 900° for 60 minutes. The sintered compacts were measured for hardness, electrical conductivity, and density. The hardness of the composites was measured using the Brinell method (with a steel ball 5 mm in diameter at a load of 250 kG) in accordance with the PN EN ISO 6506-1:2014 standard. The electrical conductivity tests were performed using the GE Phasec 3D device using the eddy current method. The density was determined by weighing the specimens in air and water using WPA120 hydrostatic scales in line with the PN EN ISO 2738:2001 standard. Microstructure analysis on the metallographic specimens were conducted using the JEOL JSM-7100F field emission scanning electron microscope fitted with OXFORD INSTRUMENTS EDS X-Max Aztec software for elemental analysis.

3. Results and discussion

3.1. Microstructural investigations

The introduction of powdered alumina particles causes a distinct change in microstructure of obtained composites. Alumina particles of various sizes are very clear in a copper matrix. Differences in a particle size result from the range of the size of the introduced alumina powder. Mixing the powders of copper and alumina led to an even distribution of the particles in the matrix. In certain areas of composites, the alumina particles are bonded into longitudinal agglomerates. Large clusters of alumina particles were not observed. Introduced alumina particles have not dissolved in the matrix in the sintering process, which is associated with high thermal resistance of alumina particles. The examples of microstructures of fabricated materials containing 5% of alumina are shown in Figure 2.

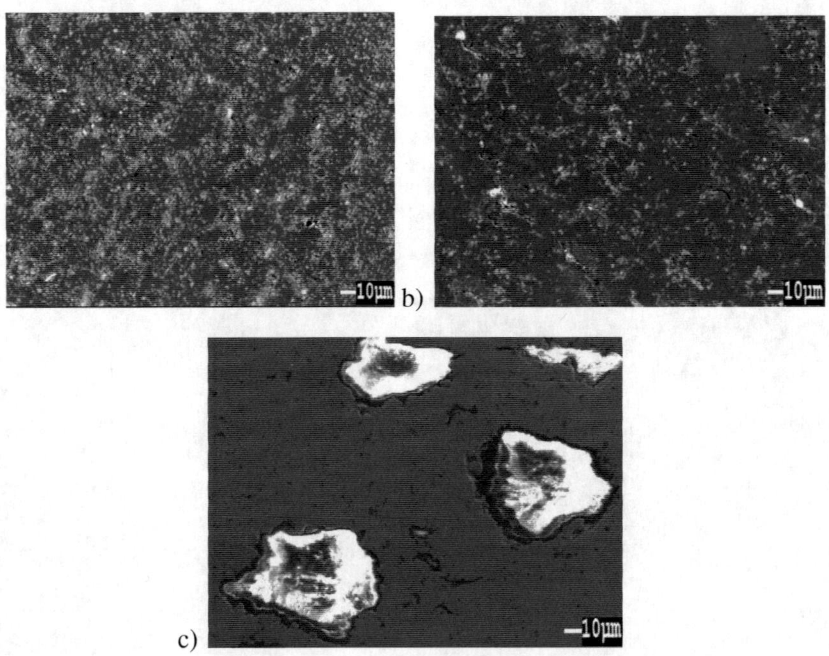

a) b) c)

Figure 2. Microstructures of sintered compacts with a SEM obtained for a) Cu+5% of alumina(3.5 µm particle size), b) Cu+ alumina (22.5 µm particle size), c) Cu +alumina (65 µm particle size)

The presence and distribution of elements in a composite containing 5 % of alumina with a particle size of 22.5 μm are presented in Figure 3. From the mapping it is clear that the major elements occurring in a fabricated composites are copper, aluminum and oxygen.

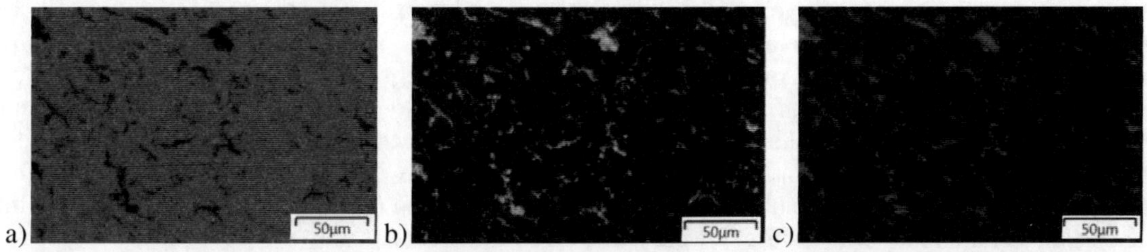

Figure 3. Distribution of elements in the micro- area of composite containing 5% of alumina with a average particle size of 22.5 μm, a) copper, b) aluminum, c) oxygen

3.2. Density and hardness measurements

The results of density and hardness measurements are presented in Tables 1-3.

Table 1. Results of the density and hardness measurements for a composites with a alumina average particle size of 3.5 μm

Material	Density (g/cm^3)	Relative density (%)	HB
Cu	8.19± 0.02	92.01	36.65± 1.5
Cu + 2.5 % of Al_2O_3	8.35± 0.04	95.26	51.14± 1.8
Cu + 5 % of Al_2O_3	7.98± 0.05	93.66	51.90± 1.3
Cu + 7.5 % of Al_2O_3	7.92± 0.03	91.57	53.08± 1.7
Cu + 10 % of Al_2O_3	7.73± 0.07	92.03	54.29± 1.8

Table 2. Results of the density and hardness measurements for a composites with a alumina average particle size of 22.5 μm

Material	Density (g/cm^3)	Relative density (%)	HB
Cu	8.19± 0.02	92.01	36.65± 1.5
Cu + 2.5 % of Al_2O_3	8.04± 0.01	91.62	46.53± 1.6
Cu + 5 % of Al_2O_3	7.97± 0.04	92.23	48.58± 1.3
Cu + 7.5 % of Al_2O_3	7.84± 0.03	92.05	50.76± 1.5
Cu + 10 % of Al_2O_3	7.51± 0.06	89.49	51.90± 1.2

Table 3. Results of the density and hardness measurements for a composites with a alumina average particle size of 65 μm

Material	Density (g/cm^3)	Relative density (%)	HB
Cu	8.19± 0.02	92.01	36.65± 1.5
Cu + 2.5 % of Al_2O_3	7.97± 0.02	90.91	45.56± 1.2
Cu + 5 % of Al_2O_3	7.68± 0.05	88.87	47.54± 1.7
Cu + 7.5 % of Al_2O_3	7.60± 0.04	89.18	49.66± 1.9
Cu + 10 % of Al_2O_3	7.41± 0.01	88.23	50.76± 1.5

The examination showed that, the introduction of alumina particles into the copper matrix increases the hardness of the materials, however, these changes are small. The increasing of Al_2O_3 content (hard

IOP Conf. Series: Materials Science and Engineering **461** (2019) 012035 doi:10.1088/1757-899X/461/1/012035

ceramic particles) with its homogeneous dispersion are the main reason for the increasing of hardness [5-8]. The highest hardness was obtained for the composite containing 10 % of alumina with average particle size of 3.5 μm which amounted 54 HB (47% higher than composite made of pure copper). Along with the increase in particle size, the hardness of composites decreases. The difference in hardness between composites containing 10% aluminum oxide with an average particle size of 3.5 μm and 65 μm was almost 4 HB. Analyzing the obtained density results, it should be stated that the introduction of alumina particles decreases density of materials. Along with the increase in the content of alumina and the increase in the size of strengthening particles the density decreases. The highest density was obtained for a composite containing 2.5 % of alumina with average particle size of 3.5 μm which amounted 8.35 g/cm^3. The obtained result is caused by filling the voids of the metal matrix with fine particles of alumina. Similar results were received by other researchers [1-3].

3.3. Electrical conductivity of the sintered compacts

The results of electrical conductivity test are shown in Figure 4.

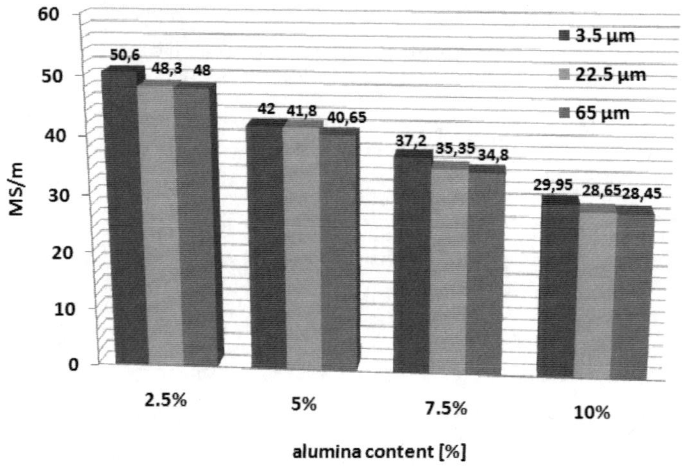

Figure 4. The results of electrical conductivity measurements of copper-alumina composites depending on amount and average particle size of alumina

The examination showed that the introduction of alumina particles into copper matrix caused decrease in electrical conductivity. This phenomenon is caused by the fact that even a small amount of alloy additives in copper causes a reduction of their electrical conductivity. Many researchers postulated that the electrical conductivity is affected by impurities. By taking into account that Al_2O_3 acts as impurities in copper matrix, so the electrical conductivity might decrease [13]. The highest electrical conductivity was obtain for a material containing 2.5 % of alumina with a average particle size of 3.5 μm which amounted 50.6 MS/m (13% less than pure copper in a solid state). Along with the increase in the content of alumina and the increase in the size of strengthening particles the electrical conductivity decreases. These results agree, to a good extent, with those reported in [14,15].

4. Conclusion

The analysis of the microstructure of the composites showed that the alumina powder particles combine into agglomerates. Before sintering, powders should be mixed thoroughly to break up the agglomerates and obtain a homogeneous mixture.

The sintering parameters were chosen correctly based on previous research and compared with the results available in the literature. The microstructural examinations showed that there were no discontinuities at the interface between the matrix and the ceramic particles. A very good bonding of alumina particles with the copper matrix was obtained, without voids, only the pores occurring in sintered metals were visible on the micrographs. Alumina particles were clearly visible in the form of irregular precipitates. The introduction of alumina particles resulted in an increase in the hardness of the composites and a decrease in the density and electrical conductivity. Higher hardness and density were observed in samples containing finer Al_2O_3 particles. An addition of Al_2O_3 and increasing the particle size of alumina as well as the amount of reinforcement increase porosity. The best properties were obtained for composites reinforced with particles of average size 3.5 μm.

5. References

[1] Konieczny M, 2012, The effect of sintering temperature, sintering time and reinforcement particle size on properties of Al-Al_2O_3 composites, *Composites Theory and Practice*, **vol.12:1**, pp 39-43.

[2] Taha M A and Zawrah M F, 2017, Effect of nano ZrO_2 on strenghtening and electrical properties of Cu-matrix nanocomposites prepared by mechanicall alloying, *Ceramics International*, **vol. 43**, pp 12698-704.

[3] Shojaeepour F, Abachi P, Purazrang K and Moghanian A H, 2012, Production and properties of Cu/Cr_2O_3 nano-composites, *Powder Technology*, **vol.222**, pp 80-84.

[4] Konieczny M, 2013, Mechanical properties of Ti-(Al_3Ti+Al) and Ti-Al_3Ti laminated composites, Composites *Theory and Practice*, **vol.13**, pp 102-106.

[5] Rajkovic V, Bozic D, Stasic J, Wang H and Jovanovic M T, 2014, Processing, characterization and properties of copper-based composites strenghtened by low amount of alumina particles, *Powder Technology*, **vol. 268**, pp 392-400.

[6] Rajkovic V, Bozic D, Devecerski A and Jovanovic M T, 2012, Characteristics of copper matrix simultaneously reinforced with nano- and micro-sized Al_2O_3 particles, *Materials Characterization*, **vol. 67**, pp 129-137.

[7] Rajkovic V, Bozic D, Popovic M and Jovanovic M T, 2009, The influence of powder particle size on properties of Cu-Al_2O_3 composites, *Science of Sintering*, **vol.41**, pp 185-192.

[8] Borowiecka- Jamrozek J, Kargul M, Borowiecki S, Engel Z, 2018, Wpływ dodatku naturalnego zeolitu na mikrostrukturę i własności mechaniczne spiekanego kompozytu o osnowie żelaza, *Rudy i metale nieżelazne*, **vol. 63:7**, pp 12-17.

[9] Martinez C, Briones F, Rojas P, Aguyilar C, Guzman D and Ordonez S, 2017, Microstructural and mechanical characterization of copper, nickel and Cu-based alloys obtained by mechanical alloying and hot pressing, *Materials Letters*, **vol.209**, pp 509-512.

[10] Zhou D, Wang X, Muransky O, Wang X, Xie Y, Yang C, Zhang D, 2018, Heterogeneous microstructure of an Al_2O_3 dispersion strengthened Cu by spark plasma sintering and extrusion and its effect on tensile properties and electrical conductivity, *Materials Science and Engineering A*, **vol.730**, pp 328-335.

[11] Rahimian M, Parvin N and Ehsani N, 2010, Investigation of particle size and amount of alumina on microstructure and mechanical properties of Al matrix composite made by powder metallurgy, *Materials Science and Engineering A*, **vol. 527**, pp 1031-38.

[12] Rahimian M, Ehsani N, Parvin N and Baharvandi H R, 2009, The effect of particle size, sintering temperature and sintering time on the properties of Al-Al_2O_3 composites, made by powder metallurgy, *Journals of Materials Processing Technology*, **vol. 209**, pp 5387-93.

[13] Hussain Z and Keong K H, 2005, Studies on alumina dispersion-strenghtened copper nanocomposite through ball milling and mechanical alloying method, *J. Teknol.*, **vol. 43**, pp 1-10.

[14] Shi Z and Yan M, 1998, The preparation of Al_2O_3- Cu nanocomposite by internal oxidation, *Appl. Surf. Sci.*, **vol. 134**, pp 103-106.

[15] Sable M J and Tripathi V K, 2011, Fabrication and study of optimum properties of copper alumina metal matrix nanocomposites for high temperature applications, *Ind. Eng. Technol.*, **vol.3:2**, pp 143-154.

IOP Conf. Series: Materials Science and Engineering **461** (2019) 012036 doi:10.1088/1757-899X/461/1/012036

Behaviour of dissimilar weld joint of steels FB2 and F during long-term creep test

J Kasl[1], D Jandová[1] and V Kanta [1]

Výzkumný a zkušební ústav Plzeň s.r.o., Tylova1581/46, 301 00 Plzeň, the Czech Republic

E-mail: kasl@vzuplzen.cz

Abstract. Creep test to the rupture of both the dissimilar weld joint made of FB2 and F martensitic steels and the base materials was carried out at temperatures ranging from 550 °C to 650 °C in the stress range from 70 to 220 MPa. Creep rupture strengths of the weld joint and the base materials were evaluated using Larson-Miller parameter. Assessment of microstructure development and changes of hardness was correlated with the creep strength. Critical zones of creep damage were determined. At lower temperatures and higher stresses the weld joint ruptured in the base material of F steel unaffected by welding, while at higher temperatures and lower stresses rupture occurred in the intercritical heated and fine-grained parts of heat affected zone of steel F. During creep at temperatures above 575 °C Laves phase precipitated in all parts of the weld joint and especially in the heat affected zones. Coarse Laves phase particles and their clusters with chromium carbides served as nucleation centres for cavities. As the fine grained heat affected zone of F steel was the softest part of the weld joint, many cavities originated and initiated causing failure of samples.

1 Introduction

There are many ongoing discussions dealing with the benefits of nuclear and steam power plants and renewable power sources in terms of economy, the environment and safety. Fossil fuel power plants cater for more than 60 % of the world's production of electricity and it is expected that they will be the most important sources of electricity in the near future in spite of the increasing role of renewable sources.

The continuous trend towards more economic electricity production together with reduced environmental pollution can be sustained by improving the thermal efficiency of power generation plants. One way to increase the efficiency of fossil power plants is to increase the temperature and pressure of the steam which finally results in the need for improved materials for the boiler and turbine design.

The class of the 9-12% Cr steels is currently used for critical components in plants operating at ultra-supercritical conditions of steam. These steels show high long-term creep strength and oxidation resistance in steam, along with ease of welding and fabrication of large forgings, castings and pipe sections. Both F and FB2 steels were developed in the under the European COST programmes [1] for the production of large scale forgings for applications with creep conditions at temperatures up to 600 °C and 620 °C respectively.

The rotor is one of the most important components of steam turbines. Large rotors for turbines of high power can be produced in the form of a solid forging or as a component welded of several parts. The welded rotors have the following advantages in comparison with solid-forged ones: application of different materials with optimal properties for different parts of a rotor at given operating conditions (properties of the rotor are gradually changed from the low pressure to the high pressure part), reduction in weight of welded parts, which can be forged more precisely than large ones, higher metallurgical accuracy, reduced heat load and faster start-up.

Doosan Škoda Power Ltd. has established a station for vertical welding of heavy components up to 135 tonnes. It, among other parts, enables automated welding of turbine rotors for single-casing turbines intended for high-temperature steam input (600 °C) as well as for low-pressure components of the highest output rating turbines. A trial dissimilar weld joint of steels COST F and COST FB2 was prepared in the conditions used for industrial welded rotor production. Specimens were machined, creep tested and analysed at the Research Testing Institute in Plzeň, Ltd. This paper deals with the results of these analyses.

2 Experiments

2.1 Materials and welding process

Dissimilar weld joints were prepared from forgings of two rings (external diameter of 600 mm and thickness of 200 mm) made of steels type COST FB 2 ((X13CrMoCoVNbN9-1-1) – the base material 1 (BM 1) and COST F (X14CrMoVNbN10-1) – the base material 2 (BM 2). Producer of both the forgings was SAARSCHMIEDE Gmbh Freiformschmiede according to the specification of Doosan Skoda Power Ltd. The quality heat treatment of forgings was 1070°C/6,5h + 570°C/12,5h + 710°C/24h and 1050°C/6h + 1100°C/6h + 570/12,5h + 720/24h for the base material 1 and the base material 2 respectively. Two filler materials were tested, namely Thermanit MTS 3 (W-CrMo91) and PSM Thermanit MTS 616 (W-ZCrMoWVNb). Welds were carried out using automated welding method 141+111 (TIG HOT WIRE) into narrow gap with internal protection by argon. The thickness of welded walls was 120 mm. Three post-weld heat treatments (PWHT) processes were applied to the weldments. On the basis of evaluation of mechanical properties, hardness and microstructure observation, the weld joint with the weld metal of THERMATIT MTS 616 and the second PWHT were chosen as the most promising variant for rotor production. For inspections of welded zones after post-heating treatments, the ultrasonic testing TOFD method was used as well as standard NDT surface inspection. The chemical compositions of the base materials COST F and COST FB2 and the filler metal used are given in Table 1.

Table 1. Chemical composition of the base materials and filler metal [w-%].

Mater.	C	Mn	Si	Cr	Ni	Mo	V	Co	W	Nb	N	B
BM1	0.13	0.34	0.08	9.6	0.17	1.48	0.2	1.32	-	0.016	0.016	0.0079
BM2	0.11	0.56	0.06	10.36	0.64	1.46	0.2	-	-	0.059	0.016	-
WM	0.11	0.42	0.3	8.87	0.57	0.56	0.18	0.15	1.49	0.051	0.018	0.0036

2.2 Sample preparation and used methods

Smooth cross-weld specimens with a length of 92 mm and a diameter of 8 mm were fabricated from the weld joint as well as from both the base materials. Long-time creep tests to the rupture of these specimens were carried out. Fracture surfaces of ruptured samples were observed using scanning electron microscope (SEM). Then specimens were cut along their longitudinal axis. Macrostructure was revealed using Villela-Bain´s reagent and location of fracture in the weldment was specified. Hardness measurement along the specimen axis was performed. Microstructure of longitudinal sections was observed using light microscopy (LM) and scanning electron microscopy (SEM). The substructure was evaluated in a transmission electron microscope (TEM) from carbon extraction replicas and thin foils. Quantitative evaluation of precipitates was performed using SEM micrographs and the image analysis program NIS–Elements.

IOP Conf. Series: Materials Science and Engineering **461** (2019) 012036 doi:10.1088/1757-899X/461/1/012036

3 Results

3.1 Mechanical properties

Integrity and mechanical properties of weld joint have been evaluated according to the welding standards EN 288-2,3. All results were satisfactory according to demands of rotor producer.

3.2 Creep tests

The creep rupture testing was carried out in air at temperatures ranging from 550 °C to 650 °C and stresses from 70 MPa to 220 MPa. The longest time to the rupture of samples so far is about 11,000 hours. A list of specimens investigated is given in Table 2. Results of creep tests as a dependence of applied stress on time to rupture are shown in Fig. 1. Obtained creep data were evaluated using the Larson-Miller parametric (LMP) equation (1):

$$P = T * (C + log\tau) \tag{1}$$

where T represents temperature given in degree Kelvin, C is a specific constant for a given material (equal 36) and τ means time to rupture in hours. Results of creep strength of the weld joint compared with the creep rupture strength data of original COST F steel [2] and creep results of the base material BM2 are graphically represented in Fig. 1.

Table 2. List of creep tested samples and some results (time to rupture in h, fracture position and LRF and SRF factors).

		Weld join		BM 2 - COST F	BM 1 - COST FB2			
Temp. [°C]	Stress [MPa]	Duration [h]	Location of fracture surface	Duration [h]	Duration [h]	Note	SRF	LRF
550	180	52 566	IC/FG HAZ F				0.97	0.45
550	200	17 626	F				0.98	0.83
575	130	39 063	IC/FG HAZ F				0.88	0.23
575	150	15 746	IC/FG HAZ F				0.98	0.37
575	200	1 354	F	799	6 406		1.03	1.59
575	220	653	F	465	3 009		1.02	2.67
600	90	21 294	FG/CG HAZ F				0.79	0.22
600	100	11 288	FG HAZ F				0.84	0.28
600	120	11 264	FG HAZ F	12 365	(41 311)	in progress	0.84	0.40
600	140	3 656	IC/FG HAZ F	9 888	(22 649)	in progress	0.99	0.56
600	160	2 712	F				1.02	0.93
600	180	577	F	586	13 626		1.09	1.65
625	80	7 137	FG/CG HAZ F	26 133	(35 056)	in progress	0.81	0.28
625	90	4 736	FG HAZ F				0.82	0.43
625	100	1 939	FG HAZ F	11 621	(23 249)	in progress	0.88	0.53
625	120	896	FG HAZ F				0.99	0.71
625	140	815	IC/FG HAZ F	372	4 985		1.01	1.04
650	70	3 552	FG/CG HAZ F	11 681			0.86	0.20
650	80	2 741	FG HAZ F (other crack)				0.89	0.52
650	100	343	FG HAZ F	394	1 410		0.94	0.98

The creep strengths of samples tested at 550°C, 575 °C, and 650 °C are in the permitted scatter band ± 20 % of the creep strength of the base material COST F. The creep strengths of the samples tested at 600 °C and 625 °C fall into the scatter band 20 % of the creep strength of the base material COST F with the exception of the sample tested at the lowest stresses, which are below this scatter band.

In order to compare welding influence on creep strength two parameters were evaluated, namely the stress reduction factor SRF

$$SRF = (\sigma_W/\sigma_{BM})_{T,t} \tag{2}$$

where σ_W and σ_{BM} are stresses needed for a given time to rupture τ at a given temperature T of the welded and non-welded material; and the life time reduction factor LRF

$$LRF = (t_W/t_{BM})_{T,\sigma} \tag{3}$$

where τ_W and τ_{BM} are the time to rupture needed for a given stress σ at a given temperature T of the welded and non-welded material BM2. Both evaluated SRF and LRF factors are shown in table 2.

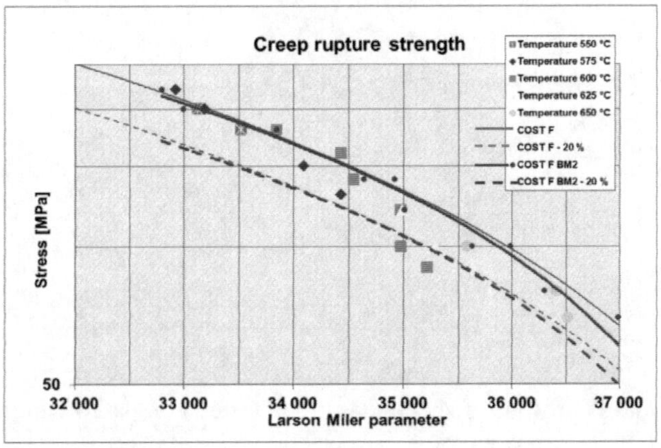

Figure 1. Creep rupture strength vs. LMP.

3.3 *Fractographic analysis*

Fractographic analysis of broken samples was performed. Firstly, it should eliminate the influence of possible defects formed during the welding process on creep results and secondly, it should find the growth mechanisms of cracks. The samples tested at lower temperatures and higher stresses failed in the base material of steel COST F unaffected by welding while those tested at higher temperatures and lower stresses ruptured in the grain refined part (FG) or in the intercritically reheated part (IC) of the heat-affected zone (HAZ) of the base material of steel COST F. Fractures in the BM occurred after short durations of creep tests. They were transcrystalline ductile with considerable macroplastic deformation (elongation about 18 %) and with the dimple morphology of the fracture surface. Fractures in the HAZ occurred by transgranular creep rupture. Elongations of these specimens were usually a few percent (figure 2).

Fractographic analysis and observation of longitudinal sections of the ruptured crept samples showed that locations of fractures depended on the creep conditions. The proper positions of fracture change from IC part of the HAZ to boundary between FG and coarse grained (CG) parts of the HAZ with increasing temperature and decreasing stress (table 2).

3.4 *Hardness profiles measurement*

Creep exposures resulted in the samples softening, which was evident from the Vickers Hardness profiles across the weld joint. The hardness of the weld metal (WM) slightly decreased, from 280 HV10 of the virgin specimen to 260 HV10 after creep at 650°C/70 MPa/3552 h. Decrease in the base materials was more significant, from 240 HV10 of the base materials to 213 HV10 of BM1 after creep at 650°C/70 MPa/3552 h and to 204 HV10 of BM2 after creep at 625°C/120 MPa/896 h. No important decrease was found in the HAZ of BM1, while a decrease from 220 to 185 HV10 occurred in the HAZ of BM2. Hardness profiles of the virgin specimen and crept specimens are shown in figure 3.

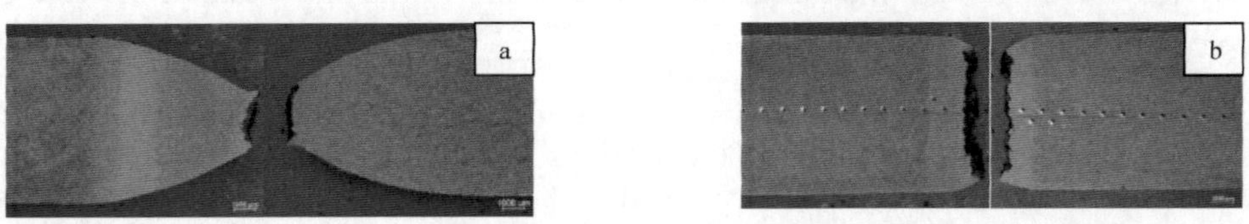

Figure 2. Macrographs of specimens creep tested at conditions: a) 575°C/200MPa/1345h and b) 600°C/80MPa/21294h.

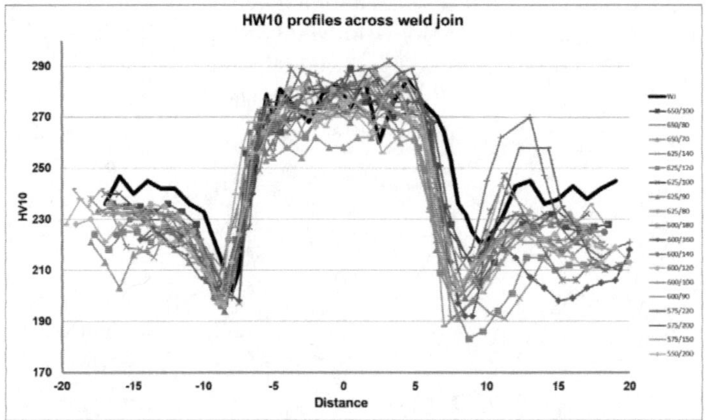

Figure 3. Vickers hardness profiles of specimens creep tested compared with not tested one (WJ).

3.5 Microstructure evaluation

The width of heat affected zones ranges from 2.5 to 3.0 mm on both sides of the weldment. Sizes of prior-austenitic grains are different in various parts of the weldment. Grain size of steel COST FB2 (about 90 μm) is significantly larger in comparison with steel COST F (about 21 μm). Average grain sizes of both the HAZs and the weld metal are comparable with grain sizes of steel F (about 20 μm). Grain sizes of coarse–grained parts of HAZs are only slightly larger compared with fine–grained parts of HAZs; from 28 to 17 μm and from 38 to 14 μm for HAZ FB2 and HAZ F respectively. Also average grain sizes in both the cast and heated parts of the weld metal are similar although the form of the grains is different.

Microstructure of the BM1 consisted of heavy tempered martensite with a small amount of δ-ferrite. Coarser prior austenitic grains contain packets of martensitic laths; some of them are subdivided into subgrains. The main contributor in suppressing the recovery of martensitic structure was a dispersion of $M_{23}C_6$ and fine vanadium carbonitride particles. However non-uniform distribution of fine vanadium carbonitrides was observed within laths. Dislocation density was about $4 \cdot 10^{14}$ m^{-2}. The WM also consisted of tempered martensite containing islands of δ-ferrite. Coarser particles of carbides rich in W, Mo, Cr, V together with some rare particles of Laves phase were observed in the WM, which is alloyed with tungsten. Dislocation density was about $5 \cdot 10^{14}$ m^{-2}. The main precipitate was $M_{23}C_6$, fine particles of V(N,C) were also present. The BM2 is fine-grained. Similar average grain size was also observed in their HAZ with only slightly higher average size in coarse-grained region. Martensitic lath structure is conserved here and laths contain subgrains only locally. On the other hand prior austenitic grains are subdivided into small equiaxed subgrains or new grains in the fine-grained region. Fine vanadium/niobium MX carbonitrides were spread within ferrite laths more often in the BM2 and also in the weld metal than in the BM1. In contrast, the density of chromium carbides was higher in the BM1 than in other parts of the weldment. Dislocation density in the BM2 was the highest, about $6 \cdot 10^{14}$ m^{-2}.

After creep testing in a temperature range from 600 to 650 °C and a stress range from 70 up to 150 MPa, cavities were detected in all parts of the weld joint. Cavitation failure in the heat affected zones on both sides of the weld joint was evident. More cavities always occurred in the HAZ of steel F and their coalescence resulted in fracture.

Some differences were found in individual zones of the weld joint, which caused premature fracture in specific zones depending on the conditions of the creep test. All structures observed corresponded to heavy tempered martensite with a high density of secondary phases. Significant differences were found in the structure recovery in the base materials, their heat affected zones and the weld metal. Small differences were found in the size and distribution of the precipitates.

Concerning the size and distribution of two of the most important secondary phases – $M_{23}C_6$ and MX particles, no important changes were found after creep tests. Intensive precipitation of Laves phase occurs in the temperature range from 600 to 625 °C. Precipitation of Laves phase occurred across the whole of the weld joint, however it was especially concentrated in the heat affected zones and the weld metal. Under the highest stress the controlling mechanism of the fracture was slip of dislocations. The grain boundaries in the HAZ pinned with $M_{23}C_6$ carbides and Laves phase particles effectively hindered the dislocation movement, and fracture occurred in BM2. Diffusion, grain-boundary sliding and dislocation climbing became more important with decreasing stress. At rather lower stresses dislocation slip and grain boundary sliding were probably equivalent processes of creep deformation. Soft small grains along previous austenite grain boundaries were deformed by slip and plastic deformation was accommodated by grain boundary sliding. In sites where soft grains were adjacent to the relatively hard lath-like base material, cavities originated as well as at grain boundaries with numerous coarse carbides.

Fractures occurred in IC HAZ, where the surface of the grain boundaries was larger than in other zones of the joint. A further decrease in stress resulted in fractures in the FG HAZ – a zone consisting of a recovered structure with soft grains separated by high angle boundaries and well developed subgrains. The slip of such grains could activate slip and consequent sliding of boundaries with jogs and coarse precipitates could originate cavities. Creep at the lowest stresses resulted in fracturing near the CG HAZ. Before creep testing, small recrystallized grains surrounded with lath-like structures with very high dislocation density occurred in this region. At high temperatures and low stresses, recovery in the regions close to the prior austenite grain boundaries played an important role. Cavities originated in these well-recovered grains, and cracks were created.

4 Conclusions

Creep testing of weld joint of COST F and COST FB2 steels designed for production rotors for fossil fuel power plants revealed sufficient creep strength, only slightly lower than that of the weaker base material steel COST F for temperatures up to 575 °C.

During creep testing the highest decrease of the hardness occurred in the intercritically reheated or in the over-tempered parts of HAZ of steel COST F where samples failed. The critical zone of the weld joint in terms of creep failure was the heat affected zone of steel COST F.

Precipitation of Laves phase and structure recovery during creep exposures at temperatures in a range from 550 to 650 °C were the main reasons for the failure. Different stages of the structure recovery and precipitation were observed in the heat affected zones in relation to the conditions of creep exposure. Consequently, contribution of individual processes taking place during creep at high temperatures – dislocation slip, climbing, grain boundary slipping and diffusion, were different in CG, FG and IC HAZ and caused a systematic shift of fracture from COST F steel to its CG HAZ. Softening of the matrix and coarsening of secondary phases as a result of Laves phase precipitation caused predominantly cavitation failure in the heat affected zone and crack propagation. Precipitation of Laves phase seems to be most significant at a temperature of 600 °C.

Acknowledgements
This work was supported by Grant TE01020118 from the Technology Agency of the Czech Republic.

References

[1] Kern T-U, Mayer K H, Donth B, Zeiler G and DiGianfrancesco A 2010 The European efforts in development of new high temperature rotor materials – COST536, in: Lecomte-Beckers J, Contrepois Q, Beck T and Kuhn B (Eds.), Materials for Advanced Power Engineering (Forschungzentrum Jülich GmbH, Jülich) pp 29–38
[2] Results of creep test of COST 522 programme; a set of internal documents available for participants of programme

IOP Conf. Series: Materials Science and Engineering **461** (2019) 012037 doi:10.1088/1757-899X/461/1/012037

Parametric study of the potential of application of high-strength steels to increasing the service fatigue life of vehicle structures

M Kepka Jr. [1] **and M Kepka**[1] **and J Tittel**[1]

University of West Bohemia, Univerzitni 8, 306 14 Pilsen, Czech Republic

kepkam@rti.zcu.cz

Abstract. The paper highlights the main impacts of HFMI technology on increasing the fatigue resistance and extending the fatigue life of joints and components welded from both conventional and high strength steels. Parametric calculations of the permissible stresses of a selected structural detail are performed, assuming its application in the bus construction, which is subjected to random loading in real operation.

1 Introduction

Steel welds are still being used in vehicle constructions. For some welded nodes, increased fatigue strength can be achieved if one of the HFMI (High Frequency Mechanical Impact) technology is applied. This is especially important for details made of high-strength steels. Of course, vehicle manufacturers are interested in whether the application of the new technology will be beneficial for real vehicle operation. Therefore, this parametric study was carried out which used the knowledge about stress spectra in the bus constructions and the recommendations of the International Institute of Welding for improving the fatigue strength of welded joints with HFMI treatment.

2 HFMI treatment for improving the fatigue strength of welded joints

Various names have been used in literature for HFMI technology: ultrasonic impact treatment, ultrasonic peening, ultrasonic peening treatment, high frequency impact treatment, etc. The working principal is identical: cylindrical indenters are accelerated against a component or structure with high frequency. The impacted material is highly plastically deformed causing changes in the material microstructure and the local geometry as well as the residual stress state in the region of impact. Weld geometry is improved and compressive residual stresses are induced. Initiations and development of fatigue cracks are prevented. Everything about the improving the fatigue strength and durability by the HFMI application is best summed up in the publication [1]. This guideline is applicable to steel structures of plate thicknesses of 5–50 mm and for yield strength ranging from 235 to 960 MPa. The publication is linked to the recommendations for fatigue design of welded joints and components [2].

The fatigue assessment of classified structural details and welded joints is based on the nominal stress range. Each fatigue strength S-N curve is identified by the characteristic fatigue strength of the detail in MPa at 2 million cycles. This value is the fatigue class (FAT). The slope of the fatigue strength S-N curves for details assessed on the basis of normal stresses is $m = 3$, if not stated expressly otherwise. The constant amplitude knee point is assumed to correspond to $N = 10^7$ cycles.

New experimental data indicate that a fatigue limit does not exist and the S-N curve should continue on the basis of a further decline in stress range of about 10 % per decade in terms of cycles, which corresponds to a slope of $m_d = 22$. For variable amplitude loading (for application of Palmgren-Miner summation) is recommended slope $m_d = 2 \cdot m - 1 = 5$. An example of a set of the resistance S-N curves is shown in figure 1.

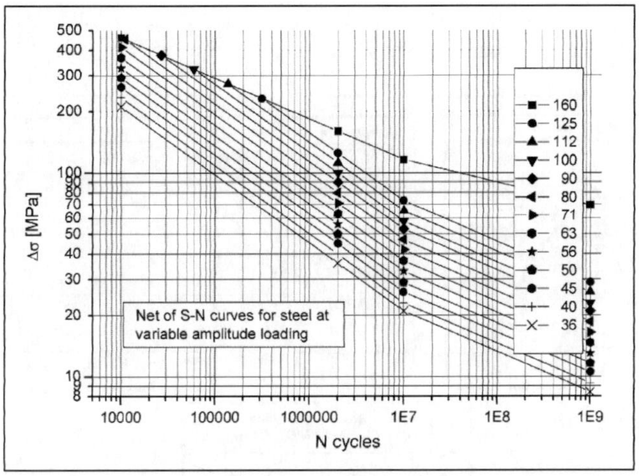

Figure 1. Resistance S-N curves at variable amplitude loading (steel, nominal stress) [2].

The most of the fatigue design methods for HFMI improved welds are based on an assumed S-N slope of m = 5, and fatigue strength improvement factor are defined at $N = 2 \cdot 10^6$. For steel with a yield strength of less than 355 MPa, HFMI processing improves to up to 4 FAT classes. For steels with a yield strength above 355 MPa, the HFMI processing is reflected by an improvement in one additional FAT class for each increase in the yield strength by 200 MPa, see figure 2.

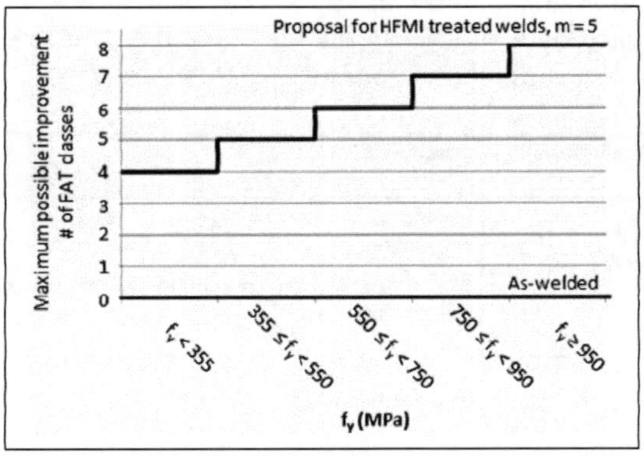

Figure 2. Maximum increase in the number of FAT classes as a function of yield strength f_y [1].

HFMI treated welds can have up to 8 FAT classes of improvement (!) depending on the material strength, welded joint geometry, etc. However, some factors limit and modify this optimistic situation. The detailed explanation is given in [1], here is just a rough summary of three important factors that are later taken into account in parametric calculations. The HFMI improvement is applied to the weld toe and is intended to increase the fatigue live of the weld threated from the view point of potential fatigue failure from the weld toe. The benefit of HFMI treatment can be claimed only for details in design Class FAT 50 to FAT 90 in the IIW notation for S-N curves. The positive effect of the technology reduces the high tension pre-stress. The stress ratio influence is expressed as a penalty with respect to the maximum increase in the number of FAT classes. In the interval ($0.15 < R \leq 0.52$) this reduction is 1 to 3 FAT classes. With the increasing stress ratio R, significant reduction is applied.

3 Parametric study

3.1 Procedure

In this study, we followed the procedure as shown in the diagram in figure 3.

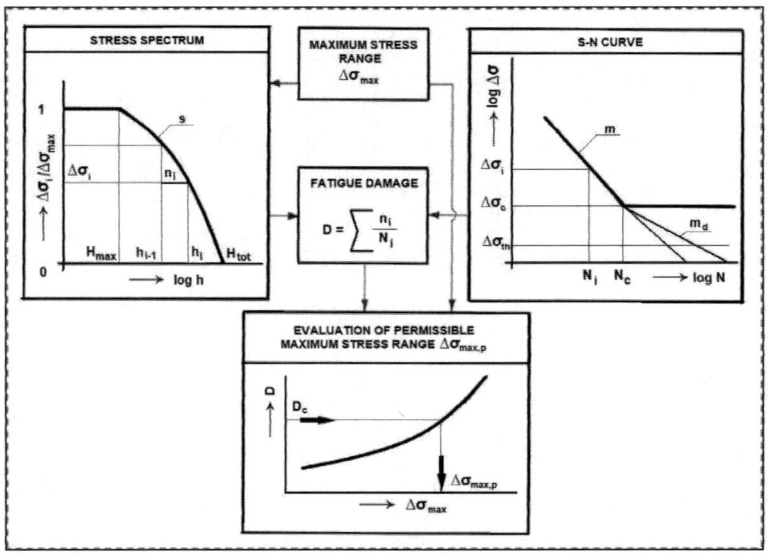

Figure 3. Evaluation of permissible maximum stress range $\Delta\sigma_{max,p}$ of design stress spectrum [5].

3.2 Structural detail and materials

For our parametric study (asked by a bus producer), we chose the structural detail that can occur in a bus construction. The selected detail is shown in figure 4 and has the FAT 71 category.

Structural Detail	Description	FAT
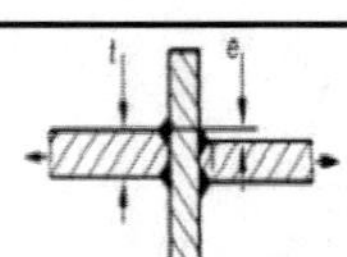	Cruciform joint or T-joint, K-butt welds, full penetration, potential failure from weld toe Misalignment < 15 % of primary plate thickness in cruciform joints	71

Figure 4. Structural detail used for parametric study [2].

We considered the following material alternatives:

- $f_y < 355$: conventional structural steel, e.g. S355J0;
- $355 \leq f_y < 550$: high-strength structural steel, e.g. WEATHERING 550W;
- $550 \leq f_y < 750$: high-strength structural steel, e.g. STRENX 700 MC.

3.3 S-N curves

We considered: maximum FAT classes improvement and 3 FAT classes reduction with respect to cycles with positive stress ratio R that may occur in random load process. The resulted S-N curves are plotted in figure 5.

- Weldable structural steel, as welded / no HFMI (+0 FAT, m = 3, $m_d = 5$);
- S355J0 + HFMI (+1 FAT, m=5, $m_d = 9$);
- WEATHERING 550W + HFMI (+2 FAT, m = 5, $m_d = 9$);
- STRENX 700 MC + HFMI (+3 FAT, m = 5, $m_d = 9$).

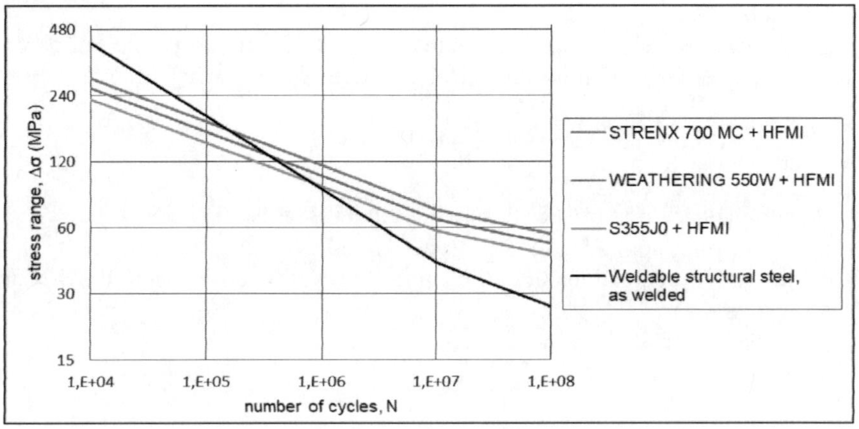

Figure 5. Considered resistance S-N curves.

From figure 5, it can be easily estimated (using exact parameters of the S-N curves precisely calculated) what extension of the fatigue life can potentially be achieved at a harmonic loading, what maximum stress range can be allowed for the required number of harmonic load cycles.

3.4 Stress spectra

At the design stage of a bus structure, no accurate data on its service loading is available. Therefore, design load (stress) spectra are employed in this parametric study. The design stress spectra were generated using the relative coordinates $\Delta\sigma i / \Delta\sigma max$ and the equation proposed in [3]:

$$h_i = H_{tot} \cdot \left(\frac{H_{max}}{H_{tot}}\right)^{\left(\frac{\Delta\sigma_i}{\Delta\sigma max}\right)^s} \quad (1)$$

$\Delta\sigma_{max}$ - maximum stress amplitude in the spectrum;
H_{max} - number of cycles with $\Delta\sigma_{max}$ amplitude in the spectrum;
H_{tot} - total number of cycles in the spectrum;
s - shape parameter of the spectrum;
h_i - cumulative frequency of cycles with an amplitude of $\Delta\sigma_i$.

Long-term monitoring of run on irregular road surfaces lead to linear distribution stress spectra with shape parameter s → 1, manoeuvres (curve riding, braking, etc.) lead to stress spectra with normal distributions and s → 2 [4]. Both types of stress spectra can occur on a bus construction.

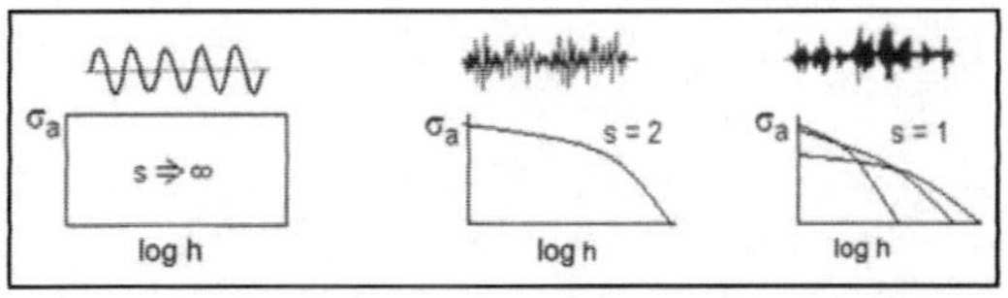

Figure 6. Characteristic shapes of design stress spectra [4].

In our parametric calculations, we considered both values s = 1 and s = 2. Further parameters of the design spectra were taken from the literature [5]. The design life was set as $L_d = 10^6$ km with the estimation of $H_{tot} = 7.2 \cdot 10^8$ cycles. The occurrence of the load cycle with the maximum stress range $\Delta\sigma_{max}$ was considered once per 100 km of run, so $H_{max} = 10^4$ cycles during the design life.

3.5 *Calculation of fatigue damage*

Fatigue damage D was calculated using the Haibach-modified version of the Palmgren-Miner rule. According to this rule, the fatigue limit state is reached when the following condition is met:

$$D = \sum_i \frac{n_i}{N_i} = D_c \qquad (2)$$

D - fatigue damage caused by the stress spectrum imposed;
n_i - number of cycles applied at the i-th level of stress with the range $\Delta\sigma i$;
N_i - limit life under identical loading (the number of cycles derived from the S-N curve at the stress range $\Delta\sigma_i$);
D_c - critical (limit) value of fatigue damage.

3.6 *Maximum permissible stress ranges*

At a constant width of the classification interval $d\Delta\sigma$, one can derive the discrete stress spectrum ($\Delta\sigma_i$ - ni). An absolute class frequency is calculated ni = hi-1 – hi and this number of cycles is assigned to the mid-point of the class (or, safely, to its upper limit) $\Delta\sigma_i$. Then, considering the S-N curve and the fatigue life accumulation hypothesis, the procedure shown in figure 3 can be applied.

The D = f($\Delta\sigma_{max}$) functions were evaluated for 2 types of design stress spectra (with shape parameter s=1 and s=2) and for 4 considered resistance S-N curves (see figure 5). The maximum permissible stress ranges $\Delta\sigma_{max,p}$ can be deduced from the functions plotted in figures 7 and 8. The maximum stress range $\Delta\sigma_{max,p}$ can be derived for the critical damage D_c. If these value $\Delta\sigma_{max,p}$ is not exceeded, the design life L_d should be guaranteed.

Estimates of permissible stress ranges $\Delta\sigma_{max,p}$ are influenced by the choice of the critical value of fatigue damage D_c. According to the original version of the Palmgren-Miner rule $D_c = 1$, but D_c can be within the wide range and non-conservative. This problem can be partially solved by recommending a value of $D_c = 0.5$. The most correct solution is to calibrate this value D_c based on a properly designed and correctly implemented experimental program (relative Palmgren-Miner).

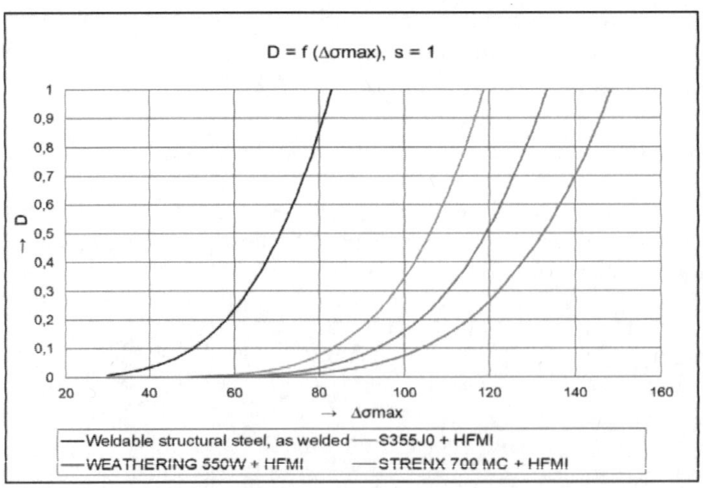

Figure 7. Functions for evaluation of maximum permissible stress range $\Delta\sigma_{max,p}$.

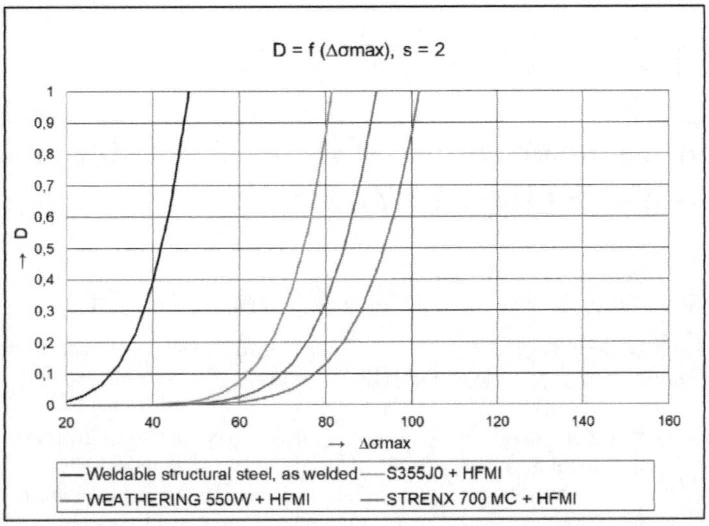

Figure 8. Functions for evaluation of maximum permissible stress range $\Delta\sigma_{max,p}$.

4 Conclusions

The study confirmed the potential of production of vehicle structures welded from high-strength steels. Taking into account HFMI process during development and production, on same load level and same lifetime, the construction can be slimmed down specifically. Considering the additional benefit of the weight advantage e.g. the achievable payload in vehicles can be increased. Using the parametric calculations with the structural detail suitable for HFMI application and theoretically made of steels with different strengths we have also proposed the procedure for determining the maximum permissible stresses at which the design lifetime can be guaranteed in case of random loading of the vehicle structure in real operation. In real practice it´s of course necessary to take into account the question of the stability of the increase the fatigue strength achieved by HFMI under the effect from real operation (for example due to the sudden overload, etc.).

Acknowledgment
This parametric study was supported by the Technological Agency of the Czech Republic, Project No. TE01020038. The support is gratefully acknowledged.

References
[1] Marquis G B and Basroum Z 2016 IIW Recommendations for the HFMI Treatment - For Improving the Fatigue Strength of Welded Joints (International Institute of Welding, Springer) p 34
[2] Hobbacher A F 2016 Recommendations for Fatigue Design of Welded Joints and Components – Second Edition (International Institute of Welding, Springer) p 143
[3] Heuler P and Klätschke H 2005 Generation and use of standardised load spectra and load–time histories, *International Journal of Fatigue* **27** (2005) 974–990
[4] J. Neugebauer, V. Grubisic, G. Fischer: Procedure for Design Optimization and Durability Life Approval of Truck Axles and Axle Assemblies (SAE Technical Paper Series 892535, 1989)
[5] Kepka M and Kepka Jr M 2018 Design, service and testing grounds stress spectra and their using to fatigue life assessment of bus bodyworks (MATEC Web of Conferences 165, 17007)

IOP Conf. Series: Materials Science and Engineering **461** (2019) 012038 doi:10.1088/1757-899X/461/1/012038

The effect of the laser spots overlap on the structure during square parts quenching of 1.7225 steel

V Kirschner, M Švec, A Škrétová, L Dvořák

VÚTS, a. s., Laser Application Center, Svárovská 619, Liberec 11, 460 01, Czech Republic, E-mail: vit.kirschner@vuts.cz

Abstract. Laser quenching is a modern technology of material processing. A high speed of the process, extremely high speeds of heating (in order of 1000 °C/s) and absence of cooling medium after heating are the biggest advantages of heat treatment by unconventional technology. Moreover, laser quenching could be in some cases only one possible way how to heat treat the material's surface. This article deals with the problematics of laser spots overlap during square parts quenching. It is necessary to set optimal distance between two neighboring tracks. If this distance is too low, the second pass of the laser can temper the previously hardened track. On the other hand, the stripe completely without heat treatment remains in the surface if a too long distance between the quenching tracks is used. The optimal distance of laser spots at quenching of 1.7225 steel was searched under present research. It was verified that at least 4 mm overlap is needed.

1 Introduction

Laser technologies are one of the modern ways of processing material. Nowadays, they are applied in a number of engineering applications from cutting, welding to heat treatment. The most advantages of laser technologies are their productivity, economy, ecology and moreover, they are often the only one alternative for the production of a particular part. One of the most used applications of lasers is in material's surface heat treatment. The component life is dependent on the quality of the material's surface because almost all fatigue cracks and material's wear start on free surface [1]. Therefore, the possibilities are being looked for an increase of material's surface layer resistance. Surface quenching is one of the possible ways how to prevent from cracks formation and premature wear [2, 3]. Moreover, the surface quenching by using laser technology has many advantages in comparison with conventional methods of thermal treatment [4]. Laser quenching reaches heating rates in an order of 1000 °C/s, thanks to that, only a thin surface layer of material will be thermally affected. No quenching medium is also necessary because the heat from the surface layer is taken away into the unheated core of a material, and thus, material self-quenching will occur. A significant advantage is also the formation of minimal undesired deformations in the material because the thermal affected area is very narrow. The width of the quenched zone can be easily modified by replacing the lens or changing of quenching parameters [5, 6]. The repeatability of the process is high because the whole process is controlled by an industrial robot [7].

When a surface quenching of square parts by means of a narrow laser spot is carried out, it will not occur a hardening of the entire surface being processed with one pass of the quenching head, but a gradual rasterization of the area to be heat-treated must be performed. During this rasterization, the distance of each quenching "tracks" must be suitably adjusted to avoid an unwanted interaction of the already quenched material with the laser beam passing in the next step by one track next to the treated surface. If the overlap of the individual spots is set too large, the first track may be accidentally tempered during the second pass of the beam and thus, it may affect the resulting mechanical properties. If, on the contrary, there is an insufficient overlap of the spots, narrow stripes will be formed between the quenched tracks in the material which will remain heat-untreated, which will again affect the resulting mechanical properties [8-10].

Adjusting the appropriate laser spot overlap degree is different for each type of material. This paper deals with the study of an appropriate degree of overlap of laser spots on steel 1.7225 (ČSN 15142).

2 Materials and experimental methods

The optimal degree of overlap of laser spots was studied on 1.7225 steel (15 142 sample). The chemical composition of the investigated material is presented in *Table 1*. The material was quenched by a Nd:YAG laser Laserline LDF 4000 with a maximal power of 4 kV and with a wavelength of 1020 – 1060 nm. The laser set was supplemented by an industrial robot KUKA KR60 HA with a repeatability of ± 0.05 mm. The sample 15 142 was quenched from 1000 °C on air, the welding speed was set on 0.003 m/s. The laser spot had a dimension of 16 x 4 mm. There were tested three variants of laser spots overlap for each material – 0 mm overlap (0), 2 mm overlap (+2) and 4 mm overlap (+4) – the scheme of the experiment is shown in *Fig. 1*. The depth of material quenching was measured from a record of hardness gradient measurement. The 550 HV1 hardness was considered for the fully quenched material. The hardness gradient was measured on a MicroMet 6020 microhardness tester at a load of 1 kg (HV1). The structure and phase compositions were studied by the Zeiss Ultra Plus and Tescan Vega SB scanning electron microscopes equipped with Oxford detectors for energy dispersive analysis (EDX). The overview structures were studied by a light optical microscope (LOM) Nikon Epiphot 200.

Table 1. The chemical composition of the investigated materials

Sample	C	Si	Mn	Cr	Ni	Mo
15 142	0.38		0.60	0.90		0.15
(1.7225 steel)	0.45	0.40	0.90	1.20	0.50	0.30

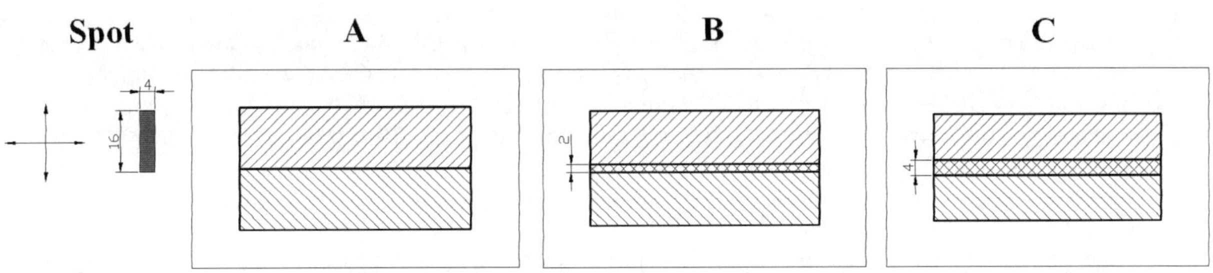

Figure 1. The schema of three investigated variants of laser spots overlap (A - 0 mm, B - 2mm, C - 4 mm).

3 Results and discussion

3.1 Unquenched 15 142 steel

The initial material used for tests of laser spots overlap was ferritic – perlitic type steel. The ferritic grains are visible as a darker background in SEM *Fig. 2*. Perlite is present in the granular form (small lighter areas in *Fig. 2*) – previously referred as coarse sorbite. There were also observed the elongated precipitates in the structure (see black needle-like precipitate in detail in *Fig. 2*). These particles were identified by EDX as manganese sulfides. The hardness of the ground material was 242 HV1.

Figure 2. Structure of ground unquenched 15 142 material (SEM, 2000x, SE, HV 20 kV, 3% Nital etchant)

3.2 The overlap of 0 mm (0)

The first tested overlap value was 0 mm. It is clearly visible from *Fig. 3* that this overlap is completely insufficient for a continuous quenching of surface parts. There is no overlap in the axis between the individual quenching spots. The width of the unquenched belt was about 3.15 mm. In comparison to the previously investigated and published materials 19312 [11] and 12050 [12], the width of this unquenched belt was larger; 0.41 mm for material 12050 and 1.6 mm for material 19312. From these data, it can be concluded that the hardenability of the material studied within this article is the lowest of all three examined steels.

Fig. 4 shows the structure of the fully quenched area. The quenching speed was not sufficient for the fully martensitic structure formation. Martensitic needles were formed only in some places where the super-critical cooling speed has been achieved. The upper bainite was the main phase in the structure. The clusters of carbides are also visible in *Fig. 4*. In the belt between the unquenched material and the fully quenched area, there was formed a transition band. The structure of this transition area is formed by ferritic grains with dimensions of 5 – 10 μm (see *Fig. 5*). These ferritic grains are soft and may adversely affect the resulting properties of the quenched layer. The effective depth of the quenched area in the axis of the laser spot is 430 μm at the used process parameters (measured by microhardness testing – see *Fig. 6*).

Figure 3. Quenched material – 0 mm overlap of laser spots (0) – LOM, 50x, 3% Nital etchant

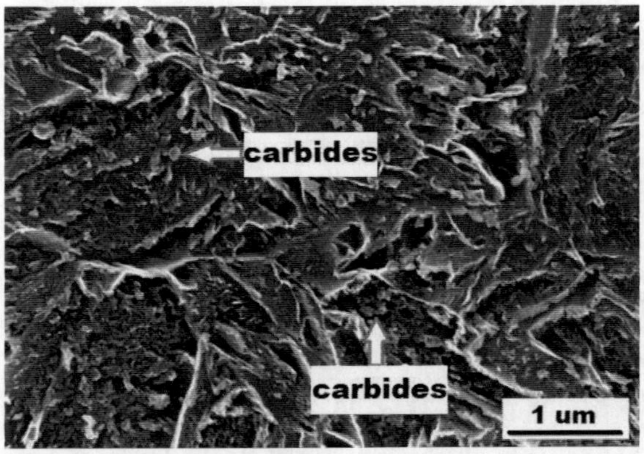

Figure 4. Fully quenched area of 15 142 steel – 0 mm overlap of laser spots (SEM, 1000x, InLens, HV 20 kV, Nital etchant)

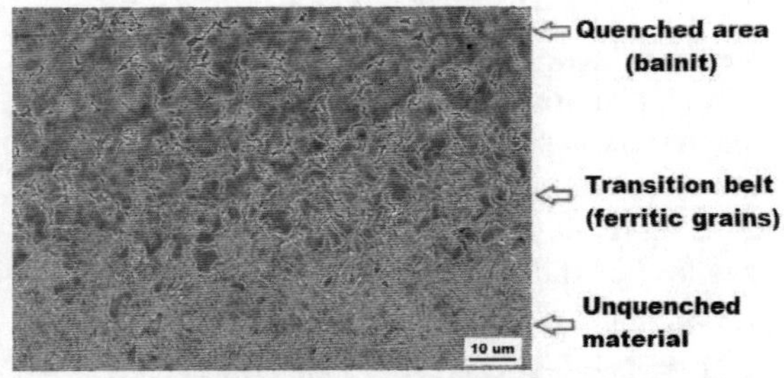

Figure 5. The transition area between fully quenched and unquenched material – 0 mm overlap of laser spots (SEM, InLens, HV 20 kV, Nital etchant)

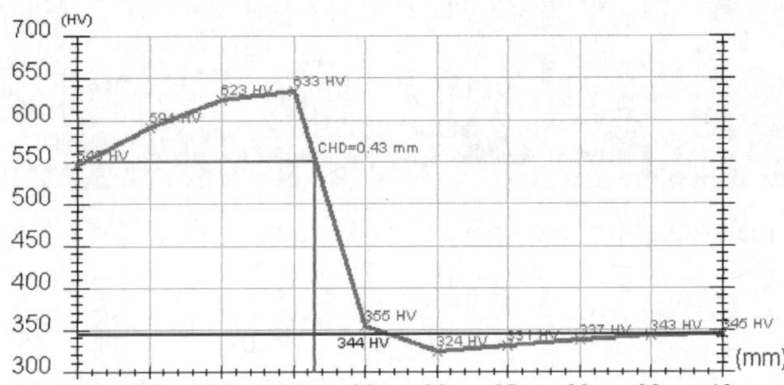

Figure 6. A record of hardness gradients of the fully quenched belt in the axis of laser spot – maximal depth of steel quenching is about 430 μm (test variant "0")

3.3 The overlap of 2 mm (+2)

When selecting the laser spot overlap of 2 mm (+2), the individual quenched tracks are still not overlapped, as shown in *Fig. 7*. The unquenched area between the individual quenching spots has a width of about 1 mm according to the image analysis of *Fig. 7*. The assumption that the hardening capability of 15142 sample is lower in comparison to the previously investigated materials [11, 12] was confirmed. The surface was completely quenched in the axis between the individual laser spots in the cases of 12050 and 19312 steels [11, 12]. The depth of these quenched belts was 413 μm for steel 12050 and 170 μm for steel 19312.

Figure 7. Quenched material – 2 mm overlap of laser spots (+2) – LOM, 50x, 3% Nital etchant

3.4 The overlap of 4 mm (+4)

Value selection of 4 mm (+4) shows some overlap, as it is visible from *Fig. 8*. However, this overlapped area was not quenched to martensite, but only to bainite state, as it was determined from the hardness gradient measurement (*Fig. 9*). The depth of bainitic area in the axis between the individual laser spots is about 140 μm. It can be summarized that the quenched belts overlapping starts at using the laser spots overlap of 4 mm (+4). On the other hand, the depth of the hardened material is too low at this overlap value. Therefore, the overlap higher than 4 mm is recommended for steel 1.7225 during square parts quenching.

Figure 8. Quenched material – 4 mm overlap of laser spots (+4) – LOM, 50x, 3% Nital etchant

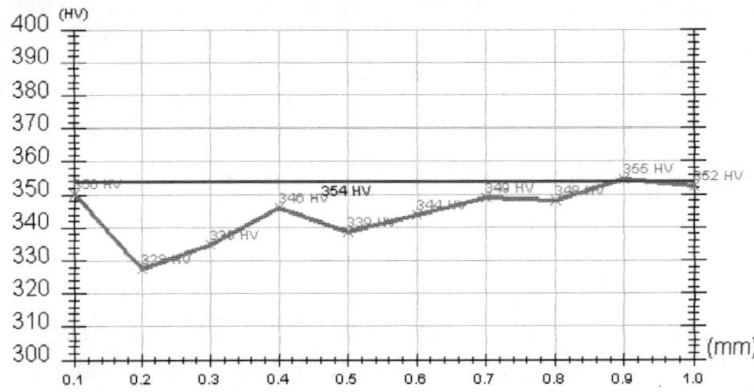

Figure 9. A record of hardness gradients in the axis between the quenched belts for variant +4

4 Conclusions

- A maximal depth of material hardening (in the axis of the laser spot) of 430 μm can be achieved for steel 1.7225 by laser quenching (hardening) using the test parameters.
- When selecting a laser spots overlap of 0 mm (0) or 2 mm (+2), the surface is not completely quenched. The unquenched belt in the axis between two laser spots has a width of 3.15 mm for variant "0" and 1 mm for variant "+2". The overlap was observed only in the case of "+4", however, the quenched depth in the axis between two laser spots was only 143 μm; moreover, this area was not created by martensite structure, but only the bainite.
- Hardening depths in the axis between the spots for each overlap variant and also for the previously tested materials [11, 12] are summarized in *Table 2*.

Table 2. The depth of the hardened layer in the axis between the spots for investigated steel 15142 (1.7225) and also for previously tested steels 12050 (1.1191) [12] and 19312 (1.2842) [11]

Overlap	Hardening depth in the axis between the spots [μm]		
	steel 15142 (1.7225)	steel 12050 (1.1191)	steel 19312 (1.2842)
0 mm	0	0	0
2 mm	0	413	170
4 mm	143 (only bainitic state)	896	580
Maximum depth of material hardenability at used laser parameters	430	1105	950

Acknowledgements
This publication was written at the VÚTS, a.s. with the support of the project LO1213 – "Excellent engineering research", as provided by the Ministry of Education, Youth and Sports of the Czech Republic in the program of "Národní program udržitelnosti" and with the support of TIP Project under the FR-TI1/604 grant. Our acknowledgment belongs to all donors that helped us to reach the results of the work.

References

[1] Veles P 1989 Mechanické vlastnosti a skúšanie kovov (Bratislava: Alfa Publishing)
[2] Zhikai CH et al. 2018 Behaviors of 40Cr steel treated by laser quenching on impact abrasive wear. In *Optics and Laser Technology* vol 103 (Amsterdam: Elsevier) pp 118-125
[3] Yuda CH et al. 2018 Influences of local laser quenching on wear performance of D1 Wheel steel. In *Wear* (Amsterdam: Elsevier) In Press
[4] Guan Y H et al. 1997 The Prediction of the Mechanical Properties of Metal during laser quenching In *Journal of Materials Processing Technology* vol 63 (Amsterdam: Elsevier) pp 614-617
[5] Zhaoyun CH et al. 2012 Microstructure and hardness investigation of 17-4PH stainless steel by laser quenching. In *Materials Science and Engineering: A* (Amsterdam: Elsevier) pp 536–541
[6] Lee J H et al. 2009 Laser surface hardening of AISI H13 tool steel. In *Trans. Nonferrous Met. Soc. China* vol 19 (Amsterdam: Elsevier) pp 917-920
[7] Martínez S et al. 2012 Laser hardening process with 2D scanning optics. In *Physics Procedia* vol 39 (Amsterdam: Elsevier) pp 309-317
[8] Qingbin L Hong L 1999 Experimental study of the laser quenching of 40CrNiMoA steel. In *Journal of Materials Processing Technology* vol 88 (Amsterdam: Elsevier) pp 77-82
[9] Yao CH et al. 2010 Study of the softening in overlapping zone by laser-overlapping scanning surface hardening for carbon and alloyed steel. In *Optics and Lasers in Engineering* (Amsterdam: Elsevier) pp 20-26
[10] Jiang J 2011 Discrete laser spot transformation hardening of AISI 01 tool steel using pulsed Nd:YAG laser. In *Surface and coatings technology* (Amsterdam: Elsevier) pp 5156-64
[11] Kirschner V, Švec M, Bruthansová R and Dvořák L 2014 The effect of the laser spots overlap on the structure during square parts quenching. In *Comat 2014*. ISBN 978-80-87294-45-1
[12] Kirschner V, Švec M, Bruthansová R and Dvořák L 2015 The effect of the laser spots overlap on the structure during square parts quenching of 1.1191 steel. In *Metal 2015*. ISBN 978-80-87294-58-1

IOP Conf. Series: Materials Science and Engineering **461** (2019) 012039 doi:10.1088/1757-899X/461/1/012039

Assessment of the predictive capabilities of commercial simulation software packages for the analysis of the C redistribution during Q&P processing

T Klein[1], M Lukas[1], M Galler[2], and G Ressel[1]

[1] Materials Center Leoben Forschung GmbH, Roseggerstraße 12, 8700 Leoben, Austria

[2] voestalpine Wire Rod Austria GmbH, Drahstraße 1, 8792 St. Peter Freienstein, Austria

E-mail: thomas.klein@mcl.at

Abstract. The C enrichment of austenite is of great importance to different kinds of transformation-induced plasticity-aided steels. Particularly, quenching and partitioning – a two-step heat treatment process, which involves the partial transformation to martensite and the subsequent C partitioning to austenite and its concomitant stabilization – has received considerable attention in recent years. The present work assesses the applicability of commercial software packages (DICTRA and MatCalc) to predict the interplay of C enrichment of austenite and concomitant C depletion of martensite. Results are additionally compared to calculations based on the "constrained carbon equilibrium" (CCE) and, moreover, to experiments conducted for two exemplary material conditions. DICTRA simulations extend the CCE calculation in terms of time dependence and allow to account for the possible displacement of the phase boundary, but lack the implications of the martensitic structure. Predictions based on MatCalc more accurately reflect experimental observations. Data gained allows for the evaluation of the predictive capabilities of commercial simulation packages corresponding to a prerequisite for the industrial implementation of novel heat treatment strategies and the required process control.

1 Introduction

The stabilization of considerable fractions of austenite via different processing routes is decisive to meet ductility requirements of advanced high strength steels (AHSS), which make use of the transformation-induced plasticity (TRIP) effect [1]. For this purpose, a suitable means is its stabilization via C enrichment by quenching and partitioning (Q&P) processing as originally introduced by Speer et al. [2], who derived a framework known as the "constrained carbon equilibrium" (CCE) for the calculation of the partitioning of C within a martensite/austenite assembly under given boundary conditions.

In the two-step Q&P process – a schematic representation of the heat treatment and the redistribution phenomena involved is shown in Figure 1 – the material is first cooled rapidly to the quenching temperature (T_Q) between martensite-start (M_s) and martensite-finish (M_f) temperatures allowing for the formation of martensite, whereby the temperature chosen allows adjusting a targeted martensite fraction. Subsequently, the material is isothermally heat treated at the partitioning temperature ($T_P \geq T_Q$) for a duration t_p with the objective of C induced stabilization of the austenite, which can then be to a large extent retained to room temperature. The mechanism of C partitioning from martensite to austenite was recently corroborated using diffraction techniques [3], thermal analysis [4] and atom probe tomography [5]. Originally, the fundamental mechanisms of C escaping from a ferrite plate were discussed in the seminal works of Mujahid and Bhadeshia [6] and Hillert et al. [7] considering the C redistribution between austenite and ferrite following its displacive formation.

This study aims at comparing the predictive accuracy of state-of-the-art commercial software packages that allow for the calculation of the C redistribution during complex heat treatments. Gained data is compared to the classical CCE treatment as well as experimental results and discussed in the frame of current literature. Thereby, we aim at highlighting the strengths of thermodynamic and thermokinetic modelling techniques, which have become decisive for material and process development and their industrial implementation.

2 Materials and methodologies

In the present study, a low alloyed steel with nominal chemical composition of Fe-2.5Mn-1.5Si-0.2C (mass percent, m.%) was investigated. The material was delivered as sheets with 4 mm thickness, whereof cylindrical specimens were fabricated using electric discharge machining. Both heat treatments were conducted using a DIL 805 L/A dilatometer from TA Instruments according to the following scheme: Heating with a rate of 2 °C/s to an austenitization temperature of 900 °C, where the material was isothermally annealed for 1800 s followed by quenching with a rate of 10 °C/s to T_Q = 290 °C. After 10 s the material was reheated with a rate of 2 °C/s to T_P = 400 °C, where the isothermal partitioning heat treatment was conducted for 120 s or 600 s followed by final cooling to room temperature again with 10 °C/s.

Optical light microscopy (OLM) was performed using an Olympus BX51 microscope. Metallographic specimens were prepared by mechanical grinding and polishing followed by an etching procedure using Klemm 1 etchant. X-ray diffraction was performed using a Bruker AXS D8 Discover diffractometer with Mo-K_α radiation in Bragg-Brentano geometry. Phase fractions present were analyzed quantitatively using Rietveld refinement with the software package Topas version 5 from Bruker AXS. The austenite C concentration was evaluated from the empirical relationship suggested by Ruhl and Cohen [9] according to equation (1):

$$a_\gamma = 3.572 + 0.033w^C + 0.0012w^{Mn} - 0.00157w^{Si} \quad (1)$$

relating the austenite lattice constant a_γ to the concentrations of substitutional elements Mn and Si and the interstitially dissolved C each denoted by w^i given in m.%. The martensite C concentration was estimated based on the lever rule.

The thermokinetic calculations of the present study were conducted in DICTRA under the local equilibrium hypothesis [10] using tcfe8 (thermodynamic) and mobfe3 (mobility) databases and MatCalc 6.0 using Mc_fe.tdb v2.059 (thermodynamic) and Mc_fe.ddb v2.011 (mobility) databases. Results presented were obtained for isothermal annealing conditions at 400 °C using Fe-0.2C (m.%) model system similarly to the simplifications proposed by Seo et al. [11], who suggested that the slow partitioning of substitutional elements at 400 °C does not affect the redistribution of C. During partitioning the formation of cementite was disregarded and phase fractions were fixed. Additionally, the influence of dislocations in the martensitic structure on C concentrations was accounted for in MatCalc using the formulations of Refs. [12–14] assuming a fixed dislocation density, which is reasonable at 400 °C [15]. Calculations were conducted using Neumann boundary conditions, i.e. closed system scenarios. Additionally, the formulations according to Speer at al. [2], i.e. the CCE condition, was used to estimate and compare the C concentrations of the involved phases.

3 Results and discussion

3.1 Experimental results

Microstructure images taken by OLM of the material conditions investigated in the present work are depicted in Figure 2. (a) shows the microstructure after 120 s and (b) shows the microstructure after 600 s of partitioning heat treatment each at 400 °C. Blue and brown specimen regions correspond to martensite (α'), whereas white areas correspond to retained austenite (γ_R). Blocky retained austenite with dimensions up to ≈ 3 µm is discernible, while fine filmy austenite present in these material conditions may not be resolvable using OLM [11]. Visually no differences in microstructure between the two specimen conditions are observable using OLM.

The austenite phase fractions evaluated using Rietveld refinement of X-ray diffractograms are similar in both material conditions, $f_\gamma(120\text{ s}) \approx 9.7$ m.% and $f_\gamma(600\text{ s}) \approx 9.8$ m.%, i.e. appear not to diminish during an extended partitioning sequence. Upon increasing the partitioning duration an increase in austenite C concentration up to 1.2 m.% is discernible, which, however, remains clearly below the maximum value of ≈ 2.02 m.% given by full redistribution from martensite into austenite using the experimental phase fraction and austenite C concentration. Thus, a substantial amount of C is trapped at dislocations and interfaces of martensite and is, therefore, not available for stabilization of the retained austenite, which is a reasonable assumption according to e.g. Ref. [16].

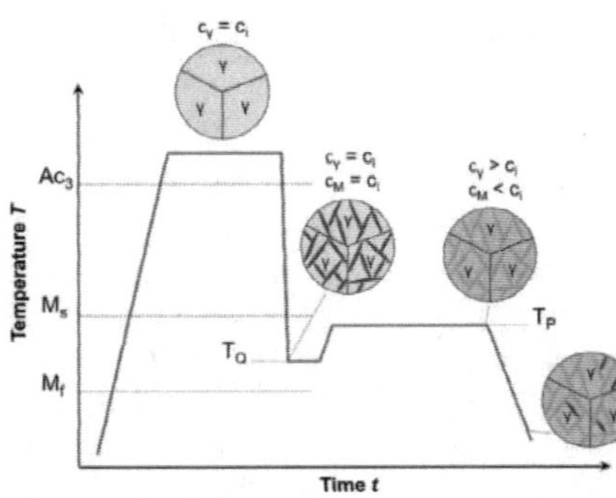

Figure 1. Schematic representation of the Q&P heat treatment and the evolution of C concentrations in the constituent phases redrawn according to Ref. [8].

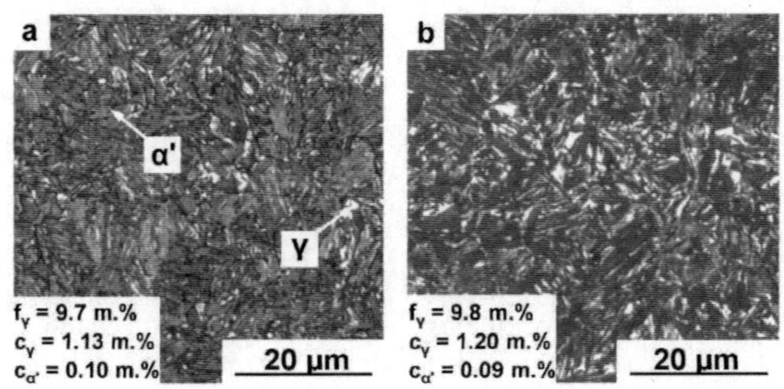

Figure 2. Microstructure images of the investigated Q&P material conditions of (a) t_p = 120 s and (b) t_p = 600 s. Apparently, morphology and phase fractions are hardly affected by the increased partitioning time, but slight variations of the C concentrations of the phases involved prevail.

3.2 Computational results

The kinetics of C redistribution across a martensite/austenite interface were calculated as depicted in Figure 3, where (a) and (b) show the DICTRA calculations and (c) and (d) correspond to the MatCalc calculations. Since C has a higher chemical potential in martensite than in austenite, the C concentration in ferrite is reduced during isothermal annealing at 400 °C [11]. Generally, the kinetics of C escaping from martensite are similarly predicted by both simulation approaches. The final concentration of the ferritic phase is obtained after 100 s using DICTRA at ≈ 0.003 m.% – a result that can be reproduced by MatCalc if trapping of C at dislocations is disregarded (see curve denoted by n.t. in (c)). If a dislocation density of 1×10^{15} m^{-3} is assumed comparing well with Ref. [15], the trapping of C inside the martensitic structure can be modelled successfully ($c_\alpha(100\text{ s}) \approx 0.06$ m.%), qualitatively matching well with the experimental results of the present work ($c_\alpha(120\text{ s}) \approx 0.10$ m.%) and literature ($c_\alpha(1000\text{ s}) \approx 0.10$ m.%) [3].

The enrichment of austenite by C, visible in Figure 3(b) and (d), shows slight discrepancies in terms of kinetics of C redistribution comparing the DICTRA and the MatCalc calculations. DICTRA predicts faster C redistribution (compare e.g. dotted lines corresponding to 1 s of partitioning). This discrepancy may result from the comparably high C concentration within the austenite at the beginning of the calculation using DICTRA in the direct vicinity of the interface to martensite and the fact that the C diffusivity in austenite strongly depends on the C concentration [11]. Also, in (d) the effect of C trapping by dislocations in the ferritic phase is discernible. Diffusion profiles without trapping are labelled by n.t. Comparing the curves for 1 s partitioning time evidences that the kinetics of C redistribution are hardly affected, while the predictions for long partitioning durations differ dramatically in terms of maximum austenite C concentrations ($c_\gamma(100\ \mathrm{s})_{\mathrm{n.t.}} \approx 1.98$ m.% and $c_\gamma(100\ \mathrm{s}) \approx 1.49$ m.%. This is due to the fact that a larger amount of C is available for austenite enrichment in the absence of trapping.

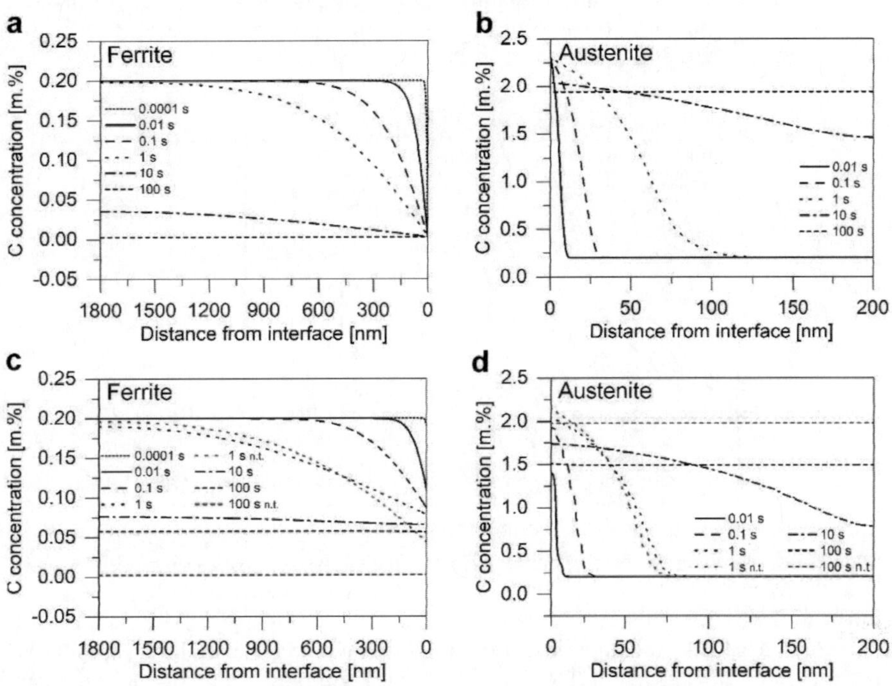

Figure 3. Results of the kinetic calculations of the C redistribution between ferrite and austenite during annealing at 400 °C. (a) and (b) correspond to DICTRA calculations, whereas (c) and (d) correspond to MatCalc calculations.

3.3 *Comparison of experiment, kinetic calculations and CCE calculation*

Results obtained by utilization of the CCE condition are depicted in Figure 4. Here, the C concentration in martensite (a) and austenite (b) are shown with respect to temperature for different martensite fractions. Comparing the values of the curve with a martensite fraction of 90% to the experimental results clearly shows that this calculation overestimates the C enrichment of austenite of $c_\gamma(600\ \mathrm{s}) \approx 1.20$ m.% in the investigated alloy and material condition, which is due to the fact that a non-negligible amount of C of about $c_\alpha(600\ \mathrm{s}) \approx 0.09$ m.% remains (inhomogeneously) in martensite. This effect can be accounted for in MatCalc as visualized in Figure 3(c). It should be mentioned that the differences between actual martensite C concentration of $c_\alpha(600\ \mathrm{s}) \approx 0.09$ m.% and the MatCalc prediction of $c_\alpha(100\ \mathrm{s}) \approx 0.06$ m.% may be rationalised by considering that by XRD a mean C concentration is determined averaging over different C concentrations present in tempered martensite and fresh martensite. Thus, the calculated C concentration is expected to closely match the C concentration of the tempered martensite.

The DICTRA calculations for long partitioning durations obtained in the present work resemble the concentrations predicted by the CCE condition (compare results of Figure 3(a) and (b) with the value the 90% martensite fraction curve depicted in Figure 4(a) and (b)). Moreover, these calculations fit well to the kinetic calculations performed by Seo et al. [11] using thermokinetic simulation and Mecozzi et al. [17] using phase field simulation. The approach implemented in DICTRA, however, falls short of accurately modelling the complex structure of martensite and the resulting implications on C concentrations in both martensite and austenite. A suitable means to account for the actual implications of the martensitic structure is the physically well-founded treatment available in MatCalc based on Refs. [12–14], whereby the effect of the presence of dislocations is treated via their effect on the C chemical potential. The presence of these traps results in a reduction of the C chemical potential within ferrite. Since C redistribution stops once the C chemical potentials equal each other in martensite and austenite, i.e. no driving force for further diffusion is available, the equilibrium C concentration in martensite is increased in comparison to a defect free phase.

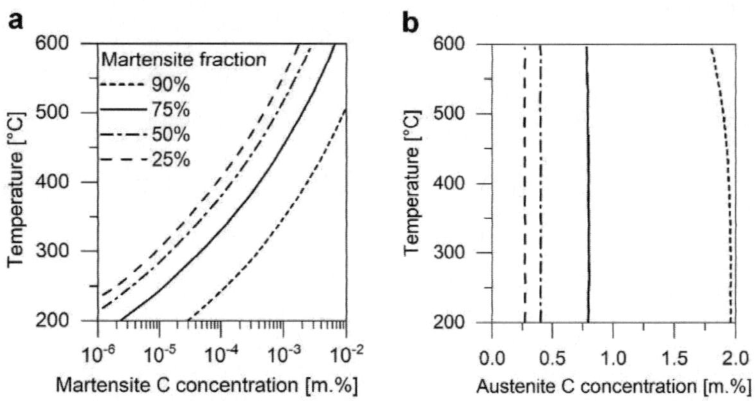

Figure 4. Dependence of (a) martensite C concentration and (b) austenite C concentration with respect to temperature for the Fe-0.2C (m.%) model system according to the CCE condition.

4 Summary and conclusion

The present work assesses the applicability of commercially available thermodynamic and thermokinetic software packages (DICTRA and MatCalc) to analyze the C enrichment of austenite and the depletion of martensite during a Q&P heat treatment. Moreover, two Q&P material conditions of a low alloyed steel are investigated experimentally. Results of kinetic calculations and experiments are compared to calculations based on the CCE condition and discussed in the frame of current literature. From the performed analyses the following conclusions can be drawn:

- Experiments conducted provide evidence for the enrichment of austenite with C during the partitioning heat treatment. The redistribution, however, clearly remains incomplete and a substantial amount of C is present within the martensite even after long partitioning heat treatment durations.
- DICTRA and MatCalc yield slightly differing kinetics of austenite C enrichment, which can be rationalised based on the dependence of the C diffusivity in austenite on the C concentration and differences in the boundary conditions applied.
- The effect of C trapping at dislocations within martensite can be accurately modelled using MatCalc resulting in increased C concentrations within martensite, while the kinetics are hardly affected.
- DICTRA calculations for long partitioning durations resemble estimations based on the CCE condition. Both generally overestimate the stabilization of austenite by C, whereas MatCalc calculations better approximate experimental results.

Acknowledgements
The authors gratefully acknowledge the financial support under the scope of the COMET program within the K2 Center "Integrated Computational Material, Process and Product Engineering (IC-

MPPE)" (Project No 859480). This program is supported by the Austrian Federal Ministries for Transport, Innovation and Technology (BMVIT) and for Digital and Economic Affairs (BMDW), represented by the Austrian research funding association (FFG), and the federal states of Styria, Upper Austria and Tyrol. The authors thank Y.V. Shan and Prof. Dr. E. Kozeschnik for fruitful discussions and support with the MatCalc calculations.

References

[1] Liu L, He B and Huang M X 2018 The Role of Transformation-Induced Plasticity in the Development of Advanced High Strength Steels *Adv. Eng. Mater.* **20** 1701083
[2] Speer J, Matlock D K, De Cooman B C and Schroth J G 2003 Carbon partitioning into austenite after martensite transformation *Acta Mater.* **51** 2611–22
[3] Clarke A J, Speer J G, Miller M K, Hackenberg R E, Edmonds D V, Matlock D K, Rizzo F C, Clarke K D and De Moor E 2008 Carbon partitioning to austenite from martensite or bainite during the quench and partition (Q&P) process: A critical assessment *Acta Mater.* **56** 16–22
[4] De Moor E, Föjer C, Penning J, Clarke A J and Speer J G 2010 Calorimetric study of carbon partitioning from martensite into austenite *Phys. Rev.* B **82** 1–5
[5] Toji Y, Matsuda H, Herbig M, Choi P P and Raabe D 2014 Atomic-scale analysis of carbon partitioning between martensite and austenite by atom probe tomography and correlative transmission electron microscopy *Acta Mater.* **65** 215–28
[6] Mujahid S A and Bhadeshia H K D H 1992 Partitioning of carbon from supersaturated ferrite plates *Acta Metall. Mater.* **40** 389–96
[7] Hillert M, Höglund L and Ågren J 1993 Escape of carbon from ferrite plates in austenite *Acta Metall. Mater.* **41** 1951–7
[8] Edmonds D V, He K, Rizzo F C, De Cooman B C, Matlock D K and Speer J G 2006 Quenching and partitioning martensite - A novel steel heat treatment *Mater. Sci. Eng.* A **438–440** 25–34
[9] Ruhl R and Cohen M 1969 Splat quenching of iron-carbon alloys *Trans. Met. Soc. AIME* **245** 241
[10] Borgenstam A, Engström A, Höglund L and Ågren J 2000 DICTRA, a Tool for Simulation of Diffusional Transformations in Alloys *J. Phase Equilibria* **21** 269–80
[11] Seo E J, Cho L and De Cooman B C 2016 Kinetics of the partitioning of carbon and substitutional alloying elements during quenching and partitioning (Q&P) processing of medium Mn steel *Acta Mater.* **107** 354–65
[12] Fischer F D, Svoboda J and Kozeschnik E 2013 Interstitial diffusion in systems with multiple sorts of traps *Model. Simul. Mater. Sci. Eng.* **21** 025008
[13] Svoboda J, Shan Y V, Kozeschnik E and Fischer F D 2013 Determination of depths of traps for interstitials from thermodynamic data: A new view on carbon trapping and diffusion *Model. Simul. Mater. Sci. Eng.* **21** 065012
[14] Svoboda J, Shan Y V, Kozeschnik E and Fischer F D 2014 Determination of depths of multiple traps for interstitials and their influence on diffusion kinetics *Model. Simul. Mater. Sci. Eng.* **22** 065015
[15] Kim B, Boucard E, Sourmail T, San Martín D, Gey N and Rivera-Diaz-del-Castillo P E J 2014 The influence of silicon in tempered martensite: Understanding the microstructure–properties relationship in 0.5 – 0.6 wt.% C steels *Acta Mater.* **68** 169–78
[16] Behera A K and Olson G B 2018 Nonequilibrium thermodynamic modeling of carbon partitioning in quench and partition (Q&P) steel *Scr. Mater.* **147** 6–10
[17] Mecozzi M G, Eiken J, Santofimia M J and Sietsma J 2016 Phase field modelling of microstructural evolution during the quenching and partitioning treatment in low-alloy steels *Comput. Mater. Sci.* **112** 245–56

IOP Conf. Series: Materials Science and Engineering **461** (2019) 012040 doi:10.1088/1757-899X/461/1/012040

Experimental-numerical approach to efficient TTT-generation for simulation of phase transformations in thermomechanical forming processes

B.-A. Behrens, A. Chugreev, C. Kock[1)]

IFUM – Institut für Umformtechnik und Umformmaschinen, Leibniz Universität Hannover, An der Universität 2, Garbsen 30823, Germany

[1)] E-mail: kock@ifum.uni-hannover.de, phone +49 511 762 2161

Abstract. Residual stresses in components are an important issue in most manufacturing processes, as they influence the performance of the final part. Regarding hot forming processes there is a great potential of defining a targeted residual stress state, due to numerous adjustment parameters like deformation state or temperature profile. In order to ensure appropriate numerical modelling of resulting residual stresses in a thermomechanical process, comprehensive material data regarding phase transformation are required. This paper presents an experimental-numerical procedure to efficiently determine time-temperature-transformation diagrams for cooling simulations after hot forming. The transformation behaviour of the steel alloys 42CrMo4 and 100Cr6 is determined by experiments as well as FE-simulations. Finally, the simulation model is validated by dilatometric experiments and metallographic investigations.

1 Introduction

In forming processes, the arising residual stresses influence the material behaviour during and after manufacturing as well as the performance of the final component [1]. This calls for a thorough consideration and evaluation of the mentioned residual stresses. Hot forming, in particular, represents not only a great challenge for numerical prediction of the resulting stress states but also a high potential for the targeted adjustment of residual stresses due to numerous influencing factors resulting from the thermomechanical process [2, 3]. During such a process, the degree of deformation after hot forming in combination with the temperature-time profile experienced by the component have an important influence [4]. In general, the polymorphic transformation behaviour of the microstructure in steel alloys is described by continuous-cooling-transformation (CCT) diagrams and time-temperature-transformation (TTT) diagrams. Thereby, CCT describe the transformation behaviour of the material during continuous cooling, whereas TTT represent the transformation behaviour after quenching and subsequent holding at a certain test temperature [5]. Such diagrams are either provided by steel manufacturers or listed, for example, in compendia for heat treatment [6]. Furthermore it is possible to calculate CCT- and TTT-diagrams on the basis of the chemical composition of the steel alloys with the software JMatPro from Sente Software [7] using the empirical equations from the work of *Kirkaldy et al.* [8] and *Li et al* [9]. However, each of these diagrams is strongly dependent on the experimental boundary conditions. For example, *Schulze et al.* [10] investigated the influence of grain growth on the transformation behaviour of a 42CrMo4 steel alloy due to heating strategy, austenitisation temperature and holding time. Another important influence on the phase transformations is the deformation state applied before cooling, which is why deformation-dependent diagrams D-CCT-diagrams and D-TTT-diagrams are recorded in literature [11, 12]. For these reasons, it is important to always use transformation diagrams specifically for the process under consideration. For numerical description of phase transformations the Koistinen-Marburger equation is widely used for diffusion-free transformations and the Johnson-Mehl-Avrami equation for diffusion-controlled transformations respectively [13]. However, these two approaches are valid only for isothermal processes, hence commercial FE-solvers like MSC.Marc require isothermal TTT-diagrams for the residual stress calculations by means of phase transformations. The experimental determination of the isothermal TTT-diagrams is very challenging, especially since they are costly and time-consuming due to the long durations of the tests. Besides phase transformations under isothermal conditions are hardly to detect by dilatation measurements. Furthermore, it is challenging to generate sufficiently high cooling rates of more than 175 K/s [14] with the quenching gases helium or nitrogen, to prevent that transformations have already taken place before the testing temperature has been reached [9, 15]. Therefore, some authors have developed mathematical approaches to calculate TTT- from CCT-diagrams [16, 17]. For the numerical investigations of residual stresses it is necessary to implement accurate information about the microstructural transformation process into the FE-software. Therefore, the aim of this work is to determine TTT-diagrams of the steel alloys 42CrMo4 and 100Cr6 with a time and cost efficient as well as precise experimental-numerical approach.

2 Investigation on phase transformation behaviour of steel alloys 42CrMo4 and 100Cr6

As explained in the introduction, there are some restrictions of getting TTT-diagrams directly from experiments. Therefore, this paper presents an approach for the determination of TTT by combination of modern software as well as the established standardised experiments. In the first step CCT-diagrams were calculated by JMatPro based on the chemical composition of the alloys (table 1).

Table 1: Chemical composition of steel alloys 42CrMo4 and 100Cr6 according to international standards [18, 19] used for material data generation with JMatPro.

	C	Si	Mn	P	S	Cr	Mo	Fe
42CrMo4	0.42	0.25	0.75	0.025	0.035	1.10	0.22	balance
100Cr6	0.99	0.25	0.35	0.025	0.015	1.475	0.10	balance

Due to batch fluctuations of investigated materials and partial shortcomings of the empirical methods in JMatPro, an experimental fitting of the calculated diagrams by experiments was necessary. Subsequently, the software tool "TTT-CCT-diagram generator" of Transvalor was used to determine the TTT- from the CCT-diagram. The determined TTT were implemented in the Software Simufact Forming 15 to simulate the experimental CCT tests. Finally, by means of the length-change-temperature-profiles and metallographic investigations the simulations were calibrated and validated.

2.1 Experimental determination of continuous cooling transformation diagrams

The experimental CCT-tests were carried out and evaluated at the Institute of Forming Technology and Machines based on the specifications of the "Deutsche Stahl-Eisen-Prüfblätter" SEP 1680 and SEP 1681 [20, 21] at the quenching-deformation dilatometer system DIL 805A/D+T (Co. TA Instruments Inc.). Cylinders with a diameter of 4 mm and a height of 10 mm were examined. The samples were first heated inductively with a heating rate of 50 K/s to the austenitising temperature of 1050 °C followed by a 10 minutes soaking time. Subsequently a cooling phase of 30 seconds to 950 °C takes place. In case of the deformation-dependent CCT-test, the specimens were upset to a degree of deformation of φ=0.6 at a temperature 950 °C and a strain rate of 1 s^{-1}. Afterwards the specimens were linearly cooled by the medium nitrogen. The investigated cooling rates together with the resulting hardness values H are listed in table 2. The obtained results were used to calibrate the CCT-diagrams calculated with JMatPro. Figure 1 shows the CCT-diagrams with and without a previous deformation for both materials. For the alloy 42CrMo4 the formation of ferritic (F), pearlitic (P), bainitic (B) and martensitic (M) phase fractions from the initial austenitic (A) phase was observed.

IOP Conf. Series: Materials Science and Engineering **461** (2019) 012040 doi:10.1088/1757-899X/461/1/012040

Table 2: Hardness values of the specimens from experimental CCT-tests.

			42CrMo4				
$\dot{\vartheta}$ **[K/s]**	30	6	3	1	0.3	0.15	0.07
H(φ=0) | H(φ=0.6) [HV1]	667|710	647|670	455|456	336|328	317|310	307|272	270|235
			100Cr6				
$\dot{\vartheta}$ **[K/s]**	30	5	2.5	1	0.5	0.15	-
H(φ=0) | H(φ=0.6) [HV1]	842|815	832|707	749|486	495|442	425|396	407|366	-

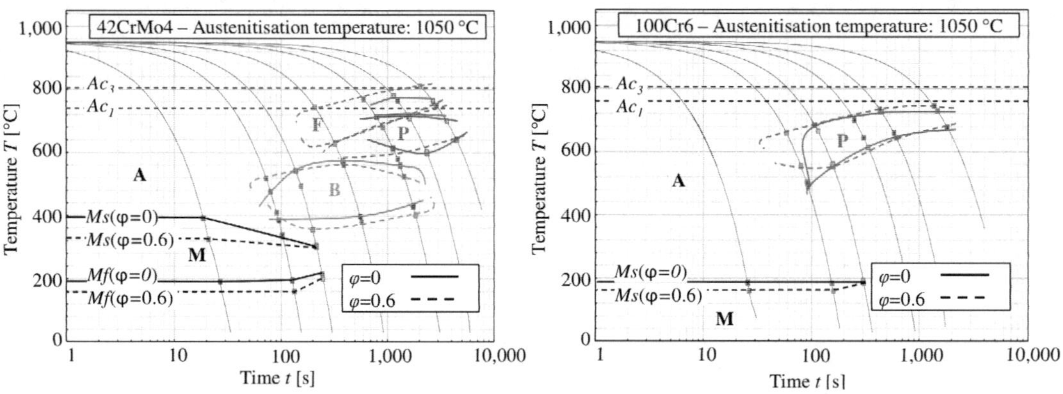

Figure 1: CCT-diagrams for the materials 42CrMo4 (left) and 100Cr6 (right) with and without deformation and an austenitising temperature of 1050 °C.

Obviously, the plastic deformation causes a decrease of the martensite-start-temperature Ms as well as the martensite-finish-temperature Mf. Furthermore, previous deformation influences the ferrite area, which is extended to lower times and temperatures. For the specimens from steel alloy 100Cr6 only the pearlitic and martensitic microstructural phases were observed. Firstly, the steel has a relatively low critical cooling rate (between $\dot{\vartheta}_{crit}$=2.5 K/s and 5 K/s) due to its high carbon content. Secondly, the bainite formation is prevented by means of the high percentage of the carbide former chromium. Thus, before a bainitic transformation can take place, all austenite grains are transformed into either martensite or pearlite grains at a corresponding cooling rate. Due to the finer grain after forming of the specimens, a shift of the pearlite area to lower times was observed. Furthermore, a reduction of Ms in the case of deformed specimens was detected for the specimen from 100Cr6 alloy. However, an Mf could not be determined since the martensitic transformation was not yet completed even at room temperature. Therefore, the samples have to be cooled further below room temperature. As evidence for the experimental CCT-tests all samples were metallographically examined and the most interesting ones are exemplarily shown in figure 2. The steel 42CrMo4 shows the formation of pure bainite in CCT-test with the cooling rate $\dot{\vartheta}$=1 K/s without previous deformation (figure 2a). In case of the CCT-test with $\dot{\vartheta}$=1 K/s and a previous deformation, ferritic and pearlitic phases are formed (figure 2b). Regarding the specimen from 100Cr6, at a cooling rate of $\dot{\vartheta}$=2.5 K/s after deformation no bainite or martensite could be found in metallographic investigations (figure 2c1). Therefore, with the kind support of the Institut fuer Werkstoffkunde (IW) raster electron microscope (REM) analyses were used to get a closer look at the phase structure near to the indentations from hardness measurements (figure 2c2). As a result, a lamellar pearlitic structure with a comparatively high hardness H=486 HV (cf. table 2) was found. The shift of the pearlitic area for the samples of 100Cr6 due to previous deformation was also observed in the metallographic investigations. In the CCT-tests with the cooling rate $\dot{\vartheta}$=5 K/s and φ=0, a purely martensitic microstructure was formed (figure 2d) whereas a in the tests with a previous forming a pearlitic-martensitic microstructure was generated (figure 2e). After calibration, the CCT-diagrams were converted to TTT-diagrams using the TTT-CCT-diagram generator.

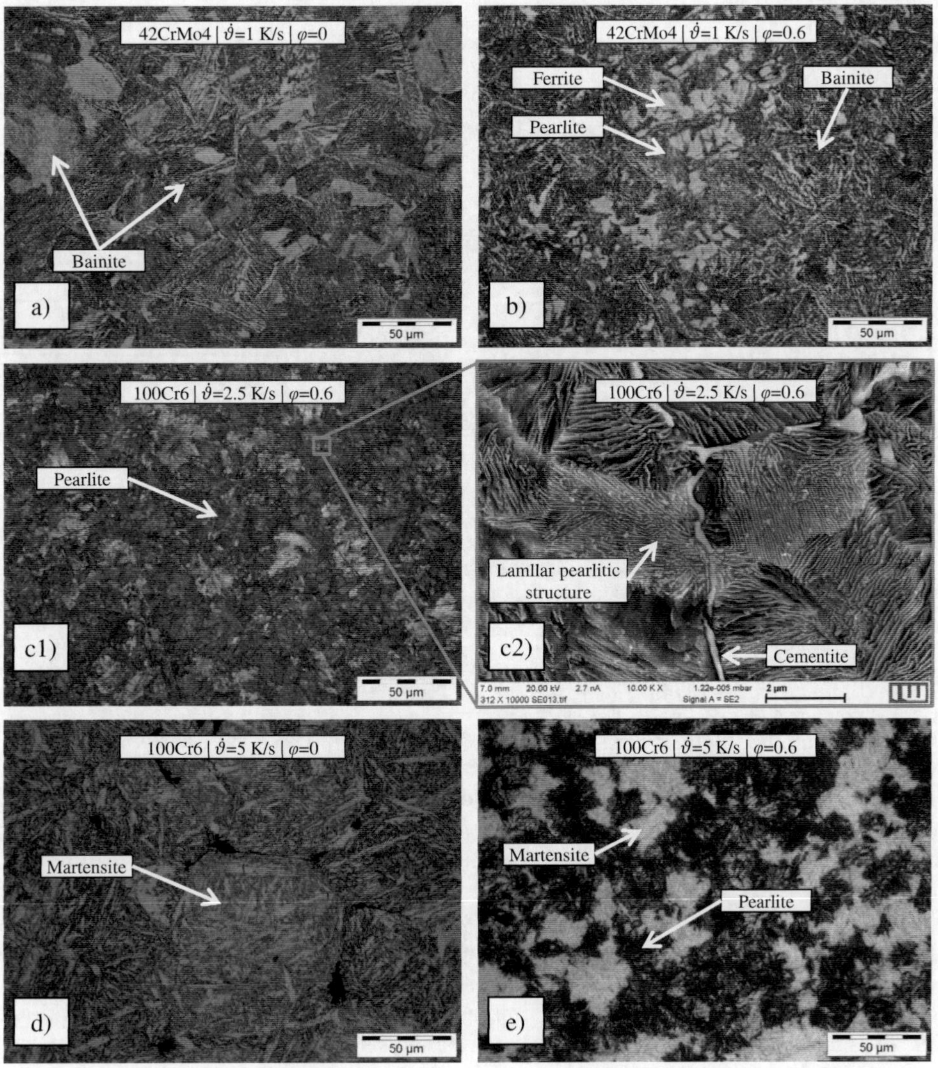

Figure 2: Microstructure of the specimens from 42CrMo4 after CCT-tests without (a) and with previous deformation (b); pearlitic phase of 100Cr6 after CCT-tests and previous deformation from metallographic (c1) and REM (c2) investigations; microstructure of the specimens from 100Cr6 after CCT-tests without (d) and with previous deformation (e).

2.2 Numerical determination of time temperature transformation diagrams

The determined TTT-diagrams were implemented in the commercial FE-software Simufact Forming 15 for numerical simulation of the CCT-tests. In order to reduce computational time rotationally symmetric 2D FE-simulations were carried out by exploiting the cylindrical shape of the specimens. The specimens have been discretised using a quad mesh with an element-edge-length of 0.05 mm. As initial configuration the thermally expanded geometry of the specimens at 950 °C with a completely austenitised structure was assumed. In the case of a CCT test with deformation, thermally conductive rigid punches with a quad mesh of 0.1 mm element-edge-length were used.

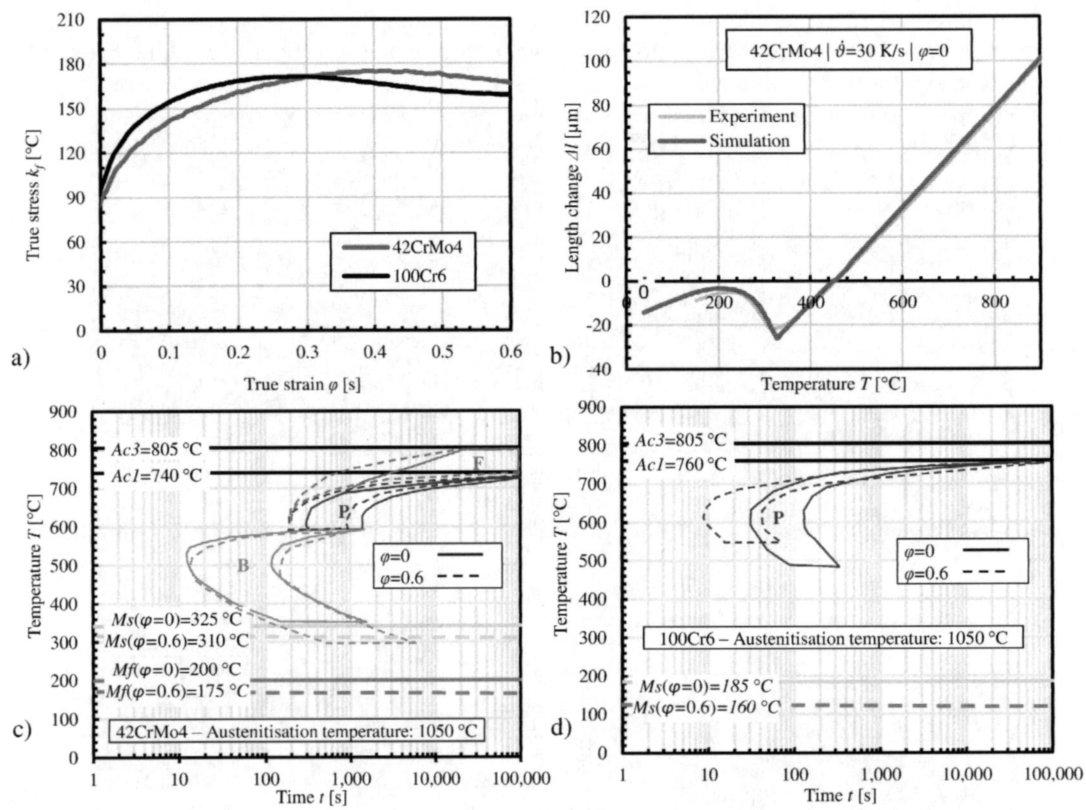

Figure 3: True stress k_f with respect to true strain φ of the investigated materials (a); comparison of experimentally measured and numerically calculated length-change Δl for the cooling rate $\dot{\vartheta}$ =30 K/s (b); TTT-diagrams with and without a previous deformation for the materials 42CrMo4 (c) and 100Cr6 (d).

The plastic deformation behaviour is described by flow curves at the relevant temperature 950 °C and the strain rate of 1 s^{-1} determined with help of the servohydraulic forming simulator Gleeble 3800-GTC from DSI company (figure 3a). As exemplarily illustrated in figure 3b the length-change Δl with respect to temperature T from the FE-simulations and the dilatometric CCT-experiments were compared. For further calibration, in an iterative procedure the TTT-diagrams were manually adjusted until the Δl-T profiles from experiment and simulation showed a good agreement. Additionally, the model was validated by a qualitative metallographic analysis of all experimentally tested specimens. Figure 3c and d show the finally determined TTT-diagrams for both materials with and without previous deformation.

3 Conclusion

In this study a cost-effective, time-saving as well as precise method for the determination of TTT-diagrams was presented for the steel alloys 42CrMo4 and 100Cr6. Initially CCT-diagrams were calculated with the software JMatPro of Sente Software based on the chemical composition of the steel alloys. These were optimised by experimental CCT-tests and subsequently converted into TTT-diagrams using the software tool "TTT-CCT-diagram generator" of Transvalor. The obtained TTT-diagrams were validated by a comparison of Δl from the FE-simulations and the dilatometric CCT-experiments as well as metallographic analyses. With this data in hand, phase transformation simulations aiming at subsequent residual stress calculation are possible.

Acknowledgement
Funded by the Deutsche Forschungsgemeinschaft (DFG, German Research Foundation) - 374871564 within the priority program SPP 2013.

References

[1] Verlinden B, Driver J, Samajdar J and Doherty RD (2007) Thermo-mechanical processing of metallic materials, Elsevier Science 11, ISBN: 9780080444970.

[2] Behrens BA and Olle P (2007) Consideration of phase transformations in numerical simulation of press hardening. Steel research international, 78(10-11), pp. 784-790.

[3] Behrens BA, Olle P, Schäffner C (2008) Process simulation of hot stamping in consideration of transformation-unduced stresses, Numisheet, Interlaken, Schweiz.

[4] Behrens BA, Bouguecha A, Bonk C, Chugreev A (2017) Numerical and experimental investigations of the anisotropic transformation strains during martensitic transformation in a low alloy Cr-Mo steel 42CrMo4, Procedia Engineering 207, pp. 1815-1820, ICTP2017, Cambridge, Great Britain.

[5] Doege E and Behrens BA (2016) Handbuch Umformtechnik, 3. Ed., Springer-Verlag, Berlin.

[6] Wever F, Rose A and Peter W (1961) Atlas zur Wärmebehandlung der Stähle Vol. 1 & Vol. 2, Verlag Stahleisen, Düsseldorf.

[7] Saunders N, Guo Z, Li X, Miodownik AP, Schillé JP (2004) The Calculation of TTT and CCT diagrams for General Steels.

[8] Kirkaldy JS and Venugopalan D (1983) Phase Transformations in Ferrous Alloys, D.A.R. Marder and J.I. Goldstein, eds., AIME, New York, NY, pp. 128-48.

[9] Li MV, Niebuhr DV, Meekisho LL, Atteridge DG (1997) A Computational Model for the Prediction of Steel Hardenability, Metall Mater Trans B, vol. 29, pp. 661-672.

[10] Schulze P, Schmidl E and Lampke T (2014) Entwicklung der Korngröße beim Erwärmen des Stahles 42CrMo4, WOMag (3).

[11] Nürnberger F, Grydin O, Schaper M, Bach FW, Koczurkiewicz B and Milenin A (2010) Steel Res. Int. 81, pp. 224-233.

[12] Behrens BA, Bouguecha A, Götze T, Moritz J, Sunderkötter C, Helmholz R and Schrödter J, (2013) Numerical and experimental analysis of the phase transformation during hot stamping process in consideration of strain-dependent CCT-diagrams, 4th International Conference on Hot Sheet Metal Forming of High-Performance Steel, Lulea, Sweden, pp. 329-336.

[13] Behrens BA, Schrödter J (2014) Numerical simulation of phase transformation during the hot stamping process, 5th International Conference on Thermal Process Modeling and Computer Simulation, Orlando, USA, S. 179-190.

[14] ASTM A1033 (2018) Standard Practice for Quantitative Measurement and Reporting of Hypoeutectoid Carbon and Low-Alloy Steel Phase, American Society for Metals, 2018.

[15] Yu Z, Gretzki T, Nürnberger F, Kästner M, Haskamp K, Bach F-W, Schaper M (2014) Wärmebehandlung, in Prozesskette Präzisionsschmieden, Bach F-W, Kerber K (Hrsg.), Springer-Verlag, Berlin Heidelberg, 2014.
Buza G, Hougardy HP, Gergely M (1990) Calculation of the isothermal transformation into two different microstructures from measurements with continuous cooling. Steel res 61(10), pp. 478–481.

[16] Rios PR (2005) Relationship between non-isothermal transformation curves and isothermal and non-isothermal kinetics. Acta Mater 53, pp. 4893–4901.

[17] DIN EN 10083-3 (2007) Steels for quenching and tempering – Part 3: Technical delivery conditions for alloy steels, Beuth-Verlag.

[18] DIN EN ISO 683-17 (2014) Heat-treated steels, alloy steels and free-cutting steels – Part 17: Ball and roller bearing steels, Beuth-Verlag.

[19] SEP 1680 (1990) STAHL-EISEN-Prüfblätter (SEP), 3rd Ed.

[20] SEP 1681, STAHL-EISEN-Prüfblätter (SEP), 2nd Ed., (1998).

On the sensitivity of the Mohr-Coulomb ductile fracture criterion in the mixed stress-strain space on varying hardening curve approximations

T Komischke and P Hora

ETH Zürich, Institute of Virtual Manufacturing, Tannenstrasse 3, 8092 Zürich, Switzerland

E-mail: komischke@ivp.mavt.ethz.ch

Abstract. In recent years the use of the Mohr-Coulomb type fracture criteria has become popular in ductile fracture prediction. In ductile fracture prediction, fracture criteria are often transformed into the mixed stress strain space formulation and applied to numerical simulations in this formulation. This paper will investigate the shape of the Mohr Coulomb fracture criterion in the mixed stress strain space as a function of the used hardening curve formulation, in particular the hardening curve exponent n that determines the slope of the hardening curve approximation for large strains. It is found that the shape of the fracture criterion significantly depends on the slope of the hardening curve. This makes a correct extrapolation of the hardening curve for large strain values essential for a correct ductile fracture prediction using the Mohr-Coulomb fracture model in its mixed stress strain space formulation.

1 Introduction

Recent studies have shown good accuracy of Mohr Coulomb type of fracture criteria in ductile fracture prediction [1-4]. In order to account for non-constant stress-states that are dominant in almost every forming process, usually a damage accumulation rule is assumed (equation (1)).

$$D = \int \frac{n}{\varepsilon_f(\eta,\theta)} D^{\frac{n-1}{n}} d\varepsilon \tag{1}$$

In this equation $\varepsilon_f(\eta, \bar{\theta})$ is the current fracture strain predicted by the Mohr-Coulomb fracture criterion. In order to calculate it, the criterion is usually expressed in its mixed stress-strain space formulation.

1.1 *The mixed stress strain space*

In accordance with Mohr's nomenclature [2] the space of stress triaxiality η, normalized Lode angle parameter $\bar{\theta}$ and fracture strain ε_f will be referred to as the "mixed stress strain space". Equations (2) and (3) give the definitions of η and $\bar{\theta}$ used in this paper, where σ_h is the acting hydrostatic stress, $\bar{\sigma}$ is the equivalent von-Mises stress and σ_i are the Cauchy stress tensor's principal stresses.

$$\eta = \frac{\sigma_h}{\bar{\sigma}} \tag{2}$$

$$\bar{\theta} = 1 - \frac{2}{\pi}\arccos\left(\frac{27}{2}\frac{(\sigma_1 - \sigma_h)(\sigma_2 - \sigma_h)(\sigma_3 - \sigma_h)}{\bar{\sigma}^3}\right) \tag{3}$$

1.2 The Mohr-Coulomb (MC) fracture criterion

The Mohr Coulomb (MC) [5] fracture criterion assumes that fracture occurs as soon as a critical weighted sum of acting shear stresses and acting normal stresses is exceeded on an arbitrary plane in the material. The MC fracture criterion is therefore given by equation (4).

$$|\tau| + c_1\sigma_n = c_2 \tag{4}$$

It can be proven [2] from considerations on Mohr's stress circles that this failure criterion is fully equivalent with equation (5).

$$(\sigma_1 - \sigma_3) + c(\sigma_1 + \sigma_3) = b \tag{5}$$

Equation (5) can be transformed to an expression of η and $\bar{\theta}$ by purely changing the stress state parametrization [4]. Equation (6) gives this formulation, where $\overline{\sigma_f}$ is the equivalent von Mises stress at which fracture occurs.

$$\overline{\sigma_f} = c_2\left[\sqrt{\frac{1+c_1^2}{3}}\cos\left(\frac{\bar{\theta}\pi}{6}\right) + c_1\left[\eta + \frac{1}{3}\sin\left(\frac{\bar{\theta}\pi}{6}\right)\right]\right]^{-1} \tag{6}$$

To transform this equation into its mixed stress strain formulation (ε_f, η, $\bar{\theta}$) the material hardening curve formulation (equation (7)) is used. In particular this hardening curve formulation will be given by equations (10) and (11) with varying parameters.

$$\overline{\sigma_f} = k(\varepsilon_f) \tag{7}$$

This paper will focus on how the material hardening curve formulation influences the shape of the Mohr-Coulomb fracture criterion in the mixed stress strain space (ε_f, η, $\bar{\theta}$).

2 Material parametrization

In this section, the material's hardening and fracture behavior are characterized. The investigated steel is a low-alloyed boron tempered steel.

2.1 Hardening behavior

To investigate the material's hardening behavior, dilatometer compression tests were carried out on cylindrical probes at a Temperature of 363K (90°C) at a strain rate of 0.01/s. The experimentally measured data were then corrected for the influence of friction by equation (8) [6].

$$\sigma_y \cong \frac{F(\varepsilon)}{\pi r(\varepsilon)^2} \frac{1}{\left[1 + \frac{2\mu r(\varepsilon)}{3h(\varepsilon)}\right]} \tag{8}$$

The Colulomb friction coefficient μ was found by measuring the probes barreling and comparing it to the barreling of the probe in numerical simulations with varying μ. The simulation with the probe barreling closest to the experimentally observed one was assumed to have the correct friction coefficient μ. Both a Hockett Sherby and a Hensel- Spittel hardening curve approach were used for parameter identification. The parameter identification was performed by minimizing square errors while at the same time fulfilling the constraint given in equation (9). This constraint ensures the equality of the experimental and approximated slope of the hardening curve in the last measured data point of ε=40%. This ensures a better agreement of the hardening curves in the extrapolated regions for large strains.

$$\left.\frac{d\sigma_{fit}}{d\varepsilon_{fit}}\right|_{\varepsilon=40\%} = \left.\frac{d\sigma_{meas}}{d\varepsilon_{meas}}\right|_{\varepsilon=40\%} \tag{9}$$

Figure 1 gives the experimental hardening curve, as well as the two approximations and their slopes.

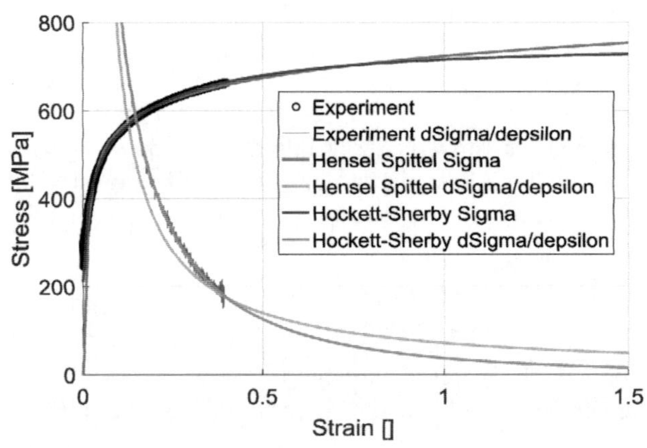

Figure 1. Experimental hardening curve and mathematical approximations

The resulting flow curve approximations are given in equations (10) (Hensel-Spittel) and (11) (Hocket-Sherby). For simplicity reasons, the flow curve is assumed as independent of temperature and strain rate. Although these effects shouldn't be neglected in actual forming processes, for this study they do not have to be taken into account.

$$\sigma_{Hensel} = A\varepsilon^n = 750.5\ \varepsilon^{0.0944}[MPa] \tag{10}$$

$$\sigma_{Hocket} = A - (A - B)\mathrm{e}^{-m*\varepsilon^n} = 741.6 - (741.6 - 195.1)e^{-3.01\varepsilon^{0.4855}}[\mathrm{MPa}] \tag{11}$$

2.2 *Fracture behaviour*

The material's fracture behavior was investigated in pure torsion tests, Torsion tests with superimposed compression/ tension as well as unnotched and notched tensile tests. Figure 2, left gives the specimen geometries used. The results are given in figure 2 right, where varying η and $\bar{\theta}$ were averaged according to equation (12).

$$\eta_{av} = \int_0^{\varepsilon_f} \frac{\eta}{\varepsilon_f} d\varepsilon \tag{12}$$

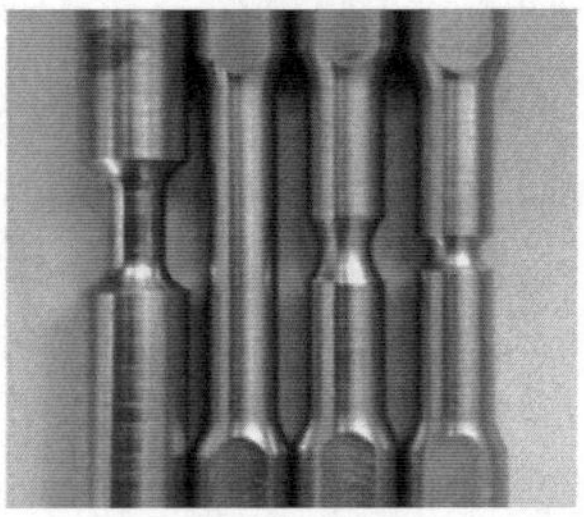

Test type	ε_f	η (Av.)	$\bar{\theta}$ (Av.)
Torsion+ compression	206%	-0.0439	-0.1492
Pure Torsion	155%	0.0	0.0
Torsion + tension	112%	0.0937	0.2115
Torsion + tension	111%	0.1147	0.3698
Tension, unnotched round bar	111%	0.5723	1
Tension, notched round bar	78.6%	0.7483	1
Tension, strongly notched round bar	64.7%	1.0357	1

Figure 2. Experimental results and specimen geometries used for calibration of parameters of the MC fracture criterion

These experimental results were then used to determine the material's Mohr-Coulomb- fracture criteria parameters. The found values are given in figure 4 right under letter c (grey background). They were identified by minimizing square errors. The resulting shape of the MC criterion in the mixed stress-strain space is given in figure 4 left, letter c.

3 Varying the hardening curve approximation

3.1 Re-Parametrization of the hardening curve approximation

In this part of the paper, the material's hardening curve will be re-parametrized. It is assumed that there is a small error in the measurement of the hardening curve. In dilatometer compression tests, this error could be caused by numerous disturbances of the measurement, such as temperature effects, non-compensation of friction effects or a small error in the determined friction coefficient μ, causing an error in the friction compensation according to equation (8). This error will be introduced into the parametrization procedure by setting the hardening exponent n to a predefined value. In order to not only investigate a small error's influence on the MC fracture criterion but also the general shape of the criterion as a function of the hardening curve, n values are varied in a wide range from n=0.04 to n=0.20. The Hocket-Sherby flow curve can be seen as the limiting case of n→0 since the slope of the hardening curve will approach zero for ε→∞ for this saturating hardening curve approximation. The parameter A that is found by minimizing square errors in function of parameter n is given in figure 3 right. Figure 3 left shows the different hardening curve approximations and the experimental result.

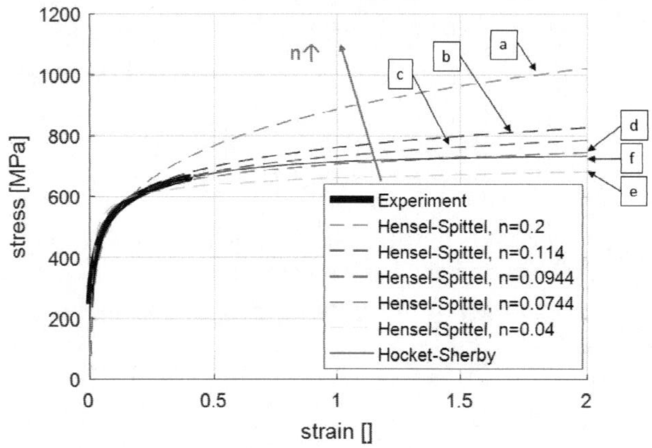

Configuration	n	A
a	0.2	891.7
b	0.1144	765.8
c	**0.0944**	**750.5**
d	0.0744	710.6
e	0.04	665.0

Figure 3. Varying hardening curve approximations

3.2 Re-Parametrization of the MC-fracture criterion

Under the varying hardening curve approximations, the Mohr-Coulomb fracture criteria's parameters have been re-identified from the experimental data set given in chapter 2.2 by minimizing square errors. The identified MC-parameters are given figure 4 right in function of the hardening curve formulation. Figure 4 left shows the resulting shapes of the MC fracture criteria in the mixed stress strain space.

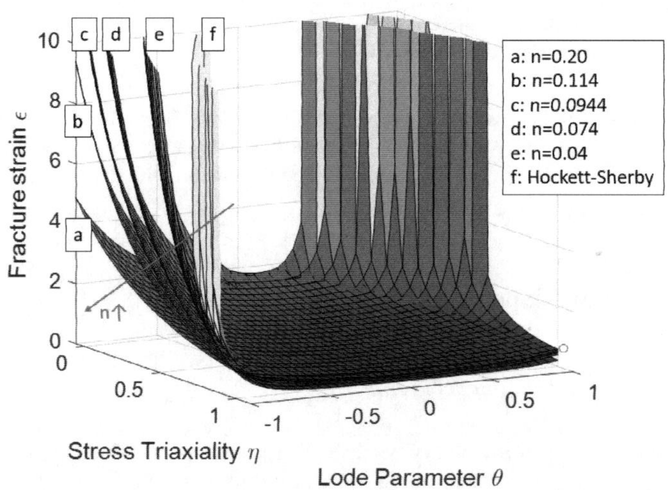

Configuration	c1	c2
a	0.1815	573.66
b	0.1377	469.99
c	**0.1286**	**455.71**
d	0.1201	462.91
e	0.1076	392.01
f	0.0976	420.76

Figure 4. Varying shapes of the MC fracture criterion in function of used hardening curve approximation

The differences in fracture strains making use of varying hardening curve formulations can be significant, depending on the stress state. Using the Hocket-Sherby flow curve approximation there are stress states, such as low stress triaxiality and high norm of the normalized lode angle parameter, under which fracture would not occur at all. These stress states are indicated in yellow in figure 5. This is due to the saturating nature of the hardening curve approximation that will never reach equivalent stresses under which fracture will occur for certain stress states.

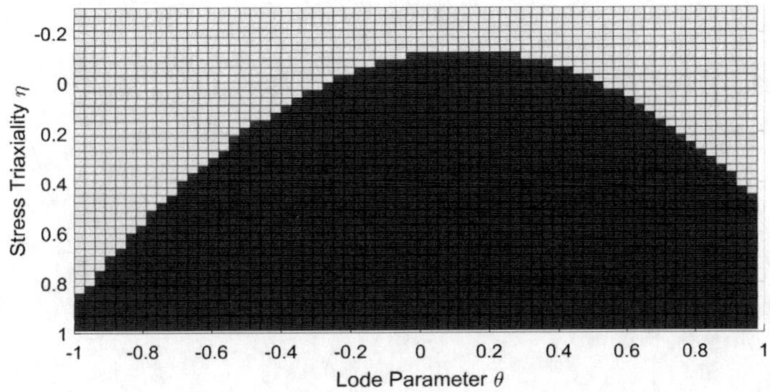

Figure 5. Stress states under which fracture wouldn't occure using the Hockett-Sherby flow curve

4 Conclusions

The following conclusion can be made from the findings in this paper:

- The hardening curve formulation and its slope for high strains changes the shape of the MC fracture criterion in the mixed stress-strain space in a considerable extent.
- When extrapolating the flow curve, special care has to be taken in order to achieve correct results for fracture strains given through the MC criterion in the mixed stress strain space.
- Since there is some uncertainty with respect to the large strain hardening of materials, there is some uncertainty with respect to the shape of the MC-criterion for high strains.
- The uncertainty with respect to large strain hardening should be minimized. One way to do so, could be the approximation of the large strain hardening behavior by hardness measurements.

References

[1] Dunand M and Mohr D 2014 Effect of Lode parameter on plastic flow localization after proportional loading at low stress triaxialities. *Journal of the Mechanics and Physics of Solids, 66*, pp 133-153.

[2] Mohr D and Marcadet S J 2015 Micromechanically-motivated phenomenological Hosford-Coulom model for prdeicting ductile fracture initiation at low stress triaxialities. *International Journal of Solids and Structures, 67-68*, pp 40-55.

[3] Komischke T, Hora P, Domani G, Plamondon M, Kaufmann R 2018 Prediction of crack induced failure phenomena in rolling operations. *Procedia Manufacturing, 15*, pp 176-184.

[4] Bai Y and Wierzbicki T 2008 A new model of metal plasticity and fracture with pressure and Lode dependence. *International Journal of Plasticity, 24*(6), pp 1071-1096.

[5] Coulomb C A 1776 Essai sur une application des regles des maximis et minimis a quelques problemes de statique relatifs a l'architecture. Memoires de l'Academie. *Royale pres Divers Savants, 7*.

[6] Siebel E, Pomp A 1927 Die Ermittlung der Formänderungsfestigkeit von Metallen durch den Stauchversuch. *Mitt. KWI Eisenforsch., 9*, pp 157- 171.

Mechanical properties of laminated CuAl10Fe3Mn2 aluminum bronze – intermetallics composites

M Konieczny

Kielce University of Technology, Department of Metals Science and Materials Technologies, Poland

E-mail: mkon@tu.kielce.pl

Abstract. CuAl10Fe3Mn2-intermetallic phases laminated composites have been fabricated through reactive bonding at 875°C in vacuum using bronze foils (0.3 mm thick) and Ti foils with different initial thickness (0.05, 0.1 and 0.15 mm). The titanium layers were completely consumed during reaction resulting in microstructure of well-bonded bronze-intermetallics composites. The mechanical properties and fracture behavior of the fabricated composites were examined under different loading conditions through tension and impact tests. The results of investigations indicated that tensile properties of the composites strongly depended on the proportion between thickness of individual bronze layers and the thickness of intermetallics. The impact tests showed that when the load was applied perpendicular to the laminates, the composites displayed superior impact toughness. The toughness increased with increasing of remaining bronze thickness. The failure during impact testing occurred by cleavage mode for the specimens loaded parallel to the laminated composites and showed extensive plastic deformation for specimens loaded perpendicularly.

1 Introduction

Since almost three decades there has been great interest in the investigation of mechanical behavior of a variety of laminated metal composites, especially from metal-metal [1], metal-ceramic [2] and metal-intermetallic [3-13] systems.

Metal-intermetallic laminated (MIL) composites have the potential to perform various functions, such as heat exchange, blast mitigation, ballistic protection, thermal management and vibration damping [5]. MIL composites combine the ductility and high toughness of metals with the higher elastic modulus, higher strength, and lower density of intermetallics. Lamination improves fracture toughness, fatigue behavior, wear, corrosion and damping capacity. It can also provide enhanced formability or ductility for normally brittle intermetallics [1]. There are two groups of commonly used production techniques of MIL composites: deposition or bonding. Sputtering or vapor deposition techniques involve atomic scale transport of the component materials. Such nano-laminated materials are typically fabricated by depositing hundreds of alternate nanoscale layers and they have received significant interest due to their extremely high strength [13]. Unfortunately deposition techniques require sophisticated manufacturing equipment and are very slow. These factors increase the cost of component production and make these composites economically unattractive. Bonding techniques, e.g. diffusion bonding, transient liquid phase bonding and reaction bonding between metal foils have many advantages. They involve relatively simple processing and the size and the number of layers that can be produced are not limited [3]. Moreover, the laminated structure of the composite allows for variations in the layer thickness and phase volume fractions of the components simply through the selection of initial foils thickness. A great number of laminated composites have been produced using Al and Ni [3, 6, 8], Ti [5, 11, 12], Nb [4] or Fe [7] foils. Previous works also revealed that titanium-intermetallic and copper-intermetallic composites can be produced by reaction that occurs at the interface of Ti and Cu [8-10]. Aluminum bronzes are corrosion-resistant alloys of copper containing from 4 to 15 % aluminum and small amounts of other metals, used to make many machine parts and tools. Alloys containing approximately 10 % aluminum are fabricated by sand casting and gravity diecasting into strong objects, including ship propellers [14]. The laminated composites with aluminum bronze matrix could be considered for structural applications because of their lower density than monolithic bronze. This kind of laminated composites has been investigated by Tsai et al. [15]. In the previous study [16] the reaction synthesis was successfully employed to fabricate laminated composites in vacuum using aluminum bronze and Ti foils. Designing laminated composites to obtain optimal mechanical requires knowledge of the composite failure mechanism. Many investigators [3-6, 17-22] have analyzed the mechanical and fracture properties of metal-intermetallic laminated composites, and proposed several useful models of crack propagation. The aim of this work is to study the mechanical properties of CuAl10Fe3Mn2 aluminum bronze-intermetallics laminated composites under different loading conditions and volume fraction of constituents.

2 Experimental procedure

For bronze-intermetallics laminated composites fabrication titanium foils were alternatively stacked between CuAl10Fe3Mn2 foils of 0.3 mm thickness that were produced by hot rolling.

Chemical composition and mechanical properties of used materials are shown in table 1.

Table 1. Chemical composition and mechanical properties of foil materials

Material	Chemical composition (at %)	YS (MPa)	UTS (MPa)	Elongation (%)
Ti	99.51 Ti, 0.09 Fe, 0.08 C, 0.07 Al, 0.18 O, 0.05 V, 0.02 N	250	320	24
CuAl10Fe3Mn2	83.36 Cu, 2.78 Fe, 10.75 Al, 1.98 Mn, 0.76 Ni, 0.29 Zn, 0.08 Sn	325	580	12

The titanium foil thicknesses were initially selected (0.05, 0.10 and 0.15 mm thick) to completely consume the metal in forming the intermetallic compounds with alternating layers of partially unreacted bronze. Foils were cut into 60 x 10 mm rectangular pieces. Any contamination was removed in a bath of 5 % HF in water. After rinsing in water, foils were stacked into laminates in an alternating sequence. A pressure of 5 MPa was employed at room temperature in a specially constructed vacuum furnace to ensure good contact between foils. Series of attempts allowed to find that a temperature of at least 870 °C was necessary for the start and rapid development of structural processes at the interface between aluminum bronze and titanium. The temperature was increased from 20 to 800°C at a heating rate of 0.25 °C/s. The samples were heated in vacuum of 0.01 Pa at 800 °C for 1h under applied 5 MPa pressure to allow diffusion bonding of the layers. After that the foils were heated to 875°C and held at this temperature for 60 minutes. The pressure was removed during this processing sequence with the purpose of eliminating possible expulsion of liquid phases. The temperature was then decreased slowly (cooling rate of 0.16 °C/s) to 700 °C and the pressure of 5 MPa was applied again. Finally, the samples were furnace-cooled to room temperature (figure 1). The laminated composites prepared from 0.05, 0.10 and 0.15 mm thick titanium foils are denoted later as samples Ti0.05 (12 layers of bronze and 11 layers of Ti), Ti0.10 (11 layers of bronze and 10 layers of Ti), and Ti0.15 (10 layers of bronze and 9 layers of Ti), respectively. After fabrication, the samples were cut using diamond blade and polished applying standard metallographic techniques.

Microstructural observations were performed using a JEOL JMS 5400 scanning electron microscope and a Nikon ECLIPSE MA 200 optical microscope. The chemical composition of the phases was determined by an energy dispersive spectroscopy utilizing a ISIS 300 Oxford Instruments.

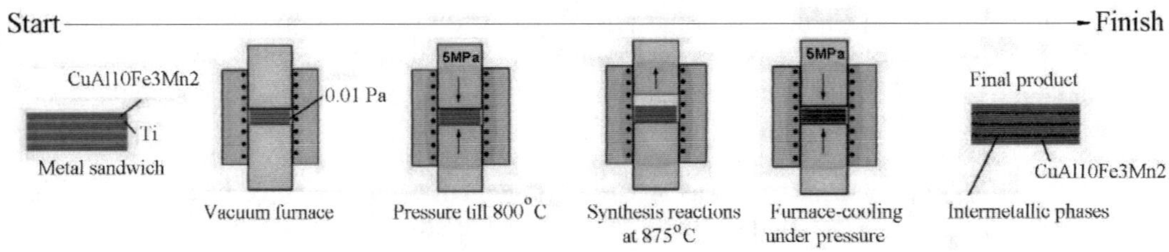

Figure 1. A schematic diagram of fabrication of CuAl10Fe3Mn2-intermetallic phases composite

Before the samples were examined with the optical microscope they had been etched to reveal grain boundaries and the structure of the intermetallic layers. Etching was performed with solution of 40 g CrO_3, 7.5 g NH_4Cl, 8 ml H_2SO_4, 50 ml HNO_3 and 1900 ml H_2O. Vickers (HV0.1) measurements were performed by Matsuzawa microhardness tester. Samples with dimensions of 4 x 8 x 50 mm were subjected to tension test on an INSTRON screw machine at a constant crosshead speed of 0.1 mm/min. The Charpy impact tests were conducted on un-notched rectangular samples with dimensions of 4 x 4 x 40 mm on a pendulum impact tester with a maximum capacity of 24.5 J.

3 Results

After fabrication, the composition of synthesized phases in the composites was determined.

The previously published results [16] showed that only three intermetallic phases were formed in the processed samples: TiCuAl, $TiCu_2Al$ and Ti_2Cu (figure 2). The predominant part of the intermetallic layers were $TiCu_2Al$ and TiCuAl phases, since a liquid front of reaction was moving into the aluminum bronze.

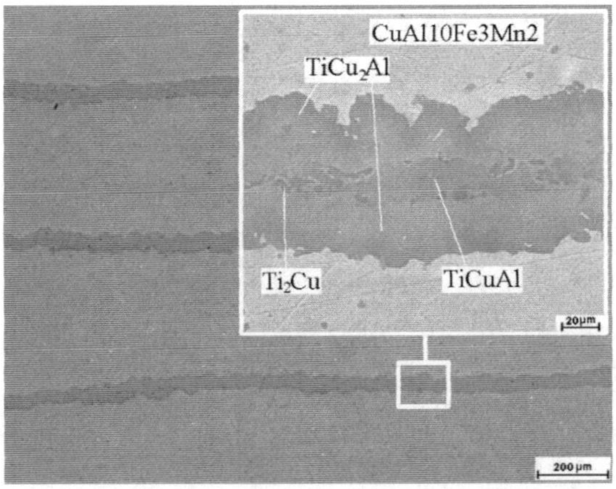

Figure 2. Microstructure of the CuAl10Fe3Mn2-intermetallic phases laminated composite

The laminated materials were well-bonded and nearly fully dense. The thickness of the intermetallic layers was dependent of the starting thickness ratio of the titanium and aluminum bronze foils. Microstructural investigations indicated also that the central zones of the formed intermetallic layers contained oxides inclusions which originated from surface oxide films on the used foils.

3.1 Microhardness measurements

Microhardness measurements were performed for used materials (for titanium grains and phases α and γ' in aluminum bronze) as well as for formed intermetallic layers.

Results of microhardness measurements are given in table 2.

Table 2. Results of hardness measurements (HV0.1)

CuAl10Fe3Mn2	Titanium	Bronze/intermetallic interface	Middle of the intermetallic layer
222±97	215±5	715±55	530±20

The maximum hardness values in the range of 660 to 770 HV were achieved at the bronze/intermetallic interface due to the presence of the $TiCu_2Al$ intermetallic phase. In the middle of the intermetallic layers, comprising mainly the mixture of TiCuAl and Ti_2Cu phases, hardness was a little bit lesser and contained in the range of 510 to 550 HV. In comparison, for materials used to produce the laminated composites, CuAl10Fe3Mn2 and titanium, the values of hardness were from two to four times lesser.

3.2 Tensile testing

Figure 3 shows the typical tensile stress-strain curves for the processed laminated composites and aluminum bronze.

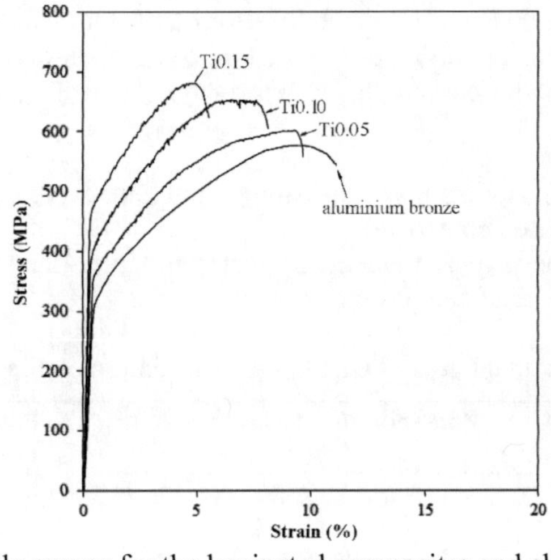

Figure 3. Tensile curves for the laminated composites and aluminum bronze

With an increase of thickness of the used titanium foils, the intermetallics layers grow, leading to an increase in the yield strength and the tensile strength of composites and a decrease in the total strain at fracture. The failure mechanism in the course of straining of the composites was carefully investigated. At the early stage of plastic straining the aluminum bronze grains were deformed by slip. No traces of plastic deformation were noticed in the intermetallic layers. Unfortunately, during deformation the crystal structures mismatch between phases makes almost impossible slip-bands slide through an alloy/intermetallic interfaces. Therefore, formation of cracks in the intermetallic layers was the characteristic feature of the prolonged deformation. The serrations in the stress-strain curves in figure 3 correspond to the formation of the multiple cracks in the intermetallic layers. The energy absorption capability of the aluminum bronze layers allowed numerous cracks to develop within each intermetallic layer before failure. With permanent increase of the crack number in the intermetallic layers the bronze layers gradually underwent the total external load. As a result the plastic flow that took place in the alloy layers was restricted to the small regions between opposite cracks in the neighboring intermetallic layers. When the number and distribution of cracks reached a critical limit, the final failure occurred by shearing fracture of the aluminum bronze layers. Of course it is obvious that strain hardening of the tensile specimens (figure 3) was produced due to plastic deformation of the aluminum bronze layers. Table 3 summarizes the tensile properties of synthesized composites. The presented results the average of three measurements for each type of composite.

Table 3. Summary of tensile properties of CuAl10Fe3Mn2-intermetallics laminated composite

Sample designation	YS (MPa)	UTS (MPa)	Elongation (%)
Ti0.05	348±5	591±4	10±0.5
Ti0.10	377±4	656±2	8±0.4
Ti0.15	462±5	683±5	6±0.5

The obtained data are consistent with the results reported by Aleman et al. [3] and Vecchio et al. [5, 12, 17]. The authors demonstrated that fracture mechanism and failure energy in metal-intermetallic laminated composites were controlled by varying the intermetallic-to-metal volume ratio. The failure characteristic described for the aluminum bronze-intermetallic phases composites is strictly consistent with previous studies of cracking and damage mechanisms in various laminated composites [2-12, 17-22]. The damage mechanisms of the unlike laminated composites during tensile testing are alike because different intermetallics formed from different constituent metals behave very similar. They normally are brittle due to the limited mobility of dislocations, and have very low surface energy resulting in little or no plastic deformation at crack tips.

3.3 Impact test

For the impact toughness measurements, two loading directions were used: one in perpendicular and another parallel to the laminated composites.

Results of impact toughness measurements are listed in table 4 and they are the average of three measurements for each type of composite.

Table 4. The impact toughness of CuAl10Fe3Mn2-intermetallics laminated composite

Sample designation	Perpendicular to layers (J/cm^2)	Parallel to layers (J/cm^2)
Ti0.05	125±3	52±2
Ti0.10	117±4	44±3
Ti0.15	96±2	31±1

For the comparison also the impact toughness for aluminum bronze was measured and it was 67 J/cm^2. Results shows that laminated composites exhibit anisotropic properties. When the load is applied perpendicular to the layers, the composites display superior impact toughness. Furthermore, it increases with increasing volume fraction of unreacted bronze. A crack initiated in the intermetallic layer did not travel across the specimen, but was arrested and deflected by the adjacent bronze layer. Further loading causes the formation of some new cracks in the next intermetallic layer. As a result a plastic strain of bronze was localized in shear bands between opposite cracks. This process was repeated until all the intermetallic layers have cracked resulting in tortuous crack propagation patch. Therefore there was one main crack and several branching smaller cracks. The failure occurred by cleavage mode showing extensive plastic deformation of bronze layers. It should be added that delamination or debonding of layers has been shown in the literature to be detrimental to the

IOP Conf. Series: Materials Science and Engineering **461** (2019) 012042 doi:10.1088/1757-899X/461/1/012042

performance of laminated composites [23, 24], but thesc failure mechanisms were not observed in any of tested samples. The absence of delamination of the bronze and intermetallic layers during mechanical testing is an indication of the excellent bonding between the layers. When the specimens were tested parallel to the layers the failure occurred by cleavage mode showing limited plastic deformation. It was evident that there was solely one main crack that travel across the specimen.

4 Conclusions

The mechanical properties and fracture behavior of the CuAl10Fe3Mn2-intermetallic phases laminated composites have been examined under different loading conditions through tension and impact test.

The results of investigations indicated that tensile properties of the composites strongly depended on the proportion between thickness of individual bronze layers and the thickness of intermetallics. The yield and tensile strength of all investigated composites were higher than for aluminum bronze, and for Ti0.15 were higher even 40 and 18 %, respectively. The impact tests showed that when the load was applied perpendicular to the laminates, the composites displayed superior impact toughness, that was even 2 times higher than for bronze.

References

[1] Wadsworth J and Lesuer D 2000 *Mater Charact* **45** 289
[2] Shen Y L and Suresh S 1996 *Acta Mater* **44** 1337
[3] Alman D, Dogan C P, Hawk J A and Rawers J C 1995 *Mat Sci Eng A-Struct* **192-193** 624
[4] Bloyer D R, Venkateswara Rao K T and Ritchie R O 1997 *Mat Sci Eng A-Struct* **239-240** 393
[5] Vecchio K S 2005 *JOM* **57** 25
[6] Wang H, Han J, Du S and Northwood D O 2007 *Mat Sci Eng A-Struct* **445-446** 517
[7] Konieczny M and Mola R 2008 *Steel Res Int* **79** 499
[8] Konieczny M 2013 *Mat Sci Eng A-Struct* **586** 11
[9] Konieczny M and Dziadon A 2007 *Arch Metall Mater* **52** 555
[10] Dziadon A, Konieczny M, Gajewski M, Iwan M and Rzączyńska Z 2009 *Arch Metall Mater* **54** 455
[11] Peng L M, Li H and Wang J H 2005 *Mat Sci Eng A-Struct* **406** 309
[12] Harrach D J and Vecchio K S 2001 *Metall Mater Trans A* **32** 1493
[13] Tixier-Boni S and Van Swygenhoven H 1999 *Thin Solid Films* **342** 188
[14] Gazda A, Górny Z, Kluska-Nawarecka S, Saja K and Warmuzek M 2010 Effect of modification and heat treatment on structure and mechanical properties of CuAl10Fe3Mn2 bronze (Krakow: Works of Foundry Research Institute vol 12) p 37
[15] Tsai H C, Higashi K and Sherby O D 1991 Proc. Int. Conf. of Advanced Composites (Warrendale: TMS) p 151
[16] Konieczny M, Mola R, Kargul M and Aniołek I 2018 Proc. of 27th Int. Conf. on Metallurgy and Materials METAL (Brno) p 161
[17] Rohatgi A, Harach D, Vecchio K S and Harvey K 2003 *Acta Mater* **51** 2933
[18] Chung D, Kim J and Enoki M 2005 *Mater Sci Forum* **475-479** 1521
[19] Xian L, Yanqing Y, Yungwang M, Bin H, Meini Y and Yan C 2007 *Scripta Mater* **56** 569
[20] Buchholz F G, Rikards R and Wang A 1997 *Int Journal of Fracture* **86** 37
[21] Rudnitskii N 2002 *Strength of Materials* **34** 612
[22] Li M and Soboyejo 2000 *Metallurgical Trans A* **31** 1385
[23] Li T, Grignon F, Benson D J, Vecchio K S, Olevsky E A, Schwarz R B and Meyers M A 2004 *Mat Sci Eng A-Struct* **374** 10
[24] Odette G R, Chao B L, Sheckherd J W and Lucas G E 1992 *Acta Metall Mater* **40** 2381

IOP Conf. Series: Materials Science and Engineering **461** (2019) 012043 doi:10.1088/1757-899X/461/1/012043

Applicability of miniature tensile test in the automotive sector

P Konopík[1], P Farahnak[1], M Rund[1] J Džugan[1], and S Rzepa[2]

[1]COMTES FHT a.s., Dobřany, Czech Republic, EU
[2]AGH University of Science and Technology, Kraków, Poland, EU

E-mail: pavel.konopik@comtesfht.cz

Abstract. The aim of this paper is to show the applicability of miniature tensile test (M-TT) for local properties determination on steel sheets for drawing and forming which are mostly used in automotive industry. Knowledge of local mechanical properties is a key factor for parts quality assurance. A comparison of the tensile test results from standard sized and miniature sized geometry has been done to assess the M-TT reliability. Various mechanical properties, such as the ultimate tensile and yield strengths the elongation, and reductions in area will be obtained. Moreover, thanks to advanced measurement techniques like digital image correlation with high resolution, strain hardening parameters and plastic strain ratio can be evaluated as well. Finally, flow stress curves are calibrated by inverse analysis in Abaqus FEM software.

1 Introduction

The tensile properties of materials, such as the ultimate tensile strength, yield strength, elongation and cross-sectional area, are very important factors for engineering design. In automotive industry, a group of cold-rolled steels for drawing and cold forming (DC01 – DC06) is widely used [1]. These steel grades have excellent formability, which facilitates cold forming operations, and are ideal for deep drawing. They are used for bending and deep drawing forming processes for applications where strength, rigidity and ductility are required.

There are guaranteed mechanical properties of delivered sheet which can be easily determined using standard tensile test. However, for the quality and reliability assessment of the final manufactured part, local properties can be crucial. The most critical are often sharp bending locations with small radiuses where residual formability and elongation can be near to the material limit.

On the other hand, due to lower friction than expected during cold forming, some areas can be deformed less than expected and thus, these locations could have lower strength. In comparison with [2], the recently-developed miniature tensile test (M-TT) is an excellent tool for local properties determination [3] or, in all cases, where only a limited amount of material is available [4-5].

The aim of this paper is to show the applicability of miniature tensile test for local properties determination on steel sheets for drawing and forming. For this purpose, standard and miniature tensile tests (M-TT) were performed in 3 orientations: 0°, 45° and 90°. Moreover, for thickness influence evaluation, miniaturized specimens were machined with several different thicknesses and material model using Abaqus FEM software will be created.

2 Testing method, M-TT

For local properties determination, miniature test pieces had to be used. The recently-developed miniature-tensile test (M-TT) technique was chosen for this purpose. The shape of the M-TT test pieces was initially derived from the small punch test (SPT). Small punch test specimens are discs 8 mm in diameter and 0.5 mm in thickness. SPT is used for determining the mechanical properties of in-service components because it only requires a very small amount of material. However, its drawback comes when interpreting the test readings, which must be converted by means of empirically-derived correlations to ordinary mechanical properties, such as yield and ultimate strength [6].

On the contrary, M-TT stress-strain curves are identical to standard ones [7], provided that several conditions are met. M-TT only requires small amounts of experimental material, and therefore enables local mechanical properties to be measured. To obtain consistent data, the test piece thickness must be ≥ 10× grain size [8]. When standard conditions [7] are met, such as accuracy of the measurement of force and extension and the required strain rate, M-TT data are identical to standard test data within a broad range of metals and their strength levels [9][10].

The initial test piece design (Figure 1a) was modified (Figure 1b) for current investigated material DC01. A modified M-TT specimen has a thickness of 0.5 mm and parallel length of 4.5 mm.

M-TT is carried out using a special test device with high load cell sensitivity. Its linear motor runs smoothly at very slow speeds to fulfill all criteria given by the standard [7]. The test piece is clamped in special flat grips which have been developed for this purpose. The axial extension is captured with a high-speed CCD camera integrated into the systems Mercury RT or ARAMIS™ that relies on digital image correlation (DIC). As the force and extension data are captured accurately, the record need not be specially processed and the evaluation can follow the standard procedure [7].

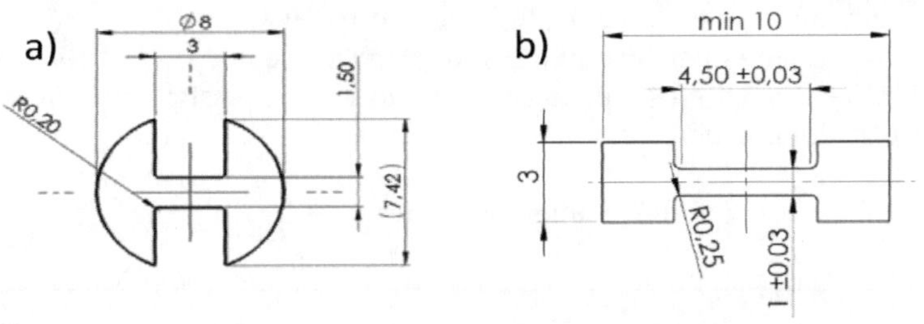

Figure 1. Geometry of an M-TT test piece; (a) initial design, (b) modified geometry for testing ductile steel sheets

3 Experiment

The uniaxial tensile tests are performed on the specimen geometries with different thicknesses 0.2, 0.5 and 1.5 mm to investigate material behavior. Specimens were taken from DC01steel sheet in three directions: longitudinal direction 0°, diagonal direction 45° and transverse direction 90°, related to the rolling direction.

Electro-mechanical testing system Mayes is used to conduct the uniaxial tests and Aramis™ system with 12 megapixel is utilized to monitor 2D strain fields with high resolution. Figure 2a illustrates the experimental setup and data acquisition rate is set at 100 Hz in testing system for clear visualizing the tensile behavior. It is noticeable that for all experiments initial lengths of extensometers are chosen 4 mm. Figure 2b represents engineering stress- strain curves of tensile tests for all considered specimens geometries in rolling direction (RD).

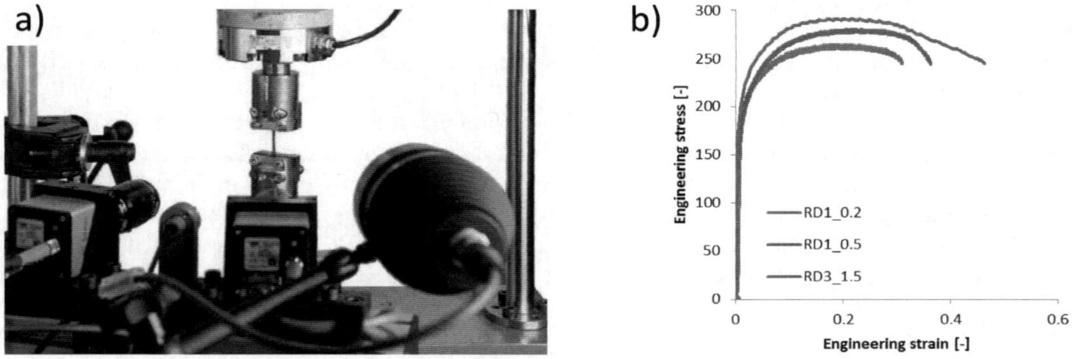

Figure 2. (a) Experimental setup, (b) engineering stress- strain curves for different thicknesses

4 Tensile test results

Tensile tests were evaluated based on the standard [7]. The measurement for the plastic strain ratio determination was performed with the use of continuous strain measurement by DIC system ARAMIS. Subsequently, r-values (i.e. plastic strain ratio) for the plastic strain range of 2-10 % were evaluated according to the standard [11].

Results of averaged values from three tests per conditions are shown in Table. 1, where YS is Yield strength, UTS is Ultimate tensile strength, A_g is plastic extension at maximum force, A is plastic elongation of the gauge length after fracture (GL for standard geometry was 50 mm and for M-TT 4 mm), Z is maximum change in cross-sectional area and $r_{2\text{-}10}$ is plastic strain ratio evaluated in the range of 2-10 % plastic deformation.

Table 1. Summary of tensile tests results

Specimen	YS MPa	UTS MPa	A_g %	A %	Z %	$r_{2\text{-}10}$ -----
M-TT_0°_0,2	154	241	15.8	34.5	57.2	2.05
M-TT_0°_0,5	185	295	18.8	53.2	83.7	2.04
M-TT_0°_1,5	186	295	20.9	56	80	2.08
Standard_0°_1,5	195	309	25.4	44.6	70.3	2.11
M-TT_45°_0,2	176	272	16.8	31.7	50.8	1.21
M-TT_45°_0,5	184	315	22.3	44.4	77.7	1.22
M-TT_45°_1,5	196	307	21.7	57.6	83.9	1.19
Standard_45°_1,5	206	323	24.4	44.8	44.7	1.21
M-TT_90°_0,2	150	231	17.9	36	53.7	2.16
M-TT_90°_0,5	166	298	23.9	54	87.2	2.28
M-TT_90°_1,5	170	287	26	59.8	84.2	2.31
Standard_90°_1,5	196	304	24.2	48.1	71.4	2.45

The evaluation of strain hardening parameters was carried out based on Hollomons equation (1):

$$\sigma = C.\varepsilon^n \tag{1}$$

where σ is the stress, K is the strength index or strength coefficient, ε is the plastic strain and n is the strain hardening exponent. In equation (1), the power law relationship between the stress and the amount of plastic strain is used. The fitting range considered for the Hollomons law parameters is between 2-10% of the plastic strain. Results are summarized in Figure 3.

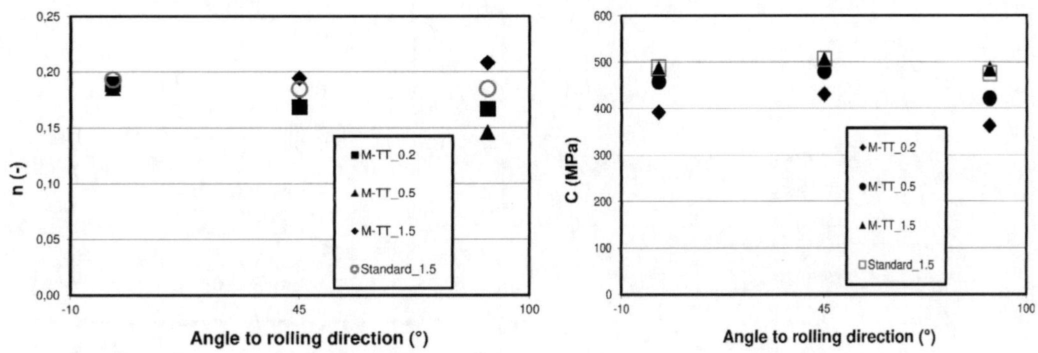

Figure 3. (a) Strain hardening exponent, (b) Strength coefficient

5 Abaqus FEM simulation

Abaqus FEM software is utilized for numerical simulation with conventional plasticity [12] based on von Mises material model. Flow curves are calibrated through inverse analysis by comparing experimental and numerical force-displacement curves and are depicted in Figure 4.

For all simulations, 3D elements are used (C3D8R according to Abaqus element library) and mesh size is chosen $0.06\ mm$ between the extensometer points and 0.15 mm for surrounding area. Moreover, 5 elements are assigned along the thickness. As it can be seen in Figure 4a, 4b and 4c, numerical simulations are in close agreement with experimental data measured in rolling direction. For measuring Lankford ratios, Gauge sections are defined in transversal direction as well as axial direction. Three experiments are performed on specimens in rolling, diagonal and transverse directions. GOM Aramis™ captures 2D displacement fields during deformation, e.g., axial and transversal extensometers, provides engineering surface strains which are then transformed into true values. Finally, thickness strain is measured based on incompressibility condition. Lankford ratio (plastic strain ratio) can be calculated from average slope of the true plastic width (ε_w^p) and thickness (ε_{th}^p) strain based on equation (2) for the specific range of equivalent plastic strain $(\bar{\varepsilon}_p)$ and are indicated in figure 4d.

$$r = \frac{d\varepsilon_w^p}{d\varepsilon_{th}^p} \qquad while\ 0.1 \le \bar{\varepsilon}_p \le 0.2 \tag{2}$$

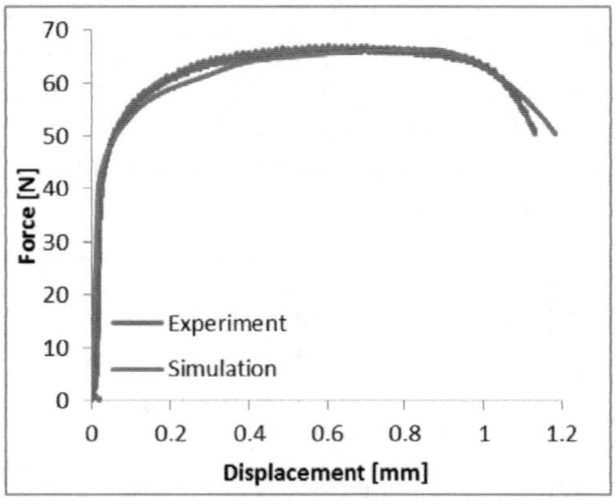

(a)

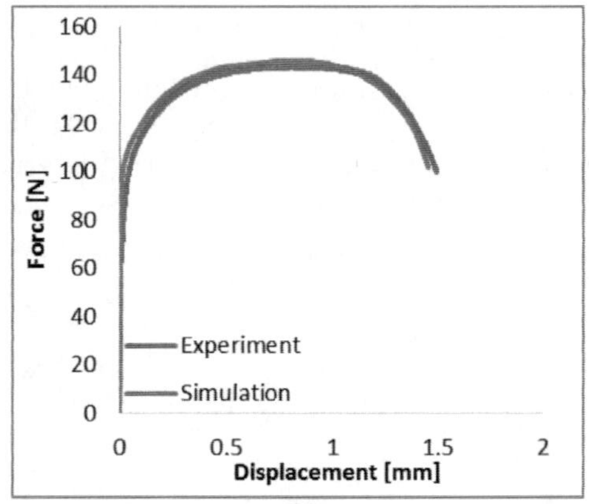

(b)

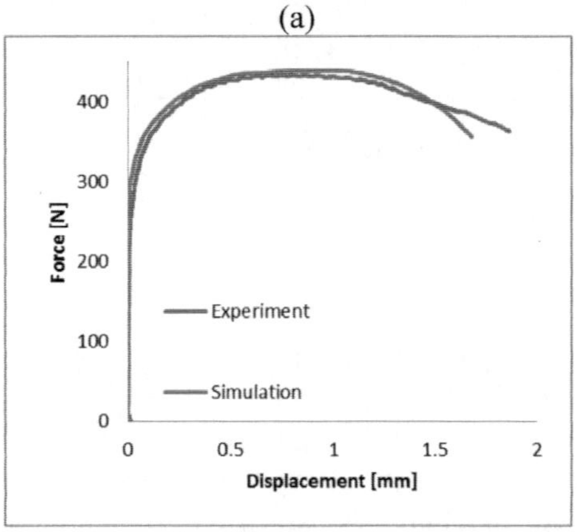

(c)

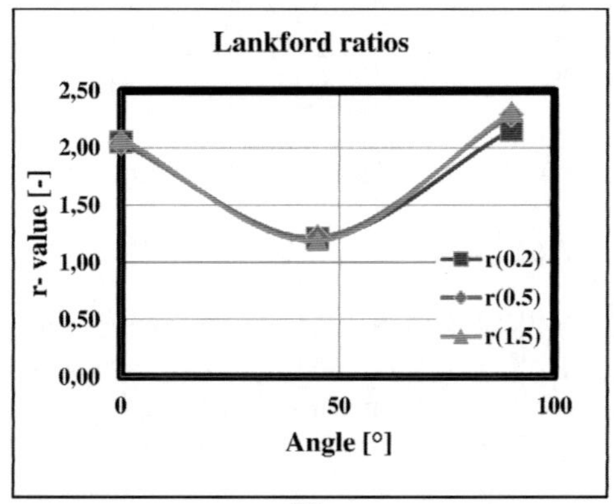

(d)

Figure 4. Engineering stress- strain curves for different thicknesses (a). for thickness 0.2 mm, (b). for thickness 0.5 mm, (c). for thickness 1.5 mm, (d). Lankford ratios

6 Results discussion

Overall, the agreement ranged from acceptable results to excellent results. The biggest differences were seen in the stress-strain curves obtained from M-TT 0.2 mm in thickness for all rolling directions, where the most significant deviation was measured in respect of UTS and corresponding C values. This can have several causes. Firstly, machining and handling of a specimen 0.2 mm thick is challenging and surface finish could possibly have a big influence. Furthermore, the specimens were taken from the middle part of the sheet where the material can be softer than material near to the surface.

However, this study indicated that values of plastic strain ratio are not dependent on specimen thickness. Trends obtained here for this parameter are in agreement with those published in [13-15].

7 Conclusions

The paper presented here successfully shows the possibility of metal sheet characterization for forming processes with the use of miniaturized tensile specimens. Generally, very good agreement was found for all considered parameters and conditions between the results attained with the use of standard and M-TT specimens. FEM simulation in Abaqus software demonstrated that data obtained from M-TT can be easily used as input data for all software requiring reliable stress – strain data and thus, local properties can be taken into count for part quality assurance.

Acknowledgements
This paper was created by project No: TH02010544, financed by the TA ČR and project No.: LO1412 - Development of West-Bohemian Centre of Materials and Metallurgy, financed by the MEYES of the Czech Republic.

References

[1] ArcelorMittal, Extract from the product catalogue, 2018, [online] https://automotive.arcelormittal.com/saturnus/sheets/ArcelorMittal%20Automotive%20product%20offer%20EN.pdf

[2] Janoušek J and Balda M 2014 Strain gage measurements for accurate yield point determination, Applied Mechanics and Materials 486, pp. 123-128

[3] Jirková J, Rubešová K, Konopík P and Opatová K 2018 Effect of the Parameters of Semi-Solid Processing on the Elimination of Sharp-Edged Primary Chromium Carbides from Tool Steel, *Metals - Open Access Metallurgy Journal* 8(9):713

[4] Dzugan J, Prochazka R and Konopik P 2017 Low cycle fatigue tests with the use of miniaturized test specimens, ASME Pressure Vessels and Piping Conference, Location: Waikoloa, HI, Date: JUL 16-20, 2017, Article Number: UNSP V01AT01A072

[5] Dzugan J, Prochazka R and Konopik P 2015 Micro-Tensile Test Technique Development and Application to Mechanical Property Determination, 6th ASTM International Symposium on Small Specimen Test Techniques Location: Houston, Date: JAN 29-31, 2014, Volume: 1576, Pages: 12-30

[6] Dobes F and Dymacek P 2016 Fracture-based correlation of uniaxial and small punch creep data *Theoretical and Applied Fracture Mechanics*, Part: A, 86, 34-38, DOI: 10.1016/j.tafmec.2016.08.020

[7] ISO 6892-1:2016: Metallic materials -- Tensile testing -- Part 1: Method of test at room temperature, 2016

[8] Kumar K, Madhusoodanan K and Rupani, B. B. 2006 Miniature Specimen Technique as an NDT Tool for Estimation of Service Life of Operating Pressure Equipment., International Conference & Exhibition on Pressure Vessel and Piping, Chennai, India, 7-9 February 2006; 92-102

[9] Rund M, Procházka R, Konopík P, Džugan J and Folgar H 2015 Investigation of Sample-size Influence on Tensile Test Results at Different Strain Rate. *Procedia Engineering*, 114, 410-415, https://doi.org/10.1016/j.proeng.2015.08.086

[10] Prochazka R, Dzugan J, Konopik P and Rund M 2018 Investigation of high-strength stainless steel using small specimen test techniques - tensile and fatigue properties, 7th International Conference on Mechanics and Materials in Design (M2D), Albufeira, Portugal, 11-15 June 2017; Eds. Gomes, J.F.S.; Meguid, S.A.; INEGI/FEUP: 2017; Paper ref. 6841, 343-354

[11] ISO 10113:2006: Metallic materials Sheet and strip determination of plastic strain ratio

[12] Farahnak P, Azinpour E, Belinha J and Cesar de Sa J 2016 The numerical non-linear analysis of elasto-plastic materials using the radial point interpolation method *International book of Material Modelling: Applications, Challenges and Research* ISBN: 978-1-53612-161-2.

[13] Gavrus A and Francillette H 2011 An Anisotropic Behaviour Analysis of AA2024 Aluminium Alloy Undergoing Large Plastic Deformations *Aluminium Alloys, Theory and Applications*, ISBN 978-953-307-244-9.

[14] Ramos G, Stout M, Bolmaro R E, Signorelli J W and Turner P 2010 Study of a drawing-quality sheet steel. I: Stress/strain behaviors and Lankford coefficients by experiments and micromechanical simulations, *International Journal of Solids and Structures* Volume 47, Issue 17, 15 Pages 2285-2293.

[15] Huh J, Huh H, Lee C S 2013 Effect of strain rate on plastic anisotropy of advanced high strength steel sheets, *International Journal of Plasticity*, Volume: 44, Pages: 23-46, 2013, DOI: 10.1016/j.ijplas.2012.11.012

IOP Conf. Series: Materials Science and Engineering **461** (2019) 012044 doi:10.1088/1757-899X/461/1/012044

Accelerated Carbide Spheroidisation and Refinement in Spring Steel 54SiCr6

Kotous J Dlouhy J Nachazelova D Hradil D

COMTES FHT a.s., Prumyslova 995, 334 41, Dobrany, Czech Republic

jakub.kotous@comtesfht.cz

Abstract. Reduction of processing time is the goal of every business. In some cases, speeding up manufacturing processes is impossible or takes place at the expense of quality. The purpose of soft annealing is to obtain globular carbides in the microstructure. However, it is one of the most time-consuming heat treatment operations. Even its optimised versions require several hours of time. „Accelerated Carbide Spheroidisation and Refinement" (ASR) is an alternative process which can be performed using induction heating or thermomechanical treatment. It only takes several minutes to complete and has a beneficial effect on resultant mechanical properties. In this experiment, the impact of various microstructures in 54SiCr6 spring steel on hardening behaviour was explored. For this purpose, specimens with two different microstructures containing spheroidised carbides were produced by ASR accelerated annealing and by conventional soft annealing. Another initial microstructure, with lamellar pearlite, was obtained by hot rolling. It was found to have a certain effect as well. Quenched and tempered microstructures were observed using electron microscopy (SEM). Their mechanical properties were measured. The effects of various initial microstructures on resultant mechanical properties after cryogenic treatment were studied as well.

1 Introduction

Soft annealing inserted prior to hardening is an important stage in the treatment of springs. Parts from spring steels are mostly subjected to alternating cyclic stresses and relaxation, which is why manufacturers strive to improve their yield strength and ultimate strength while maintaining their ductility [1]. Accelerated Spheroidisation and Refinement (ASR) is one way to produce spheroidised microstructure [2]. The classical method is soft annealing, which is lengthy due to diffusion processes involved. The material is held near the A_{c1}. Where shorter times are required, slow cycling around A_{c1} is often used. Diffusion, a slow and long process, is required for carbides to change from the lamellar to the globular form. At long annealing times, the carbides which are already globular undergo Ostwald ripening. Consequently, their particles coarsen and their thermodynamic potential decreases, resulting in lower hardness and strength [3].

The ASR process is much less time-consuming. In its accelerated annealing, it relies on rapid cycling around the A_{c1} temperature. The partial austenitization and transformations accelerate the formation of globular particles [3]. Fully spheroidised microstructure can thus be attained within minutes without coarsening. The result is uniformly-dispersed fine globular carbides.

Morphological distinctions between the products of soft annealing and ASR, particularly the particle size, are reflected in mechanical properties as well [4].

They can mislead those who evaluate the extent of carbide spheroidisation. The reason is that successful spheroidisation is usually defined as 95% of carbides being in the spheroidal form.

However, morphological criteria are rarely set in engineering practice and the only criterion for spheroidisation annealing is the maximal permissible hardness of the treated material. ASR-processed material may meet the morphological criteria but fail to meet the hardness limits imposed on soft-annealed material [5]. In still other cases, the differences between spheroidised microstructures obtained by different methods may affect the kinetics of the subsequent processes, such as quenching. The experiments reported in this paper explored the impact of the initial microstructure of 54SiCr6 spring steel on its mechanical properties and microstructure after quenching and tempering. The initial microstructures either contained globular carbides (cementite) produced by either conventional long-time soft annealing (SA) or accelerated annealing (ASR) – or lamellar cementite obtained by hot rolling (HR). In some sequences, cryogenic treatment was performed between the quenching and tempering operations. The motivation was to find whether it reduces the effects that the initial microstructure may have or whether differences between mechanical properties will be retained [6].

2 Materials and methods

The experimental steel was the 54SiCr6 grade, which is used widely for the manufacture of springs. The steel was made at COMTES FHT a.s. The chemical composition of the steel is given in table 1. It was cast in a vacuum induction furnace, then forged, hot-rolled, and air-cooled after rolling.

Table 1. Chemical composition of the experimental steel 54SiCr6.

Element	C	Si	Mn	Cr	Mo	Cu	P	S	Fe
wt. %	0.57	1.51	0.68	0.75	0.03	0.01	0.008	0.003	bal.

Carbide spheroidisation was carried out using two different sequences. The accelerated spheroidisation process (ASR) was completed using medium-frequency (fmax=17 kHz) induction heating equipment with a maximum power of 24 kW. The specimens were 20 mm-diameter bars with 130 mm length. Temperature was measured with a thermocouple welded onto the specimen. The ASR sequence comprised of three thermal cycles with follows regime: heating at 19°C/s to 820°C, holding for 15 seconds, cooling in still air (at approx. 1.5°C/s) to 725°C and holding for 5 minutes. This was followed by cooling in still air to 600 °C. The second cycle was identical. In the third cycle, the hold at 725°C was extended to 10 minutes. The final step was cooling in still air to the ambient temperature. The sequence is plotted in figure 1. The total duration of the spheroidisation stage was approximately 30 minutes.

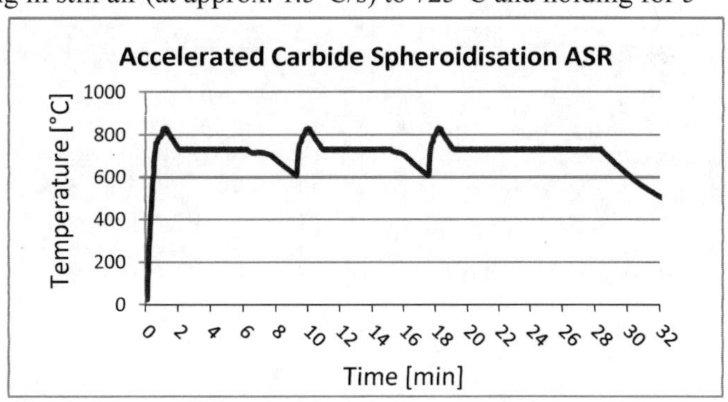

Figure 1. ASR induction heat treatment for the 54SiCr6 grade

Soft annealing (SA) was performed in an electrical air furnace. Specimens had the same size as those for the ASR sequence. In the SA sequence, the material was held just above the A_{c1}. The sequence was as follows: heating at 10°C/min to 720°C, then at 15°C/hour to 770°C, holding for 5 hours, furnace cooling at 5°C/hour to 720°C, then at 25°C/hour to 650°C, free cooling in the furnace to 400°C and then cooling in still air. The total time until removal from the furnace was approximately 27 hours. With this steel, the slow cooling at 5°C/hour in the soft annealing sequence and the isothermal hold at 725°C in the ASR sequence were necessary for preventing formation of new cementite lamellae.

Quenching and tempering was carried out in electrical air furnaces. Quenching temperatures were in the range from 810 to 890°C and the soaking time was 20 minutes. The quenching medium was oil. Tempering at 400°C/2 hours was followed by cooling in air.
Cryogenic treatment (CT) was performed in a cryogenic box. The sequence comprised cooling to -160°C over 1 hour, holding for 20 hours and reheating to 20°C in 1 hour.
The effects of the differences between initial microstructures and heat treating sequences were studied in the processed specimens using scanning electron microscopy (SEM) and mechanical testing, which involved measurement of hardness and tensile tests.

3 Results and discussion

3.1 Initial state of material

The initial microstructure after hot rolling (HR) was pearlite with hardness of 290±7 HV10. The microstructure was homogeneous, composed of a ferritic matrix and lamellar cementite (figure 2). Both spheroidised microstructures, post-SA and post-ASR, contained globular cementite particles in a ferritic matrix. In addition, there were some scarce cementite lamellae in them. Soft annealing (figure 3) produced large globular particles up to 2 μm in size. By contrast, the ASR sequence (figure 4) led to densely-dispersed small globular cementite particles (0.5 μm max.) in a ferritic matrix. The hardness after long-time soft annealing was 207 ± 1 HV10. After ASR, it was higher by 17 points, owing to a finer microstructure. Both values meet the delivery specifications for 54SiCr6 steel annealed to obtain globular cementite.

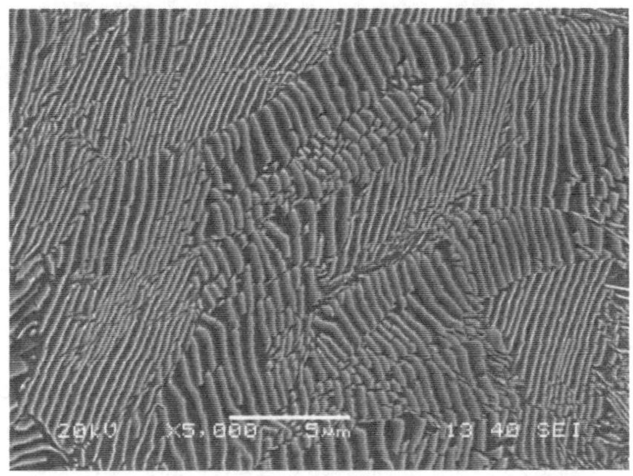

Figure 2. Microstructure after hot rolling (HR)

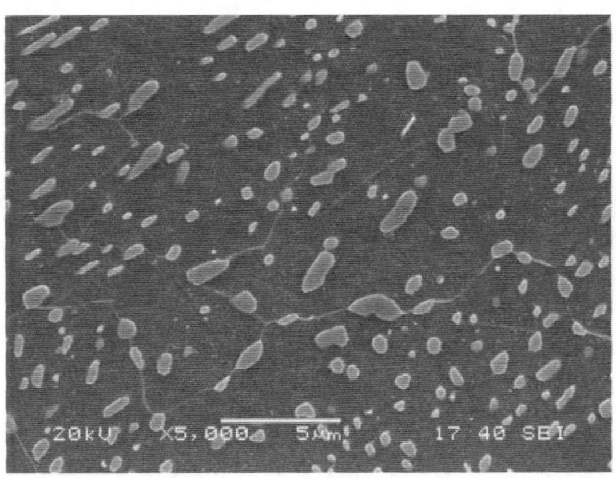

Figure 3. Microstructure after soft annealing (SA), 207 HV10

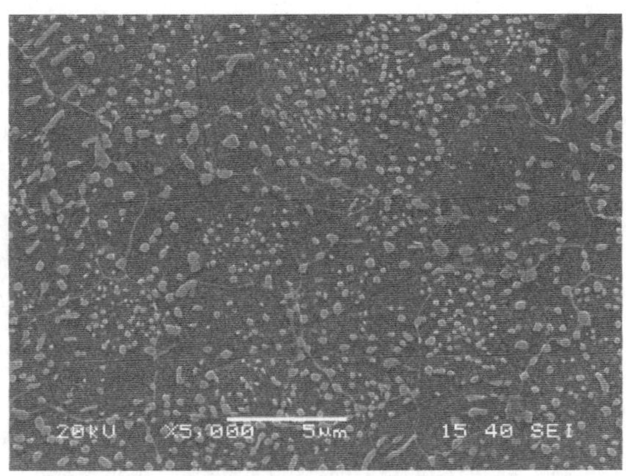

Figure 4. Microstructure after accelerated annealing (ASR), 224 HV10

3.2 Quenching and tempering

Quenching was performed from 810, 830, 850, 870 and 890°C. The austenitizing time was 20 minutes. Soft-annealed specimens, which were quenched from 810, 830 and 850°C, contained, in addition to martensite and undissolved cementite, some ferrite grains. With higher quenching temperatures, the matrix was fully martensitic. The amount of undissolved cementite decreased with increasing quenching temperatures. Yet, some undissolved globular cementite was found in the material upon quenching from 890°C. In the soft-annealed material, the largest undissolved cementite particles had a size of approx. 1000 nm. They remained in the material even after quenching from higher temperatures (figure 5) because the smaller ones had dissolved more quickly.

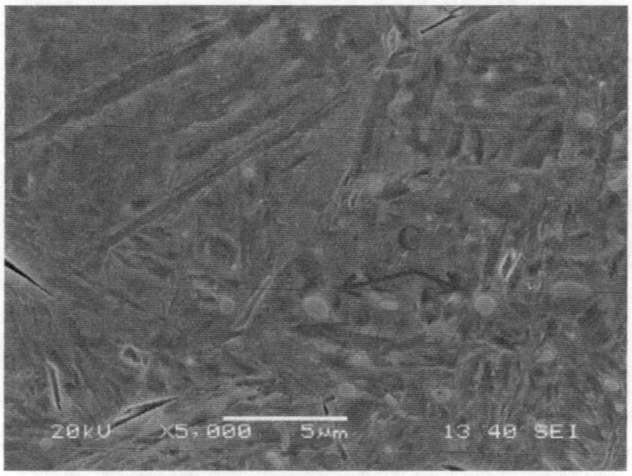

Figure 5. Quenched sample SA 870°C, 742 HV10, (C – undissolved cementite)

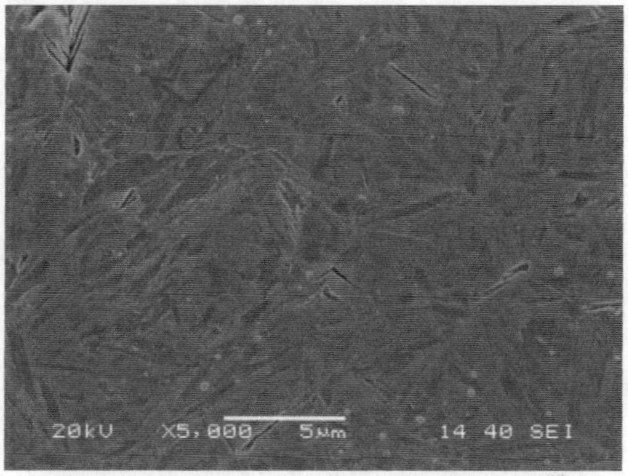

Figure 6. Quenched sample ASR 830°C, 756 HV10, (C – undissolved cementite)

Quenched ASR specimens only contained ferrite after quenching from 810°C. Fine globular cementite dissolved much more readily. After ASR, the maximum size of undissolved cementite particles was 300 nm (figure 6). Following the quench from 870°C, undissolved cementite was rare. After quenching 890°C, it was dissolved completely.

Quenching of pearlitic specimens obtained by hot rolling had similar results. Lamellar cementite austenitized even faster than the globular structure after ASR. Some remaining ferrite grains were also found after quenching from 810°C. Their amount was smaller than in the other cases. The largest undissolved cementite particles in the martensitic matrix had a size of 200 nm. A summary of the austenitizing experiment is given in table 2.

Table 2. Austenitizing (the highest T_{quench} upon which 54SiCr6 steel still contains a particular phase)

	Soft annealing (SA)	Accelerated annealing (ASR)	Hot rolling (HR)
Ferrite	850 °C	810 °C	810 °C
Undissolved cementite	890 °C	870 °C	870 °C

During tempering, cementite precipitated within and along martensite plates. Apart from the size of undissolved cementite particles, no differences were found between tempered microstructures obtained in specimens in different initial conditions.

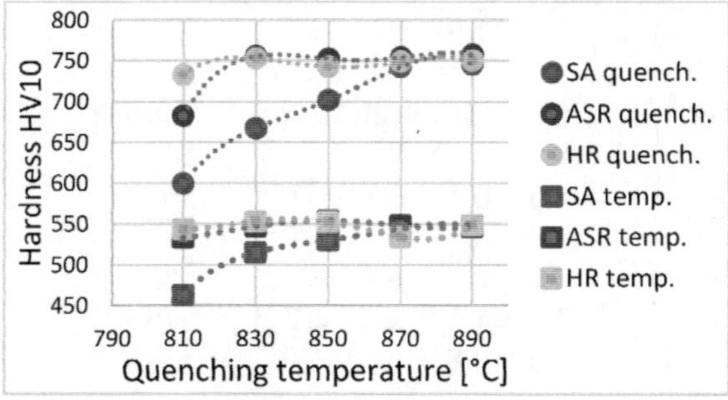

Figure 7. Hardness after quenching and tempering for 54SiCr6

Hardness is in agreement with the microstructure, as seen in figure 7. When the extent of austenitization proved insufficient (ferrite in the microstructure) in the specimens, lower hardness levels were found. Hot-rolled and ASR-treated specimens proved well-hardened even upon quenching from 830°C but the soft annealed ones required 870–890°C. The dissolution of remaining cementite particles has not led to a major increase in overall hardness. After transformation of ferrite to austenite, hardness appeared almost constant.

Maximum hardnesses were similar for all initial conditions. When quenching was appropriate, the hardness level was around 750 HV10 and the hardness upon tempering was approx. 550 HV10.

In the ASR-treated material, carbides dissolved almost as quickly as lamellar pearlite. In soft-annealed material, austenitizing was retarded, as cementite was present in large globules whose neighbourhood was depleted of carbon. Hence, slower dissolution of large particles and the need for long-distance diffusion of carbon poses greater demands on austenitizing prior to quenching. The desired properties were only attained once the quenching temperature had been raised to 870°C or higher.
The ASR process therefore combines the advantages of soft-annealed initial condition and lamellar pearlite. Thanks to fine carbides, austenitizing is rapid and high quenching temperatures are not needed. In 54SiCr6 steel, the difference in quenching temperatures is 40°C. Globular morphology of cementite guarantees the reduction in hardness which is the purpose of soft annealing.
In cryogenically-treated (CT) material, the microstructure was examined and hardness was measured as well. The sequences involving SA and quenching temperature of 870°C and ASR a HR with the quenching temperature of 830°C were selected for this purpose. However, no differences from non-CT tempered microstructures were found in these specimens using scanning electron microscopy (SEM). Despite that, their hardness levels were higher. Upon tempering, the increase was approx. 20 HV10. Hardness values after cryogenic treatment are shown in table 3.

3.3 Tensile tests
Room-temperature tensile tests were in accordance with ČSN EN ISO 6892-1. The gauge sections had a size of Ø10×60 mm. The thread in the head was M16 and the total length of the specimens was 110 mm. Specimens prepared by sequences with quenching temperatures of 830 and 870°C were chosen. They were tested in tempered condition. Results of the tensile tests are given in table 3.
The readings obtained after soft annealing and quenching from 830°C suggested that austenitization had been insufficient. With 870°C, the yield strength was 1635 MPa and the ultimate strength reached 1834 MPa. After ASR (830°C), yield strength and ultimate strength were higher by 66 MPa and 43 MPa, respectively. There was no appreciable difference between the conditions obtained by quenching from 830°C and 870°C after ASR. The ASR material quenched from 870°C had the highest elongation and reduction of area, despite its yield strength and ultimate strength being higher than in the SA material. The hot-rolled material quenched from 830°C showed an even higher yield strength and ultimate strength (by 18 MPa and 31 MPa, respectively) than the ASR material. However, its elongation and reduction of area was similar to the values upon SA. After quenching from 870°C, mechanical properties were slightly lower. This means that the microstructure with lamellar pearlite (HR) began to coarsen at higher quenching temperatures.

Table 3. Final results of tensile tests and hardness

Heat treatment	$R_{p0.2}$ [MPa]	R_m [MPa]	A_g [%]	A_5 [%]	Z [%]	HV10
SA 830 °C	1592 ± 4	1766 ± 5	3.1 ± 0.1	8.1 ± 1.0	15.3 ± 2.5	515 ± 3
SA 870 °C	1635 ± 7	1834 ± 12	3.4 ± 0.1	8.9 ± 0.2	16.5 ± 0.5	544 ± 2
SA 870 °C + CT	**1764 ± 23**	**1943 ± 19**	**2.6 ± 0.1**	**8.0 ± 0.1**	**30.5 ± 0.6**	**566 ± 2**
ASR 830 °C	1701 ± 17	1877 ± 16	3.1 ± 0.1	8.1 ± 0.9	16.4 ± 2.7	547 ± 2
ASR 830 °C + CT	**1786 ± 10**	**1957 ± 9**	**2.5 ± 0.1**	**8.4 ± 0.1**	**35.8 ± 0.6**	**567 ± 4**
ASR 870 °C	1694 ± 22	1875 ± 13	3.1 ± 0.1	9.7 ± 0.2	22.0 ± 0.5	548 ± 6
HR 830 °C	1719 ± 3	1908 ± 7	3.1 ± 0.1	8.9 ± 0.3	17.5 ± 1.1	553 ± 2
HR 830 °C + CT	**1771 ± 11**	**1951 ± 12**	**2.4 ± 0.1**	**7.0 ± 1.1**	**28.0 ± 3.7**	**575 ± 3**
HR 870 °C	1666 ± 9	1867 ± 6	3.2 ± 0.2	8.6 ± 0.8	15.9 ± 0.4	534 ± 7

Cryogenic treatment led to higher ultimate strengths and yield strengths. The increases in yield strength were 129 MPa, 85 MPa and approx. 52 MPa in materials after SA, ASR and HR, respectively. To some extent, they compensated for the differences in strength characteristics (table 3). After ASR, the ultimate strength and yield strength were slightly higher than in other conditions, including the hot-rolled one. Greater differences remained in elongation and reduction of area. The best values were found after ASR: reduction of area was 35.8%. Following soft annealing, reduction of area was 30.5%. After hardening of the hot-rolled material with lamellar pearlite, the reduction of area was no more than 28%.

The quenched and tempered ASR material probably had finer austenite grain, which was reflected in its reduction of area. This was made even more conspicuous by cryogenic treatment. Differences between strength characteristics were reduced, probably by precipitation of very fine eta carbides. Nevertheless, the smaller prior austenite grains led to better performance under triaxial load, as seen in increased reduction of area after ASR.

4 Conclusion

Effects of three different initial microstructures on mechanical properties and microstructures upon quenching, tempering and cryogenic treatment were studied in 54SiCr6 spring steel. The initial microstructures included lamellar cementite obtained by hot rolling (HR, 290 HV10), fine globular cementite after Accelerated Spheroidisation and Refinement (ASR, 224 HV10) and coarser globular cementite produced by soft annealing (SA, 207 HV10).

Austenitizing was fastest in lamellar pearlite obtained by hot rolling. It was almost matched by the rate of dissolution of spheroidised carbides upon ASR. Austenitizing was much slower in soft-annealed material where the appropriate quenching temperature proved to be 40°C higher than for the other microstructures. Maximum hardnesses upon quenching were the same for all three initial conditions. Dissolution of remaining carbides in austenite has not led to an appreciable increase in hardness appreciably.

The ASR material showed higher tensile properties than the SA one. Quenching from 870°C led to higher yield strength and ultimate strength, by 59 MPa and 41 MPa, respectively. At the same time, elongation and reduction of area were higher by 0.8 points and 5.5 points, respectively. Quenched and tempered HR material (830°C) had slightly higher strength characteristics than the ASR material. However, its elongation and reduction of area was lower, similar to SA.

Cryogenic treatment (CT) led to improved mechanical properties, while reducing the differences between yield strengths and ultimate strengths of materials with different initial conditions. This was not the case with elongation and reduction of area which still reflected the microstructures before quenching. The highest reduction of area was found in the ASR material: 35.8%. With SA, it was 5.3 points less and with HR even 7.8 points less.

Acknowledgement

This paper was created under the project "Development of West-Bohemian Centre of Materials and Metallurgy" No.: LO1412 financed by the Ministry of Education of the Czech Republic.

References

[1] Gariboldi E, Nicodemi W, Silva G and Vedani M 1994 Effect of Tempering Level on Mechanical properties and Toughness of Spring Steel *Mat. Sci. Forum* vol 163 p 107-14

[2] Dlouhy J, Hauserova D and Novy Z 2013 Accelerated Carbide Spheroidisation of Chromium Bearing Steel by Induction Heating *Adv. Mat. Res.* vol 739 p 161-6

[3] Shtansky D V, Nakai K and Ohmori Y 1999 Pearlite to austenite transformation in an Fe-2.6Cr-1C alloy *Acta Mater.* vol 47/9 p 2619-32

[4] Masek B, Jirkova H and Kucerova L 2012 Rapid Spheroidization and Grain Refinement Caused by Thermomechanical Treatment for Plain Structural Steel *Mat. Sci. For.* vol 706-9 p 2770-5

[5] Hauserova D, Dlouhy J and Kotous J 2017 Structure Refinement of Spring Steel 51CrV4 after Accelerated Spheroidisation *Arch. of Metall. and Mater.* vol 62 p 1473-77

[6] Hradil D, Duchek M, Hrbackova T 2018 Gas nitriding with deep cryogenic treatment of high-speed steel *Acta Metallurgica Slovaca* vol 24/2 p 187-93

IOP Conf. Series: Materials Science and Engineering **461** (2019) 012045 doi:10.1088/1757-899X/461/1/012045

Study of the geometrical structure of copper surface after abrasive waterjet cutting

S Spadło, D Krajcarz

Kielce University of Technology, Division of Materials Science and Munitions Technology, Al. Tysiąclecia Państwa Polskiego 7, 25-314 Kielce, Poland

E-mail: d.krajcarz@wp.pl

Abstract. The article concerns the assessment of the intersection structure of copper samples after abrasive waterjet (AWJ) cutting. The surface quality of the processed material was tested on a macroscopic and microscopic scale. The assessments of surface quality was carried out in two characteristic areas of the cutting surface. The macroscopic studies concerned the evaluation of post-treatment cutting traces, with particular emphasis on the trajectory of the stream. Microscopic study of the geometric structure of the surface included the assessment of the conditions of impact of individual abrasive grains with the surface treated by using AWJ cutting. Characteristic orientations of machining traces were observed on the treated surface. The surface of the cutting zone located at the inlet of the stream to material is characterized a better cutting quality with visible of a privileged direction of traces of the impact of abrasive grains on the machined surfaces. As a result of the decrease of the kinetic energy of the cutting jet along with material penetration the surface at the outlet of the stream from the cut material, a surface characterized by the occurrence of multi-directional traces of abrasive grains. In this area are visible traces of decohesion of the workpiece in the form of microgrooves. In the studied areas, the effect of embedding fragments of crushed grains on the surface of the processed material was observed.

1 Introduction

In recent years there has seen a growth in the use of abrasive waterjet (AWJ) cutting. At present the technology can be used to cut many different types of materials varying in properties. AWJ cutting has various distinct advantages over the other non-traditional cutting technologies including no thermal distortion, small cutting forces, high machining versatility, high flexibility and minimum stresses on the workpiece [1-3]. Because of its universality, the method can easily compete with the conventional material cutting methods [4] or non-traditional processes [5].

Erosion of the material is dependent on the hydraulic energy of the waterjet and the kinetic energy of the abrasive grains in the waterjet. The main role of water is to accelerate abrasive particles to a high velocity and to produce a high coherent jet [6]. Abrasive particles are added to water to intensify the machining process [7]. In the area close to the inlet a jet to the material, the structure is formed by microcutting. The cutting of the area further from the surface is a result of abrasive wear or mechanical erosion [8]. When the energy of the waterjet is too low, the cut surface is not a straight but a curved line down through the material [9-11].

The aim of this study was concerns the assessment of the intersection structure of copper samples after abrasive waterjet (AWJ) cutting. The surface quality of the processed material was tested on a macroscopic and microscopic scale by using the NIKON AZ100 macroscope and the JEOL JSM-

7100F scanning microscope. Further detailed research are required to better understand the processes occurring during the AWJ cutting [12] in particular concerning the kinetic energy of the abrasive grains.

2 Materials and methods

The tests were carried out on copper specimens. Figure 1 presents the views of the finished samples.

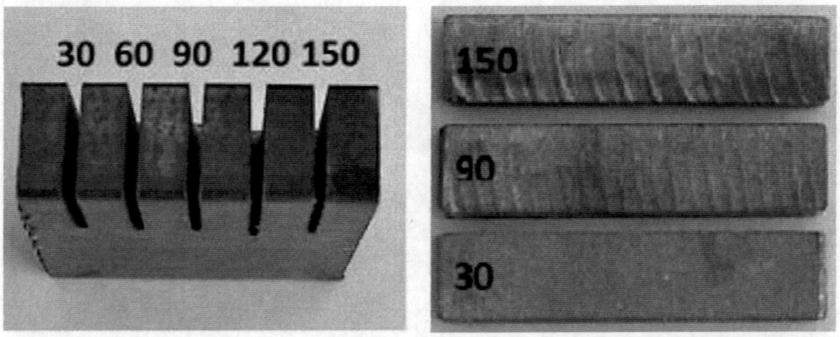

Figure 1. Views of the finished samples with marked traverse speed.

The samples were cut from copper plate with a thickness of 10 mm. The sample was cut every 5 mm The chemical composition of material is presented in table 1.

Table 1. Chemical composition of copper EN-AW CW020A.

Cu	*Bi*	O	*Pb*
≥ 99.9	≤ 0.0005	≤ 0.04	≤ 0.005

The experiments were performed at the Laboratory of Electrical Discharge Machining and Finishing of the Kielce University of Technology. The variable process parameter was traverse speed. The tests were performed at five different traverse speeds: 30, 60, 90, 120 and 150 mm/min. The constant process parameters were: water pressure (280 MPa), stand-off distance (4 mm), abrasive mass flow rate (4 g/s), water orifice diameter (0.3 mm), focusing tube diameter (1.02 mm) and average grain size (80 mesh).

3 Results and discussion

The macrostructure study was focused on the visual evaluation of kerf shape of cutting copper. The study particularly concerned in important areas of the cutting area. Figure 2 presents the shape of the slot for view of the jet entry into the cut material and the view of the jet exit from the material being cut. The pictures were taken on a NIKON AZ100 microscope. In both cases, for different traverse speeds, the top surface of the jet entering the cutting material is clearly rounded. The degree of rounding of the upper edge is affected by the coherence of the abrasive waterjet, which depends primarily on the shape of the water-abrasive nozzle and the degree of its wear.
Shape of kerf for different in two different areas presented figure 3. The presented slot of the intersection has a divergent shape. It is slightly wider at the entrance of jet material, than at the exit jet of material. The radius of rounding in the upper cutting zone is clearly visible here. The edge at the outlet of the water-abrasive stream is characterized by an almost perpendicular finish of the intersection surface to the plane of the material and the lack of greater waviness of the cut surface. An

IOP Conf. Series: Materials Science and Engineering **461** (2019) 012045 doi:10.1088/1757-899X/461/1/012045

additional advantage of the discussed edge is a very small burr formation and minimal plastic deformation. At a lower traverse speed it is possible to obtain a perfectly perpendicular cut surface to the cutting material. Cutting with a higher traverse speed will result in larger burr formation and shape errors.

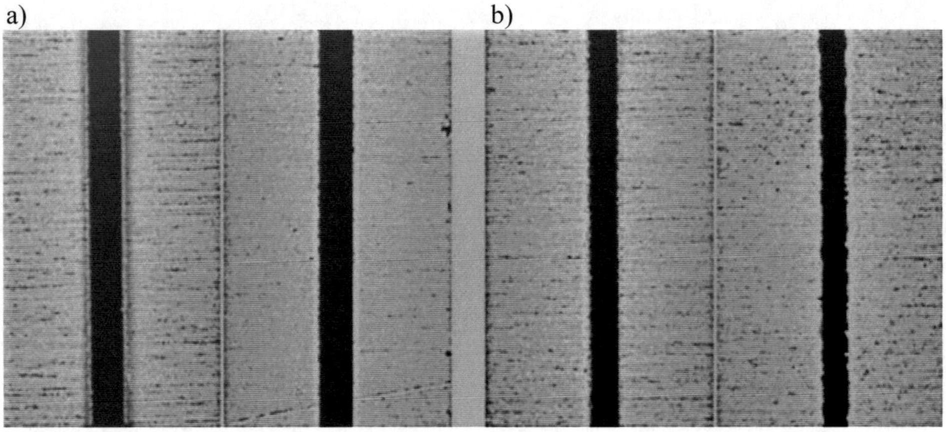

Figure 2. Kerf width of copper: a) top view, bottom view (traverse speed 30 mm/min; b) top view, bottom view (traverse speed 150 mm/min (magnification x5).

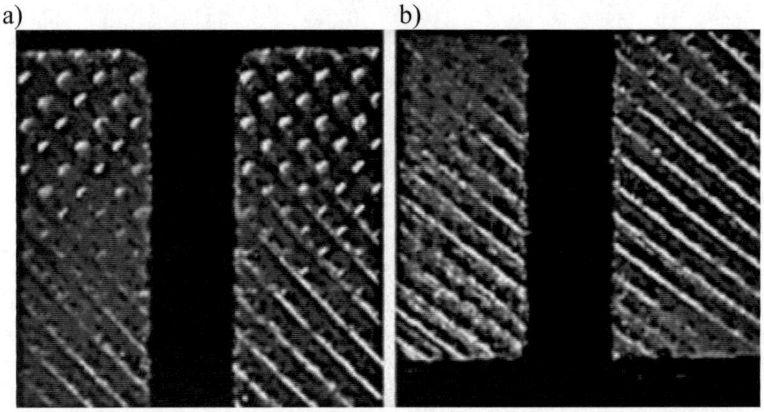

Figure 3. The shape of the intersection kerf for traverse speed 90 mm/min a) top view; b) bottom view; (magnification x10).

Evaluation of the microstructure of the intersection surface was made on a micro scale in two cutting areas: in the upper cutting zone at the inlet of the stream in the cut material (at 1 mm from jet entry) and in the lower cutting zone at the outlet of the cut material stream at (9 mm from jet entry). The tests were carried out using a JEOL JSM-7100F scanning microscope.

The geometric structure of the surface treated with a high-pressure abrasive jet is shaped mainly by the impact of a large number of accelerated abrasive grains on the surface of the workpiece. Abrasive grains are crushed as a result of impacts on the material. The kinetic energy of individual abrasive grains depends on their shape, size and angle of impact of grains on the material. The speed of grains also influences the degree of surface deformation, which decreases as a result of interactions with the work material and under the influence of aerodynamic force.

IOP Conf. Series: Materials Science and Engineering **461** (2019) 012045 doi:10.1088/1757-899X/461/1/012045

The sample cut with a feed rate of 90 mm/min was used for the microstructure study. The sample cut at this speed is characterized by an average cutting quality (Q3) with a clearly delimited smooth upper cutting zone and a rough lower cutting zone (figure 1).

Figures 4-6 shows the morphology of the cutting copper surface samples in two different areas and for different magnification (x50, x200 and x500).

a) b)

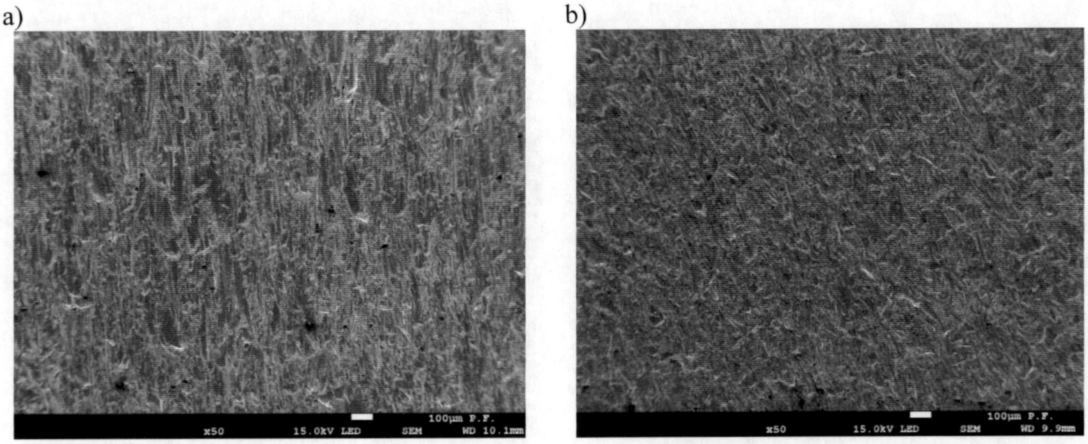

Figure 4. SEM image of AWJ cutting copper surface (magnification x50) a) surface at 1 mm from jet entry; b) surface at 9 mm from jet entry.

The application of the traverse speed 90 mm/min, water pressure 280 MPa and abrasive mass flow rate 4 g/s resulted in a relatively homogeneous microstructure cut at the inlet of the stream into a cut material. The homogeneous vertical direction of the stream's influence is visible here. This is confirmed by the microstructure pictures of the sample at higher magnification (figure 5a).

The view of the surface in the lower intersection zone is completely different, where the diagonal direction of the cutting marks dominates (figure 5b). This is related to the occurrence of the phenomenon of deviation of the erosive stream and the reflection of abrasive grains in the direction opposite to the direction of cutting. It is noticeable that the surface in the lower cutting zone is characterized by multi-directional traces of the grains moving. This proves the high ductility of copper, which in the effect of reduced kinetic energy of the cutting stream in the lower zone causes absorption of the kinetic energy of abrasive grains.

a) b)

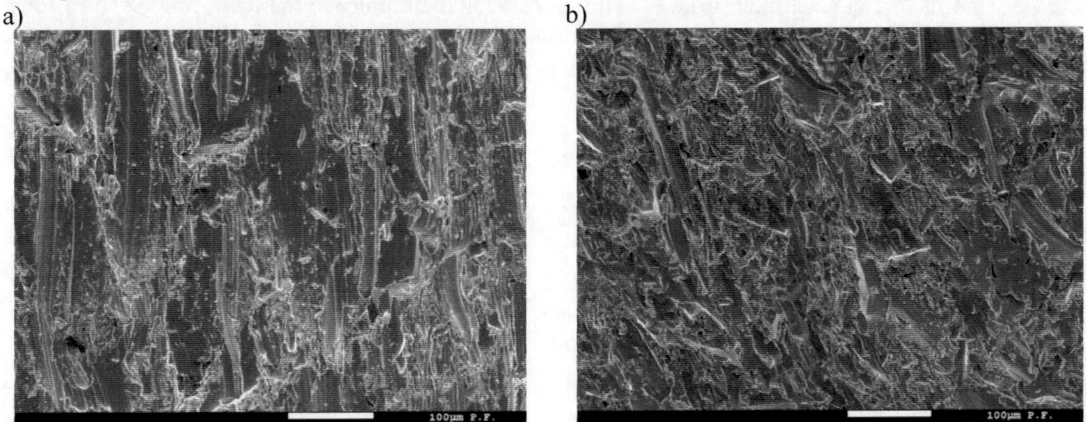

Figure 5. SEM image of AWJ cutting copper surface (magnification x200) a) surface at 1 mm from jet entry; b) surface at 9 mm from jet entry.

As a result of the interaction of numerous abrasive grains with the machined surface, the peeling of the material can be observed on the tested surfaces. As a result of the impact of the abrasive grain on the machined surface, swarf is not always formed. Sometimes, in the cutting surface especially on the deeper zone, abrasive particles cause partial detachment the weaker material layers (figure 6b). It is possible to embed fragments of crushed grains in the material.

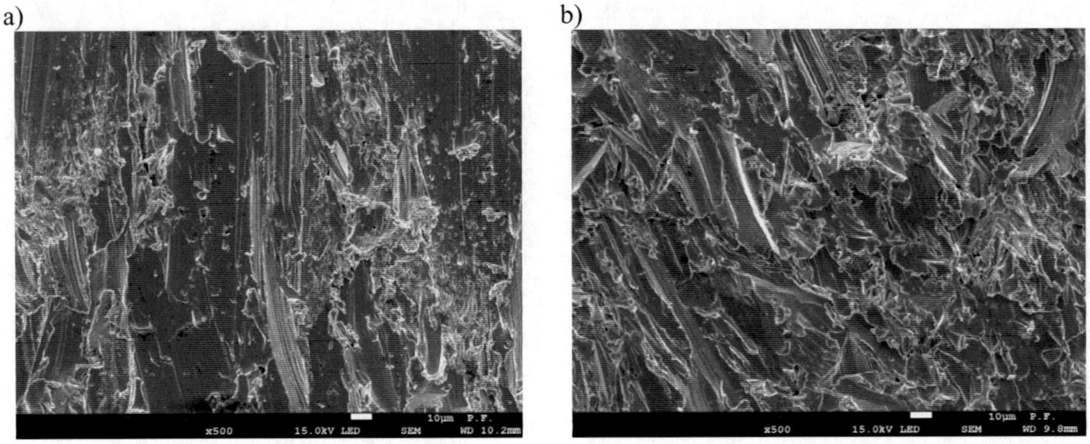

Figure 6. SEM image of AWJ cutting copper surface (magnification x500) a) surface at 1 mm from jet entry; b) surface at 9 mm from jet entry.

4 Conclusions

The experimental results obtained for copper indicate that the traverse speed of the waterjet has a significant effect on the geometrical structure. The surface quality of the processed material was tested on a macroscopic scale and concerned the evaluation of post-treatment cutting traces, with particular emphasis on the trajectory of the stream in the bottom edge of the cut. A small traverse speed causes excessive extension of kerf width in the bottom zone (jet exit). However, too much feed results in the formation of a narrow kerf of intersection and the occurrence of burr formation.

In microscopic study of the geometric structure were observed on the treated surface. Traces differ depending on the test area. The surface of the cutting zone located at the inlet of the stream to material were characterized a better cutting quality with visible of a privileged direction of traces of the impact of abrasive grains on the machined surfaces. The surface microstructure in the lower intersection zone is completely different. It is noticeable that the surface in the lower cutting zone is characterized by multi-directional traces of the grains moving. This is related to the occurrence of the phenomenon of deviation of the erosive stream and the reflection of abrasive grains in the direction opposite to the direction of cutting. The peeling was observed of the cutting surface. It was result of the interaction of numerous abrasive grains with the machined surface. Sometimes, in the cutting surface were observed fragments of crushed grains in the material.

References

[1] Akkurt A, Kulekci M K, Seker U and Ercan F 2004 Effect of feed rate on surface roughness in abrasive waterjet applications *Journal of Materials Processing Technology*, vol. 147, pp. 389-96

[2] Borkowski J, Sutkowska M, Sobczak R and Praźmo J 2007 The quality of surface cut after high-pressure abrasive water jetting, *International Symposium on Unconventional HydroJetting Technologies*, Koszalin, Poland, pp. 117-28

[3] Kovacevic R, Mohan R and Beardsley E 1996 Monitoring of thermal energy distribution in abrasive waterjet cutting using infrared thermography *Journal of Manufacturing Science Engineering*, vol. 118, pp. 555-63

[4] Spadło S, Hlavac L M, Krajcarz D and Hlavacova I M 2015 Influence Traverse Speed on Surface Quality after Water-jet Cutting for Hardox Steel, In *METAL 2015: 24th International Conference on Metallurgy and Materials*. Ostrava: TANGER, pp. 723-728

[5] Spadło S, Młynarczyk P., Depczyński W, Skowron E and Mijas R 2017 Research on the influence of electron beam welding of titanium alloy microstructure connection. In *METAL 2017: 26th International Conference on Metallurgy and Materials*. Ostrava: TANGER, pp. 1939-1944

[6] Sutowska M 2015 The quality model of the awj cutting process for aluminum type Pa4 alloy. *Archives of Mechanical Technology and Materials*, vol. 35, pp. 3-11

[7] Hashish M 1989 A Model for Abrasive - Waterjet (AWJ) Machining. *Journal of Engineering Materials and Technology*, vol.111, no 2, pp. 154-62

[8] Momber A and Kovacevic R 1998 *Principles of abrasive water jet machining*. Springer, ISBN 3-540-76239-6, Berlin

[9] Hlavac L M, Strnadel B, Kalicinsky J and Gembalová L 2012 The model of product distortion in AWJ cutting *International Journal of Advanced Manufacturing Technologies*, vol. 62, no. 2, pp. 157-66

[10] Hlavac L M, Krajcarz D, Hlavacova I M and Spadło S 2017 Precision comparison of analytical and statistical-regression models for AWJ cutting *Precision Engineering-Journal of the International Societies for Precision Engineering and Nanotechnology*, vol. 50, pp. 148-59, DOI: 10.1016/j.precisioneng.2017.05.002

[11] Krajcarz D and Spadło S 2016 Influence of the process conditions on the diameter of cylindrical holes produced by abrasive water jet cutting, *In METAL 2016: 25th Anniversary International Conference on Metallurgy and Materials.* Ostrava: TANGER, pp. 1462-1467

[12] Begic-Hajdarevic B, Cekic A, Mehmedovic M and Djelmic A 2015 Experimental Study on Surface Roughness in Abrasive Water Jet Cutting, *Procedia Engineering*, vol. 100, pp. 396-99

Influence of processing technology on fracture feature of Ti6Al4V

V Kubeš, M Losertová, J Škoda, K Konečná and F M Ijaz

VŠB - Technical university of Ostrava, Faculty of Metallurgy and Materials Engineering, Ostrava - Poruba, Czech Republic

E-mail: mlosertova@vsb.cz

Abstract. Selective laser melting is an advanced manufacturing technology providing alternative method of producing complex components directly from 3D computer models. The paper is focused on problems related with selective laser melting technology (SLM) used for preparing titanium based implants. Two series of ten Ti6Al4V samples were mechanically tested in tension and compression respectively before and after heat treatment. Annealing under argon atmosphere consisted of solution treatment at 955 °C followed by furnace cooling to 855 °C with water quenching and ageing at 600 °C with air cooling. Fractography study of fracture surfaces was performed using scanning electron microscopy and the observation was correlated with structural and mechanical properties of the SLM specimens in the condition as-built or heat treated. The failure features of as-built specimens corresponded to the microstructure with relatively high porosity, martensite needles, microcracks and residual stresses. The heat treatment resulted in the transformation of martensite to (alpha + beta) lamellar structure modifying plastic properties that corresponded to the morphology of fracture surfaces. In spite of the porosity occurrence both specimen series exhibited high mechanical properties in tension as well as in compression.

1 Introduction

Selective laser melting (SLM) is an additive manufacturing process in which successive layers of powder are selectively melted by the interaction of a high energy density laser beam. Molten and re-solidified material forms parts, while non-melted powder remains in place to support the structure. This layer-wise production technique offers some advantages over conventional manufacturing techniques such as high geometrical freedom, short design and manufacturing cycle time and made-to-order components. Layer-wise production techniques have evolved rapidly in the last 10 years and SLM has changed from a rapid prototyping to an additive manufacturing technique [1].

Consequently, the static and dynamic material properties must be sufficient to meet in service loading and operational requirements. The SLM process is characterized by rapidly solidified and non-equilibrium microstructures due to high temperature gradients. Residual stresses developed at high localized thermal gradients and very short interaction times lead to rapid volume changes. Furthermore, the change of process parameters can have a strong influence on the microstructure, density and surface quality. As a result, the mechanical properties of SLM parts can differ substantially from one another and from those produced by conventional techniques. In this respect, it is recognized that the advantages of SLM are only performed when the mechanical behavior of the final products is at least similar to conventionally produced components of the same material [2].

In recent years, much research [2 – 5] has focused on optimizing the SLM process. The SLM parts and material properties specifications were studied in order to characterize how the mechanical properties obtained with SLM may differ from the ones of conventional material and improve the quality of the resulting products. At present, Ti – based alloys can be processed by SLM technology with high reproducibility and hence low variation in material density and mechanical properties whereas the material density can be significantly affected by varying scanning parameters and scanning strategies. The effect of applying specific heat treatments of Ti6Al4V produced using SLM has been studied in order to produce a preferred final microstructure.

Although the quasi-static material properties such as tensile strength, hardness and impact toughness have been well characterized and reported to match those of conventional wrought materials, the dynamic mechanical behavior during crack initiation and propagation is more complex to explain because of critical sensitivity to the relations between fracture path, orientation, microstructure and loading conditions. Previous studies [3, 6] have shown that the residual stresses have detrimental effects on the mechanical behavior and fatigue life of SLM parts though the knowledge of failure mechanisms remains limited and consequently insufficient confidence to predict product life-time. Nevertheless, the SLM parts have great potential to meet the fatigue life requirements.

More recently, the influence of post-built heat treatment was investigated [7 – 9]. The two-step annealing of as-built products was performed at 955 °C for one hour with subsequent furnace cooling (FC) to 850 °C and water quenching (WQ) followed by ageing at 600 °C for six hours and air cooling (AC). The heat treatment of as-built material not only relieve residual stress, but also modify rapidly solidified microstructure which comprises of fine acicular α' martensite. The as-built microstructure is highly directional as a result of the imposed solidification mode during the vertical layer-by-layer build process. Hence, the purpose of the post-built heat treatment is to generate the preferred lamellar α + β equilibrium structure which provides a more desirable combination of strength and toughness. However, the maximum heat treatment temperature does not exceed the β transus temperature and the time at temperature does not result in excessive grain growth.

The present work is focused on the evaluation of the fracture surfaces in relation with as-built defects, heat treatment and mechanical properties that were determined for the as-built and annealed conditions in the previous study [9].

2 Experimental

The specimens of Ti6Al4V ELI for both tensile and compressive test were prepared by selective laser melting with vertical building orientation (in the stress axis) and specified parameters: laser power 200 W, slice thickness 30 μm, 5x5 mm islands rotated 45° one from another.

As-built products were heat treated (HT) in accordance with the schema presented in figure 1. For the metallographic study, the specimens were ground on grinding papers with grain size ranged from 600 to 2500, electro-chemically polished using Struers LectroPol-5 with A3 electrolyte and etched by Kroll's solution (6 % HF, 8 % HNO_3, 86 % H_2O) for 10 - 60 seconds. The microstructure was observed by means of Olympus IX70 microscope equipped with digital camera DP12. The porosity was measured using Image-Pro Plus software of Media Cybernetics. The mechanical tests were performed at strain rate of 0.0025 s^{-1} and room temperature by devices Walter + Bai UTM 100 and Inova Fu-0-350-170-V1 for tensile and compression tests, respectively, and the results were evaluated elsewhere [9]. The fractographic study was realized using an SEM JOEL JSM-649OLV microscope.

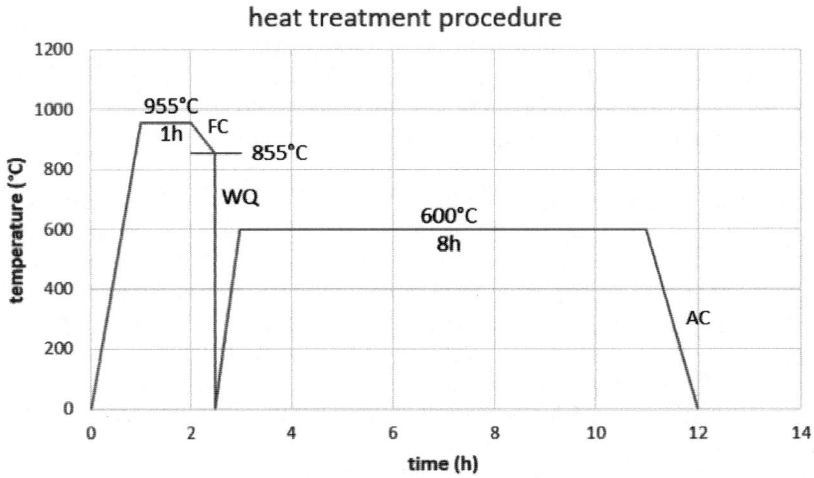

Figure 1. Diagram of two-step heat treatment (HT) of Ti6Al4V specimens before mechanical tests (where AC – air cooling, FC – furnace cooling, WQ – water quenching).

3 Results and discussion

The mechanical properties of SLM specimens in tension and compression that have been evaluated and discussed elsewhere [9], are summarized in table 1. The as-built porosity occurring in the as-built material ranged locally from 1.7 to 10.6 %. The macrostructure of vertical cut was characterized by patterns related with melted pools, while in cross section the texture corresponding to scanning strategy was observed.

Table 1. Tensile and compressive mechanical properties of Ti6Al4V (average values from 3 tests) [9].

Specimen	YS (MPa)	UTS (MPa)	$\varepsilon_{fracture}$ (%)
		Tensile test	
As-built	1056±29	1351±34	5.5±0.8
Heat treated	1098±41	1181±9	5.0±0.1
		Compression test	
As-built	1167±115	1681±74	35±10
Heat treated	1113±23	1627±28	26±1

Note: YS – yield strength, UTS – ultimate tensile strength, $\varepsilon_{fracture}$ – strain to fracture

The microstructure of as-built specimens before the HT was formed of martensite needles in the prior β phase due to rapid cooling of melted pools, as seen in figure 2a. After the two-step HT the α' martensite was transformed to α + β lamellar structure (figure 2b). The lath thickness measured ranges from 3 to 6 μm and it is depending on maximum temperature of treatment below T_β transus and cooling rate whereas their influence is increasing with increasing vicinity to T_β.

After tensile test, failures of both as-built and annealed conditions were ragged and consisted of shear and transcrystalline quasi-cleavage surfaces (figures 3 and 4). In some areas of fractures, cavities with unmelted alloy particles could be observed (figure 4b). It is clear that SLM process-induced imperfections were the driving factor at the failure initiation for both microstructural modifications on the microscopic scale. Moreover, the pores present on the specimen surfaces acted as notches localizing stresses at the loading. Generally, the Ti6Al4V alloy shows low ductility and the SLM-process providing the residual stresses and metastable microstructure can contribute to that the possibility of plasticity-induced accommodation of these elevated stresses is limited. Comparing the average mechanical values (table 1) it is evident that martensite microstructure showed rather lower yield strength (YS) but higher ultimate tensile stress (UTS) which is corresponding to more ductile feature of the fracture surfaces with cracked martensite needles rounded by ductile β phase. Unlike martensitic microstructure the lamellar α + β structure increased the YS value by 4 % but decreased the UTS by 13 %.

Macroscopically, the failure feature of the fine lamellar microstructure seemed to be very similar to martensitic one, the evident difference consisted in thickness of martensitic needles and α laths that determined the size of brittle cleavage facets. As both phases are inserted in the ductile β phase the impact of the martensite to α + β transformation on the fracture feature was less apparent.

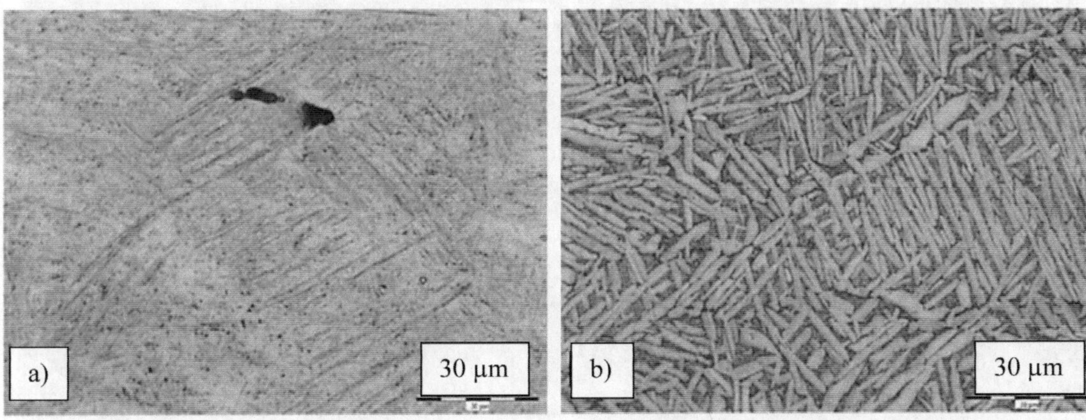

Figure 2. Microstructure of Ti6Al4V specimens produced by SLM process: a) martensite needles in prior β phase and voids (as-built before HT) and b) α + β lamellae (after HT).

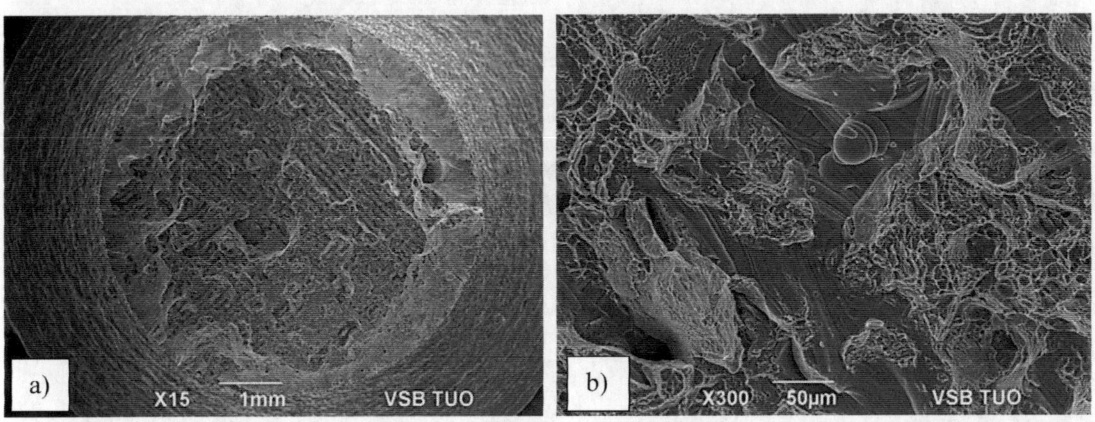

Figure 3. Tensile fracture of Ti6Al4V specimen tested in as-built condition: a) view of rugged fracture surface, b) detail of mixed failure feature with shear areas, transgranular quasi-cleavage surfaces and voids.

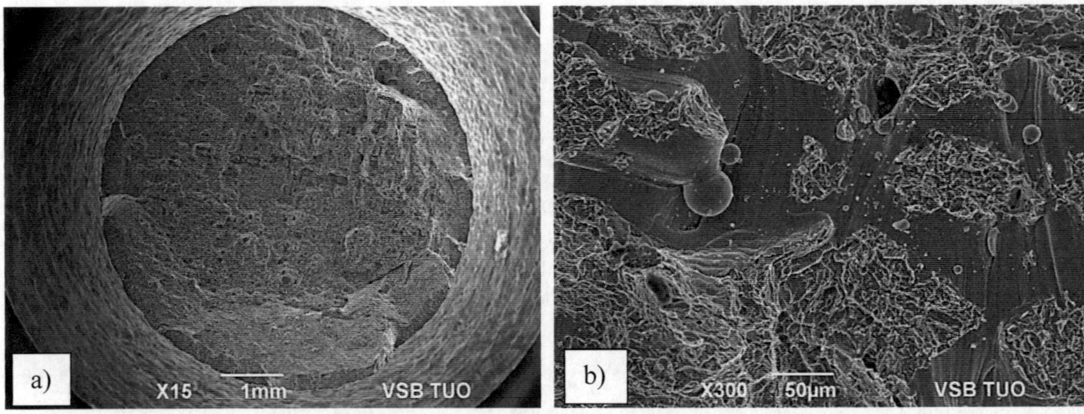

Figure 4. Tensile fracture of Ti6Al4V specimen tested in heat treated condition: a) view of rugged failure surface, b) detail of mixed fracture feature with shear areas, transgranular quasi-cleavage surfaces, voids and unmelted powder particles.

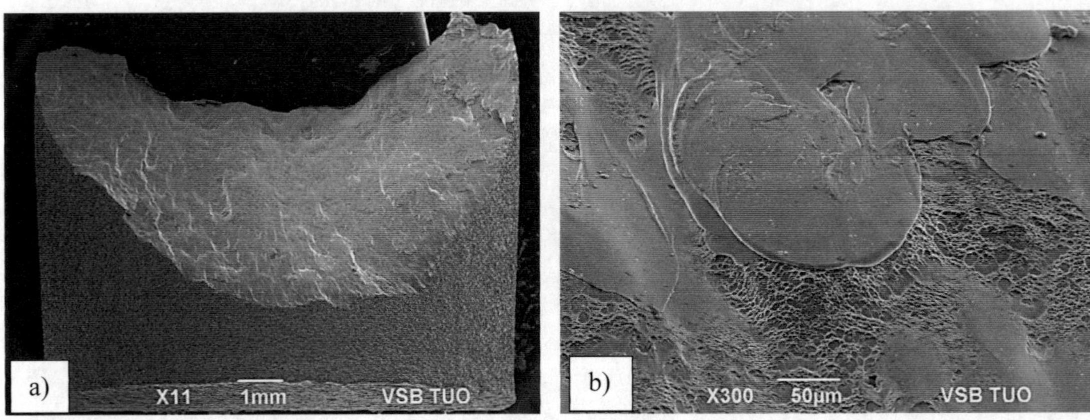

Figure 5. Compression fracture of Ti6Al4V specimen tested in as-built condition: a) view of smooth failure surface, b) detail of mixed feature of fracture with shear areas and ductile dimples.

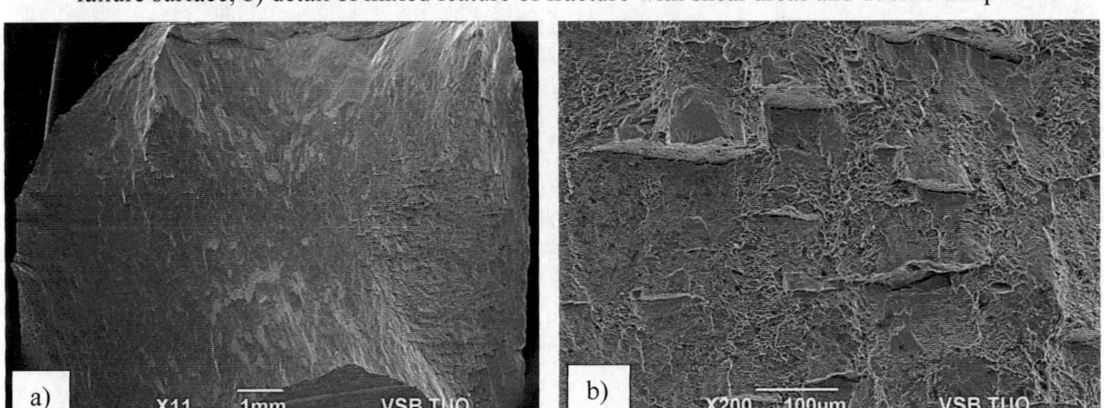

Figure 6. Compression fracture of Ti6Al4V specimen tested in heat treated condition: a) view of mixed failure feature, b) detail of steps, secondary cracks, shear areas and transgranular quasi-cleavage surfaces.

After compressive tests both as-built and HT conditions showed the failures with smooth surfaces (figures 5 and 6). The HT changed the predominant shear feature of the fracture with elongated very shallow dimples (figure 5a) to more rugged one with occurrence of secondary cracks, sharp steps, transgranular quasi-cleavage surfaces and elongated ductile dimples (figure 6b) that corresponded to microstructure changes as it was stated for tensile tests. In this case, martensitic structure in compression showed higher mechanical values by 3 % and 5 % for the YS and UTS, respectively (table 1), together with higher ductility by 35 %.

4 Conclusion

The fractographic study was performed on the specimens prepared by the selective laser melting process and tested in tensile and compressive loading. The SLM material originally consisted of fine α' martensite needles and after two-step heat treatment the microstructure transformed to α + β lamellae. The failure surfaces for both type of loading corresponded to microstructure states in accordance with very small changes in mechanical values.

Acknowledgements

This article has been elaborated in the framework of the project No. LO1203 "Regional Materials Science and Technology Centre - Feasibility Program" funded by Ministry of Education, Youth and Sports of the Czech Republic and co-financed by the European Social Fund and of the projects SGS SP2018/100 supported by Ministry of Education, Youth and Sports of the Czech Republic.

References

[1] Van Hooreweder B, Moens D, Boonen R, Kruth J-P and Sas P 2012 Analysis of fracture toughness and crack propagation of Ti6Al4V produced by selective laser melting *Adv. Eng. Mater.* **14** 1–2 92–97

[2] Thijs L, Verhaeghe F, Craeghs T, Van Humbeeck J and Kruth J-O 2010 A study of the microstructural evolution during selective laser melting of Ti–6Al–4V *Acta Mater.* **58** 9 3303–3312

[3] Kruth J-P, Badrossamay M, Aysa E, Deckers J, Thijs L and Van Humbeeck J 2010 Part and material properties in selective laser melting of metals *Proceedings of the 16th international symposium on electromachining* ISEM XVI pp 1–12

[4] Yasa E, Deckers J, Kruth J-P, Rombouts M and Luyten J 2009 Experimental investigation of Charpy impact tests on metallic SLM parts *Innovative Developments in Design and Manufacturing* 4th Int. on Advanced Research in Virtual and Rapid Prototyping Conf. proceed. (CRC Press: -Taylor & Francis Group) pp 207–214

[5] Vilaro T, Colin C and Bartout J-D 2011 As-fabricated and heat-treated microstructures of the Ti-6Al-4V alloy processed by selective laser melting *Metall.Mater.Ttrans.* A **42** 10 3190–3199

[6] Mercelis P and Kruth J-P 2006 Residual stresses in selective laser sintering and selective laser melting. *Rapid prototyping journal* **12** 5 254–265

[7] Wauthle R, Vrancken B, Beynaerts B, Jorissen K, Schrooten J, Kruth J-P and Van Humbeeck J 2015 Effects of build orientation and heat treatment on the microstructure and mechanical properties of selective laser melted Ti6Al4V lattice structures *Additive Manufacturing* **5** 77–84

[8] Vrancken B, Thijs L, Kruth J-P and Van Humbeeck J 2012 Heat treatment of Ti6Al4V produced by Selective Laser Melting: Microstructure and mechanical properties *J. Alloy. Compds.* **541** 177–185

[9] Losertová M and Kubeš V 2017 Microstructure and mechanical properties of selective laser melted Ti6Al4V alloy *IOP Conference Series: Mater. Sci. Eng.* **266** 1 23

The behaviour of Al_2O_3 precipitation at the steel-slag interface

P L Żak[1] , K. Kuglin[1] and D Kalisz[1]

[1] AGH University of Science and Technology, Faculty of Foundry Engineering, 23 Reymonta Street, 30-059 Kraków, Poland

E-mail: pawelzak@agh.edu.pl

Abstract. The paper focuses on the numerical investigation of the Al_2O_3 precipitation behaviour at the steel-slag phases interface. Several assumptions were made to simplify the process in order to prepare its description. A mathematical model of the steel-slag-precipitation system was introduced. The model base on the equation of motion, and takes into account forces balance of buoyancy force, added mass force, drag force and rebound force. Mentioned forces act in the same direction which is perpendicular with the steel-slag interface surface, but is positive or negative. The precipitation initial velocity can be calculated on the base of steel and inclusion density, liquid steel viscosity and particle radius. The numerical model based on the firstly prepared mathematical model was presented. Results of the computations show what is the precipitation radius size influence on particle movement and its assimilation to the slag.

1 Introduction

Stirring of liquid steel with neutral gas or its convection movement caused by the gravity force causes the movement of the metal as well as non-metallic precipitants. At the same time precipitants are effusing to the slag, where they can be assimilated or pushed away. Stirring force is crucial for the intensity of mentioned processes [1, 2].

The chemical composition of slag depends on the type of steel to be cast. However in the case of certain steels a considerable change of content of particular components can be expected, depending on the advancement of the casting process. The composition may change as a result of flowing off, interaction with solidification front, adsorption of inclusions and reactions on the metal/slag interface [3-8]. Adsorption and dissolution of Al_2O_3 particles by liquid slag solution can be a problem while casting, e.g. TRIP steel, mainly as far as the proper selection of the mould flux is concerned [9, 10]. Experiments performed by [9] revealed that casting steel with Al as an alloy additive resulted in a considerable change of chemical composition of slag during casting after 10-15 minutes from its beginning. A high value of Al_2O_3 absorption index was observed, and consequently lower SiO_2 content in slag, being a result of the reaction:

$$3(SiO_2) + 4[Al] = 3[Si] + 2(Al_2O_3) \tag{1}$$

Apart from this reaction the evolution of slag composition was caused by the adsorption of flowing off Al_2O_3 inclusions.

The calculation of the process of flowing off Al_2O_3 and interaction of an Al_2O_3 particle on the liquid steel/liquid slag/particle interface was performed with the use of a computer program.

IOP Conf. Series: Materials Science and Engineering **461** (2019) 012047 doi:10.1088/1757-899X/461/1/012047

2 Mathematical model

Mathematical model of precipitation behavior at the steel-slag interface basis on several assumptions:

- precipitations have a shape of sphere;
- precipitations do not change shape or volume;
- the liquid is Newtonian;
- the computational domain is isothermal;
- slag has characteristic of liquid;
- front between slag and steel is flat;
- interfacial tension between any phases is constant during computations;
- the forces which causes precipitations movement is driven by following forces [1,12,13]:
  - buoyancy force:

$$F_w = \frac{4}{3}\pi R^3(\Delta_b\, \rho_S - \rho_w)\, g \tag{2}$$

  - added Newton mass

$$F_m = \frac{2}{3}\pi R^3 \rho_S \Delta_b \frac{d^2Z}{dt^2} \tag{3}$$

  - drag force

$$F_{vis} = 6\pi R \eta_S B \frac{dZ}{dt} \tag{4}$$

  - repulsive force

$$F_{rep} = 2\pi R \sigma_{MS}\left(\frac{Z}{R} - 1 - \cos\theta\right) \tag{5}$$

where: R denotes the radius of the precipitation [m]; Δ_b is a coefficient which depends on the position of the precipitation in liquid and slag, and is given by formula:

$$\Delta_b = \frac{1}{4}\left(\frac{\rho_M}{\rho_S} - 1\right)\left(\frac{Z}{R}\right)^3 - \frac{4}{3}\left(\frac{\rho_M}{\rho_S} - 1\right)\left(\frac{Z}{R}\right)^2 + \frac{\rho_M}{\rho_S}, \tag{6}$$

ρ_S, ρ_M, ρ_w denotes density of slag, steel and precipitation respectively [kg m^{-3}]; Z is location of the particle in the slag, so it's equal 0 at the moment when particle is touching the slag, but is still all in liquid [m]; g is gravity acceleration [m s^{-2}]; η_S, η_M, η_w is dynamic viscosity of slag, steel and precipitation respectively [Pa s]; coefficient B denotes impact of location of the particle in the slag on drag force, and is given by formula:

$$B = \left(\frac{\eta_M}{\eta_S} - 1\right)\left(\frac{Z}{R}\right)^2 - 2\left(\frac{\eta_M}{\eta_S} - 1\right)\left(\frac{Z}{R}\right) + \frac{\eta_M}{\eta_S}, \tag{7}$$

$\cos\theta = \frac{\sigma_{WM} - \sigma_{WS}}{\sigma_{MS}}$ is cosine of wetting angle, which can be calculated basing on the knowledge about the values of interfacial tensions between: σ_{WM} - particle and steel, σ_{WS} – particle and slag, σ_{MS} – steel and slag [N m^{-1}].

At the beginning of the computations the center of the precipitation is in R distance from slag, at this moment start of the frame of reference is in the particle center and its location is set to be $Z = 0$. Computations runs until $Z = R$.

Analyzing process of particle flow in liquid metal at the beginning the particle which density is less than the steel density is accelerating until buoyancy and drag forces are equal. Equilibrium of those forces allows one to calculate speed of the particle:

$$v_{eq} = \frac{2}{9}R^2(\rho_M - \rho_w)\frac{g}{\eta_M} \tag{8}$$

This speed is chosen to be initial in the model as most of the particles reach this speed as their distance from slag is significant.

Finally the process is governed by ordinary differential equation, which takes into account balance of forces (2) – (5), and initial condition for Z and speed eq.(8):

$$\begin{cases} \left(\left(\frac{4}{3}\pi R^3 \rho_W + \frac{2}{3}\pi R^3 \rho_S \Delta_b\right)\frac{d^2Z}{dt^2} + 6\pi R \eta_S B \frac{dZ}{dt} + 2\pi R \sigma_{MS}\left(\frac{Z}{R} - (1 + cos\theta)\right)\right) \\ \qquad = \frac{4}{3}\pi R^3 (\Delta_b \rho_S - \rho_W) g \\ \frac{dZ}{dt}(t = 0) = \frac{2}{9} R^2 (\rho_M - \rho_W)\frac{g}{\eta_M} \\ Z(t = 0) = 0 \end{cases} \tag{9}$$

Discretization of the equation (9) was performed with the use of Runge-Kutta fourth rank scheme.

3 Results of the simulation

To analyze the movement of the Al_2O_3 particle based on the mathematical model introduced above the Dev-Cpp environment was used to prepare computer software. The computer program works in console window and as a result prepare the file, where the time, actual speed and position of the particle is stored. An exemplary calculation window has been presented in Fig. 1.

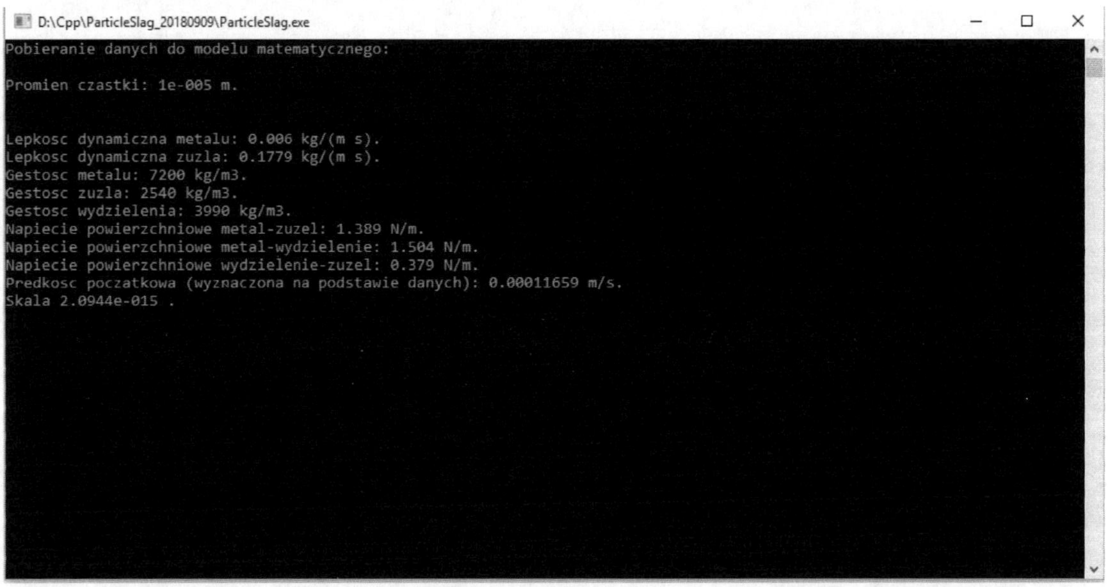

Figure 1. Window of the simulation software with some messages about uniqueness conditions of the model

The following assumptions were made in the calculations [1]: ρ_M- density of steel 7000 kg m^{-3}, ρ_I – density of inclusions Al_2O_3 3990 kg m^{-3}, ρ_S – density of slag 2750 kg m^{-3}, η_M – viscosity of steel 0.006 Pa·s, η_s – viscosity of slag 0.006 Pa·s, σ_{MS} – tension on steel/slag interface 1.389 N/m, σ_{IS} – tension on inclusion/slag interface 1.504 N/m, σ_{IM} – tension on inclusion/steel interface 0.379 N/m. The results of calculations performed for variant 1: Al_2O_3 particle radius: r=0.0001; 0.0003 and 0.0005 m, and variant 2: particles with radius r=0.000001; 0.000003; 0.000005 m have been presented in figures 2 and 3, respectively.

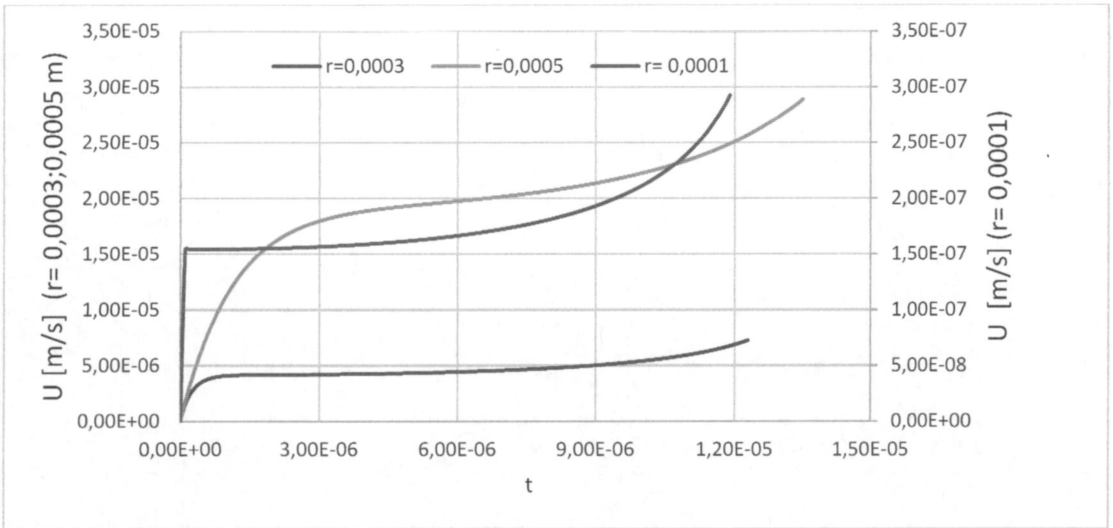

Figure 2. Speed U (m/s) of Al_2O_3 inclusion in reference to surface for particle radius $1\cdot10^{-4}$, $3\cdot10^{-4}$, $5\cdot10^{-4}$ m

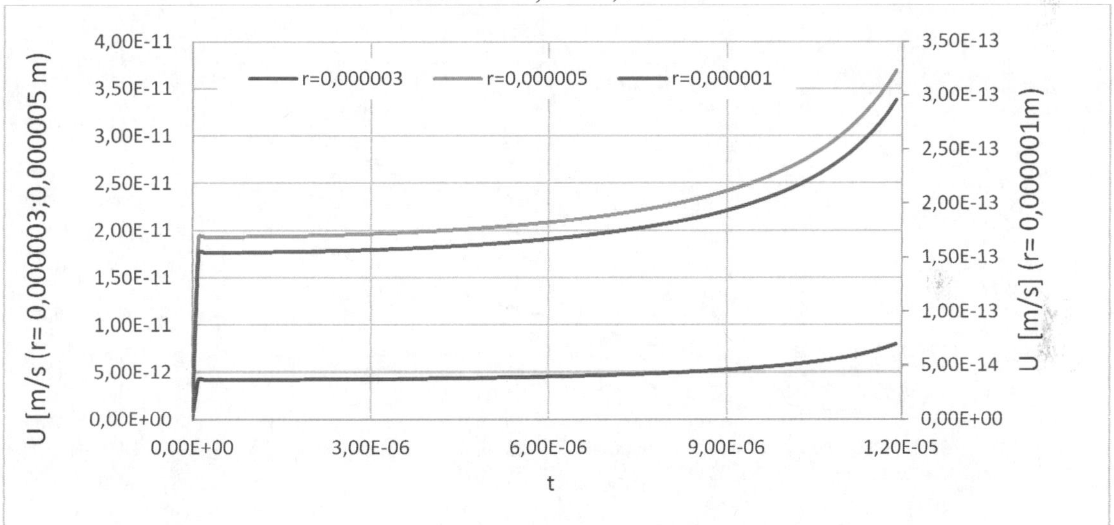

Figure 3. Speed U (m/s) of Al_2O_3 inclusion in reference to surface for particles radius $1\cdot10^{-6}$, $3\cdot10^{-6}$, $5\cdot10^{-6}$ m

A system of forces associated with the speed of free flow-off regulated by the Stokes law and force of entrainment resulting from the steel speed, act on the particle. Particles of very small radius reach low speed U values and so their removal and absorption by slag is hindered. The operation of capillary and gravity forces counteract buoyancy defined with Stokes equation. The change of placement of particle H(m) with respect to particle radius has been presented in figures 4 and 5.

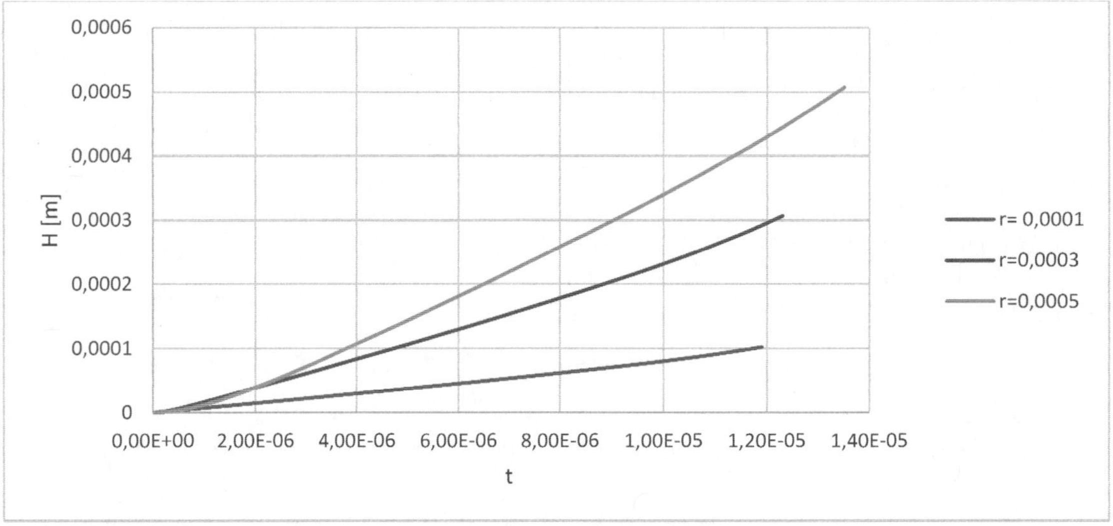

Figure 4. Change of placement H(m) of Al_2O_3 particle with respect to surface for particle radius $1\cdot10^{-4}$, $3\cdot10^{-4}$; $5\cdot10^{-4}$ m

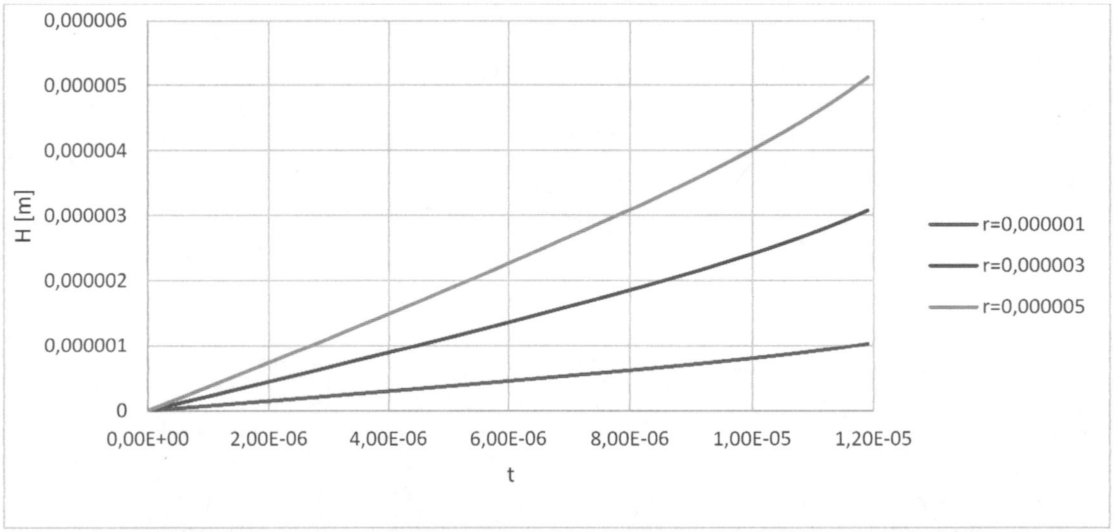

Figure 5. Change of placement H(m) of Al_2O_3 particle with respect to surface for particle radius $1\cdot10^{-6}$, $3\cdot10^{-6}$; $5\cdot10^{-6}$ m

A particle which reaches the liquid steel/slag interface may pass through this boundary and be dissolved in slag or pushed off. It was assumed in the calculations that for H(m)=0 the particle is in the initial position, for H(m) the particle is in slag.

4 Conclusions

The removal of non-metallic inclusions to slag is most importantly determined by the system of forces acting on a particle, and so physicochemical parameters, hydrodynamic conditions of the analyzed system and radius of the particle. These parameters determine the position of the particle in reference to the interface. Al_2O_3 particles have higher density than slag, therefore they stay close to the interface. Moreover the presented results refer to the behavior of a single particle, which is a considerable simplification. This type of simulation should be performed for a population of inclusions, taking into account agglomeration. The performed calculations are a starting point for proper experimental works, first with a water model of two immiscible fluids, then observations with a high-temperature microscope.

Acknowledgements
The authors acknowledgement the AGH University of Science and Technology for its financial support within grant no. 15.11.170641.

References

[1] Kalisz D 2013 Termodynamiczna charakterystyka powstawania fazy niemetalicznej w ciekłej stali [in Polish] (Krakow: Wyd. Nauk. Akapit) chapter 6 pp 163-181
[2] Taniguchi S, Kikichi A, Ise T and Shoji N 1996 Model Experiment on the Coagulation of Inclusion Particles in Liquid Steel *ISIJ Int.* vol **36** pp 117-120
[3] Cwudziński A 2011 Numerical Simulation of Behaviour a Non-Metallic Inclusions in an One-Strand Slab Tundish with Steel Flow Controll Devices *Archives of Metallurgy and Materials* vol **56**(3) pp 611-618
[4] Derda W and Staniewski I 2012 Numerical Modeling of Nonmetallic Inclusions Flowing Out during Solidification of Continuous Cast Steel Billets in Mould Zone *Materials Science Forum* vol **706-709** pp 1385-1390
[5] Michalek K, Gryc K, Socha L, Tkadleckova M, Saternus M, Pieprzyca J, Merder T and Pindor L 2016 Study of slag enrainment using ghysical modeling *Archives of Metallurgy and Materials* vol **61**(1) pp 257-260
[6] Socha L, Bazan J, Gryc K, Styrnal P, Pilka V and Piezga Z 2017 Proc. Metal 2011 (Brno), (Brno: Tanger)
[7] Kalisz D and Żak P L 2016 PSG method for simulating agglomeration of Al_2O_3 inclusion in liquid steel *Acta Physica Polonica A* vol **130** pp 157-159
[8] Kalisz D and Żak P L 2015 Modeling solute segregation and formation of non-metallic inclusion during solidification of a titanium – containing steel *Kovove Mater.* vol **53**(1) pp 35-41
[9] Blazek K, Frazee M, Yin H, Park J H and Woon S 2013 Assessment of CaO–Al2O3 Based Mold Flux System for High Aluminum TRIP Casting *ISIJ International* vol **53**(1) pp 62-70
[10] Zhang Z, Wen G, Tang P and Sridhari S 2008 SridhariThe Influence of Al_2O_3/SiO_2 Ratio on the Viscosity of Mold Fluxes *ISIJ International* vol **48**(6) pp 739-746
[11] Persson M, Gornerup M and Seetharaman 2007 Viscosity Measurements of some mould flux slags *ISIJ International* vol **10** pp 1533-1540
[12] Bouris D and Bergeles G 1997 Investigation of Inclusion Re-Entrainment from the Steel-Slag Interface *Metallurgical and Materials Transactions B* vol **29B** pp 641-649
[13] Strandh J, Nakajima K, Eriksson R and Jönsson P 2005 Solid Inclusion Transfer at a Steel-Slag Interface with Focus on Tundish Conditions *ISIJ International* vol **45**(11) pp 1597-1606

Characterization of Ti-6Al-2Sn-4Zr-6Mo alloy flow behaviour using Malas's stability criterion

O Lypchanskyi[1], K Zyguła[1], T Śleboda[1], M Wojtaszek[1], A Łukaszek-Sołek[1]

[1]AGH University of Science and Technology, Faculty of Metals Engineering and Industrial Computer Science, Av. Mickiewicza 30, 30-059 Cracow, PL

lypchans@agh.edu.pl

In the presented study the hot deformation behavior of Ti-6Al-2Sn-4Zr-6Mo (Ti-6246) alloy was investigated by isothermal hot compression tests conducted on the Gleeble 3800 thermomechanical simulator in the temperature range of 800–1100 °C and in strain rate range of 0.01 – 100 s^{-1} to true strain of 1. For determining the unstable flow regions of the investigated alloy during hot deformation, the processing maps based on the Malas's criterion of plastic flow stability were used. For characterization of the material flow during forming operations, the microstructure analysis of the processed alloy was performed and associated with the processing maps. Basing of this analysis, the results of predicting the flow instability were confirmed, what made possible to understand the mechanical behavior of Ti-6246 alloy in a wide range of strain rates and temperatures.

1. Introduction

In the modern aircraft industry α+β titanium alloys are often used for manufacturing turbine engine rotating components. These types of alloys are characterized by the ability of changing the distributions and amounts of untransformed phases by aging or other thermal treatments. The Ti-6Al-2Sn-4Zr-6Mo (Ti-6246) is α+β alloy with high tensile strength, especially in β- processed condition, used for the production of compressor disks and blades [1, 2], working at intermediate temperature in the range of about 315 to 400 °C [3]. The high content of β-stabilising elements reduces the intensity of microstructure transformation during processing, and facilitates uniform heat treatment of cross section and uniform property levels.

High strength of this alloy is observed at high strain rates, without significant strain hardening, and decreases significantly with increasing temperature [4]. The initiation of dynamic recrystallization is facilitated by increasing the strain rate and extension of the processing time, but the fully recrystallized microstructure is observed at low strain rates [5, 6]. Some researchers noticed, that the plastic deformation in the range of α phase limits the growth of β phase grains [6]. The influence of the microstructure on the tribological properties of Ti-6246 alloy is shown in [7]. It was shown, that this alloy has one of the lowest value of friction coefficient in comparison with many other types of titanium alloys.

The production of structural components made of Ti-6246 alloy often involves the use of ingots as charge material and hot die forging or isothermal forging for processing. Understanding the mechanical behavior of Ti-6246 alloy in a wide range of strain rates and temperatures is crucial in proper controlling the properties and microstructure of final products. For description of the material flow during forming operations processing maps based upon various criteria, such as Malas's stability criterion, are successfully used [8-11]. This criterion was proposed by Malas and Seetharaman [12]. This criterion – based on Gegel's criterion - assumes four inequalities (equations 1–4) containing strain rate ($\dot{\varepsilon}$) sensitivity m and $m_{criterion}$ and temperature sensitivity of flow stress s and $s_{criterion}$ as follows:

$$0 < m \leq 1 \tag{1}$$

$$m_{criterion} = \frac{\partial m}{\partial(\ln\dot{\varepsilon})} < 0 \tag{2}$$

$$s \geq 1 \tag{3}$$

$$s_{criterion} = \frac{\partial s}{\partial(\ln\dot{\varepsilon})} < 0 \tag{4}$$

The sensitivity parameter (s) of flow stress in deformation process can be defined as:

$$s = \frac{1}{T}\frac{\partial \ln\sigma}{\partial(\ln\frac{1}{T})} \tag{5}$$

Where T is temperature and σ is flow stress.

In Malas's criteria with increasing temperature and decreasing strain rate the strain hardening should increase. Besides, with increase in the temperature, the material should exhibit flow softening and the flow stress with respect to strain rate curve must have a convex character [8]. The equations (1) and (2) consider the mechanical stability, (3) and (4) the thermodynamic stability. The four criteria of stability (equations 1–4) determine the stability conditions and in the result of their overlaying the complex map can be developed. The shadowed areas on the complex processing map display the domains of instability.

2. Material and experimental procedures

Ti-6246 alloy of chemical composition presented in Table 1 was used in this study. The cylindrical specimens of 10 mm in diameter and 12 mm high were cut out from Ti-6246 alloy rod along its axis. The isothermal hot compression tests were carried out on the Gleeble-3800 thermomechanical simulator under vacuum atmosphere. The specimens were heated up at the heating rate of 2.5 °C/s and kept at given temperature for 10 s. The samples were deformed in compression at the temperatures of 800, 900, 950, 1000, and 1100 °C, and at the strain rates of 0.01, 0.1, 1, 10, and 100 s^{-1}. A graphite foil was used to reduce the friction effects. The specimens were deformed to the true strain of 1. After the plastic deformation samples were quenched in compressed air. The microstructure of the processed materials was observed on the optical microscope LEICA DM4000M.

Table 1. Chemical composition of Ti-6246 alloy (weight %)

C	Si	Mo	Al	Fe	Sn	Zr	Ti
0.01	0.09	6.18	6.13	0.09	1.8	3.8	Bal.

Figure 1 shows the microstructure of initial material used for testing, observed on optical microscope (a) and on SEM (back scattered electron SEM-BSE micrographs) (b) at transverse cross-section of Ti-6246 alloy rod. The microstructure consists of equiaxed α grains and precipitations of α phase needles in β phase matrix.

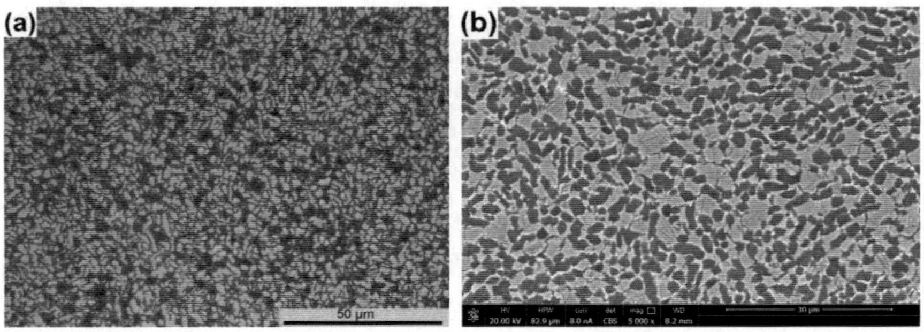

Figure 1. The optical (a) and SEM-BSE images (b) of the microstructure of the investigated Ti-6246 alloy at transverse cross-section of Ti-6246 alloy rod.

3. Results and discussion

3.1 Stress–strain behavior

True stress - true strain curves for Ti-6246 alloy were obtained in compression tests performed in temperature range from 800 °C up to 1100 °C and strain rate range of 0.01-100 s^{-1} are partly shown in Figure 2. As can be seen from the curves, higher temperature and lower strain rate reduce the flow stress. The flow stress curves (Figure 2a) exhibit a moderate to extensive flow softening at all strain rates. As can be seen in Figure 2a, the softening is less remarkable with decreasing strain rate. The curves obtained from the compression tests performed at the deformation temperature of 1100 °C (Figure 2b) are characterized by the insignificant drop in the flow stress at the beginning of hot deformation and further, almost constant, flow stress level signifying no visible flow softening mechanisms.

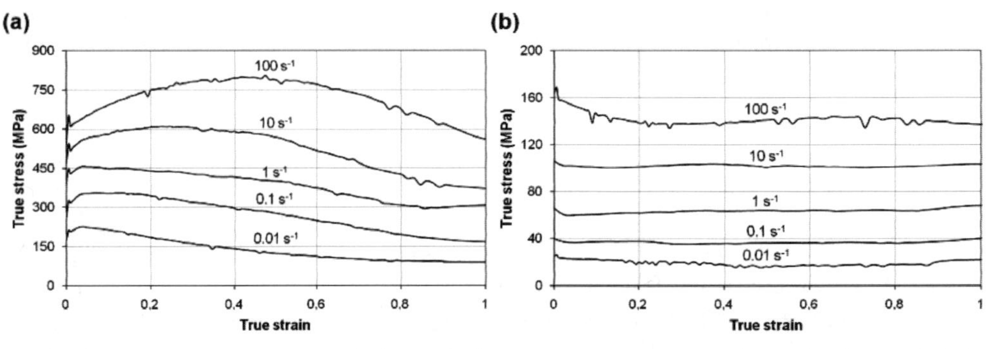

Figure 2. True stress-strain curves of Ti-6246 alloy at the temperature of: a)800 °C and b)1100 °C.

3.2 Microstructural validation and application of Malas's criterion

The contour maps of Ti-6246 alloy for the true strain of 1 at the temperature range of 800–1100 °C and for the strain rate range of 0.01 – 100 s^{-1}, describing the criteria of stability in accordance with Malas are shown in figure 3. As can be seen from the distribution of $m_{criterion}$ parameter (Figure 3a), the area of mechanical instability (shaded area) is in the range of high strain rates of deformation (from 5 s^{-1} to 100 s^{-1}) and temperature range of 800 °C – 990 °C. Figure 3b shows the contour map describing $s_{criterion}$ parameter distribution with one area of flow instability for the temperature range from 800 °C to 975 °C, and the strain rate range of 0.01 – 0.4 s^{-1}. It should be taken into account, that such structural instability areas can be characterized by the occurrence of microstructural banding or slip bands. From the distribution of temperature sensitivity of flow stress s (Figures 3c), the thermodynamic instability of the process at high temperatures (1030 – 1100 °C) and high strain rates (from 6 to 100 s^{-1}) can be noticed. The distribution of the strain rate sensitivity m (Figures 3 d) did not reveal the areas of instability in the investigated temperatures and strain rate range.

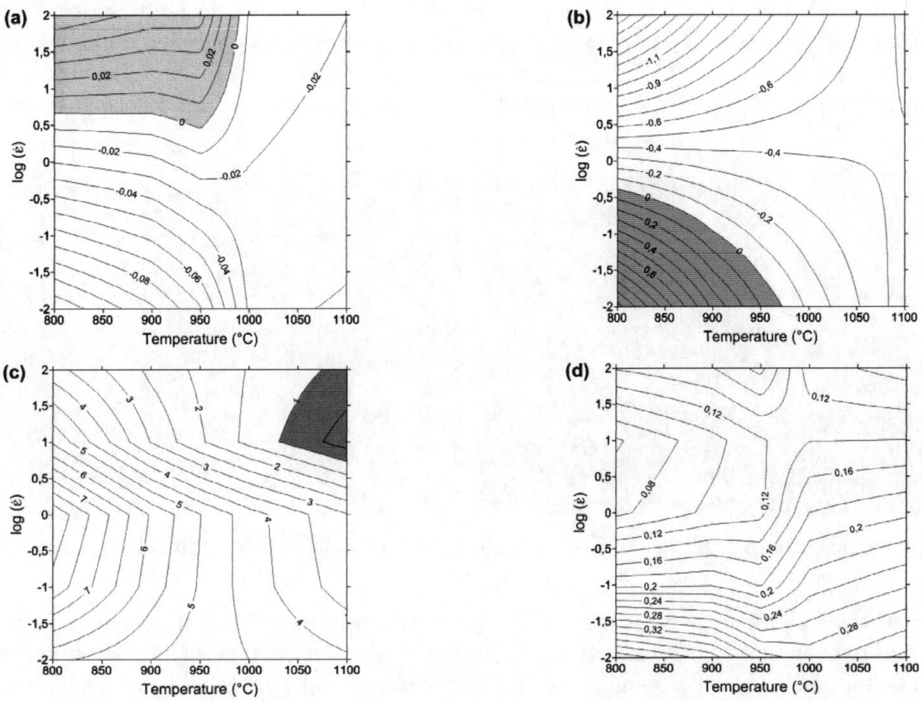

Figure 3. Contour map of Ti-6246 for the $m_{criterion}$ (a), $s_{criterion}$ (b), for the temperature sensitivity of flow stress s (c) and strain rate sensitivity m (d).

The processing map of Ti-6246 alloy created by superimposing of the four criteria of flow stability based upon Malas's stability criterion for the true strain of 1 is shown in Figure 3. This map makes it possible to identify areas providing a stable flow (white area) and the areas of flow instability (shadowed areas). In this case, there are three characteristic stability regions enabling determination of processing windows. The first window is in the range of low strain rates (from 0.01 to 0.4 s^{-1}) and high temperature range (975 °C – 1100 °C). A wide range of processing temperatures (800 – 1100 °C) and processing strain rates (0.4 – 6 s^{-1}) for the investigated alloy are characterized by the second processing window. This area is suitable for die forging on hydraulic-mechanical or hydraulic presses. The last window describes the processing parameters in narrow temperature range (from 990 °C – 1030 °C) and strain rate range of 6 – 100 s^{-1}.

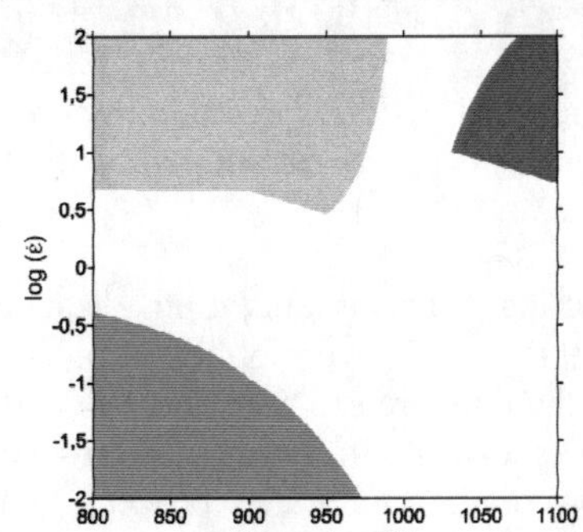

Figure 4. Processing map for Ti-6246 alloy based on Malas's stability criterion.

The typical microstructures of the specimens deformed under processing parameters corresponding to the instability area at the temperatures of 950 °C and 1100 °C, at the strain rates of 0.01 s^{-1} and 100

s^{-1} are shown in Figures 5a and 5b. As can be seen from these figures, in both cases only a partial recrystallization of the material could be observed. The microstructures consisted of the mixture of the original coarse grains and small recrystallized grains, what could result in unstable plastic flow and decreasing in the formability of the investigated alloy. Consequently, the hot deformation of Ti-6246 alloy in the instability area is not recommended.

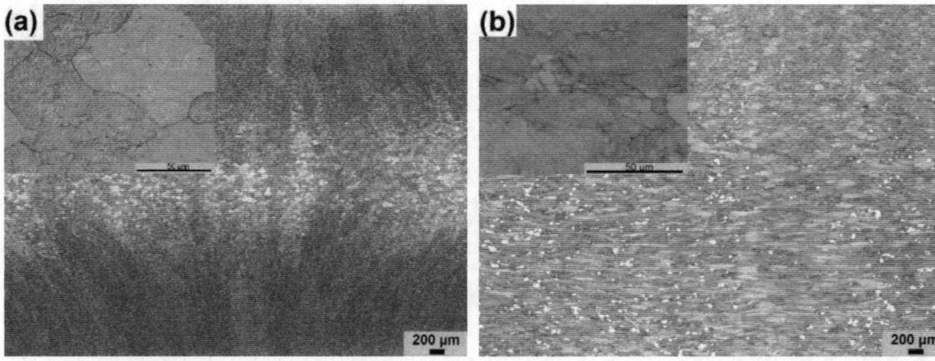

Figure 5. The microstructures of Ti-6246 alloy specimens deformed in instability area at the temperature and strain rate of: a) 950 °C /0.01 s^{-1}, b) 1100 °C /100 s^{-1}.

On the other hand, the microstructures of the specimens deformed in stability area at the strain rate of 0.1 s^{-1} and temperatures of 950 and 1000 °C (Figure 6) did not contain visible deformation defects such as internal cracks or shear bands, showing the stability and good quality of such structures. It should also be noted, that as it was also noted earlier [6], the presence of secondary precipitations of α phase in the microstructure, affects the final grain size and this fact should be taken into account in designing the hot working parameters for Ti-6246 alloy processing.

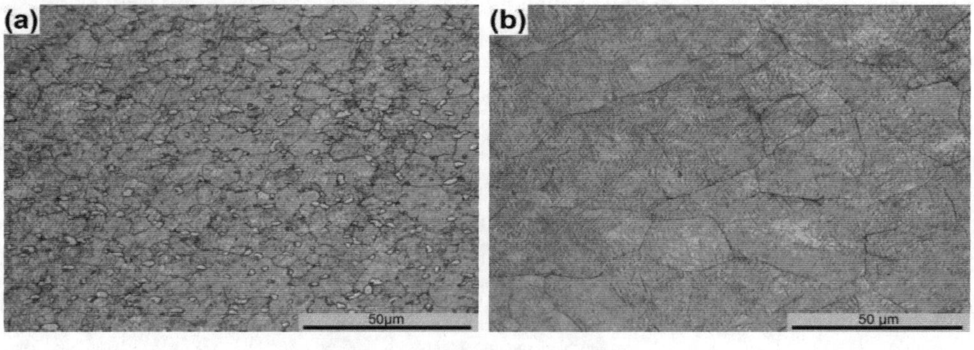

Figure 6. The microstructures of the specimens deformed at the processing parameters corresponding to the stability area - strain rate of 0.1 s^{-1} and temperatures of 950 (a) and 1000 °C (b).

4. Concluding remarks

The behavior of Ti-6246 alloy during forming operations has been studied based upon Malas's stability criterion. The true stress-strain curves were obtained in the isothermal hot compression tests performed to the true strain of 1 at the temperature range of 800–1100 °C, and at the strain rate range of 0.01 – 100 s^{-1}. The processing maps for the investigated alloy were created by superimposing four criteria of flow stability. According to the elaborated processing map (shown in Figure 4), stable processing of the investigated alloy should be assured in a wide range of the temperature and average strain rates in the range of 0.4 - 6 s^{-1}. At lower strain rates (below 0.4 s^{-1}) and at high temperatures (>950 °C), processing of this alloy should also be successful. Moreover, processing in the temperature range of 990 °C – 1030 °C and strain rate range of 6 – 100 s^{-1} should result in good quality Ti-6246 alloy products, too. The microstructure analysis confirmed the results of predicting the flow instability of Ti-6246 alloy, making possible understanding the mechanical behavior of Ti-6246 alloy at a wide range of strain rates and temperatures.

Acknowledgements
Financial support of the Polish Ministry of Science and Higher Education is gratefully acknowledged (AGH-UST statutory research project No. 11.11.110.591).

References

[1] Kuhlman G W 1994 Ti–6Al–2Sn–4Zr–6Mo in: *Materials Properties Handbook-Titanium Alloys* (ASM International) p 479
[2] Huda Z and Edi P 2013 Materials selection in design of structures and engines of supersonic aircrafts *A. Rev. Mat. and Des.* **46** pp 552–60
[3] Matthew J and Donachie Jr 2000 Titanium: *A Technical Guide 2nd Ed.* (ASM Internetional) p 369
[4] Gomon D, Hokka M and Kuokkala V-T 2012 Dynamic Compression Behavior and Numerical Modeling of Ti-6246 Alloy at Different Temperatures *Key Eng. Mat.* **527** pp 159–64
[5] Krawczyk J, Tokarski T, Łukaszek-Sołek, Dąbrowski R, Śleboda T and Lypchanskyi O 2016 Dynamic recrystallization in titanium alloys *Key Eng. Mat.* **687** pp 47–54
[6] Krawczyk J, Łukaszek-Sołek A and Dąbrowski R. 2015 The role of the deformation conditions in the evolution of the microstructure of Ti−6Al−2Sn−4Zr−6Mo alloy *Key Eng. Mat.* **641** pp 116–19
[7] Frocisz Ł, Krawczyk J, Madej M and Kopyściański M 2016 Correlation of tribological properties of titanium alloys with their microstructures *Key Eng. Mat.* **687** pp 41–46
[8] Xiong M, Weidong Z, Kaixuan W, Yunjin L and Yigang Z 2012 The investigation on the unstable flow behavior of Ti17 alloy in α+β phase field using processing map. *Mat. Sci. and Eng. A* **550** pp 131–137
[9] Wojtaszek M, Łukaszek-Sołek A, Śleboda T and Lypchanskyi O 2016 Application of Alexander-Malas criterion of plastic flow stability for estimation of PM Ti−6Al−2Sn−4Zr−6Mo alloy forging conditions *Metal 2016* pp 1577–1582
[10] Park N-K, Yeom J-T and Na Y-S 2002 Characterization of deformation stability in hot forging of conventional Ti–6Al–4V using processing maps *J. of Mat. Proc. Tec.* **130–131** pp 540–45
[11] Xiong M, Weidong Z, Yu S, Yungqing Z, Shaoli W and Yigang Z A 2010 Comparative Study of Various Flow Instability Criteria in Processing Map *Rare Met. Mat. and Eng.* **39(5)** pp 756−61
[12] Malas J C and Seetharaman V 1992 Using Material Behavior Models to Develop Process Control Strategies *J. of the Min.* **44** pp 8–13

IOP Conf. Series: Materials Science and Engineering **461** (2019) 012049 doi:10.1088/1757-899X/461/1/012049

Investigation of the composite strength of hybrid steel-steel semi-finished products manufactured by laser beam welding and friction welding

B-A Behrens[1], A Chugreev[1], T Matthias[1], S Nothdurft[2], J Hermsdorf[2], S Kaierle[2], H Ohrdes[3], J Twiefel[3], J Wallaschek[3], M Mildebrath[4] and T Hassel[4]

[1] Institute of Forming Technology and Machines (IFUM), Leibniz Universität Hannover (LUH), Garbsen, Germany
[2] Laser Zentrum Hannover (LZH) e.V., Hannover, Germany
[3] Institute of Dynamics and Vibration Research (IDS), LUH, Hannover, Germany
[4] Institute of Materials Science (IW), LUH, Garbsen, Germany

E-mail: tmatthias@ifum.uni-hannover.de

Abstract. An ever-increasing demand for more effective products with lower weight, more compact design and extended functionality has led to the use of material combinations with locally adapted properties instead of mono-materials. The Collaborative Research Centre (CRC) 1153 "Tailored Forming" at LUH addresses this topic with the aim to develop hybrid bulk-metal components on the basis of a new, tailored production process using joined semi-finished products. In contrast to existing manufacturing processes for hybrid bulk-metal components where the joining takes place during the forming process or at the end of the production chain, CRC 1153 uses tailor-made semi-finished products which are joined before the forming process. This contribution focuses on the production of hybrid semi-finished products from a steel-steel (C22.8 and 41Cr4) material combination by means of laser beam as well as friction welding. The effects of additional ultrasonic coupling into the weld-pool are presented for laser beam joining and a positive effect on the weld pool was observed. Furthermore, the resulting properties of the joining zone are investigated by uniaxial tensile tests at room temperature and metallographic analyses of the respective manufacturing processes. The friction welding proved to be a very robust process for the production of semi-finished steel-steel products.

1 Introduction and motivation

The desire for a reduced resource usage and an increased component as well as production process efficiency gives rise to an enhanced demand for adapted components and adjusted processes. These components need to offer locally customized areas for different stresses and load conditions. The use of more expensive, more difficult to process and higher alloyed steels can be reduced in combination with lower grade steels. For this approach, a joining process for the two similar materials has to take place. High material thicknesses, inconvenient component geometries and materials which are difficult to weld represent a challenge for the joining process. High carbon contents of the hardenable steel lead to hot cracking susceptibility for thermal welding of these steels. The use of ultrasound during welding is an approach to reduce this behavior due to the vibrations impact on the melt. The impact of the ultrasound amplitude on the shape of the weld reinforcement, the crack area and the hardness peak value in the weld was shown before [1]. This indicates the ultrasound assisted laser beam welding as an appropriate method for joining hot cracking susceptible steel grades in dissimilar joints. This procedure is compared with the manufacturing process friction welding in this research contribution.

1.1 Laser beam welding

Laser beam welding offers some advantages compared to other fusion welding processes. There are, for example, the localized heat input, the resulting narrow weld and heat affected zone, low thermal distortion and the good automation capability. High material thicknesses can be joined. Regarding sheet metal, stainless steels up to 20 mm with 16 kW laser power were welded. For round bars, the heat conduction and distribution is significantly different, so a transfer and comparison of the sheet metal process is not simple. The effects of ultrasound or vibration in common for welding processes were shown before for different materials as example steel [1], aluminum alloys [2-4] or nickel base alloys [5]. The results are grain refinement [2], reduced hot cracking [2] and an increase of the mechanical characteristics. Large dendrites are fewer [5], the grains are orientated more randomized and hardness levels are lower, furthermore there is an effect on the residual stress level [5]. The methods of generation of vibration are the direct introduction with tungsten plunger [6] or electrodynamic shaker [3] or indirect introduction (sonotrode [6] or piezoelectric transducer [2]).The frequencies investigated reach from low frequencies (48 Hz [5]) to ultrasound (20-40 kHz [6]). For EN AW-6082, there was shown an influence depending on the position in the waveform and amplitude of the ultrasonic with the system used [4]. The grain refinement is depending on the seam depth [3].

1.2 Friction welding

The friction welding process, which has gained much importance in recent years, is one of the press welding processes. In this process, the materials are joined together by plastic deformation. The heat required for the formation of welded joint is generated by the friction arising as a result of the rotation of one workpiece against the firmly fixed joining partner as well as due to the pressing force acting on the workpieces. Thus, this process is of particular advantage not only for the application of similar materials but also for the welding of material combinations, which cannot be joined by the application of other welding processes. Due to the temperatures just below the liquidus limit, the resulting recrystallization, the high compressive stresses as well as the resulting intensive material deformation, a very fine-grained microstructure with particularly good static and dynamic mechanical properties can be achieved. In contrast to fusion welding, almost pore free joints can be produced by this process. The relatively shorter heating time, as compared to the fusion welding, leads to a very small expansion of the heat affected zone and limits the thermally induced, undesirable structural changes of the base material only to a narrow area around the joining zone. [7]

2 Experimental setup

2.1 Materials

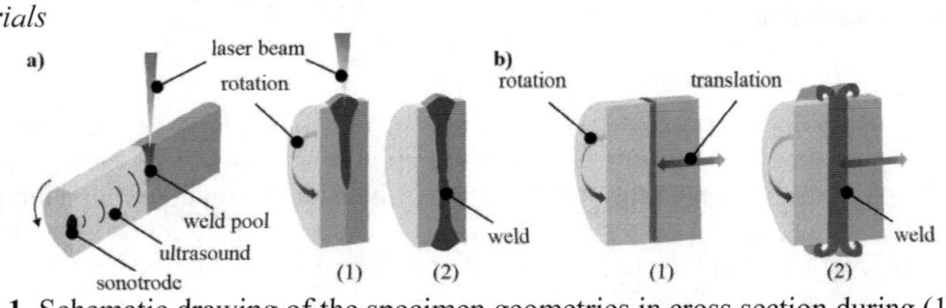

Figure 1. Schematic drawing of the specimen geometries in cross section during (1) and after (2) welding: a) laser beam welding; b) friction welding

Two steels are examined in this contribution. C22.8 (1.0460) is an unalloyed steel, the second steel 41Cr4 (1.7035) has an maximum equivalent carbon content of 0.840. 41Cr4 has a higher tensile and yield strength and is hardenable as compared to C22.8. Specimens are used in the form of round bars with a diameter of 30 mm. The joint shape is a butt weld. Figure 1 shows a schematic drawing of the sample geometries cross sections during and after welding for both processes.

2.2 *Manufacturing of semi-finished hybrid components by ultrasound assisted laser beam welding and friction welding*

2.2.1 Ultrasound system. The ultrasonic assisted welding system was developed to enable an ultrasonic assisted laser beam welding process. To ensure the desired vibration distribution, which is a longitudinal vibration at 20 kHz, sonotrodes for the specimens used are designed. The combination of the sonotrodes and the specimens lead to a longitudinal vibration resonance frequency of 20 kHz, where the connection of the specimens is located in the vibration maximum. To validate the vibration behaviour, the amplitude along the specimens is measured utilizing a laser-doppler-vibrometer with an in-plane setup. Additionally, the dependency between the vibration amplitude in the vibration maximum and the current at a second short circuited measurement transducer at the end of the vibration system is measured. The measured current at this transducer is used to control the vibration amplitude to the desired value. In addition, the vibration amplitude at the driving transducer is measured during the welding process to monitor the vibration amplitude behaviour during the process.

2.2.2 Laser beam welding system. The laser beam source used is a diode-pumped, solid-state disk laser with a maximum power output of 16 kW. The diameter of the optical is with 200 μm, the focal length of the collimation lense is 150 mm, focal length of the focusing lense is 300 mm) and the resulting focal diameter is 400 μm. The beam is not positioned on the vertex of the sample to avoid the formation of droplets by extending the time for solidification in order to realize a full penetration weld. The position of impact of the laser beam is moved rotationally by 6 mm, corresponding to the circumference of the sample, there is an angel of 20° to realize a still perpendicular angel of incident of the laser beam. Argon is used as process gas via two gas nozzles to shield the surface and to blow away the metal vapor. The samples were welded with ultrasonic vibration amplitudes of 0 μm and 3 μm. Laser power is 8.0 kW, welding speed 0.95 m/min.

2.2.3 Friction welding system. For the production of the steel-steel semi-finished products, the "Genius Plus" friction welding machine of the company Kuka, which is available at the IFUM, was used. The semi-finished products to be joined were placed in the friction welding machine in a slide and spindle clamp. A rotational movement of the spindle and a translational displacement of the slide caused both metallic semi-finished products to glow at the end faces due to the frictional heat. Both components were thus welded together by breaking the rotating semi-finished product and pressing under an increased force of the stationary semi-finished product.

3 Results and discussion

3.1 *Ultrasonic system evaluation*

To ensure that the position of the connection is at a vibration maximum, the vibration distribution along the workpieces is measured. Figure 2 a) depicts the vibration amplitude $\hat{x}$ along the workpieces. The distribution verifies that the designed sonotrodes lead to a vibration maximum in the joining zone of the two workpieces. In a further investigation, the dependency between the measured current at the sensor transducer and the vibration amplitude in the vibration maximum is measured for the pre welding condition. The results are shown in Figure 2 b). The dependency is approximately linear in the range up to 7 μm at 400 mA. The measured current used to control the vibration amplitude during the welding process. In addition, the vibration amplitude is monitored during the welding process and depicted in Figure 2 c) for an exemplary weld. It shows a nearly constant behavior, so the influence of the vibration is expected to be constant throughout the welding duration.

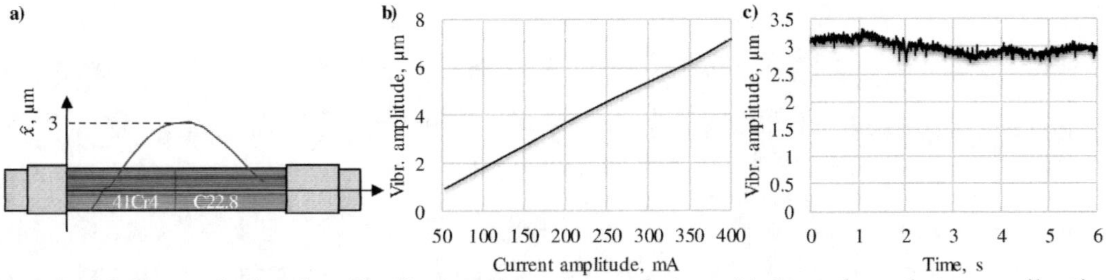

Figure 2. a) Measured vibration distribution along the workpieces b) dependency between vibration amplitude and measured current amplitude at measurement converter c) measured vibration amplitude during welding process

3.2 Metallographic analyses

Before a metallographic analysis of the three finished hybrid semi-finished products could be carried out, corresponding longitudinal sections of the respective samples were prepared. After the specimens had been set in place and then ground as well as polished, the specimens were etched in order to visualize and identify the grain boundaries as well as all microstructural constituents of the samples. Three longitudinal sections are shown in the figure 3: a) laser welding without the ultrasound system (LWwoUS); b) laser welding with the ultrasound system (LWwUS) and c) friction welding (FW).

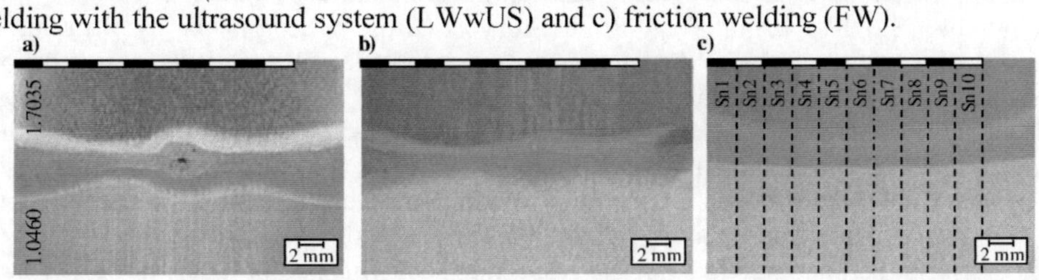

Figure 3. Longitudinal grinding of hybrid semi-finished products according to: a) LWwoUS; b) LWwUS; c) FW

The micrographs in the figure 3 show the influences of the respective manufacturing processes. The advantage of ultrasound during the welding process can be seen in micro-sections shown in figure 3 a) and b). Figure 3 a) shows clear defects, such as pores and micro cracks, in the center of the sample. With the help of ultrasound, as already described in the state of the art, the weld pool is positively influenced. This leads to an improved mixing of the weld pool, thus resulting in a significant reduction in the formation of flaws and micro cracks. After the manufacturing process, very similar heat-affected zones exist in both the methods. Due to the profile of the laser beam, these vary significantly across the cross-section of the sample. In the outer area as well as in the center of the sample, the materials are subjected to very high thermal stress. The friction welding process (Figure 3 c), on the contrary, has a very homogeneous heat-affected zone across the cross-section. This is also quite low due to the very short heating time of a few seconds. As a result, the undesirable structural change of the base materials takes place within a few millimeters around the joining zone. Micro hardness measurement is used to determine the condition of a material and its microstructure components. The hardness measurement was carried out based on the Vickers test with a four-sided diamond pyramid indent. The hardness of the three samples was respectively measured at 26 points at the edge as well as in the middle of the sample. Figure 4 shows the hardness profile depths of the three samples a) - c). The metallographic analyses can be confirmed on the basis of the hardness depth gradients. On the one hand, due to defects and micro-cracks in the center of the semi-finished product a), lower hardness values were determined. On the other hand, the undesirable structural changes of the heat-influenced zones are significantly more pronounced in semi-finished products a) and b) than in semi-finished product c). This expansion extends to 4-5 mm for laser-welded semi-finished products, whereas in the case of friction-welded semi-finished products, it is limited to 2-3 mm. The hardness values for all three processes are at a very similar level.

IOP Conf. Series: Materials Science and Engineering **461** (2019) 012049 doi:10.1088/1757-899X/461/1/012049

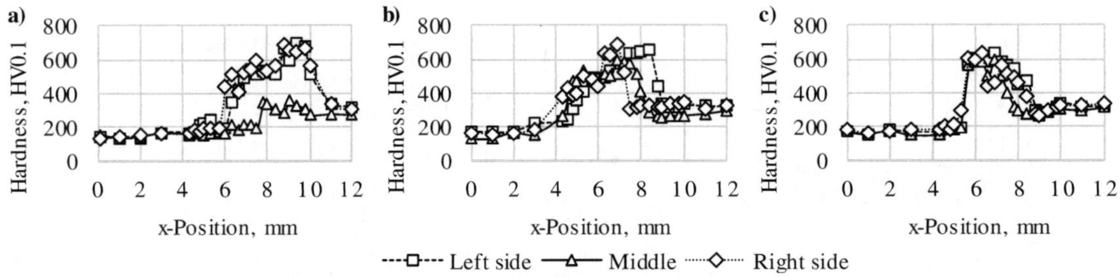

Figure 4. Hardness gradients of the hybrid semi-finished products: a) LWwoUS ; b) LWwUS; c) FW

3.3 Mechanical characteristics

In order to determine the critical thermomechanical stresses that lead to failure of the joining zone, the resulting joining zones were characterized analogous to Behrens et al. [8] and uniaxial tensile tests were performed. The determination of the bond strength of the hybrid components was initially carried out globally, i.e. the largest possible cross-section of the joining zone surface was tested by means of tensile tests. For this purpose, the global tensile specimens were prepared from the semi-finished products by machining. Due to the fact that the joining zone varies depending on the manufacturing process, local joining zone investigations were subsequently carried out. Therefore, ten small flat tensile specimens (Specimen number (Sn) 1 to 10), at every 2 mm along the cross-section of the hybrid semi-finished products, see Figure 3 c), were prepared analogous to the method described in [8] by wire erosion. With the help of the smaller test specimens, it was possible to evaluate the entire joining zone area locally. These tests were repeated thrice for each of the hybrid components. Figure 5 a) shows the results of the global bond strength. The determined composite strengths of the three different hybrid semi-finished products are on a similar level after a repetition of 5 tests. However, the friction welded semi-finished products exhibit the lowest values of scatter during the tests (Figure 5 a)).

In order to assess the bond strengths of the respective semi-finished products in further detail, the results of the local bond strengths at room temperature (exemplarily for one hybrid component) as well as those of the two mono-materials are shown below (Figure 5 b - c). To determine the mechanical properties of the joined hybrid semi-finished products, tensile specimens were prepared by wire erosion from the respective semi-finished products, as shown in Figure 3 c). Thus, it is possible to clearly describe the influence of specimen on the cross-section of the respective semi-finished products. The center of the respective semi-finished products lies between specimen numbers 6 and 7.

The results of the composite strengths of the laser-welded semi-finished products (with as well as without ultrasound) show a very similar course. Both are characterized by a strong decrease of the bond strengths in the center. In the case of LWwoUS, the bond strength decreases to 330 MPa. Whereas, with ultrasonic coupling, bond strengths of 505 MPa can be determined. The bonding strengths of hybrid components are higher than the tensile strength of the mono-material 1.0460 (see Figure 5 c). The decrease in the bond strength can be attributed to defects and micro-cracks. As described above, ultrasound is used to better mix the weld pool, thus minimizing the formation of flaws. The wave-like course of the bond strength can also be attributed to the formation of the weld seam: The thicker the weld, the greater is the bond strength. The results of the bond strength of friction-welded semi-finished products can confirm the positive properties described in section *3.2*. The very homogeneous heat-affected zone is reflected in the very uniform results of the bond strength. In this manufacturing process, there is no reduction in bond strength due to defects. In all tensile tests, with the exception of tensile tests with specimens 6 and 7, the specimens for the mono-material 1.0460 have experienced a failure. The joining zones have withstood the tests. On the contrary, the failure occurred in the joining zone for the case of the laser beam welded semi-finished products with and without ultrasonic coupling (specimens 6 and 7) due to the defects in the center. Disregarding these samples, the maximum strength of the laser welded samples is a little higher than the strength of the friction welded samples.

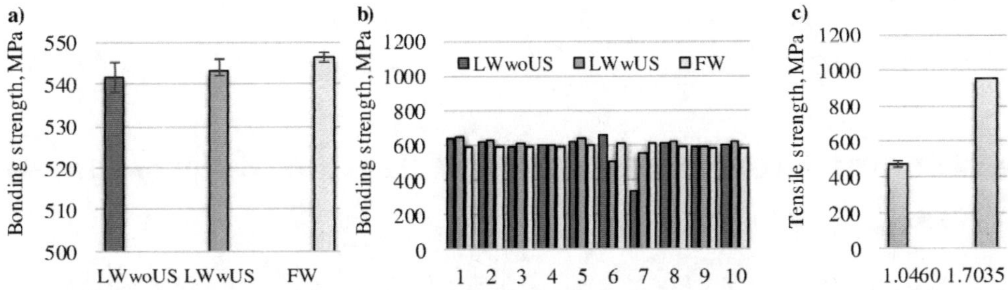

Figure 5. a) Determination of the global bond strengths; b) The bond strengths of the hybrid semi-finished products; c) tensile strengths of the mono materials according to the literature [8]

4 Summary and Outlook

In this contribution, the hybrid semi-finished products used in the CRC 1153 for the production of a hybrid shaft were characterized. The production is carried out by two different forming processes, a cross wedge rolling and a full forward extrusion and will be presented in a future contribution. With these forming processes, it is of utmost importance that the semi-finished products are not damaged or have internal defects in advance. Based on the metallographic analyses and the results of the tensile tests to determine the bond strength, various manufacturing processes have been compared with each other. The additional coupling of an ultrasound during laser beam welding had a positive effect on the weld pool, so that the previous internal flaws and micro-cracks were significantly reduced. The friction welding process proved to be a very robust process for the production of semi-finished steel-steel products. The local bond strength was very uniform and the failure always occurred in the base material of the 1.0460. Further investigations aim to the reduction of the defects in the laser welded samples with refined ultrasound usage to homogenize the strength.

Acknowledgements

The results presented in this paper were obtained from the Collaborative Research Centre 1153 "Tailored Forming" in subprojects A3 and C1 funded by the Deutsche Forschungsgemeinschaft (DFG, German Research Foundation) - 252662854.

References

[1] Nothdurft S *et al* 2018 Laser welding of dissimilar low-alloyed steel-steel butt joints and the effects of beam position and ultrasound excitation on the microstructure *J Laser Appl* **30** 032417

[2] Balasubramanian K *et al* 2011 Studies on the effect of mechanical Vibration on the Microstructure of the Weld Metal *Int. J. Eng. Sci. Technol.* **2** pp 253-256

[3] Woizeschke P *et al* 2017 Laser deep penetration welding of an aluminum alloy with simultaneously applied vibrations *Lasers Manuf. Mater. Process.* **4** pp 1-12

[4] Nothdurft S *et al* 2018 Influence of ultrasonic amplitude and position in the vibration distribution on the microstructure of a laser welded aluminum alloy *Proceedings of the International Congress on Applications of Lasers & Electro-Optics*, 14.-18.10.2018, Orlando, 2018

[5] Wu W 2000 influence of vibration frequency on solidification of weldments *Scr. Mater* 42 pp 661-665

[6] Jäger, S 2009 Nonvakuum-Elektronenstrahlschweißen mit gepulstem Strahlstrom und Ultraschallanregung (Non Vacuum Electron Beam Welding with pulsed electron beam and ultrasonic excitation), PhD Thesis, Hannover, Leibniz Universität

[7] Roder O *et al* 2005 Mixed Inconel Alloy 718 Inertia Welds for Rotating Applications – Microstructures and Mechanical Properties *In: TMS (Hrsg.): 6th International Conference on Superalloys* **718**, 625, 706 and Derivatives. Warrendale, USA 2005, pp 649-659.

[8] Behrens B-A *et al* 2018 Characterisation of the joining zone of serially arranged hybrid semi-finished components *AIP Conference Proceedings* **1960**, 040002 (2018)

Electrospinning process and characterization of PVP/hematite nanofibers

W Matysiak[1], T Tański[1], M Zaborowska[1]

[1] Institute of Engineering Materials and Biomaterials, Faculty of Mechanical Engineering, Silesian University of Technology, Konarskiego 18A, 44-100 Gliwice, Poland

E-mail: wiktor.matysiak@polsl.pl

Abstract. Iron oxide (III) or also known as hematite gained their popularity of material science engineers due to its properties, including photocatalytic activity. Nowadays, some research is being carry out to confirm the potential use of Fe_2O_3 nanomaterials, in particular nanofibers and nanowires in such devices as chemical sensors, magnetic resonances, supercapacitors and transistors. Another important property of hematite is its stability, it has been proved that it is the most durable iron oxide type. Taking into consideration the above, the aim of this work was to produce composite nanofibers with polymer matrix using poly(vinylpyrrolidone) with the doping phase of α-Fe_2O_3 (30% wt.) using combining method of sol-gel technique and electrospinning process. Final fiber mats were characterized by scanning electron microscope (SEM), to determine the influence of iron oxide nanoparticles on the morphology of PVP/hematite nanofibers and UV-Vis spectroscopy to measure the optical properties of nanofibers and determine whether such obtained fibers can be used in manufacturing process of iron oxide (III) crystal nanowires.

1 Introduction

Composite nanofibers became attractive for many industrial and economic reasons. The most important property of such fibrous composites is high surface-to-volume ratio, which is successfully used
in production of chemical sensors and dye-sensitized solar cells, where the increase in surface area plays significant role in efficiency of such devices [1-3]. When it comes to another feature, the simplicity and low cost of setup and process in general is cruel for electrospinning technique success. Compared to other nanomaterials manufacturing processes, spinning in electric field is characterized by high production rates, which makes it perfect technique for industrial scale [4]. Moreover, the great attention of nanofibers is being caused by easiness of controlling the morphology and other properties of final fibers, such as pore size, structure and other physical and chemical features [5-7].

With the dynamic development of nanotechnology and techniques of producing nanomaterials, the electrospinning process also have been developed. Nowadays, there are a few kinds of composite nanofibers. It stands out polymer/polymer fibers, polymer/nanoparticles and polymer/inorganic phase fibers and combination of these. It is not surprising, that there is potential in use of fibrous mats in many areas, for example biomedical applications, where in tissue engineering nanofibers play crucial role in matrix for tissue culture cells [8], but also fibers in nano scale are used to produce novel bandages for wound healing with antimicrobial properties [9]. The convenience in such applications is that almost every known polymer can be dissolved and used as the precursor for nanofibers process, in particular even polymers biocompatible with human body [10]. But not only biomedicine benefits from spinning in electric field technique, also the electronic industry can be revolutionized. It has been reported, that based on composite nanofibers, supercapacitors and nanogenerators can be made [11-13].

Hematite, being one of the most durable iron oxide types is widely used in metallurgy and glassmaking. But when it comes to nano scale α-Fe_2O_3, it gains some other specific features. Iron oxide (III) is typical n-type semiconductor with the width of energy band gap of 2.2 eV, chemically stable and corrosion resistant [14]. Some research has been carrying out to find its application in photocatalysis, due to visible-light responsive compound [15]. The addition of hematite nanoparticles to polymer matrix in electrospinning process results in composite nanofibers PAN/Fe_2O_3 which can be used in CO_2 gas sensors, as reported in [16]. Also, due to the increase of absorption level of PAN/hematite nanofibrous mats, such material can be used in electromagnetic radiation shields, in particular UV radiation [17].

In this paper, the Authors present the manufacturing process of PVP/α-Fe_2O_3 composite nanofibers via combining methods of sol-gel and electrospinning techniques. The doping phase of hematite nanoparticles was set to 30% (wt.) relative to polymer mass concentration in spinning solutions. To compare structure and physical properties, pure polymer PVP nanofibers have been produced. In order to investigate the influence of iron oxide (III) nanostructures on composite nanofibers, scanning electron microscopy (SEM) was used, to examine the morphology of final mats and UV-Vis spectroscopy was used to measure the absorption of produced composites. The obtained results can testify that above described technique can be the first step to produce crystal iron oxide (III) nanowires, a novel manufacturing process of making nanowires, which was reported previously [18, 19].

2 Materials and Methodology

The spinning solutions were prepared as follows: PVP (poly(vinylpyrrolidone), Mw= 1 300 000 g/mole, purity of 99%, Sigma Aldrich) was dissolved in ethanol (EtOH, purity of 99.8%, Sigma Aldrich)
to prepare polymer solution in concentration of 10% (wt.), next the suspension of α-Fe_2O_3 nanoparticles in ethanol was subjected to sonication process for 15 minutes, after which the measured amount
of polymer granulate was added into it. After mixing for 24 hours, final suspensions: PVP/EtOH, PVP/α-Fe_2O_3 (30% (wt.) were electrospinned on the FLOW – Nanotechnology Solutions Electrospinner 2.2.0 - 500 device using following parameters: the solution flow rate – 4.5 ml/h, the distance between the electrodes – 12.5 cm and the voltage between the syringe and collector – 15 kV. The as manufactured fibrous mats were left to dry in room temperature for 24 hours.

3 Results and Discussion

3.1 Morphology characterization of PVP and PVP/hematite nanofibers

Fig. 1 and 2 show SEM images of the obtained fibrous mats with EDX graph and histograms of fibers diameters distribution. The morphology analysis indicate that the applied electrospinning parameters allowed for the production of smooth and uniform in width polymer as well as composite nanofibers. Moreover, the entire length of fibers was characterized by the lack of any structural defects. The average value of PVP nanofibers diameter was 565 nm. However, the most values of fibers polymer width was in the range of 200 to 900 nm. EDX analysis confirm the presence of some characteristic chemical elements for poly(vinylpyrrolidone) nanofibers, carbon (C) and oxygen (O) and also aluminum (Al), which was the collector for produced fibrous mats.

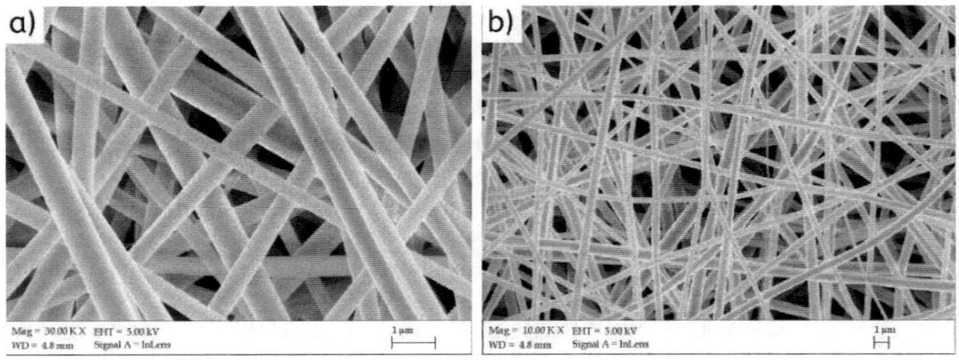

Figure 1. SEM images of PVP nanofibers: a) magnification of 30k X, b) magnification of 10k X

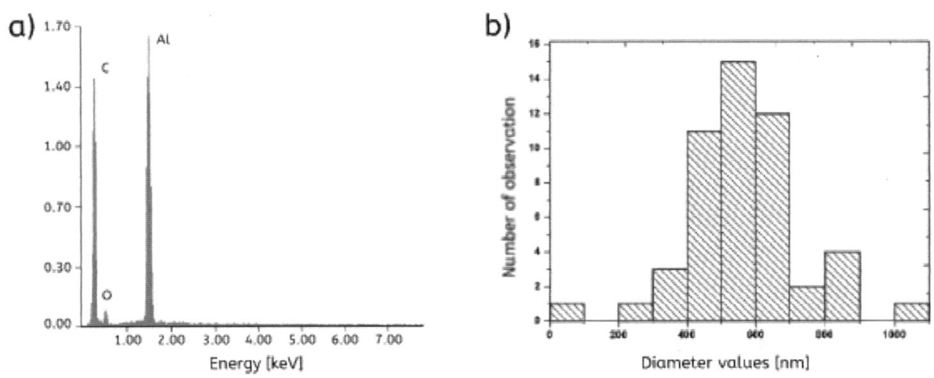

Figure 2. PVP nanofibers: a) EDX spectrum, b) histogram with fibers diameter values distribution.

Based on the histograms with diameter distribution, the addition of iron oxide (III) nanoparticles caused the decrease in fibers diameter. Scanning electron microscope images of PVP/Fe_2O_3 are shown on Fig. 3, EDX spectra and histogram with width values fibers distribution are shown on Fig. 4.

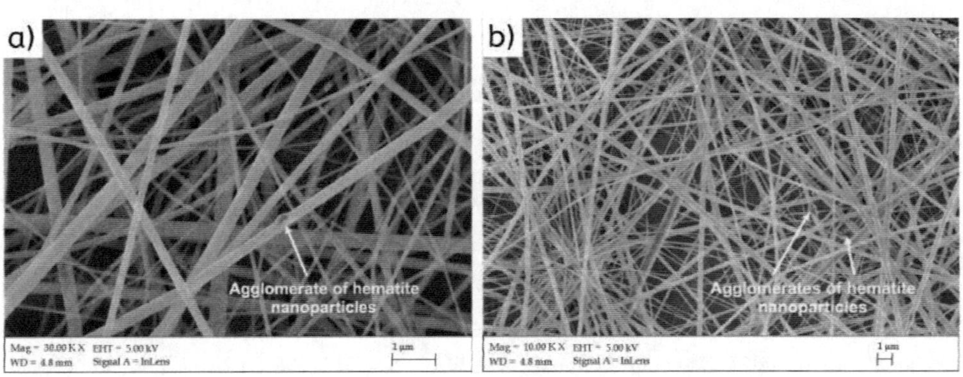

Figure 3. SEM images of PVP/α-Fe_2O_3 nanofibers: a) magnification of 30k X, b) magnification of 10k X.

The composite nanofibers are characterized by equal diameters along the entire fiber's length. There are single hematite nanoparticles aggregations (marked with yellow arrows in Fig. 3). The most important notable thing is that the values of composite fibers diameter increased in relation to polymer diameter nanofibers. It is caused by the addition of semiconductor doping phase, which changed the movement and elongated the jet of spinning solution. Basing on the histogram (Fig. 4 b)) the average value of PVP/iron oxide (III) diameter was 165 nm. The diameter values were distributed more evenly than those of polymer fibers diameters and were in the range of below 100 to 330 nm. Considering the above, addition of hematite nanoparticles to PVP nanofibers results in almost 3 times decrease in the fibers width values. The EDX spectrum of composite fibrous mats reveal four expected peaks corresponding to carbon (C), oxygen (O), which are attributed to chemical formula of polymer, iron (Fe) from doping phase and aluminum (Al) from aluminum foil, which was the collector for the obtained material.

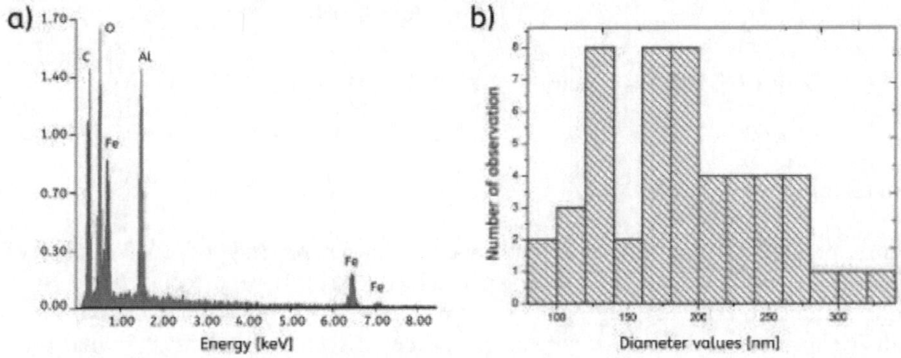

Figure 4. PVP/hematite nanofibers: a) EDX spectrum, b) histogram with fibers diameter values distribution.

3.2 UV-Vis spectroscopy

In order to determine the influence of iron oxide (III) nanoparticles on optical properties of as obtained composite nanofibers and to compare the absorption of polymer and composite nanofibers, UV-Vis spectroscopy was made (Fig. 5). Based on the absorption in the function of wavelength graphs, there can be seen slight increase in absorption level of composite nanofibers in comparison to pure polymer nanofibers. The absorption peak for PVP fibers is at 310 nm, while for PVP/hematite there is another peak at about 560 nm [17]. Such obtained results indicate, that doping phase in composite nanofibers cause higher absorption in the range of visible light wavelengths, which could be used in photocatalytic processes in such range of light [20]. Moreover, higher level of absorption in the UV radiation region indicate the use of composite nanofibers in UV light protection shields.

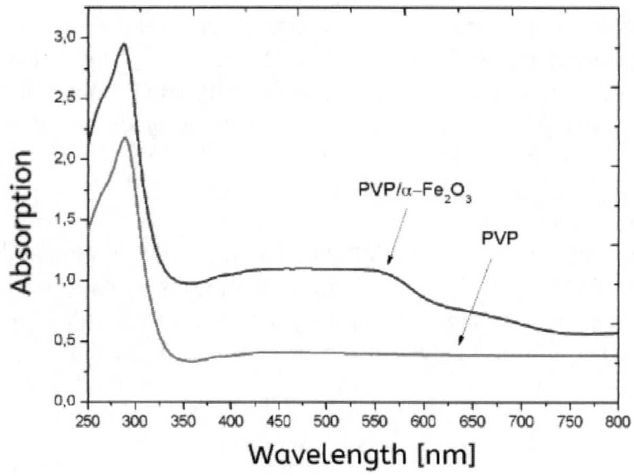

Figure 5. UV-Vis spectrum of PVP and PVP/α-Fe_2O_3 fibrous mats.

4 Conclusions

The Authors present successfully provided manufacturing process of PVP and PVP/hematite nanofibers via combining methods of sol-gel and electrospinning techniques. Based on the morphology examination, the doping phase of iron oxide (III) nanoparticles caused almost 3 times decrease in the diameter values of composite nanofibers. Such result indicates, that the composite fibrous mats were characterized by higher surface area. With the addition of hematite particles, the absorption level of nanofibers relative to pure polymer fibers increased. The obtained nanomaterial provides the possibility of producing crystalline iron oxide (III) nanowires, with great photocatalytic features.

Acknowledgements
The research presented in this article was financed by the National Science Centre, Poland based on the decision number 2016/23/B/ST8/02045.

References
[1] G Wang, T Liu, X Sun, P Li, et al. 2018 Flexible pressure sensor based on PVDF nanofiber *Sensors and Actuators A: Physical* **280** pp 319-325
[2] Y Sun, Z Wang, W Wang, G Li, et al. 2018 Electrospinning preparation of Pd@Co3O4-ZnO composite nanofibers and their highly enhanced VOC sensing properties *Materials Research Bulletin* in press, accepted manuscript
[3] F Zheng, Z Zhu 2018 Preparation of the Au@TiO2 nanofibers by one-step electrospinning for the composite photoanode of dye-sensitiez solar cells *Materials Chemistry and Physics* **208** pp 35-40
[4] Q Wei, *Functional Nanofibers and their Applications* pp 415-427
[5] T Tański, W Matysiak, B Hajduk, 2016 Manufacturing and investigation of physical properties of polyacrylonitrile nanofibre composites with SiO2, TiO2 and Bi2O3 nanoparticles, *Beilstein J. Nanotechnol.* **7** pp 1141-1155
[6] T Tański, W Matysiak, Ł Krzemiński, 2016 Analysis of optical properties of TiO_2nanoparticles

and PAN/TiO_2 composite nanofibers, *Mater. Manuf. Process.* **32** pp 1218-1224

[7] T Tański, W Matysiak, Ł Krzemiński, P Jarka, K Gołombek, 2017 Optical properties of thin fibrous PVP/SiO_2 composite mats prepared via the sol-gel and electrospinning methods, *Applied Surface Science* **424** pp 184-189

[8] Y Wang, W Cui, J Chou, S Wen, Y Sun, et al. 2018 Electrospun nanosilicates-based organic/inorganic nanofibers for potential bone tissue engineering *Colloids and Surfaces B: Biointerfaces* **172** pp 90-97

[9] M Liu, X Duan, Y Li, D Yang, et al. 2017 Electrospun nanofibers for wound healing *Materials Science and Engineering: C* **76** pp 1413-1423

[10] H Urena-Saborio, W Alfaro-Viquez, D Esquivel-Alvarado, et al. 2018 Electrospun plant mucilage nanofibers as biocompatible scaffolds for cell proliferation *International Journal of Biological Macromolecules* **115** pp 1218-1224

[11] M Nowak, M Kępińska, T Tanski, W Matysiak, P Szperlich, D Stroz, 2018 Optical properties of nanocomposite fibrous polymer mats containing SbSeI nanowires, *Optical Materials* (accepted 05.07.2018)

[12] M Nowak, T Tański, P Szperlich, W Matysiak, M Kępińska, D Stróż, B Toroń 2018 Using of sonochemically prepared SbSI for electrospun nanofibers, *Ultrasonics Sonochemistry* **38** pp 544-544

[13] X Liu, M Marlow, S Cooper, B Song, et al. 2018 Flexible all-fiber electrospun supercapacitor *Jorunal of Power Sources* **384** pp 264-269

[14] F Achouri, S Corbel, A Aboulaich, L Balan, et al. 2014 Aqueous synthesis and enhanced photocatalytic activity of ZnO/Fe2O3 heterostructures *Journal of Physics and Chemistry of Solids* **75** pp 1081-1087

[15] M L Maya-Trevino, J L Guzman-Mar, L Hinojosa-Reyes, et al. 2014 Activity of the ZnO-Fe2O3 catalyst on the degradation of Dicamba and 2,4-D herbicides using simulated solar light *Ceramics International* **40** pp 8701-8708

[16] Y Kim, E Park, D Young Lee, M Lee, et al. 2007 Electrospun nanofibrous polyacrylonitrile(PAN)/Fe2O3 membrane as CO2 gas sensor *Journal of the Korean Ceramic Society* **44** pp 194-197

[17] T Tański, W Matysiak, P Witek 2017 UV-Vis analysis of composite polyacrylonitrile/iron oxide nanoparticles thin fibrous mats *Applied Engineering Letters* **2** pp 54-59

[18] W Matysiak, T Tański, M Zaborowska 2018 Manufacturing process, characterization and optical investigation of amorphous 1D zinc oxide nanostructures *Appl. Surf. Sci.* **442** pp 382-389

[19] T Tański, W Matysiak 2018 Synthesis of the novel type of bimodal ceramic nanowires from polymer and composite fibrous mats *Nanomaterials* **8** art. no. 179

[20] L Cheng, L Liu, R Li, et al. 2017 Liquid phase deposition of α-Fe2O3/ZnO heterojunction film with enhanced visible-light photoelectrocatalytic activity for pollutant removal *Journal of The Electrochemical Society* **164** pp H726-H733

Forecasting the mean flow stress of carbon and low-alloy steels in hot rolling on a strip mill

Valeriy V Shkatov, Igor P Mazur and Victor V Shkatov

Lipetsk State Technical University, 30 Moskovskaya St., 398055 Lipetsk, Russia

E-mail: mazur_ip@mail.ru

Abstract. The effectiveness of the initial setup of the mill stands depends on the accuracy of the calculation of the mean flow stress of the rolled metal, the value of which is determined not only by the deformation parameters, but also depends on the chemical composition of the steel and the structural transformations of the rolled metal. This article proposes a mathematical model for the prediction of the mean flow stress of carbon and low-alloy steels during hot rolling on a strip mill, which allows to take into account, along with the deformation parameters, the content of chemical elements in the steel and the processes of recrystallization of the metal during deformation and during inter-pass times. The adequacy of the model was checked by comparing the results of the calculation of mean flow stress with the experimental data obtained by rolling of 100 strips 2-12 mm thick of 16 grades of carbon and low-alloy steels on a broadband mill 2000 NLMK. The influence of dynamic, metadynamic and static recrystallization on the mean flow stress of carbon and low-alloy steels in the mill stands is estimated. It is shown that taking into account the processes of recrystallization of the deformed metal can significantly improve the accuracy of the prediction of the mean flow stress.

1. Introduction

Currently, 80-90% of hot-rolled sheet steel is produced on continuous and semi-continuous broadband hot rolling mills, which is explained by higher technical and economic performance of these mills compared to other types [1]. For control and management of technological process of rolling on mills the computer equipment is now widely used.

When changing the thickness, width or chemical composition (steel grade) of the strip, it is necessary to carry out the initial setup of the mill stands, assuming the appointment of optimal deformation and speed parameters of rolling for the new working conditions of the mill. The effectiveness of the initial setup of the mill stands depends on the accuracy of the calculation of the mean flow stress of the rolled metal, the value of which is determined not only by the deformation parameters, but also depends on the chemical composition of the steel (including fluctuations in the content of elements within the steel grade), structural transformations of the rolled metal during deformation and during inter-pass times.

During repeated deformation of the strip steel in the stands of the finishing group of the mill, the following options for the development of recrystallization are possible, leading to the softening or hardening of the metal:

- softening due to dynamic recrystallization during deformation;
- softening due to metadynamic or static recrystallization during inter-pass times;

- hardening due to the summation of strain in adjacent stands in the absence or partial development of recrystallization processes during inter-pass times.

Currently, mathematical models for calculating the mean flow stress during hot rolling on strip mills are known [2-4], taking into account the influence of recrystallization processes. The main drawback of these models is the ability to predict the kinetics of recrystallization only for certain groups of steels with similar chemical composition (for example, C – Mn steels, microalloyed Nb steels), which does not allow to take into account the influence of the branded chemical composition of steels and leads to a decrease in the accuracy of prediction of the mean flow stress.

The aim of the work is to develop a mathematical model for the prediction of mean flow stress of carbon and low-alloy steels during hot rolling on a strip mill, to assess the impact of recrystallization processes (dynamic, metadynamic and static) on mean flow stress of steels.

2. Model for calculating the mean flow stress during hot rolling

During hot rolling of carbon and low alloy steels on strip mills to calculate the mean flow stress in the deformation zone (Fig. 1) the modernized Misaka equation [2, 5] is used:

$$\sigma_S = 9.8\exp\left(0.126 - 1.75[C] + 0.594[C]^2 + \frac{2851+2968[C]-1120[C]^2}{T+273}\right)\times \\ \times (0.768 + 0.137[Mn])\varepsilon^{0.21}\dot{\varepsilon}^{0.13}, \quad (1)$$

where ε is true strain; $\dot{\varepsilon}$ is strain rate, s^{-1}; T is temperature, °C; $[C]$, $[Mn]$ - carbon and manganese content in steel, % (wt.)

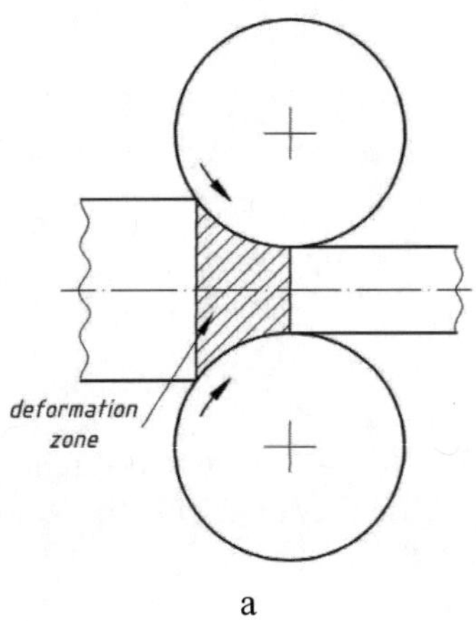

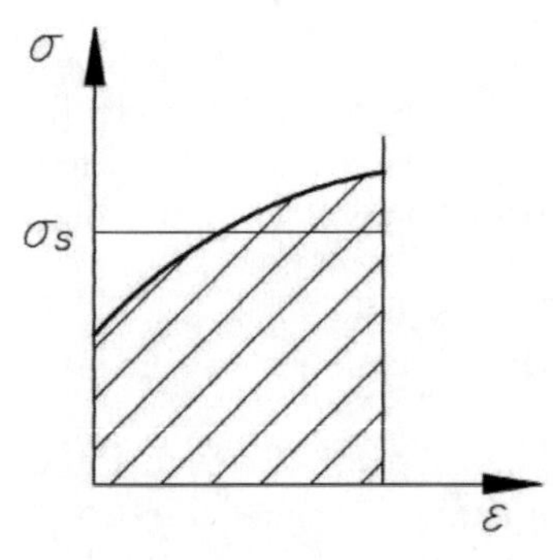

Figure 1. Scheme of deformation (a) and strain hardening (b) of metal during rolling. σ_S is mean flow stress in the deformation zone.

The processes of metal recrystallization, which are developed in the course of multiple deformation in the mill stands, impose a number of restrictive conditions on the use of equation (1). Thus, when the strip is deformed in the cage above a certain critical value of strain ε_C (depending on the steel composition, temperature and strain rate), dynamic recrystallization develops, leading to the softening of the metal during deformation. The value of mean flow stress in this case is determined not by the value of strain in the pass, but by the value of the residual metal hardening, depending on the degree of dynamic recrystallization.

When rolling the strip in the finishing group of the stands of the strip mill, the time between individual compressions varies from a few seconds to fractions of a second. Under these conditions, depending on the chemical composition of the steel and rolling parameters, several options for the development of recrystallization in the inter-stand spaces leading to the softening or hardening of the metal are possible.

If the strain in the stand is less than critical for the beginning of dynamic recrystallization ($\varepsilon < \varepsilon_c$), then the removal of the hardening in the deformed strip during the movement in the inter-stand interval occurs by static recrystallization of the deformed austenite.

If the strain in the stand is more than critical for the beginning of dynamic recrystallization ($\varepsilon > \varepsilon_c$), then the removal of the hardening in the deformed strip during the movement in the inter-stand interval occurs by metadynamic recrystallization of the deformed austenite, which develops immediately after the end of the deformation (without incubation period).

If during the time between the deformations in the i-th and (i+1)-th stands the recrystallization is not complete or has not begun, then after the deformation in the i-th stand there is a residual hardening, which is added to the deformation in the next stand. The value of the "accumulated" total strain for the calculation of the mean flow stress in the (i+1)-th stand is determined from the expression [6]:

$$\varepsilon_{i+1}^{a} = \varepsilon_{i+1} + K_a(1 - X_i)\varepsilon_i, \quad (2)$$

where X_i is degree of recrystallization (static or metadynamic) after deformation in the i-th stand;
K_a is a constant characterizing the degree of recovery development in the deformed metal.

The kinetics of static recrystallization by the Johnson–Mehl–Avrami–Kolmogorov equation was described [7]:

$$X = 1 - \exp\left[-B\left(\frac{\tau}{\tau_{0.5}}\right)^{n}\right], \quad (3)$$

where X is degree of static recrystallization, fractions;
$\tau_{0.5}$ is the time for 50% of recrystallization at given deformation parameters and the size of the initial (before deformation) austenite grain, s;
τ is the current time, s;
$B = -\ln 0.5$;
n is the coefficient.

For $\tau_{0.5}$ calculation, the mathematical description of the kinetics of static recrystallization of austenite in carbon and low-alloy steels during hot rolling was used. Along with the deformation parameters, the content of chemical elements in the steel and the grain size of austenite before deformation were taken into account [8]:

$$\tau_{0,5} = 8.591 \cdot 10^{-10} \varepsilon^{-4.09 d_0^{-0.15}} \dot{\varepsilon}^{-0.69} d_0^{1.02} \exp\left(\frac{Q_{rec}}{RT}\right), \quad (4)$$

where T is the deformation temperature, K; d_0 is the austenite grain size the before deformation, μm; R is the universal gas constant, $\mathrm{J/(mol{\cdot}K)}$.

The recrystallization activation energy Q_{rec} depends on the content of chemical elements in steel and is calculated as [8]:

$$Q_{rec} = 166620.38 - 9890.75[C] + 3828.18[Mn] + 9029.92[Si] + 17793.04[Cr]. \quad (5)$$

A common feature of static and dynamic recrystallizations is the movement of high-angle grain boundaries. The driving force for both types of recrystallizations is the excess volume energy accumulated in the process of plastic deformation [9]. Thus, the mechanism of static and dynamic recrystallizations is the same, the difference is the features of their implementation: after the completion of deformation or during deformation in conditions of continuous growth of the strain. The possibility of predicting the kinetics of dynamic recrystallization based on information about the kinetics of static recrystallization was first shown in works [10, 11].

For calculation the kinetics of metal recrystallization during deformation, the model of dynamic recrystallization of carbon and low-alloy steels was used [12, 13], based on the recalculation of the kinetics of static recrystallization (Eqs. (3)-(5)) to the conditions of continuous growth of the strain degree during deformation. To calculate the structural transformations of dynamically recrystallized metal during the subsequent inter-deformation pause, the model of metadynamic recrystallization was used [14], which was based on the same principles as for the model of dynamic recrystallization.

3. Software implementation and verification of the mathematical model

The model for calculating the mean flow stress of the rolled metal in the finishing stands of the broadband mill was implemented by OBJECT PASCAL language in the DELPHI application development environment. The following information is used as input: the chemical composition of steel, the size of the finished strip, the grain size of austenite at the entrance of the roll to the finishing group of stands, temperature and deformation-speed mode of metal rolling in the finishing group of stands. The output parameters are the degree of recrystallization of the metal in the stands and inter-stand spaces, the mean flow stress σ_S. of the rolled metal.

Checking adequacy of the model and assessing the impact of processes of recrystallization on the accuracy of the calculation of mean flow stress was performed on the experimental data that contains information about the deformation-speed parameters of rolling and the experimental values of the mean flow stress of the metal in the finishing stands of continuous broadband mill 2000 NLMK for 100 strips with a thickness of 2-12 mm of the 16 grades of carbon and low alloy steels.

4. Results and discussion

Calculations using the developed model showed that during rolling of carbon and low-alloy steels in the finishing group of stands, dynamic recrystallization can develop during the deformation of the roll in stands 1-3. Residual hardening of the metal after exiting the stand is completely removed by metadynamic recrystallization during the movement of the metal in inter-stand interval. Further deformation in the finishing stands is accompanied by static recrystallization of the deformed metal in the inter-stand intervals, the degree of development of which depends on the steel composition and rolling mode.

The results of the calculation of grain size, recrystallization of austenite and mean flow stress during the rolling of strip of a cross section 3×1500 mm of st3sp steel with a content of 0.15% C, 0.45% of Mn, and 0.20% of Si are presented in table 1, the kinetics of recrystallization of the steel is shown in figure 2.

Table 1. Grain size, recrystallized fraction and mean flow stress forecasting for St3sp grade

Finishing stand #	T (°C)	ε	$\dot{\varepsilon}$	τ_{ip} (s)	d (μm)	ε^a	ε_c	X_{dyn}	X	σ_S^{model} (MPa)	σ_S^{exper} (MPa)
1	924	0.57	9.5	4.7	50.0	0.57	0.44	0.04	1	169	163
2	914	0.54	18.5	3.0	19.9	0.54	0.40	0.06	1	186	181
3	903	0.42	30.8	2.1	17.5	0.42	0.44	–	0.96	195	205
4	874	0.41	51.7	1.4	18.4	0.43	0.55	–	0.90	223	231
5	865	0.39	84.4	1.0	18.3	0.44	0.62	–	0.83	245	250
6	837	0.33	120.8	0.8	18.0	0.41	0.76	–	0.67	272	265
7	828	0.17	105.7	–	17.9	0.29	0.77	–	0.71	255	247

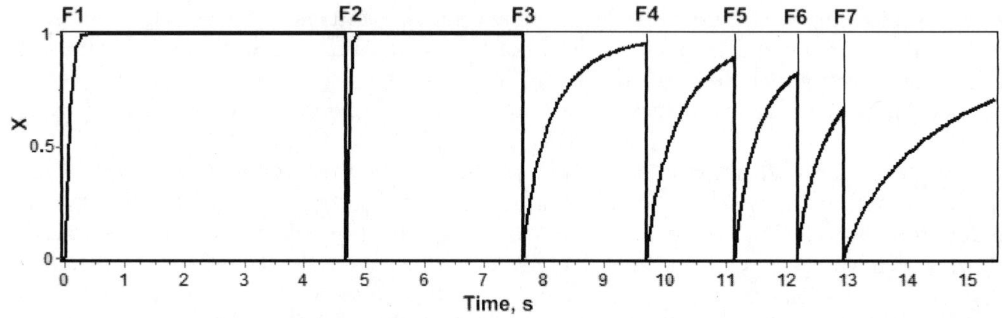

Figure 2. Variation of recrystallized fraction (*X*) during rolling a strip of st3sp steel in the finishing stands

To assess the effect of recrystallization processes on the prediction accuracy of mean flow stress, the calculation was carried out according to three schemes:

1) without taking into account of recrystallization processes of deformed austenite;
2) taking into account only the static recrystallization;
3) taking into account all types of recrystallization (dynamic, metadynamic and static).

The calculation results of mean flow stress for 100 strips are compared with the experimental data in the table.2. The use of the second scheme allowed to reduce the average relative error of the mean flow stress forecast by 1.5 times compared to the first scheme. Taking into account all types of recrystallization (the third scheme), the average relative error of the mean flow stress forecast decreased by 1.8 times.

Table 2. Results of comparing the calculated and experimental values of mean flow stress

Finishing stand #	Average relative calculation error σ_S, %		
	Scheme 1	Scheme 2	Scheme 3
1	10.49	9.25	5.40
2	5.61	4.84	3.49
3	4.18	4.04	3.71
4	4.60	3.96	3.66
5	4.95	4.66	3.14
6	8.77	4.20	4.15
7	13.53	4.79	4.92
Average by the stand	7.44	5.10	4.07

5. Conclusion

A mathematical model for calculation of mean flow stress of carbon and low-alloy steels during hot rolling on a strip mill is proposed. The influence of dynamic, metadynamic and static recrystallization on the mean flow stress of steel is estimated. It is shown that taking into account the processes of recrystallization of the deformed metal can significantly improve the accuracy of the prediction of the mean flow stress.

References

[1] Konovalov Y V 2008 A Reference Book of the Millman. Book 1: Production of Hot-Rolled Sheets and Strips (Moscow: Teplotekhnik) p. 640 (in Russia)

[2] Siciliano F and Jonas J J 2000 Mathematical modeling of the hot strip rolling of microalloyed Nb, multiply-alloyed Cr-Mo, and plain C-Mn steels Metall. Mater. Trans. A 31 511–30.
[3] Jonas J J 2000 The hot strip mill as an experimental tool *ISIJ Int.* **40** 731–8
[4] Minami K, Siciliano F, Maccagno T M and Jonas J J 1996 Mathematical modeling of the mean flow stress, fractional softening and grain size during the hot strip rolling of C-Mn steels *ISIJ Int.* **36** 1500–06
[5] Misaka Y and Yoshimoto T 1967 Formulation of mean resistance of deformation of plain carbon steel at elevated temperature *J. of the Japan Society for Technology of Plasticity* **8** 414–22
[6] Choquet P, Le Bon A and Perdrix Ch 1985 Mathematical model for prediction of austenite and ferrite microstructures inhot rolling processes *Proc. of Int. Conf. on Strengh of Metals and Alloys* vol 2 (Pergamon Press: Oxford) 1025
[7] Avrami M 1939 Kinetics of Phase Change. I General Theory. *J. Chem. Phis.* **7** 1103-1112
[8] Shkatov V V Mazur I P Kavalek A and Zhuchkova T S 2017 Model of static recrystallization kinetics of austenite in carbon and low-alloy steels during hot strip rolling *Vestnik of the Nosov Magnitogorsk State Technical University* **2** 69–74 (in Russia)
[9] Gorelik S S, Dobatkin S V and Kaputkina L M 2005 *Recrystallization of metals and alloys* (Moscow: MISIS) p. 432 (In Russia)
[10] Barnett M R, Kelly G L and Hodgson P D 2000 Predicting the critical strain for dynamic recrystallization using the kinetics of static recrystallization *Scripta mater.* **43** 365–9
[11] Barnett M R, Kelly G L and Hodgson P D 2002 Inferring dynamic recrystallization in ferrite using the kinetics of static recrystallization *Metall. Mater. Trans.* A **33** 1893-1900
[12] Shkatov V V and Shkatov M I 2007 Modeling the kinetics of dynamic recrystallization of low-alloy steels under hot deformation *Proc. Int. Conf. no Modern metallurgy of the beginning of the new millennium (Lipetsk)* vol 1 (Lipetsk: Lipetsk State Technical University) 131–6 (in Russia)
[13] Shkatov V V and Shkatov M I 2008 Predicting the critical strain for dynamic recrystallization in steels *Izvestiya vuzov. Chernaya metallurgiya* **3** 59–61
[14] Shkatov V V and Shkatov M I 2008 Mathematical modeling of the kinetics of metadynamic recrystallization of low-alloy steels *Proc. Int. Conf. no Modern metallurgy of the beginning of the new millennium (Lipetsk)* vol 1 (Lipetsk: Lipetsk State Technical University) 97–101 (in Russia)

Acknowledgments
The work was performed within the framework of the state task of the Ministry of Education and Science of Russia for Project No. 11.1446.2017 / 4.6.

Fatigue properties of SLM-produced Ti6Al4V with various post-processing processes

K Mertova[1] **and J Dzugan** [1] **and M Roudnicka**[2]

[1]COMTES FHT a.s., Průmyslová 995, Dobřany, 334 41

[2] VŠCHT Praha, Technická 5, Praha 6 – Dejvice, 166 28

E-mail: kmertova@comtesfht.cz

Abstract. Titanium and its alloys are very popular for various applications. They are the most widely used materials for biomedical applications. Titanium-based materials have shown excellent biocompatibility, as well as mechanical properties and corrosion resistance. The high fatigue life of this material is often demanded, therefore fatigue testing is necessary. Additive manufacturing (AM) is an innovative, widely investigated method for the production of metal components. AM allows the possibility of producing complex shaped components with near to net shape. Selective laser melting (SLM) is one of the most promising AM techniques. This research explores the fatigue behaviour of titanium alloy Ti6Al4V, produced by SLM. Different conditions and surface preparations for additive manufactured specimens were investigated by fatigue testing and then compared to each other. The aim of this research is to explore the influence of surface machining and processing by hot isostatic pressing (HIP) on the fatigue properties of the SLM produced components.

1 Introduction

The titanium alloy Ti6Al4V is the most frequently used Ti-based alloy; it is applied in more than half of all titanium alloys in the word [1]. It is a well-known fact that titanium and its alloys show good properties, such as low density, high specific strength, high fracture toughness, corrosion resistance and stability, and biocompatibility used for biomedical applications [1–3]. Despite its many advantages, there are also some disadvantages, such as high price of this metal and the difficulty of titanium processing. Titanium is highly reactive with oxygen, hydrogen and nitrogen at high temperatures [1,4]. Therefore, Ti has been used in applications that need proper material, despite the material cost [3].

Selective laser melting (SLM) is a type of an additive manufacturing process based on powder bed fusion. The concept of this method is based on depositing layer by layer until the whole component is built. This method allows producing from porous parts to parts with close to full density. Laser energy source melts the powder and subsequently fuses the required areas [5]. A deposition by SLM is carried out in an inert atmosphere. Post-processing including heat treatment and infiltration of printed components is also important [4].

Fatigue properties are very important values from the application point of view. The surface roughness and the presence of defects have the biggest influence on the fatigue properties of the material. The porosity and surface roughness provide stress concentration in their surroundings and can cause crack initiation. The sample roughness can be improved by surface treatment [3,6–8]. In

addition, many researchers explore minimization of the sample defects by the hot isostatic pressing (HIP) technique [9]. During the HIP treatment, products are heated at high temperatures and under high pressure. This creates plastic flow and prepares almost a fully dense structure. Generally, the lamellar or globular structure of grains α+β is desired, therefore temperature above the β-transformation is selected for the HIP processing of Ti6Al4V [10–12].
The purpose of this research was to investigate the differences in the fatigue life among machined, as deposited samples and the HIPed samples.

2 Material and Method

High cycle fatigue (HCF) properties of Ti6Al4V specimens at room temperature are investigated here.

2.1 Materials

The sample building was performed using an M2 Concept Laser machine under an argon atmosphere. The deposition was performed using the laser power of 200 W and the scanning velocity size of 1250 mm/s. The thickness of deposited layer was set to 30 μm and offset distance of 80 μm was used. The samples were produced by island scanning strategy in ZXY orientation [13]. The Ti6Al4V powder ELI grade 23 was used for deposition. A part of the specimens was subjected to recommended heat treatment to minimize residual stresses. The slow heating to 1093 K lasted 4 hours and was followed by hold at the temperature for another 1.5 hours; samples were then cooled slowly to 773 K and finally cooled to room temperature. The annealing was performed in vacuum furnace to avoid oxidation. The second part of the specimens was treated with HIP. Specimens were held for two hours under an argon atmosphere at the pressure of 1020 bar and the temperature of 1193 K. Annealed specimens and one half of HIPed specimens were machined by turning and polishing to the final surface roughness R_a = 0.2, with the direction corresponding to the building direction. The rest of the specimens was kept with the as-deposited surface.
The fatigue life of three different sample conditions was investigated, see Table 1. Each batch consisted of at least 12 samples.

Table 1 Sample designation

Sample condition	Designation
As deposited specimens with HIP treatment	AH
Machined specimens with HIP treatment	MH
Machined specimens, no HIP treatment	M

2.2 Testing and evaluation of fatigue properties

The specimens were tested under high fatigue cyclic (HCF) conditions until failure or the endurance limit of 10 million cycles was reached. The aim of this testing was to obtain the Wöhler (S-N) curve for each processing condition at room temperature. The specimen geometry is depictured in Figure 1.

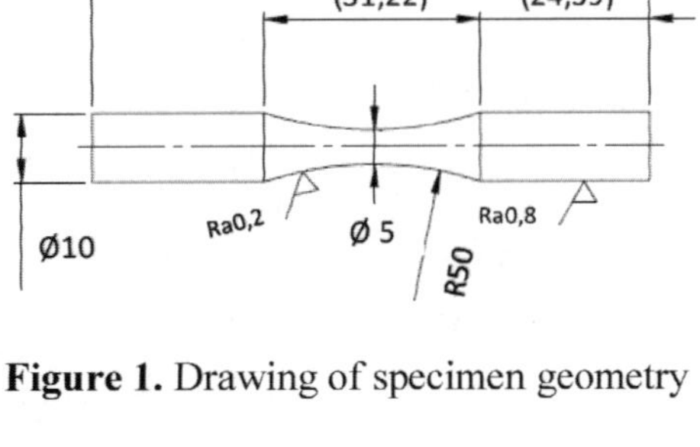

Figure 1. Drawing of specimen geometry

Figure 2. The test setup of HCF

The fatigue tests were performed in accordance with ASTM E466-7 standard under force controlled constant amplitude at room temperature in the air. The HCF tests were carried out with asymmetry ratio R = -1 at a frequency of 50 Hz. The fatigue limit σ_c was evaluated as the stress level under which the specimens survived 10^7 cycles. The testing setup is shown in Figure 2.

2.3 *Microstructure analysis*

The microstructure characterization was carried out on the microscope OLYMPUS PME3. The specimens for microstructural characterization were grinded on the silicon carbide paper and subsequently polished on the diamond paste. Final chemical-mechanical treatment was performed in colloid silica polishing suspension containing 20 vol. % of H_2O_2. Porosity was determined from micrographs using the image analysis by ImageJ.

3 Results and Discussion

Three S-N curves were constructed for different surfaces and process conditions of the titanium alloy at room temperature, as shown in Figure 5, Figure 4 and Figure 5. Orange symbols represent the specimens that withstood 10^7 cycles, blue dots represent failed samples. In the graphs fitted curve that represents the reliability of the results is plotted. Very high scattering of the results was obtained because of the random presence of defects. More considerable dispersion of the results was revealed for as deposited samples. In this case surface defects influenced the fatigue behavior and decreased the value of fatigue limits.

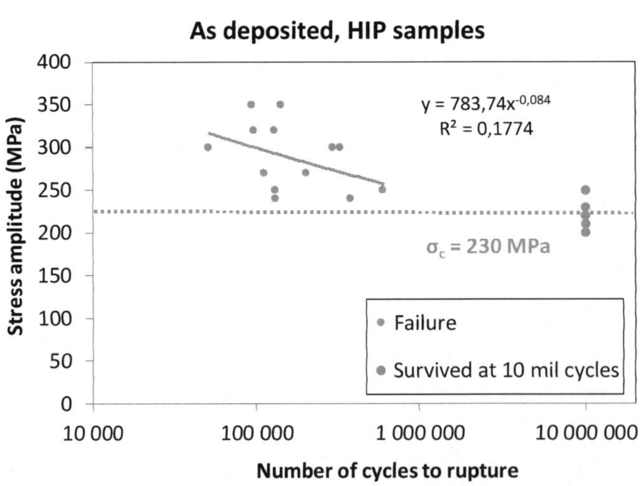

Figure 3 S-N curve of as deposited HIPed specimens

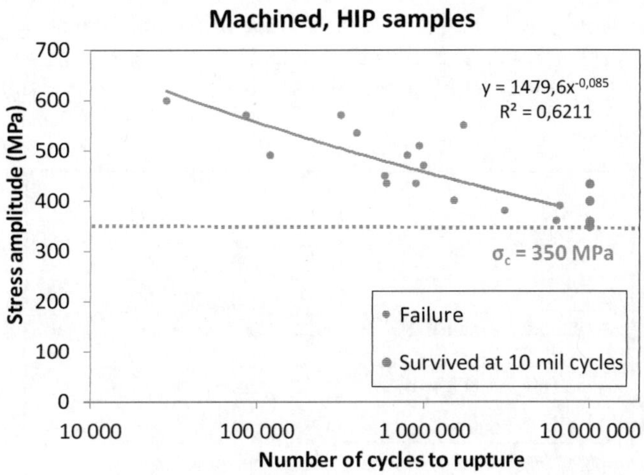

Figure 4 S-N curve of machined HIPed specimens

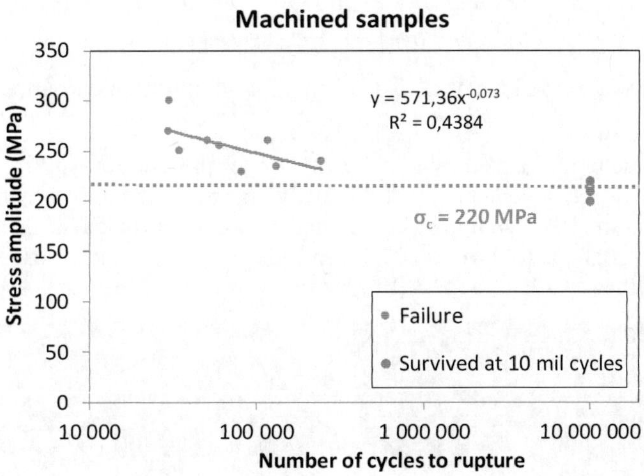

Figure 5 S-N curve of machined specimens

Table 2. Fatigue endurance limits

Sample			Fatigue endurance limit σ_c (MPa)
HIP	As deposited	AH	**230**
HIP	Machined	MH	**350**
without HIP	Machined	M	**220**

Table 2 represents fatigue endurance limits of all the tested batches. No significant change in values of fatigue limit has been obtained between the M and AH samples. However, a significant difference was observed for the HIPed and machined samples. The fatigue limit of these samples reached an increase by nearly 40 % in comparison with the others. All of these points to the fact that the large inner defects strongly affect the fatigue properties of the AM titanium alloy. At the same time, the surface treatment of the material has major impact on the results. The shape of the curves in the graph in Figure 6 demonstrates the above-described facts. Eliminating high surface roughness and inner defects plays a key role in the fatigue of the titanium alloy processed by SLM. Only removal of the surface roughness by machining is insufficient as internal defects become exposed to the surface. Conversely, eliminating porosity does not pronounce significantly if as built surface is kept.

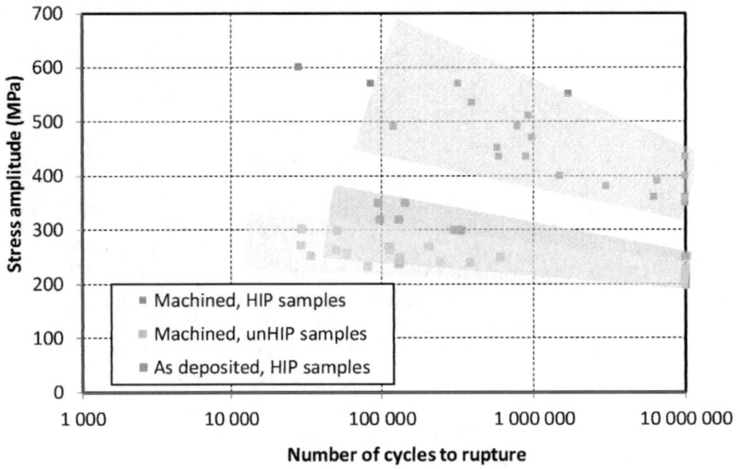

Figure 6. Comparison of all S-N curves for all investigated samples

The microstructure of the tested samples is depictured in Figure 7. In other studies, it has been shown that the sample fracture is caused by a lack of fusion or present porosity in AM products when compared to the HIP processed samples [14]. A distinctive porosity of samples without HIP treatment was found in comparison to HIPed samples. Image analysis revealed that HIP decreased the porosity significantly, from 0.67 ± 0.30 % to 0.01 ± 0.01 %. In Figure 7 can be noticed finer lamella thickness of specimens without HIP.

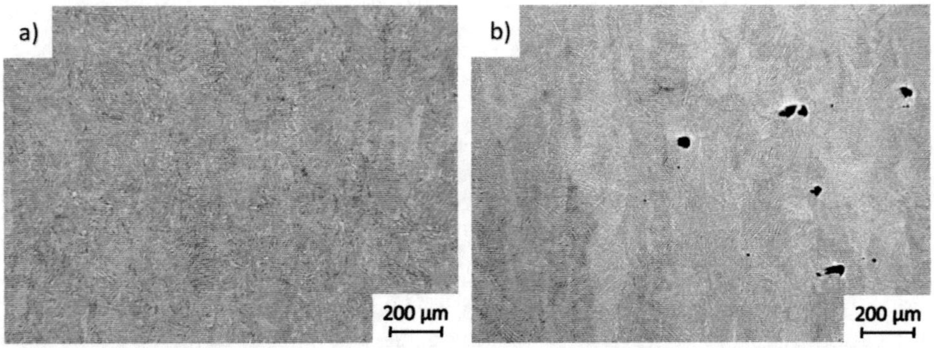

Figure 7 Microstructure of: a) HIP treated sample, b) sample without HIP treatment

4 Conclusions

Fatigue properties were investigated for three surface and processing conditions of the titanium alloy at room temperature. High scattering in the results is caused by defects of various amount, size, dimension or place in the specimens. No significant differences in fatigue life limit were observed between as deposited HIPed samples and machined samples without HIP post-processing. Additional HIP processing significantly increased the fatigue endurance limits of material. There, it can be stated that post-processing by HIP leads to effective increment of the endurance limit. The results revealed that it is need to eliminate inner pores and high surface roughness in case of as deposited samples. The defects and surface roughness are initiators of fracture during cyclic loading.

Acknowledgements
This paper was created by the project Development of West-Bohemian Centre of Materials and Metallurgy No.: LO1412, financed by the MEYS of the Czech Republic.

References

[1] Shipley H, McDonnell D, Culleton M, Coull R, Lupoi R, O'Donnell G and Trimble D 2018 Optimisation of process parameters to address fundamental challenges during selective laser melting of Ti-6Al-4V: A review *Int. J. Mach. Tools Manuf.* **128** 1–20
[2] Niinomi M, Nakai M and Hieda J 2012 Development of new metallic alloys for biomedical applications *Acta Biomater.* **8** 3888–903
[3] Romero C, Yang F and Bolzoni L 2018 Fatigue and fracture properties of Ti alloys from powder-based processes - a review *Int. J. Fatigue* **117** 407–19
[4] Srinivas M and Babu B S 2017 A Critical Review on Recent Research Methodologies in Additive Manufacturing *Mater. Today Proc.* **4** 9049–59
[5] Yap C Y, Chua C K, Dong Z L, Liu Z H, Zhang D Q, Loh L E and Sing S L 2015 Review of selective laser melting: Materials and applications *Appl. Phys. Rev.* **2**
[6] Dumas M, Cabanettes F, Kaminski R, Valiorgue F, Picot E, Lefebvre F, Grosjean C and Rech J 2018 Influence of the finish cutting operations on the fatigue performance of Ti-6Al-4V parts produced by Selective Laser Melting *Procedia CIRP* **71** 429–34
[7] DebRoy T, Wei H L, Zuback J S, Mukherjee T, Elmer J W, Milewski J O, Beese A M, Wilson-Heid A, De A and Zhang W 2018 Additive manufacturing of metallic components – Process, structure and properties *Prog. Mater. Sci.* **92** 112–224
[8] Dzugan J, Prochazka R, Rund M, Podany P, Konopik P, Seifi M and Lewandowski J J 2018 Effects of thickness and orientation on the small scale fracture behaviour of additively manufactured Ti-6Al-4V *Mater. Charact.*
[9] Greitemeier D, Palm F, Syassen F and Melz T 2017 Fatigue performance of additive manufactured TiAl6V4 using electron and laser beam melting *Int. J. Fatigue* **94** 211–7
[10] Benedetti M, Torresani E, Leoni M, Fontanari V, Bandini M, Pederzolli C and Potrich C 2017 The effect of post-sintering treatments on the fatigue and biological behavior of Ti-6Al-4V ELI parts made by selective laser melting *J. Mech. Behav. Biomed. Mater.* **71** 295–306
[11] Fleck C and Eifler D 2010 Corrosion, fatigue and corrosion fatigue behaviour of metal implant materials, especially titanium alloys *Int. J. Fatigue* **32** 929–35
[12] Shamsaei N and Simsiriwong J 2017 Fatigue behaviour of additively-manufactured metallic parts *Procedia Struct. Integr.* **7** 3–10
[13] ASTM WK49229 2017 Standard Guide for Orientation and Location Dependence Mechanical Properties for Metal Additive Manufacturing Work in Progress
[14] Günther J, Krewerth D, Lippmann T, Leuders S, Tröster T, Weidner A, Biermann H and Niendorf T 2017 Fatigue life of additively manufactured Ti–6Al–4V in the very high cycle fatigue regime *Int. J. Fatigue* **94** 236–45

AZ31 Magnesium Alloy Subjected to Severe Plastic Deformation at Different Temperatures: Microstructure and Hardness

J Michalski[1], J Till[1], J Petrů[1], S Rusz[2], O Hilšer[2] and J Vontorová [3]

[1]VSB - Technical University of Ostrava, Faculty of Mechanical Engineering, Department of Machining, Assembly and Engineering Technology, Ostrava, Czech Republic, EU.

[2]VSB - Technical University of Ostrava, Faculty of Mechanical Engineering, Department of Mechanical Technology, Ostrava, Czech Republic, EU.

[3]VSB - Technical University of Ostrava, Faculty of Metallurgy and Materials Engineering, Department of Chemistry, Ostrava, Czech Republic, EU.

E-mail: jakub.michalski.st@vsb.cz

Abstract. This paper is focuses on study of a hot extruded AZ31 magnesium alloy processed by severe plastic deformation method TCAP (Twist Channel Angular Extrusion) at different temperatures. In order to characterize processing temperatures and the resulting structure and properties, the specimens were processed by 4 passes at 250°C, 230°C and 200°C. Microstructure and mechanical characterization was performed by Vickers hardness testing and metallographic analysis on an optical microscope.

1 Introduction

It is well established that grain size has a significant influence on the strength (hardness) of materials based on the Hall-Petch relationship (1) [1,2].

$$YS(HV) = \sigma_o + k \cdot \frac{1}{\sqrt{d_{AVE}}} \quad (1)$$

where ***YS (HV)*** is yield stress and hardness, respectively. $\boldsymbol{\sigma_o}$ is the stress necessary for overcoming of Peierls-Nabarr friction stress, resistance of dissolved foreign atoms, resistance of precipitates from solid solution and lattice defects, $\boldsymbol{k}$ is the constant, the measure of which is the value of shearing stress necessary for release of accumulated dislocations and $\boldsymbol{d_{AVE}}$ is the average grain size.

It follows from the equation (1) that material yield value increases with decreasing grain size. This phenomenon is a driving force for research and development of high-strength structural materials, particularly of steels. It turns out that refining of grain can lead to increased drawability of metallic materials. On condition of identical strengthening mechanism refining of grains down to the level of nanometers can bring enormous increase in material strength.

Severe plastic deformation (SPD) is well known as an effective process in refining the grain size of metals and alloys to the nanometer range [3,4]. Several techniques have been presented as SPD methods, such as accumulative roll bonding (ARB) [5], twist extrusion (TE) [6] and friction stir processing (FSP) [7], but the most common methods are equal channel angular pressing (ECAP) [8] and high-pressure torsion (HPT) [9].

Interest in magnesium alloy has increased in recent years due to its high specific strength and low density compared to other structural metals and alloys. Weight reduction is an important factor in fuel economy, so magnesium alloys are attractive for use in the automotive [10]. As a consequence of its hexagonal close-packed (HCP) crystal lattice and limited slip systems, magnesium alloy lack ductility at ambient temperatures [11]. At room temperature, deformation can be achieved by two dominant mechanisms: slip in the basal plane and mechanical twinning [12,13]. Magnesium alloy has three slip systems in the basal plane, of which two are independent, but in order to achieve homogenous deformation without cracks developing, five independent slip systems are required [14]. For this reason, the homogenous deformation of magnesium at room temperature without cracking is difficult and this limits its use in many applications.

1.1 Equal Channel Angular Pressing

A typical flowchart of this technology is shown in **Figure 1**. ECAP allows deformation of a metal sample by shear without changing its dimensions. A die comprises two channels with the same cross section, which intersect each other and create a bend. When extruding a sample through the die, the shear strain occurs in this place. The process can be repeated to achieve intensive plastic strain and a finer grain. The real deformation in one pass depends on the angle between axes of both the channels and to a less degree also on the bending radius. A sample can be rotated along its longitudinal axis between the particular passes, thus creating various technological routes of ECAP [15]. The technological route influences the process of refinement of grains and their shapes significantly [16].

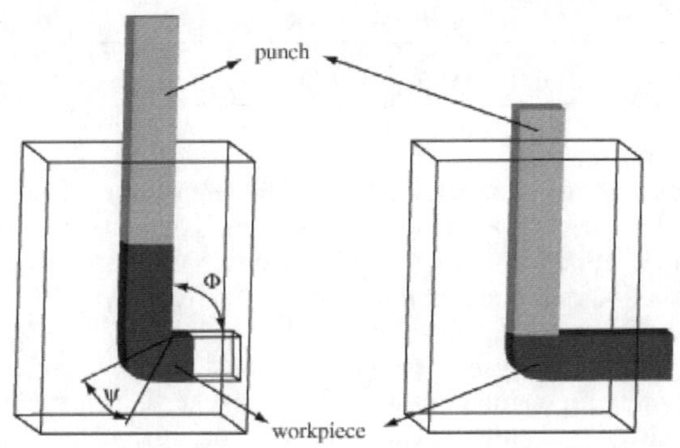

Figure 1. The principle of Equal Channel Angular Pressing (ECAP) process [16]

1.2 Twist Channel Angular Pressing

The TCAP (twist channel angular pressing) is a method for increasing the efficiency of SPD. The horizontal part of the TCAP channel is manufactured into a helix shape with an angle of lead $\gamma = 30°$, the TCAP die geometry thus combines the conventional ECAP method and the TE [17] method. During extrusion of an experimental sample with the use of a tool with modified geometry, the refining starts with shearing strain originated during the passes through the deformation zone, which is at the place of transition from the vertical position to the horizontal position of the TCAP extrusion channel. This part of the process is identical with the conventional ECAP process. After the sample passes through the deformation zone, the sample is then extruded into the helix. The effect of the helix in the horizontal part of the channel lies in the fact that the sample in the first stage is decelerated, which means it encounters a back pressure (BP). For continuation of the forming process, it is necessary to increase the extrusion force and this leads to an increase of strain intensity [18].

a) **b)**

Figure 2. Schematic illustration of ECAP die channel: convention ECAP geometry (a); ECAP geometry with helical horizontal channel, i. e. TCAP (b)

2 Experimental material and procedures

The experiments were concentrated on the influence of severe plastic deformation method TCAP on the grain refinement and tensile properties of hot extruded (at 400°C with ratio of 22) magnesium alloy AZ31.

2.1 Experimental material

The investigated material was supplied by the Institute of Non-ferrous Metals in Gliwice, Department of Light Metals (OML Skawina), Poland. The chemical composition of the experimental alloy is presented in **Table 1**.

Table 1. Chemical composition of AZ31 magnesium alloy

Weight [%]	Fe	Si	Mn	Ni	Cu	Mg	Zn	Al
AZ31	0.005	0.100	0.500	0.005	0.050	94.740	1.200	3.100

2.2 Experimental procedures

Due to the limited formability of investigated alloy at low temperature (below 200°C), the extruded alloy AZ31 was deformed by the TCAP method at a different temperature (200°C, 230°C and 250°C) for obtaining maximal grain refinement of deformed structure. TCAP processing was realized with the constant deformation rate of 40 mm/min and the samples were rotated by 90° around the longitudinal axis (Bc processing route).

The samples for metallographic analysis were cold mounted in resin, ground with silicon carbide papers and polished in cloth with diamond paste to a mirror-like finishing. The polished surface was etched with a solution of 5% vol. nitric acid and 95% vol. ethanol and observed on an optical microscope.

The hardness tests (HPO 350 testing machine) according to Vickers HV_5 (load 50 N for 20 s) were carried out on etched metallographic samples.

3 Experimental results

3.1 Microstructure evolution

Microstructure of the investigated alloy in the as-cast condition is shown in **Figure 3a-b**. The microstructure consists of the solid solution of aluminium in magnesium (α-phase), surrounded by formations smaller amount of $Mg_{17}Al_{12}$ intermetallic (β-phase), nearly continuously distributed in the interdendritic areas along grain boundaries.

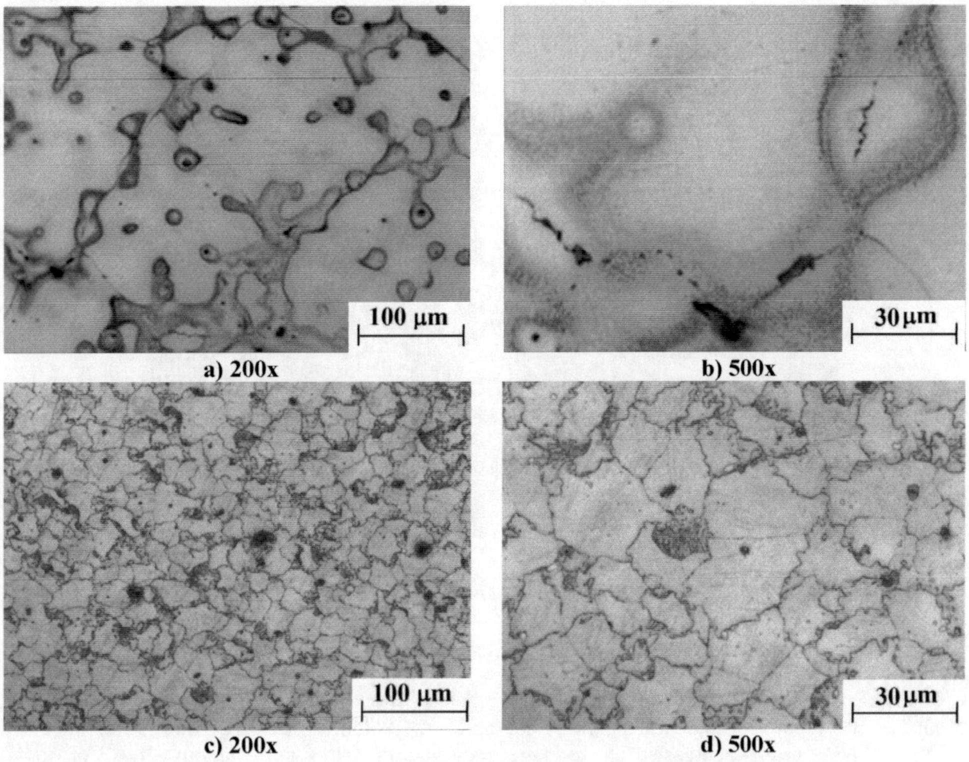

Figure 3. OM microstructure of the investigated alloy in two different conditions: as-casted (a-b) and hot extruded (c-d)

The microstructures of the casted and extruded AZ31 alloy are shown in **Figure 3c-d**. The initial microstructure is bimodal containing large grains elongated in the extrusion direction and smaller grains. An advantage of using the samples in the extruded initial state is the refinement of the cast grain having an average size of ~ 120 μm to a size of ~ 23 μm. The initial hot extrusion has a positive influence on the β-phase precipitates dissolution. It can be assumed that hot extruded alloy will have a higher plasticity and during TCAP processing will be no cracking of the specimens. We can assume that by using the initially extruded state of the samples, the number of TCAP passes necessary for the achievement of the desired refinement and homogeneity of the structure, can be reduced.

Taking into the melting temperature of the investigated alloy is ~ 625°C [19], recrystallization during deformation at 200°C is referred to as low temperature dynamic recrystallization. Obviously, deformation at those conditions, i. e. a relatively high temperature that is below the typical ductile transition temperature (225-250°C) necessary for activating pyramidal slip, caused mechanical failure.

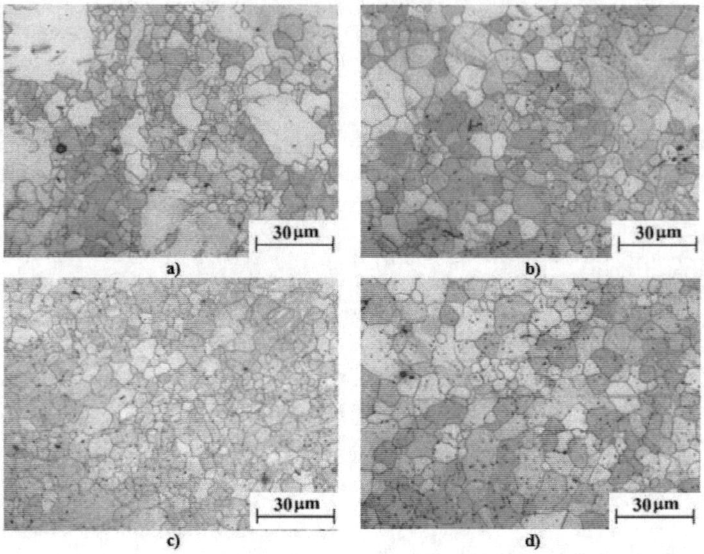

Figure 4. OM microstructure of the investigated alloy processed by TCAP at 250°C: 1 pass (a); 2 passes (b); 3 passes (c) and after 4 passes (d)

Figure 4a-d shows the OM images of AZ31 magnesium alloy subjected to TCAP method. After 1 pass of TCAP at 250°C, the microstructure is more heterogenous with very fine grains below 9 µm. With further TCAP processing at the same temperature, the volume fraction of fine grains increases while that of large grains decreases. The average grain size is further homogenous after 3 passes of TCAP (4.8 µm). Some grains are refined to a few micrometers with little residual strain whereas other large grains undergo distortion with trace of plastic deformation. This heterogeneity is believed to result from the different orientations of the grains in relation to the shear direction imposed by TCAP.

The results show a positive influence of TCAP processing realised at lower temperature on the Vickers hardness values and distribution (standard deviation in **Table 2**). After the 3 passes, the hardness increases (compared to the initial state) by approximately 22% (samples processed by TCAP at 250°C), by approximately 29% (230°C) and 38% (200°C), respectively.

Table 2. Vickers hardness with the standard deviation of the AZ31 magnesium alloy subjected to TCAP prcessing (in HV_5)

Processing method	**250°C**	**230°C**	**200°C**
Initial state (hot extruded)	55±2.6	55±2.6	55±2.6
1 pass TCAP	59±1.8	62±2.1	68±1.9
2 passes TCAP	63±1.7	67±1.9	72±1.3
3 passes TCAP	66±0.9	71±1.0	76±0.6
4 passes TCAP	62±1.3	65±1.7	73±0.9

4 Conclusions

The hot extruded (EX) AZ31 magnesium alloy was subjected to severe plastic deformation method twist channel angular pressing (TCAP) using up to 4 passes at different temperatures (250°C, 230°C or 200°C). The following conclusions have been drawn:

- The effect of TCAP processing on the microstructure and the hardness of hot extruded AZ31 magnesium alloy was studied by means of optical microscopy (OM) and hardness testing.

TCAP is successfully used to achieve a reasonably homogenous ultra-fine grained (UFG) structure for the AZ31 alloy. With increasing number of passes, the initial coarse-grained structure has been transformed gradually into a very fine-grained microstructure with an average grain size of 4.8 μm (sample deformed at 250°C), 2.9 μm (230°C) and 1.7 μm (200°C), respectively.

- The TCAP processing leads to relatively slightly increasing of Vickers hardness values. With increasing number of passes, the standard deviation (inhomogeneity factor) decreases to minimum.

- It was observed that temperature plays a key role in the effectiveness in the grain refinement of magnesium alloy.

References

[1] Vinogradov A and Estrin Y 2018 Analytical and numerical approaches to modelling severe plastic deformation. *Prog. in Mater. Sci.* **95** pp 172-242.
[2] Vinogradov A, Serebryany V N and Dobatkin V 2017 Tailoring microstructure and properties of fine grained magnesium alloys by severe plastic deformation. *Adv. Eng. Mater.* **20** 22 p.
[3] Fadei A, Farahafshan F and Sepahi-Boroujeni S Spiral equal channel angular extrusion (Sp-ECAE) as a modified ECAE process. *Mater. and Design* **113** pp. 361-368.
[4] Máthis K, Kover M, Stráská J, Trojanová Z, Džugan J and Halmešová K Micro-tensile behavior of Mg-Al-Zn alloy processed by equal channel angular pressing (ECAP). *Materials* **11** 11 p.
[5] Trojanová Z, Drozd Z, Lukáč P, Minárik P, Džugan J a Halmešová K Amplitude-dependent internal friction in AZ31 alloy sheets submitted to accumulative roll bonding. *Low Temp. Phys.* **44** pp. 966-972.
[6] Faregh S S H, Hassani A Stress and strain distribution in twist extrusion of AA6063 aluminium alloy. *Inter. J. Mater. Form.* **11** pp. 175-184.
[7] Luo X C, Zhang D T, Zhang W W, Qiu C and Chen D L Tensile properties of AZ61 magnesium alloy produced by multi-pass friction stir processing: Effect of sample orientation. *Mater. Sci. Eng. A* **725** pp 398-405.
[8] Horn T D, Silbermann Ch B, Frint P, Wagner M F-X and Ihlemann J Strain localization during equal-channel angular pressing analyzed by finite element simulations. *Metals* **8** 18 p.
[9] Alsubaie S A, Huang Y and Langdon T G Hardness evolution of AZ80 magnesium alloy processed by HPT at different temperatures. *J. Mater. Res. Technol.* **6** pp 378-384.
[10] Wang X J, Xu D K, Wu R Z, Chen X B, Peng Q M, Jin L, Xin Y C, Zhang Z Q, Liu Y, Chen X H, Chen G, Deng K K and Wang H Y What is going on in magnesium alloys? *J. Mater. Sci. and Techn.* **34** pp 245-247.
[11] Mordike B L and Ebert T Magnesium: Properties-applications-potential. *Mater. Sci. Eng. A* **302** pp 37-45.
[12] Watanabe H, Tsutsui H, Mukai T, Kohzu M, Tanabe S and Higashi K Deformation mechanism in a coarse-grained Mg-Al-Zn alloy at elevated temperatures. *Int. J. Plasticity* **17** pp. 387-397.
[13] Koike J, Kobayashi T, Mukai T, Watanabe H, Suzuki M, Maruyama K and Hihashi K The activity of non-basal slip systems and dynamic recovery at room temperature in fine-grained AZ31B magnesium alloy. *Acta Mater.* **51** pp. 2055-2065.
[14] Al-Samman T and Gottstein G Achieving exceptionally high strength in Mg-3Al-1Zn-0.3Mn extrusions via suppressing intergranular deformation. *Acta Mater.* **160** pp 97-108.
[15] Figuereido R B and Langdon T G Principles of grain refinement in magnesium alloys processed by equal-channel angular pressing. *J. Mater. Sci.* **44** pp 4758-4762.
[16] Stráská J, Janeček M, Čížek J, Stráský J and Hadzima B Microstructure stability of ultra-fine grained magnesium alloy AZ31 processed by extrusion and equal-channel angular proessing (EX-ECAP). *Mater. Char.* **94** pp 69-79.
[17] Kocich R, Greger M, Kursa M, Szurman I and Macháčková A Twist channel angular pressing (TCAP) as a method for increasing the efficiency of SPD. *Mater. Sci. Eng. A* **527** pp 6386-6392.
[18] Hilšer O, Rusz S, Szkandera P, Čížek L, Kraus M, Džugan J and Maziarz W Study of the microstructure, tensile properties and hardness of AZ61 magnesium alloy subjected to severe plastic deformation. *Metals* **8** 16 p.

The influence of relative rolling reduction of sheet metal on the changes of deep drawing forces in multi drawing steps for cylindrical drawpiece without collar made from EN AW-1050A aluminium

T Miłek and K Głowacki

Kielce University of Technology, Faculty of Mechatronics and Mechanical Engineering, al. Tysiąclecia Państwa Polskiego 7, 25-314 Kielce, Poland

E-mail: tmatm@tu.kielce.pl

Abstract. The paper presents experimental results of investigations on deep drawing process of cylindrical drawpieces without collar in four drawing steps (single draw and three redraws). The research involved the use of EN AW-1050A aluminium blanks with diameter D = 66 mm. The material was prepared by rolling with different relative rolling reductions (z = 5%; 10%; 15% and 20%, respectively) at rolling mill *DUO-100* and blanking process. In investigations of deep drawing, experimental force waveforms as the function of displacement for different relative rolling reductions of sheet metal made from aluminium were obtained.

1 Introduction

The indirect application of the drawing force during each draw is a characteristic of all deep drawing processes. The force necessary to cause deformation is not applied directly to the deformation zone, but to the punch which transmits it to the bottom of the cylindrical drawpiece, and then it is transmitted by the cup walls to the deformation zone [1].

Cold deep drawing of cylindrical drawpieces made from aluminium and aluminium alloys and the design of the tooling have been reported in some studies [2-8]. Those covered both experimental and computer modeling investigations. Moshksar et al. [2] examined the effect of the punch and die profile radii on the drawing load and formability of aluminum sheet metal. Based on the experimental data obtained, the maximum punch force for aluminium cup drawing is proposed. Chałupczak et al. [3] discussed the influence of cold work ratio of EN-AW 1050A aluminium on selected parameters in deep drawing. In their studies, they highlights that more than 12 % cold work ratio of tested aluminum sheets improve the quality of drawing. By means of both numerical simulation (LS-DYNA3D) and experiment, using two typical parts in aircraft made from 2B06 aluminium alloy (China brand), Lang et al. [4] proposed optimization of the key parameters of hydromechanical deep drawing process. Dwivedi et al. [5] presented a review on the deep drawing parameters and identified directions for future research of process. They presented result of study formed the successful aluminium alloys (AA 1200) cup by conical die. Tenner et al. [6] numerically and experimentally analyzed dry deep drawing process of AA 6014 aluminium. In paper [7], Hattali et al. showed recent technological advances of sheet metal forming processes. Fischer et al. [8] proposed system and control algorithm which are able to determine the part quality based on the draw-in directly after the first drawing step for kitchen sink.

The paper presents experimental results for deep drawing process of cylindrical drawpieces without collar made from EN-AW-1050A in four drawing steps. The investigations aimed to determine impact of the relative rolling reduction of sheet metal on the changes of forces. The degree of deformation of material was defined as the relative rolling reduction z [9]. The relative rolling reduction z was given by formula [9]:

$$z = \frac{s_0 - s_1}{s_0} 100\% \quad (1)$$

where:

s_0 – thickness of material (sheet) before deformation,
s_1 – thickness of material (sheet) after deformation.

2 Methodology

The material for experimental investigations of deep drawing were EN AW-1050A aluminium blanks whose diameter is 66 mm (D) and thickness are s_0 = 1 mm; 0.95 mm; 0.90 mm; 0.85 mm and 0.80 mm (which corresponded to relative thickness s_0/D = 0.0151; 0.0144; 0.0136; 0.0129 and 0.0121, respectively).

First, a belts of sheet with thickness s_0 = 1 mm, width b_s = 72 mm and length l_s = 300 mm were rolled with different relative rolling reductions (z = 0%; 5%; 10%; 15% and 20%, respectively). The cold longitudinal rolling process was conducted at laboratory rolling mill *DUO-100*. The rolling mill has two mill rolls at diameter equals 102 mm. The relative rolling reduction of sheet z = 0% denotes deformation for starting material without rolling. Next, the blanking process was performed at the *ZD 100* testing machine with the blanking die. *ZD100* testing machine was modified by LABORTECH firm. The machine is compliant with metrological requirements for Class 1 and was calibrated as per PN-EN ISO 7500-1:2005. Finally, the deep drawing process was conducted at a special stand which included the following:

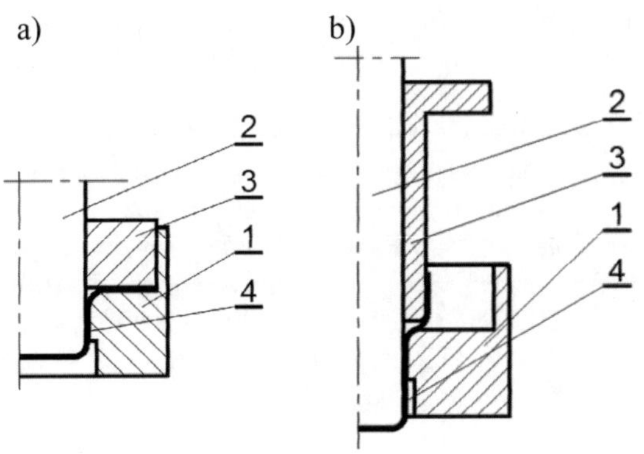

Figure 1. Diagrams of the main parts of deep drawing tools used in investigations: a) tooling for single drawing (where 1- die, 2 – punch, 3 – blank-holder, 4 – drawpiece); b) tooling for redrawing (where 1- die, 2 – punch, 3 – pilot sleeve, 4 – drawpiece).

- A laboratory press-forming die for deep drawing of cylindrical drawpiece equipped with replaceable dies and punches. Figure 1 shows diagram of the main parts of the tool and table 1 presents dimensions of punches and dies used in investigations (where d_s denotes the diameter of the punch, d_m – the diameter of the die, r_s – the radius of the punch and r_m – the radius of the die, respectively);
- LabTest5.20SP1 testing machine (LABORTECH firm) with 20 kN force. Machine was calibrated by PN-EN ISO 7500-1:2005 and meets the metrological requirements for class 0.5;
- A computer stand with Test&Motion software (LABORTECH) to measure forces and displacements.

Figure 1 shows a schematic setup of two drawing steps: first draw and first redraw. The first draw produces cylindrical drawpiece from a flat blank. During each redraw the diameter of the cup is decreased while its height is increased. In experimental investigations, the blank-holder for first drawing was used, because the condition $\frac{s_0}{D} \le 0.02$ from literature [10] has been met.

Table 1. Dimensions of punches and dies used in investigations (mm)

	Drawing	*First redrawing*	*Second redrawing*	*Third redrawing*
$\mathbf{d_s}$	35.5	26.5	21.5	17.5
$\mathbf{d_m}$	38	29	24	20
$\mathbf{r_s}$	5	3	3	3
$\mathbf{r_m}$	5	3	3	3

EN AW-1050A aluminum was selected as the testing material in these investigations due to its good formability and wide applications in industry [11,12]. The mechanical properties of the material were determined by static tensile testing [13]. Tensile test was conducted on LabTest5.20SP1 testing machine. The stress characteristics were given as a function of the relative strain $\Delta l/l_0$ (where Δl denotes the punch displacement and l_0 – initial length of the measuring part of the sample, respectively) for different relative rolling reductions of material (z) after rolling (figure 2).

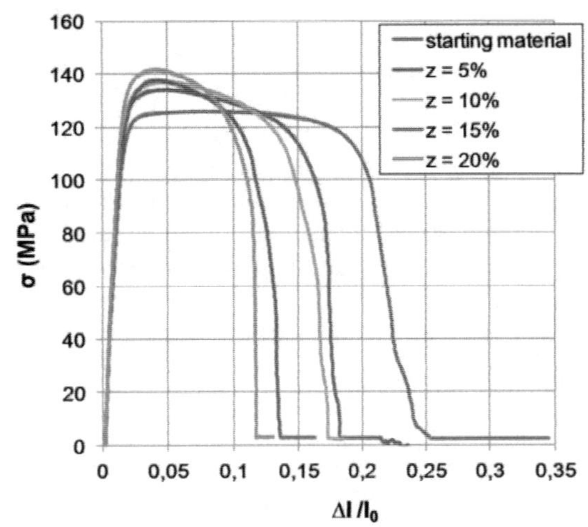

Figure 2. Stress vs. relative dispacement $\Delta l/l_0$ obtained for samples made from EN-AW 1050A aluminium for different relative rolling reductions z in tensile test.

The mechanical properties of material were defined by tensile strength R_m and percentage elongation after fracture A. The effect of the relative rolling reductions z on tensile strength R_m obtained for samples made from EN-AW 1050A aluminium is shown in figure 3. The data shown in figure 3 suggest that the relative rolling reductions have a considerable influence on the values of tensile strength R_m calculated on the basis of the static tensile test. As can be seen from figure 3, an increase in the relative rolling reduction z from 0% to 20% causes an increment in the tensile strength R_m of material. The values of R_m obtained for ratios z = 0% (starting material without rolling), z = 5%, z = 10%, z = 15% and z = 20% were: 126 MPa; 134 MPa; 137 MPa; 138% and 142 MPa, respectively. The relative increase in the tensile strength R_m for the specimens at z = 0% (material before deformations) and z = 20% was approx. 13 %.

The effect of the relative rolling reduction z on percentage elongation after fracture A obtained for samples made from EN-AW 1050A aluminium is shown in figure 4. The values of A obtained for relative rolling reductions z = 0% (starting material without rolling), z = 5%, z = 10%, z = 15% and z = 20% were 18.3 %; 14.8 %; 13.6 %; 10.2 % and 7.4 %, respectively. The percentage elongation after fracture A decreased as the ratio z rose.

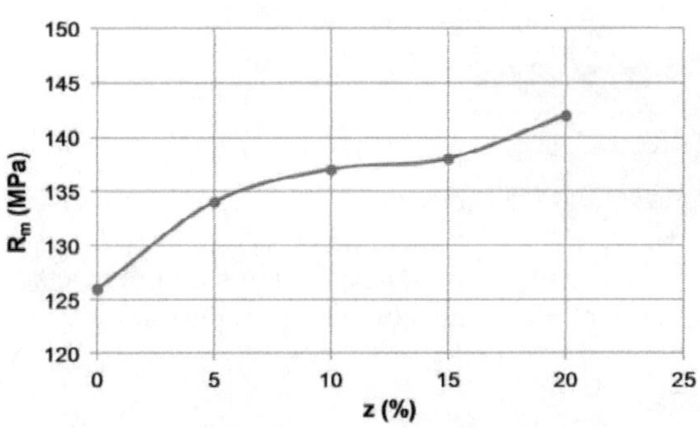

Figure 3. The effect of the relative rolling reduction of material z on tensile strength R_m for samples made from EN-AW 1050A aluminium.

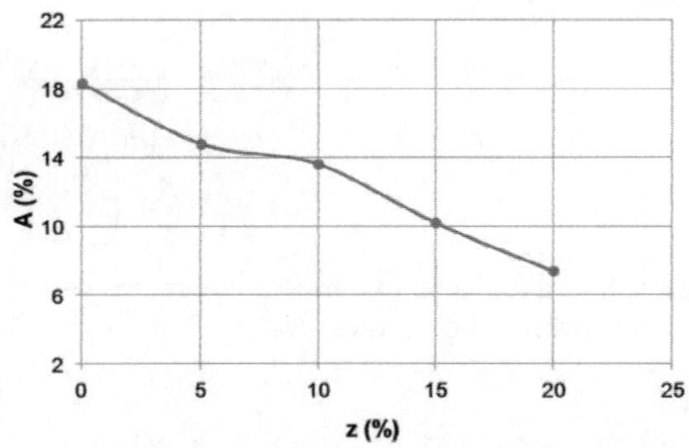

Figure 4. The effect of the relative rolling reduction of material z on percentage elongation after fracture A for samples made from EN-AW 1050A aluminium.

The results indicate that change of mechanical properties (increase of tensile strength R_m and decrease of the percentage elongation after fracture A) of samples made from EN-AW 1050A is related to strain hardening of material after cold deformation (longitudinal rolling).

3 Results and analysis

In the experimental investigations, cylindrical drawpieces in four drawing steps were formed. For the assumed coefficients of deep drawing successful tests were conducted (table 2).

Table 2. Total coefficients of deep drawing assumed in investigations for four drawing steps

m_1	m_2	m_3	m_4
0.57	0.76	0.82	0.83

In table 2, total coefficients of deep drawing (m_n) for each drawing step is given by a formula (2) [10]:

$$m_{1..n} = \frac{d_n}{d_{n-1}} \tag{2}$$

where:

d_n – diameter of drawpiece (in the middle of thickness) i.e., for drawing d_n equals d_1 as a diameter of drawpiece obtained for first step of drawing, (mm)

d_{n-1} – a diameter of initial material i.e., for drawing d_0 equals D as a diameter of the blank and for redrawing d_{n-1} is diameter of drawpiece (in the middle of thickness) from previous drawing step (mm).

Examples of cylindrical drawpieces at $d_4 = 19^{\pm 0.1}$ mm and $h_4/d_4 = 2.54 \div 2.81$ (where d_4 denotes a diameter of drawpiece in the middle of thickness for third redrawing and h_4 denotes minimum height of drawpiece for third redrawing) obtained in last step (at coefficient of third redrawing - $m_4 = 0.83$) of experimental investigations are presented in figure 5.

The investigations produced force profiles in deep drawing of EN-AW 1050A aluminum drawpieces for different relative rolling reductions of sheets (figure 6). Force waveforms as the function of displacement for different drawing steps and the relative rolling reductions of sheet metal made from aluminium were very similar in character.

Figure 5. Examples of cylindrical drawpiece without collar made from EN-AW 1050A aluminium at coefficient of deep drawing $m_4 = 0.83$ with different wall thickness obtained in experiment.

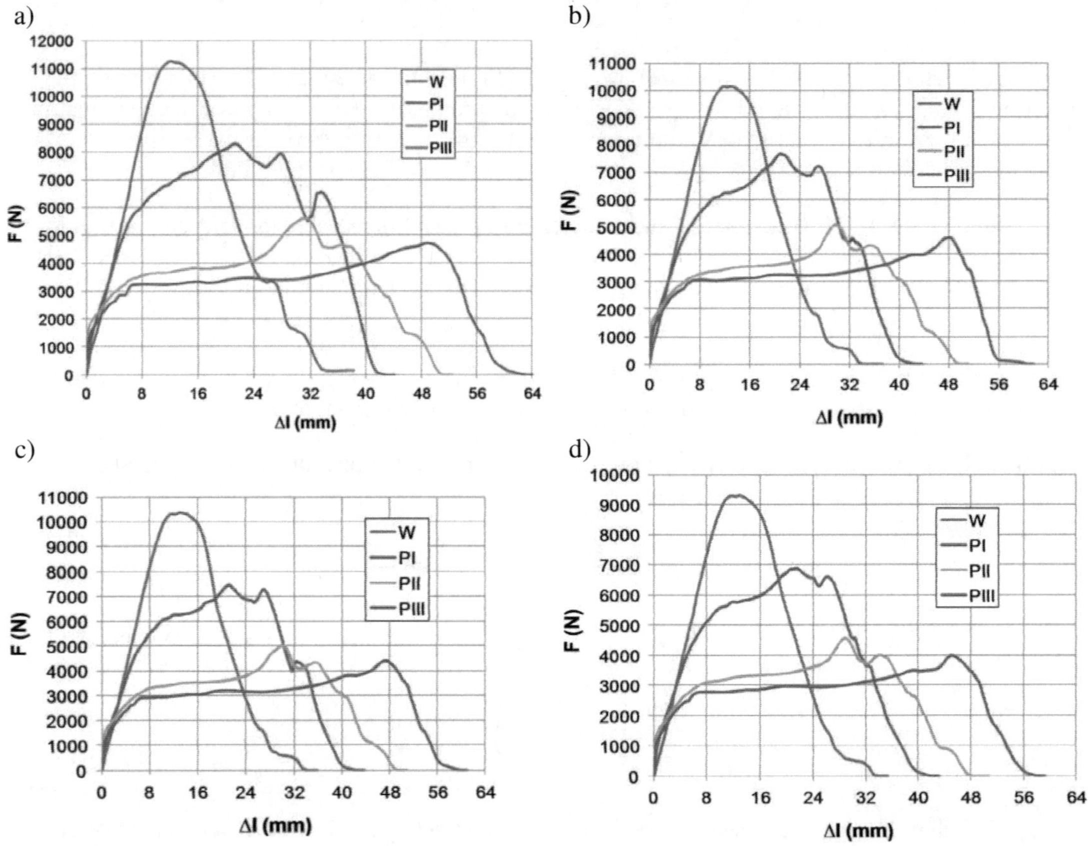

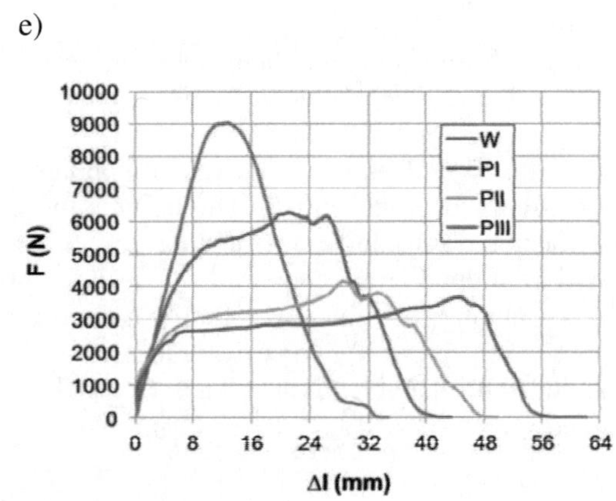

Figure 6. The force (F) vs. displacement (Δl) obtained for different relative rolling reductions of aluminium sheets in deep drawing process of cylindrical drawpieces without collar in four drawing steps (where: W - single drawing, PI – first redrawing, PII – second redrawing, PIII - third redrawing):

a) starting material at $s_0 = 1$ mm
b) $z = 5\%$ at $s_0 = 0.95$ mm,
c) $z = 10\%$ at $s_0 = 0.90$ mm,
d) $z = 15\%$ at $s_0 = 0.85$ mm,
e) $z = 20\%$ at $s_0 = 0.80$ mm.

Investigations have revealed that during deep drawing process, the maximum values of force change depending on the displacement of the punch -Δl (or height of drawpiece) for subsequent drawing steps (figures 6). Displacement of the punch -Δl_1 at which forces obtain maximum values in relation to the total distance -Δl during deep drawing were: $\Delta l_1/\Delta l$ = 0.3÷0.45 for first drawing (*W*), $\Delta l_1/\Delta l$ = 0.5÷0.6 for first redrawing (*PI*) and second redrawing (*PII*) and $\Delta l_1/\Delta l$ = 0.7÷0.8 for third redrawing (*PIII*), respectively. This analysis confirmed the study conclusions regarding changes of deep drawing forces in multi drawing steps for cylindrical drawpieces without collar made from DC01 steel, Cu-ETP copper, CuZn37 brass and EN-AW1050A aluminum for relative rolling reductions z = 0%; 12% and 24% [3].

4 Conclusions

The following conclusions were drawn from investigations into deep drawing of EN-AW 1050A aluminium drawpieces for different relative rolling reductions of sheets:

1. It was possible to conduct deep drawing of aluminium cylindrical drawpieces in four drawing steps without intermediate annealing for assumed total coefficients of deep drawing (m_n), both for small and high relative rolling reduction of sheet. It is confirmed by successfully performed tests for drawpieces with a large slenderness at z = 0% and z = 20%.
2. Force waveforms as the function of displacement for different drawing steps and relative rolling reductions of sheet metal made from aluminium were very similar in character.
3. The relative rolling reduction of sheet have a considerable influence on the values of tensile strength R_m and the percentage elongation after fracture *A*. An increase in the relative rolling reduction *z* from 0% to 20% causes an increment in the tensile strength R_m of material. The percentage elongation after fracture *A* decreased as the ratio *z* rose.

References

[1] Lange K 1985 Handbook of metal forming (McGraw-Hill Book Company)

[2] Moshksar M M and Zamanian A 1997 Optimization of the tool geometry in the deep drawing of aluminium *Journal of Materials Processing Technology* **72** Issue 3 pp 363-70

[3] Chałupczak J and Miłek T 1999 Influence of cold work ratio on selected parameters in deep drawing *Met. Form.* **3** pp 5-10 (in Polish)

[4] Lang L et al. 2009 Investigation into hydromechanical deep drawing of aluminum alloy—Complicated components in aircraft manufacturing *Materials Science and Engineering: A* **499** Issues 1–2 pp 320-24

[5] Dwivedi R and Agnihotri G 2017 Study of deep drawing process parameters *Materials Today: Proceedings* **4** Issue 2 Part A pp 820-26

[6] Tenner J et al. 2017 Numerical and experimental investigation of dry deep drawing of aluminum alloys with conventional and coated tool surfaces *Procedia Engineering* **207** pp 2245-50

[7] Hattalli V L and Srivatsa S R 2018 Sheet metal forming processes – recent technological advances *Materials today: Procedings* **5** pp 2564-74

[8] Fischer P et al. 2018 Experiences with inline feedback control and data acquisition in deep drawing *Procedia Manufacturing* **15** pp 949-54

[9] Marciniak Z 1971 Limit strains in deep drawing process of sheet metals (Warsaw: WNT) chapter 7 pp 189-225 (in Polish)

[10] Golatowski T 1984 Design of deep drawing process and press-forming dies. Selected problems (Warsaw: Warsaw University of Tachnology) p 21 (in Polish)

[11] N-EN 573-3:2014-02. Aluminium and aluminium alloys. Part 3: Chemical composition and types of products

[12] Davis J R 1993 Aluminum and Aluminum Alloys (Ohio: ASM International)

[13] Metallic materials – Tensile testing. Part 1: Method of test at room temperature. ISO 6892-1:2009 (E)

Selected properties of electro-spark deposition on carbon steel using the Alloy 400 electrodes

P. Młynarczyk[1, 3], S. Spadło[1], J. Bartos'[2]

[1] Kielce University of Technology, Faculty of Mechatronics and Mechanical Engineering, Al. Tysiąclecia Państwa Polskiego 7, 25-314 Kielce, Poland

[2] Kielce University of Technology, Faculty of Civil Engineering and Architecture, Al. Tysiąclecia Państwa Polskiego 7, 25-314 Kielce, Poland

[3] E-mail: piotrm@tu.kielce.pl

Abstract. The paper presents a study of surface layers produced by electro-spark deposition (ESD) using Alloy 400 electrode on the carbon steel. In the electro-spark deposition process, the electrode is the anode and the work piece is the cathode. The layers were investigated with metallographic methods. Microscopic examination was carried out to examine the structure of formed layers. Image analysis methods were used to observe the cross-section of the layer. For diffusion observations, ESD analyzes were performed on the cross-section of the produced layer. Scanning electron microscope (SEM) and energy-dispersive X-ray spectroscopy (EDX) analysis was conducted to characterize the microstructure and composition of the coating. The results of investigations showed that there is a possibility of obtaining the satisfying quality superficial layer on the carbon steel using Alloy 400 electrode.

1 Introduction

Electrical Discharge Machining (EDM) is a controlled metal-removal process that is used to remove metal by means of electric spark erosion [1-4]. The EDM is often included in the "non-traditional" or "non-conventional" group of machining methods together with processes such as electrochemical machining (ECM) [5-8], water jet cutting (AWJ) [9], laser cutting [10], micro-welding, etc. The EDM is undeniably classified as a removal process. However, the EDM process can also be used as a surface treatment method; then it is called ESD or EDA – electrical discharge alloying.

Electro-spark deposition (ESD) [11-15] is particularly suitable for repairing small and shallow defects, but it is not suitable for large defects, since the process is slow and the maximum thickness of the coating is about 2 mm. ESD can also be considered as a process to increase the wear and the erosion resistance of small surface. Many applications require the components to have excellent surface performance, such as corrosion resistance and wear resistance as well as the build up of worn or damaged areas on parts. To meet these requirements, some components are built with specific materials, compromising other properties and cost.

The electric motor generate electrode movement, preventing the constant contact between the electrode and substrate. During ESD process, the release of capacitor energy will generate a high temperature plasma arc between the electrode tip and the substrate. The electrode material is ionized by the plasma arc and the molten electrode material is transferred onto the substrate [16-20].

The short duration of the electrical pulse allows for an extremely rapid solidification of the deposited material. The low heat input eliminates deleterious thermal stresses or changes in metallurgical structure, otherwise known as the HAZ, allowing it to be considered for heat-sensitive applications.

The paper is focused on electro-spark deposition (ESD) as a "non-conventional" method of applying layers. The aim of research is to study the effects of ESD process to modify carbon steel using an Alloy 400 as the hot electrode, with the same discharge voltage but different capacitance capacitances.

2 Materials and methods

In the experiment, the coatings were electro-spark deposited with using an Alloy 400 electrode (the anode). Carbon steel was chosen as the base material (the cathode). The diameter of the electrode was 3 mm. The equipment used for electro-spark deposited was specially designed for the experiment. Reaching our current experience, based on previous research, following parameters were assumed to be optimal for ESD technique: voltage U = 600 V, capacitor volume C = 150 μF and 250 μF, frequency of the spark-intervals F = 50 Hz. Schematic of the ESD process is shown in figure 1.

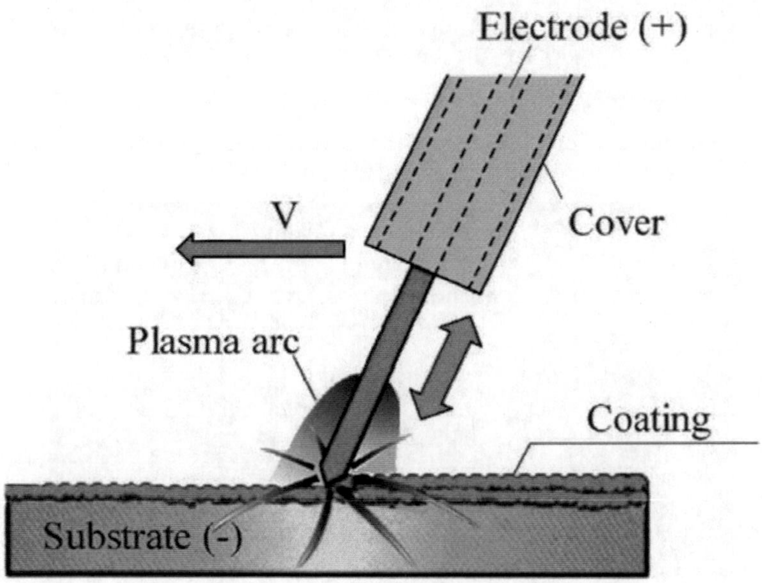

Figure 1. Scheme of the ESD process.

The chemical composition of alloy 400 electrode is shown in table 1. The chemical composition of used materials allows for good adhesion of the applied layer. Based on previous experience, the values of machining parameters have been selected.

Table 1. The chemical composition of alloy 400 electrode

element	Ni	Cu	Fe	C	Mn	S	Si
%	63.0 min	28.0-34.0	2.5 max	0.3 max	2.0 max	0.024 max	0.05 max

Samples before applying the layers were weighed to observe the growth of the material. After ESD treatment, the surface roughness was measured. During the preparation process for the obtained layer were cut to smaller pieces using AWJM machine to avoid thermal influences and mounted in resin (Technovit 5000) and prepared for microscopic analysis with a STRUERS automatic polishing machine. The final polishing was performed using a 0.05 μm Al_2O_3 suspension. After proper polishing the layer was subjected to observation (figure 2,3). The preparation of the specimens for the OM and SEM examinations did not include etching.

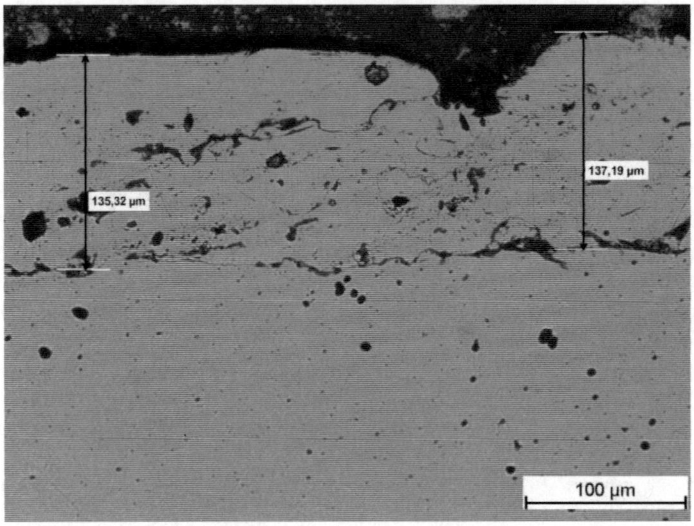

Figure 2. Micrograph image of cross-section of the surface alloyed after the ESD process, 600 V, 150 μF, frequency 50Hz, mag 200x.

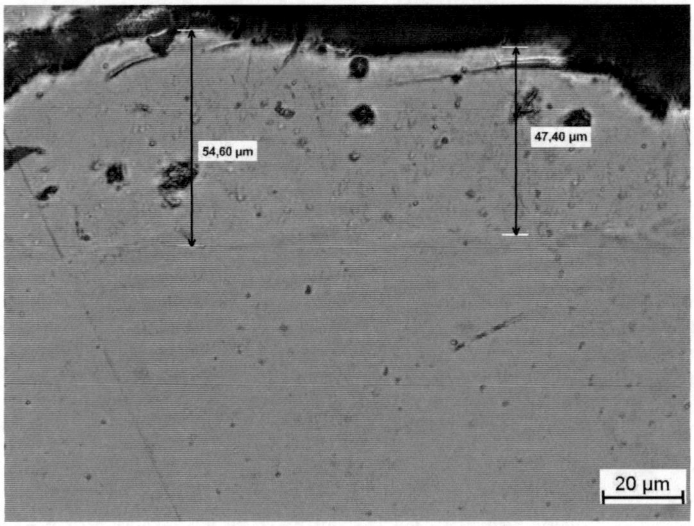

Figure 3. Micrograph image of cross-section of the surface alloyed after the ESD process, 600 V, 250 μF, frequency 50Hz, mag 500x.

Analysis of the geometric structure after ESD were performed using optical profilometer Talysurf CCI Lite, using the TalyMap Platinum software, and Gaussian filter 0.8 mm (ISO 4287). The tests were performed on scanning profilometer which can not only give visual information on the shape of discharge traces but also enables measurement of trace geometry and give information on volume of pits and flashes.

The analysis of the ESD process shows that relationships between the discharge energy and surface roughness. Surface roughness increases when the discharge energy whas higher (higher electric charge on the capacitor), from Sa 3,9 μm (lower electric charge on the capacitor) to 8,7 μm.

3 Test methods

The microstructures of the alloyed layers were analysis was conducted using a JEOL JSM 7100F (field emission system) scanning electron microscope. The chemical composition of the specimens was studied by means of an EDS spectrometer Oxford X-MAX.

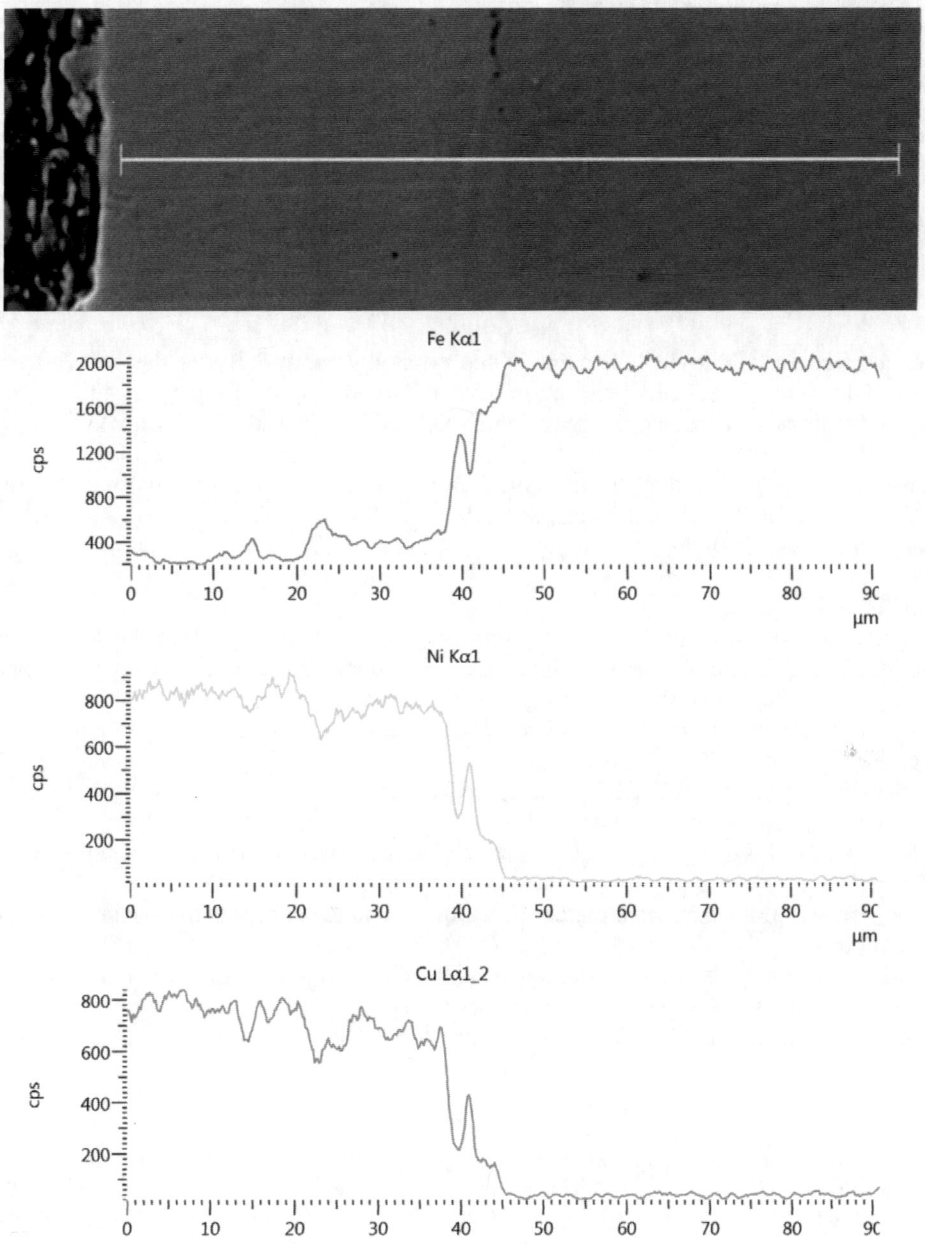

Figure 4. The specimen cross-sections SEM image and linear distribution elements in cross-section of 150 µF covered sample.

Chemical composition tests for surface layer have been carried out using X-ray micro-analysis. Table 2 shows analysis of chemical composition in points, and we can observe the diffusion of the electrode in to material.

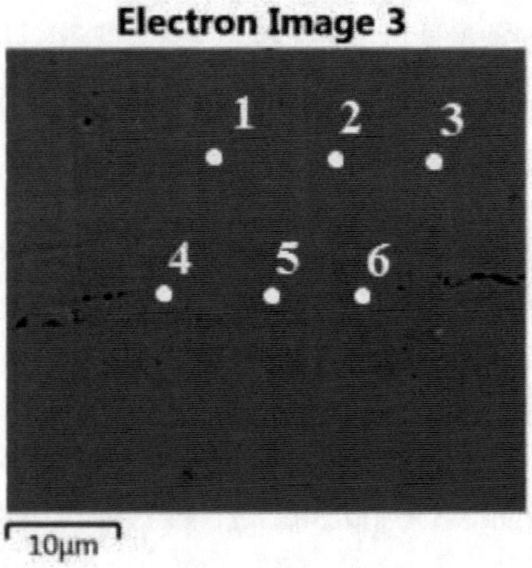

Table 2. Composition (in weight %) of the surface layer

Spectrum Label	Fe	Ni	Cu	Total
1	1.98	64.77	32.23	98.98
2	1.41	65.60	32.95	99.96
3	1.02	65.16	32.85	99.03
4	68.29	21.82	8.66	98.77
5	64.55	24.88	9.80	99.23
6	73.18	19.02	7.15	99.35

Transfer advantages of alloying elements is the possibility of modifying the selected structural elements, and as a result reducing production costs and improving the properties of the material. Microhardness tests were carried out by using INNOVATEST Nexus 4504 with Vickers indenter. Applied load was 0.1961 N for 15 s. The indentations were positioned at regular intervals in the transverse direction across the surface layer, from the diffusion zone up to the base metal. The microhardness of the alloyed layers was $HV_{0.02} = 140$, the fusion zone $HV_{0.02} = 260$ and the base material $HV_{0.02} = 210$.

4 Conclusion

The investigations into electro-spark deposition process using an Alloy 400 electrode shown that:

- increasing of the capacity of the capacitors causes the increase discharge energy in between electrode gape and having an influence over a larger machining area,
- the surface layer subjected to the deposition process contains chemical components of the hot electrode (Ni – 65 %, Cu – 32 %),
- thickness recast layer was from 50 to 130 micrometers with using lower discharge energy (lower electric charge on the capacitor),
- X-ray analysis shows an increase of the hot electrode elements on the obtained layer, and show the diffusion zone 5-8 μm,
- the hardness of the deposited material was lower than the substrate material but the diffusion zone was higher,
- surface roughness increases with increasing the electric charge of the capacitors, but the thickness of the superficial layer was bigger at lower capacity

The results of investigations showed that there is a possibility of obtaining the satisfying quality superficial layer on the carbon steel using an Alloy 400 electrode.

References

[1] Lipiec, P., Machno, M., Skoczypiec, S. The Experimental Research on Electrodischarge Drilling of High Aspect Ratio Holes in Inconel 718, *PROCEEDINGS OF 21ST INTERNATIONAL ESAFORM CONFERENCE ON MATERIAL FORMING (ESAFORM 2018)*, Book Series: AIP Conference Proceedings, Volume: 1960, Article Number: UNSP 100009, Published: 2018

[2] Schumacher, B. After 60 years of EDM the discharge process remains still disputed. In: *Proc. XIV. ISEM*, vol. 149, Edinburgh, UK, 2004, pp. 376-381.

[3] Świercz, R., Oniszczuk-Świercz, D. Experimental investigation of surface layer properties of high thermal conductivity tool steel after electrical discharge machining. *METALS*, vol. 7, Issue: 12, Article Number: 550, Published: DEC 2017

[4] Świercz, R., Oniszczuk-Świercz, D. Influence of electrical discharge pulse energy on the surface integrity of tool steel 1.2713. In *METAL 2017: 26th International Conference on Metallurgy and Materials*, Ostrava: TANGER, 2017, pp. 1450-1455

[5] Davydov, A.D., Kozak, J. Physico-chemical principles of electrochemical discharge machining. *Surface Engineering and Applied Electrochemistry*, vol. 3, (1991), pp. 3-13

[6] Grabowski, M., Skoczypiec, S., Wyszynski, D. A study on microturning with electrochemical assistance of the cutting process, *MICROMACHINES*, Volume: 9, Issue: 7, Article Number: 357, Published: JUL 2018

[7] Paczkowski, T., Sawicki, J. Experimental studies of vibration of the technological device for electrochemical machining of curvilinear surfaces, *ENGINEERING MECHANICS 2017,* Book Series: Engineering Mechanics, pp.: 746-749, Published: 2017

[8] Paczkowski T., Zdrojewski J.: Monitoring and control of the electrochemical machining process under the conditions of a vibrating tool electrode, *JOURNAL OF MATERIALS PROCESSING TECHNOLOGY*, (2017), Volume 244, pp.: 204-214

[9] Krajcarz, D. Comparison Metal Water Jet Cutting with Laser and Plasma Cutting. In *24th DAAAM International Symposium on Intelligent Manufacturing and Automation*, 2013, Book Series: Procedia Engineering, 2014, Vol. 69, pp. 838-843

[10] Mucha, Z., Widlaszewski, J, Kurp, P. et al. Mechanically-assisted laser forming of thin beams. In *Laser Technology 2016: Progress and Applications of Lasers,* Book Series: Proceedings of SPIE, 2016, vol. 10159, Article Number: 101590U

[11] Gangadhar, A., Sunmugam, M.S., Philip, P.K. Surface modification in electrodischarge processing with a powder compact tool electrode. *Wear* 143, (1991), pp. 45–55.

[12] Młynarczyk, P., Krajcarz, D., Bańkowski, D. The selected properties of the micro electrical discharge alloying process using tungsten electrode on aluminum. *12th International Scientific Conference of Young Scientists on Sustainable, Procedia Engineering*, vol. 192, pp. 603-608, Published: 2017

[13] Młynarczyk, P., Spadło, S. The Analysis of the Effects Formation Iron - Tungsten Carbide Layer on Aluminum Alloy by Electrical Discharge Alloying Process. In *METAL 2016: 25th Anniversary International Conference on Metallurgy and Materials*, Ostrava: TANGER, 2016, pp. 1109-1114

[14] Nowicki, B., Dmowska, A., Podolak-Lejtas, A. A new method of investigating crater and flash made by individual discharge using scanning profilometers. *WEAR*, 2011, vol. 271, Issue: 3-4, pp. 634-634

[15] Wang, M.W., Pan, R., Li, S., Zhao, X.J., Zhu, Z., Zhang, L.W. Microstructure of electro-spark deposition WC coating on BT20 titanium alloy. RARE METAL MATERIALS AND ENGINEERING, vol. 43, Issue: 2, pp. 361-363

[16] Dudek, D., Stefańczyk, K., Skowron, E. An analysis of the properties of chromium - based coatings deposited by TIG welding. In *METAL 2017: 26th International Conference on Metallurgy and Materials*, Ostrava: TANGER, 2017. pp: 1246-51

[17] Lee, H. G., Simao, J., Aspinwall, D.K., Dewes, R.C., Voice, W. Electrical discharge surface alloying. *J. Mater. Process Technol.*, 2004, 149(1-3), pp. 334-340

[18] Salmaliyan, M., Ghaeni, F.M., Ebrahimnia, M. Effect of electro spark deposition process parameters on WC-Co coating on H13 steel. *SURFACE & COATINGS TECHNOLOGY*, vol. 321, pp. 81-89, DOI: 10.1016/j.surfcoat.2017.04.040

[19] Spadło, S., Depczyński, W., Młynarczyk, P. Selected properties of high velocity oxy liquid fuel (hvolf) - sprayed nanocrystalline wc-co infralloy(tm) s7412 coatings modified by high energy electric pulse, *METALURGIJA,* vol: 56 Issue: 3-4 pp: 412-414, Published: JUL 2017

[20] Spadło, S., Młynarczyk, P. Analysis of the mechanical interactions of the filament brush electrode on the formation of the surface roughness. In *METAL 2016: 25th Anniversary International Conference on Metallurgy and Materials*, Ostrava: TANGER, 2016, pp. 1169-1174

Peculiarities of microstructure and mechanical behavior of VT8M-1 alloy processed by rotary swaging

***Iu M Modina, A V Polyakov, G S Dyakonov, T V Yakovleva, A G Raab, I P Semenova**

[1]Institute of Physics of Advanced Materials, Ufa State Aviation Technical University, 12 K. Marx street, Ufa, 450008, Russian Federation

E-mail: modina_yulia@mail.ru

Abstract. This research is focused on the study of the relationship between microstructure and mechanical behavior of two-phase Ti alloy VT8M-1 (Ti-5.7Al-3.8Mo-1.2Zr-1.3Sn) in the ultrafine-grained (UFG) state, which is produced by rotary swaging (RS) as compared to equal-channel angular pressing (ECAP). It is shown that both processing techniques enable forming a UFG structure and enhanced mechanical properties in the alloy. Distinguishing features of the UFG structure produced by RS are conditioned by different stress-strained states in the alloy in comparison with ECAP. The relationship between structure, strength and ductility of the alloy processed by RS and ECAP is discussed.

1 Introduction

Two-phase Ti alloys are widely used as materials for aircraft structures and engines due to their high specific strength and corrosion resistance. VT8M-1 alloy is a two-phase Ti alloy, which has higher heat resistance and thermal stability as compared to popular Ti-6Al-4V alloy [1,2]. Parts of gas-turbine engines (GTE) are exploited in high-load states in the conditions of corrosion, erosion and thermal impacts, therefore enhanced requirements are imposed on them in today's conditions of engineering development. It is the combination of high strength, ductility, fatigue resistance, fracture toughness and creep rupture at exploitation temperatures, which determine the performance of structural materials.

Formation of ultrafine-grained states in metals and alloys with the help of different severe plastic deformation (SPD) techniques is one of the most promising approaches to improve strength and ductility; it considerably increases the fatigue endurance limit and other performance properties [3-5].

By comparing the results of grain structure refinement in Ti and its alloys via different SPD techniques [6-10], one should note that the deformation scheme, temperature and applied forces impact greatly the peculiarities of the formed UFG structure and mechanical properties. For example, the advantage of high pressure torsion (HPT) [6] over other SPD techniques is that the use of this treatment allows forming grains of a nanometer size (less than 100 nm), but this technique can be implemented on small-sized samples. Equal-channel angular pressing (ECAP) enables producing bulk billets with larger grains (over 200 nm) [8].

Combination of SPD with conventional plastic deformation techniques [8] and creation of new approaches and treatment techniques [8,11] are promising directions in SPD development and implementation of commercial technologies for producing UFG materials. Rotary swaging (RS) is one of new commercial techniques, which consists in high-frequency deformation of metals with anvils according to the quasi-scheme of all-round compression. Rotary swaging consists in periodic compressions of a rod and reduction its diameter by split dies which simultaneously move along radial, rotational and axial directions relative to the axis of the rod [9]. An important advantage of RS against ECAP is the possibility of upscaling, which is connected with the increase in geometrical sizes (diameter and length) of produced samples. The combination of the two approaches is possible, which will allow creating new technologies for producing UFG materials on a commercial scale.

This work deals with the studies on the peculiarities of UFG structure formation and mechanical behavior of Ti alloy VT8M-1 processed by RS. The aim of this study is to reveal the peculiarities of grain structure refinement in the alloy, as compared to ECAP, and, as a consequence, their impact on its mechanical behavior.

2 Material and experimental procedure

In order to carry out investigations, VT8M-1 (VSMPO-AVISMA Corporation, Verkhnaya Salda, Russia) alloy with the following composition Ti–5.7Al–3.8Mo–1.2Zr–1.3Sn–0.16Fe (wt. %) was used. The β-transus temperature (TPT) in the alloy is 980 ± 5°C. A VT8M-1 rod with a diameter of 70 mm and 1000 mm long was subjected to heat treatment (HT) by heating of the alloy to a temperature of 940°C, subsequent quenching in water, annealing for an hour at 700°C, and cooling in air to obtain a globular-lamellar structure. The ultrafine-grained state was formed via two different techniques. 4 ECAP passes via the B_C route at a temperature of 750°C (e=2.7 with a strain rate of 4 mm s^{-1}) were applied to produce rods with a diameter of 30 mm (Figure 1a), the angle of channels intersection was ψ=120°. A rod with a diameter of 70 mm was successively processed by rotary swaging to diameters of 60, 50, 40, 30 mm with intermediate annealing at 750°C (e ~ 1.7, the strain rate was higher than 300 mm s^{-1}) (Figure 1b).

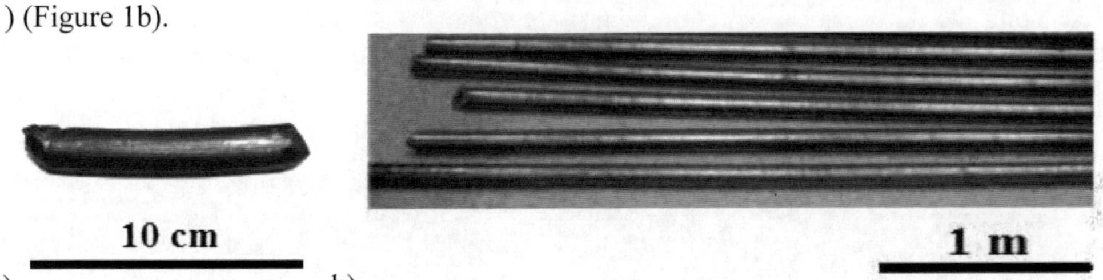

a) b)

Figure 1. VT8M-1 rods: (a) after ECAP; (b) after rotary swaging.

The microstructure characterizations were carried out using scanning electron microscopy (SEM) and transmission electron microscopy (TEM). The TEM micrographs were obtained using a JEOL JEM-2100 microscope operating under an accelerating voltage of 200 kV. Thin foils for TEM were prepared by spark cutting of plates with a thickness of 0.8-1 mm, mechanical thinning with subsequent electrolytic polishing at subzero temperatures. The average grain size was estimated according to the obtained images according to GOST 5639-82 by "Method of counting grain boundary intersections" with the help of the special software "Graine-size". The volume fraction was evaluated with the help of the "Image J" software. The stress-strain curves were recorded using an initial strain rate of 1.0 x 10^{-3} s^{-1} with an Instron universal testing machine. Three samples were tested for each state.

3 Results and their discussion

3.1 Microstructure of VT8M-1 in the as-received state

The initial structure of the VT8M-1 alloy rod (Figure 2a) is a mixed globular-plate structure. The average size of the primary α-phase is 5 μm, and its fraction in the alloy is 50%, the thickness of α lamellas is 0.2 μm.

To ensure the conditions of plastic deformation of hard-to-deform Ti alloys, preliminary heat treatment (HT) was performed with the aim to form a duplex microstructure consisting of a primary α-phase not over 25% and a thin plate α+β structure. This approach to grain structure refinement efficiency enhancement was successfully applied earlier to Ti-6Al-4V [12]. The microstructure after HT (Figure 2b) is a globular platelike structure, which consists of lamellas of the α-phase with a thickness of 0.12 μm, divided by β-phase interlayers and 25% of a globular primary α-phase with an average size of globules of 2.7 μm.

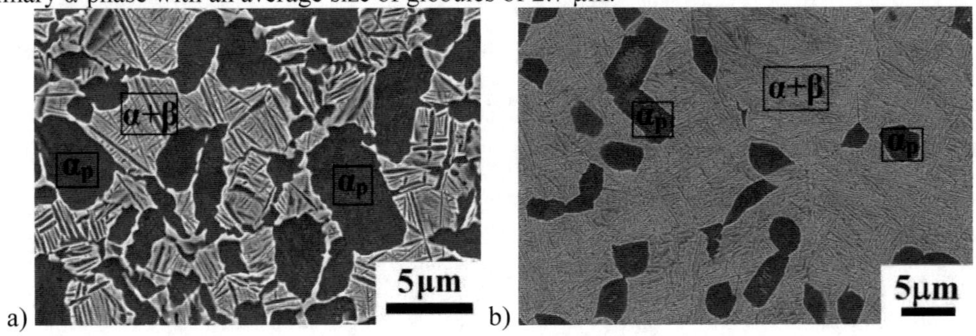

Figure 2. Microstructure of the VT8M-1 alloy: (a) in as-received (b) after HT.

3.2 Microstructure of VT8M-1 after SPD processing

As it follows from the scanning electron microscopy data (Figure 3), after ECAP and RS the plate component of VT8M-1 practically totally transforms into a globular one. The average size of globular grains of the primary α-phase after both deformation techniques is about 3 μm. Part of grains is elongated along the deformation direction. Their length in the structure achieves 7 μm after ECAP (Figure 3a), whereas after RS the shape of grains of the primary α-phase was close to an equi-axed one (Figure 3b).

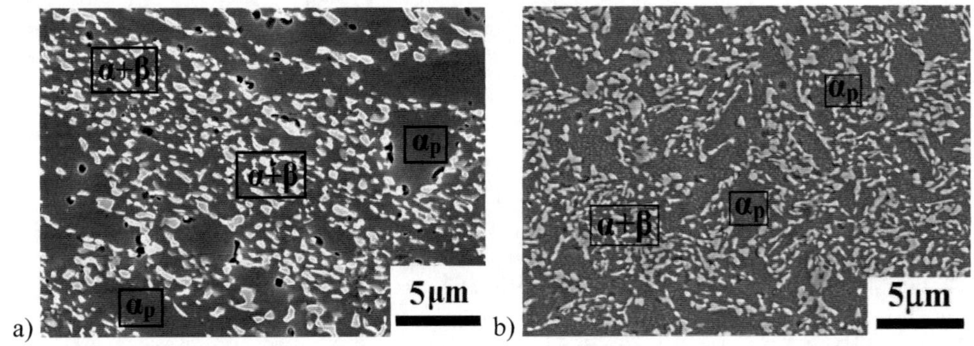

Figure 3. Microstructure of the VT8M-1 alloy: (a) after HT+ECAP (b) after HT+RS.

As it is seen in the images of the structure taken on TEM (Figure 4), the deformation of structure components occurred non-uniformly. In particular after 4 ECAP passes the average size of grains and subgrains after ECAP was about 0.48 μm (Figure 4a). Weakly deformed fragments of α-phase plates were found in the microstructure, which testified to incompleteness of globularization of the plate component [14]. New low-angle dislocation boundaries form inside the primary globular α-phase, separate fragments with a size from 0.5 to 1 μm are observed (Figure 4b). According to the results of TEM, the average size of subgrains in the structure of HT+RS samples as a result of deformation and fragmentation of plates of the α-phase was much lower and varied from 0.2 to 0.3 μm (Figure 4c). The grains of the primary α-phase as well as after ECAP were fragmented by new subgrain boundaries (Figure 4d).

The differences in the UFG structure formed in the alloy after ECAP and RS are evidently connected with different stress-strain states in the billet, which are conditioned by such parameters as rate and accumulated degree of strain. The alloy's microstructure produced at the initial stages of ECAP processing at T=750°C [13] indicates the development of spheroidization of the lamellar constituent. This process was accompanied by the fragmentation of lamellae by shear bands or by the formation of dislocation boundaries [14]. The deformation by RS led to the formation of a more homogeneous refined structure as compared to the ECAP processing (Figures 4a and 4c). The high deformation rate during RS (75 times higher than in the ECAP process), apparently, may promote the development of dynamic recrystallization, which is confirmed by the type of the formed structure (Figures 4a and 4c).

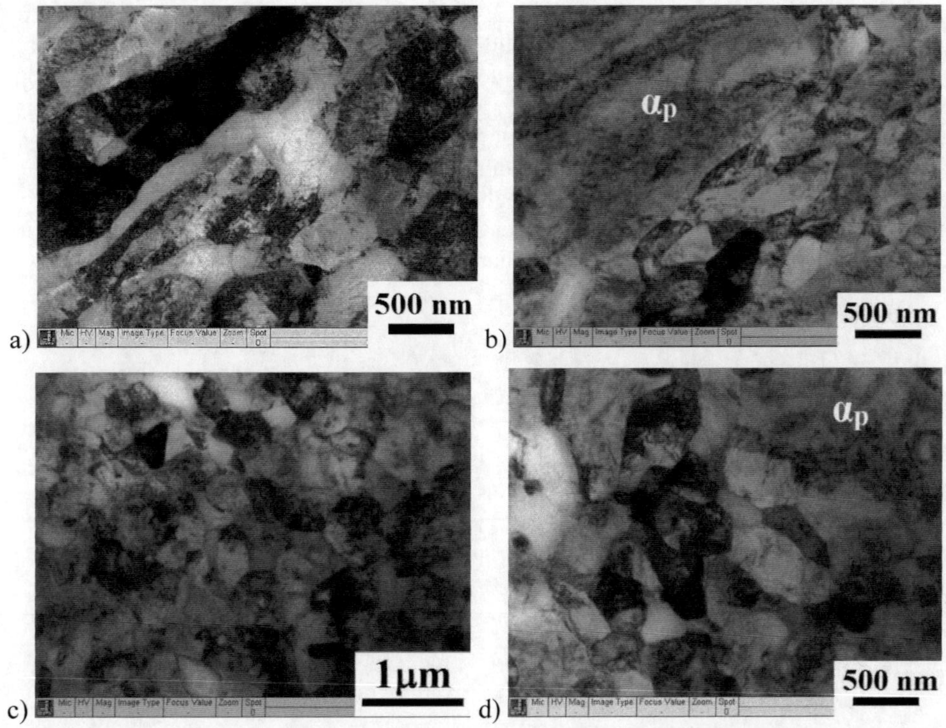

Figure 4. TEM images of the VT8M-1 alloy microstructure: (a) after HT+ECAP; (b) weakly deformed primary α-phase after HT+ECAP; (c) after HT+RS; (d) weakly deformed primary α-phase after HT+RS.

3.3 *Mechanical properties of VT8M-1 after SPD processing*

It is seen from Table 1 that the grain structure refinement in the alloy achieved by RS and ECAP techniques increases the ultimate tensile strength of the VT8M-1 alloy. It should be noted that RS leads to more considerable hardening (from 1030 to 1220 MPa) as compared to ECAP (from 1030 to 1160 MPa) (Table 1). The difference in the UTS values of the investigated states is primarily conditioned by a more homogeneous and fine structure in the alloy after rotary swaging (0.25 and 0.48 μm for RS and ECAP, respectively).

Table 1. Mechanical properties of the VT8M-1 alloy before and after SPD-processing.

State	Average grain size,	Average size of primary α-	UTS, MPa	YS 0.2, MPa	Elong., %	Uniform elongation, %

	D, μm	phase, μm				
As-received	5±1	5±1	1030	960	14.5	5.0
HT+ECAP	0.48±0.08	2.7±0.2	1160	1090	9.9	1.2
HT+RS	0.25±0.05	3.0	1220	1140	11.0	3.8

Moreover, judging by the shape of tensile curves, RS changes the mechanical behavior of the alloy against ECAP (Figure 5). In particular, the alloy after RS demonstrates uniform elongation of about 4% in comparison with ECAP, which was 1.2 %, the total relative elongation was practically the same and was found in the error range (11 and 10%, respectively) (Table 1). The higher value of uniform elongation in the alloy after RS as compared to the ECAP-processed condition can be attributed to a lower dislocation density inside grains/subgrains (as it can be seen from Figure 4c), and also to a more equilibrium state of the boundaries.
As it is known the uniform elongation value during tension indirectly points to the material ability for strain hardening and tendency to strain localization [15]. First of all it is connected with fracture toughness of material, i.e. the ability to resist brittle failure, which is one of the important properties of structural materials applied in aviation [16].

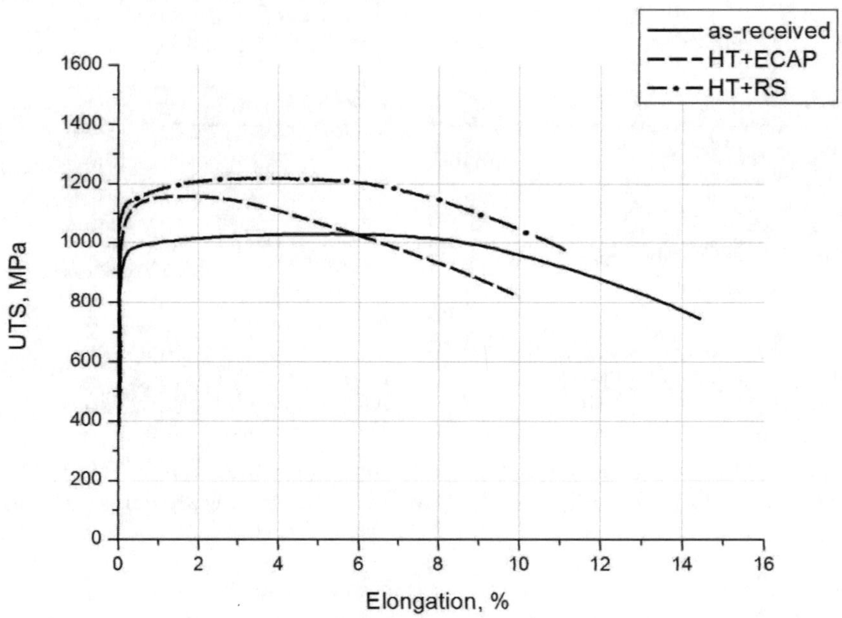

Figure 5. Typical tensile mechanical curves of Ti alloy VT8M-1 in different states.
Thus, the rotary forging ensures enhanced strength and ductile properties in the VT8M-1 alloy against ECAP processing due to formation of a UFG structure with a size of grains/subgrains of less than 0.3 μm. This technique is the most efficient way of producing rods in commercial scales as compared to ECAP. Alongside, the combination of these techniques could be the basis of creating a new technology for producing high-strength rods from Ti alloys with an increased set of exploitation properties, which is the subject of our future studies.

4 Conclusions

The results of the comparative studies of the VT8M-1 alloy processed by ECAP and rotary swaging at a temperature of 750°C enables stating that:

1. The RS treatment allows forming a bulk (α+β) UFG structure with an average size of grains/subgrains of 0.25 μm in the VT8M-1 rod unlike ECAP, as a result of which an average size of the α-phase achieved approximately 0.5 μm. In both cases during deformation primary grains of the α-phase retained their initial sizes (~3μm) with formation of subgrain boundaries. The grain and subgrain size reduction after RS as compared to ECAP is probably connected with a very high strain rate, which suppresses the static recrystallization at high temperatures.
2. The formation of the UFG structure with a finer grain size during RS as compared to ECAP (0.25 and 0.48 μm, respectively) results in a higher ultimate tensile strength (1220 and 1160 MPa, correspondingly).
3. The UFG VT8M-1 alloy after RS demonstrates a higher uniform elongation as compared to the samples produced by ECAP (4 and 1%, respectively), which can be attributed to a lower dislocation density inside grains/subgrains, and also to a more equilibrium state of grain boundaries.

Acknowledgements
The work on the investigation of the ECAP-processed samples was supported by the Russian Science Foundation (grant No. 18-79-10158), and the study of the alloy after rotary swaging was conducted within the scope of the State Assignment (grant No 11.1235.2017).

References
[1] Veiga C, Davim J P and Loureiro A J R 2012 *Rev. Adv. Mater. Sci.* **32** 133
[2] Antashev V G, Nochovnaya N A, Pavlova T V and Ivanov I 2007 (Entsiklopedicheskii Spravochnik, VIAM) http://viam.ru/public/files/2006/2006-204686.pdf (in Russian).
[3] Meyers M A, Mishra A and Benson D J 2006 *Prog Mater Sci.* **51** 427
[4] Valiev R Z, Estrin Y, Horita Z, Langdon T G, Zehetbauer M J and Zhu Y T 2015 *Mater Res. Lett.* **4** 1
[5] Valiev R Z, Estrin Y, Horita Z, Langdon T G, Zehetbauer M J and Zhu Y T 2006 *JOM* **58** (4) 33
[6] Sergueeva A V, Stolyarov V V, Valiev R Z and Mukherjee A K 2001 *Scripta Mater.* **45** 747
[7] Zherebtsov S V, Salishchev G A, Galeyev R M, Valiakhmetov O R, Mironov S Yu and Semiatin S L 2004 *Scripta Mater.* **51** 1147
[8] Semenova I P, Raab G I and Valiev R Z 2014 *Nanotechnologies in Russia* **9** (5-6) 311
[9] Klimova M, Boeva M, Zherebtsov S and Salishchev G 2014 *IOP Conf. Ser.: Mater. Sci. and Eng.* **63** 012070
[10] Belyakov A, Kimura Yu, Adachi Yo and Tsuzaki K 2004 *Mater. Trans.* **45** (9) 2812
[11] Valiev R Z and Langdon T G 2011 *Metal. Mater. Trans.* A. **42** 2942
[12] Semenova I P, Saitova L R, Raab G I, Korshunov A I, Zhu Y T, Lowe T C and Valiev R Z 2006 *Mater. Sci. Forum* **503–504** 757
[13] Semenova I P, Dyakonov G S, Raab G I, Grishina Yu F, Huang Y and Langdon T G 2018 *Adv. Eng. Mater.* **20** 1700813
[14] Semiatin S L, Seetharaman V and Weiss I 1999 *Mater. Sci. Eng. A.* **44** 257
[15] Gupta R K, Mathew C and Ramkumar P 2015 *Fronti.Aerosp.Eng.* **4** (1) 1
[16] Pippan R and Hohenwarter A 2016 *Mater. Res. Lett.* **4** (3) 127

Identification of the elasto-viscoplastic parameters for a thermoplastic polymer by instrumented indentation

A. Mokhtari [1] and N. Tala Ighil [1]

[1] Research Center in Industrial Technologies (CRTI), P.O.Box 64, Cheraga 16014, Algiers, Algeria

a.mokhtari@crti.dz; n.tala-ighil@crti.dz

Abstract. The indentation test is a simple, fast and reliable tool that allows the determination of the materials mechanical properties from experimental load-penetration curves using the inverse computation methods. Through this approach and using the Berkovich indenter, the creep and elasto-plastic properties of the polymers were estimated. Simulations of the elasto-viscoplastic behaviour of the studied polymers under nanoindentation tests were performed. A finite element analysis was carried out to simulate the mechanical behaviour of polymers which can be defined by the Young's modulus E and the parameters (K and n) that describe the materials hardening for large deformations. The obtained functions from the numerical simulations were validated by nanoindentation and compression tests for the studied polymers.

1 Introduction

Thermoplastic polymers have many applications in various industrial sectors due to their good corrosion resistance, low cost, low density, low energy consumption and low thermal conductivity. These materials are easy to manufacture by different techniques such as injection moulding, compression and thermoforming. Objects of different shapes (tubes, profiles, films, fibres and fabrics) can be made with these polymers. Their mechanical properties are relatively good, especially since they can easily be modified by copolymerization, by polymer blending, or especially by adding fibres.
The theory of indentation developed by Oliver and Pharr in 1992 [1] made possible the calculation of the Young's modulus and the hardness of a material from the discharge curve. This technique has the advantage of being simple to implement and allows testing of small volumes of material. However, obtaining the elastic and plastic properties of the materials is difficult due to the mechanical complexity of these tests. Numerous studies combining numerical simulation and experimental measurements are currently emerging and propose to link the force-displacement curves with the intrinsic properties of materials. Recently, Dao et al. [2] have established, from a finite element analysis, dimensional functions connecting the force-displacement curves measured during the Berkovich (or Vickers) indentation to the Young's modulus and to the metal flow stress. Bucaille et al. [3] have shown that by using more acute indenters, precision is increased in the determination of plastic parameters.
The authors [2, 3] defined a representative deformation independent of the indented material and related to the geometry of the indenter. It is difficult to faithfully reproduce the viscoelastic and viscoplastic behaviour of polymers with a reasonable number of parameters [4, 5]. In order to use indentation test measurements to characterize a polymer, it is more prudent to use simple two-parameter models (K and n), such as empirical Norton Hoff's model to simulate behaviour of polymers

during indentation. On the other hand, their calculation technique is relatively heavy and requires for the experimenter the use of finite element software.
In this paper, a simple and practical method is presented to predict the curing exponent of a material from an inverse analysis. Numerical simulations corresponding to 24 combinations of the parameters (E, ν, σ_y, n) were carried out making it possible to simulate the viscoplastic behaviour of polypropylene (*PP*) under an instrumented nanoindentation test. Using the results of dimensional analysis and simulations, two dimensionless functions Π_1 and Π_2 have been established to link the parameters of the charge-discharge indentation curve to the flow stress σ_r of the material. The three mechanical properties (E, σ_y, n) extracted from these functions were examined and validated by nanoindentation and compression tests for the polypropylene polymer PP.

2 Finite element simulation of indentation test

For the finite element simulation of indentation tests, a rigid conical indenter with a half-included tip angle of 70.3° is applied on an infinite half-space. The numerical model used for the nanoindentation test simulation is shown in figure 1. A two-dimensional axisymmetric model included in the commercial finite element code ABAQUS [6], was used. The size of the sample test represents ten times the radius of the contact region, so that the sample can be considered as an infinite half space. The sample was modelled using a 4-node quadrilateral element (CAX4R) with a bilinear approximation. The discretized rectangular domain is composed of a fine mesh of 4900 elements near the contact region and a coarser mesh of 1031 elements further from the contact region. A convergence analysis was performed to assess the dependence of the results with the numerical model. The indenter has been modelled as a rigid body and the contact has been modelled as frictionless [3].

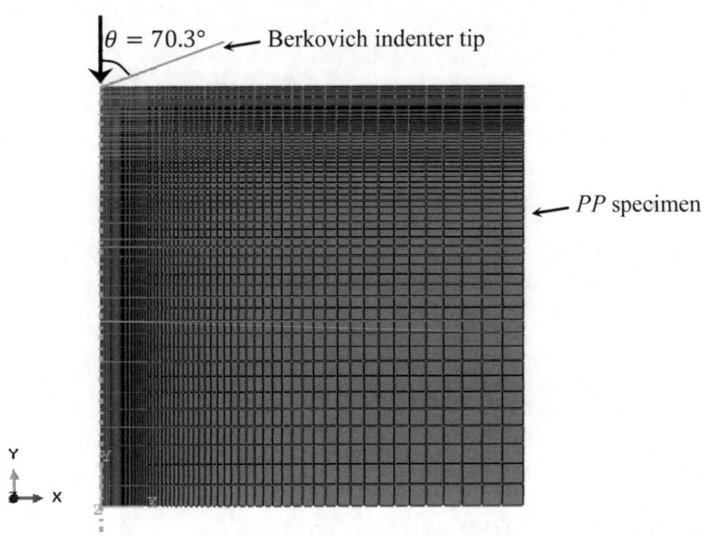

Figure 1. Axisymmetric finite element model to simulate the nanoindentation test with a rigid conical indenter.

The polymer rheology has been modelled considering that the total deformation is the sum of the elastic and plastic deformation. The elastic behaviour is modelled by the Young modulus E and Poisson coefficient ν. The power law strain-stress curve of the engineering metals in the uni-axial compression state (figure 1) can be assumed as [7]:

$$\begin{cases} \sigma = E\varepsilon & for\ \ \varepsilon \leq \frac{\sigma_y}{E} \\ \sigma = K\varepsilon^n & for\ \ \varepsilon \geq \frac{\sigma_y}{E} \end{cases} \tag{1}$$

Where K is strength coefficient and n is the hardening coefficient. The continuity of the curve in equation 1 requires $K=\sigma y(E/\sigma y)$. Therefore, v, E, n, and σ_y are independent parameters describing the viscoplastic behaviour of the material. A wide range of material parameters were considered, each parameter takes 4 values and the combination of these parameters leads to 256 combinations. In order to simplify the problem, some combinations were eliminated by fixing a few parameters (for example, Young's modulus and Poisson's ratio were fixed to 2.189 GPa and 0.4, respectively). Parametric analysis was conducted for 24 cases only and the hardening coefficient n varies from 0 to 0.5. For a given geometry of a sharp indenter (half-angle of a sharp indenter α is fixed), the loading portion of the curve can be expressed by:

$$P = P(h, E_r, \sigma_r, n) \tag{2}$$

Where σ_r is called the representative stress. During the indentation charging phase by a Berkovich indenter, the normal force applied to the indenter is proportional to the h squared penetration (h^2), that for a homogeneous material and in the context of the Continuum mechanics:

$$F = Ch^2 = \sigma_r h^2 \Pi_1\left(\frac{E_r}{\sigma_r}, n\right) \tag{3}$$

Where Π_1 is a dimensionless function. The loading curvature C is independent of the indentation depth, but depends on material properties. It has been verified for each simulation that the equation 3 perfectly represents the load-unload curve. The unloading stiffness dF/dh during the unloading step is related to viscoelastic-plastic properties. For fixed half-angle α of a sharp indenter, the unloading stiffness is given by:

$$S = \left.\frac{dF}{dh}\right|_{h=h_m} = E_r h_m \, \Pi_2\left(\frac{\sigma_r}{E_r}, n, \frac{\eta}{E_r t}\right) \tag{4}$$

Where S is the contact stiffness. The effect of creep $\frac{\eta}{E_r t}$ can be removed by replacing the measured contact stiffness S, with the true elastic contact stiffness S^* [9]. Then, equation 4 is written as:

$$\frac{S^*}{E_r h_m} = \Pi_3\left(\frac{\sigma_r}{E_r}, n\right) \tag{5}$$

This two dimensionless equations Π_1 (load) and Π_3 (unload) [2], connect C and S to Young's modulus E_r and to the stress σ_r corresponding to a representative strain ε_r. This strain is the strain accumulated beyond the elastic limit:

$$\varepsilon = \varepsilon_y + \varepsilon_r \tag{6}$$

Where ε_y is the representative strain at the end of the elastic domain.

3 Results

The results of the nano-indentation tests on the PP with the maximum load of 100 mN and for three hold times of 30s, 300s and 600s are shown in figure 2a. The specimens was loaded to 100mN at a rapid rate ($4 mN.s^{-1}$), held at constant peak load for 30s (test 1), 300s (test 2) and 600s (test 3) and unloaded quickly ($-4 mN.s^{-1}$). During unloading, an elastic return of 2350 nm is recorded which corresponds to 44.75% of the maximum displacement of the indenter. It is observed from figure 2 that the polypropylene has a very high capacity to deform under constant load (creep). The analysis of these curves reveals that the increase of the hold time under maximum load P_{max} is responsible of the increase in the contact depth h_c (figure 2b) and therefore in the projected contact area A_p. This will influence the stiffness S of the polymer and therefore the Young's modulus decreases with the elongation of the creep phase.

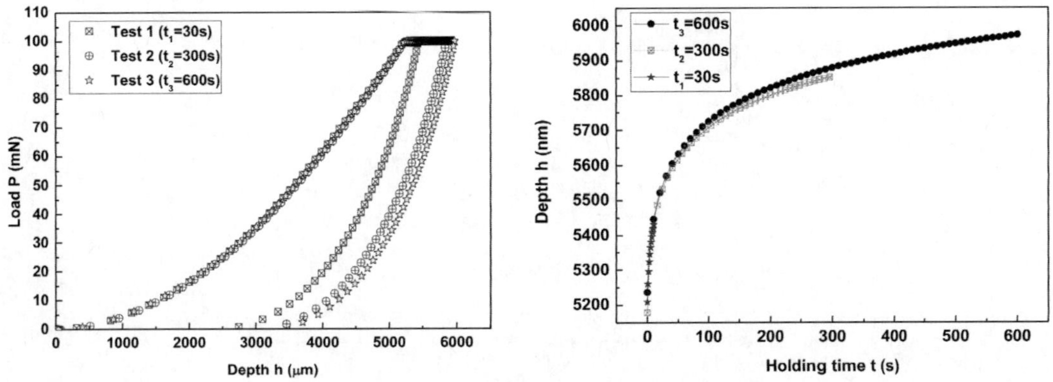

(a) (b)

Figure 2. Evolution of (a) indentation load-displacement (P-h) responses and (b) indentation depth of polypropylene PP for three different holding times.

Consequently, an overestimation of Young's modulus could occur if the load holding time is short. In this case, the Oliver-Pharr method [1] is used to determine the elastic modulus E_r, The Feng and Ngan method [10] can calculated the Young's modulus E_r* from the measured true contact stiffness S*, using the following equation:

$$E_r^*(corrected) = \frac{\sqrt{\pi}}{2}\frac{S^*}{\beta\sqrt{A_p^*}} \quad (7)$$

with $h_c(corrected) = h_{max} - \varepsilon * P_{max} * (\frac{1}{S} + \frac{\dot{h}}{|\dot{P}|})$

where A_p* is the corrected contact area, S* is the true elastic contact stiffness, $\dot{h}$ is the creep rate (*dh/dt*) at the end of holding and $\dot{P}$ is the unloading rate (*dP/dt*) at the beginning of unloading. ε is a constant related to the indenter shape ($\varepsilon = 0.72$ for sharp indenters).

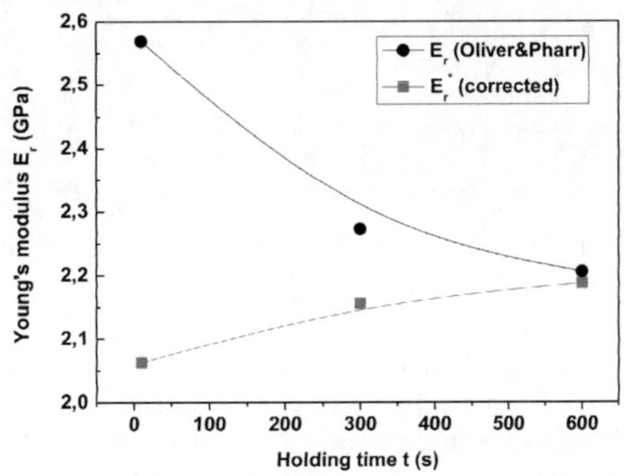

Figure 3. Variation of the corrected Young's modulus with the holding time.

Figure 3 shows a comparison between the Modulus obtained by the Oliver Pharr method and the corrected modulus obtained by the Feng and Ngan method. It can be noted that the effect of the holding time on the reduced module E_r is very important for holding time values less than 230s and less important (small correction) for values exceeding this limit. The Feng and Ngan method permits to stabilize the module for a value of 2.189 GPa.

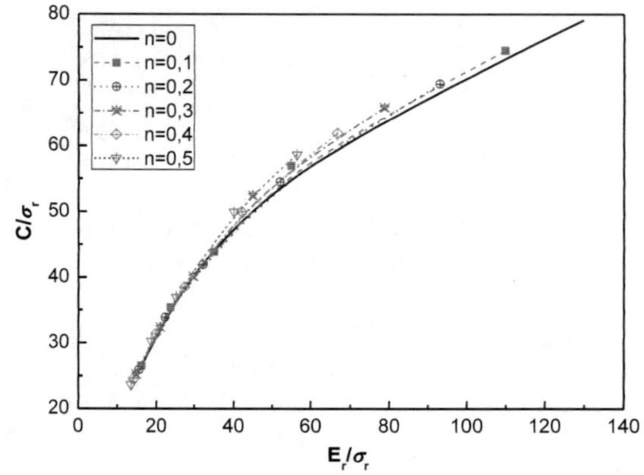

Figure 4. Variation of the ratio C/σ_r with the parameter E_r/σ_r for representative strain ε_r=4.3%.

Figure 4 shows the variation of the ratio C/E_r as a function of the parameter E_r/σ_r. It is noted that for the Berkovich pyramid, the dimensionless function Π_1 is independent of the hardening coefficient n for a representative strain of 4.3%. The points resulting from the results of the numerical computation are then interpolated by a second degree polynomial according to the parameter E_r/σ_r. The dimensionless function Π_1 is easy to build and the equation is given by:

$$\frac{C}{\sigma_r} = 0.66\left(ln\left(\frac{E_r}{\sigma_r}\right)\right)^2 + 20.63\ln\left(\frac{E_r}{\sigma_r}\right) - 36.49 \quad (8)$$

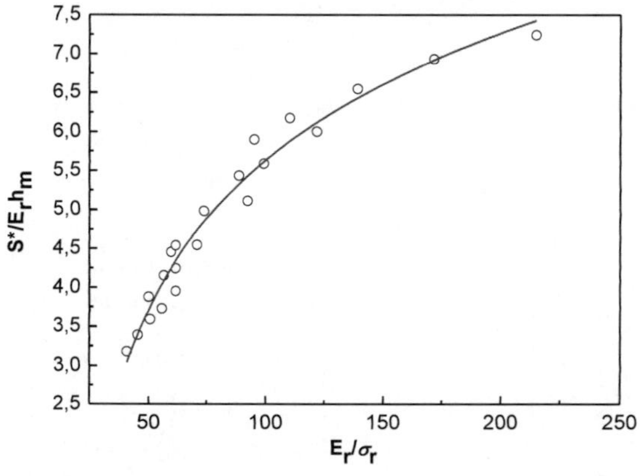

Figure 5. Variation of the ratio S^*/E_rh_m with the parameter E_r/σ_r for representative strain ε_r=4.3%.

In figure 5, the S^*/E_rh_m ratio was plotted against the parameter E_r/σ_r. It can be seen that all the points of the digital data lie approximately on a single curve, which means that the relationship between the S^*/E_rh_m ratio and the E_r/σ_r parameter is insensitive to the hardening coefficient n. The dimensionless function Π_3 obtained from the curve of figure 5 is given by:

$$\frac{S^*}{E_rh_m} = 2.63ln\left(\frac{E_r}{\sigma_r}\right) - 6.52 \quad (9)$$

Polypropylene has a great ability to deform under a constant load (creep). The Young's modulus thus decreases with the extension of the creep phase. As a result, overvaluation of the mechanical properties could occur if the hold time is short. The method of Feng and Ngan [7] allows correcting the Young modulus calculated by the classic method Oliver and Pharr [1].

IOP Conf. Series: Materials Science and Engineering **461** (2019) 012057 doi:10.1088/1757-899X/461/1/012057

4 Conclusion

This study allows the determination of the creep and elasto-plastic properties of a thermoplastic polymer. For this, two dimensionless functions have been employed that permit to link the indentation measurements charge-displacement to the intrinsic properties of the studied material. This method was applied to a thermoplastic polymer (polypropylene), performing nanoindentation tests at a constant loading rate. The Young modulus Er can be determined uniquely from a single force-depth curve of loading using the revised Oliver-Pharr method presented by Tang and Ngan [8]. To verify the uniqueness of the inverse analysis to evaluate σ_r and n, a simple and practical method proposed by Dao et al. [2] was adopted. For a representative strain of 4.3%, the computed stress by indentation is very close to the compression stresses [10]. The possibilities of using multiple indenters of different angles to obtain force-depth relationships should also be investigated both experimentally and theoretically.

References

[1] Oliver W C and Pharr G M 1992, An Improved Technique for Determining Hardness and Elastic- Modulus Using Load and Displacement Sensing Indentation Experiments, Journal of Materials Research, **7**, p 1564–1583

[2] Dao M, Chollacoop N, VanVliet K J, Venkatesh T A and Suresh S 2001, Computational Modeling of the Forward and Reverse Problems in Instrumented Sharp Indentation, Acta Materialia, **49**, p 3899–3918

[3] Bucaille J L, Stauss S, Felder E, and Michler 2003, Determination of plastic properties of metals by instrumented indentation using different sharp indenters, Acta Materialia, **51**, p 1663

[4] Van Melick H, van Dijken H, Den Toonder J, Govaert L, Meijer H 2002, Near-surface mechanical properties of amourphous polymers, Philosophical Magazine A, **82**, p 2093

[5] Bucaille J L, Felder E, Hochstetter G 2002, Identification of the viscoplastic behavior of a polycarbonate based on experiments and numerical modeling of the nanoindentation, Journal of Materials Science, **37**, p 3999

[6] Abaqus Theory Manual, version 6.9-2. Dassault Systèmes, 2009.

[7] Feng G and Ngan A H W 2002, Effects of creep and thermal drift on modulus measurement using depth-sensing indentation, Journal of Materials Research, **17**, p 660

[8] Tang B, Ngan A H W 2003, Accurate measurement of tip-sample contact size during nanoindentation of viscoelastic materials, Journal of Materials Research, **18**, p 1141-1148

[9] Peng G, Feng Y, Wen D, Zhang T (2014) An instrumented indentation method for evaluating the effect of hydrostatic pressure on the yield strength of solid polymers. J Mater Res, **29**, p 2973–2981.

[10] D. Vgenopoulos, J.Sweeney, C.A. Grant, G.P. Thmpson, P.E. Spencer, P. Caton-Rose, P.D. Coates, Nano-indentation analysis of oriented polypropylène: Influence of elastic properties in tension and compression, Polymer, (2018), doi:10.1016/j.polymer.2018.07.080.

Characteristics of AlSi11-AM60 bimetallic joints produced by diffusion bonding

R Mola, M Cieślik and K Odo

Kielce University of Technology, Faculty of Mechatronics and Mechanical Engineering, Al. Tysiąclecia Państwa Polskiego 7, 25-314 Kielce, Poland, UE

E-mail: rmola@tu.kielce.pl

Abstract. This paper demonstrates that the diffusion bonding method can be used to join AlSi11 alloy to AM60 alloy. The microstructure of the bonding zone in such joints was examined using optical and scanning electron microscopy. Its chemical composition was determined through energy dispersive X-ray microanalysis. The experimental results show that the microstructure of the bonding zone is dependent on the parameters of the diffusion bonding process. In an AlSi11-AM60 bimetallic joint fabricated at 420 °C, the bonding zone had a thickness of about 50 μm and it was composed of Mg-Al intermetallic phases with irregularly distributed Mg_2Si phase areas. When the heating temperature was 440 °C, the bonding zone was about 150 μm thick. Mg_2Si was the predominant phase and it was distributed more regularly over the Mg-Al phase matrix. The microhardness measurements of this joint showed that the bonding zone containing hard phases had much higher hardness than either of the alloys joined.

Keywords: magnesium alloy, aluminum alloy, diffusion bonding, microstructure, intermetallic phases, microhardness

1 Introduction

Lightweight alloys based on Al or Mg have been widely used in the automotive, aerospace, machine and defense industries for several decades. The materials are characterized by high strength-to-weight ratio (Rm/ρ). This property makes them desirable wherever it is necessary to reduce the structure weight in order to reduce energy and fuel consumption and, consequently, greenhouse gas emissions [1].

When Al alloys are compared with Mg alloys, differences in the physical and mechanical properties can be noticed. For instance, Al alloys have higher density, lower specific strength, lower machinability, and lower damping ability than Mg alloys. However, some properties of Al alloys are clearly superior to those of Mg alloys, e.g. higher tensile strength at elevated temperature, higher creep resistance, and higher plastic deformability at room temperature. Mg alloys, on the other hand, are characterized by low impact strength as well as much lower resistance to corrosion and abrasion than Al alloys [1].

The demand for lightweight and strong components has led to increased research into the fabrication of bimetals being a combination of these alloys. Because of their unique properties, Mg-Al bimetals are likely to become attractive materials, desirable in various industries. It is expected that Mg-Al bimetallic elements will have high strength while maintaining their low weight. The techniques used to join Mg alloys to Al alloys are: welding [2-4], soldering [5,6], friction welding [7,8], explosive welding [9], diffusion bonding [10-16], metal forming [17-20], and casting [21-23].

Diffusion bonding, also called diffusion welding, is a common method used to join both similar and dissimilar metals or alloys [24-26]. It is often used when traditional methods fail. The principal mechanism in this process is interdiffusion of atoms across the interface. The basic parameters of diffusion welding are heating temperature, contact pressure, and contact time. The process is carried out in a vacuum or inert gas environment.

This article analyzes the structure of the bonding zone in an AlSi11-AM60 bimetallic joint produced by diffusion bonding. The observations were conducted using optical and scanning electron microscopy. The chemical composition of the bonding zone was determined through quantitative EDS analysis. The microhardness of the bonding zone was also measured.

2 Experimental procedure

The chemical composition of the alloys joined is shown in table 1.

Table 1. Chemical composition of the alloys joined.

Alloy	Chemical composition, in wt.%							
	Al	Mg	Si	Fe	Zn	Mn	Cu	Ti
AlSi11	balance	0.15	10.97	0.26	0.01	0.43	0.15	0.07
AM60	5.97	balance	0.031	0.0023	0.035	0.269	0.0018	-

The procedure to prepare the materials for diffusion bonding was as follows. The specimens (20 x 10 x 50 mm) were cut from AlSi11 and AM60 ingots. The mating surfaces were first ground using up to 800 grit SiC abrasive papers, then cleaned with ethyl alcohol and finally dried. The diffusion bonding process was carried out in a Czylok vacuum furnace (figure 1 (a)). The AlSi11 and AM60 pieces were pressed against each other, heated from room temperature to a predetermined temperature for 30 min., kept at that temperature for another 30 min., and cooled back to room temperature. Two heating temperatures were used: 420 °C and 440 °C. During the heating process, the AlSi11 and AM60 specimens were under a pressure of 5 MPa. Figure 1 (b) shows a schematic diagram of the diffusion bonding process.

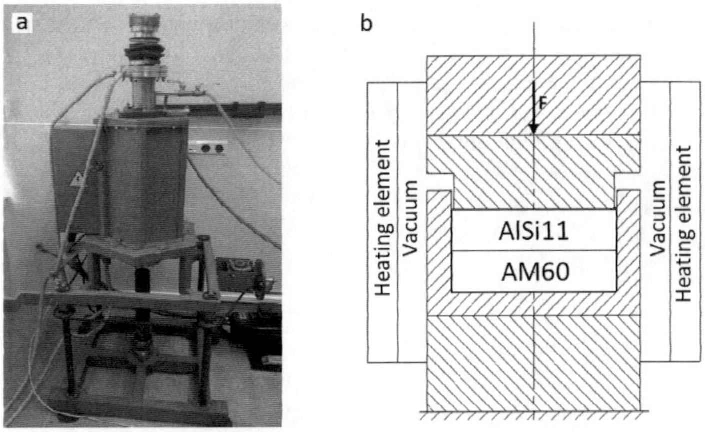

Figure 1. (a) Vacuum heat treatment furnace with an integrated pressing unit, (b) schematic diagram of the formation of a bimetallic joint using the diffusion bonding technique.

Metallographic examinations were conducted to analyze the structure of the bonding zone formed between the two alloys. A Nikon ECLIPSE MA200 optical microscope and a JEOL JSM-5400 scanning electron microscope were used. The chemical composition of the bonding zone was determined by means of an Oxford Instruments Link ISIS-300 energy dispersive spectroscopy (EDS) system coupled with a scanning electron microscope (SEM). The microhardness tests were performed at a load of 100 g using a MATSUZAWA MMT micro Vickers hardness tester.

3 Results and discussion

The microstructure of the bonding zone in the AlSi11-AM60 bimetallic joints fabricated at two different temperatures is shown in figure 2. Figure 2 (a) shows the AlSi11-AM60 joint produced at 420 °C. As can be seen, the bonding zone is thin (about 50 μm) with a lighter matrix and some darker areas. The darker areas are adjacent to the Si crystals occurring in the microstructure of the AlSi11 alloy close to the bonding zone. As observed from figure 2 (b), applying a temperature of 440 °C while keeping the other parameters unchanged resulted in the formation of an AlSi11-AM60 bimetallic joint with a much thicker bonding zone (about 150 μm). The area closer to the AlSi11 alloy was lighter and thin (about 40 μm thick) while the area adjacent to the AM60 alloy was darker and approximately 110 μm in thickness.

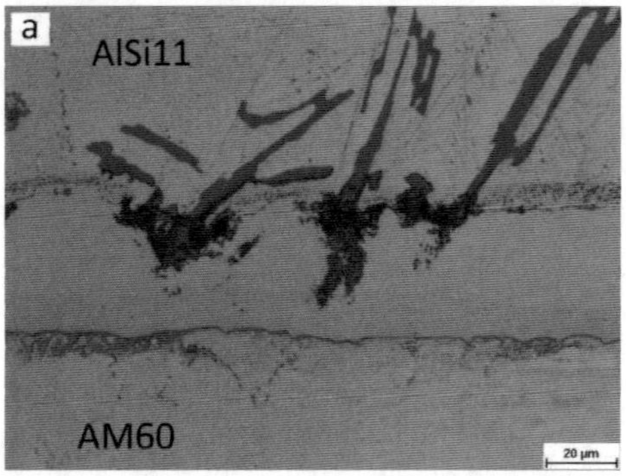

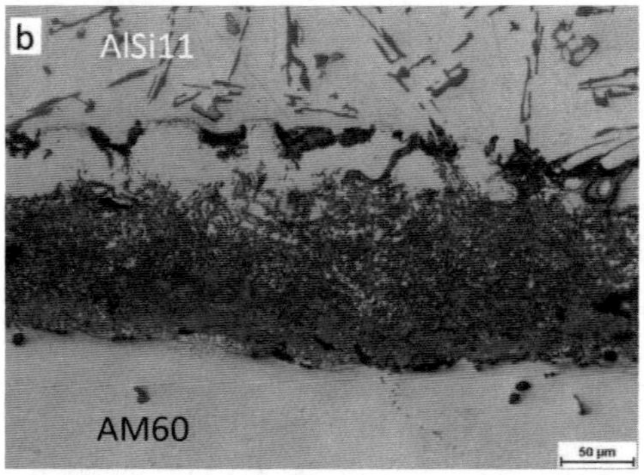

Figure 2. OM images of the microstructure of the bonding zone in the AlSi11-AM60 joints fabricated by diffusion bonding at: (a) 420 °C, (b) 440 °C.

The structure of the bonding zone in the AlSi11-AM60 joints was studied using a scanning electron microscope. Figure 3 shows the bonding zone obtained at 420 °C. The EDS results for points 1-3 (figure 3) are given in table 2. The chemical composition of the lighter area of the bonding zone closer to the AlSi11 (analysis at point 1) indicates the presence of an Al_3Mg_2 intermetallic phase. The Mg to Al ratio in the lighter area of the bonding zone closer to the AM60 alloy (analysis at point 2) suggests the $Mg_{17}Al_{12}$ intermetallic phase. The irregularly distributed darker areas adjacent to the Si primary crystals in the AlSi11 alloy have a high content of Si. The EDS data suggest the Mg_2Si phase (analysis at point 3).

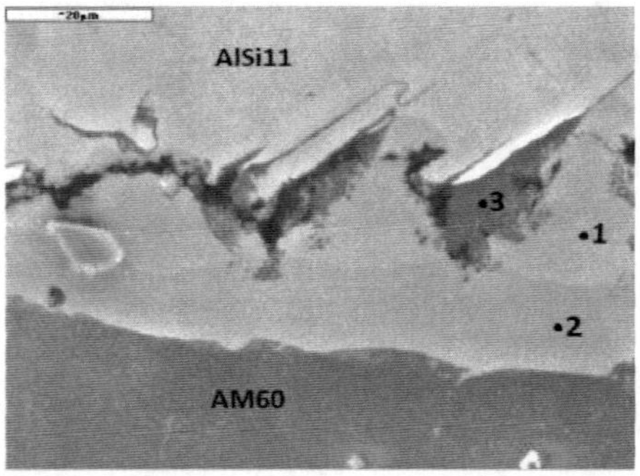

Table 2. Results of the quantitative EDS analysis corresponding to the points in figure 3.

Point	Mg		Al		Si	
	% wt.	% at.	% wt.	% at.	% wt.	% at.
1	38.42	40.92	61.58	59.08	-	-
2	58.97	61.47	41.03	38.53	-	-
3	64.02	67.27	-	-	35.98	32.73

Figure 3. SEM image of the microstructure of the AlSi11-AM60 joint fabricated by diffusion bonding at 420 °C.

Figure 4 illustrates details of the microstructure of the joint fabricated at the higher temperature (440 °C). The EDS results obtained at points 1-6 (figure 4) are provided in table 3. Figure 4 (a) shows the microstructure of the bonding zone close to the AlSi11 alloy. As can be seen, there are gray phase particles in the lighter matrix. The elemental analysis of the lighter matrix close to the AlSi11 alloy (point 1) revealed the presence of the Al_3Mg_2 intermetallic phase. The chemical composition of the lighter matrix closer to the central area of the bonding zone (point 2) is different; it is the $Mg_{17}Al_{12}$ intermetallic phase. From the elemental analysis of the gray particles in the lighter matrix (point 3) it is clear that they are the Mg_2Si phase. Figure 4 (b) illustrates the microstructure of the bonding zone adjacent to the AM60 alloy. This area has a more uniform structure. The gray phase particles (point 4), which are regularly distributed over the lighter matrix, are the Mg_2Si phase. The EDS data obtained for the lighter area of the bonding zone adjacent to the AM60 alloy (point 5) suggest the $Mg_{17}Al_{12}$ intermetallic phase. As can be seen, light particles are present in the bonding zone both close to the AlSi11 and the AM60. Those occurring closer to the AM60 alloy (point 6) indicate a multi-component phase consisting mainly of Mn and Al but also Si and Mg. The EDS analysis shows that the white multi-component phase observed close to the AlSi11 alloy contains also Fe.

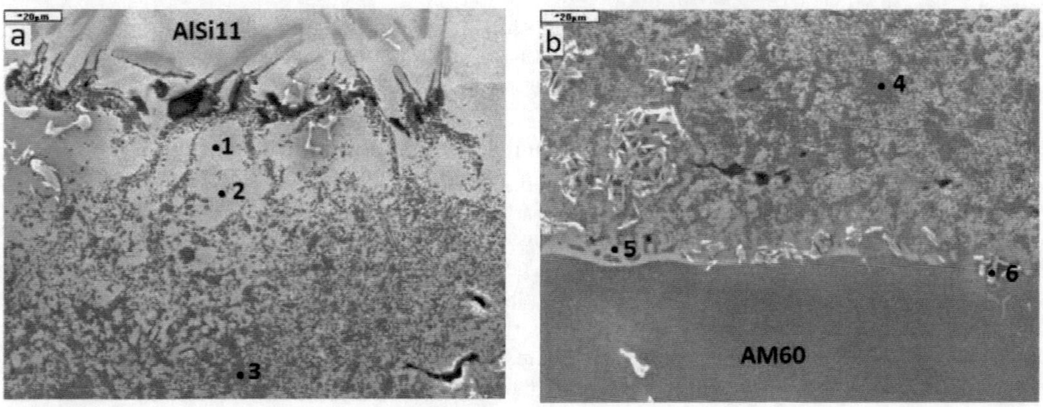

Figure 4. SEM images of the microstructure of the AlSi11-AM60 joint fabricated by diffusion bonding at 440 °C: (a) area close to the AlSi11, (b) area close to the AM60.

Table 3. Results of the quantitative EDS analysis corresponding to the points marked in figure 4 (a) and (b).

Point	Mg		Al		Si		Mn	
	% wt.	% at.	% wt.	% at.	% wt.	% at.	% wt.	% at.
1	37.68	40.16	62.32	59.84	-	-	-	-
2	57.51	60.03	42.49	39.97	-	-	-	-
3	61.18	64.43	4.9	4.65	33.92	30.92	-	-
4	63.42	66.7	-	-	36.58	33.3	-	-
5	65.21	67.53	34.79	32.47	-	-	-	-
6	3.14	5.05	33.9	49.18	1.34	1.87	61.62	43.9

Figure 5 shows indentations left after Vickers tests for the joint fabricated at 440 °C. The microhardness of the AlSi11 varied from 60 to 63 HV, while that of the AM60 ranged from 63 to 65 HV. The microhardness of the bonding zone in the lighter area closer to the Al alloy, where the predominant phase was Mg-Al, reached 227-247 HV. The microhardness of the darker area observed closer to the AM60 alloy was 287-308 HV; the higher values were due to the presence of the hard Mg_2Si phase.

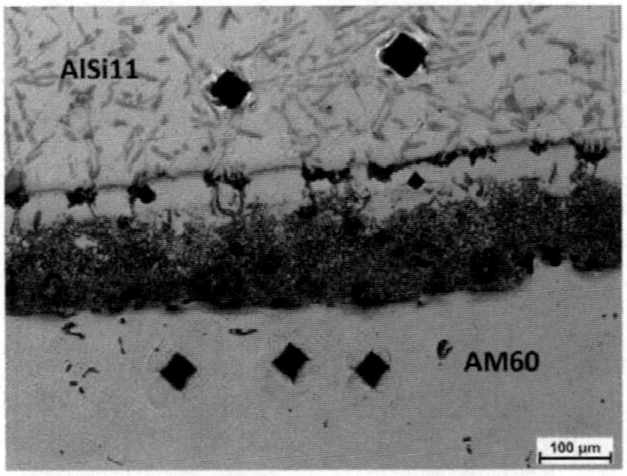

Figure 5. OM image of the joint fabricated at 440 °C with indentations left after the Vickers microhardness tests.

4 Conclusions

The experimental results showed that the microstructure and thickness of the bonding zone in the AlSi11-AM60 joints under study were dependent on the heating temperature. The joint fabricated at 420 °C had a thickness of about 50 µm and it was composed of Mg-Al intermetallic phases with irregularly distributed Mg_2Si phase areas. When the heating temperature was 440 °C, the bonding zone had a thickness of about 150 µm and two main areas could be distinguished. The lighter area closer to the AlSi11 was about 40 µm thick; it consisted of Al_3Mg_2 and $Mg_{17}Al_{12}$ intermetallic phases and irregularly distributed Mg_2Si phase particles. The darker area adjacent to the AM60 was much thicker (about 110 µm) and the predominant phase was Mg_2Si. The microhardness tests conducted for that joint revealed that the hardness of the bonding zone was several times higher than that of either of the alloys joined.

References

[1] Dziadoń A and Mola R 2013 Magnesium-trends of development of mechanical properties *Metal Forming* **24** 253-77
[2] Liu L, Liu X and Liu S 2006 Microstructure of laser-TIG hybrid welds of dissimilar Mg alloy and Al alloy with Ce as interlayer *Scripta Materialia* **55** 383-6
[3] Borrisutthekul R, Miyashita Y and Mutoh Y 2005 Dissimilar material laser welding between magnesium alloy AZ31B and aluminum alloy A5052-0 *Science and Technology of Advanced Materials* **6** 199-204
[4] Liu L, Ren D and Liu F 2014 A Review of dissimilar welding techniques for magnesium alloys to aluminum alloys *Materials* **5** 3735-7
[5] Wang Z, Wang H and Liu L 2012 Study on low temperature brazing of magnesium alloy to aluminum alloy using Sn–xZn solders *Materials & Design* **39** 14-9
[6] Liu L M, Tan J H, Zhao L M and Liu X J 2008 The relationship between microstructure and properties of Mg/Al brazed joints using Zn filler metal *Materials Characterization* **56** 479-83
[7] Somasekharan A C and Murr L E 2004 Microstructures in friction-stir welded dissimilar magnesium alloys and magnesium alloys to 6061-T6 aluminum alloy *Materials Characterization* **52** 49–64
[8] Sato Y S, Park S H C, Michiuchi M and Kokawa H 2004 Constitutional liquation during dissimilar friction stir welding of Al and Mg alloys *Scripta Materialia* **9** 1233-6
[9] Mróz S, Stradomski G, Dyja H and Galka A 2015 Using the explosive cladding method for production of Mg-Al bimetallic bars *Archives of Civil and Mechanical Engineering* **15** 317-23
[10] Zhao L M and Zhang Z D 2008 Effect of Zn alloy interlayer on interface microstructure and strength of diffusion-bonded Mg–Al joints *Scripta Materialia* **58** 283-6

[11] Liu P, Li Y, Geng H and Wang J 2005 A study of phase constitution near the interface of Mg/Al vacuum diffusion bonding *Materials Letters* **59** 2001-5

[12] Somekawa H, Hosokawa H, Watanabe H and Higashi K 2003 Diffusion bonding in superplastic magnesium alloys *Materials Science and Engineering* **339** 328-33

[13] Li Y, Liu P, Wang J and Ma H 2007 XRD and SEM analysis near the diffusion bonding interface of Mg/Al dissimilar materials *Vacuum* **82** 15-9

[14] Liu P, Li Y, Geng H and Wang J 2006 Investigation of interfacial structure of Mg/Al vacuum diffusion-bonded joint *Vacuum* **80** 395-9

[15] Zhang J, Luo G, Wang Y, Shen and Q Zhang L 2012 An investigation on diffusion bonding of aluminum and magnesium using a Ni interlayer *Materials Letters* **83** 189-91

[16] Dziadoń A, Mola R and Błaż L 2011 Formation of layered Mg-eutectic composite using diffusional process at the Mg-Al interface *Archives of Metallurgy and Materials* **56** 677-84

[17] Zhang X P, Castagne S, Yang T H, Gu C F and Wang J T 2011 Entrance analysis of 7075 Al/Mg–Gd–Y–Zr/7075 Al laminated composite prepared by hot rolling and its mechanical properties *Materials & Design* **32** 1152-8

[18] Liu X B, Chen R S and Han E H 2009 Preliminary investigation on the Mg-Al-Zn/Al laminated composite fabricated by equal channel angular extrusion *Journal of Materials Processing Technology* **209** 4675-81

[19] Zhang X P, Yang T H, Castagne S and Wang J T 2011 Microstructure; bonding strength and thickness ratio of Al/Mg/Al alloy laminated composites prepared by hot rolling *Materials Science and Engineering* **528** 1954-60

[20] Binotsch C, Nickel D, Feuerhack A and Awiszus B 2014 Forging of Al-Mg compounds and characterization of interface *Procedia Engineering* **81** 540-5

[21] Mola R, Bucki T, Dziadoń A 2017 Microstructure of the bonding zone between AZ91 and AlSi17 formed by compound casting *Archives of Foundry Engineering* **17** 202-6

[22] Hajjari E, Divandari M, Razavi S H, Emami S M, Homma T and Kamado S 2011 Dissimilar joining of Al/Mg light metals by compound casting process *Journal of Materials Science* **46** 6491-9

[23] Hajjari E, Divandari M, Razavi S H, Emami S M, Homma T and Kamado S 2012 Characteristics and mechanical properties of Al 413/Mg joint in compound casting process *Metallurgical and Materials Transactions A* **43** 4667-77

[24] Konieczny M and Szwed B 2016 Effect of cooling rate on titanium and stainless steel joints performed with copper interlayer METAL 2016: 25rd International Conference on Metallurgy and Materials (Brno) 1456-61

[25] Shao X, Guo X, Han Y, Lu W, Qin J and Zhang D 2015 Characterization of the diffusion bonding behavior of pure Ti and Ni with different surface roughness during hot pressing *Materials & Design* **65** 1001-10

[26] Aleman B, Gutierrez L and Urcola J J 2013 Interface microstructures in diffusion bonding of titanium alloys to stainless and low alloy steels *Material Science and Technology* **9** 633-41

Study hydrogen embrittlement and determination of E110 fuel cladding mechanical properties by ring compression testing

H K Namburi [1], L Ottazzi [1, 2], M Chocholousek [1], G Lomonaco [2], P Gavelova [1], J Krejci [3]

[1] Centrum výzkumu Řež s. r. o., Husinec-Řež, 252 63 Czech Republic

[2] GeNERG-DIME/TEC, University of Genova, via all'opera Pia 15/A, 16145 Genova, Italy

[3] UJP PRAHA a.s., Praha – Zbraslav, Czech Republice, EU

E-mail: nab@cvrez.cz

Abstract. Zirconium based alloys are commonly used as material for fuel claddings in the light water reactors. Claddings act as first metallic barriers against loss of fission products during the nuclear power plant operation, intermittent storage or final dry storage. The integrity of claddings is always critical issue during reactor operation and during storage of spent fuel. In this work, ring compression testing method developed was applied to study hydrogen embrittlement, to evaluate the stress–strain behaviour and hoop fracture properties of E110 (Zr-based) fuel claddings. The results show that the collapse load and the tensile strength values depend strongly on hydrogen concentration. In particular, tensile strength experiment data shows significant change in its trend after reaching the maximum hydrogen solubility limit.

1 Introduction

Zirconium based fuel claddings serve as barrier between the pellets and the fuel rod environment, avoiding the release of fuel or fission products during service into the reactor core or cooling system or after service into storage containments. Zirconium and its alloys have a significant property, a very low neutron absorption cross section, 30 times less than iron. In addition, zirconium alloys also possess good corrosion resistance, mechanical strength and are relatively resistant to radiation damage [1-3]. During the reactor operation, claddings accumulate hydrogen due to oxidation and are subjected to degradation resulting from their exposure to high temperature, high pressure and irradiation. When the hydrogen content in zirconium alloys exceeds the terminal solid solubility limit, diffused hydrogen atoms react with the zirconium matrix and formation of zirconium hydrides occurs [3]. The amount of hydrogen increases with the increase in the fuel burn-up and reactor operation time reactor cycles [4]. Possible degradation mechanisms of fuel claddings include creep, hydrogen embrittlement (reduction in ductility) are reported in detail in and delayed hydride cracking [5-12] etc. Under the influence of internal pressure, temperature, hydrogen content claddings may undergo failure. Therefore the study on mechanical integrity of the fuel cladding tubes is very important.

IOP Conf. Series: Materials Science and Engineering **461** (2019) 012059 doi:10.1088/1757-899X/461/1/012059

When the hydrogen solubility limit in zirconium based alloys exceeds, the excess amount of it results in the formation of zirconium hydride as precipitates. These precipitates have been shown to be less ductile than the surrounding zirconium alloy matrix, so can have deleterious effects on the mechanical properties. It has been demonstrated that these hydrides may embrittle the cladding and reduces its mechanical strength. The resulting embrittlement depends on the concentration and geometrical distribution of the hydrides [16-20]. Current study is focused on evaluating the mechanical properties in the hoop direction of the zirconium based fuel cladding (E110) with different hydrogen contents (as received, 120, 283, 1500 wt. ppm) using the Ring Compression Testing technique at RT-150-400 °C. And furthermore, to determine the best practices for testing irradiated claddings in hot-cells to perform cladding embrittlement studies.

2 Materials and Methods

2.1. Materials and Methods

Specimen examined in this study was fabricated from E110 alloy of tubular section, 10 mm long with outer diameter of ~ 9.1 mm and wall thickness of ~ 570 μm provided by UJP Prague. The chemical composition of E110 cladding tube was 99 % Zr and 1 % Nb with minor impurities of Fe and O. Metallography was performed before mechanical testing, to determine hydrides distribution in pre-hydrogenated specimens. Ring compression testing is performed in the present study in order to obtain the mechanical properties of tubular samples in hoop direction with various hydrogen content at different hydrogen content as in Table 1 using Zwick Z250 test machine. The thermo-mechanical cycle is shown in Figure 1 and Zwick machine with flat compression holders are shown in Figure 2. Test was performed in displacement control mode at 0.5 mm/min.

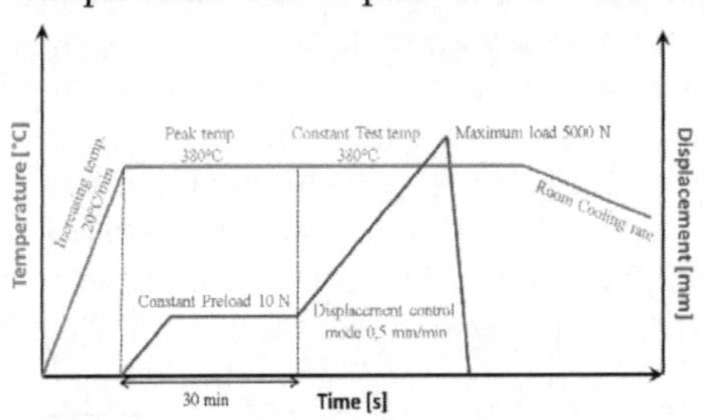

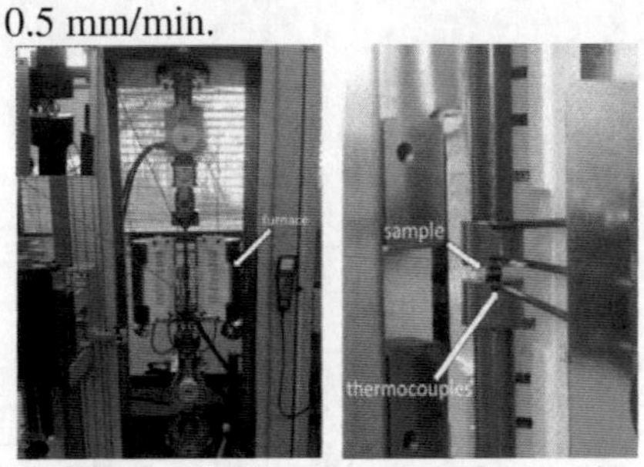

Figure 1. Thermo-mechanical cycle during RCT & **Figure 2.** Ring compression testing machine : (a) Furnace and compression heads; (b) Thermocouples inserted into furnace close to specimen during the RCT

Table 1. Specimen test matrix for ring compression testing to study hydrogen embrittlement in E110 fuel claddings

Hydrogen content (ppm)	Test Temperatures (°C)	Lengths (mm)
As -received	RT-150-400 °C	9.90 - 10.40
120	RT-150-400 °C	9.90 – 10.50
283	RT-150-400 °C	9.90 – 10.30
1500	RT-150-400 °C	9.90 – 10.20

3 Results and Discussion

3.1. Microstructural investigations

Transverse (cross-sectional) and longitudinal metallographic images from the pre-hydrogenated specimens are shown in Figure 3. Optical examination of the hydrogenated metallographic specimens revealed the presence of hydride platelets aligned in circumferential direction. These are called circumferential hydrides formed on the circumferential-longitudinal planes and were in the form of long chains. The average length of macro-hydrides in the circumferential and longitudinal direction was about 150 -182 μm. Figure 4 shows the 3D re-construction of the observed hydrides on the circumferential-longitudinal plane. On the right side of this figure, the habit planes of the circumferential hydrides in hexagonal zirconium lattice are shown.

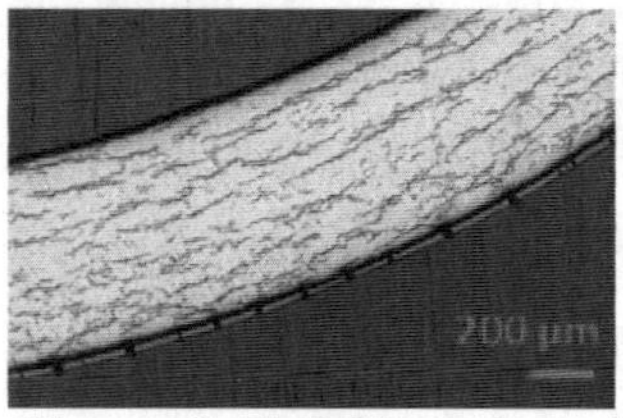

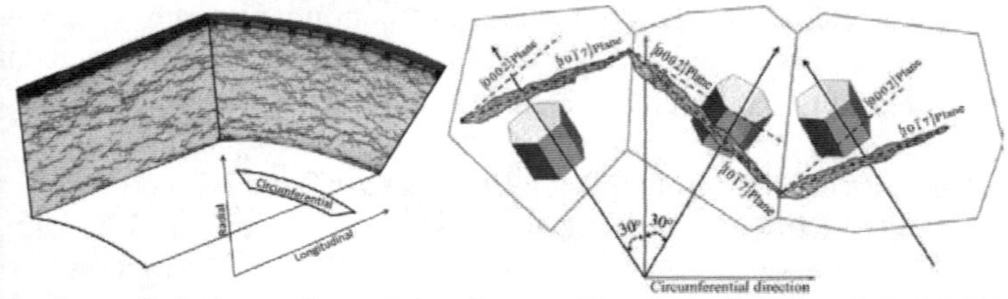

Figure 3. Hydrides oriented in the radial-circumferential orthographic plane & **Figure 4.** (a) 3D reconstruction of hydrides orientation in the creep tested hydrided E110 cladding; (b) schematic of hydride formation on habit plane [21]

3.2. Mechanical investigations

Load – displacement curves obtained from ring compression testing at various levels of hydrogen content tested at RT-150-400 °C are shown in Figure 5. On the recorded load-time curve, three areas was highlight to show the macroscopic behavior of the sample during RCT. Figure 6 represents the deformation stages of cladding during the ring compression test with varying loading at constant temperature. Initially there was a linear relationship between the stress and the strain. The next stage, known as plastic deformation zone, where drastic change of slope is noticeable. During RCT, it is known that sample underwent plastic deformation after initial elastic deformation. As shown in Figure 7, the collapse force P_0, where large plastic deformation occurred, can be obtained directly from the load-displacement curve by getting the intersection of the lines extended from elastic and plastic region, respectively. After extracting collapse force from the load-displacement curve, one of targeted properties of the material, ultimate tensile stress (UTS), could be estimated by the equations provided from previous works [22-23], which were based on theoretical models of plastic theory and several basic mechanical theories. Thus, the dimensions of sample as in Figure 7(b), including outer radius *R*, thickness *t* and length *L*, are important factors in these equations which aim to convert collapse force into UTS. Due to the complexity of the stress distribution in the sample, the ring is assumed to be perfectly plastic, which means no working hardening occurred in the whole process of RCT. With the plastic theory, the correlation between collapse stress σ_0 and plastic moment M_p can be obtained by

$$M_p = 1/4\ \sigma_0\ t^2\ L \tag{1}$$

Where, M_p is defined as the moment, at which the entire cross section is under its yield stress. Furthermore, based on the law of conservation of energy, the external virtual work done by P_0 is known to be equal the internal virtual work done by M_p. Thus, the collapse force can be defined as

$$P_0 = (4\ M_p) / R = (\sigma_0\ t^2\ L) / R \tag{2}$$

For working hardening material, a similar correlation can also be established and be used to get the collapse stress σ_0, as shown below:

$$\sigma_0 = \alpha\ P_0\ R / (t^2\ L) \tag{3}$$

Note that the constant α equals 0.866 in this work because the value of α depends upon the lengths of sample. Collapse stress is calculated from RCT, and then can be linearly correlated to the UTS of tensile test through the following coefficients

$$R_m = \sigma_0 / K_{UTS} \tag{4}$$

The load-displacement curves obtained from the RCT are shown in Figure 5, basically demonstrates two different types of results. It is evident from that the samples with lower wt. ppm H without hydrides have the point of contact between the upper and lower face (where there is the huge change of slope) higher than the other concentrations. This indicates that the samples with a higher hydrogen concentration were less ductile, due to the precipitation of δ-hydrides in the sample. The tests carried out at room temperature exhibits large cracks at the equatorial azimuths. These cracks initiate at the outer surface of the ring and propagate toward the inner surface. The crack depth in the sample with 120 and 283 wt. ppm is about half of the cladding thickness and the whole thickness in the samples with more than 1500 ppm. It may be observed that higher the hydrogen content, the smaller the number of steps in the main crack. This is consistent with easier propagation in the axial direction of the samples. Furthermore, low H content, ductile fracture occurs by formation and growth of voids. For the specimens with high H content brittle areas appeared, indicating fracture through hydrides.

The tests developed from room temperature to the set temperature of 150 °C, showed a different tensile strength compared to the tests a room temperature. In this case cracks were never observed until 1500 ppm, where the high amount of hydrides caused the loss of ductility and the formation of crack in the load-displacement curve close to 7 mm and 8 mm. The samples with 120 and 283 wt. ppm recovered their ductility as temperature increased. In these results, we can suppose two possibilities: precipitated zirconium hydrides may become ductile at 150 °C or a Zr metal matrix may regain enough ductility to accommodate the large plastic deformation even with the brittle hydride precipitates (the solubility of the zirconium matrix at this temperature is about 10 ppm). However, many studies at 150 °C suggest that the hydrides are still brittle at that temperature but the ductility of matrix improves with the increase of temperature. In addition, specimens regain their ductility at relatively high temperature above 250 °C, regardless of hydride orientation, supports the matrix recovery theory. As would expected, the load-displacement curves at 400 °C show that all the specimens can deform substantially without cracking. This implies that the Zr-matrix recovers its ductility enough to bear a large plastic deformation at 400 °C. This indicates that at this high temperature, common in the dry storage, these hydrides cannot act as a crack initiation site. At this temperature the hydrogen solubility is about 210 wt. ppm that means the specimen with 120 wt. ppm did not present hydrides, so the whole amount of hydrogen presented in the specimen was dissolved in the α-Zr. The tensile strengths R_m, which could be obtained by converting the collapse loads with Equations. (3) - (4) are summarized in Table 2.

Figure 8, shows the tensile strength behavior of E110 fuel claddings tested with various hydrogen contents at different test temperatures. The experimental results showed that this test is very sensitive to low hydrogen concentration, for 120 wt. ppm of hydrogen produce a noteworthy effect in the load vs. displacement curves. The analysis was performed to identify the fracture behavior and compare the as-received with the pre-hydrided samples at the main temperature that the cladding undergoes during its whole life. In all cases the tests showed that the increase of hydrogen leads to a decrease of fracture strain or loss of ductility. Moreover, at 400 °C and high hydrogen concentrations the E110 exhibits an almost ideal elastic-plastic behavior. A remarkable increase of the ultimate tensile strength is observed for the samples with 120 wt. ppm, whereas for the concentrations in excess of 120 wt. ppm, the UTS increases slowly.In the tests with the presence of circumferential δ-zirconium hydrides into the samples parallel to applied stress causes an increase of tensile strength due to the presence of harder (always more brittle) particles within the microstructure. The hydrides absorb more load than Zr-matrix causing the increase of the tensile strength. On the other hand the tests developed with 189 wt. ppm demonstrated the no presence of hydrides owing to the solubility (about 216 wt. ppm) at 380 °C. That means that the hydrogen was all in solid solution.

Table 2. Tensile strength (R_m) at RT-150-400 °C with different hydrogen content

Hydrogen content (ppm)	Average R_m at RT (MPa)	Average R_m at 150 °C (MPa)	Average R_m at 400 °C MPa)
0	379.31	264.98	166.52
120	434.99	310.84	192.76
283	444.35	321.26	192.19
1500	492.35	323.60	237.12

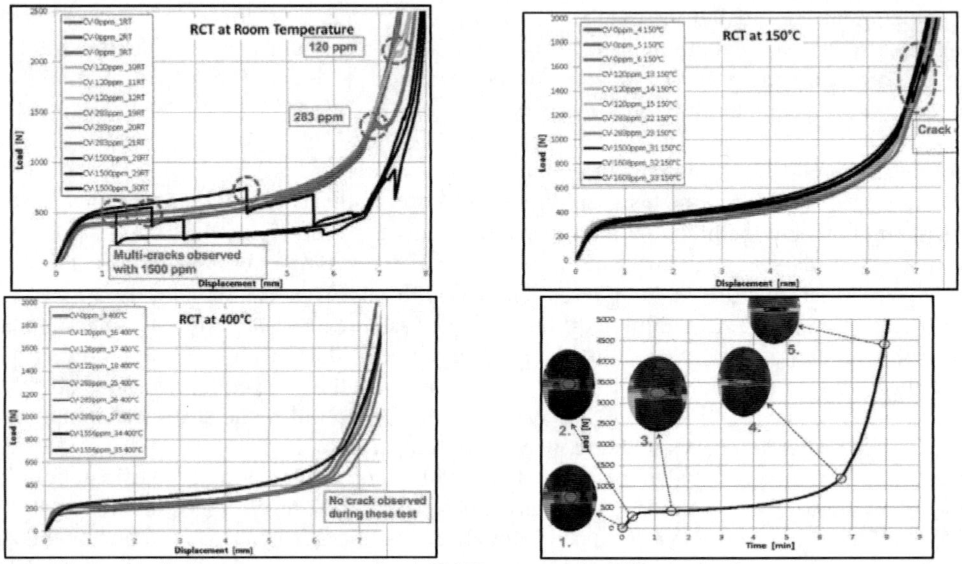

Figure 5. Load-displacement curves from RCT tests with different hydrogen concentration at RT-150-400 °C and **Figure 6.** Stages of cladding tube deformation during the ring compression testing at room temperature

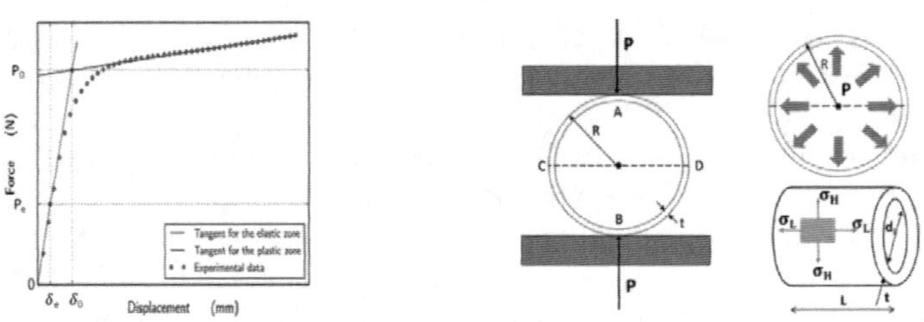

Figure 7. (a) Load-displacement curve of ring compression test; (b) Schematic of test specimen during ring compression testing

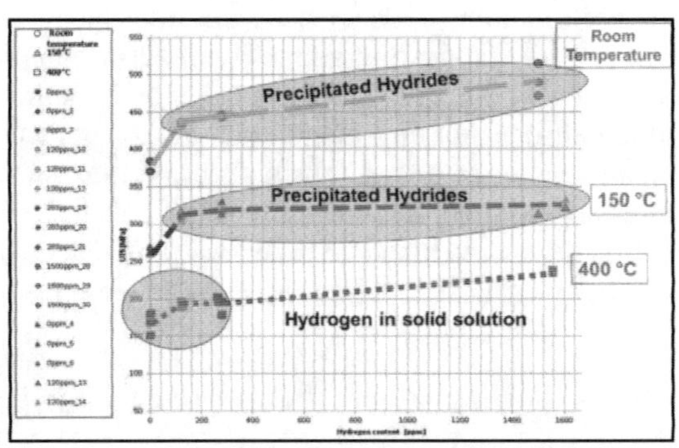

Figure 8. Tensile strength at RT-150-400 °C with different hydrogen concentrations

4. Conclusions

The experimental results showed that RCT test is very sensitive to low hydrogen concentration in test sample. The presence of δ-hydrides in the matrix causes an increase of tensile strength due to its harder properties. RCT test is an interesting test method to determine mechanical properties in hoop direction of tubular samples. Its cost-reducing and the low material consumption gave the test a real potential.

References

[1] C. Lemaignan.; Arthur T. Motta. Zirconium Alloys in Nuclear Applications. Materials Science and Technology: A Comprehensive Treatment 1994, vol. 10, pp. 1-51.

[2] Konings. Rudy. Zirconium Alloys: Properties and Characteristics. Comprehensive Nuclear Materials 2012, vol. 2, pp. 217–232.

[3] H. Okamoto. H-Zr (Hydrogen-Zirconium). Journal of Phase Equilibria and Diffusion 2006, no. 27, pp. 548-549.

[4] D. Kook Review of spent fuel integrity evaluation for dry storage. Nuclear Engineering and Technology 2013, vol. 45, pp. 115-124.

[5] T. Alam. A review on the clad failure studies. Nuclear Engineering and Design 2011, pp. 3658–3677.

[6] S. Suman. Hydrogen in Zircaloy: Mechanism and its impacts. International Journal of Hydrogen Energy 2015, pp. 5976–5994.

[7] F. Feria. ; L.E. Herranz . Creep assessment of Zry-4 cladded high burnup fuel under dry storage. Progress in Nuclear Energy 2011, pp. 395-400.

[8] R.N. Singh. Stress-reorientation of hydrides and hydride embrittlement of Zr–2.5 wt% Nb pressure tube alloy. Journal of Nuclear Materials 2004, no. 325, pp. 26–33.

[9] Hygreeva K. Namburi.; K. S. Valance.; J. Bertsch. Delayed hydride cracking in Zircaloy-2 fuel cladding tubes. TopFuel Reactor Fuel Performance 2012.

[10] Y. S. Kim. Precipitation of reoriented hydrides and texture change of α-zirconium grains during delayed hydride cracking of Zr-2.5Nb pressure tube. Journal of Nuclear Materials 2001, no. 297, pp. 292-302.

[11] R. N. Singh. Influence of temperature on threshold stress for reorientation of hydrides and residual stress variation across thickness of Zr–2.5Nb alloy pressure tube. Journal of Nuclear Materials 2006, no. 359, pp. 208–219.

[12] K. Kese Hydride re-orientation in Zircaloy and its effect on the tensile properties. SKI report 98:32 1998, pp. 49.

[13] Hyun-Gil Kim. The effects of creep and hydride on spent fuel integrity during interim dry storage. Nuclear Engineering and Technology 2010, vol. 42, no. 3, pp. 249-258.

[14] K.Kese Threat of hydride re-orientation to spent fuel integrity during transportation accidents: Myth or reality? In Proceedings of the 2007 International LWR Fuel Performance Meeting, San Francisco, 2007. pp.464-471.

[15] Hyun-Jin Cha. The effect of hydrogen and oxygen contents on hydride reorientations of zirconium alloy cladding tubes. Nuclear Engineering Technology 2015, no. 147, pp. 746-755.

[16] S.Arsene.; J. B. Bai. Effet de la microstructure et de la temperature sur la transition ductile-fragile des Zircaloy hydrurés, PhD dissertation, Ecole Centrale Paris, 1997.

[17] J. Desquines.; B. Cazalis.; Bernernaudat.; Ch.Poussard.; X. Averty.; P.J. Yvon. ASTM Int. 2 (6), 2005, Paper ID JAI 12465.

[18] G.W. Parry .; W.Evans. Nucleonics. 1964, no. 22, pp. 65.

[19] A.C.Wallace.; G.K Shek.; O.E. Lepik. ASTM Spec. Tech. Publ 1989, no. 1023, pp. 66–88.

[20] K.Linga.; I.Charit. Texture development and anisotropic deformation of Zircaloys. Progress in Nuclear Energy 2006, no. 48, pp. 325-359.

[21] S. Suman. Hydrogen in Zircaloy: Mechanism and its impacts. International Journal of Hydrogen Energy 2015, no. 40, pp. 5976 – 5994.

[22] V.Busser.; M.C.Baietto-Dubourg.; J.Desquines.; C.Duriez.; J.P.Mardon. Mechanical response of oxidized Zircaloy-4 cladding material submitted to a ring compression test. Journal of Nuclear Material 2009, no. 384, pp. 87-95.

[23] Y.T.Reddy.; S.R.Reid. On obtaining material properties from the ring compression test. Nuclear Engineering and Design, 1979, no. 52, pp. 257-263.

IOP Conf. Series: Materials Science and Engineering **461** (2019) 012060 doi:10.1088/1757-899X/461/1/012060

Determination of susceptibility of cast iron with a predetermined chemical composition to shape properties and microstructure through bainitic transformation

P. Nawrocki[1], D. Myszka[2],

[1] Institute of Precision Mechanics, Duchnicka 3, 01-796 Warsaw, Poland
[2] Warsaw University of Technology, Faculty of Production Engineering, Narbutta 85, 02-524 Warsaw, Poland

piotr.nawrocki@imp.edu.pl

Abstract. The problem of nanostructuring cast iron is hardly ever reported on scientific reports. Most of the research is undertaken in terms of steel nanostructurization. Bhadeshia states that to obtain a fragmented matrix of steel, the carbon content in the matrix must be at the level of 0.6% C. It should be noted that the carbon in the ferrite-pearlite matrix meets this condition, because the content of this element in the matrix is within the range of 0.7-0.9%. The authors of this work have attempted to evaluate the ability to shape the properties and microstructure of ductile cast iron as a result of not fully completed bainitic transformation, compared to the standard process of obtaining ADI cast iron. The tested material was characterized by a higher degree of alloying elements compared to the basic chemical composition of ADI cast iron. The aim of the work was to maximize the fragmentation of the cast iron microstructure by selecting a low temperature of isothermal transformation, i.e. 40°C above Ms temperature in the lower bainite range. The JMatPro program was used to determine the isothermal transformation temperature. Heat-treated cast iron was subjected to dilatometer tests to compare the results obtained from the computer program, microstructure research and hardness tests. Heat-treated cast iron was subjected to dilatometer tests to compare the results obtained from the computer program, microstructure research and hardness tests.

1 Introduction

Nowadays nearly 70% of the world-wide production of castings is made of cast iron [1]. The problem of this casting alloy includes issues related not only to the theory but also to the practice of obtaining a good casting with conscious shaping of the structure [2]. A special role in the development of gray iron castings is made of cast iron with the precipitation of nodular shape graphite, commonly called ductile iron. The process of obtaining ductile iron is based on spheroidization and inoculation of the Fe-C alloy with a much higher carbon content compared to its other kinds [3].

The morphology of austempered ductile iron matrix microstructure largely depends on the duration of the isothermal transformations. If the transformation will last for a very short time, the cast iron matrix will be almost entirely martensitic, with a small amount of ferrite fraction. The longer the transition time is, the ferrite content in the matrix increases in the form of austenite-separated plates. Austenite stabilizes with increasing time. This results in the formation of a full ausferritic matrix in cast iron. Bainitic microstructure is obtained after long periods of isothermal hold [4].

IOP Conf. Series: Materials Science and Engineering **461** (2019) 012060 doi:10.1088/1757-899X/461/1/012060

There is a growing trend in the literature of obtaining the finest microstructure (nanostructure) of steel by means of suitably selected thermal processes [5-8]. The problem of nanostructuring cast iron is hardly ever reported in scientific reports.
The authors of this work have attempted to evaluate the ability to shape the properties and microstructure of ductile cast iron as a result of different degrees of completion of bainite transformation, compared to the standard process of obtaining ADI cast iron.

2 Materials and methods

The subject of the research was ductile iron, with chemical composition (Table 1) selected based on statistical analysis of the database.

Table 1. Chemical composition of ductile cast iron (wt. %).

C	Si	Mn	Mo	Ni	Cu	Mg
3,55	2,5	0,35	0,55	1,90	0,5	0,05

The distinctive feature of the studied material was increased content of Mo, Ni and Cu elements.
The microstructure of the obtained cast iron in the as-cast state is presented in Figure 1.

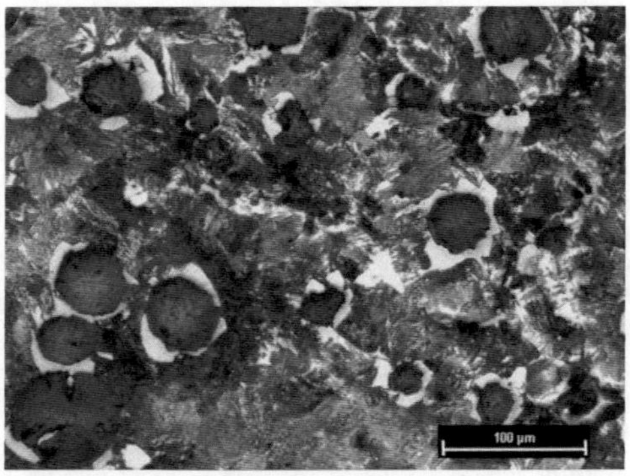

Figure 1. Microstructure of the obtained Mo-Ni-Cu cast iron, mag. x200 (etched with nital)

The obtained castings were cut into smaller samples in the shape of rectangles of 15x15x10mm, which were subject to a heat treatment processes. A diagram of the processes is shown in Figure 2.

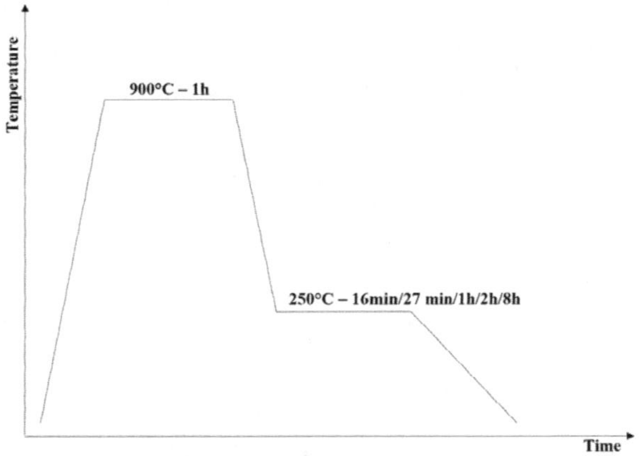

Figure 2. Schemes of conducted heat treatment process of Mo-Ni-Cu ductile iron.\

The maximum refinement of the cast iron microstructure was assumed by selecting a low temperature of isothermal transformation, i.e. 40°C above M_s in the bottom bainite range. The JMatPro program was used to determine the isothermal transformation temperature. An example chart from the program is shown in Figure 5. On the basis of dilatometric plots (Figures 3) the temperature of 250°C and the isothermal holding time were determined, i.e. 16 min, 27 min and 8h, corresponding to 30%, 50% and 100% bainite transformation, respectively. In addition, characteristic temperatures were defined, i.e. A_{c1}, A_{c3}, M_s. Due to the fact that analysis of SEM images after thermal processes with isothermal hold of 16min and 27min did not indicate achieving 30% and 50% of the bainite transformation, the times were extended to 1h and 2h.
For heat treated material microstructure observation was performed using an electron microscope. Hardness tests were carried out. Observations of metallographic specimens etched with nital were performed using the JEOL JSM-IT100 scanning electron microscope at magnifications in the range of x1000 and x5000. The HBW surface hardness tests were performed using an Alpha hardness tester, performing 3 readings on each of the samples.

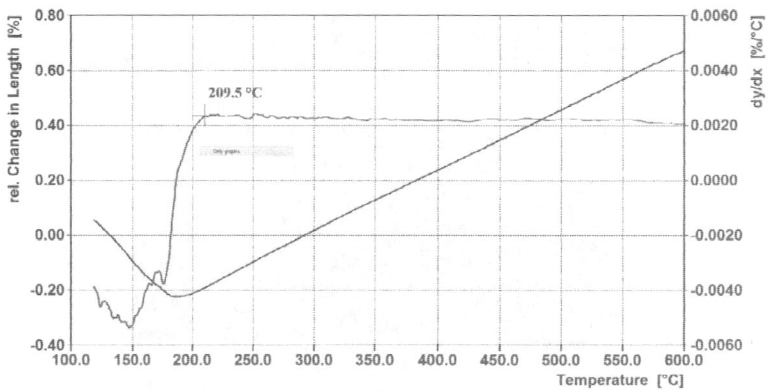

Figure 3. Dilatometric diagram for the tested Mo-Ni-Cu cast iron shows M_s temperature.

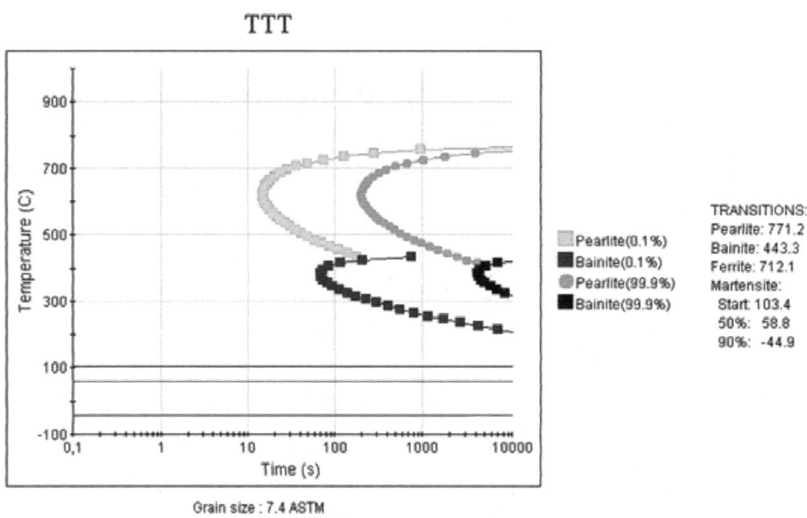

Figure 4. Example of TTT diagram of tested Mo-Ni-Cu cast iron.

IOP Conf. Series: Materials Science and Engineering **461** (2019) 012060 doi:10.1088/1757-899X/461/1/012060

3 Results and discussion

3.1 *Microstructure*

Figures 5 shows the obtained cast iron microstructure after thermal processes, for the applied isothermal holding times.

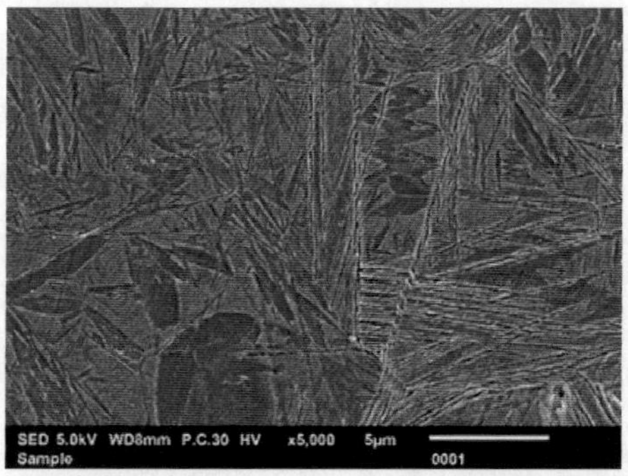

A)Isothermal treatment – 16 min, mag. x5000.

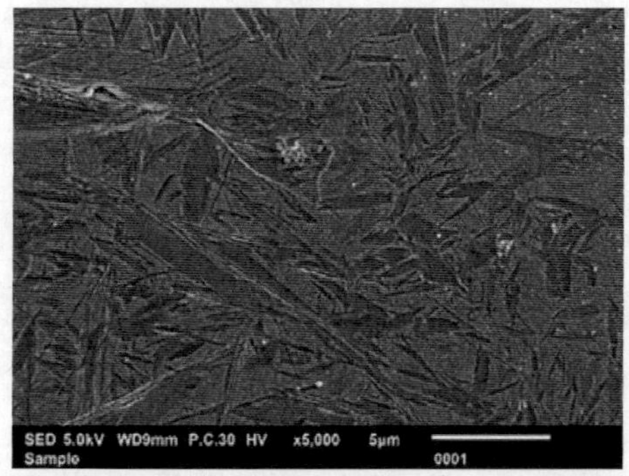

B)Isothermal treatment – 27 min, mag. x5000.

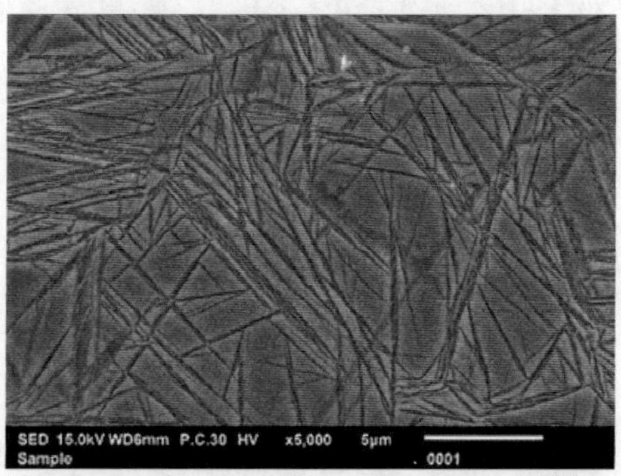

C)Isothermal treatment – 1h, mag. x5000.

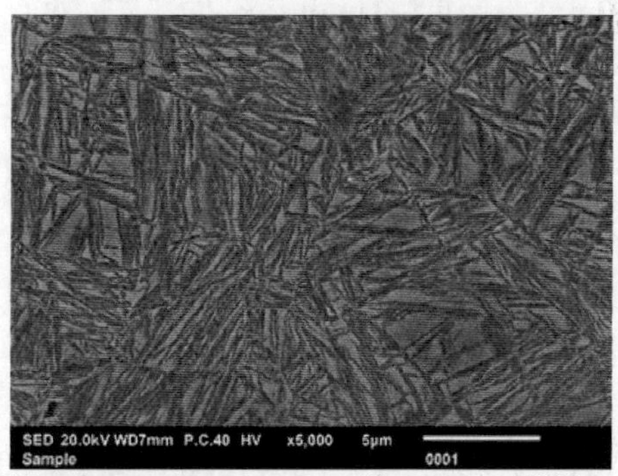

D)Isothermal treatment – 2h, mag. x5000.

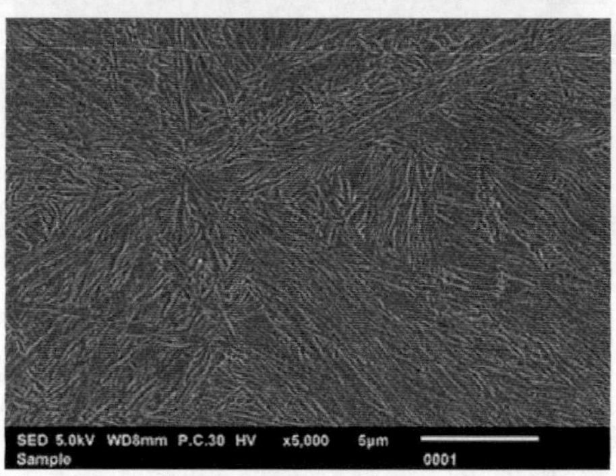

E)Isothermal treatment – 8h, mag. x5000.

Figure 5. Microstructure of the tested cast iron after various modes of heat treatment (etched with nital).

Analysis of SEM images presented in Figures 5A-B confirms the very short time of isothermal transformation. The studied ductile iron is characterized by an entirely martensitic matrix.
The obtained microstructure of cast iron after isothermal transformation after 16 and 27 minutes shows that no bainitic transformation occurs in the percentages assumed from the dilatometer plots, as

mentioned in the previous chapter of the article. Observation of the microstructures indicates a thick plate martensitic structure with the local occurrence of ferrite plates, but without any noticeable differences between the two lowest times of isothermal strength. After 1h of isothermal transformation in the tested cast iron, austenitic microstructure characterized by large blocks of austenite was observed (Figure 6C).

Continued transformation in 2h influenced on the stabilization of austenite in the matrix of cast iron and at the same time increasing the proportion of bainite in the structure (Figure 6D).

The time of 8h of isothermal treatment affected the occurrence of 100% bainite transformation, which is confirmed by the presented microstructure in Figure 5E.

3.2 *Hardness*

Table 2 presents the obtained hardness results of the tested cast iron after thermal processes. In general, it can be concluded that there is no large dispersion between subsequent measurement results.

Table 2. Results of ductile iron hardness after thermal processes.

	Hardness [HBW]				
	900°C-1h 250°C-16min	**900°C-1h 250°C-27min**	**900°C-1h 250°C-1h**	**900°C-1h 250°C-2h**	**900°C-1h 250°C-8h**
1	415	420	444	337	485
2	420	438	451	337	495
3	426	438	457	329	499
Average	420	432	451	334	493

With the progressive bainitic transformation an increase in the hardness of the tested samples is observed. After 8 hours of transformation, the average hardness of 493 HBW was obtained, which corresponds to 51 units on the Rockwell scale. The relatively high hardness after the lowest isothermal transformation times can be explained by the occurrence of the martensite structure. It is problematic to explain the low hardness value for the sample after isothermal quenching after 2 hours. Most likely, this fact is due to the lack of heterogeneity of the sample material. Therefore, it is required to repeat the hardness tests and heat treatment processes. Figure 6 presents a summary of mean values for the tested cast iron samples.

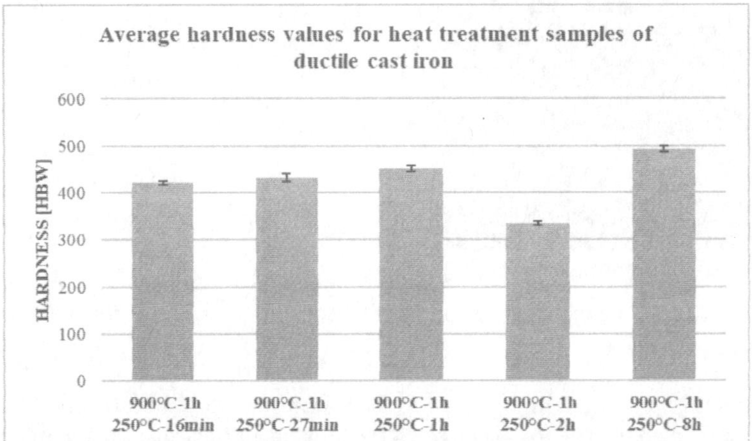

Figure 6. Average hardness values for heat treatment samples of Mo-Cu-Ni ductile cast iron for the applied isothermal hardening times.

4 Summary

The conducted research indicates it is possible to shape the properties and microstructure of ductile iron with an increased content of Ni and Cu, using different times of isothermal quenching. Short-term isothermal processes lead to the formation of martensitic structures with ferrite separation in the form of plates.

With austenitizing at 900°C and a low temperature of isothermal quenching, further formation of a ferritic-austenitic matrix with or without carbide precipitates (bainite) can be observed with the progressive bainitic transformation. To determine the nature of the resulting bainite, studies should be carried out using transmission electron microscopy. The JMatPro program and dilatometer tests proved to be a useful tool in the design of thermal processes. Nevertheless, each of the assumed values should be verified in detail by appropriate tests.

References

[1] 51th Census of World Casting Production 2017 A Modern Casting Staff Report.
[2] Podrzucki Cz 1991 Cast Iron Structure Properties Application vol.1 ZG STOP (Cracow).
[3] Myszka D Kaczorowski M and Tybulczyk J 2003 Ausferritic ductile cast iron directly isothermally hardened Foundry Research Institute Publishing (Cracow).
[4] Myszka D 2004 Technological aspects of deformation transformation in ferritic ductile cast iron Scientific work of the Warsaw University of Technology vol.265 Warsaw University of Technology Publishing House (Warsaw).
[5] Skołek E Wasiak K and Świątnicki W 2015 Structure and properties off the carburised surface layer on 35CrSiMn5-5-5 steel after nanostruturization treatment *Materials and technology* 49 pp 933-939.
[6] Garcia-Mateo C Caballero F G and Bhadeshia H K D H 2003 Acceleration of Low-temperature Bainite *ISIJ International* 43 vol.11 pp 1821–1825.
[7] Caballero FF G Bhadeshia H K D H Mawella K J A Jones D G and Brown P 2002 Very strong low temperature bainite *Materials Science and Technology* 18 vol.3 pp 279–284.
[8] Caballero F G, Bhadeshia H K D H 2004 Very Strong Bainite *Current Opinion in Solid State and Materials Science* 8 vol.3–4 pp 251–257.
[9] Myszka D Skołek E and Wieczorek A 2014 Manufacture of toothed elements in nanoausferritic ductile iron *Archives of Metallurgy and Materials* vol. 59 pp 1217-1221.
[10] Putatunda S K Panneerselvam S and Alshwigi M 2015 Development of nanostructured austempered ductile cast iron (ADI) 28th ASM heat treating society conference 20-22 October USA: ASM International pp 71-75.

IOP Conf. Series: Materials Science and Engineering **461** (2019) 012061 doi:10.1088/1757-899X/461/1/012061

Identification of lattice defects beneath fracture surfaces of cold-drawn pearlitic steel fractured in elastic/plastic regions in the presence of hydrogen

K Nishiyama[1], T Manabe[2], D Hirakami[2] and K Takai[3]

[1] Graduate Student, Sophia University, 7-1 Kioi-cho Chiyoda-ku Tokyo 102-8554, Japan
[2] Nippon Steel & Sumitomo Metal, 20-1 Shintomi, Futtsu, Chiba Prefecture 293-8511, Japan
[3] Faculty of Science and Technology, Department of Engineering and Applied Science, Sophia University, Tokyo, Japan

Corresponding author's e-mail address: kei175348@yahoo.co.jp

Abstract The kinds of lattice defects formed beneath fracture surfaces were analyzed using thermal desorption analysis (TDA) for cold-drawn pearlitic steel that was fractured in the elastic/plastic regions in the presence of hydrogen. A sample thickness of 0.3 mm made it possible to obtain two separate peaks: a low temperature peak and a high temperature peak. Tensile tests and constant load tests were conducted under various hydrogen charging conditions to prepare samples with different elongations and fracture times. Samples of 0.3 mm in thickness beneath fracture surfaces were charged with hydrogen to determine the kinds of lattice defects and hydrogen desorption profiles were obtained using TDA. For samples that were fractured in the plastic region, a broad hydrogen desorption profile continuing to around 180 °C was obtained compared to samples without hydrogen charging and straining. Since this broad desorption profile was disappeared by aging at 200 °C, it probably corresponded to desorption from vacancies. In contrast, for samples that were fractured in the elastic region, this broad desorption profile was not obtained. These findings indicate that one of the factors causing hydrogen embrittlement in the plastic region is probably vacancies, whereas other factors cause it in the elastic region.

1 Introduction

Quasi-cleavage fracture modes are reported as representative hydrogen embrittlement fracture modes in cold-drawn pearlitic steel. Hydrogen-enhanced lattice defects (HELDs) are considered to be one of the mechanisms causing quasi-cleavage fracture. As a result, various approaches are taken to clarify the mechanism of hydrogen embrittlement with respect to lattice defects. In a previous study, the formation of hydrogen-enhanced strain-induced lattice defects was not observed when cold-drawn pearlitic steel with hydrogen was fractured in the elastic region, whereas when it was fractured in the plastic region, the formation of such defects was observed [1]. However, since that analysis of the states of hydrogen present was conducted under a diffusion-limited condition of hydrogen, only a single peak was obtained by TDA. Furthermore, it made it impossible to separate hydrogen-enhanced strain-induced lattice defects. In this study, we aimed to separate lattice defects formed in the elastic plastic regions by bringing the analysis conditions close to the thermal desorption rate-limiting condition of hydrogen and setting the TDA measurement position just below the fracture surface of specimens.

IOP Conf. Series: Materials Science and Engineering **461** (2019) 012061 doi:10.1088/1757-899X/461/1/012061

2 Experimental

2.1 Materials
Specimens were prepared of cold-drawn pearlitic steel corresponding to JIS SWRS82B. Specimens were cold-drawn from 13 to 6 mm in diameter, after austenitization at a temperature of 880 °C for 600 s and isothermal transformation at a temperature of 530 °C for 120 s. Then, specimens were aged at a temperature of 300 °C for 600 s. The chemical composition of the specimens is shown in Table 1.
The microstructure was pearlite, consisting of the ferrite (α) phase and cementite (θ) elongated in the drawing direction.

2.2 Specimen preparation
Obtaining multiple peaks of the hydrogen desorption profile was conducted by thinning specimens and bringing them close to the thermal desorption rate-limiting condition of hydrogen. First, specimens were cut to thicknesses of 0.3, 0.5, 1, 3 and 5 mm, respectively. Then they were polished using emery papers (# 800 to 2000 grit) and charged with hydrogen by immersion in a 5 mass % NH_4SCN solution kept at a temperature of 30 °C until they were saturated. After hydrogen charging, hydrogen desorption profiles were obtained by TDA.

2.3 Evaluation of relationship between the amount of lattice defects and distance from fracture surface
In order to evaluate the amount of lattice defects in fractured specimens, the relationship between the amount of lattice defects and the distance from the fracture surface was examined in specimens fractured in the plastic region with hydrogen in tensile testing. The surfaces of the specimens were polished using emery papers (# 800 to 2000 grit). Specimens were electrochemically charged with hydrogen in a 0.1 N solution of NaOH + 5 $g{\cdot}L^{-1}$ NH_4SCN at a current density of 30 $A{\cdot}m^{-2}$ and a temperature of 30 °C. Charging was performed for 144 h. This was the time required for the hydrogen concentration at the center of the specimens to reach the equilibrium. After hydrogen charging, tensile testing was conducted at a crosshead speed of 0.055 $mm{\cdot}min^{-1}$. Figure 1 shows the results of tensile testing and Fig. 2 is a schema showing how samples were cut out from the fracture surface. The samples were cut to 0.5 mm in thickness were numbered from (1) to (4) in order from the fracture surface and these samples were aged at 30 °C for 7 days to degas the contained hydrogen. These samples were then polished to 0.3 mm and tracer hydrogen was charged in a 5 mass % NH_4SCN solution at 30 °C for 1 day. Hydrogen is trapped in lattice defects in the material. As a result, it is possible to evaluate the amount of lattice defects in the material by charging hydrogen after degassing the contained hydrogen. The hydrogen that is charged to investigate the amount of lattice defects is called "Tracer Hydrogen." Tracer hydrogen content of each sample was measured by TDA. This content corresponds to the amount of lattice defects.

Table 1. Chemical composition of specimen material (mass %)

C	Si	Mn	P	S
0.84	0.25	0.73	0.017	0.005

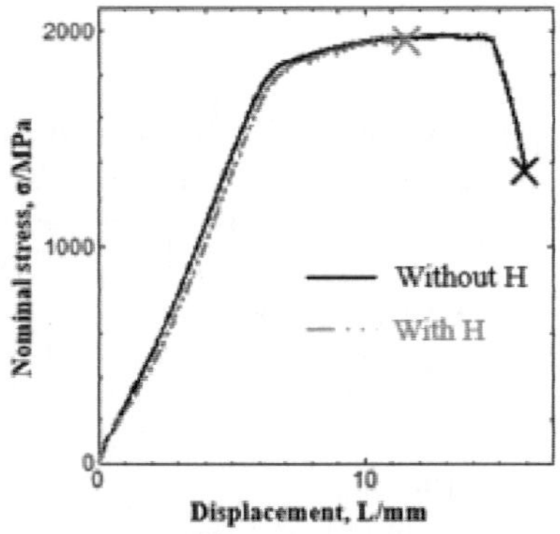

Fig. 1. Nominal stress– displacement curves of specimens with/without hydrogen

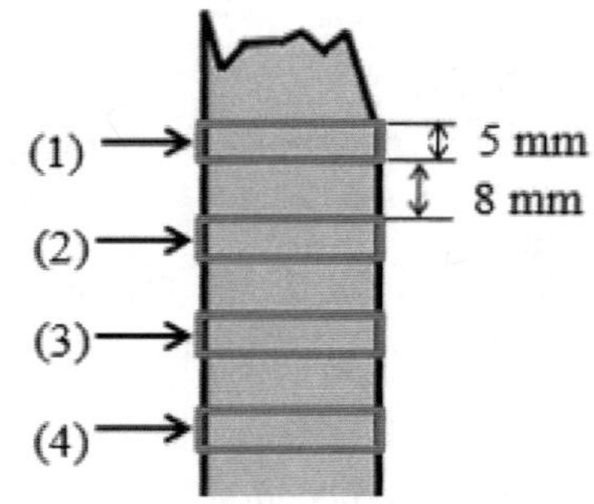

Fig. 2. Schema showing how specimens were cut out from the fracture surface

2.4 Comparison of tracer hydrogen desorption profiles of specimens showing different fracture strain by tensile tests with hydrogen

In the same experimental procedure as in section 2.3, only the hydrogen charging conditions were changed to obtain specimens with different fracture strain. The specimens were electrochemically charged with hydrogen in a 0.1 N solution of NaOH + 3.5, 4.0 and 5.0 $g{\cdot}L^{-1}$ NH_4SCN at a current density of 30 and 50 $A{\cdot}m^{-2}$ and a temperature of 30 °C. Fig. 3 shows the nominal stress-displacement curves obtained by tensile testing. These specimens were cut to a thickness of 0.5 mm just below the fracture surface and aged at 30 °C for 7 days to degas the contained hydrogen. The samples were then polished to 0.3 mm and tracer hydrogen was charged in a 5 mass % NH_4SCN solution at 30 °C for 1 day. The tracer hydrogen content of each sample was measured by TDA.

2.5 Comparison of tracer hydrogen desorption profiles of specimens showing different fracture times by constant load tests with hydrogen

Constant load tests were carried out at $0.4\sigma_B$ (stress lower than the proportional limit) and $0.9\sigma_B$ (stress higher than the proportional limit) corresponding to 0.2% proof stress in the presence of hydrogen, and the hydrogen desorption profiles just beneath the fracture surfaces were compared. Here, σ_B represents the tensile strength. First, the surfaces of the specimens were mechanically polished using emery papers (# 800 to 2000 grit). For constant load tests at a stress level lower than the proportional limit, tests were conducted at a load stress of $0.4\sigma_B$ in a 20 mass% NH_4SCN solution at a temperature of 50 °C without pre-charging hydrogen. After dividing the fracture time of this experiment(immersion) by the sane fracture time, the fracture time ratio is obtained as unity. For constant load tests at a stress level higher than the proportional limit, hydrogen pre-charging was conducted in a 0.1 N NaOH solution with 1.5 to 3 $g{\cdot}L^{-1}$ NH_4SCN at a current density of 30 $A{\cdot}m^{-2}$ and a temperature of 30 °C until the specimens were saturated with hydrogen. Then constant load tests were carried out at a load stress of $0.9\sigma_B$. After dividing two fracture times of this experiment(electrolysis) by the fracture time of the previous experiment(immersion), the fracture time ratios are obtained as 0.7 and 7.5. These specimens were cut to a thickness of 0.5 mm just beneath the fracture surface and degassed at 30 ° C for 7 days to remove the contained hydrogen. They were mechanically polished to 0.3 mm and tracer hydrogen was then charged in a 5 mass % NH_4SCN solution at 30 °C for 1 day. The tracer hydrogen content of each sample was measured by TDA.

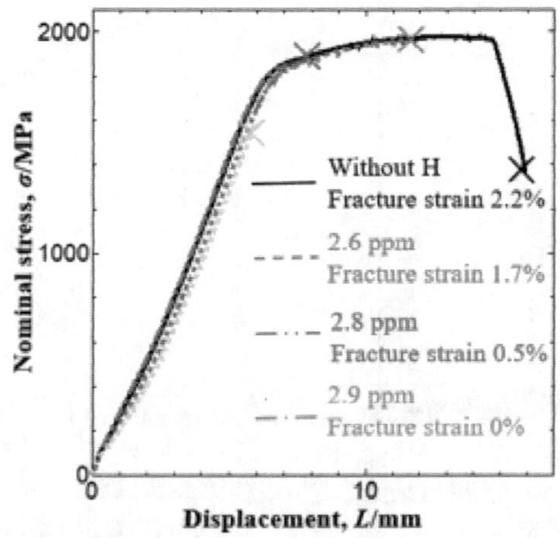

Fig. 3. Nominal stress– displacement curves for various strain amounts

2.6 Identification of lattice defect corresponding to high temperature shoulder

Obtaining multiple peaks in the TDA profile was performed on specimens to which strain was applied in the presence of hydrogen. Polishing, hydrogen charging and tensile tests were carried out under the same conditions as in section 2.3, and specimens having the same fracture strain as in section 2.3 were obtained. Two samples were cut out to 0.5 mm from just below the fracture surfaces and aged at 30 °C for 7 days to degas the contained hydrogen. One sample was subjected to aging at 200 °C for 2 h, whereas the other one was not. It was reported that vacancies are disappeared by aging treatment at 200 °C. As a result, it is possible to identify vacancy peak by comparing those with and without aging. Then these two samples were polished to 0.3 mm and tracer hydrogen was charged in a 5 mass % NH_4SCN solution at 30 °C for 1 day. Tracer hydrogen content of each sample was measured by TDA.

3 Results and Discussion

3.1 Hydrogen desorption profiles of samples with different thickness

Fig. 4 shows the TDA profiles of each sample with a different thickness charged with hydrogen under the conditions in section 2.2. The peak temperature and desorption end temperature decreased as the sample thickness was reduced. The sample thickness of 0.3 mm in particular made it possible to obtain two separate peaks: a low temperature peak and a high temperature peak. This result indicates that the condition of the sample approached the thermal desorption rate-limiting and it was determined to use the thickness of 0.3mm for obtaining multiple peaks in this study.

3.2 Relationship between tracer hydrogen content and distance from fracture surface

Fig. 5 shows the relationship between the tracer hydrogen content and the distance from the fracture surface. Numbers (1) -(4) correspond to those in the schema in Fig. 2. The closer the samples were to the fracture surface, the more tracer hydrogen content was obtained by TDA. This result suggests that the amount of lattice defects increased as the measurement position was moved closer to the fracture surface. Based on this result, samples just below the fracture surfaces were used in the sections 2.4, 2.5, and 2.6 since these samples contained a sufficient amount of lattice defects at the time of fracture.

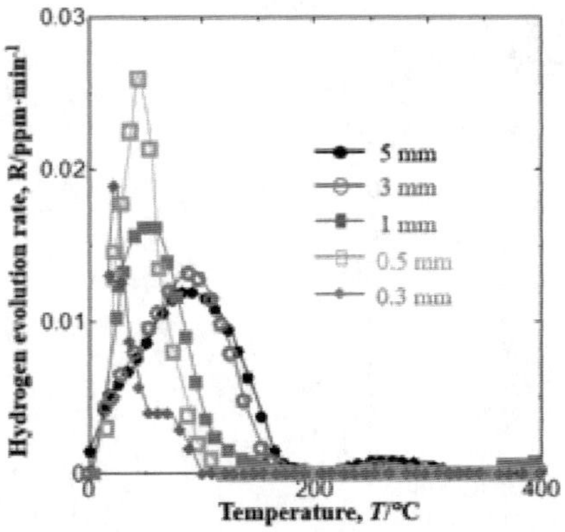

Fig. 4. TDA profiles of samples with a thickness of 0.3 mm to 5 mm

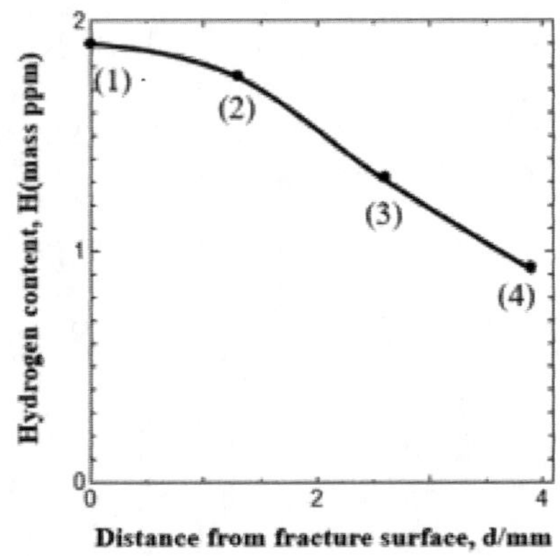

Fig. 5. Tracer hydrogen content at each distance from fracture surface

3.3 *Comparison of tracer hydrogen desorption profiles of samples cut out from just below fracture surface with different fracture strain*

Fig. 6 shows the tracer hydrogen desorption profiles of samples cut out from just below the fracture surface with different fracture strain with hydrogen. For the specimens that fractured in the elastic region, no increase in tracer hydrogen content was observed when compared with a specimen without hydrogen and straining. In contrast, for the specimens that fractured in the plastic region, the tracer hydrogen content increased and a broad hydrogen desorption profile appeared that continued to around 180 °C.

3.4 *Comparison of tracer hydrogen desorption profiles of samples cut out from just below fracture surface with different fracture time*

Fig. 7 shows the hydrogen desorption profiles of unstrained samples that were fractured in the plastic region in tensile testing and in the elastic region in constant load test. In constant load test, there was little change in the profile and hydrogen content when compared with the hydrogen desorption profile of the unstrained sample. In addition, compared to the sample that was fractured in the plastic region in tensile testing, there was no broad shoulder on the high temperature side, as shown in section 3.3. The hydrogen desorption profiles of the unstrained samples that were fractured at different time are shown in Fig. 8. Loading in the plastic region increased the hydrogen content more than for the unstrained sample. Furthermore, the shoulder on the high temperature side became broader as the time to be fractured was increased.

3.5 *Obtaining multiple peaks in TDA profile of samples fractured in the plastic region with hydrogen*

Fig. 9 shows the tracer hydrogen desorption profiles of "unstrained", "H + strain" and "H + strain (200 °C aging)" samples. Here, H + strain is a sample that the strain was applied in the presence of hydrogen, H+strain (200 °C aging) is the one that strain was applied in the presence of hydrogen then aged at 200 °C. Compared to the unstrained sample, the tracer hydrogen content increased for the "H + strain" and "H + strain (200 °C aging)" samples. In addition, the shoulder on the high temperature side, which was generated by applying plastic strain in the presence of hydrogen, decreased as a result of aging at 200 °C and the desorption end temperature was the same as that of the "Non-strain" sample. These results suggested that the amount of lattice defects increased by applying the strain in the presence of hydrogen and defects corresponding to the shoulder at the high temperature that disappeared upon aging at 200 °C was the vacancy peak. This implies that vacancies are involved in the fracture in the plastic region in the presence of hydrogen.

4 Conclusion

The kinds of lattice defects formed beneath fracture surfaces were analyzed using TDA for cold-drawn pearlitic steel specimens that were fractured in the elastic/plastic regions in the presence of hydrogen.
(1) At a thickness of 0.3 mm, the condition of the samples approached the thermal desorption rate-limiting of hydrogen. As a result, two separate peaks could be obtained from TDA profiles: the low temperature peak and the high temperature peak.
(2) It was found that the amount of lattice defects was higher as the measurement position of the specimens was closer to the fracture surface. Hence, for the measurement of the amount of lattice defects of fractured specimens, the measurement position was set just below the fracture surface.
(3) After comparing TDA profiles samples with different fracture strain and time, following observations were made: (i) In tensile testing, amount of lattice defects was not increased in elastic region and shoulder appeared on high temperature side in plastic region. (ii) In constant load tests, amount of lattice defects was not increased in elastic region, however, it was increased slightly in plastic region. There is tendency for defects to increase with the increased time to be fractured.
(4) The sample that was fractured in the plastic region in tensile testing with hydrogen displayed the broad shoulder on the high temperature side in the tracer hydrogen desorption profile, which disappeared upon aging at 200 °C. As a result, the types of defect, which corresponded to the shoulder, was presumably the vacancies.

References

[1] R. Konno, T. Manabe, D. Hirakami and K. Takai : *Tetsu-to-Hagané*, **104** (2018) 36.

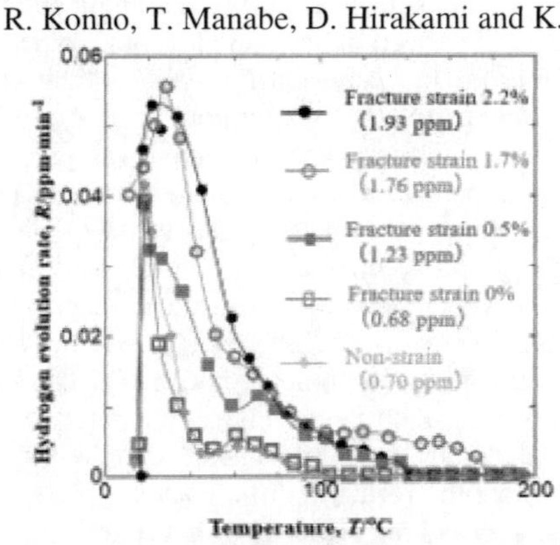

Fig. 6. TDA profiles of unstrained samples fractured at various strain levels by SSRT

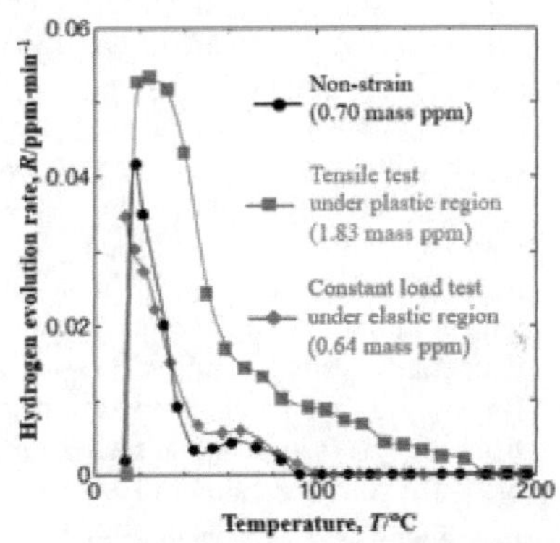

Fig. 7. TDA profiles of unstrained samples fractured by tensile testing and constant load testing

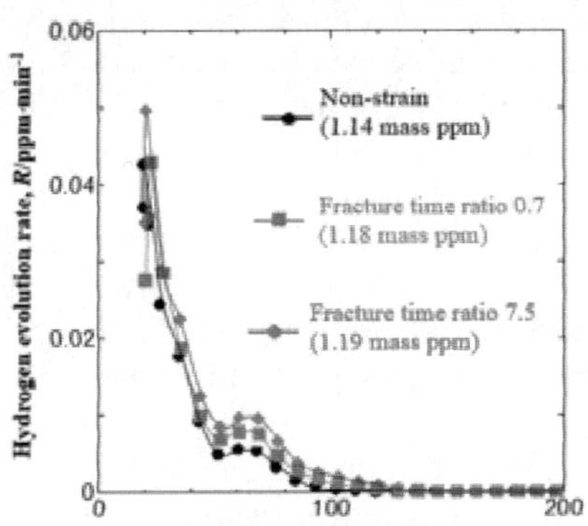

Fig. 8. TDA profiles of unstrained samples fractured in fracture time ratio of 0.7 and 7.5

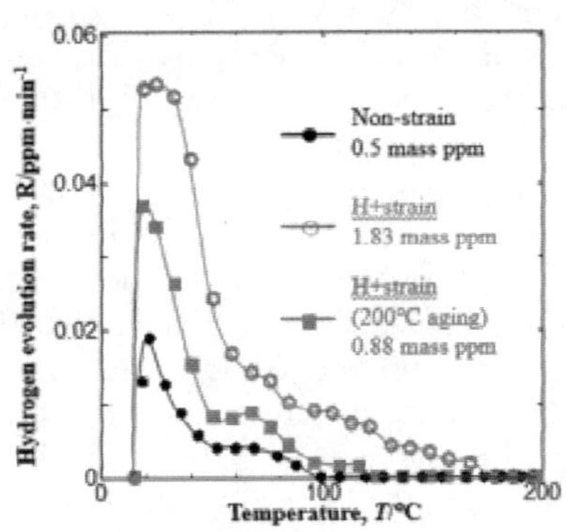

Fig. 9. TDA profiles of unstrained samples fractured at 1.7% strain by SSRT with/without aging at 200 °C

IOP Conf. Series: Materials Science and Engineering **461** (2019) 012062 doi:10.1088/1757-899X/461/1/012062

Changes in the substructure of tempered martensitic steel during the application of cyclic elastic stress in the presence of hydrogen

M Ohori[1], T Chiba[1], Y Matsumoto[1], H Suzuki[2] and K Takai[2]

[1]Graduate School of Science and Technology, Sophia University, 7-1 Kioi-cho, Chiyoda-ku, Tokyo 102-8554, Japan
[2]Faculty of Science and Technology, Department of Engineering and Applied Sciences, Sophia University, Tokyo, Japan

Corresponding author's e-mail address: m-ohori-3wr@eagle.sophia.ac.jp

Abstract. Changes in the substructure of tempered martensitic steel during the application of cyclic elastic stress in the presence of hydrogen have been investigated. Three types of specimens were prepared. The first type was subjected to cyclic elastic stress (strain rate: 3.0×10^{-4} s^{-1}, number of cycles: 70000, and stress level: 0.8±0.1 σ_B) without hydrogen. The second type was subjected to cyclic elastic stress with hydrogen. The third type was prepared by annealing the second type at 200 °C for 2 h. These specimens are denoted here as [(n)-cycled], [H+(n)-cycled] and [H+(n)-cycled+annealed] specimens, respectively. Tracer hydrogen was used as a probe for clarifying changes in the specimen substructure. After tracer hydrogen charging, hydrogen states present in the specimens were examined using low-temperature thermal desorption spectrometry (L-TDS). Tracer hydrogen desorption in the high-temperature range from the [H+(n)-cycled] specimen increased compared with that from the [(n)-cycled] specimen. In addition, tracer hydrogen desorption in the high-temperature range from the [H+(n)-cycled+annealed] specimen was reduced by annealing at 200 °C to the level of the [(n)-cycled] specimen. As a result, it is inferred that the formation of vacancy type defects was enhanced under the application of cyclic elastic stress in the presence of hydrogen.

1 Introduction

Hydrogen is trapped at various trapping sites in steels. The degree of influence of hydrogen on hydrogen embrittlement differs among the different kinds of trapping sites. In previous studies, the formation of vacancy-type defects in the vicinity of the fracture surface of specimens that were fractured under constant elastic tensile stress with hydrogen has been reported for tempered martensitic steels [1]. Furthermore, it has also been reported that plastic deformation with hydrogen increased the formation of vacancy-type defects [2]. Vacancy-type defects have been reported to cause degradation of mechanical properties such as ductility loss. However, steels are subjected not only to constant elastic stress but also to cyclic stress in actual environments. Accordingly, it is imperative to clarify the changes in the substructure of steels during the application of cyclic elastic stress in the presence of hydrogen. In terms of fatigue behavior, the effects of hydrogen on fatigue crack propagation [3] and cyclic hardening/softening [4] under cyclic loading with hydrogen have been reported. On a contrary, there have been few reports regarding the effects of hydrogen on substructure changes before the occurrence of damage such as crack formation. Correspondingly, in this study, we

IOP Conf. Series: Materials Science and Engineering **461** (2019) 012062 doi:10.1088/1757-899X/461/1/012062

investigated changes in the substructure of tempered martensitic steel during the application of cyclic elastic stress with hydrogen.

2 Experimental

2.1 Materials

The chemical composition of the tempered martensitic steel specimens used in this study is listed in Table 1. The specimens were induction quenched at 960 °C and tempered at 360 °C. The tensile strength (σ_B) and proof stress of 0.2% were 1471 MPa and 1384 MPa, respectively. Figure 1 displays the specimen shape having a diameter of 5 mm and a gauge length of 30 mm.

2.2 Hydrogen precharging conditions

Prior to hydrogen precharging, the specimen surface was mechanically polished using emery papers (#800, 1000 and 2000). The specimens were charged electrochemically with hydrogen at a current density of 10 $A{\cdot}m^{-2}$ in a 4 $g{\cdot}L^{-1}$ N NaOH aqueous solution kept at 30 °C and containing a NH_4SCN additive of 1 $g{\cdot}L^{-1}$. The charging time was 72 h to reach an equilibrium hydrogen concentration at both the surface and the center of the specimens. These charging conditions provided the equilibrium hydrogen concentration of approximately 2.6 mass ppm in the specimens.

2.3 Preparation of specimens subjected to cyclic elastic stress

Three types of specimens subjected to cyclic elastic stress with/without hydrogen were prepared. The applied cyclic elastic stress was a triangular wave at 0.8±0.1 σ_B. The strain rate during cyclic elastic stress was 3.0×10^{-4} s^{-1}. The numbers of cycles (n) were 700, 7000 and 70000. The first type, denoted as the [(n)-cycled] specimens, was subjected to cyclic elastic stress at 30 °C without hydrogen. The second type, denoted as the [H+(n)-cycled] specimens, was subjected to cyclic elastic stress at 30 °C with hydrogen. The third type, denoted as the [H+(n)-cycled+annealed] specimens, was obtained by annealing the second type at 200 °C for 2 h. The second and third specimen types were precharged with hydrogen and then cyclic elastic stress was applied concurrently with hydrogen charging for maintaining the hydrogen content of the specimens.

2.4 Tracer hydrogen charging and hydrogen analysis

Figure 2 shows a schematic diagram of the applied cyclic elastic stress and subsequent L-TDS. After the application of cyclic elastic stress, the specimens were aged for 5 days in air at 30 °C to completely degas the absorbed hydrogen. The specimens were then cut to 1 mm in gauge length for hydrogen analysis. The deformation layer on the specimen surface of all three types was removed by mechanical polishing followed by chemical polishing, which reduced the specimen thickness to 0.2 mm. Hydrogen was used as a tracer to clarify the changes in the specimen substructure during the application of cyclic elastic stress with hydrogen. Tracer hydrogen was charged into the specimens by immersion for 24 h in a 0.5 $g{\cdot}L^{-1}$ NH_4SCN solution kept at 30 °C. The specimens were then immediately immersed in liquid nitrogen to prevent the release of hydrogen. States of hydrogen were examined using low-temperature thermal desorption spectroscopy (L-TDS), which can heat specimens from lower temperatures than conventional TDS. The specimens were then mounted on the heating stage of the L-TDS equipment and heated at a rate of 1 °C/min from –200 to 200 °C.

Table 1. Chemical composition of specimens (mass %)

C	Si	Mn	P	S
0.34	0.27	0.80	0.012	0.008

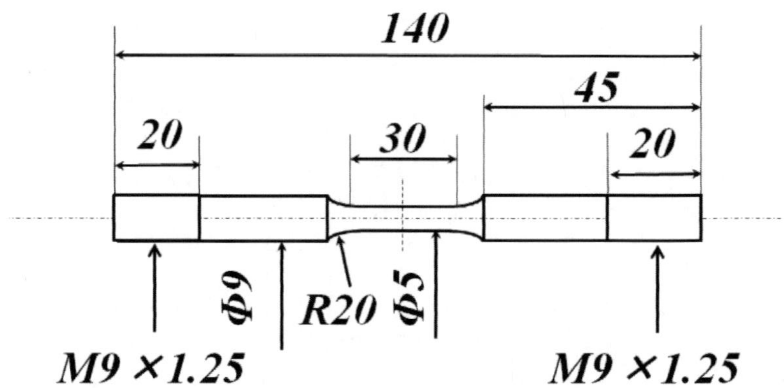

Figure 1. Geometry and dimensions (in mm) of the specimens subjected to cyclic elastic stress

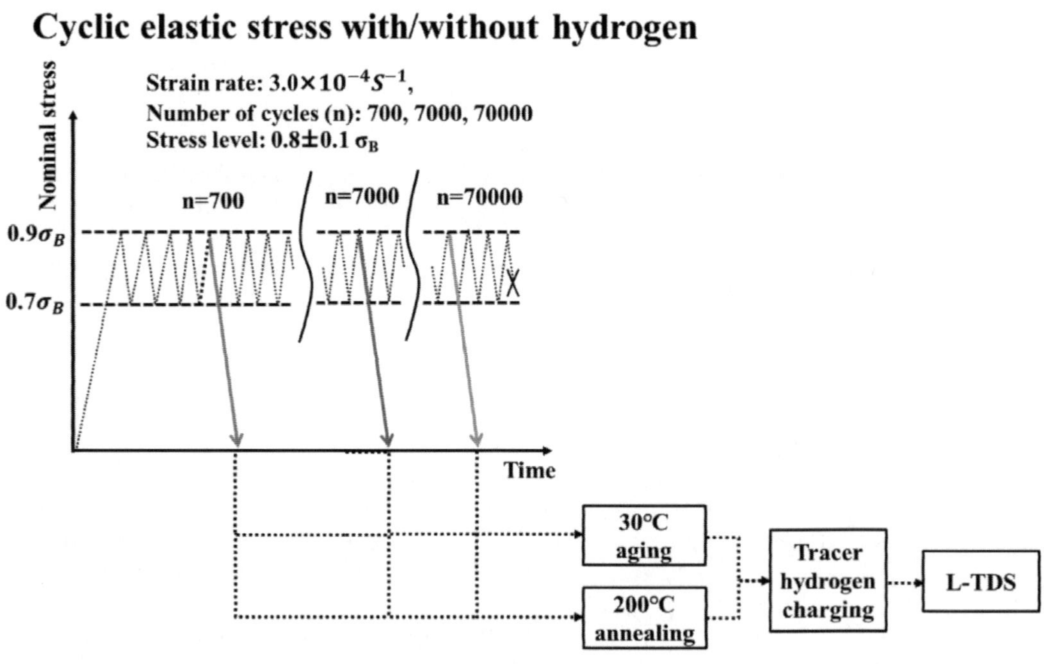

Figure 2. Schematic diagram of the applied cyclic elastic stress and subsequent L-TDS

3 Results

3.1 Tracer hydrogen desorption spectra of the [(n)-cycled] specimens

Figure 3 shows the tracer hydrogen desorption spectra of the [(n)-cycled] specimens, after tracer hydrogen charging under the same conditions as indicated in Section 2.4. The tracer hydrogen of the specimens firstly desorbed at approximately -50 °C and disappeared at approximately 60 °C in all the tracer hydrogen desorption spectra. The figure shows no significant difference among the specimens with regard to the states of hydrogen. For the [(n)-cycled] specimens subjected to cyclic elastic stress without hydrogen, the influence of the number of cycles on the states of tracer hydrogen was small.

3.2 Tracer hydrogen desorption spectra of the [H+(n)-cycled] specimen

Figure 4 displays the tracer hydrogen desorption spectra of the [H+(n)-cycled] specimens, after tracer hydrogen charging under the same conditions as indicated in Section 2.4. The tracer hydrogen of the [H+700-cycled] specimen firstly desorbed at approximately -50 °C and disappeared at approximately 60 °C. The tracer hydrogen desorption behavior of the [H+700-cycled] specimen was similar to that of the [(n)-cycled] specimens. However, the tracer hydrogen of the [H+7000-cycled] and [H+70000-cycled] specimens firstly desorbed at approximately -50 °C and disappeared at approximately 100 °C as shown in Figure 4. The disappearance temperature of the tracer hydrogen desorption spectra of the [H+7000-cycled] and [H+70000-cycled] specimens shifted from 60 °C to 100 °C in comparison with that of the [(n)-cycled] specimens. Tracer hydrogen desorption in the high-temperature range from the [H+7000-cycled] and [H+70000-cycled] specimens increased in comparison with that from of the [(n)-cycled] specimens. Additionally, one new peak appeared at approximately 60 °C. These findings indicate that application of cyclic elastic stress in the presence of hydrogen formed new lattice defects with a higher number of cycles.

3.3 Tracer hydrogen desorption spectra of the [H+(n)-cycled+annealed] specimens

Figure 5 depicts the tracer hydrogen desorption spectra of the [H+(n)-cycled+annealed] specimens, after tracer hydrogen charging under the same conditions as indicated in Section 2.4. The tracer hydrogen of the specimens firstly desorbed at approximately -50 °C and disappeared at approximately 60 °C in all the tracer hydrogen desorption spectra. The tracer hydrogen desorption behavior of the [H+(n)-cycled+annealed] specimens was similar to that of the [(n)-cycled] specimens. Tracer hydrogen desorption in the high-temperature range from the [H+(n)-cycled+annealed] specimens was reduced by annealing at 200 °C to the level of the [(n)-cycled] specimens in comparison with that of the [H+(n)-cycled] specimens. These findings indicate that the new lattice defects formed by application of cyclic elastic stress in the presence of hydrogen were annihilated by annealing at a temperature of 200 °C.

3.4 Effects of the number of cycles on tracer hydrogen desorption spectra of three types of specimen

Figure 6 shows the effects of the number of cycles of cyclic elastic stress on the tracer hydrogen contents of the [(n)-cycled], [H+(n)-cycled] and [H+(n)-cycled+annealed] specimens. The tracer hydrogen contents of the specimens were calculated from the area of the tracer hydrogen desorption spectra shown in Figures 3, 4 and 5. The tracer hydrogen contents of the [700-cycled], [7000-cycled] and [70000-cycled] specimens were 0.088, 0.092 and 0.093 mass ppm. The tracer hydrogen contents of the [H+700-cycled], [H+7000-cycled] and [H+70000-cycled] specimens were 0.099, 0.104 and 0.108 mass ppm. The tracer hydrogen contents of the [H+(n)-cycled] specimens increased in comparison with those of the [(n)-cycled] specimens. The tracer hydrogen contents of the [H+700-cycled+annealed], [H+7000-cycled+annealed] and [H+70000-cycled+annealed] specimens were 0.091, 0.094 and 0.096 mass ppm. The tracer hydrogen contents of the [H+(n)-cycled+annealed] specimens were similar to those of the [(n)-cycled] specimens. As shown in Figure 6, the influence of the number of cycles on the tracer hydrogen contents of the [(n)-cycled] and [H+(n)-cycled+annealed] specimens was small. In contrast, the tracer hydrogen contents of the [H+(n)-cycled] specimens increased with an increasing number of cycles.

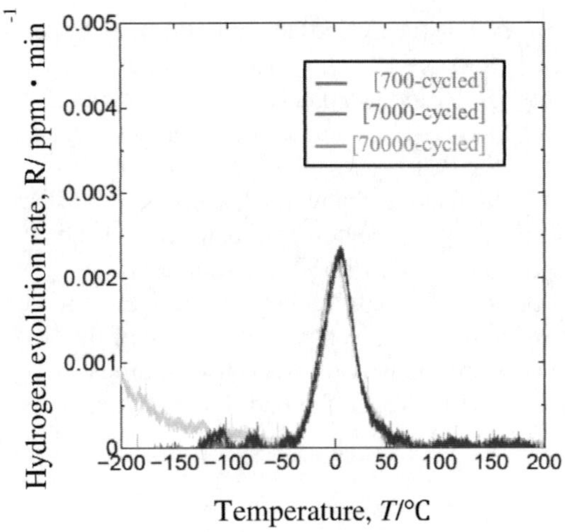

Figure 3. Tracer hydrogen desorption spectra of the [(n)-cycled] specimens

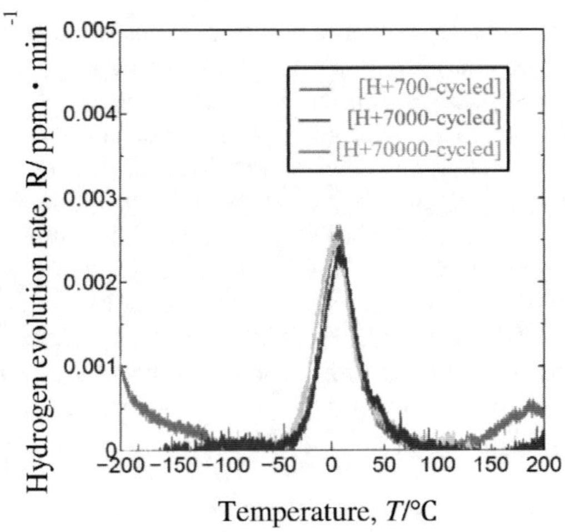

Figure 4. Tracer hydrogen desorption spectra of the [H+(n)-cycled] specimens

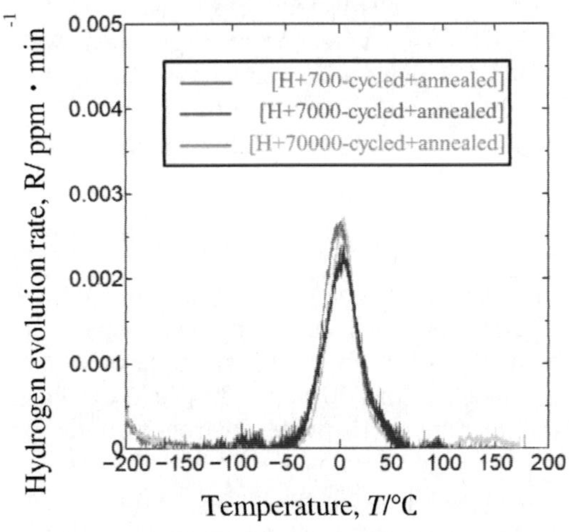

Figure 5. Tracer hydrogen desorption spectra of the [H+(n)-cycled+annealed] specimens

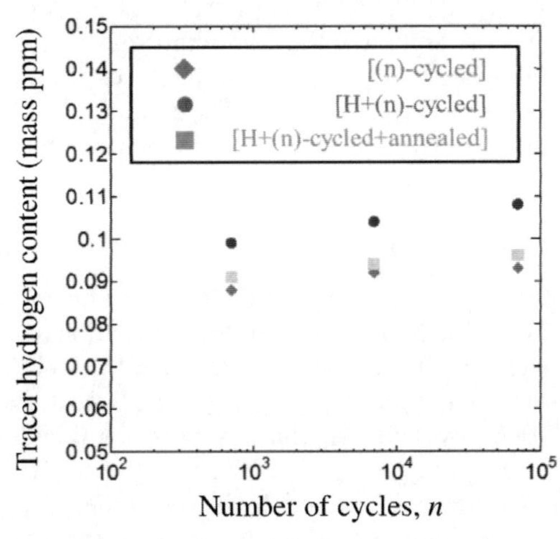

Figure 6. Effects of the number of cycles of cyclic elastic stress on the tracer hydrogen contents of three types of specimens.

4 Discussion

The hydrogen desorption temperature obtained from the tracer hydrogen desorption spectra provides an indication of the binding energy between hydrogen and trapping sites. In addition, the tracer hydrogen content measured by L-TDS provides an indication of the number of lattice defects in steel when hydrogen is charged until saturation under the same conditions. The tracer hydrogen desorption spectra in Figures 3 and 4 show that tracer hydrogen was desorbed in the high-temperature range from the the [H+(n)-cycled] specimen but not from the [(n)-cycled] specimens. In addition, the tracer hydrogen content of the [H+(n)-cycled] specimens increased in comparison with that of the [(n)-cycled] specimens even though tracer hydrogen charging was carried out until saturation under the same conditions. These findings suggest that changes occurred in the substructure of the tempered martensitic steel specimens during cyclic elastic stress with hydrogen such as the formation and accumulation of lattice defects. However, tracer hydrogen desorption in the high-temperature range from the [H+(n)-cycled+annealed] specimens was annihilated by annealing at a temperature of 200 °C as shown in Figure 5. Furthermore, the tracer hydrogen contents of the [H+(n)-cycled+annealed] specimens also decreased to the level of the [(n)cycled] specimens as shown in Figure 6. The tempered martensitic steels used in present study contained dislocations, cementite interfaces and grain boundaries as hydrogen trapping sites. The binding energies between hydrogen and dislocations, cementite interfaces and grain boundaries were 20-30 [5], 19-23 [6] and 10-27 [7] kJ mol^{-1}, respectively. It has also been reported that lattice defects annihilated at 200 °C were vacancy-type defects for medium-carbon steels [8], pure iron [9] and tempered martensitic steel [2]. Taking into consideration of all these factors, it is inferred that the formation of vacancy-type defects was enhanced under the application of cyclic elastic stress in the presence of hydrogen.

5 Conclusions

Changes in the substructure of tempered martensitic steel during the application of cyclic elastic stress in the presence of hydrogen were investigated using L-TDS. The results obtained in this study can be summarized as follows.

(1) The tracer hydrogen desorption behavior of the specimens subjected to cyclic elastic stress without hydrogen showed no significant difference with a higher number of cycles.
(2) Tracer hydrogen desorption in the high-temperature range from the specimens subjected to cyclic elastic stress with hydrogen increased with a higher number of cycles.
(3) Tracer hydrogen desorption in the high-temperature range from the specimens subjected to cyclic elastic stress with hydrogen was reduced by annealing at 200 °C to the level of the specimens subjected to cyclic elastic stress without hydrogen. The formation of vacancy-type defects was enhanced under the application of cyclic elastic stress in the presence of hydrogen.

References

[1] Doshida T et al (2012) Enhanced lattice defect formation associated with hydrogen and hydrogen embrittlement under elastic stress of a tempered martensitic steel *ISIJ International* **52.2** 198-207
[2] Saito K et al (2015) *CAMP-ISIJ* **28** 935
[3] Birenis D et al (2018) Interpretation of hydrogen-assisted fatigue crack propagation in BCC iron based on dislocation structure evolution around the crack wake *Acta Materialia* **156** 245-253
[4] Uyama H et al (2006) Effects of hydrogen charge on microscopic fatigue behaviour of annealed carbon steels *Fatigue & Fracture of Engineering Materials & Structures* **29.12** 1066-1074
[5] Hirth JP (1980) Effects of hydrogen on the properties of iron and steel *Metallurgical Transactions A* **11.6** 861-890
[6] Hagi H (1994) Effect of interface between cementite and ferrite on diffusion of hydrogen in carbon steels *Materials Transactions* **35.3** 168-173
[7] Kaneko M et al (2013) *CAMP-ISIJ* **26** 978
[8] Nagumo M et al (2002) Function of hydrogen in intergranular fracture of martensitic steels *Philosophical Magazine A* **82.17-18** 3415-3425
[9] Takai K et al (2008) Lattice defects dominating hydrogen-related failure of metals *Acta Materialia* **56.18** 5158-5167

Possibilities of Assembling of Processing Maps by Utilizing of an Artificial Neural Network Approach

P Opěla[1], I Schindler[1], S Rusz[1], P Kawulok[1], R Kawulok[1], V Ševčák[1]

[1] VSB – Technical University of Ostrava, Ostrava, Czech Republic, EU

E-mail: petr.opela@vsb.cz

Abstract. The processing maps (i.e. power dissipation maps superimposed over instability maps) can be used as a very convenient tool in case of an optimizing of hot forming processes. In this research, processing maps of C45 medium-carbon steel were assembled on the basis of an experimental flow stress dataset. This dataset was acquired via series of uniaxial hot compression tests in the temperature range of 1173 K – 1553 K and the strain rate range of 0.1 s^{-1} – 100 s^{-1}. In addition, a predicted flow stress dataset was created with use of an artificial neural network approach – it allowed extending of the experimental dataset with additional temperature levels. The experimentally compiled processing maps have been subsequently enhanced by this additional dataset to encourage the overall information capability. The results have showed that the predicted dataset was useful to reveal additional instability regions in the experimentally assembled processing maps.

1 Introduction

In recent years, it has been showed that so-called processing maps are a very useful tool for an optimizing of hot forming processes. These maps simply represent the areas of temperature, strain rate and strain which are appropriate to use in a forming process. They, of course, also denote the areas at which the forming process becomes to be precarious [1, 2]. The processing maps are compiled by superimposing of power dissipation maps over instability maps. These maps are usually assembled by means of experimental flow stress data which are acquired by e.g. uniaxial hot compression or torsion tests. These tests are performed at given combinations of temperatures and strain rates – covering a presumed range of concrete forming process. However, the limited amount of the experimental flow stress data can lead to an inferior accuracy of the assembled processing maps. Nevertheless, this issue can be overcome by utilizing of a predicted flow stress dataset [3]. Flow stress prediction is usually performed by so-called flow stress models. These models contain different parameters (auxiliary variables), e.g. peak point coordinates, hardening and softening exponents. These parameters are dependent on the thermomechanical conditions (i.e. temperature, strain rate), so they need to be related to these thermomechanical circumstances [4]. In this work, in the first place, power dissipation and instability maps of the C45 medium-carbon steel have been compiled on the basis of an experimental flow stress dataset. These maps were then enhanced by a predicted flow stress dataset (assembled with use of an artificial neural network approach). The aim of this research is to determine the influence of the additional flow stress dataset on completeness of the assembled processing maps.

IOP Conf. Series: Materials Science and Engineering **461** (2019) 012063 doi:10.1088/1757-899X/461/1/012063

2 Experimental flow stress dataset

Experimental flow stress dataset of the C45 medium-carbon steel was acquired by means of series of uniaxial hot compression tests. The entire experimental procedure was in detail described previously in [5]. The tests were performed at the deformation temperatures (T) of 1553 K, 1473 K, 1373 K, 1273 K, 1173 K and the strain rates ($\dot{\varepsilon}$) of 0.1 s^{-1}, 1 s^{-1}, 10 s^{-1}, 100 s^{-1}, upon the true strain (ε) up to 1.0. Afterward, twenty flow curves (i.e. dependence of true flow stress on true strain at given thermomechanical conditions) were assembled on the basis of this experimental dataset.

3 Predicted flow stress dataset

3.1 Hot flow stress models

The experimentally compiled flow curves were described via known flow stress models. Cingara and McQueen's model [6] was used to describe these curves in the strain range of $0 \leq \varepsilon \leq \varepsilon_p$ (i.e. up to the peak point) – see equation (1). A modified version of this model [7] was then applied to describe the curves in the strain range of $\varepsilon_p \leq \varepsilon$ (i.e. beyond the peak point) – see equation (2):

$$\sigma = \sigma_p \cdot \left[\frac{\varepsilon}{\varepsilon_p} \cdot \exp\left(1 - \frac{\varepsilon}{\varepsilon_p}\right)\right]^c \tag{1}$$

$$\sigma = \sigma_{ss} + \left(\sigma_p - \sigma_{ss}\right) \cdot \left[\frac{\varepsilon}{\varepsilon_p} \cdot \exp\left(1 - \frac{\varepsilon}{\varepsilon_p}\right)\right]^s \tag{2}$$

In the equations (1) and (2), ε (-) is true strain and σ (MPa) is corresponding flow stress. The subscript p indicates the coordinates of maximum flow stress level (so-called peak point) and subscript ss corresponds to the initiation of steady-state flow. The variables c (-) and s (-) represent the work-hardening exponent and softening exponent, respectively [6, 7].

3.2 Description of the flow stress models parameters via an artificial neural network approach

The flow stress models parameters (i.e. ε_p, σ_p, σ_{ss}, c and s) are highly dependent on the temperature and strain rate level – so they need to be related to these thermomechanical circumstances before use in the models (1) and (2) [6, 7].

As the very first step, the experimental values of these parameters have to be obtained. The values of ε_p, σ_p and σ_{ss}, can be easily deducted at each combination of temperature and strain rate from the experimentally compiled flow curves. The experimental values of work-hardening exponent, c, can be obtained via regression analysis of the logarithmic form of equation (1), i.e. $\ln(\sigma / \sigma_p)$ vs. $\ln(\varepsilon / \varepsilon_p) + 1 - \varepsilon / \varepsilon_p$. Similarly, it is possible to achieve the values of work-softening exponent, s, from the logarithmic form of equation (2), i.e. $\ln[(\sigma - \sigma_{ss}) / (\sigma_p - \sigma_{ss})]$ vs. $\ln(\varepsilon / \varepsilon_p) + 1 - \varepsilon / \varepsilon_p$ [6, 7].

As the second step, an artificial neural network approach was utilized to mathematically describe the obtained experimental values of the above mentioned parameters. A Multi-Layer Feed-Forward Artificial Neural Network (ANN) with the Back-Propagation (BP) learning algorithm is the most used type of ANN in case of the material modeling [3]. This type of ANN was used to create five neural networks – one network for each examined parameter. Architecture of the generated networks was relatively simple. Each network connects the input variables (temperature and strain rate) with the output variable (one of the examined parameters) via set of artificial neurons arranged in three layers, see figure 1.

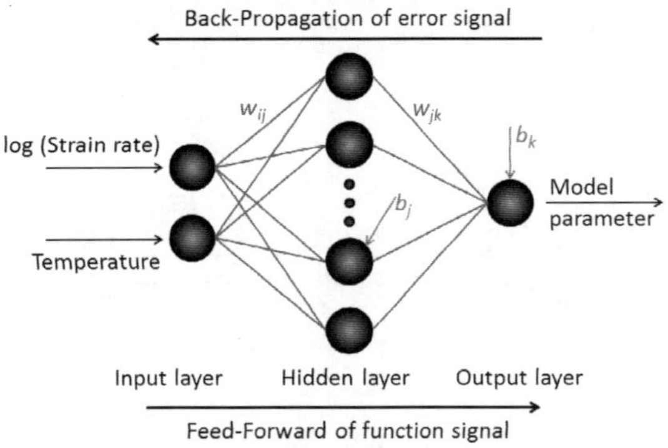

Figure 1. Schematic illustration of the used artificial neural network architecture.

The neurons inside of the neural network are connected via synaptic weights – w_{ij} (connecting input and hidden layer) and w_{jk} (connecting hidden and output layer). Each neuron of the hidden and output layer is at the same time connected with a unique bias value (b_j, b_k). This weights and biases must be then properly set to get the correct response of the network (provided by a network training process) [3]. Detailed specifications of the architecture of the created networks are presented in table 1. Appropriate number of neurons in the hidden layer of given network is clearly shown in table 2. Comparison of the experimental and predicted flow curves can be seen in figure 2.

Table 1. Architecture of the assembled artificial neural networks.

Network architecture parameter	**Settings**
Network type	Multi-Layer
Propagation of function signal	Feed-Forward
Learning (training) type	With supervisor
Learning (training) algorithm	Back-Propagation of error signal (BP)
Performance function	Mean Square Error (MSE)
Minimization algorithm of performance function	Levenberg-Marquardt (LM)
Hidden layer transfer function	Hyperbolic tangent sigmoid (tansig)
Output layer transfer function	Linear (purelin)

Table 2. Appropriate number of neurons in the hidden layer of given network.

Network intended for the parameter	ε_p	σ_p	σ_{ss}	c	s
Ideal hidden layer neuron number	5	3	3	8	6

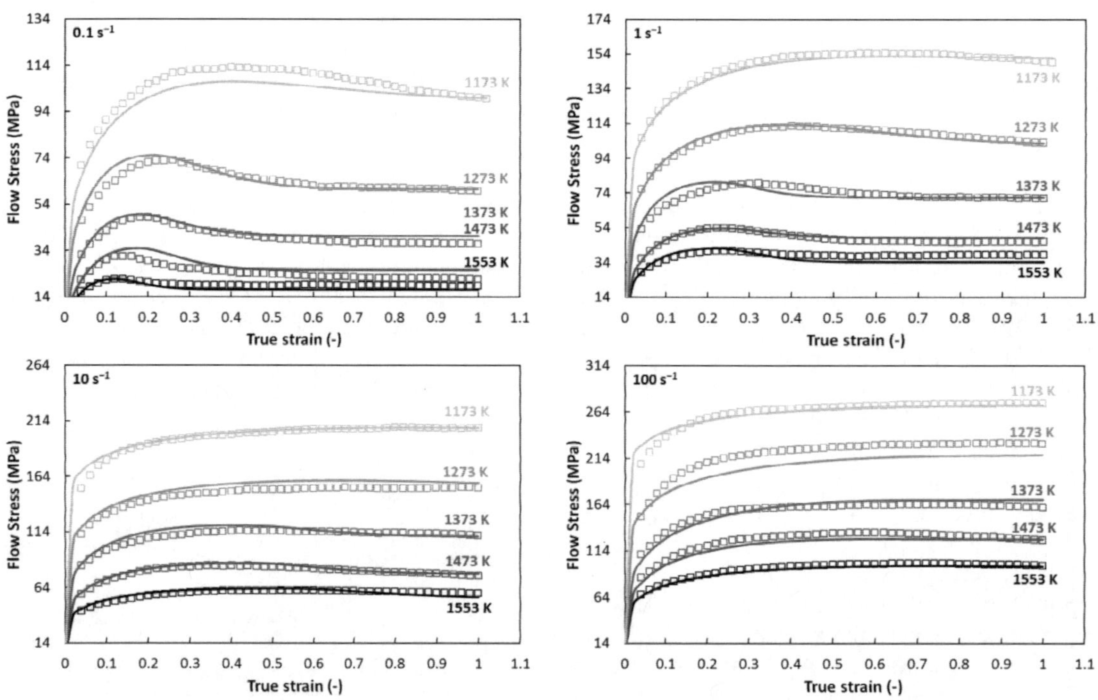

Figure 2. Experimental flow curves (squares) and predicted flow curves (lines) of the steel C45.

It can be seen, the flow tress level of the examined medium-carbon steel is highly dependent on temperature and strain rate – flow stress decreases significantly with an increase of temperature and decrease of strain rate. The course of the flow stress is also highly influenced by a combination of the thermomechanical parameters – see e.g. observation of peak point or steady-state flow in dependence on temperature and strain rate level. It can be noted, the observed flow stress behavior was possible sufficiently described by means of the above mentioned models (1) and (2) in combination with the ANN approach.

4 Processing maps

In this investigation, processing maps of the examined medium-carbon steel were assembled on the basis of so-called Dynamic Material Model (DMM) [8 – 10]. According to the DMM, the material passing through thermoplastic deformation can be considered as a power dissipater [11]. The instantaneous power per unit volume, P (J), which is dissipated by the formed material consist of two complementary parts – G (J) and J (J). The first one is the power dissipated as a consequence of temperature increase with respect to plastic deformation. The second one is the power consumed by metallurgical processes, e.g. dynamic recovery (DRV) and recrystallization (DRX) or defect formation [1, 2, 12]. The power dissipation can be quantified on the basis of so-called efficiency of power dissipation, η (-) [8]:

$$\eta = \frac{2 \cdot m}{m+1} \tag{3}$$

In equation (3), m (-) is the strain rate sensitivity index which determines the correlation between the G content and J co-content – it can be expressed as [1]:

$$m = \frac{dJ}{dG} = \frac{\dot{\varepsilon} d\sigma}{\sigma d\dot{\varepsilon}} = \frac{d \ln \sigma}{d \ln \dot{\varepsilon}} \tag{4}$$

In equation (4), σ (MPa) is the instantaneous flow stress at given strain and temperature level, and $\dot{\varepsilon}$ (s^{-1}) is the strain rate [1] – the relevant data can be obtained from the flow stress dataset. It should be noted, the data points of ln σ vs. ln $\dot{\varepsilon}$ should be fitted by cubic spline [11]. A graphical representation of the efficiency of power dissipation with respect to temperature and strain rate is known as so-called power dissipation maps [1, 12]. These maps simply reveal various domains which correlate with specific microstructural processes (e.g. DRV, DRX) [11]. Each microstructural process is associated with the specific η-range, e.g. η-values in the range of 30 – 50 % are typical for the DRX process and η-values above of 60 % are then attributed to superplasticity [12]. In order to detect the regimes of flow instability during hot deformation, a continuum criterion was proposed by Kumar [13] and Prasad [14] on the basis of Ziegler's [15] principle of maximum rate of entropy production. In accordance with this criterion, flow instability can occur if [1]:

$$\xi(\dot{\varepsilon}) = \frac{\partial \ln\left(\frac{m}{m+1}\right)}{\partial \ln \dot{\varepsilon}} + m \leq 0 \tag{5}$$

In equation (5), $\xi(\dot{\varepsilon})$ is the parameter of flow instability. The variation of this parameter with temperature and strain rate is then known as flow instability map [1]. The above mentioned processing maps are then given by superimposing of the flow instability maps over the power dissipation maps [11].

With use of the above described approach, the processing maps of the medium-carbon steel C45 were assembled on the basis of two flow stress datasets. The first one is given by the experimental procedure. So, the experimental processing maps were compiled with use of five temperatures (1553 K, 1473 K, 1373 K, 1273 K and 1173 K) and four strain rates (0.1 s^{-1}, 1 s^{-1}, 10 s^{-1}, 100 s^{-1}). The second one is obtained on the basis of the mathematical description of the experimental dataset – when the flow stress was predicted at additional temperature levels. So, the predicted processing maps were compiled with use of ten temperatures (1553 K, 1523 K, 1473 K, 1423 K, 1373 K, 1323 K, 1273 K, 1223 K, 1173 K and 1073 K) and four strain rates (0.1 s^{-1}, 1 s^{-1}, 10 s^{-1}, 100 s^{-1}). The processing maps were compiled at true strains of 0.2, 0.4, 0.6 and 0.8. Figure 3 shows the processing maps created altogether on the basis of experimental and predicted flow stress data. The contours clearly show the efficiency of power dissipation, η, in percent. As mentioned above, the η-values in the range of 30 – 50 % are typical in case of DRX process [12]. Based on this assumption, it can be concluded, the examined steel is undergoing through the DRX which is, of course, gradually evolved by an increasing strain level.

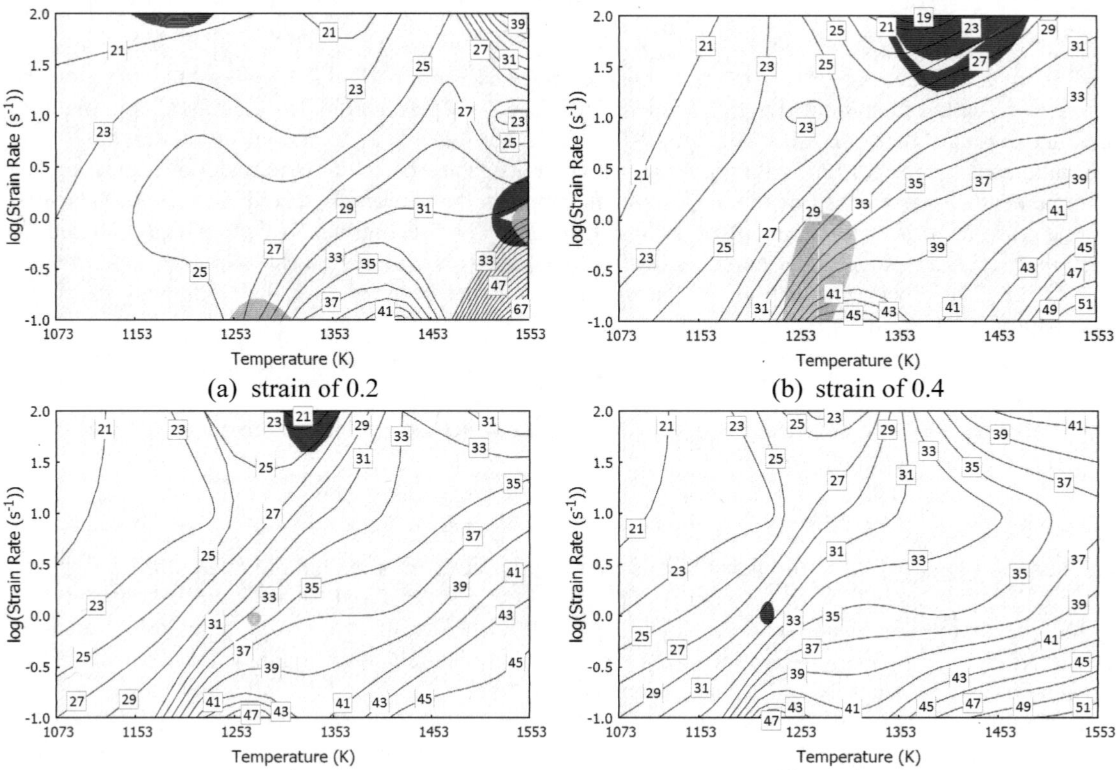

(a) strain of 0.2 (b) strain of 0.4

(c) strain of 0.6 (d) strain of 0.8

Figure 3. Processing maps of the steel C45; experimental instability (light gray areas) and predicted instability (dark grey areas).

It should be noted, the high value of power dissipation efficiency does not mean the necessarily unproblematic forming conditions. So, the flow instability criterion should be taken into account. The shaded areas represent the flow instability domains. Note there are two types of instability domains. The first one (light gray areas) corresponds to the instability domains which were revealed on the basis of the experimental flow stress dataset. The second one (dark grey areas) then relates to the instability domains which were achieved on the basis of the predicted flow stress dataset. It can be seen, there are two experimentally revealed instability domains at strain of 0.2. The first one is nearby the temperature of 1253 K and around the strain rate of 0.1 s^{-1}. The second one is around of 1553 K and 1 s^{-1}. This second domain is, however, somewhat extended by the predicted flow stress dataset. In addition, the predicted dataset revealed another instability domain near to the temperature of 1153 K and strain rate of 100 s^{-1}. Two distinct areas of instability flow are clearly visible at the strain of 0.4. The first one was calculated on the basis of the experimental dataset in the vicinity of 1253 K and the strain rate range of 0.1 s^{-1} – 1 s^{-1}. The second one was revealed on the basis of predicted dataset in the temperature range of 1353 K – ca 1500 K and strain rate range of ca 17 s^{-1} – 100 s^{-1}. Quite different situation occurs at the strain of 0.6. The experimental flow stress dataset reveals very insignificant instability area (close to 1253 K and 1 s^{-1}). Nevertheless, the predicted dataset allowed revealing of instability area at the temperature range of 1300 K – 1353 K and the strain rate range of 32 s^{-1} –

100 s^{-1}. Only one tiny instability area can be then observed at the strain of 0.8 – revealed just on the basis of the predicted dataset. Based on the above performed analysis can be stated, the experimental flow stress dataset can be insufficient in case of revealing of instability flow areas. The additional (predicted) flow stress dataset can be then used to reveal other areas of instability flow, so the assembled processing maps could be hypothetically more accurate. This hypothesis, however, should be proved by metallographic analyses.

5 Conclusion

Experimental flow curves of the C45 medium-carbon steel were mathematically described by means of two flow stress models in combination with an artificial neural network approach. Additional flow curves were then predicted to extend the experimental dataset. The experimental and predicted datasets were then used to compile processing maps of the examined steel. Results have showed the additional (predicted) dataset is helpful to reveal higher amount of flow instability areas. This fact should lead to the greater information capability of the assembled processing maps. Nevertheless, the metallographic analyses should be performed to confirm this assumption.

Acknowledgements

This paper was created on the Faculty of Metallurgy and Materials Engineering in the Project No. LO1203 "Regional Materials Science and Technology Centre – Feasibility Program" funded by Ministry of Education, Youth and Sports of the Czech Republic.

References

[1] Gao X, Jiang Z, Wei D, Jiao S and Chen D 2014 *Key Eng. Mater.* **622-623** 330–39
[2] Kliber J 2016 *Materials Physics and Mechanics* **25** 16–21
[3] Quan G-Z, Zou Z-Y, Wang T, Liu B and Li J-Ch 2017 *High Temp. Mater. Processes* **36** 1–13
[4] Gronostajski Z 2000 *J. Mater. Process. Technol.* **106** 40–44
[5] Opěla P, Schindler I, Kawulok P, Vančura F, Kawulok R, Rusz S and Petrek T 2015 *Metalurgija* **54** 469–72
[6] Cingara A and McQueen H J 1992 *J. Mater. Process. Technol.* **36** 31–42
[7] Opěla P, Schindler I, Kawulok P, Vančura F, Kawulok R and Rusz S 2016 Proc. 25th Anniversary International Conference on Metallurgy and Materials (Brno) (Ostrava: Tanger Ltd) p 458
[8] Prasad Y V R K, Gegel H L, Doraivelu S M, Malas J C, Morgan J T, Lark K A and Barker D R 1984 *Metall. Trans.* A **15** 1883–92
[9] Gegel H L, Malas J C, Doraivelu S M and Shende V A 1987 Metals Handbook vol 14, (ASM International) p 417
[10] Alexander J M 1989 Modelling Hot Deformation of Steels, ed J G Lenard (Berlin: Springer-Verlag) p 101
[11] Quan G-Z, Zhao L, Chen T, Wang Y, Mao, Y-P, Lv W-Q and Zhou J 2012 *Mater. Sci. Eng.* A **538** 364–73
[12] Sonnek P and Petruželka J 2001 Proc. 10th International Metallurgical and Materials Conference (Ostrava) (Ostrava: Tanger Ltd)
[13] Kumar A K S K 1987 Master's Thesis (Bangalore: Indian Institute of Science)
[14] Prasad Y V R K 1990 *Indian J. Technol.* **28** 435–51
[15] Ziegler H 1963 Progress in Solid Mechanics vol 4, (New York: John Wiley and Sons) p 93

IOP Conf. Series: Materials Science and Engineering **461** (2019) 012064 doi:10.1088/1757-899X/461/1/012064

Influence of the steel rolling direction on the mechanical properties and distribution of the local stress in front crack

R Pała[1,1], S Lipiec[1,2], I Dzioba[1,3]

[1)] Faculty of Mechatronics and Mechanical Engineering, Department of Machine Design, Kielce University of Technology, Al. 1000-lecia PP 7, 25-314 Kielce, Poland

[1)] rpala@tu.kielce.pl, [2)]slipiec@tu.kielce.pl, [3)]pkmid@tu.kielce.pl

Abstract. In this article presented the results of studies of strength characteristics (yield strength- σ_y, ultimate strength σ_{uts}, power exponent- *n)* and fracture toughness (J_R, J_{IC}) determined on the specimens of steel S235 taken in the rolling and perpendicular direction. Standard strength characteristics are similar for the specimens in both directions. However, there occurs an essential difference in fracture toughness characteristics for the specimens from different directions. To determine (explain) factors that cause this phenomenon performed some metallographic studies and the analysis of local stress and strain distributions in front of the crack.

1 Introduction

Low-carbon structural steel is a widely used material in structural elements and in construction. The termo-mechanical process causes the grains to elongate along the direction of rolling and the segregation of non-metallic inclusions. Increased amount of impurities and poor selection of plastic processing parameters intensifies the phenomenon of segregation, clusters of inclusions form, which cause anisotropy of mechanical properties and especially fracture toughness [1, 2, 3, 5, 6, 7]. Figures 1a and 1b show examples of microstructure pictures for specimens cut out in two perpendicular directions with visible inclusions. The specimens for fracture toughness tests were taken in the T-L and L-T directions according to the ASTM standard [4].

a) b)

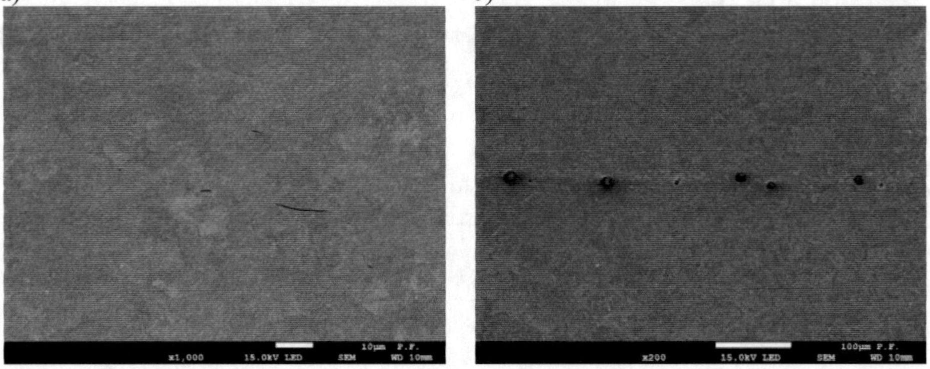

Figure 1. Steel microstructure with visible inclusions for the direction: a) T-L, b) L-T.

2 Experimental studies

Specimens for determination of strength characteristics for a uniaxial tensile test with a diameter of 5 mm and a 25 mm extensometer base [8] were made directly from tested SENB specimens cut from a 12 mm thick plate in the T-L, L-T directions.

The determined mean values of nominal strength characteristics are similar for two directions (table 1). The nominal stress-strain diagrams (σ_n-ε_n), have different trend after reaching the ultimate strength value, $\sigma_{uts,}$ (when neck is formed) (figure 2a). Differences were obtained in relative necking levels (Z), and particular in the fracture stress values (σ_f) for the two analyzed directions of specimen cutting.

Table 1. Nominal values of mechanical characteristics

	σ_y	σ_{uts}	σ_f	Z	A_5
	MPa	MPa	MPa	%	%
T-L	336	539	1070	66	26.9
L-T	335	542	1300	76	27

Fracture toughness was determined on three-point bent specimens with one-edge notch, SENB, with dimensions B=11 mm, W=22 mm, a_0/W≈0.5, cut in the T-L and L-T orientation. The control and registration system made it possible to record force signals, specimen deflections at the point of load, crack opening displacement, potential changes. The obtained signals courses allowed to compute the crack growth by both the drop potential technique and unloading compliance technique.

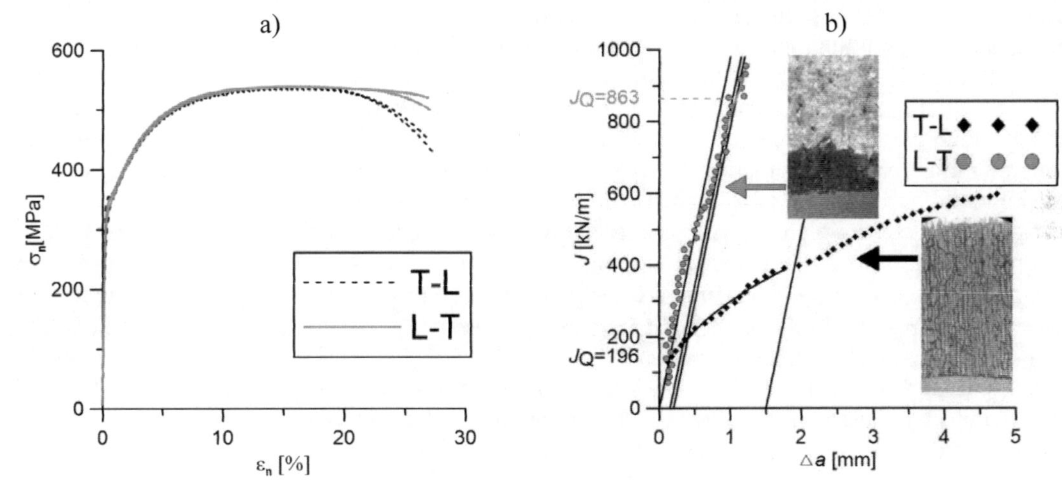

Figure 2. a) diagrams σ_n-ε_n for the analyzed directions, b) J_R curves with the views of the crack extension obtained by the change unloading compliance technique.

The resistance fracture toughness curve for the L-T direction runs significantly above the J_R curve for the T-L direction (figure 2b). Large differences in the critical values of J_Q fracture toughness were also noted. For specimens of both analyzed directions the condition for the minimum thickness B_c (equation 1) was not met, which ensures the dominance of plane strain in front of the crack (table 2). Thus, the determined critical values of J integral cannot be considered as J_{Ic}, but only as J_Q values [4]. The J_Q value was determined at the intersection of the offset line with the appropriate J_R curve.

$$B_c = 25\frac{J_{\mathrm{Q}}}{\sigma_{\mathrm{y}}} \tag{1}$$

where: σ_y is the value of yield strength of the material at the test temperature, J_Q is the critical fracture toughness value obtained for the tested specimen.

Table 2. Values of fracture toughness and critical thickness for plane strain domination

	J_Q	σ_y	B	B_c	$B > B_c$
	kN/mm	MPa	mm	Mm	-
T-L	863	336	11	64.4	NO
L-T	196	335	11	14.6	NO

3 Numerical computations

The material constitutive relationships obtained as a result of the uniaxial tensile test, for the purpose of numerical computations, should be properly calibrated. This procedure was described in [12] (in print). Due to the analysis of the results in the direct proximity of the crack tip, the constitutive relationships were subjected to the calibration process [9, 10, 11, 12, 13]. As a result of numerical computations, distributions of stresses and strains in front of the crack were determined. The model of the three-point SENB specimen was used for computations, which was used during fracture toughness tests (figure 3). The computations were carried out in the Abaqus ver. 6.12-2 [14]. In the model the crack front was modeled as an arc with a radius of 0.12 mm. The specimen in the thickness direction was divided into 11 layers. In the model used 8-node three-dimensional finite elements. The arrangement of the mesh of elements was compacted as the crack front approached. In the definition of the boundary conditions of the computations, the possibility of displacement of the specimen surface to the crack front in the x direction, the area in the specimen axis from the y direction was blocked, and the lower roll was completely immobilized. In the computations the displacement of the upper roll which pressed on the specimen was controlled. The displacement value corresponded to the moment of initiation of the subcritical crack, determined as a result of experimental studies. The results presented from the axis of the numerical specimen model.

In the figure 4a presented a comparison of distributions of stress components in front of the crack, for both compared directions of specimen cutting. Higher levels of the stresses were recorded for specimen L-T direction. For the opening stresses (σ_{22}), the difference in the maximum is at the level of nearly 6.5%. For the σ_{11} component, the difference is approx. 4%, while for the σ_{33} component it is 3.5%. In each case, for a material with a direction L-T, the occurrence of the local maximum of stresses is shifted from the front of the crack, in relation to the material with the direction T-L. Much larger differences in data were obtained for the distributions of effective strains in front of the crack. (figure 4b). In this case, the effective strain values for L-T direction specimen were about twice as high.

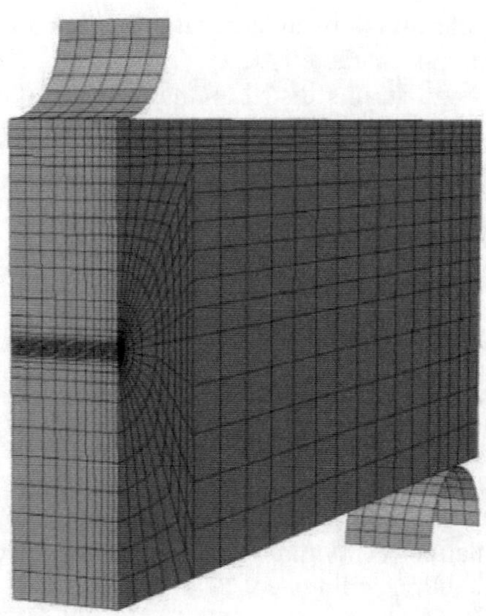

Figure 3. Modeled specimen type SENB.

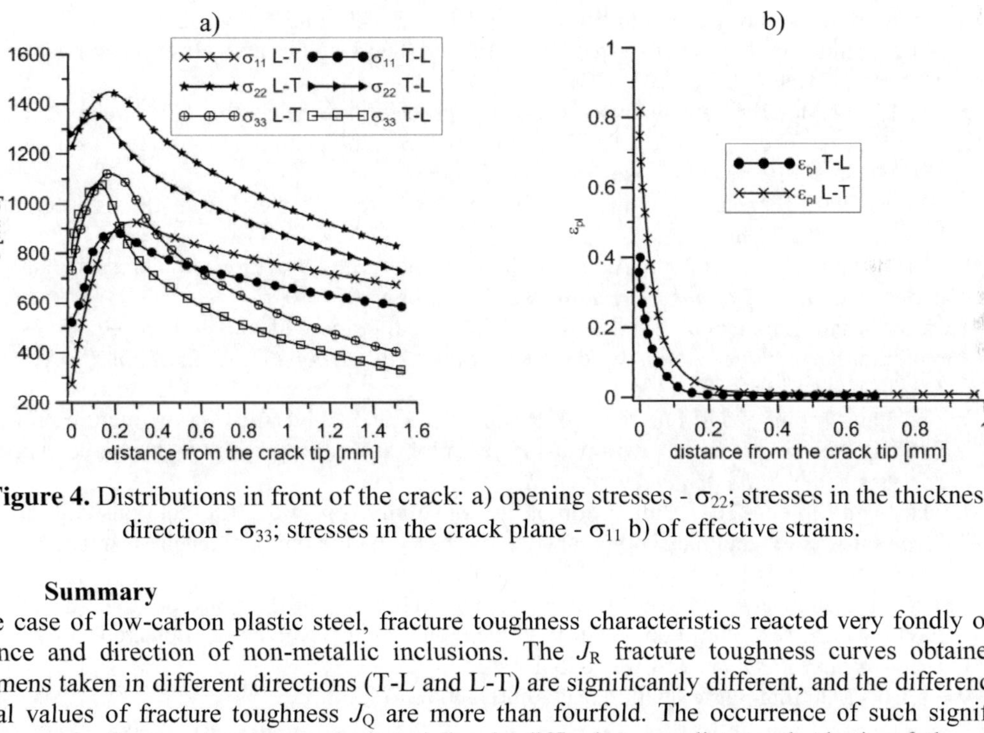

Figure 4. Distributions in front of the crack: a) opening stresses - σ_{22}; stresses in the thickness direction - σ_{33}; stresses in the crack plane - σ_{11} b) of effective strains.

4 Summary

In the case of low-carbon plastic steel, fracture toughness characteristics reacted very fondly on the presence and direction of non-metallic inclusions. The J_R fracture toughness curves obtained on specimens taken in different directions (T-L and L-T) are significantly different, and the differences in critical values of fracture toughness J_Q are more than fourfold. The occurrence of such significant differences in fracture toughness characteristics is difficult to predict on the basis of the results obtained in the uniaxial tensile test. The courses of nominal diagrams (σ_n-ε_n) are similar to the value of σ_{uts}. The differences are only on the section corresponding to the formation of the neck.

When the necking formed in the tensile tested specimen, the process of strengthening and damaging the material in 3D stress-strain state takes place. At the test to determine fracture toughness characteristics, the same processes take place in front of the tip during crack growth. So, comparing the nature of the courses of tensile diagrams at the stage of neck formation, we can conclude what material has higher fracture toughness characteristics.

Quantitative analysis of stresses and strains distributions in front of the crack tip at the moment of subcritical crack initiation obtained on the basis of numerical computations of modeled SENB specimens indicates higher stress and strain values in L-T samples as compared to T-L. This means that when the subcritical crack starts, the level of J integral is higher in L-T specimens, which is consistent with the experimental results.

Acknowledgements

The research was partially financed by: National Science Centre, Poland (No. 2017/25/N/ST8/00179), the Ministry of Science and Higher Education of Poland (No. 01.0.08.00/2.01.01.01.0008)

References

[1] Jaroniek M and Niezgodziński T 2014 Interaction of cracks in the case of lamelar cracks Material XXV Fatigue and Fracture Mechanics Symposium (Bydgoszcz) pp
[2] Pała R and Dzioba I Evaluation of fracture toughness of the specimens containing delamination cracking AIP Conference Proceedings vol 1780 pp 1-7
[3] Anderson TL 2008 Fracture Mechanics (New York)
[4] ASTM E1820-09 2011 Standard Test Method for Measurement of Fracture Toughness Annual book of ASTM standards pp 1070-1118.
[5] Kocanda D and Hutsaylyuk V and Slezak T and Torzewski J and Nykyforchyn H and Kyryliv V 2012 Fatigue crack growth rates of S235 and S355 steels after friction stir processing *Materials Science* 726 pp 203-2100
[6] Dzioba I 2012 Modeling and analysis of fracture process in ferritic steels (TU Kielce) p 242
[7] Dzioba I 2011 The influence of the microstructural components on fracture toughness of 13HMF steel Materials Science 47(5) pp 357-364
[8] ASTM E8 2003 Standard Test method for tension testing of metallic materials International West Conshohocken
[9] Bai Y and Wierzbicki T 2008 A new model plasticity and fracture with pressure and Lode dependence *International Journal of Plasticity* 24 pp 1071-1096
[10] Neimitz A and Gałkiewicz J and Dzioba I 2018 Calibration of constitutive equations under conditions of large strains and stress triaxiality *Archives of Civil and Mechanical Engineering* 18 pp 1123-1135
[11] Dzioba I and Lipiec S and Furmańczyk P and Pała R 2018 Investigation of fracture process of S355JR steel in transition region using metallographic, fractographic tests and numerical analysis *Acta Mechanica et Automatica* 12 pp 145-150
[12] Dzioba I and Lipiec S 2018 Calibration of the constitutive equations for materials with different levels of strength and plasticity characteristic based on the uniaxial tensile test data COMAT 2018 Recent Trends in Structural Materials (in print)
[13] Depreński Ł and Seweryn A 2011 Experimental Research into Fracture of EN-AW 2024 and EW-AW 2007 Aluminum Alloy Specimens with Notches Subjected to Tension *Experimental Mechanics* 51 pp 1075-1094
[14] Abaqus 6.12 Getting Started with Abaqus Interactive Edition.

OBTAINING OF LONG-LENGTH RODS WITH ULTRAFINE-GRAINED STRUCTURE BY THE RADIAL-SHEAR ROLLING

A B Naizabekov [1], S N Lezhnev [1], A S Arbuz [2] and E A Panin [2]

[1] Rudny Industrial Institute, Rudny, Kazakhstan,
[2] Karaganda State Industrial University, Temirtau, Kazakhstan

E-mail: cooper802@gmail.com

Abstract. Improving the quality of metal products by modifying the structure is one of the main trends in materials science. However, the main problem of such materials is the small final size of the workpieces and low technological production. The paper presents data of the experimental production of ultra-fine-grained structure in a long rod of steel grade AISI 321 by radial-shear rolling. With a grain size of 300-600 nm, the mechanical properties and micro-hardness increase by more than 2 times while having a heterogeneity across the bar section.
Keywords: radial-shear rolling, ultrafine-grained structure, stainless steel, mechanical properties, microhardness, intensive plastic deformation, transmission electron microscopy

1 Introduction

Currently, progress in the production of new metal materials is focused on a deep change in the internal structure of the metal. The most effective way to improve the structure and properties of the metal is severe plastic deformation (SPD). Thus, bringing the metal microstructure into an ultra-fine-grained (UFG) state opens up great prospects in the field of materials engineering [1-2].

In the process of grinding the microstructure, with increasing the proportion of intergranular boundaries in the volume of the workpiece, along with changes in the mechanical properties of the material, its other physical properties also change, such as electrical resistance or radiation resistance [3]. The latter phenomenon is based on the fact that the grain boundaries serve as the drain surface for radiation defects of the grain crystal lattice, which are mass formed due to neutron irradiation in the reactor core. A large specific number of grain boundaries in UFG-structural materials significantly increase the number of such surfaces. The use of SPD will provide a new class of materials for nuclear power not only with improved mechanical properties, but also with increased resistance to neutron irradiation. This problem is very relevant, because a significant share in the cost of nuclear power plants is the replacement and disposal of structures operating in the core. The extension of the life cycle of such elements by manufacturing them from UFG materials can justify the relatively high cost of such products and make them commercially attractive, which in general is of high relevance for the UFG products market.

However, here the difficulties associated primarily with the lack of a reliable way to obtain a sufficient number of such materials with sufficient dimensions for commercial products. The most frequently used and studied methods, such as equal-channel angular pressing (ECAP) with good microstructure quality, have critical limitations on the length of the resulting workpiece, which significantly complicates its industrial application [1]. The method of multi-axis forging allows you to get a relatively large workpiece. On the other hand it is very time-consuming and has its limitations. The problem of obtaining a long billet with UFG-structure is solved by means of ECAP-Conform and rolling-pressing, but only for non-ferrous metals and alloys. The method of accumulated roll bonding (ARB) for sheet materials is promising, but also still has a number of disadvantages for commercial use.

2 Experimental part

Radial-shear rolling (RSR) allows to obtain UFG structure of the long-length workpieces, but also has its disadvantages associated with the homogeneity of the structure and properties of the cross section of the workpiece (from the edge to the center). This method was invented by S. P. Galkin in "MISIS" [4-5]. The difference from the usual cross-screw rolling, which is used in the tube rolling, is that a single round bar is rolled here along a three-roll scheme with increased feed angles of the rolls, which contributes to the intensification of shear deformation and non-monotonicity of the metal flow under the predominance of the hydrostatic compression scheme [1]. Such conditions are favorable for the formation of UFG structure, especially at low temperatures of deformation.

Despite the lack of unevenness of the cross-section structure, this method has great potential, because unlike other methods it is as simple, efficient and technological as possible. And this drawback may be insignificant in the case of production of products having the shape of a pipe (for example, tubes of fuel elements of the reactor). It adds relevance to the question of obtaining UFG structure on the materials used in the nuclear power industry.

For the experiment, austenitic stainless steel AISI-321 (0.08% C, 17-19% Cr; 9-11% Ni; 2% Mn; 0.8% of Si; 0.5 to 0.7% Ti, all in. wt. %) was chosen, since it is the main structural material of nuclear power. In particular, it is used for the manufacture of tubes of fuel elements of the most advanced fast neutron reactors. This steel is also used for the production of equipment operating in high temperature and aggressive environments (heat exchangers, pipes, parts of furnace and reactor valves, spark plug electrodes).

The experiment was conducted at Rudny Industrial Institute on the radial-shear rolling mill in-house marked 14-40. The mill is designed for hot and warm deformation of bars from almost any metal materials, including low-plastic, cast, composite and powder blanks. A special feature of the mill is the construction of the working stand of increased rigidity, which is a massive cylinder with grooves for the rolls cassettes. For expansion of assortment of mill two types of cassettes with rolls of a diameter of 90 mm and rollers with a diameter of 46 mm were used. Power of motors is 3×7.5 kW. Since one of the conditions for the formation of UFG structure is a reduced temperature of deformation, the temperature of the experiment 800 °C as the lowest technically possible value for rolling this steel at the mill 14-40 was chosen. The same temperature in the production of UFG structure on this steel by ECAP method in work [6] was used. The initial blank with a diameter of 30 mm and a length of 300 mm should be rolled in several passes to a diameter of 13 mm with an extraction equal to 5.

Before the experiment to determine the deformation parameters, in the software complex DEFORM-3D (SFTC, USA) finite element simulation of the process with the above conditions was carried out. The number of finite elements was 50 000, heat transfer and friction parameters - recommended for hot deformation. Geometric and speed parameters of the model corresponding to the real process, except for the length of the workpiece, which was significantly reduced, because the real mill, unlike the model, successfully rolled short workpiece is technically difficult.

3 Results and discussion

The results of the model calculation are shown in figure 1. According to the obtained results, the degree of deformation of the peripheral zone of the sample by the end of rolling exceeds 15, while the axial zone receives the degree of deformation about of 5-6. Such deformation is already considered to be large and can lead, under other favorable conditions, to the formation of UFG structure. However, the flow of metal in this zone is laminar, which should lead to the formation of highly elongated narrow grains such as rolling texture.

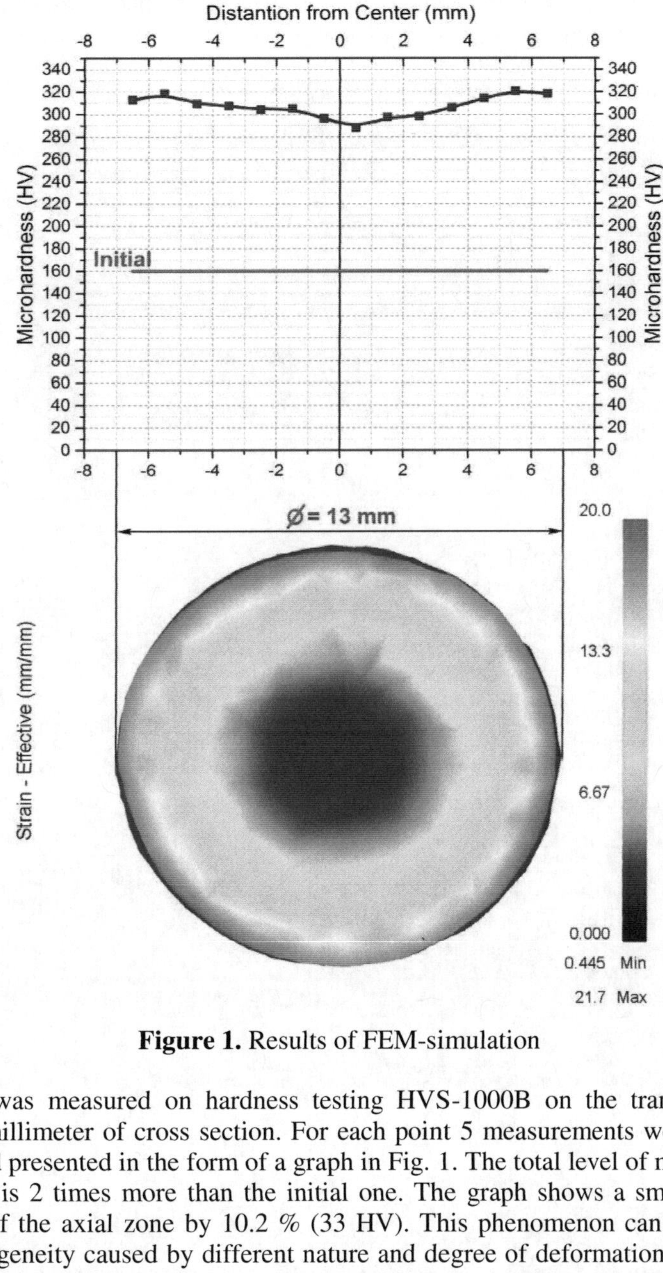

Figure 1. Results of FEM-simulation

Microhardness was measured on hardness testing HVS-1000B on the transverse microsection through every millimeter of cross section. For each point 5 measurements were made, the results are averaged and presented in the form of a graph in Fig. 1. The total level of microhardness is 290-320 HV, which is 2 times more than the initial one. The graph shows a smooth decrease in the microhardness of the axial zone by 10.2 % (33 HV). This phenomenon can be explained by the structural heterogeneity caused by different nature and degree of deformation of different areas of the workpiece. The microhardness drop in the center is also correlated with the calculated distribution of the degree of deformation in the cross section of the model below the graph.
Further, according to the described scheme, a physical experiment on the mill 14-40 was carried out. Before rolling, samples were taken from the rod to determine the original structure and properties. The initial structure is equiaxed grains of size 60-80 microns, tensile strength is 493 MPa, Vickers microhardness is 160HV (load is 9.87 N, exposure is 15 sec), which generally corresponds to the known data [6-7].

Heating was carried out in a tubular furnace Nabertherm R 170/1000/13 to 800 °C with a heating time of 30 minutes. Then the heated billet was rolled in series in several passes to a diameter of 13 mm, after which it was intensively cooled with water. After rolling, from deformed rod in the longitudinal direction at a distance of half the radius from the center were made flat samples 30 x 3 x 0,3 mm for tensile testing on the machine Instron-1195. According to the test results of 5 samples, tensile strength is 1073 MPa, which is more than 2 times higher than the original.

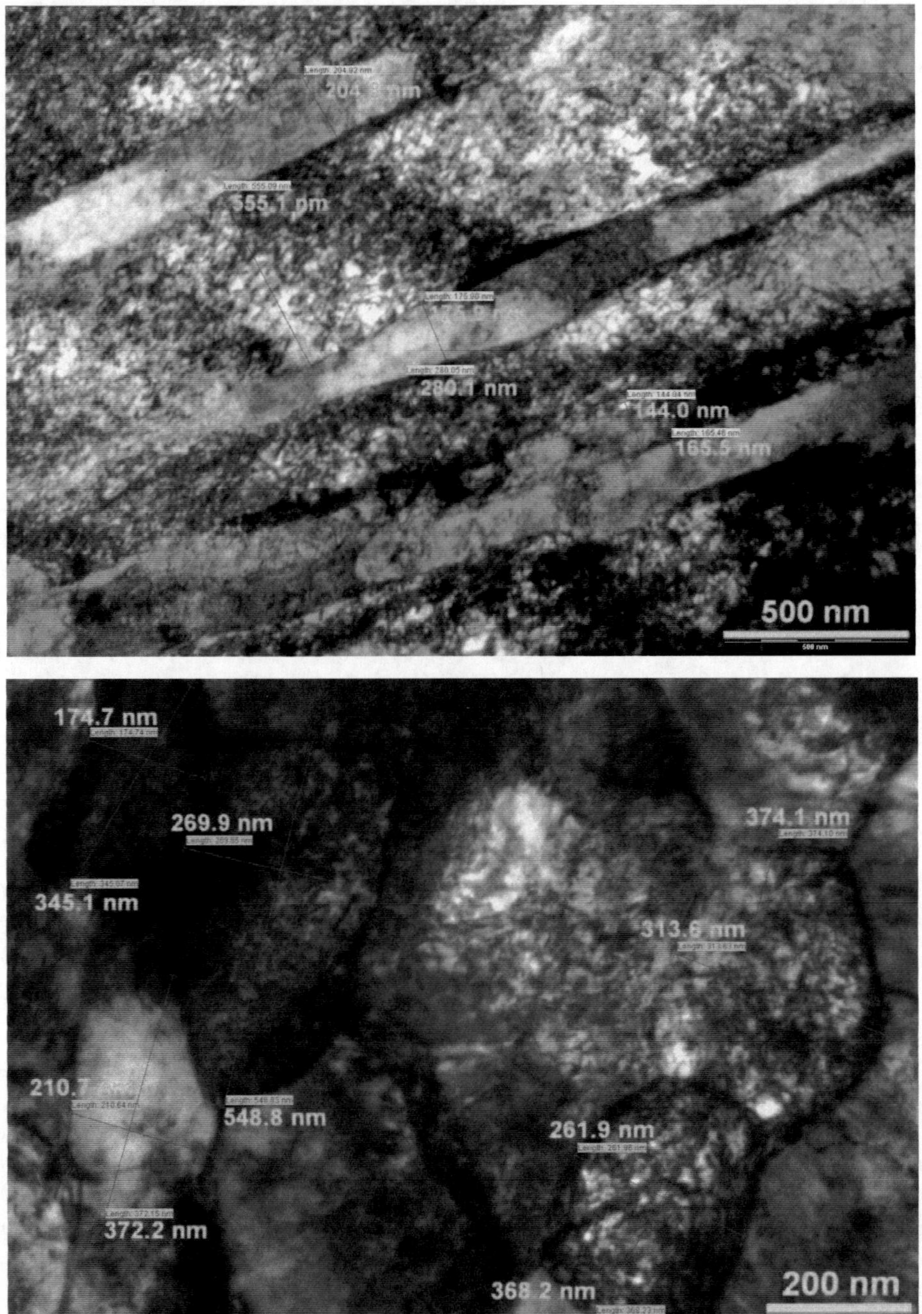

a)
b)

Figure 2. Results of experiment: a – axial zone, b – peripheral zone

From the plates cut out along the longitudinal section of the rod, samples were also prepared for the study of the microstructure of the axial and peripheral zones on the transmission electron microscope JEM-2100. Photos of typical areas are shown in Fig. 2. The images show that in the peripheral zone formed equiaxial UFG structure with grains of 300-600 nm, while in the axial zone formed a completely different type of structure – a texture with elongated grains of 200-600 nm and a length of several microns. The resulting structure corresponds to theoretical expectations, and correlates with the structure of the obtained by other methods of SPD under similar conditions [6]. However, it is important to note that in this case, a long steel rod UFG was obtained at a lower cost.

References

[1] Valiev R Z, Islamgaliev R K and Alexandrov I V 2000 *Prog. Mater. Sci.* **45(2)** 103-89.
[2] Koch C C, Ovidko I A, Seal S and Veprek S 2007 Structural nanocrystalline materials: Fundamentals and applications (Cambridge, Cambridge University Press) p 181.
[3] Maksimkin O P, Gusev M N, Tsai K V, Yarovchuk A V, Rybalchenko O V, Enikeev N A, Valiev R Z and Dobatkin S V 2015 *Phys. Met. Metallogr.* **116(12)** 1270-8.
[4] Patent of Russian Federation № 2293619 Method of screw rolling Galkin S P 2007 bulletin 5.
[5] Galkin S P 2014 *Steel in Translation* **44(1)** 61-4.
[6] Tsay K, Arbuz A, Gusseynov N, Nemkaeva R, Ospanov N and Krupenkin I 2016 *J. Chem. Technol. Metall.* **51(4)** 385-92.
[7] Rybalchenko O V Effect of severe plastic deformation on the structure, mechanical and service properties of steel 08H18N10T. Thesis for the degree of candidate of science Moscow 2014 p.107

IOP Conf. Series: Materials Science and Engineering **461** (2019) 012066 doi:10.1088/1757-899X/461/1/012066

Additively Manufactured CP-Ti (Grade 2) Single Strut Size Effect of Mechanical Response Under Building Direction

E. Pehlivan [1], M. Daniel [2] and J. Dzugan[3]

1 Department of Mechanics, Biomechanics and Mechatronics, Czech Technical University, Prague, Czech Republic, eren.pehlivan@fs.cvut.cz

2 Matej.Daniel@fs.cvut.cz and 3 jdzugan@comtesfht.cz

Abstract. The open cell structure is commonly used in the medical device industry and consists of small struts connected to each other. Relatively new additive manufacturing technique "SLS" has been used in this study. This approach allows to create a sufficiently small specimens to represent each strut in a commercially available orthopaedic porous structure. Pure titanium Grade 2 (CP-Ti grade 2) powder has been used during the experiments because of higher resistance to corrosion and it is one of the most biocompatible metal. The mechanical response of each strut affects the general behaviour of the porous system which influences stress shielding and bone attachment. In this study, small strut size effect and building direction of mechanical response have been investigated in order to provide vital data for upcoming rhombic dodecahedron structure FEM research

1. Introduction

Materials are vital for medical application and supposed to have a specific parameters to achieve their role. Metal materials are widely used in the medical application and their properties should be suitable to survive in human or animal body [1][2]. Commercially available pure titanium has the high specific strength which makes them good for medical application. Also, it can deliver excellent biocompatibility [3][4][5][6]. Surface coatings and porous architecture designs have been developed to improve the biological fixation of the implant due to increasing subtract surface [7]. Rapid prototyping allows creating titanium porous parts with mechanical response near to bone like material. New rapid prototyping technologies, like selective laser melting (SLM) and selective electron beam melting (SEBM) can produce porous titanium parts with multifaceted geometries and organized inner architectures [8]. 3D printed porous structure does not only provide low density, also lower stress shielding which may help to bone remodeling [9]. Cellular materials are also known as a porous structure and it acts like trabecular bone. SLS applications have been used with different unit cell shape such as cubic, octahedron, rhombic dodecahedron. Several types of research and commercial product use rhombic dodecahedron for biological and mechanical response close to the trabecular bone. Therefore, rhombic dodecahedron consists of 12 identical rhombic faces with 24 struts [10]. Each strut mechanical performance would affect the complex 3D porous structure and there is a lack of literature regarding the subject. In this work, additively manufactured pure titanium (grade 2) small strut has been investigated. The size of the small samples was chosen according to real rhombic dodecahedron unit cell in commercially available implants.

2. Material and Test Method

2.1 Material

The chemical composition of CP-Ti can be found in ***Table 1***. During specimen preparing Arcam CP-Ti Grade 2 powder has been used. The particle size is between 45 to 100 μm [11]. ***Table 2***. Shows the mechanical properties of CP-Ti grade 2. It has been expected that additively manufactured material is supposed to be higher yield strength than reference material data [12].

Table 1 Chemical composition of titanium grade 2 [11]

	C	Fe	O	N	H	Ti
Arcam Grade 2 CP-Ti	0,005%	0,05%	0,19%	0,004%	0,0009%	Balance
CP-Ti Grade2, Required*	<0,08%	<0,3%	<0,25%	<0,03%	<0,015%	Balance

*ASTM F67 (Unalloyed Titanium for Surgical Implant Applications)

Table 2 Mechanical properties of titanium grade 2[11]

	Yield Strength (σ_{YS})	Ultimate Tensile Strength (UTS)	Elongation	Reduction of Area
Arcam Grade 2 CP-Ti	540 MPa	570 MPa	21%	>20%
CP-Ti Grade2, Required*	275 MPa	345 MPa	55%	>30%

*ASTM F67 (Unalloyed Titanium for Surgical Implant Applications)

2.2 *Designing struts and SLS* process

Plate tensile specimens were designed with SolidWorks CAD software (SolidWorks Corp., Dassault Systemes, Concord, MA, USA) and imported to Materialize (Leuven, Belgium) software for final orientation to run in ConseptLaser the M2 Cusing (ConceptLaser, Lichtenfels, Germany) 3D printer. The thickness of the samples are designed to be 0.5mm however, their printed parameters might be slightly different due to the accuracy of the printing machine. The length of the samples is adjusted with 22mm and 7 mm thickness on the machine grip part. Parameter "a" has been changed the range of 0.15mm to 4.2mm and 74 samples were printed in order to reach a fine result, ***Figure 1***.

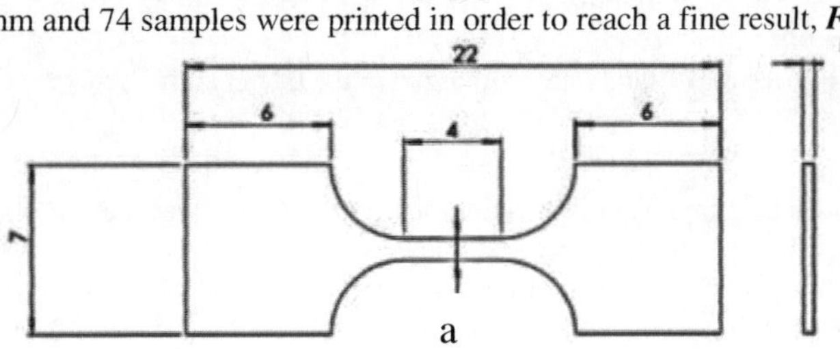

Figure 1 Technical drawing of small tensile specimens (mm)

The cross-sectional area was calculated by multiple with parameter "a" and "b". All samples were printed perpendicular to the platform surface (z-direction). Cross-sectional area (S_0) of the samples are up to 2.4 mm^2. 46 samples were printed orientation of ZXY and 27 samples were built with YZX orientation, ***Table 3***. Illustrates the number of printed samples with building orientation respectively.

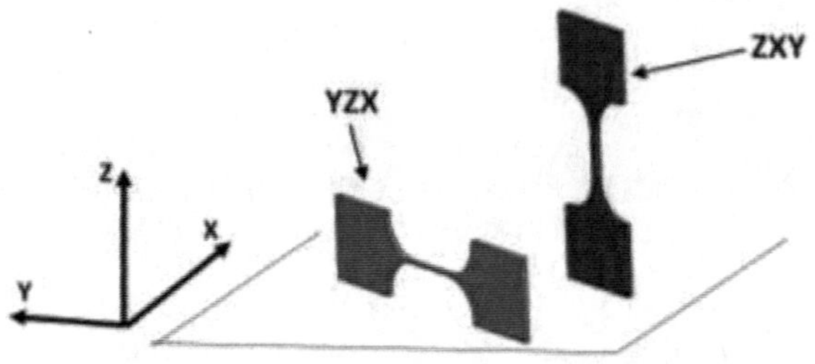

Figure 2 Sample printing direction on the platform

Table 3 Specimen numbers and orientations

Orientation	Number of samples
ZXY	46
YZX	27

During additive manufacturing, ConseptLaser the M2 cusing (Lichtenfels, Germany) 3D printer has been used. Fundamental building parameters are, building chamber is not pre-heated, the laser beam power is set at 200W, the scan speed is 7m/s, the layer thickness is 20μm and 75μm offset distance. The scan strategy used was Concept Laser's patented 'island' scan strategy in which each layer is divided into 5 mm × 5 mm square islands and exposed in a zig-zag fashion. Scan vectors in each island are exposed at 90◦ with respect to the neighboring island and alternating layers. A shift of 1 mm with respect to the previous island layer is used to limit porosity build-up [13], ***Figure 3***.

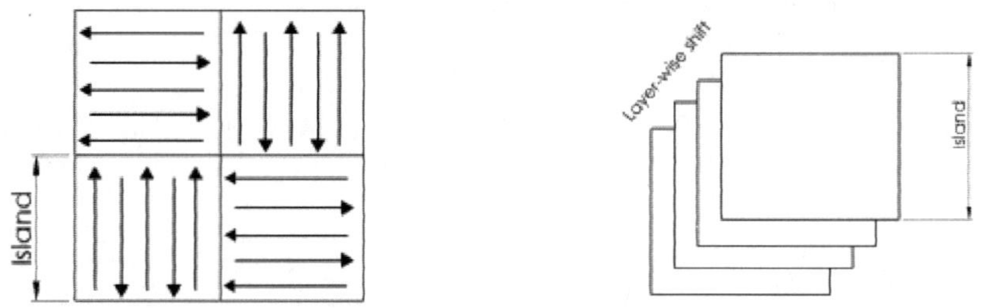

a) b)

Figure 3 a) Island hatch strategy and b) Island shift strategy

2.3 *Test Method*

The tensile test has been accomplished in room temperature. And it was carried out with the use of a small-sized linear drive-based testing system with a capacity of 5 kN using valid calibration certificate [14]. The test has been completed regarding the ASTM E8 requirements, initial displacement rate set as 0.2 mm/min. The tensile direction is perpendicular da printing layers. ***Chyba! Nenalezen zdroj odkazů.*** In order to maintain traceability, each sample kept in the pocket for the further surface analysis. Additively manufactured CP-Ti parts have tested without any chemical and mechanical post process.

2.4 *Geometrical investigation*

Since the tests sample has the significant small cross-sectional area, the accuracy of the SLS machine performance worth to be investigated. The powder size is approximately 50 μm therefore the smallest specimen would contain very few melted powder particle if we compare to specimens have bigger cross-section area. As a result of that, it may affect the general mechanical properties or elongation, further argument will be carried out in discussion part. Even minor unmelt particles would lead to a weak point around fracture areas. Scanning electron microscope (SEM) (Tescan VEGA-3 LMU) and RedLux (RedLux Ltd. Southampton, UK) machines have been used to investigate surface accuracy. Although during the tensile test laser measurement have been done, the load bearing area of the samples were validated with different aspects.

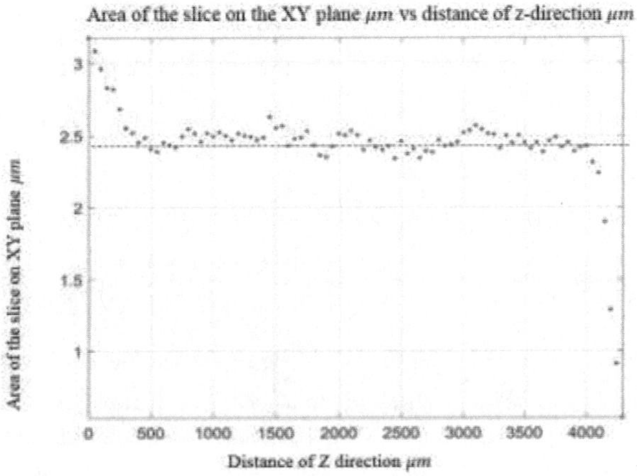

Figure 4 RedLux cross-sectional measurement on the XY plane. Result of sample with ZXY printing orientation.

After the tensile test, the samples were analysed with the RedLux measurement. RedLux is non-contact coordinate measuring machine (CMM) which provides 3D capture. It is mounted on an anti-vibration platform, the machine is unaffected by environment and gives on-screen results for analysis as soon as the measurement is completed. The machine has a white light confocal sensor, which of allows for analysis relevant surfaces without touching them and therefore potentially affecting the extent of the damage of the small fragile sample [15]. Scan performed with z-direction and it created helix on XY plane. The helix slice thickness is 0.05mm so, it contains a considerable number of cross-section area. Each section was projected onto the XY plane, and an envelope was constructed using the convhull function in MATLAB software environment, ***Figure 6*** illustrates the closed cross-sectional area. In ***Figure 4***, the points represent cross-sectional areas on the XY plane.

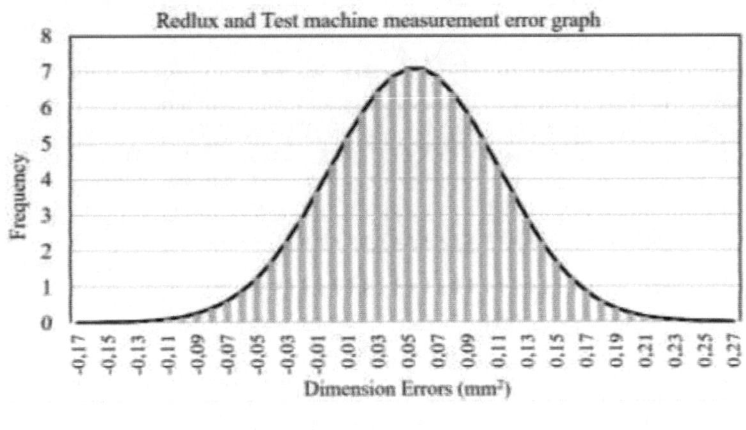

Figure 5 Redlux and tensile test machine measurement of cross-section area and dimension errors vs frequency

The method can provide also the statistical accuracy of geometry as well. Moreover, the closed loop result presents a load bearing area. Each sample has been investigated by the mentioned method and

IOP Conf. Series: Materials Science and Engineering **461** (2019) 012066 doi:10.1088/1757-899X/461/1/012066

compared with the test result. In order to understand the geometrical accuracy, tensile test machine and Redlux machine measurement of cross-section area have compared, ***Figure 5*** shows statistical difference between two methods to validate load bearing area.

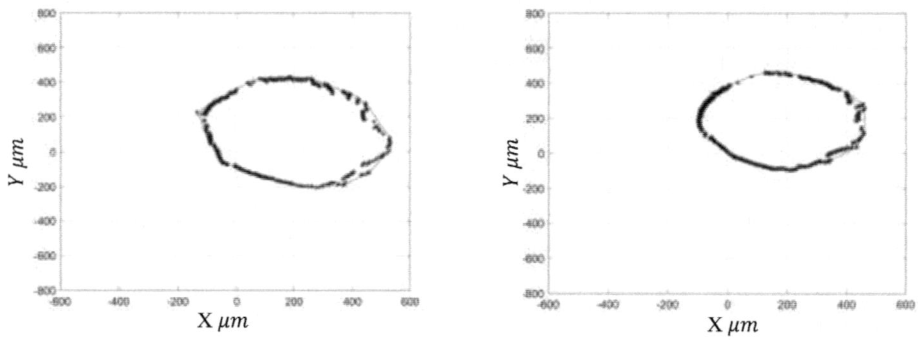

Figure 6 Single slice cross-section area on the XY plane of the sample printed with ZXY orientation

ZXY and YZX building orientation surfaces have been investigated with scanning electron microscope (SEM) Tescan VEGA-3 LMU. Melt pool may affect the building properties. Besides, melt pool is affected by thermal properties such as heating-cooling cycle [16]. These parameters are also considered during the investigation.

3. Result

The cross-sectional area and surface analysis have been carried out in order to identify the difference between additively manufactured samples and cad files. Cad design and laser scan micrometers gives comparable values of cross-sectional area. While So is getting smaller, accuracy of SLS technique is decreasing. As it can be said that cross-sectional area is becoming more circular than rectangular. Additively manufactured small samples contain unpredictable mechanical response while the cross-sectional area has smaller diameters such as less than $1.5mm^2$. It has been mentioned previously that samples are grouped in two according to building orientation. ***Table 3*** illustrates samples number were built with ZXY and YZX orientation. Sample with ZXY orientation yield strength is generally higher than samples with YZX orientation. This affect can be noticed that strongly while So is getting smaller.

4. Discussion

3D printed technology does not only provide design freedom, it can also offer good mechanical response and lightweight structure. The open cell structure is commonly used in the medical device industry and it consists of small struts connected to each other. The mechanical response of each strut affects the general behavior of the porous system. This mechanical performance may affect the stress shielding in the medical application. During the experiment small sample mechanical response is significantly different in the cross-section area smaller than $0.5mm^2$ and this is the range of complex 3D structure strut thickness. Moreover, the finer microstructure can produce good mechanical performance and the mechanical response of AM samples may even have better performance than wrought or machine titanium [17][18]. This is the promising feature can improve by post treatment method such as heat treatment of HIP (Hot Isostatic Process) application.

5. Conclusion

As a result of conducting this research, the following findings have been concluded:

- The size effect of building orientation occurs until certain cross-section area. Particularly, the yield strength distribution of the samples is considerable differences in this section with small cross-section area, $So<1.5mm^2$ respectively. While the cross-section area is getting bigger than $1.5mm^2$, the trend of yield strength reaches a stable result regardless of building orientation.

- The result of yield strength changes with the cross-section diameter respectively, this may influence the design aspect. The porous structure is used in the medical device has small struts due to the open cell diameters and it helps to allow bone ingrowth [19].
- In addition, the paper demonstrates that additively manufactured pure grade 2 titanium parts tend to have higher yield strength than conventionally manufactured pure titanium grade 2 since SLS can provide finer microstructure.

References

[1] D. Bombač, M. Brojan, P. Fajfar, F. Kosel, and R. Turk, "Review of materials in medical applications," *RMZ – Mater. Geoenvironment*, vol. 54, no. 54, pp. 471–499, 2007.

[2] C. N. Elias, J. H. C. Lima, R. Valiev, and M. A. Meyers, "Biomedical applications of titanium and its alloys," *Jom*, vol. 60, no. 3, pp. 46–49, 2008.

[3] R. K. Quinn and N. R. Armstrong, "Electrochemical and Surface Analytical Characterization of Titanium and Titanium Hydride Thin Film Electrode Oxidation," *J. Electrochem. Soc.*, vol. 125, no. 11, p. 1790, 1978.

[4] N. Schiff, B. Grosgogeat, M. Lissac, and F. Dalard, "Influence of fluoride content and pH on the corrosion resistance of titanium and its alloys," *Biomaterials*, vol. 23, no. 9, pp. 1995–2002, 2002.

[5] B. Salvador-Culla and P. E. Kolovou, "Keratoprosthesis: A review of recent advances in the field," *J. Funct. Biomater.*, vol. 7, no. 2, 2016.

[6] R. Adell, B. Eriksson, U. Lekholm, P. I. Branemark, and T. Jemt, "Long-term follow-up study of osseointegrated implants in the treatment of totally edentulous jaws.," *Int. J. Oral Maxillofac. Implants*, vol. 5, no. 4, pp. 347–359, 1990.

[7] J. D. Bobyn, G. J. Stackpool, S. A. Hacking, M. Tanzer, and J. J. Krygier, "Characteristics of bone ingrowth and interface mechanics of a new porous tantalum biomaterial," *J. Bone Jt. Surg.*, vol. 81, no. 5, pp. 907–914, 1999.

[8] L. Mullen, R. C. Stamp, W. K. Brooks, E. Jones, and C. J. Sutcliffe, "Selective laser melting: A regular unit cell approach for the manufacture of porous, titanium, bone in-growth constructs, suitable for orthopedic applications," *J. Biomed. Mater. Res. - Part B Appl. Biomater.*, vol. 89, no. 2, pp. 325–334, 2009.

[9] A. Chamay and P. Tschantz, "Mechanical influences in bone remodeling. Experimental research on Wolff's law," *J. Biomech.*, vol. 5, no. 2, pp. 173–180, 1972.

[10] S. Babaee, B. H. Jahromi, A. Ajdari, H. Nayeb-Hashemi, and A. Vaziri, "Mechanical properties of open-cell rhombic dodecahedron cellular structures," *Acta Mater.*, vol. 60, no. 6–7, pp. 2873–2885, 2012.

[11] Arcam EBM system, "Arcam," 2008.

[12] B. Vrancken, L. Thijs, J. P. Kruth, and J. Van Humbeeck, "Heat treatment of Ti6Al4V produced by Selective Laser Melting: Microstructure and mechanical properties," *J. Alloys Compd.*, vol. 541, pp. 177–185, 2012.

[13] G. M. Ter Haar and T. H. Becker, "Selective laser melting produced Ti-6Al-4V: Post-process heat treatments to achieve superior tensile properties," *Materials (Basel).*, vol. 11, no. 1, 2018.

[14] J. Dzugan *et al.*, "Effects of thickness and orientation on the small scale fracture behaviour of additively manufactured Ti-6Al-4V," *Mater. Charact.*, no. February, 2018.

[15] F. Out, W. Artificial, and J. Fail, "Case Study : Rush University Medical Center Figuring Out Why Artificial Joints Fail Rush University Medical Center study examines artificial implants Customer Benefit : Understanding failure , and success," vol. 44, no. 0.

[16] B. Cheng and K. Chou, "Melt Pool Geometry Simulations for Powder-Based Electron Beam Additive Manufacturing," *Proc. Solid Free. Fabr. Symp.*, pp. 644–654, 2013.

[17] H. Li, L. Zheng, H. Zhang, S. Li, H. Zhan, and S. Gong, "Effects of solidification parameters on microstructures of Ni3Al based single crystal," *Procedia Eng.*, vol. 27, pp. 1187–1192, 2012.

[18] J. Alcisto *et al.*, "Tensile Properties and Microstructures of Laser- Formed Ti-6Al-4V," vol. 20, no. 2, pp. 203–212, 2011.

[19] S. Lee *et al.*, "Potential Bone Replacement Materials Prepared by Two Methods," *MRS Proc.*, vol. 1418, pp. mrsf11-1418-mm06-02, 2012.

Modelling of interface stress transfer in graphene monolayer nanocomposites under static extension load

T Petrova[1,3], E Kirilova[1], W Becker[2], J Ivanova[3]

[1]Institute of Chemical Engineering, Bulgarian Academy of Sciences, Acad. G. Bonchev str. Bl.103, 1113 Sofia, Bulgaria

[2] Institute of Structural Mechanics, TU Darmstadt, Franziska-Braun-Str. 7, L5|01 347a, 64287 Darmstadt, Germany

[3] European Polytechnical University, 2300 Pernik, Bulgaria

Corresponding author e-mail: tancho66@yahoo.com

Abstract. An investigation on the interfacial stress transfer in nanocomposite between a monolayer of graphene and a common size polyethylene terephthalate (PET) matrix under extensional traction applied to the matrix is presented. The constitutive law describing the behaviour of the interface is assumed to be a bi-linear cohesive one. Three zones are defined along the interface: elastic, cohesive and delaminated. The system of ordinary differential shear lag equations together with boundary, symmetric and continuity conditions is solved consequently at each zone and the respective solutions are fitted at common points. The values of these common points – the ends of the lengths of the elastic, cohesive and delaminated zones are determined using the solutions of two nonlinear algebraic system equations about them. All results are illustrated in figures and discussed in detail. A comparison of the results obtained with the model considered and experimental data is done and confirms that for the structure of a length 21.8 nm the interface delamination is not observed. The implemented method offers a simple tool to understand the nature of the interface behaviour and evaluation of the damaging of the nanostructure.

1 Introduction

Graphene is a monolayer of honeycomb lattice of carbon atoms discovered in 2004. The research communities have shown a lot of interest in this novel material and its unique properties. As a consequence, the publications on graphene have dramatically increased in recent years. Graphene is also one of the stiffest (modulus ~1 TPa) and the strongest (strength ~100 GPa) material with thermal conductivity (5000 W/mK). Based on these exceptional properties, graphene has found its applications in various fields such as sensors, electrodes, solar cells, energy storage devices and nanocomposites. Therefore, graphene-polymer nanocomposites have demonstrated a great potential to serve as next generation functional or structural materials.

Over the last decades, a variety of numerical and analytical models has been developed to investigate the behavior of monolayer graphene nanocomposites. Recently, the shear lag method which is 1-D analytical method has been widely used in many engineering problems related with interface stress transfer and delamination [1-14].

From mathematical point of view, the shear-lag model is the simplest of 1-D analytical model, used for the stress transfer analysis in unidirectional composites. This model is based on a simple differential equation which relates the fiber axial stress to the interfacial shear stress. It is not applicable at higher volume fractions due to the significant interactions of fibers. Also, the model cannot predict changes of axial stress and strain distributions in the radial direction. As a result, due to these limitations, the shear-lag model cannot provide reliable predictions for the composite properties, and thus, its application for short fiber composites has been limited overtime.

In the present work the bi-linear shear stress is implemented in shear lag analysis for a monolayer graphene/PET nanocomposite. The representative model unit consists of two layers (graphene hosted on substrate of PET is considered). The substrate is loaded by axial traction on extension. As a consequence the statements as well as the boundary and continuity conditions are given in tractions. Three zones are defined along the interface: elastic, cohesive and delaminated. The respective solutions are fitted using continuity conditions. The system of two nonlinear algebraic equations gives a possibility to find the lengths of the elastic and cohesive zone. The obtained results are illustrated in figures and discussed. A comparison with experimental results [13] shows that the used method is sufficiently good to give a quick prognosis for the damaged composite.

2 Statement of the problem

The representative model unit of the structure and the interface shear stress behaviour are presented in Figure1(a) and (b), respectively, where the layer A is traction free and the layer B is loaded by an extensional traction T. The parameters $h_A << h_B$, E_A, E_B are the layers thicknesses and elastic moduli, respectively. The modified shear lag analysis will be applied to the considered structure. The interface I works only on shear, bending is neglected. The coordinate system Oxz is placed in the middle of the structure with a length $2l$. Due to the symmetry, only half of the structure will be considered.

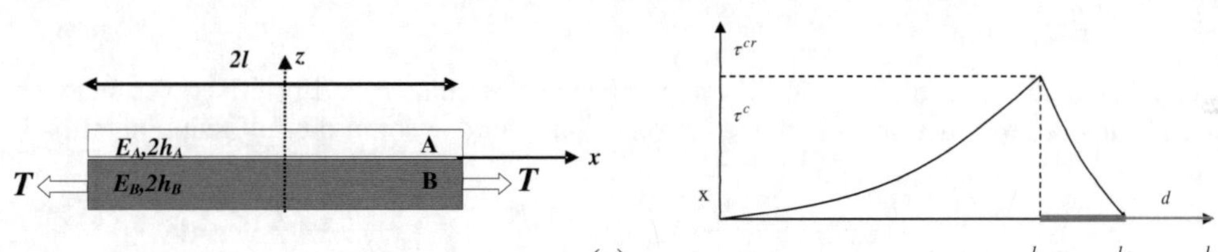

(a) (b)

Figure 1. Representative unit of graphene monolayer nanocomposite (a) and interface shear stress behaviour (b)

The following shear lag equations for the model hold:

$$\frac{dT_A}{dx} = 2h_A \frac{d\sigma_A}{dx} = \tau_I \quad ; \quad \frac{dT_B}{dx} = 2h_B \frac{d\sigma_B}{dx} = -\tau_I \tag{1}$$

$$T_A(x) = 2h_A\sigma_A \quad ; \quad T_B(x) = 2h_B\sigma_B ; \quad \varepsilon_A(x) = u'_A \quad ; \quad \varepsilon_B(x) = u'_B ; u_I = u_A - u_B$$

For interface shear stress τ_I we assume:

$$\tau_I = \begin{cases} G_I \dfrac{u_I}{h_I} = K_t u_I \quad for \quad u_I \le u_I^c = \dfrac{\tau_I^c}{K_t} \\ \tau_I^c - \dfrac{K_s}{h_I}\left(u_I - u_I^c\right) = \tau_I^c - K_s\left(u_I - u_I^c\right) \quad for \quad u_I > u_I^c \; ; \; u_I^f = \tau_I^c\left(\dfrac{1}{K_t} + \dfrac{1}{K_s}\right) \\ 0 \quad for \quad u_I \ge u_I^f \end{cases} \tag{2}$$

Here interface thickness is h_I, and K_t, K_s are effective elastic shear and softening moduli of the interface bonding layer. The plate B is under extension $T = 2h_B E_B \varepsilon_0$, plate A is free. The interface I is also assumed linear elastic, working only on shear. During the monotonically increasing traction T and due to the assumed interface stress behaviour in equation (2), three zones can be defined along the length of the interface I (see Figure1(b)). Along the first zone the elastic behaviour for the interface is valid. The interface remains continuous until reaching the critical value of interface shear stress $\tau_I = \tau_I^c$ at both ends of the structure. The length of the pure elastic zone is l_e.

Let the length of the softening zone be l_d. The softening zone follows the cohesive law for the interface shear stress, where $\tau_I > \tau_I^c$. This zone will be called "cohesive" (damaged) zone. The proper continuity conditions should be stated to match different stress and displacement regimes at transition point l_e with length ($l_d - l_e$) between the elastic and softening zones.

The third zone consists of pure delamination of the interface and then $\tau_I = 0$. At the same time, the proper continuity conditions at transition point l_d (length of this zone) should be stated.

In all three zones the stress interface behaviour is determined, satisfying the BC and proper continuity conditions. Finally, the lengths of both transition points l_e and l_d are calculated. The boundary, symmetry and continuity conditions for the solutions in three zones are:

Elastic zone - $0 \le x \le l_e$	**Softening zone -** $l_e \le x \le l_d$	**Delamination zone -** $l_d \le x \le l$	
$T_A^e(l_e) = 0$	$T_A^s(l_e) = 0$	$T_A^d = 0$	
$\tau_I^e(l_e) = -\tau_I^c$	$T_A^s(l_d) = 0$	$T_B^d = T$	(2.1)
$\tau_I^e(0) = u_I^e(0) = 0$	$\tau_I^s(l_d) = 0$	$\tau_I^d = 0$	
$T_B^e(l_e) = T$	$T_B^s(l_e) = T$		

For simplicity, the analytical solutions in all three zones are calculated and briefly given below only for interface shear stresses, tractions and deformations, taking into account the following notations:

$$\lambda_e^2 = \left(\frac{1}{2h_A E_A} + \frac{1}{2h_B E_B}\right) K_t \; ; \quad \rho = \frac{2h_A E_A}{2h_B E_B} \; ; \quad \lambda_s^2 = \left(\frac{1}{2h_A E_A} + \frac{1}{2h_B E_B}\right) K_s$$

$$F = \left(\frac{1}{2h_A E_A} + \frac{1}{2h_B E_B}\right)\left(1 + \frac{K_s}{K_t}\right)\tau_I^c = \frac{\lambda_s^2}{K_s}\left(1 + \frac{K_s}{K_t}\right)\tau_I^c \tag{2.2}$$

Elastic zone: $0 \le x \le l_e$: $\quad \dfrac{d^2 u_I}{dx^2} = \lambda_e^2 u_I$

Solutions: $\tau_I^e(x) = -\tau_I^c \dfrac{\sinh(\lambda_e x)}{\sinh(\lambda_e l_e)}$; $\quad T_A^e(x) = -\dfrac{\tau_I^c}{\lambda_e}\dfrac{\cosh(\lambda_e x)}{\sinh(\lambda_e l_e)} + \dfrac{T\rho}{1+\rho}$; $\quad \varepsilon_A^e(x) = T_A^e(x)/(2h_A E_A)$ (3)

Softening zone: $l_e \le x \le l_d$: $\quad u_I'' + \lambda_s^2 u_I = F$

Solutions: $\tau_I^s(x) = -\dfrac{\rho}{1+\rho} T\lambda_s \sin(\lambda_s(l_d - x))$; $T_A^s(x) = \dfrac{\rho}{1+\rho} T[1 - \cos(\lambda_s(l_d - x))]$; $\varepsilon_A^s(x) = T_A^s(x)/(2h_A E_A)$ (4)

Delamination zone: $l_d \le x \le l$

Solutions: $\tau_I^d(x) = 0$; $T_A^d(x) = 0$; $\varepsilon_A^d(x) = 0$ (5)

For determination of zones lengths $(l_e),(l_d)$ it is required that: $\tau_I^s(l_e) = -\tau_I^c$ and $T_A^e(l_e) - T_A^s(l_e) = 0$, or:

$$l_d = l_e + \frac{1}{\lambda_s} arctg\left[\frac{\lambda_e}{\lambda_s}\coth(\lambda_e l_e)\right];\quad T = \frac{\tau_I^c}{\lambda_e}\frac{1+\rho}{\rho}coth(\lambda_e l_e)\left[1+\left(\frac{\lambda_e}{\lambda_s}tanh(\lambda_e l_e)\right)^2\right]^{1/2} \tag{6}$$

3 Numerical example and discussion

Material properties and geometry of the considered nanostructure in Figure 1 are taken from [13] and are presented in Table 1. For pure elastic zone over the whole graphene length of 21.8 μm it can be obtained an additional equation (7) for model strain, derived from last equation in Equations (3) at $l_e = l$ (the same result is obtained in [11]):

$$\varepsilon_A^e(x) = \frac{T\rho}{2h_A E_A(1+\rho)}\left[1 - \frac{\cosh(\lambda_e x)}{\cosh(\lambda_e l)}\right] \tag{7}$$

This equation is used to make a comparison is between our results for this strain and experimental data in [13,14] for graphene length of 21.8 μm. The comparison is presented in Figure 2 and the model results fit well with experimental data, both for two different external uniaxial loads (ε_0=0.4% and 0.6%, which correspond to applied tractions T =42 and 63 kN/m, respectively).

Table 1. Material properties and geometry of considered structure

E_A, TPa	E_B, GPa	K_t TPa/m	K_s, TPa/m	τ_I^c MPa	$2h_A$ nm	$2h_B$ m	$2l$ μm
1	3	74	1.596	0.5	0.35	0.35e-04	21.8;60

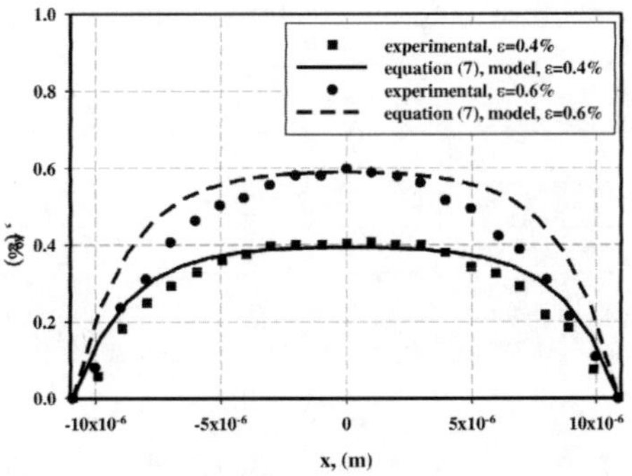

Figure 2. Comparison between experimental data in [13] and results obtained from equation (7) for the strain in graphene along the graphene length 21.8 μm only, at two different external uniaxial loads (ε_0=0.4% and 0.6%).

In Figure 3(a), the axial strain $\varepsilon_A(x)$is shown for five values of the external tractions in all three zones – elastic, softening and delamination (equations (3)÷(5)), and in Figure 3(b) the behaviour of the interfacial shear stress in all three zones at the same external loadings is shown. For simplicity, in Figure 3(a) and 3(b) only the essential part of the length $2l$=60 μm is presented.

It can be seen that increasing the load, maximal values of$\varepsilon_A(x)$decrease and delamination increases. This also confirms the non-linear cohesive shear lag model observations in [13] in the sliding region for strain in graphene. The character of the curves for $\tau_I(x)$ changes as the applied load increases (at

Figure 3(b)) – up to 1% there is still a small plateau in the middle of graphene length, like in [13], after 2% the bi-linear cohesive behaviour is well visible. The cohesive behaviour described in equations (3÷5) occurs at graphene length 60 μm after 0.025%, at which the obtained lengths of elastic and cohesive zones are 5.1883e-07m and 1.546e-05m, respectively. At loads less than 0.025% no real roots $(l_e),(l_d)$ of equation (6) are obtained.

In Table 2 the results obtained from equation (6) for the lengths $(l_e),(l_d)$ are given at different loading tractions for graphene length of 60 μm.

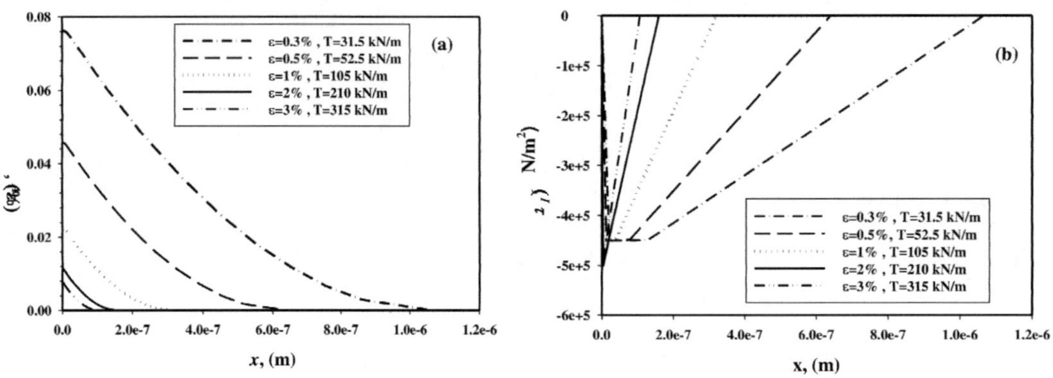

Figure 3. The axial strain $\varepsilon_A(x)$ in graphene along the graphene length, for various external uniaxial tractions (a), and interfacial shear stress $\tau_I(x)$ along the graphene length, for various external uniaxial tractions (b)

Table 2. Model results for $(l_e),(l_d)$ at graphene length 60 μm.

T (kN/m)	31.5	52.5	105	210	315
ε_0 (%)	0.3	0.5	1	2	3
l_e(m)	2.2579e-8	1.3525e-8	6.7582e-9	3.3785e-9	2.2523e-9
l_d (m)	1.0674e-6	6.4012e-07	3.1998e-7	1.5998e-7	1.0665e-7

4 Conclusions

In this work the modelling of interfacial stress transfer in a nanocomposite between graphene monolayer and a common size polyethylene terephthalate (PET) matrix under extensional traction applied to the matrix is presented. The constitutive law describing the behaviour of the interface is assumed to be a bi-linear cohesive one. Three zones are defined along the interface: elastic, cohesive and delaminated. The system of ODE shear lag together with boundary, symmetry and continuity conditions is solved consequently at each zone and the respective solutions are fitted at common points. The values of these common points – the ends of the lengths of the elastic, cohesive and delaminated zones are determined using the solutions of two nonlinear algebraic system equations about them. A comparison of the model strain results in graphene and experimental data of [13] is done and confirms that for the structure of a length 21.8μm the interface delamination is not observed. For graphene length of 60 μm after parametric analysis it was found, that increasing the external load, the maximal values of strain at the centre of graphene for all three zones solutions decrease and delamination increases. At graphene length of 60 μm the interfacial shear stress also follows the predicted cohesive bi-linear behaviour, and delamination increases with applied load increasing.

Acknowledgements: This study is carried out by the financial support of DFG project "Shear lag analysis by a cohesive bilinear model for the interface stress transfer in a graphene monolayer nanocomposite", BE 1090/46-1, 2018.

References

[1] Abedian A, Mondali M, Pahlavanpour M 2007 Basic modifications in 3D micromechanical modeling of short fibre composites with bonded and debonded fibre end *Computational Materials Science* **40** 421

[2] Agarwal B D and Broutman L J 1980 *Analysis and Performance of Fibre Composites* (New York:John Wiley and Sons)

[3] Agarwal BD, Lifsitz JM, Broutman LJ 1974 Elastic plastic finite element analysis of short fibre composites *Fibre Science and Technology* **7** 45

[4] Glavinchevski B, Piggott M 1973 Steel disc reinforced polycarbonate *Journal of Materials Science* **8** 1373

[5] Jiang G and Peters K 2008 A shear-lag model for three-dimensional, unidirectional multilayered structures *International Journal of Solids and Structures* **45** 4049

[6] Hsueh CH 1994 A two-dimentional stress transfer model for platelet reinforcement, *Composites Part B: Engineering* **4**1033

[7] Cannilloa V, Pellacania GC, Leonelli C, Boccaccini AR 2003 Numerical modelling of the fracture behaviour of a glass matrix composite reinforced with alumina platelets *Composites:Part A* **34** 43

[8] Lusti HR, Hine PJ, Gusev AA 2002 Direct numerical predictions for the elastic and thermoelastic properties of short fibre composites *Composites Science and Technology* **62** 1927

[9] Narin JA 1997 On the use of shear-lag methods for analysis of stress transfer in unidirectional composites *Mechanics of Materials* **26** 63

[10] Cox HL 1952 The elasticity and strength of paper and other fibrous materials *British Journal of Applied Physics* **3** 72

[11] Mroz K and Mroz Z 2015 Analysis of delamination and damage growth in joined bi-linear systems *Geomechanics for Energy and Environment* **4** 4

[12] Fattahi AM and Mondali M 2014 Theoretical study of stress transfer in platelet reinforced composites *Journal of Theoretical and Applied Mechanics* **52** 3

[13] Guo G and Zhu Y 2015 Cohesive shear lag modeling of interfacial stress transfer between a monolayer graphene and polymer substrate *Journal of Applied Mechanics* **82** 031005-1

[14] Jiang T, Huang R, Zhu Y 2013 Interfacial Sliding and Buckling of Monolayer Graphene on a Stretchable Substrate *Advanced Functional Materials* **24** 396

Beta phase stability of Ti-35Nb-6Ta-7Zr-0.7O beta titanium alloy

D Preisler[1*], J Stráský[1], K Bartha[1], J Džugan[2], M Janeček[1]

[1] Department of Physics of Materials, Charles University, Ke Karlovu 5, Praha 2, Czech Republic
[2] COMTES FHT, a.s., Průmyslová 995, Dobřany, Czech Republic

* preisler.dalibor@gmail.com

Abstract. Phase stability of metastable beta Ti alloy Ti-35Nb-6Ta-7Zr-0.7O (wt.%) in a beta solution treated condition was studied during annealing at 400°C – 700°C for 1 h – 16 h. Resulting microstructure was characterized by scanning electron microscopy; the samples were also characterized by microhardness measurements. The microhardness did not change in any of the annealed condition. The reason follows from the microstructural observations, the α phase precipitation did occur only at the grain boundaries and only at higher temperatures / longer annealing times. Thus, the β phase stability is high and this fact can be used for designing annealing treatments or the thermo-mechanical processing route.

1 Introduction

Titanium has two main crystallographic structures, hcp α phase and bcc β phase. In pure Ti, α phase is present at low temperatures and transforms into β phase at 882°C. By adding the so-called β stabilizing elements (Nb, Mo, Ta, etc.), pure β phase can be achieved in metastable form by quenching from the β region (β solution treated condition). The metastable β Ti alloys in β solution treated condition can be aged at sub-transus temperatures to promote precipitation of the α phase [1].

Recently developed Ti-35Nb-6Ta-7Zr-0.7O (wt.%) (TNTZO) [2] has promising properties for implant manufacturing – low Young's modulus of 80 GPa, high strength over 1000 MPa and contains only biocompatible elements [3,4]. The low Young's modulus is attributed to single β phase condition [5,6] and therefore α phase precipitation should be avoided in final product. High strength is caused by interstitial strengthening by oxygen [7,8].

TNTZO is a high oxygen variant of Ti-35Nb-6Ta-7Zr metastable β alloy [6]. Oxygen is α stabilizing element in terms of increasing β transus temperature [9] (the β transus temperature of TNTZO alloy was not measured, but solution treatment at 820°C / 2h was successful). On the other hand, the effect of oxygen on the precipitation, especially concerning nucleation, of α phase particles in metastable β alloys is unclear.

Currently the most used Ti alloy for implant production is α+β alloy Ti-6Al-4V (wt.%) [10], with extra low interstitial content (ELI) that causes good ductility simultaneously with high strength. In the comparison with Ti-6Al-4V ELI, the TNTZO alloy has similar strength, lower Young's modulus and does not contain toxic vanadium.

Preparation of the β Ti alloys often comprises of hot working (hot forging or hot rolling) [2,7,11–13] or cold working combined with subsequent annealing [14–16]. It is therefore valuable to study the kinetics of α phase precipitation in this alloy to successfully design the thermal / thermo-mechanical processing route to avoid the precipitation (and connected rise of Young's modulus).

2 Materials and Methods

The studied material with nominal composition Ti-35Nb-6Ta-7Zr-0.7O was cast at company Retech Systems LLC by sequential pour melting procedure. Chemical composition shown in Table 1 was analyzed by manufacturer by Inductive Coupled Plasma Atomic Emission Spectroscopy for analyzing content of metals while content of other elements was determined by Combustion Infrared Absorption. A small rod of diameter 10 mm was machined from the cast ingot and beta solution treated at 820°C/2h in pure Ar atmosphere in a quartz tube. The rod was water quenched immediately after finishing the annealing by breaking the quartz tube in water. This condition will be denoted as ST (solution treated).

Table 1. Chemical composition of Ti-35Nb-6Ta-7Zr-0.7O alloy after casting.

Element	Ti	Nb	Zr	Ta	O	N	C
Composition (wt.%)	Bal.	35.6 ± 0.2	7.35 ± 0.04	6.0 ± 0.1	0.71 ± 0.01	0.012 ± 0.002	0.010 ± 0.001

After solution treatment, the rod was cut into slices 3 mm thick and aged at temperatures 400°C, 500°C, 600°C and 700°C for 1 h, 4 h and 16 h. The samples were water quenched immediately after ageing. To prevent oxidation, the ageing at 400°C and 500°C was done in preheated inert salt baths (potassium nitrate and sodium nitrate mixture). The samples were put directly into salt bath preheated to ageing temperature. Ageing at 600°C and 700°C was done in preheated tube, which was evacuated immediately after inserting the sample.

The samples were ground on SiC papers up to 2400 grit and vibratory polished using Vibromet polisher with suspensions Alumina 0.3 μm and 0.05 μm for at least 6 h and Colloidal Silica 0.05 for at least 3 hours. Vickers microhardness measurements were done using automatic Qness Q10 hardness tester with load 0.5 kgf and load time 10 s. Scanning electron microscopy (SEM) observations using back-scattered electrons (BSE) were done at FEI Quanta 200F, equipped with an energy dispersive X-ray spectroscopy (EDS), with accelerating voltage 15 kV.

3 Results and Discussion

3.1 Microhardness measurements

Table 2 shows results from microhardness measurements of aged samples. Within the statistical error, all aged conditions have the same microhardness. The values from aged samples also do not differ from ST condition having microhardness 321 ± 8 HV. Achieved values are comparable to the microhardness of previously prepared TNTZO alloy in laboratory environment [2] and are consistent with the high strength of TNTZO alloy [17]. For comparison with other metastable beta Ti alloy, in Timetal LCB (Ti–6.8Mo–4.5Fe–1.5Al in wt.%) the microhardness grows by 100 HV, compared to ST condition, already after ageing at 400°C / 1 h [18]. The microhardness changes in Timetal LCB are attributed to homogeneous distribution of ω and α phases.

Table 2. Microhardness of Ti-35Nb-6Ta-7Zr-0.7O alloy annealed at 400°C – 700°C for 1h – 16h.

Microhardness (HV 0.5)				
Anenaling time	Annealing temperature			
	400°C	500°C	600°C	700°C
1 h	327 ± 7	330 ± 10	332 ± 6	333 ± 11
4 h	335 ± 7	332 ± 5	331 ± 6	333 ± 9
16 h	332 ± 6	332 ± 4	332 ± 8	332 ± 4

3.2 SEM observations

Figure 1 shows SEM images of selected conditions at low magnification. All conditions show very large grains with size of approx. 1 mm. There is also porosity in the material and dendritic chemical inhomogenieties, both were created during solidification during casting. The concentration variations of elements with respect to bulk did not exceed 3 wt. %, as proven by EDS. Neither of these features was changed much during ageing. We can observe, that the α phase does not precipitate at the dark areas of chemical inhomonieties – areas with higher amount of Ti. On the other hand, there are α lamellae grown at grain boundary (GB) visible at image of 700°C / 16 h aged condition as black areas.

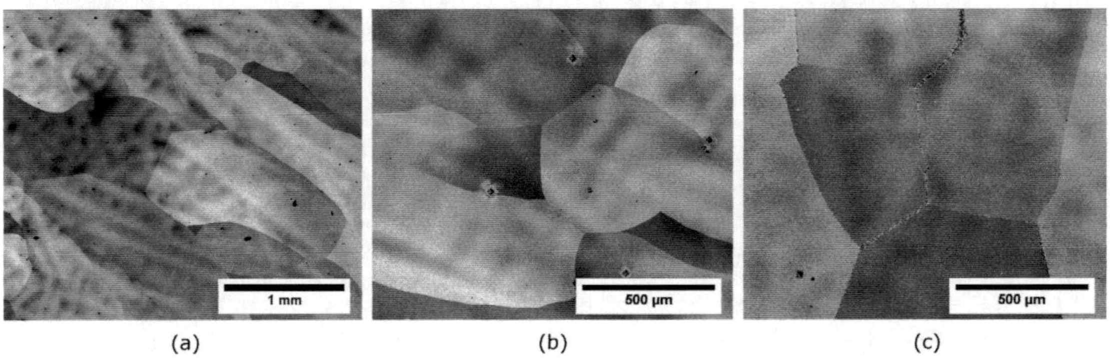

Figure 1. SEM images of overall microstructure (a) ST condition, (b) 600°C / 1h ageing, (c) 700°C / 16 h ageing. Small square indents originate from microhardness measurements.

Thorough analysis was conducted to determine, at which conditions the GB α starts to precipitate. Figure 2 shows GBs and triple points of selected aged conditions. No α phase precipitation was observed at ageing temperature 400°C. At 500°C, the GB α phase was found only on several GBs after the longest annealing time 16 h. The ageing at 600°C / 1 h was not long enough for α precipitation, same as lower temperatures with shorter annealing times. Figure 2 (a) shows, that even at high magnification, there is no GB α. After 4 h at 600°C, several grain boundaries with α phase were found, similar as in the 500°C / 16 h aged sample. After ageing at 600°C for 16 h, the GB α phase starts to form lamellae growing from the grain boundary into neighboring grains. Similar, but smaller lamellae are observed after 1 h ageing at 700°C. Prolonged ageing at 700°C leads to presence of GB α at majority of grain boundaries and growth of lamellar α phase towards the interior of β grains.

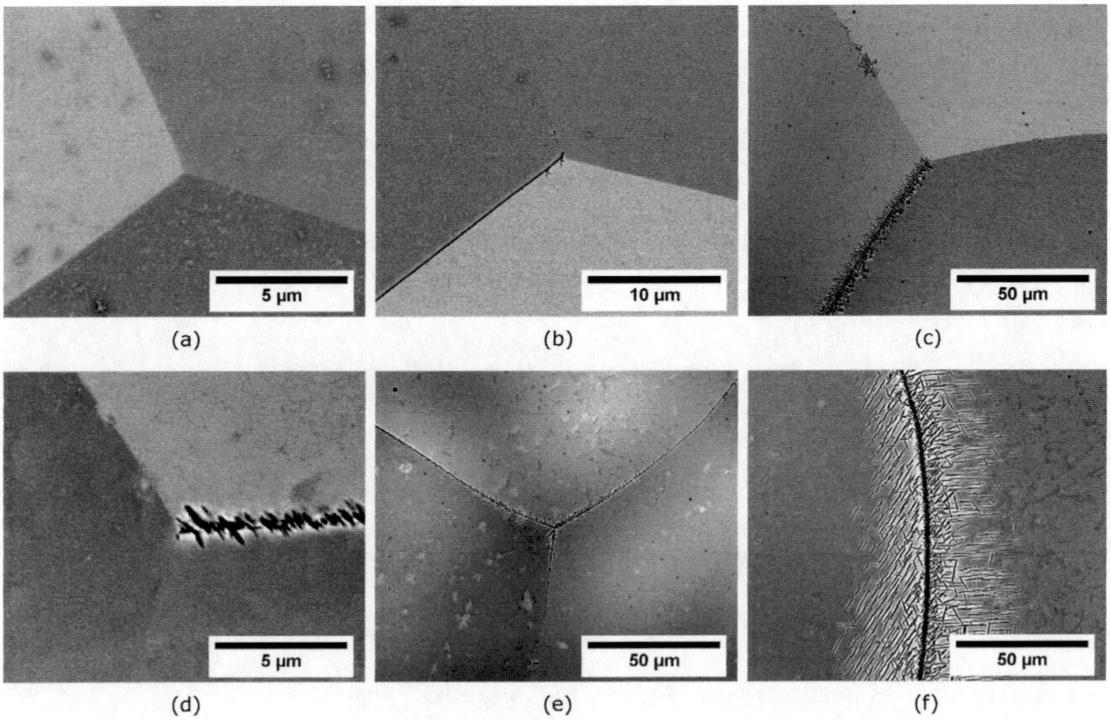

Figure 2. Selected triple points and grain boundaries in TNTZO aged at 600°C for (a) 1 h, (b) 4 h, (c) 16 h and at 700°C for (d) 1 h, (e) 4 h, (f) 16 h.

Limited or no precipitation of the α phase in aged samples explains lack of microhardness variations between the conditions. Typical size of the microhardness indent is 50 µm and therefore the distant precipitates do not affect the overall microhardness despite the α phase precipitates are harder than β matrix. The material therefore exhibits β phase stability up to relatively high temperatures. The consequence is that water quenching is not needed to prevent formation of α phase during cooling after using thermo-mechanical processing or any thermal treatment. The downside is that, for practical purposes, this material cannot be strengthened by α phase precipitation in sufficiently short time. This however, is not much limiting as the desired use of this alloy is in biomedical endoprostheses, where the lowest Young's modulus is desired and α phase would cause the Young's modulus to rise.

In comparison with other β Ti alloys, the TNTZO has rather high β phase stability. In Ti–29Nb–13Ta–4.6Zr alloy, the α and ω were found after 48 h long ageing at 350°C – 600°C and 300°C – 400°C, respectively. On the other hand, Ti–39Nb–13Ta–4.6Zr with much 10% more Nb had to be aged for 12 days to achieve any α phase precipitation [19]. In the Ti-27Nb-2Ta-3Zr alloy, the α phase precipitates at 500°C already after 0.5 h on GB and after 2 h inside grains [20]. To compare with oxygen-containing alloys, the amount of α phase is raised [12,21] by oxygen content.

It is interesting that similar alloy Ti–35Nb–7Zr–5Ta–0.68O with lower amount of Nb (by 0.6%) and Ta (by 1%) shows α phase precipitates both on GB and in grain interiors when aged at temperatures 538°C and 598°C for 8 h [21]. It must be noted that all alloys in the literature were subjected to hot forging before annealing treatment, while in this study, as-cast material was used. Considering that the preferential sites for heterogeneous α phase nucleation are grain boundaries and dislocations, the precipitation is limited to the GBs in the coarse-grained as-cast material. It was already reported that microstructural condition significantly affects alpha phase precipitation [22].

4 Conclusion

The β titanium alloy with the composition Ti-35Nb-6Ta-7Zr-0.7O (wt.%) after solution treatment and ageing in the temperature range 400°C – 700°C for 1 h – 16 h was characterized by SEM and we can draw following conclusion.

- There is no α phase present in the aged material inside the grains, not even at chemical inhomogenieties, but only on grain boundaries.
- Grain boundary α phase starts to grow after ageing at 500°C / 16 h or 600°C / 4 h. Further ageing yields α lamellae growing from grain boundary into grain.
- Microhardness does not change with ageing, as there was only GB α phase present in aged samples.
- Due to high β phase stability, water quenching from β region may not be required, unless α phase precipitation is not substantially enhanced by microstructural refinement.

Acknowledgements

Financial support by the Ministry of Industry and Trade, project no. FV20147 is greatly acknowledged. JS gratefully acknowledges Czech Science Foundation (project 17-20700Y) for financial support. DP acknowledges financial support by the Grant Agency of Charles University (project no. 1530217).

References

[1] Lütjering G and Williams J C 2003 *Titanium* vol 2 (Springer)
[2] Stráský J, Harcuba P, Václavová K, Horváth K, Landa M, Srba O and Janeček M 2017 Increasing strength of a biomedical Ti-Nb-Ta-Zr alloy by alloying with Fe, Si and O *J. Mech. Behav. Biomed. Mater.* **71** 329–36
[3] Eisenbarth E, Velten D, Müller M, Thull R and Breme J 2004 Biocompatibility of β-stabilizing elements of titanium alloys *Biomaterials* **25** 5705–13
[4] Kopova I, Stráský J, Harcuba P, Landa M, Janeček M and Bačákova L 2016 Newly developed Ti–Nb–Zr–Ta–Si–Fe biomedical beta titanium alloys with increased strength and enhanced biocompatibility *Mater. Sci. Eng. C* **60** 230–8
[5] Tane M, Nakano T, Kuramoto S, Hara M, Niinomi M, Takesue N, Yano T and Nakajima H 2011 Low Young's modulus in Ti–Nb–Ta–Zr–O alloys: Cold working and oxygen effects *Acta Mater.* **59** 6975–88
[6] Ahmed T and Rack H 1999 Low modulus biocompatible titanium base alloys for medical devices, US patent no. 5,871,595.
[7] Geng F, Niinomi M and Nakai M 2011 Observation of yielding and strain hardening in a titanium alloy having high oxygen content *Mater. Sci. Eng. A* **528** 5435–45
[8] Nakai M, Niinomi M, Akahori T, Tsutsumi H and Ogawa M 2009 Effect of Oxygen Content on Microstructure and Mechanical Properties of Biomedical Ti-29Nb-13Ta-4.6Zr Alloy under Solutionized and Aged Conditions *Mater. Trans.* **50** 2716–20
[9] Welsch G, Boyer R and Collings E W 1993 *Materials Properties Handbook: Titanium Alloys* (ASM International)
[10] Rack H J and Qazi J I 2006 Titanium alloys for biomedical applications *Mater. Sci. Eng. C* **26** 1269–77
[11] Wei Q, Wang L, Fu Y, Qin J, Lu W and Zhang D 2011 Influence of oxygen content on microstructure and mechanical properties of Ti–Nb–Ta–Zr alloy *Mater. Des.* **32** 2934–2939
[12] Qazi J I, Rack H J and Marquardt B 2004 High-strength metastable beta-titanium alloys for biomedical applications *JOM* **56** 49–51
[13] Hou F Q, Li S J, Hao Y L and Yang R 2010 Nonlinear elastic deformation behaviour of Ti-30Nb-12Zr alloys *Scr. Mater.* **63** 54–7

[14] Kim J I, Kim H Y, Hosoda H and Miyazaki S 2005 Shape Memory Behavior of Ti–22Nb–(0.5–2.0)O(at%) Biomedical Alloys *Mater. Trans.* **46** 852–7
[15] Guo S, Meng Q, Zhao X, Wei Q and Xu H 2015 Design and fabrication of a metastable β-type titanium alloy with ultralow elastic modulus and high strength *Sci. Rep.* **5**
[16] Elmay W, Laheurte P, Eberhardt A, Bolle B, Gloriant T, Patoor E, Prima F, Laille D, Castany P and Wary M 2010 Stability and elastic properties of Ti-alloys for biomedical application designed with electronic parameters *EPJ Web Conf.* **6** 29002
[17] Preisler D, Václavová K, Stráský J, Janeček M and Harcuba P 2016 Microstructure and mechanical properties of Ti-Nb-Zr-Ta-O biomedical alloy *Met. 2016 25rd Int. Conf. Metall. Mater. Ostrava TANGER* 1509–13
[18] Šmilauerová J, Janeček M, Harcuba P, Stráský J, Veselý J, Kužel R and Rack H J 2017 Ageing response of sub-transus heat treated Ti–6.8Mo–4.5Fe–1.5Al alloy *J. Alloys Compd.* **724** 373–80
[19] Li S, Hao Y L, Yang R, Cui Y and Niinomi M 2002 Effect of Nb on Microstructural Characteristics of Ti-Nb-Ta-Zr Alloy for Biomedical Applications *Mater. Trans. - MATER TRANS* **43** 2964–9
[20] Wang L, Lin Z, Wang X, Shi Q, Yin W, Zhang D, Liu Z and Lu W 2014 Effect of Aging Treatment on Microstructure and Mechanical Properties of Ti27Nb2Ta3Zr β Titanium Alloy for Implant Applications *Mater. Trans.* **55** 141–6
[21] Qazi J I, Marquardt B, Allard L F and Rack H J 2005 Phase transformations in Ti–35Nb–7Zr–5Ta–(0.06–0.68)O alloys *Mater. Sci. Eng. C* **25** 389–97
[22] Gatina S, Semenova I, Janecek M and Strasky J 2014 Effect of high pressure torsion on the aging kinetics of β-titanium Ti-15Mo alloy *IOP Conf. Ser. Mater. Sci. Eng.* **63** 012068

Mechanical property of thin walled selective laser melted parts and the effect of heat treatment

S Raghavan[1*], N Soh[1], L J Hao[1], R Muthu[1], H K Rafi[2], N A Khan[1], J Dzugan[3]

[1]Advanced Remanufacturing and Technology Centre, 3 Cleantech loop, Singapore 637143
[2]Underwriter Laboratories Singapore Pte Ltd. 1 Fusionopolis Walk , Singapore 138628
[3]COMTES FHT a.s., Prumyslova 995, 334 41 Dobrany, Czech Republic

Corresponding author: raghavans@artc.a-star.edu.sg

Abstract. The variability in mechanical properties in selective laser melted (SLM) maraging steel parts was captured by testing miniature size tensile coupons that were extracted from different locations in the part. Tensile testing provides better mechanical response to defects such as pores and inclusions as compared to hardness testing. The effect of heat treatment on the tensile property variation with different section thickness was also studied. It was observed that tensile property of miniature samples was more sensitive to porosity defects in maraging steel samples. The elongation values in the tensile specimens exhibited more variability as the section thickness of the part was reduced. This was due to the presence of more pore defects in thin wall sections. Heat treatment did not influence the tensile properties with respect to the wall thickness except for an increase in the strength due to hardening.

1 Introduction

Additive manufactured parts have the potential to incorporate complex part designs with variable section thickness. This variability in section thickness can also lead to variation in mechanical properties; hence it is essential to characterize the mechanical properties at different locations within the part to capture any variability. Previously, hardness measurements were routinely employed [1] to capture the location specific mechanical properties. However, hardness testing does not completely capture the mechanical properties since, defects formed such as porosity and inclusions affect only the tensile properties. To obtain location specific tensile properties in the part, it may be necessary to machine out miniature samples and perform tensile testing. Tensile testing of miniature samples have been attempted by various researchers in the recent past and have been compared with standard specimen tensile results [2-5]. The important feature of miniature sample testing, is it can assess the quality of printed parts and be used as a part qualification procedure. Any variation in mechanical properties within part or batch to batch variability can be captured and reported.

In the current research an additive manufactured part of maraging steel, with different sectional thickness were fabricated in order to assess wall thickness effect on mechanical properties, as these may vary with deposition thickness [6]. Miniature samples were machined out to perform tensile testing in the as-built and heat treated condition. The variation in tensile properties as a function of part section thickness were analysed, and the tensile properties of miniature samples were compared and benchmarked with ASTM size tensile samples that were also extracted from the different wall thickness of the part.

2 Materials and Methods

Different wall thickness ranging from 1 mm to 10 mm was printed in a part using Concept Laser machine with maraging steel alloy from OEM supplied powders. The process parameters used were also OEM recommended which was uniform for the whole part as given in Table 1. The part design with different wall thickness and the orientation with respect to build direction is shown in Figure 1. Miniature and ASTM size samples were extracted from different wall thickness in the part as shown in Figure 1. Both transverse and lateral oriented (along the build plane and perpendicular to build plane) miniature samples were machined out for tensile testing. For ASTM size samples only lateral oriented samples were tested for different wall thickness. The 1 mm thick wall ASTM sample was not tested due to slight distortion. The dimensions of the tensile samples are illustrated in Figure 2 as shown below. The larger sized samples were machined as per ASTM E8 specification.

Table 1. Process parameter setting for Surface and contour region

Layer thickness (μm)	Region	Laser Power (W)	Scan Speed (mm/s)	Spot Size (μm)	Trace distance (mm)
25	Surface	260	950	125	0.11
	Contour	90	150	50	-

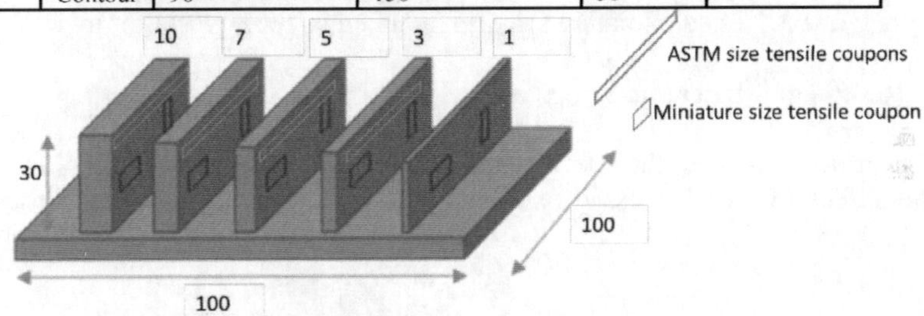

Figure 1. 3D printed part with different wall thickness. Also shown the sample orientations extracted for miniature and ASTM size samples. All dimensions in mm.

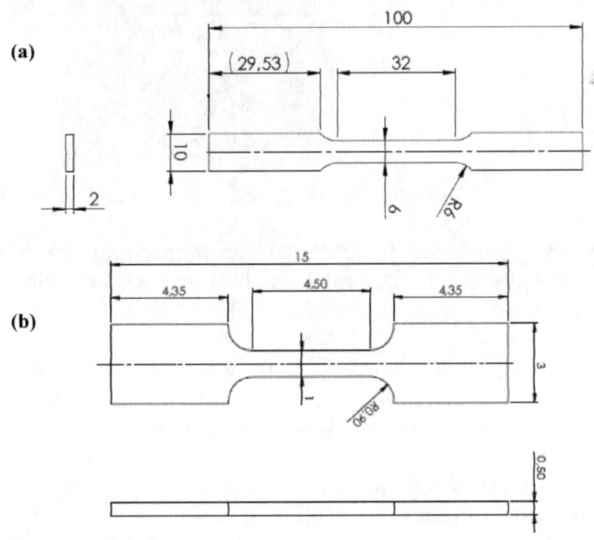

Figure 2. (a) Standard tensile coupon, dimensions as per ASTM E8 specifications (b) Tensile coupon dimensions of miniature samples. All dimensions in mm.

The miniature sample dimensions shown in Figure 2 (b) were 15 mm long, 3 mm wide and 0.5 mm thick. Such small size specimens enabled a study of location specific tensile properties in a complex part. Three samples for each orientation (lateral and transverse) were extracted for the miniature samples and one sample as per ASTM E8 specification, for the different wall thickness.

The samples were tested using a special testing apparatus for miniature tensile samples. The load cell has a capacity of 5kN and the deformation was measured by means of optical extensometer based on digital image correlation.
The fracture surface of the tensile tested samples were examined in a scanning electron microscope (SEM) to identify defects and the cause of crack initiation. The samples were then mounted in an epoxy resin and fine polished as shown in Figure 3, in order to measure the Vickers micro-hardness and examine the porosity near the fracture location.

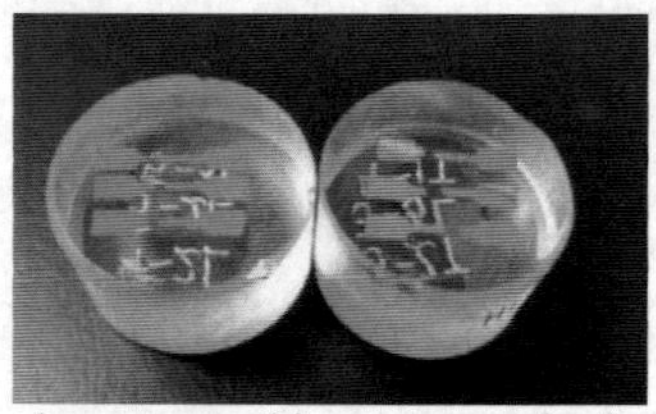

Figure 3. Tested miniature samples mounted in epoxy resin for microstructure analysis

3 Results and Discussion

The printed part with the different wall thickness is shown in Figure 4. Dimensional measurements of the different wall thickness were observed to lie within 2% of the actual prescribed dimensions.

Figure 4. 3D printed part with different wall thickness.

The tensile strength (yield and ultimate tensile strength) for different samples in the as-printed and heat-treated samples were extracted from different wall thickness as shown in Figure 5. Also shown are the tensile properties of the ASTM standard samples extracted from different wall thickness.
From Figure 5 it can be observed that the yield strength of miniature samples were higher than that of standard ASTM samples, whereas such a difference was not observed in the tensile strength values for both as-printed and heat-treated conditions. No clear trends were observed for strength variation as a function of wall thickness in the as-printed condition, which would indicate there is minimal microstructural variation with different wall thickness for maraging steels. Figure 6, shows the microstructure observed using SEM for 1 mm and 10 mm wall thickness for the as-printed part. A cellular morphology was observed in both images for the two wall thickness indicating high cooling rates which is typical for selective laser melting process. The spacing between cells was calculated by counting number of cells in a given area (as highlighted in Figure 6 (a) and (b)), and the spacing(λ) is related to number of cells in an area (N_a) by the equation $\lambda = 1/(N_a)^{0.5}$ [7]. The cellular spacing (N_a) calculated from multiple images was found to be 1.24 ± 0.73 μm for 1 mm wall thickness and 1.14 ± 0.16 μm for 10 mm wall thickness, clearly indicating no difference in the scale of microstructure for maraging steel.

The tensile strength values for the miniature samples matches well with ASTM standard samples which indicates there is no change in strength due to size effect. It has been reported [7] that the tensile strength values remain stable for different thickness until a critical ratio of sample thickness to grain size (t/d) is reached, whereby the results starts to deviate. This critical (t/d) ratio is material dependent and for SS304 material, the strength and ductility reduces below a sample thickness of 0.147 mm [8]. For maraging steel miniature samples with the thickness of 0.5 mm, has not reached the critical limit and does not deviate from the ASTM samples.

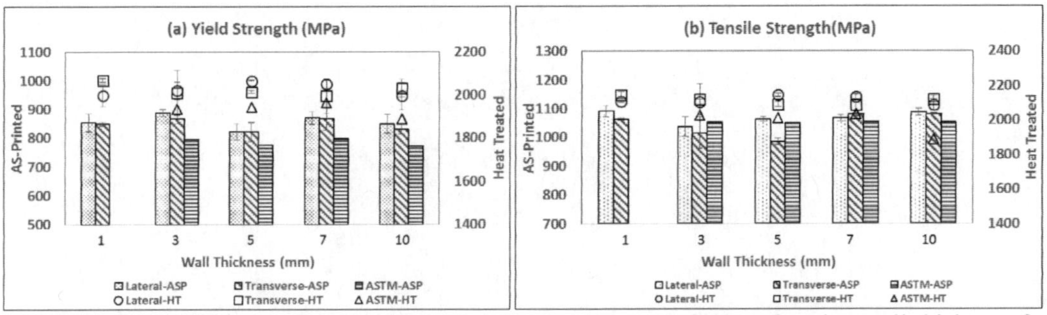

Figure 5. Tensile properties ((a) Yield and (b) Tensile strength) as a function wall thickness for miniature samples and ASTM standard samples extracted from the as-printed and heat treated part.

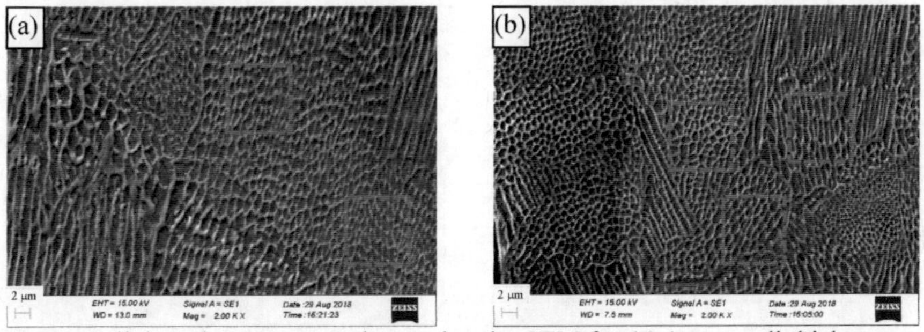

Figure 6. Cellular microstructure observed under SEM for (a) 1 mm wall thickness and (b) 10 mm wall thickness. Boxes highlighted to calculate cellular spacing.

Maraging steels are precipitation hardening steels, whereby the alloy is heat-treated to allow the precipitation of intermetallic phases which improves the strength significantly. A standard recommended heat treatment for maraging steels is 490°C, held for 6 hours followed by fast cooling was carried out in a vacuum furnace for the part. It is clear that the strength of the samples have increased by more than 100% after heat treatment when compared to the as-printed part. The increase in strength was offset by reduction in elongation from 12% for as-printed samples to 3% for heat- treated samples as seen in Figure 7.

From Figure 5, it was also observed that the lateral oriented samples showed slightly higher strength as compared to transverse oriented samples. This is due to the columnar grain texture developed during 3D printing which yields anisotropic behaviour in terms of strength. Such similar observations were also found by other researchers [9-10].

The elongation percent exhibited more variation especially for thin wall thickness samples as seen in Figure 7. The ductility of the samples were mainly influenced by the pore defects present in the part. The tensile properties of miniature samples are more sensitive to defects in the sample since the relative size of the defect to sample size is higher in miniature samples when compared to the larger ASTM sized samples. It is clear from Figure 7 that 3 mm, 5 mm and 1 mm samples exhibited large variance in elongation. The biggest pores in the samples from different wall thickness were characterised as shown in Table 2. A trend was observed in the presence of large pores as a function of wall thickness as seen in Table 2. The samples from thin walls have larger sized pores as compared to samples from thick walls which possibly explains the larger variance in ductility values in thin section samples. The influence of pore size on ductility in relation to sample size is also reported by other researchers [11-12]. Boyce et al. [11] reported a maximum pore size in relation to tensile sample thickness that did not affect the ductility when compared to pore-free samples.

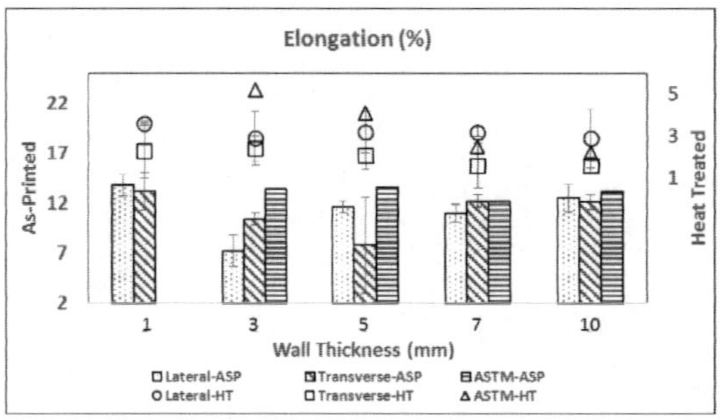

Figure 7. Elongation (%) as a function wall thickness for miniature and ASTM standard samples from the as-printed and heat-treated part.

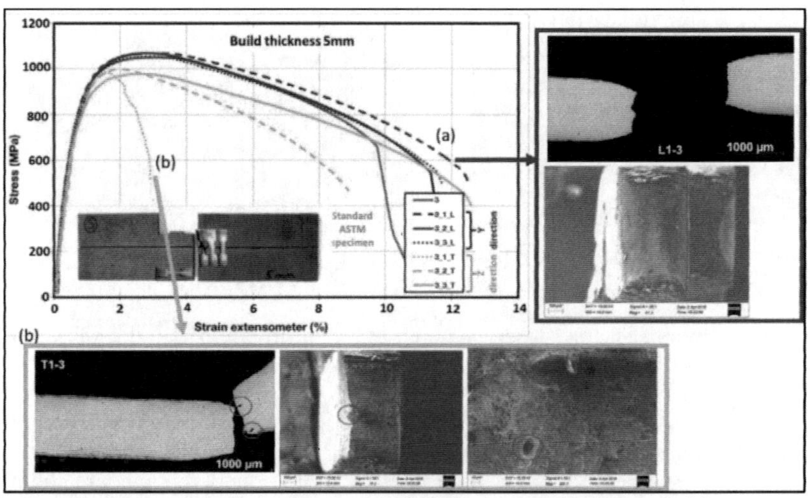

Figure 8. Tensile flow curves from 5 mm wall thickness as-printed part and fracture surface of samples that have failed with low elongation (b) and exhibited good elongation (a).

Table 2. Largest pore sizes observed in the as-printed miniature samples.

sample	Biggest pore size (μm)			
	1	2	3	average
1mm	40.56	34.06	27.77	34.13
3mm	31.91	20.32	20.3	24.18
5mm	21.61	16.89	13.37	17.29
7mm	22.29	17.07	15.73	18.36
10mm	18.25	14.1	13.67	15.34

The low elongation value observed in the tensile curve of the sample correlates well with the defects that were seen on the fracture surface. In Figure 8, the tensile curves of 5 mm samples from the as-printed part and the fracture surfaces were observed for two samples 3_1_L and 3_1_T, marked (a) and (b) respectively. Sample (b) which was oriented transverse to build plane exhibited poor strength and elongation as seen in Figure 8. The fracture surface of the sample (b), revealed presence of numerous pore defects (highlighted by the red circles), that potentially caused early failure. The fracture surface of sample (a) which was oriented laterally to build plane, exhibited highest elongation, and did not have any obvious pore defects as seen in Figure 8(a).

4 Summary

The following points can be concluded from the current study:

1. The tensile values for the miniature samples of gauge length 4.5 mm correlates well with ASTM size samples which had a gauge length of 32mm. The miniature samples exhibited good repeatability in both tensile and yield strengths.
2. The samples tested from thin walls (1 mm, 3 mm and 5 mm) in the part exhibited more variability in elongation as compared with samples from thicker walls (7 mm and 10 mm). This was due to the thin walls having larger sized pore defects as compared to the thicker walls.
3. No trends were observed for strength values as a function of wall thickness. This was possibly due to the microstructure of maraging steel not varying with thickness of the part. However this variation is material dependent, and it has been observed that Ti6Al4V [1] exhibited a variation of approx. 15% in hardness between the section thickness.
4. Miniature tensile sample testing provides an effective way to measure localized mechanical behaviour of the part and compares well with ASTM standard tests.
5. Miniature tensile testing also achieves cost savings because a small extended portion in the part can be created and machined off to obtain tensile properties. This is more representative, than printing separate larger coupons and test for tensile properties.

References

[1] Tan XP, Kok YH, Tan YJ, Vastola V, Pei QX, Zhang YW, Tor SB, Leong KF and Chua CK, 2015, *Journal of Alloys and Compounds*, **Vol. 646**, pp. 303-309.

[2] Dzugan J, Procházka R, and Konopik P, 2015, *Small specimens Test Techniques*: **Vol 6** ASTM International, pp 12-30

[3] Dzugan J, Sibr M, Konopík P, Procházka R and Rund M, 2017, *IOP Conference Series: Materials Science and Engineering,* **Volume 179**, conference 1

[4] Rund M et al. , 2015, Procedia Engineering Investigation of Sample-size Influence on Tensile Test Results at Different Strain Rates, **Vol. 114** (Amsterdam: Elsevier science bv) pp 410-15

[5] Procházka R, Džugan J and Kövér M, 2015, Arch. of Mat. Sci. and Eng. Miniature specimen tensile testing of AZ31 alloy processed by ECAP, Volume 76, Issue 2 (Gliwice: International OCSCO World Press) pp 134-9

[6] Dzugan J, Prochazka R, Rund M, Podany P, Konopik P, Seifi M and Lewandowski J J 2018 Effects of thickness and orientation on the small scale fracture behaviour of additively manufactured Ti-6Al-4V *Mater. Charact.*

[7] Andrew J. Elliott. “Directional solidification of large cross-section Ni-base superalloy castings via liquid-metal cooling”. PhD Thesis, University of Michigan, Ann Arbor, MI, 2005

[8] Kundan Kumar, Arun Pooleery , Madhusoodanan K , Singh RN, Arnomitra Chatterjee , Dutta BK, Sinha RK, 2016, *Materials Science and Engineering A*, **Vol 675**, pp 32-43

[9] Mang Ni, Chao Chen, Xiaojun Wang, Pengwei Wang, Ruidi Li, Xiaoyong Zhang, Kechao Zhou, 2017, *Materials Science and Engineering A*, **Vol 710**, pp 344-351

[10] Kok Y, Tan XP, Wang P, Nai MLS, Loh NH, Liu E, Tor SB, 2018, *Materials and Design*, **Vol 139**, pp 565-586

[11] Boyce BL, Reu PL, Robino CV, 2006, *Metallurgical and Materials Transactions A*, **Vol 37A** , pp 2481-2492

[12] Agnew SR, Elliott BR, Youngdahl CJ, Hemker KJ, Weertman JR, 2000, *Materials Science and Engineerng A*, **Vol 285**, pp 391-396

IOP Conf. Series: Materials Science and Engineering **461** (2019) 012070 doi:10.1088/1757-899X/461/1/012070

The influence of niobium content and initial microstructure of steel on the occurrence of Lüders band at the start of the plastic flow during cold deformation

S Rešković, I Jandrlić, T Brlić

University of Zagreb Faculty of Metallurgy, Sisak, Croatia

E-mail: reskovic@simet.hr

Abstract. In this paper, the results of a research on influence of niobium and initial microstructure on start of plastic flow of a material and formation of Lüders bands during cold plastic deformation in niobium microalloyed steel are presented. Research was carried out on hot rolled strips from low carbon steel and three steels with the same base chemical composition but with different amount of niobium: 0.035% Nb, 0.048% Nb, 0.060% Nb. Niobium microalloyed steel strips are rolled with selected parameters of thermomechanical treatment and additionally heat treated with the goal of achieving inhomogeneous microstructure. Static tensile testing was performed on samples, while simultaneously being recorded with thermographic camera. The results of thermographic measurements were analysed with software package Irbis 3 professional. The initial microstructure is tested for all samples. During the start of plastic flow in all microalloyed steel samples, with fine-grained homogeneous microstructure, inhomogeneous deformations are created. These were not found on samples which do not contain microalloying element niobium. In area of inhomogeneous deformations, Lüders bands are formed and propagate through the deformation zone. Lüders band was not found in microalloyed steels with coarse grain. The tests show that the addition of micro-alloying element niobium affects Lüders bands formation but the content of niobium in steel has no effect on the appearance of the Lüders bands. Additionally, it was found that the initial microstructure has a great influence on Lüders bands occurrence.

1 Introduction

In recent years, inhomogeneous deformations are intensively investigated at the beginning of the plastic material flow during cold deformation [1]. It has been found that they are associated with the occurrence of the Lüders bands [2-4].

They have not yet been given a realistic explanation of why Lüders bands occurs and which are the influential parameters for their occurrence and propagation. Some authors associate them with precipitates of alloying and micro-alloying elements [4-6]. Other authors associate them with the initial microstructure, primarily with grain size and homogeneity of microstructure [5-10]. The influence of strain rate is also extensively investigated [4,10-12]. Today, thermography has found its application [13-14] of studies of stress changes at the appearance and propagation of the Lüders bands.

This paper presents the results of detailed research on behavior hot-rolled strip of thickness 3 mm with different contents of niobium and different initial microstructure at the beginning of plastic flow during cold deformation.

2 Materials and methods

Four types of steels were selected for this study: low carbon steel and three steels with the same base chemical composition but with different amount of micro-alloying element niobium 0.035%, 0.048% and 0.060%, Table 1.

Table 1. Chemical composition of steels [wt.%]

	C	*Mn*	Si	*P*	S	*Al*	*Nb*	*N*
1	0.13	0.77	0.18	0.010	0.019	0.020	0.000	0,0080
2	0.09	0.75	0.05	0.018	0.014	0.020	0.035	0.0081
3	0.12	0.78	0.18	0.011	0.018	0.020	0.048	0.0080
4	0.13	0.70	0.14	0.018	0.014	0.040	0.060	0.0081

In this research, samples from hot-rolled strip with thickness of 3 mm with different contents of niobium and different initial microstructures were taken. Steels with 0.00% and 0.048% of niobium are rolled in industrial conditions in 3 mm thick strip of hot rolled steel and 0.035% and 0.060% Nb steels are rolled in semi-industrial conditions with thickness of 3 mm. Thermomechanical treatment was carried out on all niobium microalloyed steels whit resulted in a fine-grained ferrite-pearlite microstructure. Steels 2 and 4 have been cooled at various rates after the thermomechanical treatment in order to achieve changes in microstructure homogeneity and grain size. Steel 3 was heated at 900 °C and air-cooled. The static tensile test was performed on the Zwick 50 kN at stretching rate of 5 mm/min.

3 Results and Discussion

The obtained diagrams of Force-Elongation curves are shown in Figure 1.

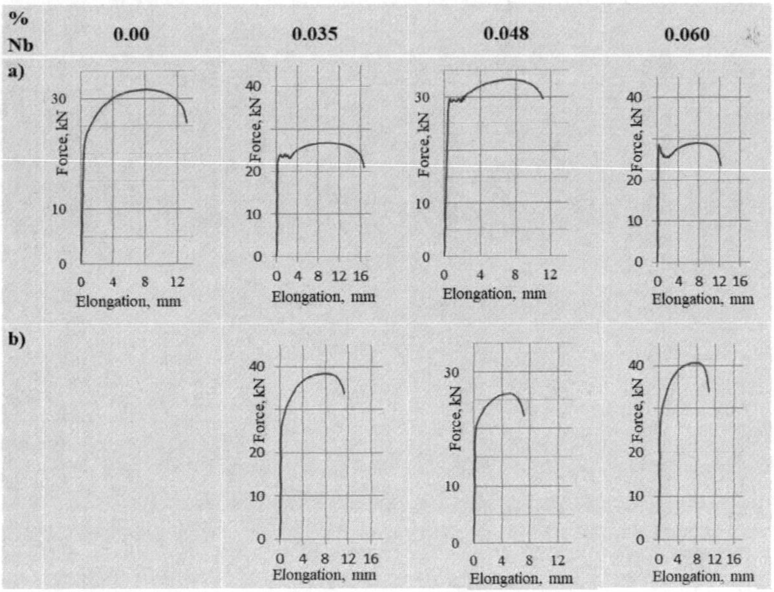

Figure 1. The appearance of inhomogeneous deformations during static tensile tests

It is clearly apparent that the steel without niobium did not have pronounced transition from elastic to plastic region and no inhomogeneous deformations at the beginning of the plastic flow. Samples

containing the different content of micro-alloying element niobium and homogeneous fine-grained microstructure have a clearly pronounced proportionality limit (Rp) and have the appearance of inhomogeneous deformations at the beginning of the plastic flow. Steels with the same base chemical composition and different initial microstructure show different behaviors.
Metallographic microanalysis of rolled strips was performed. The obtained results are shown in Figure 2.

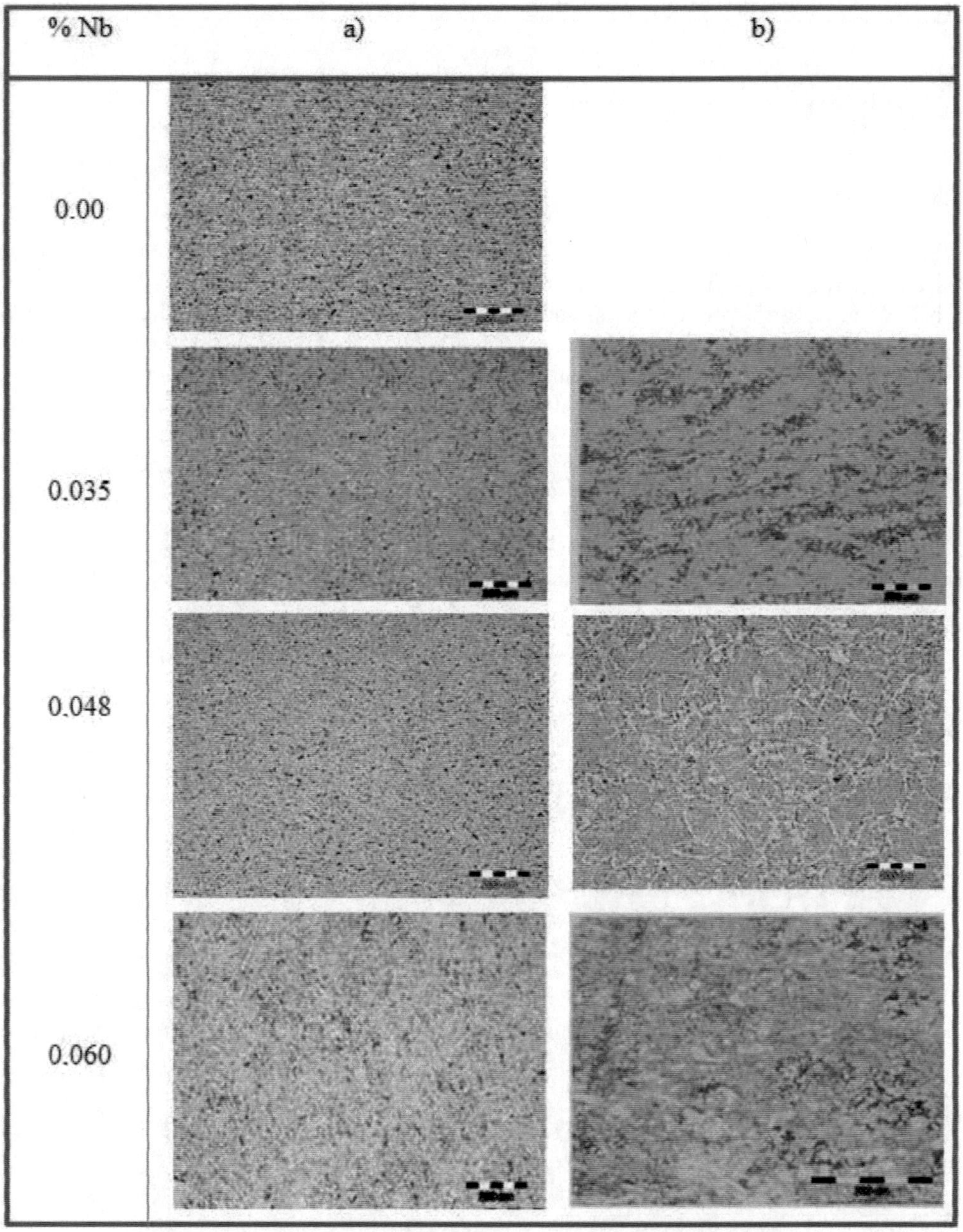

Figure 2. The initial microstructure of investigated steels:
a) After thermomechanical treatment and air-cooled
b) After rolling cooled at different rates

Figure 2 clearly shows that hot rolled strips (a) have a homogeneous fine-grained microstructure. The steel strips (b) with 0.035% and 0.060% Nb, are cooled on different temperatures and at different cooling rates. The steel strips (b) with 0.048% Nb, which was subsequently heated and cooled, have a coarse-grained inhomogeneous microstructure.

Thermographic measurements were also carried out simultaneously with static tensile test. Black matte coating Lechsys 29141 was applied on samples for thermographic tests.

Thermographic measurement was performed using Vario CAM® M82910 thermographic camera. The analysis of the measured results was carried out by Irbis 3 Professional software package. Stress changes have been established during deformation. Figure 3 shows thermograms of all tested samples.

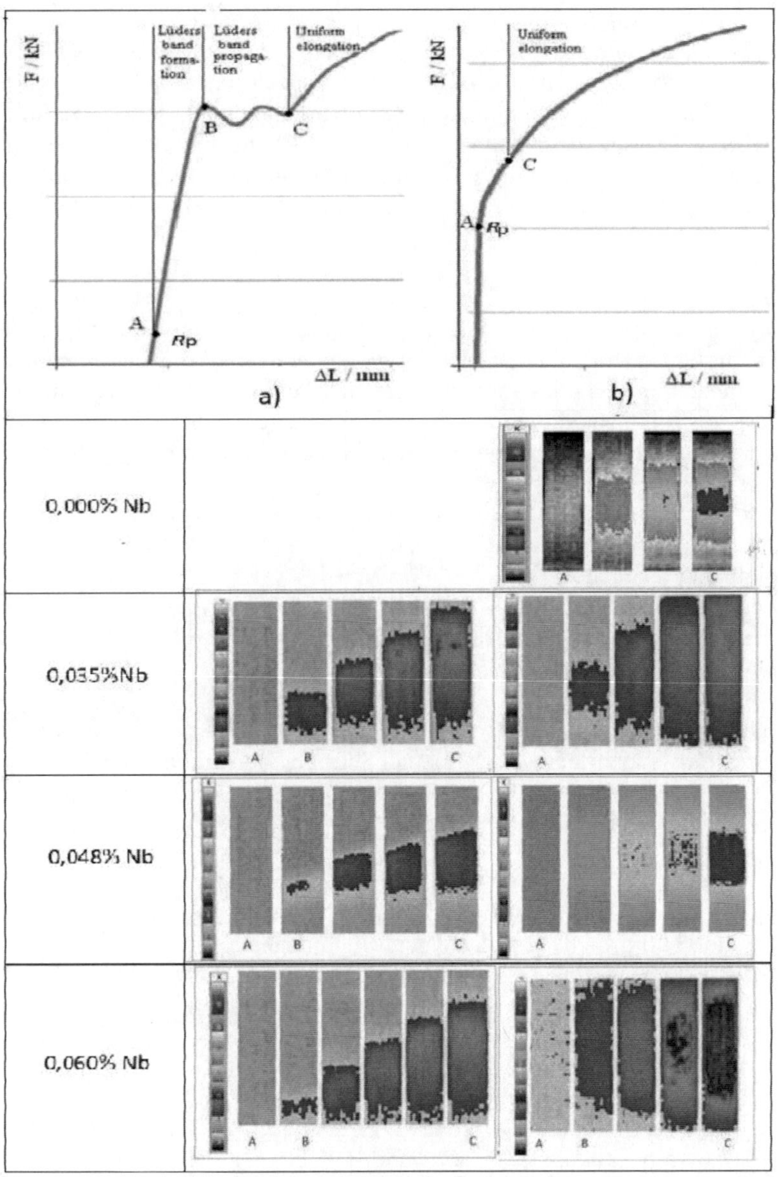

Figure 3. Thermograms of tested samples:
a) with the appearance of Lüders band, b) without appearance of Lüders band

Thermographic analysis has shown that in the area of non-homogeneous deformations are formed Lüders bands. In all three niobium microalloyed steels with homogeneous fine-grained microstructure (a), after reaching proportionality limit, point A, Figure 3, Lüders band formation and propagation begins from one end of the sample to the other end of the sample. The Lüders band is formed up to point B and propagates through the deformation zone, point C. Then, in the formed deformation zone, plastic deformation takes place until reaching tensile strength (Rm).
In the low carbon steel without niobium and all steels with niobium but having with a inhomogeneous microstructure, Figure 3, (b), no Lüders band were formed. After reaching proportionality limit, stresses are concentrated in the deformation zone and increase with the stretching force increase, from point A to C Figure 3.

Changed content of niobium has no effect on the appearance of the Lüders bands.
The reason why no niobium microalloyed steels has no occurrence Lüders band is related to the size and distribution of niobium precipitates. It has been shown that in the steel micro-alloyed with 0.048% Nb, Figure 3 a) size is niobium precipitates is 10 nm and the arranged in strings, while niobium precipitates in steel with the same content of niobium which was subsequently heated, Fig. 3 b) are 60nm [6-11].

4 Conclusion

The investigations have shown that the Lüders bands appear in the niobium micro-alloyed steel at the beginning of the plastic flow in cold deformation. For steel of the same base composition but without niobium and with homogeneous fine-grained microstructure, Lüders bands have not formed.
In the low-carbon steel, the addition of the micro- alloyed element niobium leads to the appearance of Lüders bands, but the content of the niobium in the steel does not affect the appearance of the Lüders bands.
It was also found that niobium microalloyed steels with inhomogeneous microstructure at the beginning of plastic flow during cold deformation, Lüders bands have not formed. This is related to the size of niobium precipitates.
Further research on microstructures, crystal grain orientation and size and distribution niobium precipitates, is needed to explain forming and propagation of the Lüders bands with cold deformation niobium microalloyed steel.

Acknowledgements
This work has been fully supported by the Croatian Science Foundation under the project number IP-2016-06-1270.

References

[1] Nagarajan S, Narayanaswamy R and Balasubramaniam V 2013 Study on local zones constituting to band growth associated with inhomogeneous plastic deformation, *Mater. Lett.* **105** 209–212
[2] Srinivasan N, Narayanaswamy R and Venkatraman B 2013 Advanced imaging for early prediction and characterization of zone of Lüders band nucleation associated with pre-yield microstrain, *Mater. Sci. Eng. A* **561** 203–211
[3] Srinivasan N, Raghu N and Venkatraman B 2013 An insight into lüders deformation using advanced imaging techniques *J. Mater. Eng. Perform.* **22** (2013) **10** 3085–3092
[4] Brlić T, Rešković S, Vodopivec F and Jandrlić I 2018 Lüders bands at the beginning of the plastic flow of materials, *Metalurgija* **57** (2018) **4** 357–359
[5] Emadoddin E, Akbarzadeh A and Daneshi G.H 2007 Correlation between Luder strain and

retained austenite in TRIP-assisted cold rolled steel sheets, *Mater. Sci. Eng. A* **447** (2007) **1-2** 174–179

[6] Rešković S and Jandrlić I 2013 Influence of niobium on the beginning of the plastic flow of material during cold deformation, *The Scien. Wor. J.* **2013** (Article ID 723725)

[7] Johnson DH, Edwards M R and Chard-Tuckey P 2015 Microstructural effects on the magnitude of Luders strains in a low alloy steel, *Mater. Sci. Eng. A* **625** 6–45

[8] Hyun Ryu J, In Kim J, Kim H S, Oh C S, Bhadeshia H K D H and Suh D-W 2013 Austenite stability and heterogeneous deformation in fine-grained transformation induced plasticity assisted steel, *Scripta Mater.* **68** 933–936

[9] Tsuchida N, Masuda H, Harada Y, Fukaura K, Tomota Y and Nagai K 2008 Effect of ferrite grain size on tensile deformation behavior of ferrite-cementite low carbon steel, Materials *Mater. Sci. Eng. A* **488 (2008) 1-2** 446–452

[10] Tsuchida N, Tomota Y, Nagai K and Fukaura K 2006 A simple relationship between Lüders elongation and work-hardening rate at lower yield stress, *Scripta Mater.* **54** 57–60

[11] Rešković S, Jandrlić I and Vodopivec F 2016 Influence of testing rate on Lüders band propagation in niobium microalloyed steel *Metalurgija* **55** (2016) **2** 157–160

[12] H.B. Sun H B, Yoshida F, Ohmori M and Ma X 2003 Effect of strain rate on Lüders band propagating velocity and Lüders strain for annealed mild steel under uniaxial tension, *Mater. Lett.* **57** 4535–4539

[13] Wang X G, Crupi V, Guo X L and Zhao Y G 2010 Quantitative thermo-graphic methodology for fatigue assessment and stress measurement *Int. J. Fatigue* **32**, 1970–1976

[14] Louche H and Chrysochoos A 2001 Thermal and dissipative effects accompanying Lüders band propagation, *Mater. Sci. Eng. A* **307** 15–22

The effect of heat treatment on the microstructure and mechanical properties of 3D-printed AlSi9Cu3Fe alloy

M Roudnicka[1,2], **D Dvorsky**[1,2] **and D Vojtech**[1]

[1]Department of Metals and Corrosion Engineering, University of Chemistry and technology Prague, Technicka 5, 166 28 Prague 6, Czech Republic

[2]Institute of Physics, Academy of Sciences of the Czech Republic (AS CR), Na Slovance 1999/2, 182 21 Prague 8, Czech Republic

E-mail: michaela.roudnicka@vscht.cz

Abstract. In this paper, we characterize AlSi9Cu3Fe alloy prepared by selective laser melting (SLM); a 3D printing technique based on the localized melting of a powder material by a laser beam in thin successively deposited layers. The layer-wise building-up resulted in an anisotropic and hierarchical microstructure of this alloy. High cooling rates brought a very fine microstructure, which contributed to high material strength, but also residual stresses, which are undesirable. Residual stresses can be reduced by various heat treatments. We carried out stress-relief and T6 heat treatments and studied their effect on the microstructure and mechanical properties. T6 heat treatment brought a significant increase in tensile yield strength.

1 Introduction

Al–Si–Cu alloys have been employed for various engineering applications due to their low density and good strength, especially for replacement of cast iron and steels in automobile parts [1]. Alloys based on Al-Si system are well known for their good castability due to eutectic transformation. However, conventional manufacturing via casting, working and welding is limited in terms of component geometry. A possible solution for the production of complex parts is the application of additive manufacturing (AM) technologies.

Due to processing materials layer upon layer by localized heat source, additively manufactured parts have usually different features from traditionally produced parts. Between irradiated volume and its surroundings, there are great temperature gradients. Melt pools thus solidify under cooling rates reaching up to 10^6 °C/s [2]. Under such rates, microstructure evolution can become completely distinctive. Generally, very fine microstructures are obtained [3] but phase composition and other microstructural features can vary depending on specific materials and specific processing conditions.

Selective Laser Melting (SLM) is one of the most promising AM techniques. Although it is one of the most thoroughly studied and advanced techniques, there are some weak sides we have to deal with. Because SLM belongs among powder-bed techniques, SLM parts suffers poor surface quality as particles of surrounding loose powder adhere to their surface. Therefore, even though SLM enables direct production of net parts, some additional treatment is often required to obtain high-quality surface. Otherwise, high surface roughness strongly limits material fatigue life [4]. Another

weak side of SLM manufactured metals is often their low plasticity. Due to steep temperature gradients (10^4 °C/mm) in melt pools, large thermal stresses remain in materials [5].
Retained internal stresses can be minimized already before actual production by setting of the SLM process. The first step is to select an appropriate scanning strategy. Osakada and Shiomi [6] concluded that the thermal stress increase with longer laser distance. Therefore, it is advisable to build parts in segments (e.g. chess-board scanning strategy). Mercelis et al. [7] also found that stresses are higher in the building direction than in the perpendicular one and so residual stresses increase with the height of SLM parts. Hence, orientation of parts can be considered before production too. In some SLM machines that enable preheating of the substrate, temperature gradients can be lowered and so the residual stresses, however, at the expense of yield strength [8].
After the manufacture, residual stresses can be reduced by heat treatment. Before the removal from the build substrate, most SLM producers recommend to subject final parts to a so-called stress-relief treatment to avoid possible deformations [7, 9]. For example, material data sheets for AlSi10Mg alloy by EOS or Renishaw suggest 300 °C/2 h. Another possibility is to apply different modes of heat treatment and tune microstructures and mechanical properties according to requirements given by a particular application. T6 heat treatment is commonly used to increase the strength of conventionally manufactured Al-Si alloys containing Cu or Mg. During solution treatment at a higher temperature, large intermetallic particles dissolve and alloying elements redistribute homogeneously. During artificial aging at a lower temperature fine precipitates are formed [10]. When the material is loaded, these homogeneously distributed precipitates become efficient obstacles for dislocation movement. For the here-studied AlSi9Cu3Fe alloy containing 3 wt.% of copper, T6 heat treatment might be thus applied to induce precipitation strengthening by Cu-rich precipitates [11].
So far, there are not many references in the literature to the preparation and characterization of AlSi9Cu3Fe alloy by AM. Therefore, we studied and characterized this alloy prepared by SLM. We subjected it to two conventional heat treatments – stress-relief and T6 - and described their effect on microstructural changes and mechanical properties.

2 Materials and Methods

2.1 Specimens preparation

Samples of AlSi9Cu3Fe alloy (chemical composition given in table 1) were prepared by means of metallic 3D printing by NETME Centre in Brno. Selective laser melting (SLM) technology was applied (SLM Solution 280HL machine). Fine gas-atomized powder of AlSi9Cu3Fe alloy (LPW, $d_{average}$ = 40 µm) was compacted by localized melting of a 400W laser beam. The laser beam melted the powder material, deposited in 50 µm thin layers, according to a chess board scanning strategy, with a hatching distance between adjacent scan lines set to 150 µm. The scanning speed of the laser beam was 1300 mm/s. The working chamber was supplied with an argon atmosphere to prevent the powder oxidation.

Table 1. Chemical composition of the SLM AlSi9Cu3Fe alloy.

wt.%	**Al**	**Si**	**Cu**	**Fe**	**Mg**	**Zn**	**P**
SLM	bal.	8.9±0.1	3.1±0.1	1.2±0.1	0.3±0.02	0.01	0.01

SLM provided a set of samples intended for mechanical testing, directly in the form given by the input 3D CAD model (dog-bone samples). All samples were printed in vertical orientation. As their surface was covered with adherent particles of surrounding powder, sand-blasting was applied to smoothen the surface.
One third of the tested samples was kept in the as-built form. The rest of the samples were subjected to heat treatment (HT) in a muffle furnace Martinek MP 05. Two common HT regimes were applied: (1) stress-relief HT to remove residual stresses, and (2) T6 HT to induce precipitation of θ″ hardening phase. Stress-relief HT was carried out at a temperature of 300 °C for 2 hours. For

T6 HT, samples were solution treated at 520 °C/6 h, water-quenched and artificially aged at 160 °C/10 h.

2.2 *Characterization of microstructural changes*

For each material state (as-built, stress-relieved, T6), microstructure was studied on longitudinal metallographic sections. After etching in 5% HF, microstructures were observed by light metallographic microscope OLYMPUS PME 3 and scanning electron microscope TESCAN VEGA-3 LMU. Scanning electron microscopy (SEM) was completed by EDS analysis (Oxford instruments INCA 350 analyzer). Phase composition was determined by X-ray diffraction (XRD, PANalytical X'Pert PRO with Cu anode, High Score Plus software).

2.3 *Mechanical testing*

On the prepared metallographic section, Vicker's hardness HV1 (1 kg load, 10 s persistence) was measured using a Future-Tech FM-700 microhardness tester. Average hardness was evaluated statistically from 10 measurements. Uniaxial tensile tests were carried out on a universal loading machine LabTest 250SP1- VM. All measurements were carried out under a constant deformation rate of 0.001 s^{-1}.

3 Results and Discussion

3.1 *Microstructure of 3D-printed AlSi9Cu3Fe alloy*

Gradual localized melting of powder material repeatedly in many successive layers is responsible for the formation of distinctive microstructure of the AlSi9Cu3Fe alloy. At the moment when the beam is focused on powder layer (interaction time in ms), it melts a certain volume of the material. As the beam is moved away in its vector, the melt solidifies. As a result, we can observe solidified melt pools in the microstructure (figure 1a-b). Their width and depth reaches about 200-250 μm, which approximately corresponds to the size of the laser beam spot. The depth of melt pools exceeds the thickness of a powder layer by a factor of 4-5, which is important for ensuring good interconnection between layers.

In higher magnification, one can see that the microstructure is formed by very fine cellular substructure (figure 1c-d). Inside melt pools, cells are finer (figure 1c) and elongated in one direction. Their width and length was determined by image analysis as 1.0±0.1 μm and 3.7±0.8 μm, respectively. Such fineness is the result of very high cooling rates [2]. The elongation and directionality of cells is determined by temperature gradients in melt pools. From the hottest center of a melt pool, the heat is radially conducted away by already solidified material or surrounding powder. Because cellular substructure is very homogeneous in the whole material, it is hard to distinguish grains. Nevertheless, slight differences in cell morphologies and optical contrast between particular grains make it possible to discern columnar grains growing from pool borders towards their centers (indicated by dashed lines in figure 1b). At melt pool borders, cells are larger and more rounded (figure 1d), which is caused by lower undercooling at the interface between a melt pool and its surroundings [12].

Comparable to AlSi10Mg alloy prepared by SLM [13, 14], in AlSi9Cu3Fe alloy, the intercellular network is formed by eutectic Si and the cells by oversaturated solid solution α-Al. EDS quantified the chemical composition of the solid solution and showed 5.4±0.5 wt.% of Si, 3.2±0.4 wt.% of Cu and 1.1±0.2 wt.% of Fe. However, the quantification is distorted by surrounding Si network which interferes with the interaction volume of the EDS analysis. Therefore, it can be expected that the actual content of Si in the solid solution is lower. Further measurement in transmission mode of electron microscopy is planned to obtain more accurate evaluation. Except from the two main constituting components, XRD also revealed the presence of small amount of Al_2Cu phase (2 vol.%).

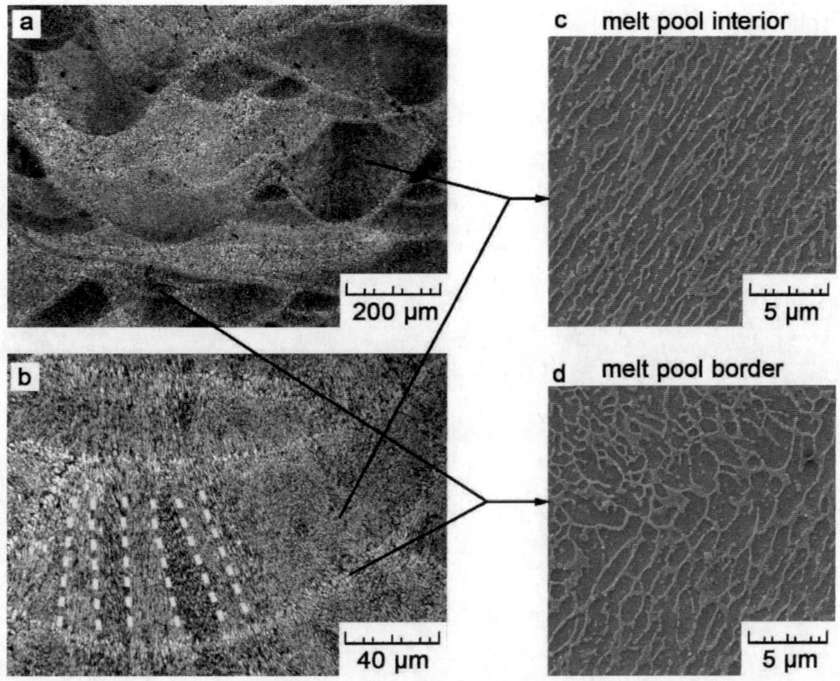

Figure 1. Microstructure of the SLM AlSi9Cu3Fe alloy: (a) melt pools, (b) melt pool interiors (dashed lines indicate grains), (c) cellular structure inside melt pools and (d) cellular structure at melt pool borders.

3.2 Microstructural changes after heat treatment

Microstructural changes after applied heat treatments are depicted by micrographs in figure 2. Microstructure after stress-relief HT (figure 2b,e) does not apparently differ from the as-built microstructure (figure 2a,d). Only slight coarsening of Si network and its initial disruption into idiomorphic particles was observed. Increased temperature enhanced Si diffusion from the oversaturated solid solution. Otherwise, additively built macrostructure remained unchanged. After T6 HT, microstructure homogenized (figure 2c,f); melt pool borders vanished and the Si network completely disintegrated. High temperatures during solution treatment led to Si diffusion and formation of large Si particles (1-3 µm in diameter). Also, needlelike Al_5FeSi phase formed (figure 2f). Subsequent quenching again resulted in an oversaturated solid solution α(Al)′ with concentrations of alloying elements higher than their maximal equilibrium solubility. Therefore, at subsequent aging, oversaturated solid solution α(Al)′ decomposed into coherent GP zones followed by precipitation of fine θ″ phases. Within the scope of this work, it was not possible to visualize the precipitates but their presence is highly anticipated as the course of T6 heat treatment is the same as for castings [15]. In Al-Si-Cu alloys, GP zones consisting of localized concentrations of Cu atoms form at room temperature already. Above 100 °C, GP zones dissolve and are replaced by θ″ phases. With longer aging time, θ″ phase transforms into the metastable θ′ semi-coherent with the matrix. Finally the stable incoherent equilibrium phase θ (Al_2Cu) is formed. Maximum strength is obtained when the highest fraction of θ″ phase is present [10].

By comparison of images a-c in figure 2, it is apparent that a significant amount of small pores appeared after T6 HT. Most probably, the porosity formed due to the melting of Al_2Cu phase during solution treatment at 520 °C. By image analysis, the porosity was assessed to represent 2 vol.%, which corresponds to the amount of Al_2Cu phase detected in the material in the as-built state. For Al-Si-Cu alloys, there is a risk of the incipient melting of Cu-rich phases at temperatures higher than 495˚C. On the other hand, solution treatment at temperatures of 495˚C or less is not capable of maximizing the dissolution of the copper-rich phases, nor is it able to modify the silicon particle morphology sufficiently [16]. Gautier et al. [17] showed that the best combination of tensile strength and ductility is obtained with solution treatment at 515°C.

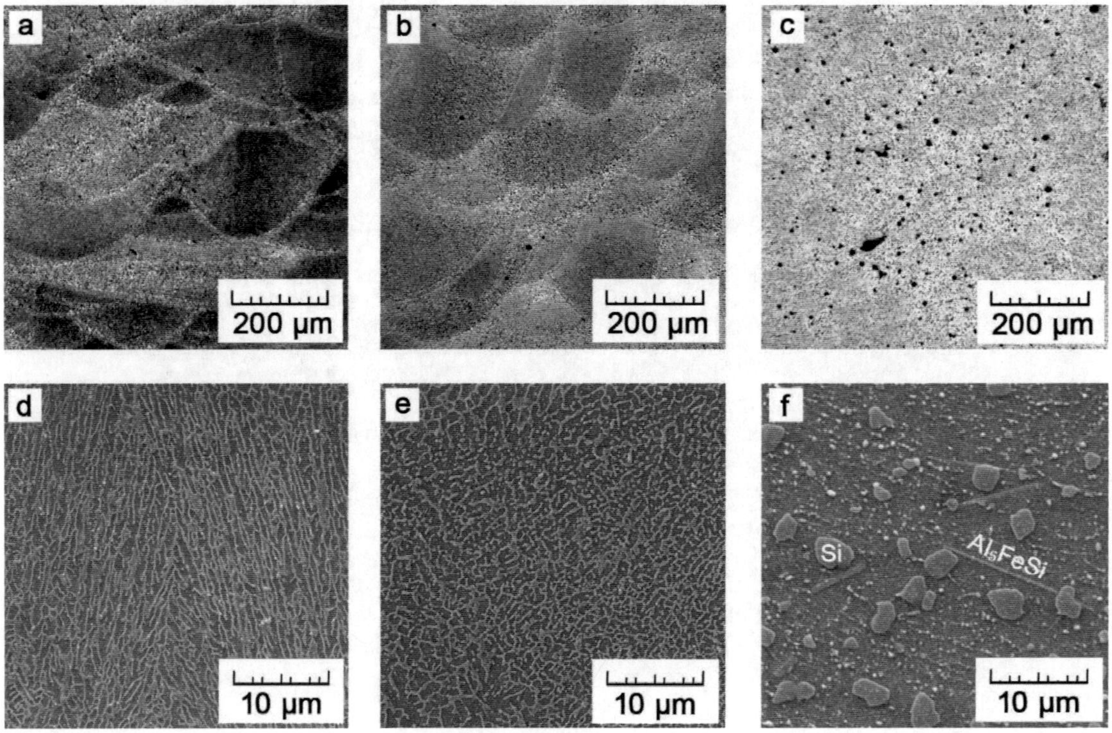

Figure 2. Microstructures of the SLM AlSi9Cu3Fe alloy: (a,d) as-built, (b,e) stress-relieved and (c,f) T6.

3.3 Mechanical properties

Changes in mechanical properties related to heat treatment are described by chart in figure 3.

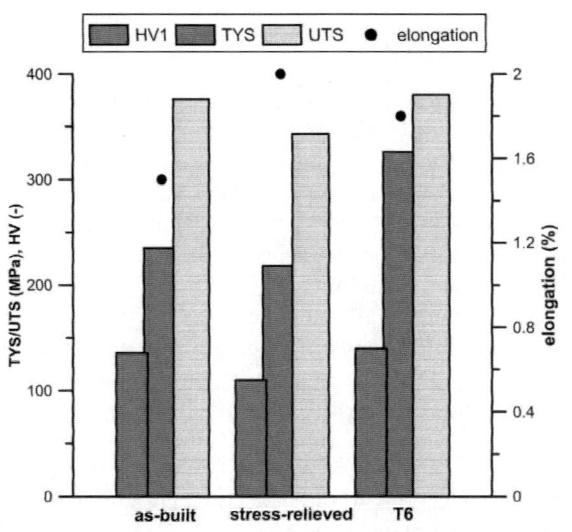

Figure 3. Changes of mechanical properties of the SLM AlSi9Cu3Fe alloy after heat treatment.

Vickers hardness do not show significant differences. Stress-relief HT relieved residual stresses and subtly coarsened the Si network, which resulted in a slight decrease in hardness (from 136±4 to 110±3 HV1). After T6 HT, hardness remained almost unchanged. Hardening brought by precipitation of θ″ phase was counterweighted by softening due to the break-up of Si network forming fine cellular microstructure and significant coarsening of Si particles. Ultimate tensile strengths (UTS) show a similar trend. In addition to the described features, the increase in UTS in T6 condition was also limited by porosity.

The effect of precipitation strengthening by θ″ phase is more pronounced in the values of tensile yield strength (TYS). After T6 HT, about 40% increase in TYS was registered. Elongation remained at a comparable level. Negative effects of precipitation strengthening, brittle Al_5FeSi phase and porosity on elongation were offset by an increase in plasticity due to the break-up of Si network.

Compared to SLM-fabricated AlSi10Mg alloys, where the T6 HT resulted in decrease in tensile strength due to the coarsening of Si precipitates [12, 13], in AlSi9Cu3Fe alloy, this treatment have shown to be contributory in terms of TYS.

4 Conclusion

Two heat treatments of the AlSi9Cu3Fe alloy prepared by 3D printing technique SLM have been studied. Stress-relief did not significantly change the microstructure, only relieved residual stresses resulting from high cooling rates during SLM. After T6 heat treatment, characteristic 3D-printed microstructure completely vanished. During solution treatment and subsequent quenching, fine cellular substructure disintegrated and Si and Al_5FeSi phases formed in the matrix of oversaturated α(Al)′ solid solution. Aging led to the precipitation of fine θ″ precipitates, bringing a significant increase in tensile yield strength.

Acknowledgements

Authors wish to thank the Czech Science Foundation (project no. P108/12/G043) and specific university research (MSMT No 21-SVV/2018) for the financial support of this research.

References

[1] Okayasu M, Takeuchi S and Ochi T 2017 *Int. J. Cast Metals Res.* **30** 217-225
[2] Brandl E, Heckenberger U, Holzinger V and Buchbinder D 2012 *Mater. Design* **34** 159-169
[3] Wang L Z, Wu J J, Huang X F and Hong X F 2018 *Opt. Laser Technol.* **107** 89-98
[4] Olakanmi E O 2013 *J. Mater. Process. Technol.* **213** 1387-1405
[5] Li Y and Gu D 2014 *Addit. Manuf.* **1-4** 99-109
[6] Osakada K and Shiomi M 2006 *Int. J. Mach. Tools Manuf.* **46** 1188-1193
[7] Peter M and Jean-Pierre K 2006 *Rapid Prototyp. J.* **12** 254-265
[8] Buchbinder D, Meiners W, Pirch N, Wissenbach K and Schrage J 2014 *J. Laser Appl.* **26** 012004
[9] Sun S, Brandt M and Easton M 2017 Laser Additive Manufacturing: Materials, Design, Technologies and Applications ed M. Brandt (Cambridge: Woodhead Publishing) chapter 2 p 55-77
[10] Sjölander E and Seifeddine S 2010 *J. Mater. Process. Technol.* **210** 1249-1259
[11] Mondolfo L F 1976 Aluminum alloys: structure and properties (London: Butterworths)
[12] Li W, Li S, Liu J, Zhang A, Zhou Y, Wei Q, Yan C and Shi Y 2016 *Mater. Sci. Eng. A* **663** 116-125
[13] Fousova M, Dvorsky D, Michalcova A and Vojtech D 2018 *Mater. Charact.* **137** 119-126
[14] Zhou L, Mehta A, Schulz E, McWilliams B, Cho K and Sohn Y 2018 *Mater. Charact.* in press
[15] Kang H, Kida M, Miyahara H and Ogi K 1997 *J. JFS* **69** 828-834
[16] Samuel A M, Doty H W, Valtierra S and Samuel F H 2013 *Int. J. Cast Metals Res* **26** 354-363
[17] Gauthier J, Louchez P and Samuel F 1995 *Cast Metals* **8** 91-106

IOP Conf. Series: Materials Science and Engineering **461** (2019) 012072 doi:10.1088/1757-899X/461/1/012072

Investigation of hold period influence on residual fatigue live in steam turbine shaft design

M Rund[1], R Procházka[1], J Džugan[1], P Konopík[1], M Nesládek[2], J Jurenka[2], M Růžička[2], P Měšťánek[3]

[1]COMTES FHT a.s., Průmyslová 995, Dobřany 33441

[2]Czech Technical University in Prague, Faculty of Mechanical Engineering, Technicka 4, 16607 Prague 6, Czech Republic

[3]Doosan Škoda Power s.r.o., Tylova 1/57, 30128 Pilsen, Czech Republic

E-mail: martin.rund@comtesfht.cz

Abstract. Growth of share renewables in the field of electric power supplier and a lack of mechanism to cover shortage, force standard supplier of energy to stabilize the electrical grid. This situation is then demonstrated by increase of hot-start and cold-starts. But there is a lack for robust tool for qualitative assessment of residual life of the turbine shaft and for design of such a component. The goal of this paper is to present influence of the holding period on low-cycle fatigue (LCF), thermo-mechanical fatigue (TMF) and the so called Damage Operator Approach (DOA) by Nagode, which were used as a new approach for design of turbine shaft in Doosan Škoda Power.

1 Introduction

The energy industry is nowadays subjected to greater demands. One of them is growing share of renewables in gross final energy. In comparison, in 2004 was the share of renewable in gross final energy consumption 8.5%, compared with 17.0% in 2016 [1]0. Goal from EU is to have a 20% share in 2020. In Czech republic the share of renewables was 10.3% in the beginning of the year 2018 [1]. However, the fluctuating nature of the renewables creates instability in the energy systems. Because a lack of system, which would cover "energy fluctuations", these energy instabilities has to be in large measure compensated by thermal power sources. As results of this, the operating conditions are changing, particularly in number of startups and shutdowns, and at the same time, is here demand on acceleration the whole process of startups and shutdowns.

Turbine shaft are designed according to the number of required *cold-starts* cycles and *hot-starts* cycles. Usual number of *cold-start* is 50-250 cycles and 6000-8000 cycles of *hot-starts*, respectively. Where the *cold-start* is defined as a start from almost zero speed and cooled state, in comparison to the *hot-start,* where starts begin from warmed state, these cycles are mostly related to the fluctuation of the power supply.

Above mentioned, this brings questions about residual life of several decades running turbine shaft and needs for new attitudes in turbine shaft design.
The standard attitudes for derivations of material criterion are based usually on low cycle fatigue (LCF). Limitations of LCF is because of isothermal nature of LCF and because the damage mechanism in turbine shaft is thermo-mechanical (TMF). In literature is therefore possible to find new attitudes in LCF testing [3][4][5].
The life accessing for such a complexly loaded component is not possible without finite element method (FEM), which is very challenging task, because of combination damage process, where the most significant are fatigue, creep and oxidation [6][7]. Nagode et al. [8] suggested new the so-called Damage Operator Approach (DOA) which proposed computation process for estimating fatigue damage of component subjected to thermal and mechanical loading. Aim of this paper is to introduce experimental work and partial results carried out within the FLEXTURBINE project. Namely, experimental work dealing with comparison of Young's modulus measured by standard tensile test and dynamic modulus attitude. Subsequently, problem of the hold period on LCF tests results is discussed. TMF test experiment used for verification of the DOA model is outlined.

2 Experimental work

For estimation of the basis material proprieties the tensile tests at room and elevated temperature were carried out. From the nature of tensile tests it is not in all cases appropriate to take the results of the Young's modulus. The Young's modulus is fundamental element used for deriving other parameters, whose underestimation can lead in mismatch in results. Therefore, the dynamic modulus was measured and evaluated according to ASTM E1876-09, where the modulus is obtained based on the impulse excitation technique analyzing the vibration of a test sample or component after it was impulse excited. The material subjected to testing was a chrome-molybdenum creep resistant steel applicable to steam turbine shafts. Comparison of Young's modulus resulted from tensile tests and dynamic modulus is presented in **Figure 1**.

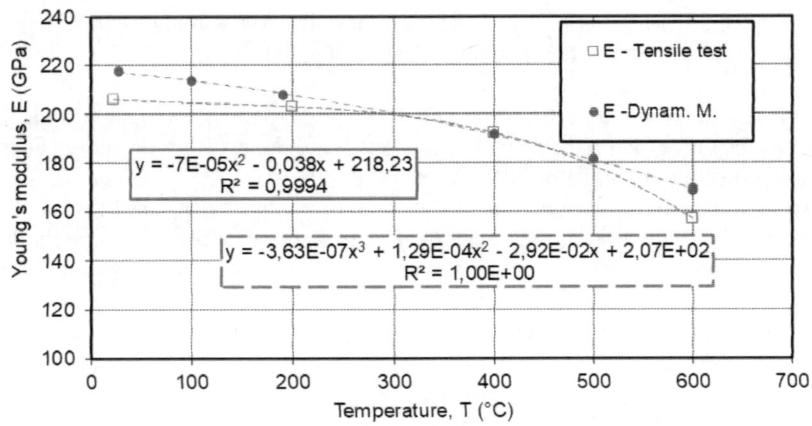

Figure 1. Temperature dependence on Young's modulus

As was presented above, the standard attitude for life prediction is based on standard LCF test. Low cycle fatigue is covered by standard e.g. ASTM E606. But loading of the turbine shaft is rather out of description listed in the standard, from the view of velocity/frequency of loading. In a fact, in a real shaft loading regimens with long hold period between the individual loading cycles can be found. Effect of the hold period on LCF results wasn't know and therefore investigation of this effect were investigated.

2.1 *Low-cycle fatigue*

Fatigue damage and life time prediction under LCF is described by Manson-Coffin and Basquin curve, see equation (1).

$$\varepsilon_a = (\sigma_f'/E_{eff})(2N_f)^b + \varepsilon_f'(2N_f)^c \quad (1)$$

Where ε_a is a strain amplitude, N is number of cycles to failure, σ_f' and b are the fatigue strength coefficient and exponent, ε_f' is fatigue ductility coefficient and c is the fatigue ductility exponent. Standard LCF tests were carried out according to the standard ASTM E606 in temperature range 20-600°C.

Firstly, the specimen size was designed in that manner to avoid buckling during cyclic compressive load increases. In **Figure 2** is presented the final specimen size which was used for all tests and results from simulation of the buckling at 600°C. LCF tests were carried out on servo-hydraulic testing machine MTS 810 with load capacity 250kN at room and elevated temperatures. In the second case, the split three-zone tube furnace Mellen was employed for heating samples up to 600°C.

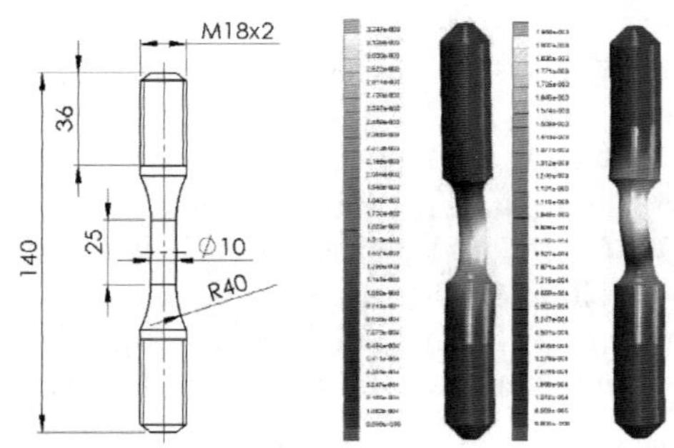

LEFT RIGHT

Figure 2. LEFT Geometry of the specimen; RIGHT Simulation of the Buckling, F=50kN, E=170GPa, T=600°C

One of the key parameters of LCF tests is number of cycles to initiation. One of the attitude to evaluate this, which comes from the standard, is to evaluate the cycles to initiation as intersection of the envelope Force-Count and a horizontal line which lie in the thirty percent of the maximal force. According to our investigation, this attitude lead to underestimating of the results. Therefore new methodology for evaluation of the cycles to initiation was proposed. This methodology is based on finding of the saturated part of the Force-Count curve, middle of the interval is evaluated as cycles to saturation (C_{SAT}), cycles to initiation (C_{IN}) are evaluated to the intersection point of parallel line to the saturated part of the Force-count curve which lie below the saturated part in distance of five percent, and Force-count curve.

2.2 *Low-cycle fatigue with hold period 10 min*

For investigation of the hold period effect the LCF test were carried out, but 10 minutes delay between individual cycles. During the holding period constant value of deformation was hold. Test were carried out on servo-hydraulic testing machine Inova, deformation of the specimens were measured by means of High temperature extensometer Epsilon model 3549.

Demonstration of the Force – Deformation course is shown in **Figure 3**. All tests were carried out with asymmetry coefficient R = 0 and strain amplitude 0,7%. This test takes roughly 7 days to reach a rupture. In comparison, LFC test with the same parameters takes 20 minutes. The longer-term exposure will give preference to further degradation process, oxidation. Metallographic analysis reveal, that the hold period effect cause growth of radial crack in the specimen with sharper manner, than in standard LCF. Influence of the atmosphere and temperature on the oxidation process will be done within the future work.

Comparison of metallographic evaluation from specimens after LFC and LCF with hold period is in **Figure 4**. The hold period effect cause a decrease of the number of the cycles to initiation because of above mentioned reason. Comparison LCF and LCF with hold period results in **Figure 5**.

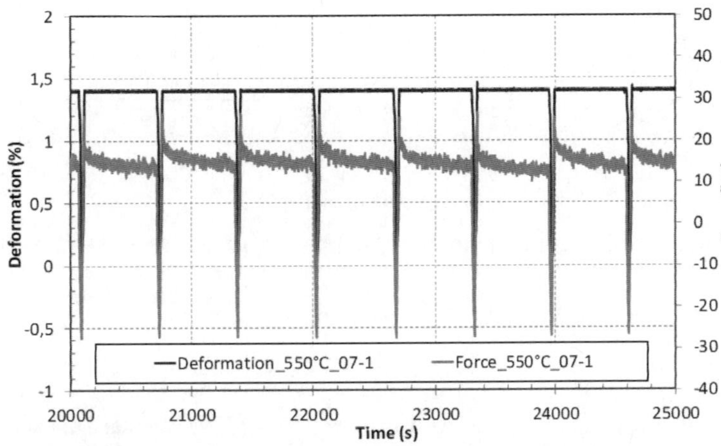

Figure 3. Demonstration of the course of the force and deformation during LCF with hold period at 550°C and strain amplitude 0,7%, R=0

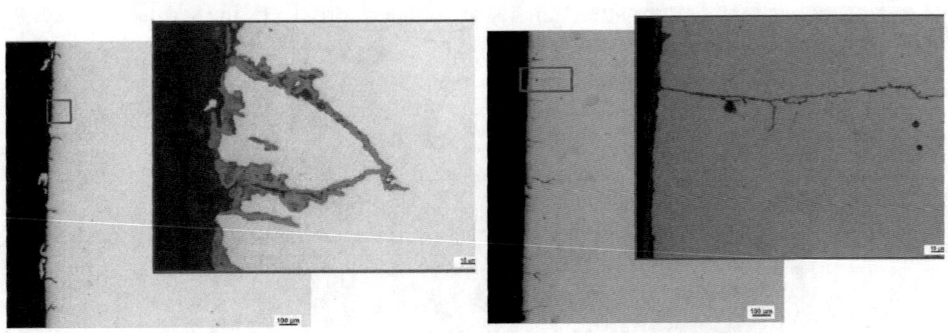

LEFT RIGHT

Figure 4. LEFT radial crack length in specimen after LCF test; **RIGHT** radial crack length in specimen after LCF with hold period

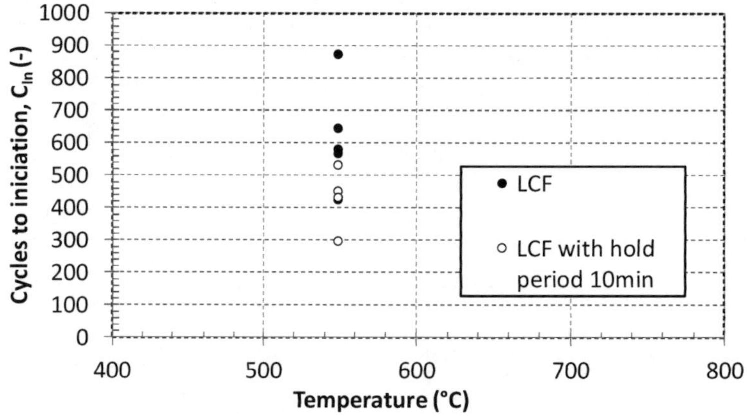

Figure 5. Relation between cycles to initiation and temperature for LCF and LCF with hold period; R0, deformation amplitude 0,7%

2.3 Thermo-mechanical fatigue

Design of the tests and specimens was derived from ASTM E2207 standard. Tests are carried out on servo-hydraulic axial-torsional testing machine. DIC system Mercury provides the control signal for the machine in axial and torsional direction. Heating of the specimens realized by resistive heating system. Infrared camera was used for evaluation of temperature distribution field. It was found that only 5mm is possible to consider as temperature homogeneity area. Therefore, L_o of the virtual extensometer was set to be 4 mm. Experimental setup is presented in **Figure 6**.

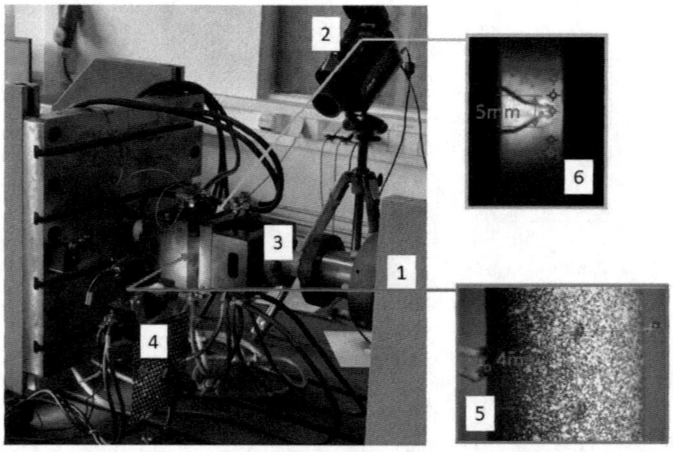

Figure 6. Setup for TMF experiment, 1- Servo-hydraulic axial torsional machine; 2- Infrared camera; 3 – resistive heating grips; 4 - DIC system; 5 – detail of the specimen from DIC; 6 – detail of the specimen from infrared camera

TMF experiments were carried out under strain and temperature control. Whereas the temperature was in- or out-off-phase, respectively. The cycle asymmetry coefficient R-1 was used for new fatigue tests, where the maximum axial strain ε_a 0.005 and maximum temperature T 600 °C was in phase (0°) or out-off phase (180°), as document **Figure 7**.

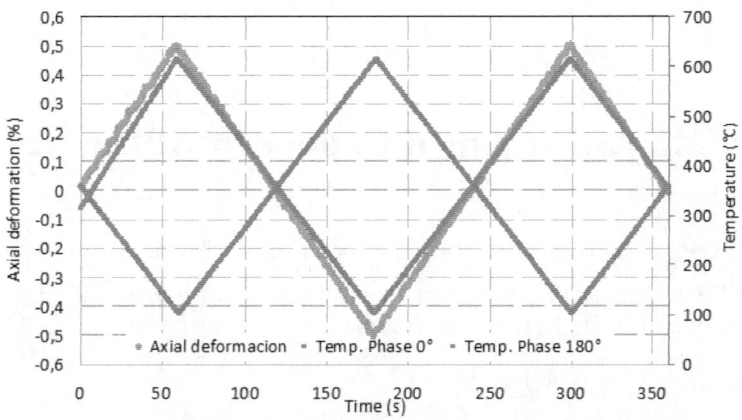

Figure 7. Time dependence of axial strain and temperature in phase (0) and out of phase (180)

3 Conclusion

This paper briefly presents the experimental program carried out in the scope of FLEXTURBINE project. Influence of measurement technique on results of Young's modulus was presented. It was found that hold period effect significantly LFC test results. The long term exposure leads to involvement of other damage mechanism such as oxidation. Simultaneously attitude for verification of the DOA by means of TMF experiment was preformed.

Acknowledgements

This project has received funding from the European Union's Horizon 2020 research and innovation program under the Grant Agreement No. 653941

References

[1] Consumption of renewable energy doubled between 2004 and 2016,"Available: https://ec.europa.eu/eurostat/statistics- explained/index.php/ Renewable_energy_statistics# Renewable_energy_produced_in_the_EU_increased_by_two_thirds_in_2006-2016.,

[2] M. Nesládek, M. Bartošák, J. Jurenka, et al. Thermo-Mechanical Fatigue Prediction of a Steam Turbine Shaft. MATEC Web Conf 165, 2018.https://doi.org/10.1051/matecconf/ 201816522016.

[3] Beck T, Pitz G, Lang K.-H, Löhe D. 1997 Thermal-mechanical and isothermal fatigue of IN 792 CC, Materials Science and Engineering: A, https://doi.org/10.1016/S0921-5093(97)00281-5

[4] Malpertu, J.L, Rémy, L. 1990 Influence of test parameters on the thermal-mechanical fatigue behavior of a superalloy, Metall and Mat Trans A, https://doi.org/10.1007/BF02782419

[5] Dzugan J, Prochazka R, Konopik P 2017 Low cycle fatigue tests with the use of miniaturized test specimens PROCEEDINGS OF THE ASME PRESSURE VESSELS AND PIPING CONFERENCE, 2017, VOL 1A

[6] Neu R W and Sehitoglu H 1989. Thermomechanical Fatigue, Oxidation and Creep: Part I. Damage Mechanisms. Metallurgical Transactions A 20A, 1755-1767

[7] M. Nesládek, J. Jurenka, M. Lutovinov, M. Růžička, P. Měšťánek, J. Džugan, An assessment of thermo-mechanically induced fatigue damage of a steam turbine shaft, Procedia Structural Integrity, Volume 7, 2017, Pages 190-197, ISSN 2452-3216, https://doi.org/10.1016/j.prostr.2017.11.077

[8] Nagode M, Angler F L and Hack M 2011 A Time-Dependent Damage Operator Approach to Thermo-mechanical Fatigue of Ni-resist D-5S. Int J Fatigue 33(5):692–99.doi:https://doi.org/10.1016/j.ijfatigue.2010.11.009

Refining Structure of Copper by SPD process

J Švec[1], P Szkandera[1], S Rusz[1], O Hilšer[1] and J Petrů

[1]VSB – Technical University of Ostrava, Faculty of Mechanical Engineering, 17. listopadu 15, 708 33 Ostrava – Poruba

E-mail: stanislav.rusz@vsb.cz

Abstract. This research concerned the whole production of UFG materials, using Severe Plastic Deformation (SPD). These new technologies for production of semi-finished products with ultra-fine grained structure differ from conventional technologies. One of them is new type of equipment DRECE (Dual Rolling Equal Channel Extrusion), designated for obtaining UFG structure and increase of mechanical properties in strip of sheet. The main goal of the paper is a review of current achieved results obtained by processing of Cu based materials by DRECE method, the equipment of which was put in service towards the end of 2009 at the Department of mechanical technology, Faculty of mechanical engineering, VŠB - Technical University of Ostrava. Cu-based materials – technical pure Cu (99.3% purity) – is used as experimental material. Experiments with use of the formed structural material - copper was made on the forming device in order to achieve grain refinement in the strip of sheet (Cu 99,3% purity) with dimensions 58 x 2 x 1000 mm The paper describes results of determination of an influence of number of passes on resulting mechanical and micro-structural properties of the processed Cu-based materials.

Keywords: severe plastic deformation, DRECE method, copper, mechanical properties, heat treatment, structure

Introduction

In many technical processes of forming the deformation is substantially greater than conditions at the tensile test. These new activities demonstrated at the beginning of the nineties that it was possible to manufacture nano-crystalline metallic materials by very high plastic deformation at low homological temperatures. It is possible to achieve in ductile metallic materials at the tensile test a deformation from 30% to 70% [1-2]. Obtaining of nano-crystalline structures requires typical magnitudes of deformation of the order from 100 to 1000%. High deformation at comparatively low homological temperatures is an efficient method for manufacture of ultrafine grained (UFG) massive materials [3]. New technologies, which use high deformation for obtaining the fine-grained structure is described namely by Valiev [4]. This research concerned the whole production of UFG materials, using Severe Plastic Deformation (SPD). Several types of SPD technologies serving for production of UFG metals were developed already at the beginning of the nineties. One of them is new type of equipment called DRECE (Dual Rolls Equal Channel Extrusion), designated for obtaining UFG structure in a strip of sheet.

1 Experimental materials and procedures

A sheet from commercial pure aluminum was used at the very beginning of experimental works [5]. This material was used for practical solving of the first technological problems with jamming and high warp of the processed sheets. Copper-based materials were selected for next experiments after performed literature search and after obtaining of the binary phase diagrams (in the case of alloys). Chemical composition of copper is shown in Table 1

Table 1. Chemical composition of copper (wt%)

Sn	Al	As	Pb	Sb	Fe	Cu
0.15	0.05	0.1	0.1	0.08	0.05	Rest

Sheets of dimensions (58x2x1000 mm) were used as an input material. The sheets were bought in commercial trade network in as cold-rolled state, without previous heat treatment, in chemical purity corresponding to the relevant standard. Forming equipment based on process CONFORM, modified for sheet forming. During 2009 a prototype of this equipment was put into trial operation at the working site of the VSB - Technical university of Ostrava [5-6]. Fig. 1 shows a principle and overall view of the prototype of this equipment. It consists of the following main parts: gear of the type Nord with electric drive, disc clutch, feed roller and pressure rollers with regulation of thrust, forming tool made of the steel grade Dievar. Strip of sheet is fed into the working space and it is pushed by the feed roller with help of pressure rollers through the forming tool without change of its cross section. Next pass is realized after rotation of sheet by an angle of 180°. Repeated plastic deformation realised in this manner brings substantial refinement of structure. During the trial operation first experiments were made followed by their evaluation. On the basis of these works some modifications of design were proposed [6].

a) b) c)

Figure 1 a) Principle of DRECE method, b) Forming equipment, c) Detail of the equipment

Both experimental materials were formed by SPD process at ambient temperature without previous heat treatment and selected samples operative heat between individual passes. The sheets were rotated between individual passes by 180° around their longitudinal axis [5]. Forming processed sheets were subjected to metallographic evaluation and basic mechanical properties were determined.

The extruded samples of used materials after completion of selected passes were then cut from sheets into individual series for preparation of individual testing specimens for mechanical tests and metallographic evaluation, which were marked by symbols Cu-X, where X expresses the number of passes (X = 1-6 or 10). The sample of initial state is marked with the symbol 0. Mechanical properties of studied samples were tested by tensile test and by Vickers hardness method. Metallographic analysis of structure was made on light microscope and with use of TEM.

2 Description and discussion of achieved results

Mechanical properties of studied samples were tested by Vickers hardness method on the HPO 250 testing device. Average values of hardness from five measurements were calculated. Tensile test was performed at ambient temperature on the Inova TSM 50 testing machine.

2.1 Hardness test

Results of Vickers hardness test for the samples of copper are shown in Table 2.
As it can be seen from Table 2 these values rapidly increased from the 1st to the 4th pass. After the 5th pass these values increased slightly and reached even the value of 119 HV5 in the case of Cu-10.

Table 2 Average values of hardness of Cu in the samples

Number of passes	Hardness HV5
Cu-0	84
Cu-2	98
Cu-4	101
Cu-6	108
Cu-8	109
Cu-10	119

As it can be seen from Table 2 for samples without heat treatment, these values rapidly increase from the 1st to the 4th pass. After the 4th pass the value of hardness stays nearly the same. It may be assumed from this dependence, that the biggest increase of hardness caused by dislocation strengthening in the course of plastic deformation occurs till the 4th pass and subsequent passes do not contribute substantially to further increase of strengthening. Values of hardness are in the cases of the 2nd and the 4th pass slightly lower than those of initial state, and in the case of the 6th pass they are higher than in initial state. These values of hardness are also increasing with an increasing number of passes.

2.2 Tensile tests

Tensile test of the samples was made on the testing machine Inova TSM 50 at the strain rate $1.10^4 s^{-1}$. Results of tensile test of the copper samples are given in Table 3 and they are shown in Figure 2.

Table 3 Influence of SPD process on mechanical properties

Number of passes	$R_{p0.2}$ [MPa]	R_m [MPa]	A80 [%]
Cu-0	198	264	37.3
Cu-2	252	284	12.6
Cu-4	255	311	8.7
Cu-6	247	324	6.8
Cu-8	231	308	6.4
Cu-10	223	315	4.4

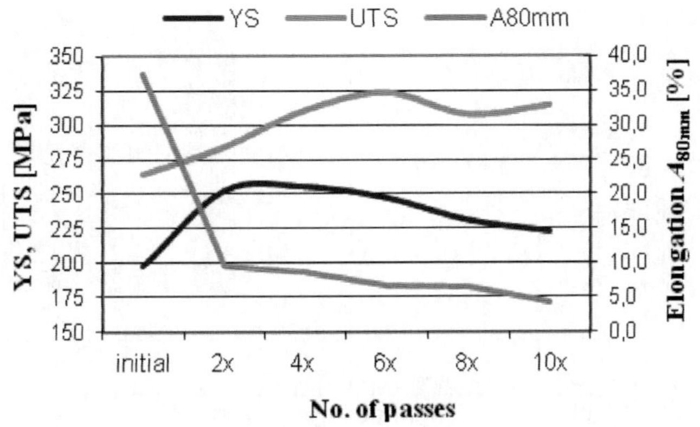

Figure 2 Tensile test

It is obvious from the results of tensile test on copper samples that the biggest increase of yield strength occurs after the 2nd pass and that during the next passes it gradually slowly decreases, while the strength limit gradually increases till the 6th pass and it stagnates during subsequent passes. These results are reflected in considerable drop of ductility after the 2nd pass to the value of 12.6%. During next passes it decreases slowly to the value of 4.4% after the 10th pass. These results correspond with dependence of the number of passes on the hardness values given in the previous chapter. Results of tensile test of samples Cu (99,3 % purity) without heat treatment are given in Tab. 3 and they are shown in Figure 2. As it can be seen from this figure the yield stress and ultimate tensile stress after DRECE processing are increased till the 4th pass, while the elongation is decreased. After the 6th pass, similarly as in the case of copper, these values slightly decrease. These results are also in agreement with determination of hardness values given in the previous chapter.

2.3 Metallographic analysis

For investigation of structure a metallographic analysis was made on light microscope NEOPHOT 2. Initial state was compared with the state after selected number of passes.

More precise analysis was made on TEM microscope. After usual metallographic preparation the samples were electrolytically etched. Figures 3 and 4 show the microstructure of selected samples of copper. Microstructure of initial state of the Cu samples is shown in Figure 3a. This microstructure consists of grains in agreement with the fact that this material was formed before the forming process. Grain size reached the value G4 according to the ASTM. Microstructure of the sample after the 4th, 8th and 10th pass through the DRECE tool is shown in Figure 3 b, c, d.

Refining of grains after each pass was only small as it can be seen in Figure 3d showing the result after last passes. Grain size decreased on the value G5-G6 according to the ASTM. More information by TEM analysis was acquired as it can be seen in Figure 4. This picture shows increasing dislocations' density and presence of slip band. This fact is in agreement with the hardness measurement. In further work, for the exact analysis of the structure, it is necessary to analyze a larger number of samples on TEM and SEM.

IOP Conf. Series: Materials Science and Engineering **461** (2019) 012073 doi:10.1088/1757-899X/461/1/012073

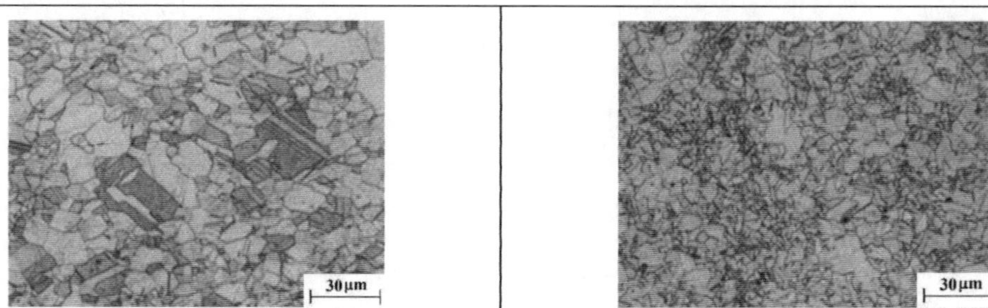

a) **b)**

Figure 3. Structure of pure Cu: Initial state (a) and after 10 DRECE passes (b)

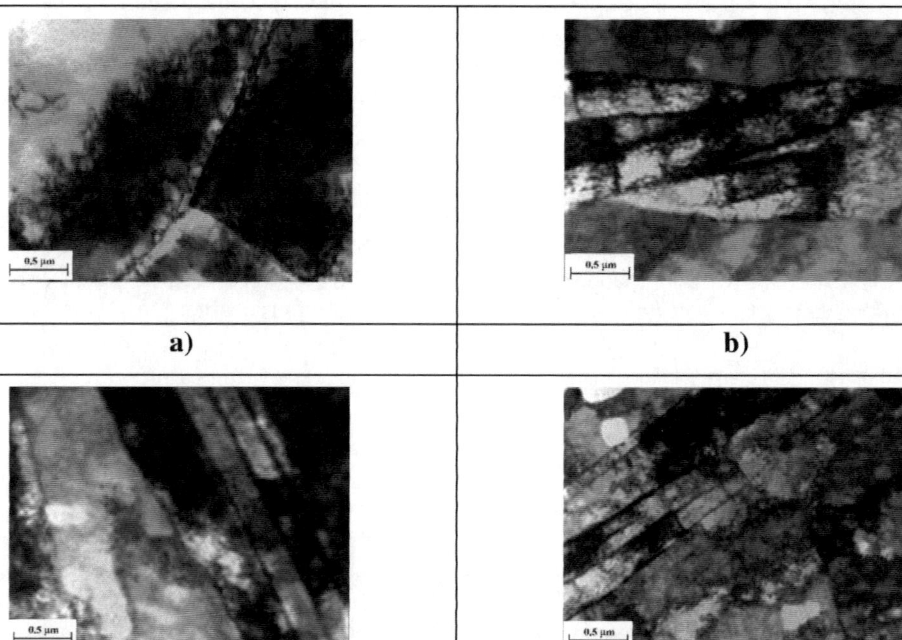

a) **b)** **c)** **d)**

Figure 4. TEM images of investigated material: initial state (a); after 4 passes (b); after 8 passes (c) and after 10 passes (d)

3 Summary

The device has been designed as a prototype equipment for production of UFG structure in a strip of sheet made of non-ferrous metals with subsequent possibility of deformation, and also in a strip of steel sheets with thickness of 2 mm. Creation of UFG structure in the strip of sheet is closely connected to the design of suitable geometry of the forming tool, appropriately dimensioned power unit and control system enabling setting of various values of peripheral velocities. It is also necessary to optimise the pressure force and thus also pressure on the pressure rollers, so that only extrusion occurs, but not the rolling. From the viewpoint of the forming parameters higher number of passes will bring considerable strengthening of the formed material. The forming equipment is at the stage of verification and future works will verify the influence of technological parameters on the increase of efficiency of SPD process for obtaining the UFG structure in non-ferrous metals. According to the level of the obtained results of extrusion of examined materials it is possible to state that the equipment is fully functional. It may be assumed from dependence of mechanical properties on the number of passes, that the biggest increase of hardness caused by dislocation strengthening in the course of plastic deformation occurs till the 4th pass and subsequent passes do not contribute substantially to further increase of strengthening. The results show the necessity of next experiments especially in terms of for the analysis the changing of electrical parameters (electro-conductivity).

Acknowledgments

The authors would like to acknowledge gratefully the Ministry of Education, Youth and sports of Czech Republic for its support to the project "Creation of an international team of scientist and participation in scientific networks in the sphere of nanotechnology and unconventional forming metal", CZ.1.07/2.3.00/20.0038, the project Pre-seed Materials IA3, Production technology of a sheet strip with an ultra- fine grain structure, No. CZ.1.05/3.1.00/14.0320 and project FRVS, IRP No. 239/18 Innovation of teaching subject - Forming

References

[1] GUTKIN, M.YU., OVIDKO, I.A. AND PANDE, C.S. Theoretical models of plastic deformation processes in nanocrystalline materials, *Reviews on Advanced Materials Science,* September 2001, vol. 2, p. 80 – 102.

[2] KOBAYASHI, M. et al. Research and Development of Superplastic Materials „Recent Progresses and Future Prospects", *Metallurgical Transactions 18A*, 1987, p. 685-695.

[3] RUSZ, S., ČÍŽEK, L., MICHENKA, V., DUTKIEWICZ, J., SALAJKA, M., HILŠER, O., TYLŠAR, S., KEDROŇ, J., KLOS, M.: New type of device for achievement of grain refinement in metal strip. *Advanced Materials Research.* Vol. 1127 (2015), Switzerland: Trans Tech Publications, p. 91 - 97.

[4] VALIEV, R. Recent developments of severe plastic deformations techniques for processing bulk nanostructured materials, *Materials Science Forum A* 579 (2008) p. 1–14.

[5] RUSZ, S., KLYSZEWSKI, A., SALAJKA, M., HILŠER, O., ČÍŽEK, L. and KLOS, M. Possibilities of application methods DRECE in forming of non-ferrous metals, *Archives of Metallurgy and Materials*, Vol. 60, 2015, ISSUE 4, p. 3011-3015, DOI: 10.1515/amm-2015-0481

[6.] HILŠER, O., SALAJKA, M. and RUSZ, S. Study of the mechanical properties of steel and selected types of non-ferrous alloys after application of the DRECE process. In NANOCON 2015: 7th International Conference on Nanomaterials-Research & Application. Ostrava: Tanger, 2015, p. 163-167.

IOP Conf. Series: Materials Science and Engineering **461** (2019) 012074 doi:10.1088/1757-899X/461/1/012074

Atomistic Simulation of Heavily Plastic Deformation of Pearlitic Steel: Nano-sized Processes of Wiredrawing and Indentation

K. Saitoh [1], K. Oda [2], K. Yoshida [2], M. Takuma[1], Y. Takahashi[1] and T. Sato [1]

1 Department of Mechanical Engineering, Faculty of Engineering Science, Kansai University, 3-3-35 Yamate-cho, Suita, Osaka, 564-8680 Japan

2 Graduate School of Science and Engineering, Kansai University

E-mail: saitou@kansai-u.ac.jp

Abstract. We conduct molecular dynamics (MD) simulations of pearlitic steel in which iron and carbon atoms are included. The paper aims at clarifying micro- or nano-sized mechanism of plastic deformation of pearlitic steel in wiredrawing process. By using embedded atom method (EAM) potential, we construct a nano-sized computational wiredrawing model of pearlitic steel where lamellae structures of ferrite and cementite phases are attached to each other and it is subjected to conventional drawing condition using a hollow die. Basically, diffusion of carbon atoms around the ferrite-cementite interface is observed during plastic deformation. In order to recognize probable mechanisms by which carbon atoms migrate from cementite to ferrite, dislocation emission and its movement are observed in detail. Besides, the plasticity in pearlitic steel is also discussed in this research by using nano-indentation model where a nano-sized spherical indenter compresses lamellae structures of cementite and ferrite. Both wiredrawing and nano-indentation computational models provides new insight concerning atomistic mechanism occurring in heavily plastic deformation of pearlitic steel.

1 Introduction

Wiredrawing is a conventional plastic working method and has a long history, effectively producing a wide variety of engineering and metallic materials especially with wire, rod and pipe shapes. Recently, the demand for small-sized wire is more required in industry, so that the cutting-edge technique of making very thin wire is required and now it has come down to just several micrometers in diameter[1]. Though there have been developed a tremendously strong wire (fiber) from polymer material, the strongest wire-shaped material which has sufficient strength and reliability and fits to industrial use for infrastructure or transport equipment is still made of steel. Among them, pearlitic steel wire fabricated by wiredrawing technology is noteworthy and is recognized as the strongest one. The reason why the pearlitic steel is strong is believed that it has, in its microstructure, a combination of very thin fibrous structures where two different metallic phases, that is, ferrite (α-Fe) and cementite (Fe_3C), become neatly aligned along the wire direction during drawing process. However, how those two phases become narrower and thinner retaining the thin lamellae shape during heavily plastic process like wiredrawing is not well understood so far. Experimentally, 3DAP (three-dimensional atom probe) technique or TEM (transmission electron microscopy) has been utilized to try to capture the change of those structures, but

dynamic arrangement of atoms or stoichiometric change, including phase transition and crystalline defects, induced by heavily plastic deformation during wiredrawing has not clearly perceived at all. Thus, atomistic simulation method is and will be playing an important role. For example, atomistic behavior of interface between ferrite and cementite can be simulated[2], and an assessment of working principle as for producing for nanometer-sized wiredrawing and materials' flow are investigated[3]. There have been some researches for atomistic modelling of pearlite steel[4][5], but the heavily plastic deformation like wiredrawing has not been fully investigated. The topic of the research is the effect of interface geometry (mutual relationship of crystalline orientation between cementite and ferrite) and energy on dislocation behaviour as well as diffusivity of carbon atoms. The corresponding experiment should be nano-indentation[6], in which very small indenter is pressed into the surface of specimen and any plastic deformation as well as elastic response there is observed from obtained force-depth curve.
In this study, computational model of pearlitic steel is investigated by using precise EAM potential and molecular dynamics (MD) method along above context. Heavily plastic deformation of pearlitic steel are configured by modelling a conventional wiredrawing condition and more basic testing setup of nano-indentation. Nano-indentation model will be capable of investigating atomic behavior of crystalline defects during plastic deformation more in detail.

2 Theory and method

2.1 Potential functions for ferrite and cementite phases (for iron and carbon atomic system)

In this study, EAM (embedded atom method) potential for iron(Fe) and carbon(C) atomic system, in which many-body interaction between neighbor atoms is incorporated as well as primitive pair-wise interaction, is used[7]. There should be some possible choices in EAM potentials' family (including modified EAM) for Fe-C system[8]. Among them, different two formulations of EAM potential are adopted for the present wiredrawing and nano-indentation models, respectively, here. It means that direct and quantitative comparison is difficult between two MD results, but it is certainly realistic that each results are reliable and precise. The EAM potential is generally formulated as:

$$\Phi = \sum_{i=1}^{N}\{F(\rho_i) + \sum_{j=1,j>i}^{N} V(r_{ij})\}, \quad (1)$$

where Φ is total potential energy of the system, $F(\,)$ is called embedded energy function, ρ_i is electron density on the site of i-atom, $V(\,)$ is a function of pairwise interaction and r_{ij} is distance between i- and j-atoms. Here, we use two EAM potentials. One is obtained by Ruda et al.(Ruda-EAM)[9] and it is constructed from well-known Mendelev II (EAM) potential which is just for Fe-Fe interaction. It is reformulated to become also including C-C and Fe-C interactions. The characteristic of Ruda-EAM potential is that a certain degree of condensation of carbon atoms often into cluster form is reproducible. Another one is proposed by Hepburn et al.(Hepburn-EAM)[10]. Technically speaking, Hepburn-EAM potential is already integrated into LAMMPS software[11] we are using. It is based on a first principle (*ab initio*) calculation. The details of expressions of these two EAM potentials are omitted here.

2.2 Molecular dynamics modelling of pearlitic steel under heavily plastic deformation

2.2.1 Modelling of Pearlitic steel. Pearlite structure existing in carbon steel is composed of two crystal phases, ferrite and cementite, each of which are often called α and θ phases, respectively. The former has the body-centred cubic (bcc) lattice unit, and the latter has the orthorhombic crystal lattice unit. The interface between α and θ phases is not fully determined to any unique relationship. By theoretical consideration of crystalline matching based on experimental facts, there has been found three principal relationships between two crystal parts, namely, Bagaratsky(BA), Isaichev(IS) and Pitch-Petch(PP) relations as shown in figure 1. In this study, those three relations are compared. As for the nano-indentation model, at first, BA interface relation is scrutinized in this paper.
The lattice constants for α and θ phases are checked by NPT-ensemble MD at 300 K with fully periodic lattice. $a_\alpha = 0.286$ nm (one for ferrite: α) and a_θ=0.437 nm, b_θ= 0.514 nm and c_θ= 0.650 nm (three for cementite: θ) are obtained and those values are used in arranging cementite in pearlite models.

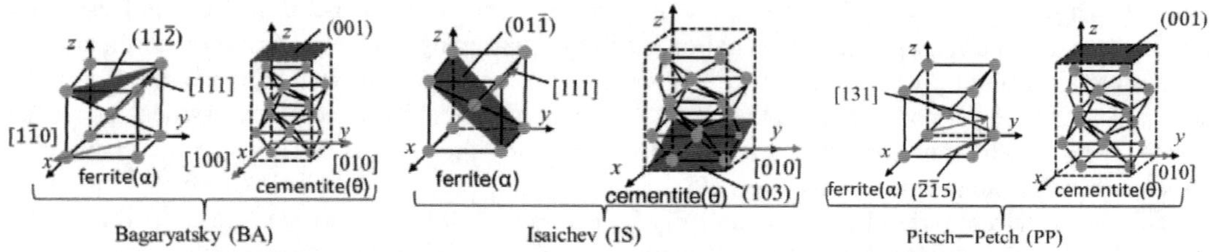

Interface type	Direction	Plane
Bagaryatsky (BA)	$[100]_\theta$ // $[1\bar{1}0]_\alpha$ and $[010]_\theta$ // $[111]_\alpha$	$(001)_\theta$ // $(11\bar{2})_\alpha$
Isaichev (IS)	$[010]_\theta$ // $[111]_\alpha$	$(101)_\theta$ // $(01\bar{1})_\alpha$
Pitsch-Petch (PP)	$[010]_\theta$ 2.6 deg. from $[131]_\alpha$	$(001)_\theta$ // $(2\bar{1}\bar{5})_\alpha$

Figure 1. Three principal relations of crystalline orientation for the interface in pearlitic steel.

2.2.2 Computation model for wiredrawing process. This model assumes drawing process of nano-sized pearlitic steel wires. As shown in figure 2, the wire material is pushed through a hollow die with conical shape, by moving two edge regions with velocity constraint. The shape of die is designed as to mimic an actual large-sized conventional drawing die, which furnishes die-semi-angle of 7.0 degrees. The wire part has 7~8 nm in diameter. Since the atoms in the interface region across wire and die parts mutually exerts too strong interaction (attractive) force in non-treated setup, it is found that a certain relief of interatomic interaction has to be applied there so that the wire is smoothly drawn through the die hole. Calculation conditions of this drawing model are summarised in table 1. Interface plane between α and θ phases is placed in parallel to the drawing direction. The potential function for this model is Ruda-EAM[9].

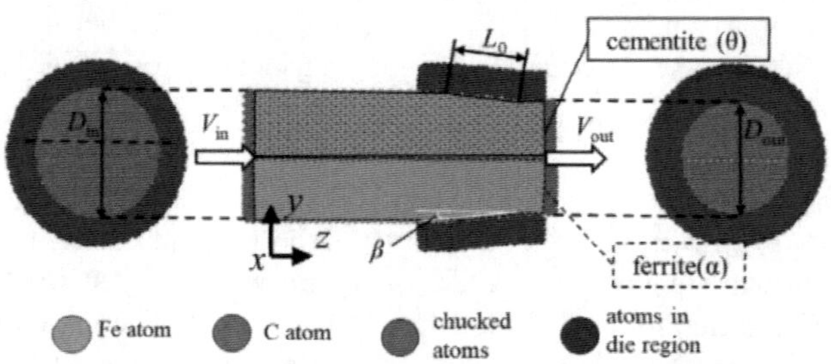

Figure 2. Molecular dynamics(MD) model for nano-sized wiredrawing of pearlitic steel.

Table 1. Computation conditions for wiredrawing MD model.

property	unit	value(s)
Cell size in *x, y,* and *z* directions	nm	9.8796, 9.8796, 16.7592
Inlet diameter D_{in}	nm	6.8796
Length of die land L_0	nm	4.12
Die semi-angle β	deg.	6.0
Moving velocity V_{in} , V_{out}	m/s	38.3, 50.0
Time increment × steps	fs	2.0 × 20000

2.2.3 Computation model for Nano-indentation process. For the purpose of investigating heavily plastic deformation around α-θ interface, a computation model of nano-indentation testing is configured. The nano-indentation experiment is well known as an actual nano-sized material testing method to see hardness or plasticity process, in particular, for thin film state by using very hard indenter. In this study, α and θ phases with very narrow thickness are stacked as shown in figure 3 and are pushed from one side by a spherical indenter. Naturally, we will be comparing different directions of indenter push, but one arrangement in which landing of indenter takes place on the surface of α phase should become the most critical condition, because many dislocations will be emitted and be contained with relative ease inside the α part. The point of mechanisms is that how dislocations are transferred from α to θ, dragging carbon atoms around their interface. Computation conditions of this nano-indentation model is shown in table 2. The thickness of each lamellae structures should be several nm in this model, which is inevitable due to the limitation of computation size. The discussion on size effect of lamellae structure in pearlite is still open and is left to further studies, but the present model provides us important insight when nano-sized lamellae exists in the pearlite. Potential function for this model is Hepburn-EAM[10].

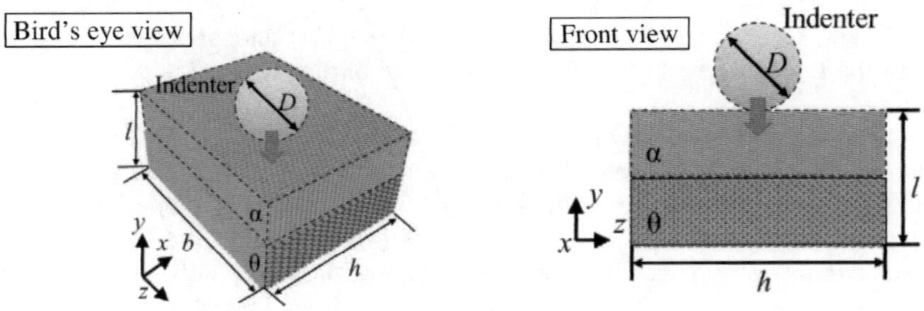

Figure 3. Molecular dynamics(MD) model for nano-indentation of pearlitic steel.

Table 2. Computation condition of nano-indentation testing performed by MD.

property	unit	value(s)
The number of atoms	-	141120
Sizes l, h, b	nm	7.34, 12.47, 16.06
Type of α / θ interface	-	Bagaryatsky(BA): $(001)_\theta$ // $(11\bar{2})_\alpha$
Indenter diameter D	nm	10.0
Pushing speed V_z	m/s	200.0
Temperature	K	300
Time increment × steps	fs	1.0 × 100000

2.3 Dislocation mechanism and its detection method

In MD simulation, from perpetual output of position and velocity of atoms, many analyses can be conducted. Analysis of lattice defects is important to understand plasticity occurring inside MD cell. Recently, DXA method (Dislocation eXtraction Algorithm) is developed by researchers and the dynamic behavior of dislocations in crystalline lattice are successfully obtained from MD results[12]. (The DXA analysis operation is already integrated into visualization software called "Ovito"[13]).

3 Results and discussions

3.1 Behaviour of pearlite in wiredrawing process model

Figure 4 shows an example of wiredrawing model of pearlite material. As the wire is drawn and its atoms go through the die, the α phase easily exhibits plastic deformation but the θ phase gets including some carbon clusters. This mechanism is understood as a kind of phase separation toward binary phase coexisting monoatomic C and Fe regions. The diameter of carbon atomic cluster approximately has grown until 1.5~2 nm. The reason why stoichiometric Fe_3C phase exhibits such phase separation is that Ruda-EAM potential function used in this wiredrawing MD model has energetically more stable state for C-C bonding than that for the initial Fe-C bonding. This behavior does not confirmed when using only pairwise interaction, which means that the nature of EAM potential function is responsible for.

These MD results predict that experimentally a segregation process of carbons will be induced by heavily plastic deformation.

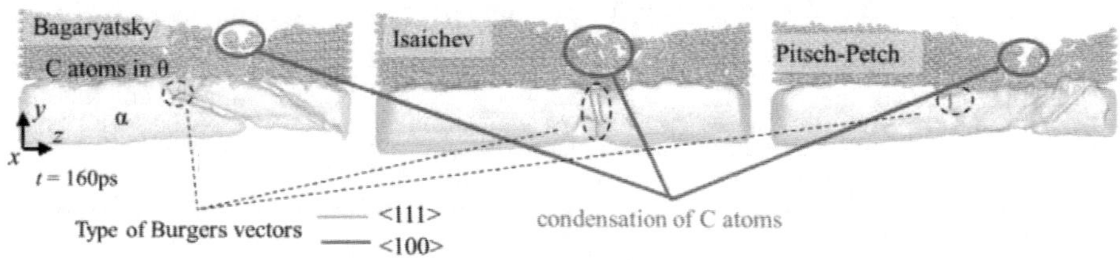

Figure 4. Results of dislocation analysis (by DXA method[12][13]) and observed condensation of carbon atoms, comparing three different crystalline orientations of the interface in pearlite.

3.2 Behaviour of pearlite in nano-indentation process model

Figure 5 shows the results obtained by nano-indentation process model. In figure 5(a), the relation between force acting on the indenter in the pushing direction and the depth of indentation. In the first part of the curve, it exhibits a quadratic elastic response. After then, it shows several local drops of force value, which mean accumulation of plastic deformation. Figure 5(b)~(d) show dislocation behavior mainly in α phase. In this simulation, the indenter enters in the specimen from the side of α phase which is displayed in upper part in the figure. In fact, after structural relaxation process to adjust atomistic strain occurring around the α-θ interface, a kind of lattice defects called "misfit dislocations" emerges. As shown in figure 5(b), these dislocations seem relatively immobile because of conserving nature of Burgers vectors. Then, as shown in figure 5(c), indenter digs into α phase and soon dislocations are emitted from the contact area of indenter. It is observed that these dislocations overhang from the surface, forming almost dislocation loops. However, as shown in figure 5(d), those overhanging dislocations must interact with the α-θ interface and a part of them is likely to be adsorbed into θ phase. It is understood that this adsorption behavior tends to occur due to the very narrow width of α phase. Note that dislocations produced by indentation also interact with misfit dislocations left in MD cell and their interaction often lead to perfectly immobile dislocation. Unfortunately, dislocations in θ phase is not clearly defined, so they are not detected at this point. The detection method are expected to be developed in future studies.

4 Conclusion

In order to understand atomistic behaviour in wiredrawing of pearlitic steel, molecular dynamics simulation of nano-sized wiredrawing and nano-indentation are performed. Since precise EAM potentials are used, realistic behavior of atoms and lattice defects are obtained. For wiredrawing model, a possibility of condensation of carbon atoms up to about 2 nm in cluster size are discovered. It depends on the formulation of EAM potential. For nano-indentation model, emission of dislocations induced by indentation are reproduced. Interaction mechanism between emitted dislocations and misfit dislocation is successfully observed. Unfortunately, dislocation in cementite phase is not detected due to under investigation, but some dislocations in the ferrite phase emitted by indenter are probably absorbed inside the cementite phase.

Acknowledgement
This work was supported by JSPS KAKENHI Grant Number JP16K05994(Grant-in-Aid for Scientific Research (C)) and Nippon Steel & Sumitomo Metal Corporation. The authors would like to acknowledge those supports here.

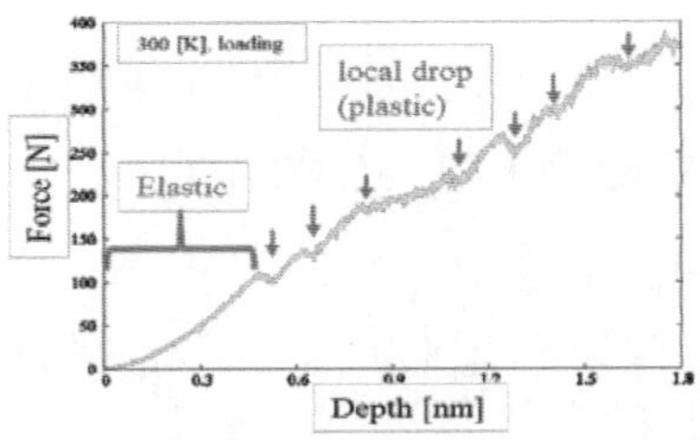

(a) Force-depth curve.

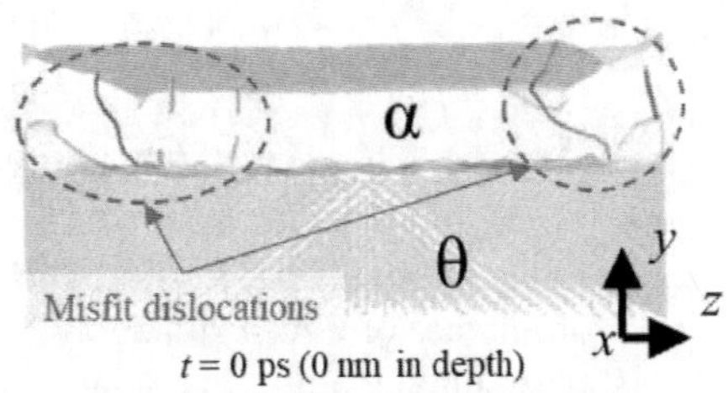

(b) Dislocation: part 1 (at the beginning).

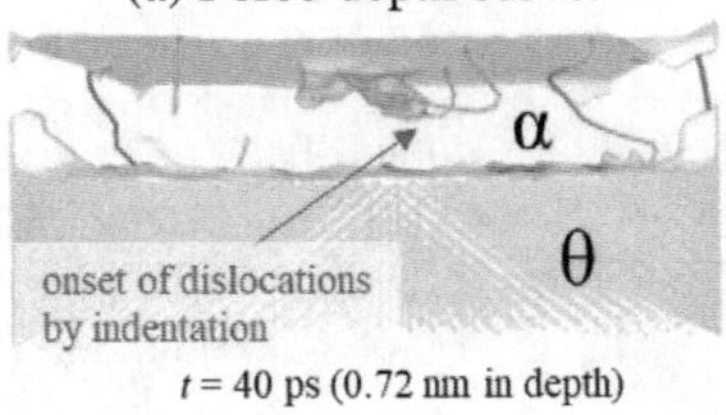

(c) Dislocation: part 2 (onset of defects).

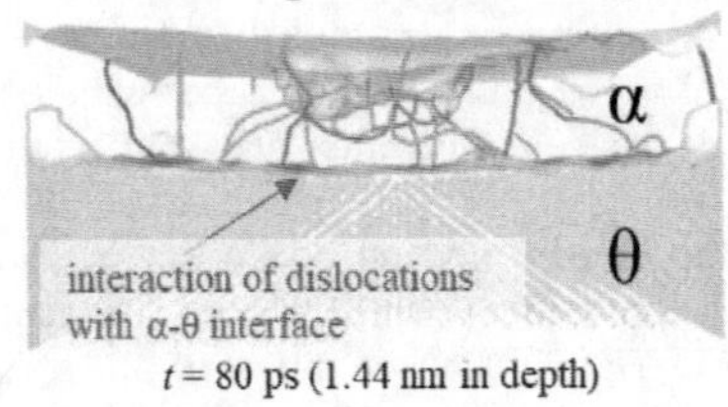

(d) Dislocation: part 3 (dislocation interaction).

Figure 5. Results of pearlite nano-indentation process model: (a) force-depth curve obtained during indentation process; (b),(c) and (d)visualized dislocations and carbon atoms remaining in the cementite phase (dislocation lines are detected by the DXA method[12][13]).

References

[1] Zelin M 2002 Microstructure evolution in pearlitic steels during wire drawing *Acta Mater.* **50** 4431-47

[2] Levchenko E V, Evteev A V, Belova I V and Murch G E 2009 Molecular dynamics simulation and theoretical analysis of carbon diffusion in cementite *Acta Mater.* **57** 846-53

[3] Saitoh K, Sameshima Y and Daira S 2014 Nano-scale modelling and simulation of metal wiredrawing by using moleculardynamics method *World J. Nano Sci. Eng.* **2** 46498

[4] Guziewski M, Coleman S P and Weinberger C R 2018 Atomistic investigation into the mechanical properties of the ferrite-cementite interface: The Bagaryatskii orientation *Acta Mater.* **144** 656-65

[5] Saitoh K, Yoshida K, Oda K, Sato T, Takuma M and Takahashi Y 2018 Molecular dynamics study on nano-sized wiredrawing:possible atomistic process and application to pearlitic steel wire *IOP Conf. Ser.: Mater. Sci. Eng.* **307** 012039

[6] Nix W D and Gao H 1998 Indentation size effects in crystalline materials: A law for strain gradient plasticity *J. Mech. Phys. Solids* **46**, 411-25

[7] Daw M S and Baskes M I 1984 Embedded-atom method: Derivation and application to impurities, surfaces, and other defects in metals *Phys. Rev. B*, 29, 6443-53

[8] Interatomic Potentials Repository (by NIST) *https://www.ctcms.nist.gov/potentials/system/C-Fe/*

[9] Ruda M, Farkas D and Garcia G 2009 Atomistic simulations in the Fe-C system *Comp. Mater. Sci.* **45** 550-60

[10] Hepburn D J and Ackland G J 2008 Metallic-covalent interatomic potential for carbon in iron *Phys. Rev. B* **78** 165115

[11] Plimpton S 1995 Fast parallel algorithms for short-range molecular dynamics *J. Comp. Phys.* **117** 01-19

[12] Stukowski A and Albe K 2010 Dislocation detection algorithm for atomistic simulations *Model. Simul. Mater. Sci. Eng.* **18** 025016

[13] Stukowski A 2010 Visualization and analysis of atomistic simulation data with OVITO - the Open Visualization Tool *Model. Simul. Mater. Sci. Eng.* **18** 015012

IOP Conf. Series: Materials Science and Engineering **461** (2019) 012075 doi:10.1088/1757-899X/461/1/012075

Shape Setting in super-elastic NiTi ribbon

S. Samal, E. de Prado, O. Tyc and P. Sittner
Institute of Physics of Czech Academy of Science, 182 21 Prague 8, Czech Republic
Corresponding author's e-mail address. E-mail: samal@fzu.cz

Abstract. The various application of shape memory alloys requires setting of NiTi ribbon in particular form. Shape setting is accomplished by constraining the NiTi ribbon on a mandrel that undergoes heat treatment. Surface annealing of super-elastic NiTi ribbon was performed shape inside the furnace. The duration of furnace temperature was kept fixed for 2 h. The sample was held in bend position for annealing the martensitic phase created on the surface of NiTi ribbon. The aim is to anneal the martensitic phase on the upper surface region and introduce the compressive stress on the edge of the alloy. The induced surface strain determines the volume fraction of martensitic formed on the sample in the flat position at room temperature. The shifting of the peak and evolution of new peaks as function of the temperature on the annealed NiTi ribbon at the flat position was observed. X-ray diffraction was carried out on the shape set bended sample at room temperature and constrained flat position (both outer and inner surface). The bended NiTi ribbon annealed at 350° C shows a mixture of austenitic phase and plastic deformation region that confirms from the broadening of some peaks. The residual stress arises from the plastically induced periphery region and the elastic core region that allows the NiTi ribbon returns partially to original flat shape. However, fully shape set was achieved on super-elastic NiTi ribbon annealed at 400 and 450 °C.

1. Introduction

NiTi ribbon belongs to the category of shape memory alloys (SMA) that exhibit unique nonlinear deformations due to the phase transformation under change of temperature or stress. The alloy returns back to the original shape after removal of stress and reverse of temperature. This unique feature in shape memory alloys is due to the transformation of the martensitic to the austenite phase. The specific characteristics are based on the initial condition of the alloy in respective of the austenite or martensitic phase and thermomechanical treatments. The phase transformation allows NiTi SMA the ability to recover strains on the order of 5 to 10 %. To understand the mechanism of stresses generated during annealing processing in NiTi wires were annealed by laser heating [1] process, this study has been arising in a new subject of interest. As the localization effect of laser limits the understanding in the very thin wire of diameter 100 μm, our approach has been shifted on the shape memory effect of NiTi ribbon in bending form, which was annealed in a furnace. The aim is brief understanding of the functional graded NiTi plate annealed in bending form at the the uniform temperature inside the furnace of fixed temperature. NiTi uses shape memory effect to generate mechanical motion usually force and displacement with a change of temperature or the applied force. Our aim shifted from laser annealed sample to furnace annealed sample for a better understanding the role of martensitic phase form during annealing on the bended sample and its conversion and transformation from martensitic to austenite through R phase. The appearance and disappearance of various diffracting peaks of the phase in the annealed ribbon depend on bended or flat position. The stresses generated during bending and unbending of the annealed sample and its formation can be explained on the basis of the volume fraction of the phase. The role of coupling between martensitic and plasticity on the super-elastic NiTi ribbon as the function of annealing temperature is the subject of curiosity [2].

The aim of this work is to induce surface stresses induced during annealing the martensitic phase on the surface of the bended NiTi ribbon. The approach is the combination of bending stress for the generation of martensitic phase on the surface and then annealed to lock the martensitic phase on the surface [3]. The various annealed temperature determines the volume fraction of martensitic remained on the surface of the sample [4]. Here we annealed the NiTi ribbon at various strains to shape set the alloy. The annealed samples were examined using X-ray Diffraction at bended shape set form and then on flat position at both top (outer) and bottom (inner) area. The aim is to study the tensile and compressive stresses generated in the annealed flat ribbon at room temperature. The effect of annealing temperature and its role on the shape memory effect of the NiTi ribbons is well explained on the localization of the volume fraction of phase transformation. The investigation was carried out on the surface of NiTi ribbon due to the phases of austenite and martensitic to respective internal surface stresses.

2. Materials and Methods

A commercial NiTi plate of composition Ti-50.7 Ni (IC = 1348, flat annealed, ϕ_d: 0.10 mm) was considered in this study. Ribbon samples with a thickness of 0.1 mm were bended around the mandrel of 8 mm diameter followed with annealing to maintain the shape setting in the furnace of fixed temperature for 2 h. Figure 1 (a) displays the schematic diagram for the NiTi ribbon annealed at various strains on position with the mandrel. The NiTi ribbon of various samples was annealed at a temperature of 350, 400, 450° C for shape setting. The change in curvature after shape set sample at three different temperatures is shown in Figure 1 (b). The shape memory effect induces the sample to return back to the original shape at a lower temperature of 350 °C. So the unrecovered strain is less influence on the bending the sample in original form. However, the sample annealed at high temperature shows the more pronounced effect of shape set and less effect of shape memory. Figure 1 (c) shows the schematic diagram of NiTi ribbon during annealing and quenching position. Annealing position of NiTi ribbon shows on the stage of the mandrel. The on axis shows quenching in position angle for lower annealing temperature. However, off axis shows quenching position lower to annealing position for an increase in annealing temperature. The Differential scanning calorimetry (DSC) was performed on the reference (virgin sample) and annealed sample to observe the phase transition, during heating and cooling. The annealed sample was extensively analysed by X-ray diffractometer (Panalytical X′Pro) in shape set bending form, un-bending (flat position) with outer and inner position by a detector (Lineal X′Celerator detector). Finally, Rietveld analysis was performed to calculate the volume fraction of martensitic formed during annealing. The peaks are identified on the bending and unbending position as the function of annealed temperature. The sample cooled down in room temperature in an air medium. The XRD analysis for all the bended samples was carried at room temperature in both bend and flat position.

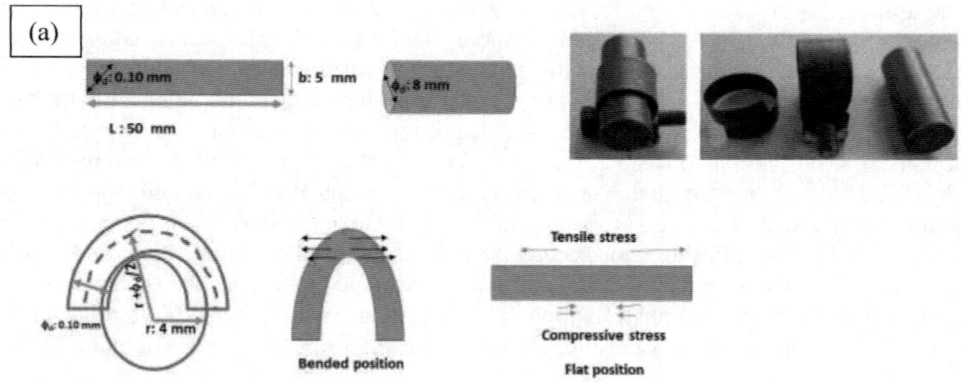

Figure 1. (a) Schematic diagram of the concept for shape set annealed NiTi ribbon.

The distance from the mandrel to the center of the NiTi ribbon $D = r + \phi_d/2$ (1)

Therefore, the strain at the center point is

$$\varepsilon = \frac{\emptyset_d/2}{\emptyset_d/2 + r} \tag{2}$$

the maximum bending strain $|\varepsilon_{max}| = \frac{\emptyset_d}{\emptyset_d + 2r} \cong \frac{\emptyset_d}{2r}$ (3)

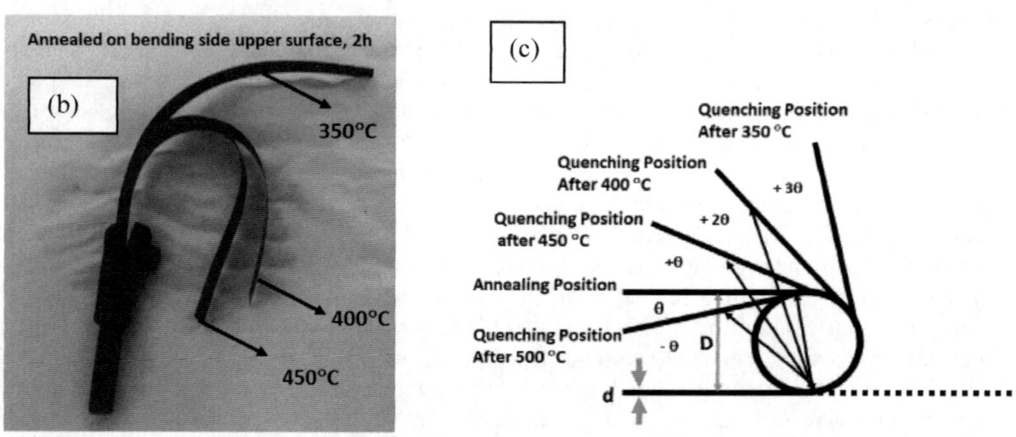

Figure 1. (b) Shape set annealed sample at various temperature (c) schematic diagram of NiTi ribbon position as a function of temperature.

3. Results and discussion

Figure 2 shows the DSC of annealed NiTi samples in the heating and cooling temperature zone of -100 to 100 °C at the scanning rate of 3 K/min. The peak of the annealed sample shows the evolution of austenite peaks around 35 °C, however martensitic peak is more prominent at low temperature in the annealed sample. The sample annealed at 500 °C shows the evolution of martensitic peaks, due to the accumulation of internal stresses in the NiTi ribbon alloy. The cooling curve of the NiTi ribbon annealed at 500 °C, shows the evolution of M_f and M_s in the martensitic phase. However, on decreasing annealed temperature, it has been observed that M_f

and M_s decreases, becomes flat and finally vanishes at a lower temperature of 350 °C. Austenite peak arises at 35 °C, that coincides with an annealed temperature of 350 and 400 °C. The shift happens on the annealed sample at 450 (A_s~ 30° C) and 500 °C (A_s~ 25° C) for NiTi ribbon. The shape set of NiTi ribbon of thickness 0.10 mm was performed in furnace annealing bending position. The furnace temperature was maintained at 350° C for 2 h. Similarly, other sets of shape set for NiTi ribbon was performed at temperatures of 400 and 450° C for 2 h (Figure 1 b). The XRD analysis was performed on the annealed alloy at bend position. The CoKα is the source of radiation with a wavelength of 1.79Å was used for measurement of profiles with a diffractometer (Panalytical X' Pro).

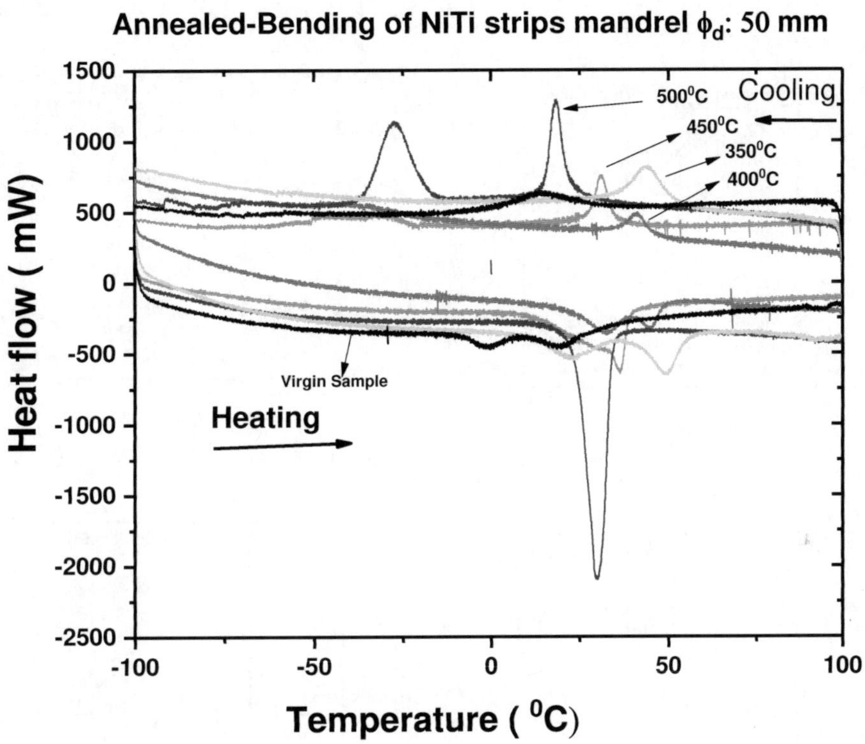

Figure 2. DSC profile of annealed NiTi ribbon from -100 to 100 °C.

The shape set, the annealed sample was examined for possible phases at room temperature by X-ray diffraction method Figure 3 (a-f). The annealed NiTi ribbon was examined at the bend, and flat condition at the outer and inner position for annealing at 350, 400, 450° C at bend position for 2h. At bend position, the NiTi ribbon shows austenite phases mostly at 350° C that corresponds towards higher temperature annealed at 400°C, which deconvulated into R-phase peak structure [5-6]. The intensity of the peak decreases and complete conversion towards R-phase at 450° C annealed shape set NiTi ribbon. However, on constrained the annealed shape set NiTi ribbon in to a flat position, R-phase, and martensitic peaks arise from the surface of the sample [7-8]. The surface strain of the shape set annealed bent sample is calculated and presented as below in Table 1.

Table 1. The surface strain of NiTi ribbon annealed on the mandrel at various temperature.

Temperature	Annealed diameter	Original mandrel diameter	Surface strain
350	7 mm	8 mm	0.125×10^{-3}
400	6.5 mm		0.188×10^{-3}
450	6.0 mm		0.250×10^{-3}

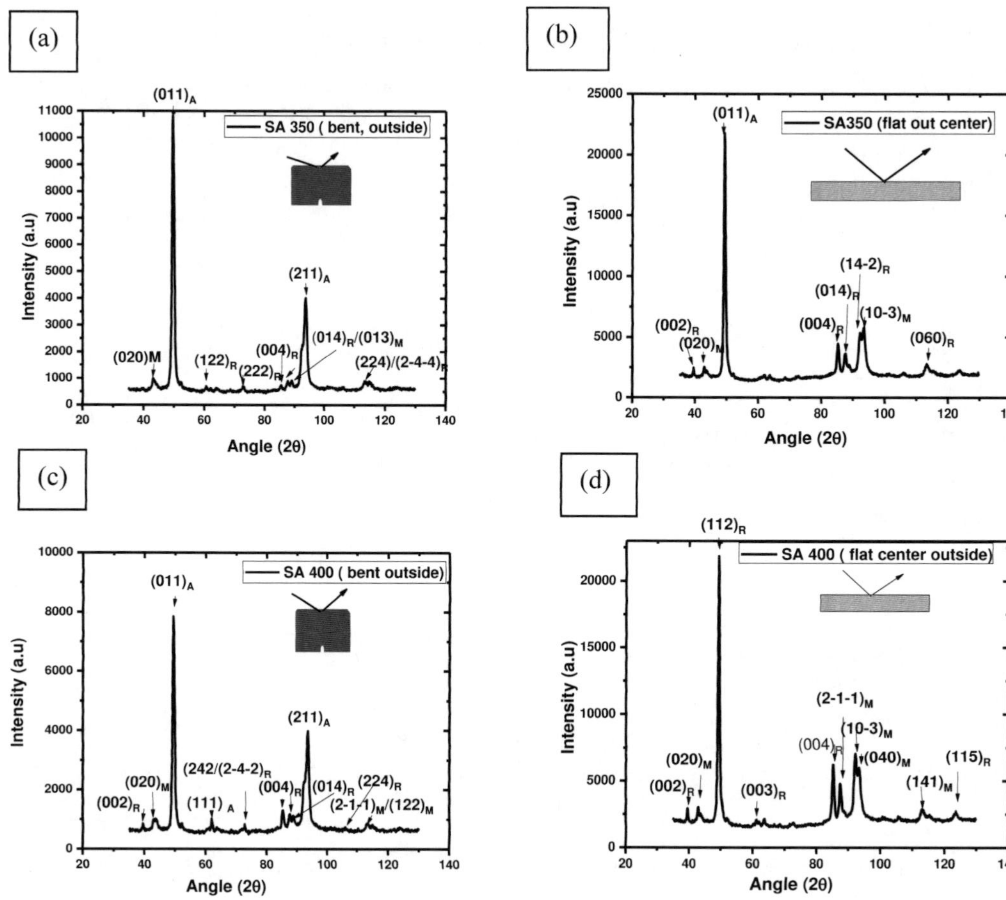

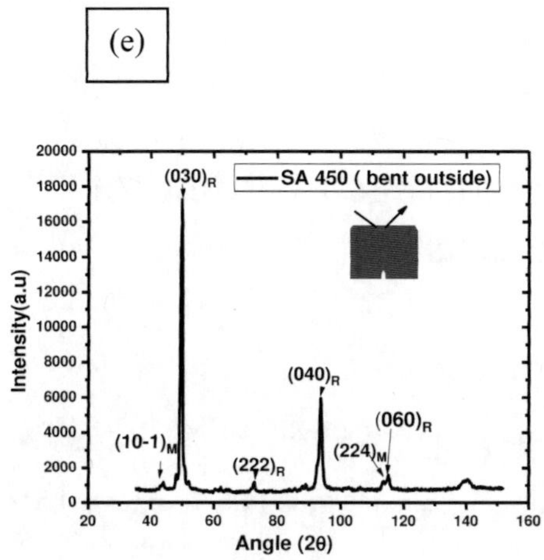

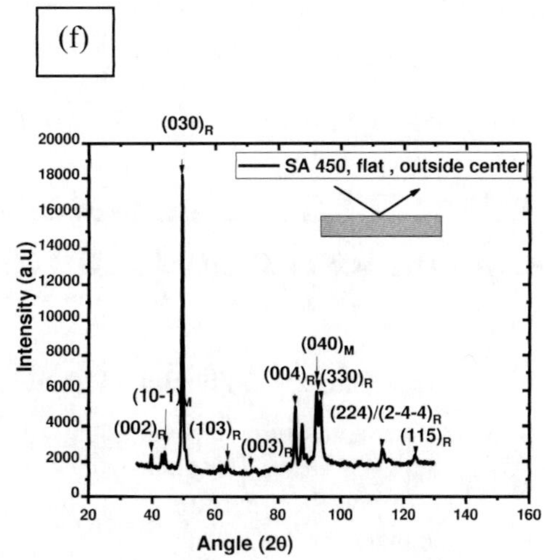

Figure 3 (a-f) XRD pattern of various annealed NiTi ribbon shape set to sample in bend and flat position at room temperature (annealed at 350, 400 and 450° C).

The partial gradient of mechanical properties observed earlier has now confirmed from the behaviour and diffraction pattern of the surface annealed super-elastic NiTi ribbon [9].

4. Conclusion

The shape set was successfully achieved on NiTi ribbon at a higher temperature of 400 and 450° C. However at 350° C, the NiTi ribbon displays the partial shape set due to the elastic nature core region overcomes plastically surface region and returns back to the original form. The residual stress induces reverse motion towards elasticity, as a result, the surface strain has less effect on the bended shape of the sample. However when NiTi ribbon annealed at 400 and 450° C, the surface strain increases, as a result, the sample maintains the bended form. When annealed at 350° C, it acquires a partially bent shape due to the tensile and compressive internal stresses introduced via the low temperature shape setting process into both surfaces of the ribbon during the heat treatment.

References

[1] Samal et al S 2018 *IOP conference series of proceedings*, DOI: 10.1088/1757-899X/362/1/012007

[2] Sittner et al P 2018 On the coupling between martensitic transformation and plasticity in NiTi: experiments and continuum based modelling, *Progress in Materials Science* 98, 249-298.

[3] Lavvafi H, Lewandowski J.R, Lewandowski J.J 2014, Flex bending fatigue testing of wires, foils, and ribbons, *Materials Science & Engineering A* 601, 123-130.

[4] Meng Q, Yang H, Liu Y and Nam T 2012 Compositionally graded NiTi plate prepared by diffusion annealing, *Scripta materialia* 67, 305-308.

[5] Vokoun D, Hu C.T 2002 Two-way shape memory effect in Fe-28.8 at.% Pd melt-spun ribbons, *Sripta Materialia* 47,453-457.

[6] Wu Z, Vokun D, Leu C & Hu C 2017, A two way shape memory study on Ni rich NiTi shape memory alloy by a combination of the all-round treatment and the R phase transformation, *JMEPEG* 26, 5801–5810.

[7] Pilch J, Sittner P, Heller L, Delville R, Malard B, Curfs C 2010, *Proc. of SMST*.

[8] Novák V, Šittner P, Pilch J and Delville R 2009, *ESOMAT*, 03009.

[9] Samal S, Heller L, Brajer J, J Manak 2018, The gradient of hardness and Young's modulus over the cross section of laser annealed NiTi wires, *Esomat*.

IOP Conf. Series: Materials Science and Engineering **461** (2019) 012076 doi:10.1088/1757-899X/461/1/012076

Dependence of trapping states on the hydrogen content of high-carbon ferrite-austenite dual-phase steel

D Sekine[1], Y Sakiyama[2], T Omura[2] and K Takai[3]

[1]Graduate Student, Sophia University, 7-1 Kioi-cho, Chiyoda-ku, Tokyo 102-8554, Japan
[2]Nippon Steel & Sumitomo Metal, 20-1 Shintomi, Futtsu, Chiba Prefecture 293-8511, Japan
[3]Faculty of Science and Technology, Department of Engineering and Applied Science, Sophia University, Tokyo, Japan

Corresponding author's e-mail address: daikimifami@gmail.com

Abstract States of hydrogen present in high carbon ferrite (α)-austenite (γ) dual phase steel were analyzed under various hydrogen charging conditions using thermal desorption analysis (TDA). Specimens including 0%γ and 10%γ in α were prepared. The 10%γ specimen consisted of bainitic ferrite and thermally stable retained austenite. When the 10%γ specimen were charged with hydrogen under high fugacity condition, the amount of hydrogen rapidly increased, the end temperature of hydrogen desorption in TDA profile increased, the hydrogen degassing time at 30 ° C increased, and the trap activation energy increased from about 30 kJ/mol to 47 kJ/mol. This 47 kJ/mol value is in excellent agree with the activation energy of hydrogen diffusion in γ phase. These results probably indicated that hydrogen is not absorbed in γ phase but α phase under lower fugacity conditions, i.e., lower amount of hydrogen. In contrast, hydrogen is absorbed in not only α phase but also γ phase under higher fugacity conditions, i.e., higher amount of hydrogen.

1 Introduction

High carbon ferrite-austenite dual-phase (α/γ) steel is a high strength steel with a tensile strength of about 1540 MPa class and excellent ductility, so it has been used for automobile parts. For weight reduction in automobiles, high strength of α/γ steels have been developed. However, since high-strength steels are susceptible to hydrogen embrittlement, it is necessary to understand hydrogen embrittlement in high carbon α/γ dual-phase steel.

There are many reports on hydrogen embrittlement of materials containing retained γ. For example, it was reported that the diffusion rate of hydrogen inside the retained γ phase is slow[1)] and that the phase transformation of retained γ due to heat or hydrogen charging affects hydrogen embrittlement susceptibility[2)]. However, few studies have focused on the trapping states of hydrogen. In this study, we aimed to explicate the role of retained γ in hydrogen embrittlement through evaluating the trapping states of hydrogen present in high carbon α/γ dual-phase steel containing 10% retained γ.

2 Experimental

2.1 Materials

As a high-carbon α/γ dual-phase steel, 10% γ specimens containing 10 vol % thermally stable retained γ in the α phase were prepared. Japanese Industrial Standards (JIS) SCM 435 with no retained austenite (0% γ specimens) was used as a comparative material. The chemical components of the specimens are shown in Table 1. Tensile strength was about 1540 MPa and about 1300 MPa, respectively. In the hydrogen analysis, flat test pieces of 10 × 10 × 0.2 mm, 10 × 20 × 0.5 mm, and 10 × 20 × 1.0 mm were used. Figure 1 shows a phase map of the 10% γ material obtained by electron back-scattered diffraction (EBSD). Red indicates the α phase and green the retained γ phase. The matrix of 10% γ specimens is composed of bainitic ferrite, ferrite, pearlite which are bcc structure. In this study, the 0% γ specimens using the high carbon model steel rather than the practical steel has a tempered martensite single-phase structure.

2.2 Hydrogen charging

To investigate the trapping states of hydrogen at various hydrogen content, hydrogen charging was carried out under five conditions using a cathodic electrolysis method. The conditions were (i) a current density of 50 $A \cdot m^{-2}$ in a 0.1 N NaOH aqueous solution, (ii) a current density of 100 $A \cdot m^{-2}$ in a 0.1 N NaOH aqueous solution, (iii) a current density of 100 $A \cdot m^{-2}$ in a 2.5 $g \cdot L^{-1}$ NH_4SCN + 0.1 N NaOH aqueous solution, (iv) a current density of 100 $A \cdot m^{-2}$ in a 5 $g \cdot L^{-1}$ NH_4SCN + 0.1 N NaOH aqueous solution, and (v) a current density of 100 $A \cdot m^{-2}$ in a 7.5 $g \cdot L^{-1}$$NH_4SCN$ + 0.1 N NaOH aqueous solution. Hydrogen charging time was respectively (i) 48 h, (ii) 48 h, (iii) 96 h, (iv) 120 h, and (v) 96 h. Hydrogen charging was conducted until the hydrogen concentration at the surface and the center of the specimens became constant and reached an equilibrium state.

X-ray diffraction (XRD) was conducted to measure the change in retained γ content in the 10% γ specimens during hydrogen charging and heating. Hydrogen charging was carried out under the condition of (iv). Heating was carried out at 100 ° C, 200 ° C and 300 ° C.

2.3 Hydrogen analysis and evaluation of activation energy

The thermal desorption method is a method of measuring the desorption rate of atoms and molecules desorbed by heating a sample that adsorbed or occluded atoms or molecules, and analyzing the adsorption and occlusion states. In this study, thermal desorption analysis (TDA) was conducted at a heating rate of 100 $°C \cdot h^{-1}$.

The activation energy was calculated to evaluate the dominant factor of hydrogen traps. Hydrogen charging was conducted for the 10% γ specimens under conditions (i) to (iv). TDA was conducted at a heating rate of 50-200 $°C \cdot h^{-1}$. Based on the peak temperature of the profile obtained, activation energy values were calculated under the various conditions using the equation proposed by Choo and Lee[3)].

Table 1 Chemical composition of 0% γ and 10% γ steel (mass %).

	C	Si	Mn	P	S	Al	Cr	Mo
γ 0%	0.35	0.24	0.79				1.09	0.15
γ 10%	0.79	2.42	0.51	0.008	0.002	0.036		

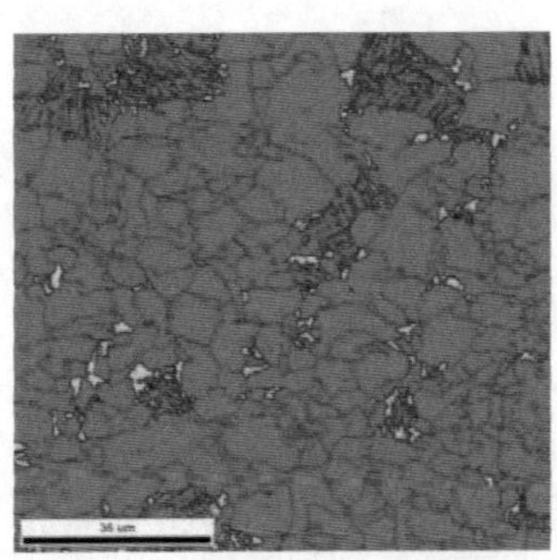

Fig. 1 Phase map of specimen obtained by EBSD: red represents α and green represents retained γ. (Online version in color.)

2.4 Effect of retention time after hydrogen charging
The effect of the retention time after hydrogen charging on the TDA profile was investigated to evaluate the trapping states of hydrogen present in retained γ utilizing low hydrogen diffusivity of retained γ. Hydrogen charging was conducted using 0% and 10% γ specimens under the conditions of (ii) and (iv). After degassing the specimens for 0 to 48 h in an environment of 30 °C, TDA was conducted at 100 °C/h.

3 Results and Discussion

3.1 Role of retained austenite in trapping states of hydrogen present in α/γ dual-phase steel
Figure 2 shows the change in retained γ content in the 10% γ specimens during hydrogen charging and heating. The results showed that there was no change in the retained γ content in the 10% γ specimens after hydrogen charging and heating.

Figure 3 shows the change in the hydrogen content of the 0% γ and 10% γ specimens charged with hydrogen under the conditions of (iv) described in section 2.2. It took more time for the hydrogen content to reach the equilibrium state for the 10% γ specimens than for the 0% γ specimens, and the hydrogen content of the former was larger than that of the latter.

Figure 4 shows the TDA profile of the 10% γ specimens with the different thicknesses and hydrogen-charged under the conditions of (iv) described in section 2.2. The result shows that the peak temperature and the release end temperature shifted to a lower temperature as the specimen thickness became thinner. Therefore, the influence of retained γ on the trapping states of hydrogen was found to be an increase in the hydrogen content and a decrease in the hydrogen diffusion rate. The result in Fig 4 also indicates that the hydrogen present in retained γ was not non-diffusible because the peak of non-diffusible hydrogen was not retained at high temperature side as the specimen thickness became thinner.

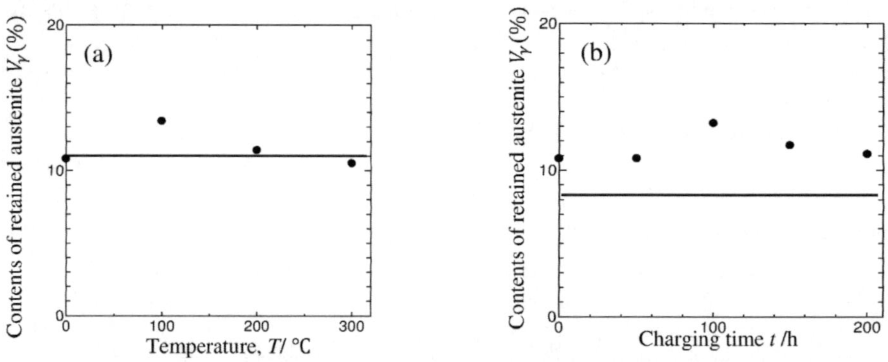

Fig. 2 Contents of retained austenite in 10%γ specimens (a)heated and (b)hydrogen charged at 100 $A{\cdot}m^{-2}$in 0.1N NaOH solution adding 5 $g{\cdot}L^{-1}$ NH_4SCN analyzed by XRD.

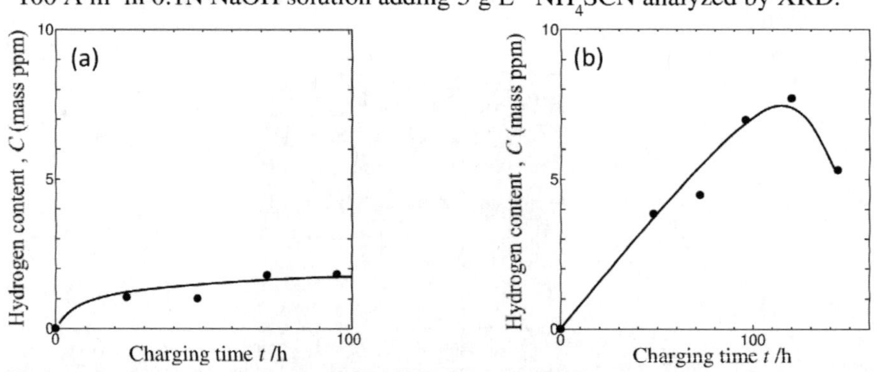

Fig. 3 Hydrogen absorption behavior of (a)[0%γ] and (b)[10%γ] specimens hydrogen charged at 100 $A{\cdot}m^{-2}$ in 0.1N NaOH solution adding 5 $g{\cdot}L^{-1}$ NH_4SCN.

3.2 *Hydrogen penetration into retained γ in high carbon α/γ dual-phase steel*

Figure 5 shows the TDA profiles of the 0% γ and 10% γ specimens charged with hydrogen under the conditions of (i), (ii), (iv) and (v) described in section 2.2. In the 0% γ specimens, there was no change in either the peak temperature or the release end temperature for the different hydrogen charging conditions. On the other hand, both the peak temperature and the release end temperature of the 10% γ specimens shifted to a higher temperature, and the hydrogen occlusion amount sharply increased. These results suggest that the trapping states of hydrogen changed in the high carbon α/γ dual-phase steel as the hydrogen content increased.

Figure 6 shows the relationship between the hydrogen content and the activation energy of the 10% γ specimens charged with hydrogen under the conditions of (i) to (iv) described in section 2.2. The activation energy values of the 10% γ specimens were about 30 kJ / mol when the hydrogen content was 2.5 ppm or less and about 47 kJ / mol when it was 7.7 ppm. The value of the activation energy at high hydrogen content corresponds to the activation energy of diffusion in the γ phase.[4)] Accordingly, it is suggested that the low hydrogen diffusion of retained γ under the conditions of (iv) was the rate-controlling for hydrogen desorption of the 10% γ specimens. That is, it indicates that hydrogen entered retained γ under the condition of a high hydrogen content of 7.7 ppm.

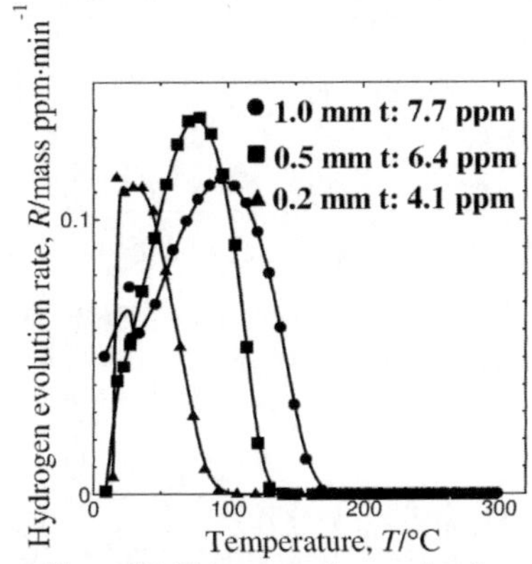

Fig. 4 TDA profiles of 10% γ specimens hydrogen charged at 100 $A \cdot m^2$ in 0.1N NaOH solution with 5.0 $g \cdot L^{-1}$ NH_4SCN and having a thickness of (●) 1.0 mm, (■) 0.5 mm, and (▲) 0.2 mm.

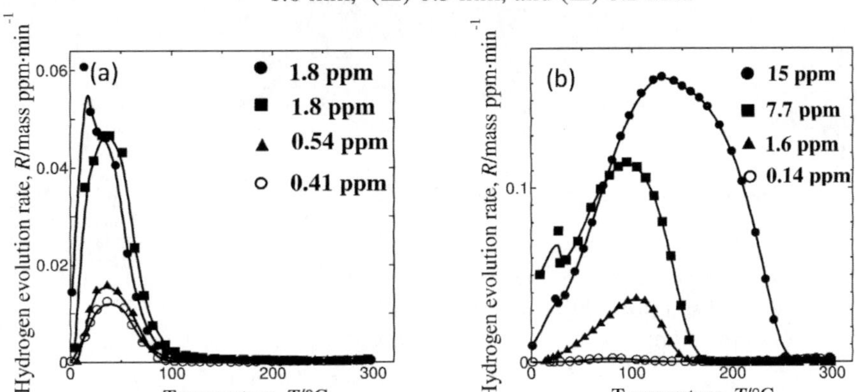

Fig. 5 TDA profiles of (a) 0% γ and (b) 10%γ specimens hydrogen charged at 100 $A \cdot m^{-2}$ in 0.1N NaOH solution with (●) 7.5 $g \cdot L^{-1} NH_4SCN$,(■) 5.0 $g \cdot L^{-1} NH_4SCN$, (▲) 0 $g \cdot L^{-1} NH_4SCN$ and (○) 50 $A \cdot m^{-2}$ in 0.1N NaOH solution with 0 $g \cdot L^{-1}$ NH_4SCN having a thickness of 1.0 mm.

Subsequently, the test results of hydrogen desorption behavior from the 0%γ and 10%γ specimens in an atmospheric environment at 30°C are shown in Fig 7 (a) and (b). Hydrogen charging was conducted under the conditions of (ii) described in section 2.2. Hydrogen desorption time was consistent for both specimens. These results indicate that most of the hydrogen in the high carbon α/γ dual phase steel was trapped in the α phase under the condition of a low hydrogen content.

Figure 7 (b) and (c) shows the results of hydrogen desorption behavior from the 10%γ specimens in an atmospheric environment at 30°C charged hydrogen under the conditions of (ii) and (iv) described in section 2.2. In the 10% γ specimens, hydrogen penetration into retained γ occurred as the hydrogen content increased and it was found that a long desorption time was required.

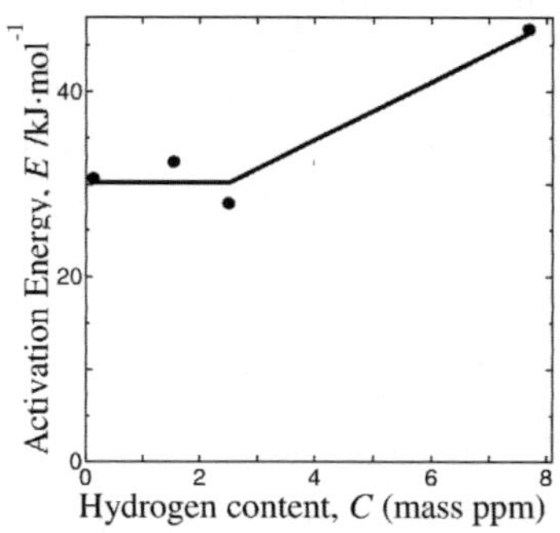

Fig. 6 Activation energy of hydrogen desorption as a function of the hydrogen content of 10%γ specimens.

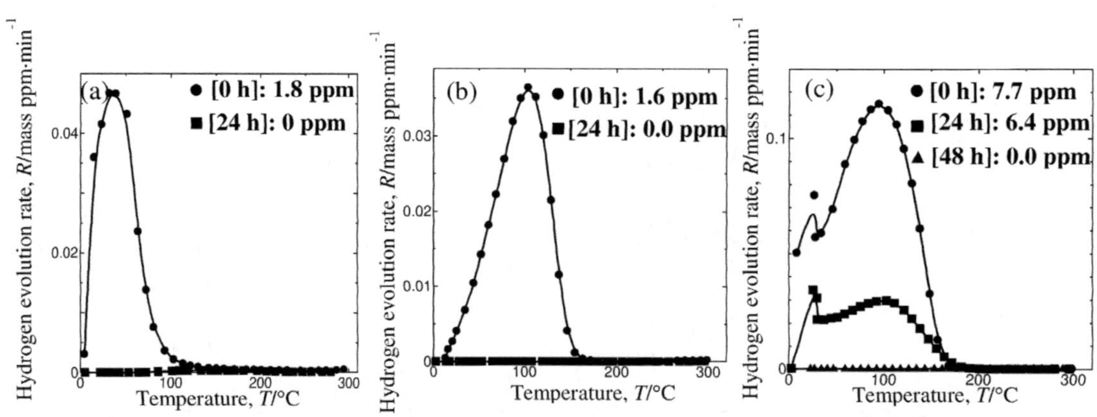

Fig. 7 TDA profiles of (a) 0%γ and (b)10%γ specimens hydrogen charged at 100 A·m^{-2} in 0.1N NaOH solution and 10%γ specimens hydrogen charged at 100 A·m^{-2} in 0.1N NaOH solution adding (c)5.0 g·L^{-1} NH_4SCN and degassed at 30°C for 0 h(●),24 h(■), 48 h(▲) having a thickness of 1.0 mm .

IOP Conf. Series: Materials Science and Engineering **461** (2019) 012076 doi:10.1088/1757-899X/461/1/012076

4 Conclusion

The hydrogen content in high carbon α/γ dual-phase steel was investigated by thermal desorption analysis (TDA), and the following findings were obtained.

(1) Hydrogen content in the 10% γ specimens having retained γ was larger than that in the 0% γ specimens, and a long desorption time was required. This indicates that retained γ had the effect of increasing the hydrogen content and delaying hydrogen diffusion.
(2) Under both low and high hydrogen content conditions, no change occurred in either the peak temperature or the release end temperature of the TDA profile of the 0% γ specimens. In contrast, both the peak temperature and the release end temperature of the 10% γ specimens shifted to a higher temperature. In addition, the activation energy was about 30 kJ/mol when the hydrogen content was 2.5 ppm or less and about 47 kJ/mol when it was 7.7 ppm. This implies that in high carbon α/γ dual-phase steel, hydrogen penetration into retained γ occurs only under a high hydrogen content.
(3) As the thickness of the specimens became thinner, the TDA profile of the 10% γ specimens moved to a lower temperature. This result suggests that hydrogen desorption of the 10% γ specimens was rate-limited by low hydrogen diffusion under a high hydrogen content.
(4) Hydrogen charging was carried out on 0% and 10% γ specimens under the conditions of (ii) (a current density of 100 $A{\cdot}m^{-2}$ in a 0.1 N NaOH aqueous solution) and degassed at 30 °C. As a result, both specimens completely released hydrogen at 24 h. Therefore, the bcc phase was the rate controlling for hydrogen desorption of the 10% γ specimens at low hydrogen content.

References

[1] D. Perez Escobar, K. Verbekena, L. Duprezc and M. Verhaegea, Materials *Science and Engineering A*, **vol. 551** (2012) 50.
[2] T. Nagase, T. Ito, Y. Nishimura, H. Suzuki and K. Takai et al Tetsu-to-Hagané, **vol. 102** (2016) 534.
[3] W. Y. Choo and J. Y. Lee *Metall. Trans. A*, **vol. 13** (1982) 135-140.
[4] K. Takai, K. Murakami, N. Yabe, H. Suzuki and Y. Hagihara, *J. Japan Inst. Metals*, **vol. 72**, No. 6 (2008), pp. 448-456

IOP Conf. Series: Materials Science and Engineering **461** (2019) 012077 doi:10.1088/1757-899X/461/1/012077

Ultra-fine grained β-type TNZT ELI alloy with high strength and low elastic modulus

A V Polyakov[1], I P Semenova[1], E Ivanov[2], R Z Valiev[1]

[1] Ufa State Aviation Technical University, Institute of Physics of Advanced Materials,12 K. Marx str., Ufa, 450008, Russian Federation

[2] Tosoh SMD Inc., 3600 Gantz Road, Grove City, OH 43123, USA

E-mail: alex-v.polyakov@mail.ru

Abstract. This work deals with the improvement of mechanical properties of medical alloy Ti-34Nb-7Zr-4.6Ta ELI (TNZT) by severe plastic deformation (SPD) techniques at temperatures below the temperature of the onset of phase transformations in order to maintain a low modulus of elasticity. For the formation of a nanostructured state, high pressure torsion (HPT) at room temperature and equal channel angular pressing (ECAP) at 200 °C, that is, below the onset of diffusion ω-transformation, were used. It is shown that HPT leads to the formation of the UFG structure in the alloy with the grain size of about 110 nm without changing in the alloy phase composition. The ECAP processing allowed forming a subgrain structure with an average size of grains/subgrains of 500 nm and an ultimate tensile strength of 745 MPa, which exceeded the ultimate tensile strength after the standard aging in the (α+β) area (686 MPa). The issues of the effect of SPD on changes in the structure, phase transformations in the alloy and ways of maintaining a low modulus of elasticity in this alloy are discussed.

1 Introduction

Titanium and its alloys are the best ones to be applied in medicine as materials for long-term application implants due to their excellent biocompatibility, corrosion resistance and high specific strength [1]. However, the elastic modulus of the bone, in which implants from such materials are usually integrated, is from 30 GPa and below [2]. Therefore, β- and near β Ti alloys have been recently the most promising ones. For example, the elastic modulus of Ti-15Mo alloy is 80 GPa, whereas α-Ti or (α+β) alloys Ti-6Al-4V and Ti-6Al-7Nb have an elastic modulus over 110 GPa [3,4]. A large difference between the elastic modulus of the implant material and the bone could lead to stress shielding, bone resorption and, as a consequence, implant loosening and need for repeated surgery [1]. Alongside, the alloys of this system have a relatively low mechanical strength, which usually does not exceed 550 MPa [4]. It is known that modern implants are produced from Ti alloys, the strength of which is over 1000 MPa, which positively effects the service lifetime of items [1]. Thus, the strength enhancement in β-Ti alloys with retention of the ductility reserve and low elastic modulus is an urgent problem.

The lowest elastic modulus is demonstrated by the Ti-Nb-Ta-Zr (TNZT) system alloys, the value of which can achieve 50 GPa depending on the content of elements in the alloy [3,4]. TNZT alloys are biocompatible, as they have only biotolerant elements in their composition. Alongside, Ti alloys of the TNZT system can have different contents of alloying elements, greatly impacting the phase transformations and elastic modulus. For example, it is known that Zr is a neutral element, however, it significantly impacts phase transformations in the alloys of the Ti-Nb system [5]. In particular, Zr suppresses formation of martensite α′ and α″ phases, and it can play the role of β-stabilizer in the presence of other strong β-stabilizers (e.g. Nb) [5,6]. Therefore, the choice of the alloy composition of this system is an important aspect for achieving the required properties. In this work the research object is medical alloy Ti-34Nb-7Zr-4.6Ta ELI with an extremely low content of oxygen (224 ppm) and enhanced content of Zr as compared to the alloy studied in [7,8]. This alloy is the stable β-phase alloy with an elastic modulus of about 55 GPa [4].

A strengthening of β-Ti alloys by conventional thermomechanical treatment (TMT) usually result in simultaneous increase of strength and elastic modulus due to precipitation of secondary ω and α-phases [4,9]. The severe plastic deformation (SPD) techniques, allow forming nanocrystalline and ultrafine-grained (UFG) structures in metals and alloys, which could cardinally change their physical and mechanical properties [7,10]. SPD can be performed at relatively low temperatures, lower than phase transformation temperatures. For example, formation of a nanostructured state in β-Ti alloy by HPT at room temperature allows increasing the strength with retention of a low elastic modulus as reported in [8]. Besides, the grain boundary area has lower values of the elastic modulus as compared to the grain volume, therefore the increase of the extension of non-equilibrium boundaries in the microstructure after SPD can reduce the total elastic modulus [11]. Two SPD techniques - high pressure torsion (HPT) and equal channel angular pressing (ECAP) - were used in order to form a UFG state in it. The aim of this study was to investigate the peculiarities of microstructure changes in the alloy, their impact on the mechanical behavior of the alloy.

2 Experimental methods

The material for the research was the stable β-alloy Ti-34Nb-7Zr-4.6Ta ELI (oxygen content 224 ppm) obtained by electron beam melting. In the initial (as recieved) state the billets of the alloy were machined only. Discs with a diameter of 10 mm and 1.6 mm thick were subjected to HPT in the press with grooved anvils under a pressure of 6 GPa at room temperature. The number of turns was 1, 5 and 10. The rods with a diameter of 10 mm and 80 mm long were subjected to ECAP in the die-set with the angle of channels intersection of 120° at the temperature of 200 °C. The Vickers microhardness measurements were performed on the Duramin Struers unit under a load of 0.5 kg with a dwelling time of 10 sec. At least three indentations were carried out per point of the length.

The microstructures of the specimens were analyzed by transmission electron microscopy (TEM) on the JEOL JEM 2100. The foils were cut out using electrical discharge machining, and mechanically thinned for electro-polishing using the TenuPol-5 Struers facility. Solution of 5% perchloric acid, 35% butanol, and 60% methanol was used at a polishing temperature within the range from –20 to –35 °C. The diffraction patterns were taken from the areas having diameters of 120 and 240 nm.

Cylindrical samples with a gauge length of 15 mm and diameter of 3 mm (ISO 6892) were subjected to tensile testing on the Instron 5982 tensile machine with the strain rate $\acute{\varepsilon} = 10^{-3}$ s^{-1} at room temperature. At least 3 samples were tested for each state.

3 Results and their discussion

3.1 HPT

In the as-received state the structure of TNZT samples consisted of coarse grains of the β phase, the size of which was from 100 to 250 μm (Figure 1a). The TEM images of the structure display well-resolved dislocations, which points to the fact that internal areas of β-grains have a very low dislocation density (Figure 1b). The identification of diffraction patterns confirms the presence of one β-phase (Figure 1b).

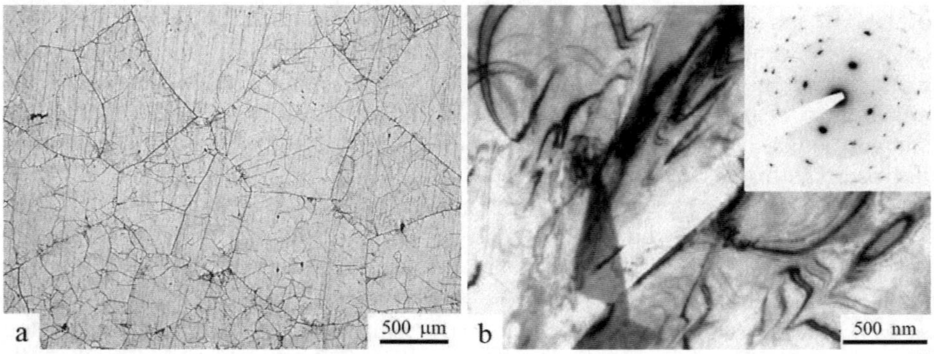

Figure 1. SEM-image (a) and TEM image (b) of the microstructure of TNZT alloy samples in the as-received state.

High pressure torsion after 1 turn results in considerable refinement of β-grains (Figure 2a). Judging by the dark-field images of the microstructure, the average size of subgrains achieves 150 nm. The microdiffraction pattern points to multiple elements with different crystallographic orientation of atomic planes. This is testified by spots located along the concentric circles.

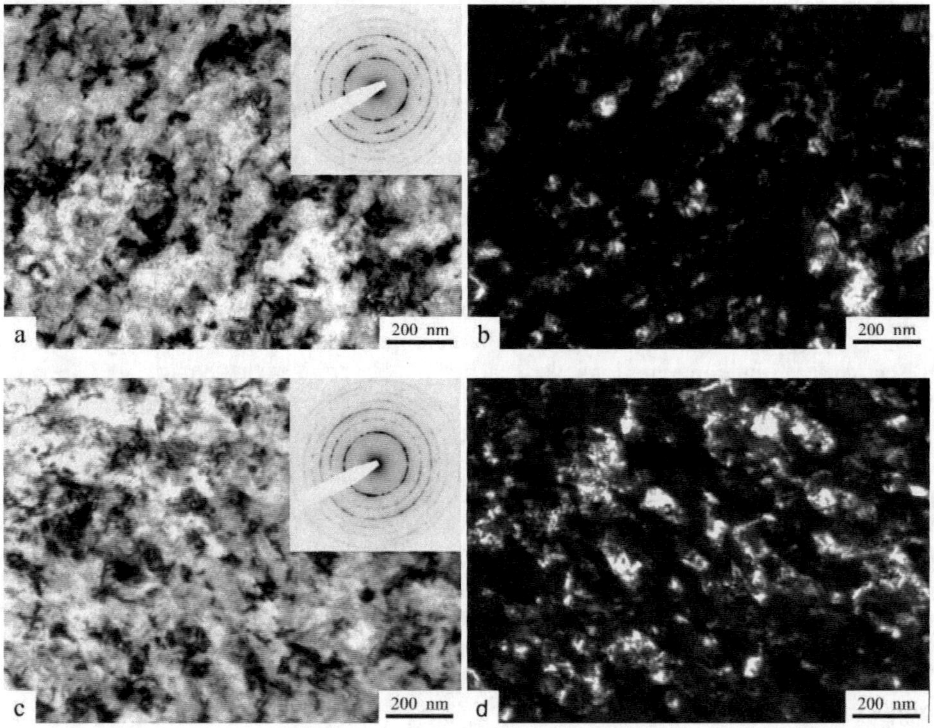

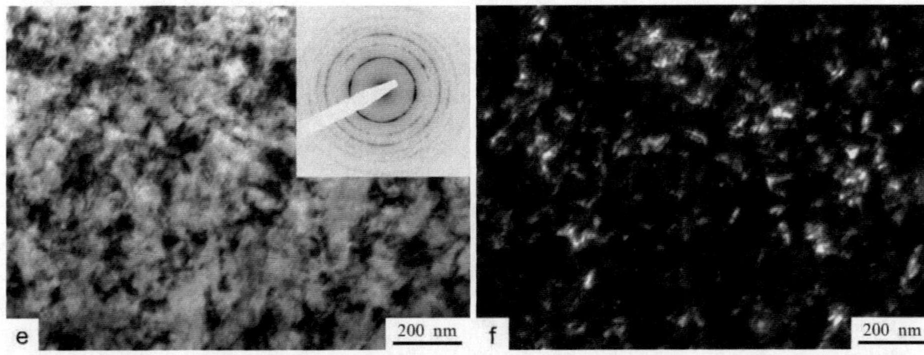

Figure 2. TEM-images of TNZT alloy sample after 1 (a), 5 (b) and 10 (c) HPT turns at 6 GPa and RT, where (a), (c) and (e) – bright field, (b), (d) and (f) – dark field images.

The deformation by 5 and 10 turns results, on the one hand, in additional reduction of grain/subgrain sizes to 130 and 110 nanometers, respectively (Figure 2b, c). On the other hand, the structure becomes more homogeneous, which is testified by the contrast in the dark-field images and more uniform distribution of spots on the concentric circles in the diffraction patterns. It should be noted that no additional spots from secondary phase precipitates were observed in the diffraction patterns. Alongside, precipitation of ω-phase particles is possible in the alloy of the given system under high pressure (6 GPa), which is induced by severe deformation [5,11]. Probably, lack of its spots in the diffraction patterns after HPT is connected with implementation of reverse transformation $\omega \rightarrow \beta$, as reported in [11]

Figure 3 displays the change of the microhardness over the radius of discs from the center to the edge. The highest increment of the microhardness from 176±5 to 236±17 HV was observed after 1 turn due to a cardinal decrease of the size of β-grains to 150 nm. The microhardness value after 5 and 10 turns increases slightly and is 253±11 and 257±10 on the average, respectively (Figure 3). It is seen from Figure 4 that on the periphery of samples the microhardness is higher than that in the center, which is conditioned by the HPT deformation scheme and typical of different materials subjected to this type of deformation [7,12]. It should be noted that with the number of turns increasing, i.e. accumulated strain degree, the HV values over the disc section are leveled (Figure 3). Such behavior was observed in near-β Ti-15Mo alloy processed by HPT at room temperature [12].

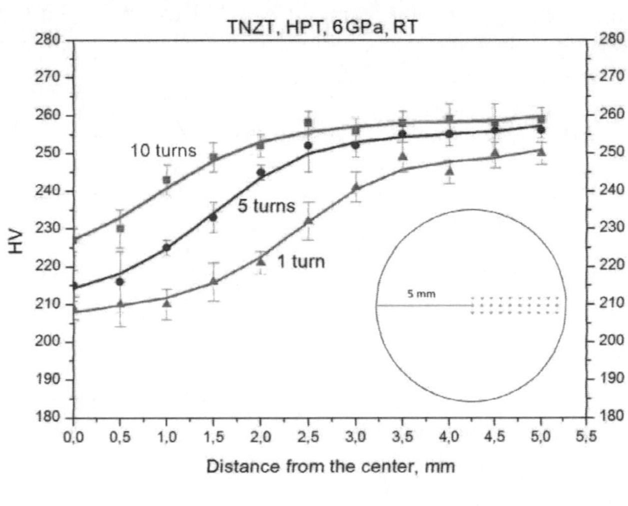

Figure 3. Distribution of microhardness over the TNZT sample radius subjected to HPT to a strain degree corresponding to 1, 5 and 10 turns.

3.2 ECAP

As it is known, the ECAP technique allows producing bulk billets with a UFG structure [7,10]. Alongside this deformation scheme applied to Ti alloys usually suggests processing at elevated temperatures, which can lead to β→ω and β→α transformation and, as a consequence, to increase of the elastic modulus. Therefore, in this work the alloy was processed by ECAP at 200°C, which was lower than the temperature of the start of diffusion precipitation of ω- and α-phases [5].

The study of coarse-sized TNZT samples with a diameter of 10 mm shows that after 6 ECAP passes a microstructure forms that is different from the HPT one (Figure 4). The TEM images display a weakly pronounced grain-subgrain structure, which is testified by the image of diffraction patterns with grouped position of spots. The shape of subgrains is non-equiaxed and elongated especially in the longitudinal section of the billet (Figure 4b). Judging by the dark-field images of the structure, the average size of grains/subgrains was 500 nm.

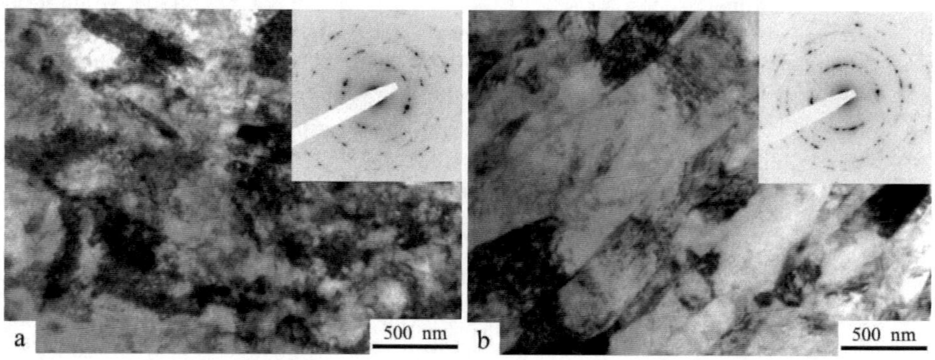

Figure 4. TEM images of the TNTZ alloy microstructure after 6 passes of ECAP at 200°C, where (a) cross section; (b) longitudinal section.

Table 1 lists the mechanical properties of the studied alloy samples in the initial state and after ECAP. As a result of grain structure refinement the ultimate tensile strength increased from 550 to 745 MPa, and the relative elongation changed insignificantly (from 20 to 15%). The mechanical properties achieved by ECAP are higher than the properties of the alloy after standard heat treatment that includes ageing at temperatures of secondary α-phase precipitation [4].

Table 1. The mechanical properties of TNZT alloy.

	Phase composition	UTS, MPa	YS, MPa	Elongation, %	HV
Initial TNZT	β	550	530	20	176
ECAP	β	745	700	15	214
ST+ aging [4] 427°C/8hours	β+α	686	630	17	-

In this case one could expect that one-phase β-structure in the alloy hardened by ECAP will enable retaining the low elastic modulus unlike two-phase β+α structure produced by standard ageing [4]. As it is known the elastic modulus value in conventional coarse-grained materials depends first of all on the phase composition [13]. Severe plastic deformation of the material ensures not only an ultrafine grain size, long extension of grain boundaries, but also introduces high elastic stresses, possible precipitation of ω-phase particles induced by severe deformation [14]. Therefore, application of the known phase rule [13] to estimate the elastic modulus will be not correct for nanostructured and ultrafine-grained materials. Thus, the interrelation between the UFG structure formed in the studied alloy and strength and elastic modulus require more careful studies, our further efforts will be aimed in that direction.

4 Conclusions

The results of the performed studies of the Ti-34Nb-7Zr-4.6Ta ELI alloy with the initial β-structure subjected to processing by different SPD techniques enable stating that:

1. HPT processing at room temperature led to the formation of a UFG structure with an average grain size about 110 nm. The microhardness of the UFG alloy was 257 HV, as compared to the initial coarse-grained state where the microhardness was 176 HV.

2. ECAP processing at a temperature of 200 °C resulted in the formation of a predominantly subgrain structure with an average grain/subgrain size of 500 nm. The UFG alloy had a tensile strength of about 745 MPa, which exceeds that of the counterpart (686 MPa) after standard aging.

3. During the analysis of diffraction patterns taken from the TEM images of the alloy microstructure after HPT and ECAP, no additional spots from ω-phase were observed. It gives grounds to expect the retention of a low elastic modulus in the UFG state of the alloy. However, this is the subject of future studies.

Acknowledgments

This work was supported by RFBR grant № 18-32-00585.

References

[1] Brunette D M, Tengvall P, Textor M, Thomsen P eds 2001 Titanium in Medicine (Berlin Heidelberg: Springer-Verlag) pp 1-1019

[2] Rho J Y, Ashman R B, and Turner C H *J. Biomech.* **26**(2) 111 https://doi:10.1016/0021-9290(93)90042-d

[3] Niinomi M 1998 *Mater. Sci. Eng. A* **243**(1-2) 231 https://doi:10.1016/s0921-5093(97)00806-x

[4] Qazi J I, Rack H J and Marquardt B 2004 *JOM* **56** 49 https://doi:10.1007/s11837-004-0253-9

[5] Tang X, Ahmed T and Rack H J 2000 *J. Mater. Sci.* **35**(7) 1805 https://doi:10.1023/a:1004792922155

[6] Abdel-Hady M, Fuwa H, Hinoshita K, Kimura H, Shinzato Y and Morinaga M 2007 *Scr. Mater.* **57**(11) 1000 https://doi:10.1016/j.scriptamat.2007.08.003

[7] Estrin Y, Vinogradov A 2013 *Acta Mater.* **61** 782 https://doi.org/10.1016/j.actamat.2012.10.038

[8] Xie K Y, Wang Y, Zhao Y, Chang L, Wang G, Chen Z, Cao Y, Liao X, Lavernia E J, Valiev R Z, Sarrafpour B, Zoellner H and Ringer S P 2013 *Mater. Sci. Eng. C* **33**(6) 3530 https://doi:10.1016/j.msec.2013.04.044

[9] Nakai M, Niinomi M and Oneda T 2011 *Metall. Mater. Tran. A* **43**(1) 294 https://doi:10.1007/s11661-011-0860-3

[10] Valiev R Z, Estrin Y, Horita Z, Langdon T G, Zehetbauer M J and Zhu Y T 2015 *Mater. Res. Let.* **4**(1) 1 https://doi:10.1080/21663831.2015.1060543

[11] Wang Y B, Zhao Y H, Lian Q, Liao X Z, Valiev R Z, Ringer S P, Zhu Y T, Lavernia E J 2010 *Scr. Mater.* **63**(6) 613 https://doi:10.1016/j.scriptamat.2010.05.045

[12] Janeček M, Čížek J, Stráský J, Václavová K, Hruška P, Polyakova V, Gatina S and Semenova I 2014 *Mater. Charact.* **98** 233 https://doi:10.1016/j.matchar.2014.10.024

[13] Kim H S, Hong S I and Kim S J 2001 *J. Mater. Process. Technol.* **112**(1) 109 https://doi:10.1016/s0924-0136(01)00565-9

[14] Gatina S, Semenova I, Leuthold J and Valiev R 2015 *Adv. Eng. Mater.* **17**(12) 1742 https://doi:10.1002/adem.201500104

IOP Conf. Series: Materials Science and Engineering **461** (2019) 012078 doi:10.1088/1757-899X/461/1/012078

POWER INCREASE OF SUPERSONIC JETS IN OXYGEN CONVERTER

V O Sinelnikov[1], D Kalisz[1], P L Żak, R D Kuzemko[2]

[1] AGH University of Science and Technology, 23 Reymonta str., Cracow 30-059 Poland, EU

[2] PSTU Pryazovsky State Technical University, 7 Universytets'ka str., Mariupol, 87555 Ukraine

E-mail: victoriano@online.ua

Numerical studies targeted to increasing power supersonic jets in oxygen converter during slag splashing process have been presented in this paper. The calculations are based on a special design of the gas-cooled lance, in which the structure was simplified, as compared with a water-cooled lance. Importantly, the sprayed slag with nitrogen and magnesium-containing (MgO) powder were used in the study. The maximum heating temperature in lance nitrogen was equal to 493 °C. The work was preceded by laboratory studies conducted for convertor slag received in one of the European plants. The analysis of the computation results shows that the cooling of the lance with gas leads to the heat recuperation, which increases the temperature of the injected mixture, so that the kinetic energy of the outflowing jet in the converter increases about 3.5 times, and the slag is splashed higher than in the case of a conventional water-cooled lance.

1 Introduction

Higher resistance of refractory lining of oxygen converters considerably limits the environmental effect of steel making technology by reducing dust emissions accompanied by the piling of waste refractory material after fixing furnace lining. For instance, in Mariupol, Ukraine there are two biggest metallurgical plants in that part of Europe, which evidently have a negative influence on the ecological situation there. Only in 2013 282,000 Mg gaseous pollutions were emitted to the atmosphere, which is about 0.6 Mg per citizen [1]. A considerable part of post-production waste consisted of slag from metallurgical technological processes and waste refractory lining. These are integral elements of the steel production process and cannot be eliminated. Over the last 100 years a few waste disposal sites were established in the vicinity of Mariupol where millions tones of slag and used refractory lining were discharged. The present environmentally-friendly policy obliges steel producers to stop degrading the environment and limit the amount of produced and deposited metallurgic waste. According to the present waste management trends, the closed circuit waste management models are recommended [2].

The slag splashing method applied after the liquid steel melting in an oxygen converter to generate a protective layer in the refractory lining of the converter is a radical way of saving resources and efficient way of limiting environmental emissions. With this technology the resistance of the refractory lining can be increased 5-10 times. The amount of generated waste can be reduced by almost the same order. However the practical implementation of this technology in the steel works requires certain investments aided by simulations of the slag splashing process with informatic tools. The interaction of high power (~4 MW) jets with liquid slag has been preliminarily modeled by various authors [3,6]. Erosion of refractory lining is caused by chemical corrosion due to its contact with liquid metal bath and slag. Another impact is on the part of mechanical damage done while loading metal waste. The cracks and microcracks facilitate penetration of liquid metal bath into the pores, which additionally deteriorates the situation. A powdered MgO is added to increase the adhesiveness of slag to the refractory lining, and protect against the corrosive activity under the influence of FeO, which is the main component of slag. Of course, more complex forms of Ca_2SiO_4, $CaSiO_3$, $FeSiO_3$, Fe_2SiO_4, $Ca_2Si_2O_6$ and $Ca_4Si_2O_8$ may occur and the presence of MgO favors the formation of complex silicates: $CaO{\cdot}MgO{\cdot}SiO_2$ (monticellite), $2CaO{\cdot}SiO_2$ (dicalcium orthosilicate). After insulfflating MgO to slag, its alkality increases and consequently the silicate structures are disrupted. Therefore, for increasing the efficiency of the slag splashing method, 10-13% MgO is introduced in the form of baked or melted or powdered materials; for blowing – powdered materials are used. This type of additive comes from refractory materials post-production or used refractory materials [7-9].

Only few scientific works point to the existence of important premises for the realization of research and experiments in this subject. The slag splashing technology can be improved by increasing the power of media jets flowing out of the lance and decreasing dissipative processes in dispersed jets directed on liquid slag. The efficiency of the slag splashing method can be significantly improved if the nitrogen + MgO powder mixture is preheated before being administered to the nozzles. In this solution the design of lance will be modified into a heat exchanger, where the heat carrier, i.e. N_2+MgO recuperates heat. In this way the heat which was previously lost through the converter's gas outlets to the atmosphere as a thermal pollution, can be partly utilized. Mariupol with its catastrophic environmental conditions needs more than any other European city new technological solutions for its two biggest steel works in Ukraine.

Liquidation of plants generating contaminations control of waste and industrial emission – these and many other environmental recommendations, are strictly regulated by the Kyoto protocol [7]. Meeting these guidelines is possible by working out 'clean technologies' and closed circuit waste management technologies. Among such activities are, among others, supersonic slag splashing with single or two-phase jets. It is conditioned by the lance nozzle size and may total up to 4 MW. The principle of this technology is that a 7 – 12 mm protective layer is introduced on the refractory lining of an oxygen converter so the melting process takes place in this layer. In this process the modified converter slag is given high-energy nitrogen jets or MgO enriched nitrogen through the lance. The total power of jets ejected from the nozzle to the converter (350 Mg) with the slag splashing method, totals to a few MW. If this energy is correctly used, the durability of the lining can be increased several times [5-6].

2 Mathematical model

The calculations were based on a mathematical model already presented in [3,5], where o – isentropic flow deceleration; 1 – nozzle exit, g – converter's gas cavity, x – control section of the jet, n – normal physical conditions.

The added mass was calculated with formula [3]

$$g = m_g / m_1 , \tag{1}$$

where:

$m_1 = \rho_1 w_1 F_1$ – mass consumption by each of 5 nozzles [kg/s];

m_g – mass consumption to be obtained by supersonic jet after leaving the nozzle [kg/s];

ρ_1, w, F_1 – density [kg/m^3], speed [m/s] and area of exit section in the nozzle [m^2].

IOP Conf. Series: Materials Science and Engineering **461** (2019) 012078 doi:10.1088/1757-899X/461/1/012078

Authors [5] describe in detail the method of calculating the supersonic jet making use of 12 mathematical equations. In [5] parameter g was described with

$$g = \frac{2\bar{r}_{\max}\bar{x}}{Ds} C_i \left(1 - C_i^2\right)^{1/2} \left(I_{1R} - I_{2R}\right) \tag{2}$$

where:

C_i – the number of Crocco = $C_i = \sqrt{1 - \left(1 + \frac{k-1}{2} M_i^2\right)^{-1}}$

where σ – similarity parameter characterizing level of jet turbulence.
$\Sigma = 12 + 2.58\ M_i$,
and:
M_i – Mach number,
k – parameter characterizing thermal capacities.
D – coefficient, on the basis of which the dependence between M_1 and n can be established $D = \left(\frac{k-1}{2}\right)^{1/2} M_1 n^{\frac{k+1}{2k}}$.
$\bar{r}_{\max}$ – radius of the first bubble in supersonic jet [3,5].
In equation (2) and in the calculation results $\bar{x} = l / r_1$ – length of jet for which calculations were performed to entry nozzle diameter ratio;
The I_{1R} and I_{2R}, values calculated for the conditions of cool jet (t_o = 30°C) and hot gas (t_g = 1500°C) are presented in [5].
If speed of ejected gases w_g is neglected, the energy conservation law can have the form:

$$m_1 c_{p1} T_O + m_g c_{p_g} T_g = \left(m_1 + m_g\right) c_{px} T_x + \left(m_1 + m_g\right) w_x^2 / 2 \times 10^3, \tag{3}$$

where:
T_0 – thermodynamic temperature [K]; c_p – isobaric thermal capacity [J/K]; w – averaged flow rate [m/s].
Hence, temperature along distance x:

$$T_x = c_{p_1} T_1 + \frac{a_1 m_1 w_1^2}{2} + g c_{p_g} T_g - a_x (1 + g) \frac{w_x^2}{2} \tag{4}$$

where:
d_1 – the diameter of the nozzle the exit section [mm].

$$w_x = \left[w_1 + \frac{p_g (n-1)}{r_1 w_1} \right] \frac{1}{(1+g) b} \tag{5}$$

Thermal capacity c_{pg} will depend on temperatures T_g and T_o.

$$c_{px} = \sum c_i g_i = c_{p_1} g + c_{p_g} g_g \tag{6}$$

$$c_{p_g} = c_{p_{N_2}} = \frac{k}{k-1} R_{N_2} \tag{7}$$

N_X– kinematic energy of supersonic jet, which can be calculated with:

$$N_X = V_n r_n (1 + g) w_x^2 / 2 \times 10^3 \quad (8)$$

3 Results and discussions

Authors of this paper modeled the process of jetting a medium to the liquid slag. This issue was investigated on the basis of a system of differential and algebraic equations of motion and heat exchange in a gas-powder flow, in a definite cross-section (at a given distance) and analysis of impact of the effect of carrier gas heating on the power of the gas/powder jet flowing into liquid slag of the oxygen converter.

The most interesting results of numerical calculations were obtained for variable MgO powder content μ. Computer simulation (**figure 1**) showed that the powder additive significantly affects the efficiency of the slag splashing process. For instance, for the jet length $x = 20$, $\mu = 0.2$, the gas temperature in the converter was 455°C, and power of supersonic jet 359 kW. With the increase of μ *to* 1, gas temperature in the converter and jet power increase to 555°C and 445 kW, respectively.

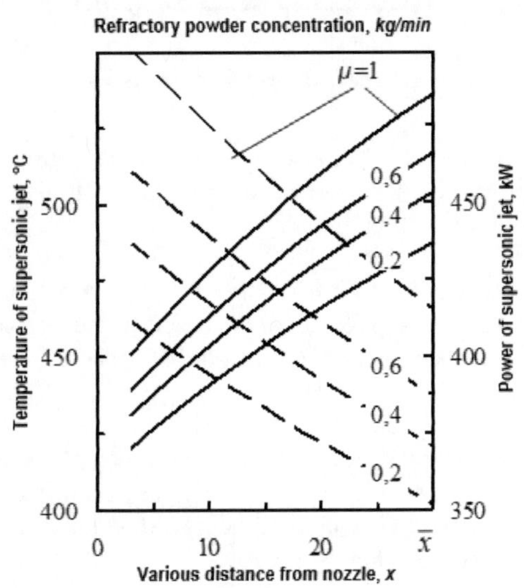

Figure 1. The influence of different concentration refractory powder on the temperature (—) and power supersonic jet (– –) at various distances from nozzle x.

The plot in **figure 2** shows that, with the increase MgO powder content μ, the added mass insignificantly increases, whereas relative power, i.e. ratio of jet power at a distance x to jet power on the nozzle inlet (N_x/N_l) considerably lowers. At a distance $x = 20$, for MgO powder content $\mu = 0.2$, the added mass amounted to 0.18 and relative power to 0.61. At the same distance, with increasing $\mu = 1$, th added mass increases to 0.2, whereas relative power decreases to 0.36.

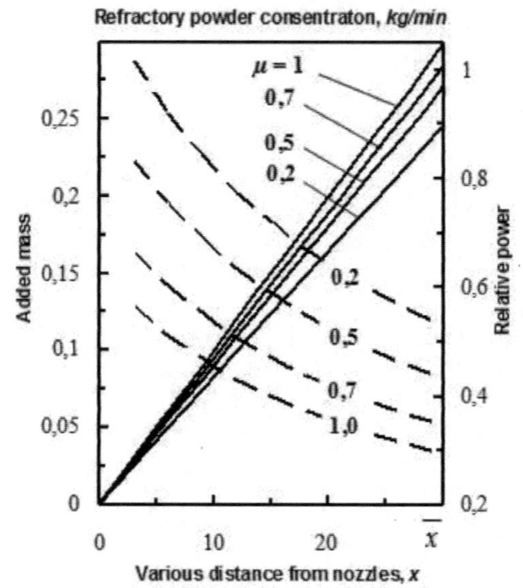

Figure 2. Dependence added mass (—) in cavity oxygen converter and different concentration refractory powder (– –) on relative power at distance *x*.

4 Conclusions

1. Heating of nitrogen in the lance prior to introducing it to the supersonic nozzes and increasing gas temperature in the convrter lead to a significant increase of power of supersonic jet when splashing slag in the oxygen converter. For increasing 2-3 times the jet power from the lance and increasing the efficiency of the *slag splashing process*, the lance has to be gas-cooled, with the possible heating of nitrogen or its mixture with MgO to at least 400°C.

2. Adhesion of the slag layer for the refractory lining significantly improves when a strictly defined chemical composition of slag is provided, as it guarantees obtaining given viscosity and formation of low- and high-melting phases. This can be achieved when the participation of MgO in the splashed slag is on a level of 10-13% [7].

References

[1] Sinelnikov V O 2012 Scientific reports of the winners of the 2nd round of the all-Ukrainian competition of students scientific works in Physics (Sumy) pp 72-79

[2] Sinelnikov V O, Kuzemko R D, Kalisz D 2018 Int. Conf. The Circular Economy – Rational Use of Raw Materials (Cracow) p 26

[3] Sinelnikov V O, Kalisz D, Kuzemko R D 2017 Influence of hydrodynamic and temperature conditions on the efficiency of the slag splashing method *Journal of Machine Constraction and Maintance. Problemy eksploatacji* **105:2** 93-98

[4] Kuzemko R D, Sinelnikov V O, Mastykash V, Kalisz D 2018 Analysis of the flow of gas/powder mixture in the oxygen converter lance nozzles *Journal of Machine Constraction and Maintance. Problemy eksploatacji* **108:1** 111-117

[5] Kharlasnin P S, Kuzemko R D, Sinelnikov V O 2015 Influence of different factors and physical impacts on the power of flowing supersonic jet during slag spraying in the converter *Theoretical and practical solutions of mineral resources mining: «New developments in mining engineering 2015»* (London) pp 597-602

[6] Kharlasnin P S, Kuzemko R D, Protopopov E V, Feiler S V, Kharin A K 2015 Adding slag a supersonic jet in an oxygen converter *Steel in Translation* **45:2** 100-104

[7] Sinelnikov V O, Kalisz D 2016 Influence the FeO content on slag viscosity at his spraying.

Increase the life of the refractory lining *Glass and Ceramics* **73:3-4** 144-148
[8] Kalisz D, Sinelnikov V O, Kuglin K 2018 Investigation of physicochemical properties of slag splashed on the lining of an oxygen converter *Refractories and Industrial Ceramics* **58:5** 463-468
[9] Yildirim I Z, Prezzi M 2011 Chemical, mineralogical, and morphological properties of steel slag, Hindawi Publishing Corp., *Advances in Civil Engineering* **2011** 1-13

Surface electrolytic-plasma polishing of Ti-6Al-4V alloy with ultrafine-grained structure produced by severe plastic deformation

***M K Smyslova, D R Tamindarov, N V Plotnikov, Iu M Modina, I P Semenova**

Ufa State Aviation Technical University, 12 K. Marx str., 450008, Ufa, Russia

E-mail: sem-ip@yandex.ru

Abstract. The paper deals with the studies on electrolytic-plasma polishing (EPP) of the surface of Ti-6Al-4V alloy with ultrafine-grained (UFG) structure produced by equal channel angular pressing. The electrolyte temperature range and working shift voltages of EPP to ensure the polishing effect in Ti-6Al-4V with UFG structure are discussed. The results of the studies on topography and chemical composition of the alloy surface in coarse grained (CG) and UFG states by scanning electron microscopy are presented. The impact of EPP duration on the surface roughness and metal removal rate of conventional and UFG alloy are demonstrated. The perspectives of the EPP process applied to a surface treatment of medical implants and items for mechanical engineering are discussed.

1 Introduction

Ti alloys are widely used in medicine, mechanical engineering and aircraft. All these application areas impose high requirements to service life of items and, therefore, to quality of their surface. Polishing techniques with employment of felt discs with abrasives, chemical and electrochemical polishing techniques are used presently in order to reduce surface roughness. However the polishing with felt discs with abrasives is restricted to flat surfaces with a large bending radius. Chemical and electrochemical polishing is connected to necessity to use concentrated acids. All the described techniques are characterized by low productivity and technology complexity. Today one of the promising techniques of surface treatment is electrolytic-plasma polishing (EPP) [1-3]. This technique is most widely used to polish items from stainless steels [4-7]. Parts processed by this technique have a surface roughness up to *Ra* 0.01 μm and microscopic defects as stripes and a defect layer with foreign inclusions are removed from the surface. In works [4, 7] surface polishing by EPP is shown to lead to increase of fatigue strength and corrosion resistance of stainless steels. The authors of [6] give examples of effective use of EPP to polish medical items from different alloys, including Ti ones. It is shown that the EPP process can be used for small-sized and complex geometry parts. Alongside, the regularities of EPP of Ti alloys and its impact on physical and chemical state of the processed surface have not been studied sufficiently.

The recent experiments on EPP of Ti alloys in [7-9] show that physical and chemical processes in them are similar to those taking place in chrome steels during polishing in solutions of inert salts. In particular, depending on processing regimes (shift voltage and electrolyte temperature), during EPP of two-phase Ti-6Al-4V alloy, conditions are created for surface oxidation, heating and polishing. However, as compared to steels, oxidation in Ti alloys occurs more intensively with formation of thin but rather dense oxide films, which obstructs electrochemical reactions directed at

IOP Conf. Series: Materials Science and Engineering **461** (2019) 012079 doi:10.1088/1757-899X/461/1/012079

surface smoothing. This impacts the physical and chemical state of the item surface. Moreover fluoride-salt-based electrolyte for polishing does not settle the problem of oxidation resistance, which requires studies on electrochemical behavior of Ti alloys during EPP.

As it is known Ti alloys have different types of microstructure depending on thermomechanical treatment. Microstructures differ in ratio of β- and α-phases, their shape and sizes, state of grain boundaries, etc. and, therefore, in electrochemical activity [10, 11]. Comparative analysis of EPP processes and further surface formation in Ti-6Al-4V with ultrafine-grained (UFG) structure produced by severe plastic deformation are of special interest. The recent decade studies show that formation of bulk UFG structure in Ti alloys allows increasing its service and technological properties (strength, fatigue resistance, superplasticity, etc) [10], which is urgent for promising items in medicine and mechanical engineering.

The main goal of the present work is to study the features of EPP and physical and chemical state of the surface of two-phase Ti-6Al-4V alloy with regular globular-plate and ultrafine-grained structure produced by equal channel angular pressing (ECAP).

2 Experimental methods

Ti-6Al-4V alloy of following chemical composition in wt.% (Ti- base, Al – 6.6 %; V – 4.9 %; Zr – 0.02 %; Si – 0.033 %; Fe – 0.18 %; C – 0.007 %; O – 0.17 %) is used to perform the research. The initial coarse-grained (CG) structure of the alloy is mixed globular plate structure consisting of primary α-phase grains (the average size is about 15 μm, the fraction of which is about 30%), and zones with (α+β) thin-plate structure (Figure 1 a). Billets with a diameter of 20 mm and 100 mm long are subjected to ECAP on the die-set with the angle of channels intersection of 120° via route B_c at T=650°C, 6 passes. The total accumulated strain after ECAP is the equivalent strain degree ε~3. In the microstructure of ECAP-processed billets low-deformed grains of primary α-phase with the average size of about 10 μm are observed (Figure 1b), and in the zones with α+β structure UFG structure is formed, the average size of grains/subgrains of α and β-phases is about 350 nm (Figure 1c).

a) 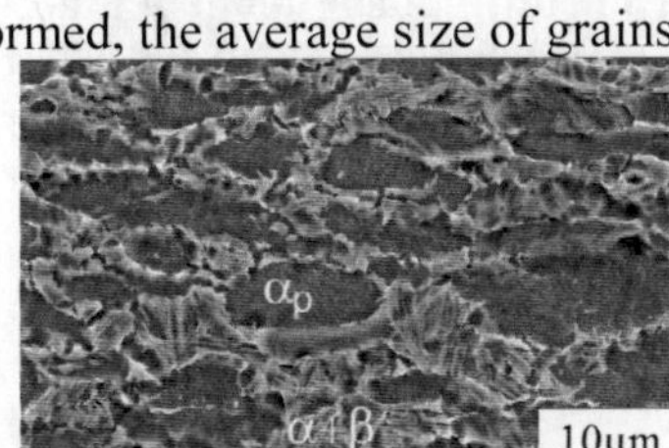

b) 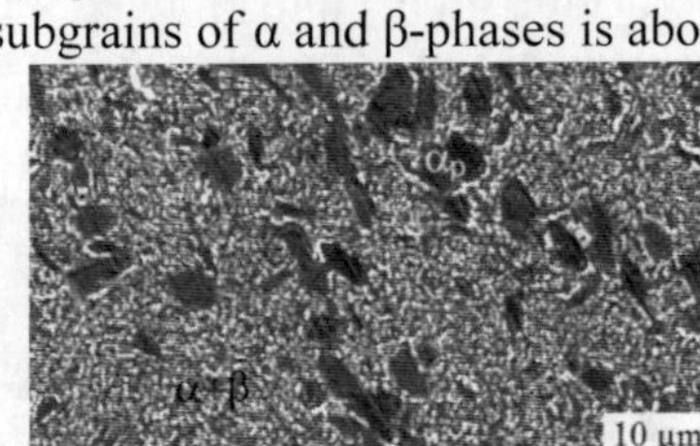

c) 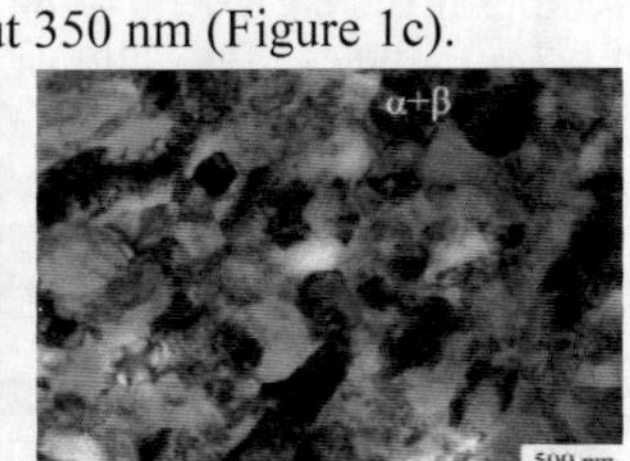

Figure 1. Microstructure of Ti-6Al-4V alloy with coarse-grained globular-plate structure (a); ultrafine-grained structure (b, c).

The EPP of samples is implemented in electrolyte composition applied to conventional Ti-6Al-4V [8], including from 4 to 6 wt. % of hydroxylamine hydrochloride and from 0.7 to 0.8 wt.% of sodium or kalium fluoride [8]. The samples are dipped in the bath of the EPP-50EM set to a depth of 50 mm from the electrolyte surface. The bath walls are cathodes; the treated samples are anodes (Figure 2). The voltage is supplied before the samples are dipped in the electrolyte. In these conditions when the parts are dipped in the electrolyte, the electrolyte starts boiling intensively round the anode, thin steam-gas shell forms (Figure 2).

The studies on the peculiarities of EPP applied to two-phase Ti-6Al-4V with CG and UFG structures include:

- the impact of shift voltage on EPP applied to alloy with UFG structure at 260 and 280 V and the electrolyte temperature 90+5°C;
- the impact of EPP duration on roughness change and metal removal rate.

The surface state and chemical composition are studied with the help of a scanning electron microscope (SEM) JEOL-6490VL with an X-ray microanalyzer Oxford Instruments. The roughness is measured with the help of a profilometer SurfTest SJ-210 (Mitutoyo Corporation). The rate of metal removal from the surface during EPP is estimated through measuring the thickness of samples with the help of a caliper CD-15AX (Mitutoyo Corporation).

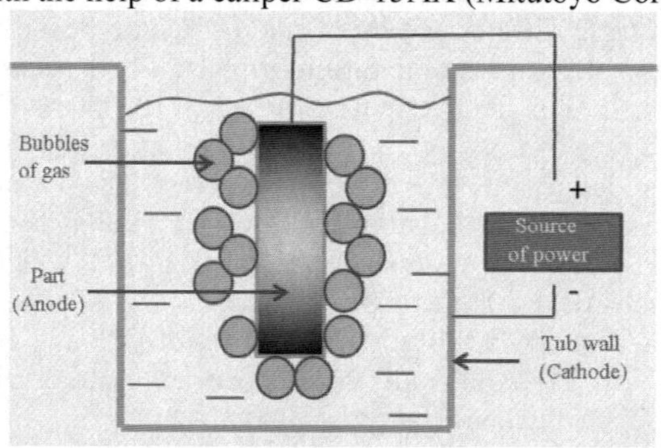

Figure 2. Scheme of electrolyte-plasma polishing and steam-gas shell formation.

3 Results

3.1 Influence of voltage during EPP on the polishing effect of samples surface

In this work it was experimentally established that for an UFG alloy the polishing effect was observed at a temperature T = 90 + 5 ° C as well as for a CG alloy. Therefore, in the course of further studies this parameter did not change.

As is known that the process of EPP for a conventional Ti-6Al-4V alloy is usually carried out at a shift voltage of 280 V [8]. The polishing process of UFG samples surface at this voltage led to the etching of the surface and the formation of etch pits (Figure 3a). The shift voltage reduction up to 260 V for UFG alloy allows achieving highly polished effect to reach the surface of the mirror (Figure 3 b and c).

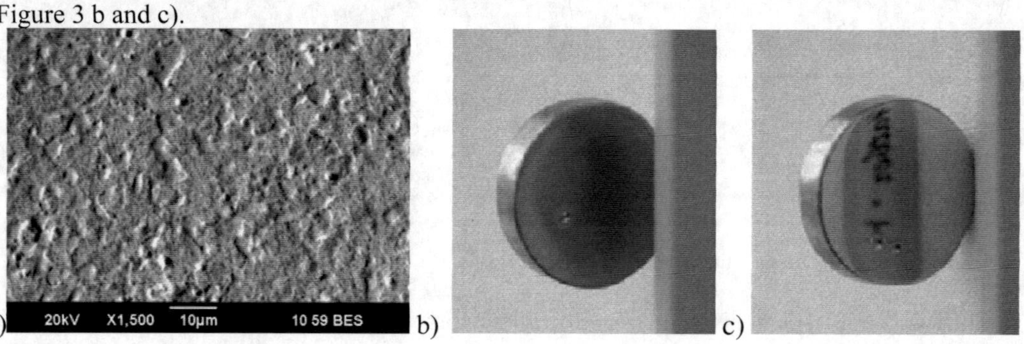

Figure 3. SEM image of surface topography (a) in UFG Ti-6Al-4V samples after EPP via the regimes of CG alloy processing (U=280 V ,T=90 °C); samples surface after EPP and a voltage: a) 280V, b) 260V.

3.2 The impact of EPP duration on chemical composition and surface roughness

The results of the studies on chemical composition of the surface of UFG Ti-6Al-4V after mechanical polishing and after EPP treatment are displayed in Table 1.

It is seen from Table 1 that electrolytic-plasma polishing in the electrolyte containing fluoride components results in some change of chemical composition of the surface as compared to conventional mechanical polishing. In particular, alongside with alloying elements fluoride ions are observed in the sample surface. It is known that fluohydric acid HF is in majority of solutions applied for etching of Ti alloys. The Ti surface treatment by fluohydric acid solutions disturbs the stability of initial oxide film TiO_2 and results in its subsequent dissolution. Oxide layer that forms

IOP Conf. Series: Materials Science and Engineering **461** (2019) 012079 doi:10.1088/1757-899X/461/1/012079

after chemical etching is more homogeneous in thickness and composition as compared to initial layer [12].

Table 1. Chemical composition of sample surface with UFG structure after mechanical polishing and after EPP (wt. %).

State	Main components and impurities (averaged values in wt. ± 0.1%)				
	Ti	Al	V	Fe	F
Mechanical polishing	89.11	5.60	4.50	0.32	-
EPP	88.82	5.99	5.08	0.40	0.48

The results of measurement the parameter *Ra* in the EPP process as a function of the polishing time for the Ti-6Al-4V alloy with a CG and UFG structure are shown in Figure 4. It can be seen that a decrease in the surface roughness with a UFG structure from *Ra* 0.2..0.25 µm to *Ra* 0.05 µm occurs in 2 minutes, whereas similar values of *Ra* for a CG alloy were achieved only after 4 minutes of polishing. Further increase of polishing time does not lead to significant change of roughness: a maximally achievable value of the *Ra* is 0.040...0.045 µm for both CG and UFG alloys.

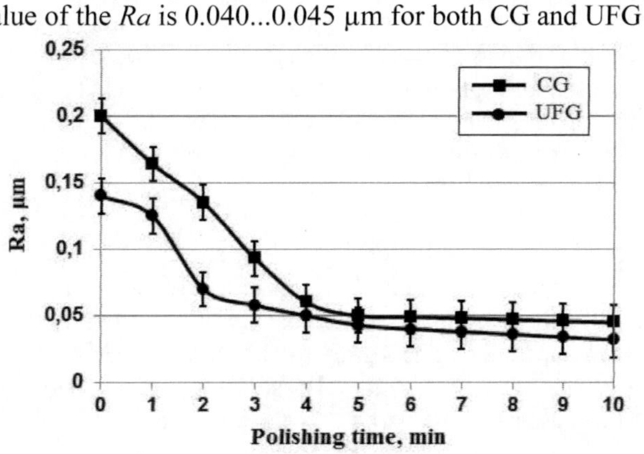

Figure 4. The surface roughness of Ti-6Al-4V samples with UFG and CG structure during EPP against polishing duration.

3.3 Impact of EPP duration on rate of metal removal from the surface

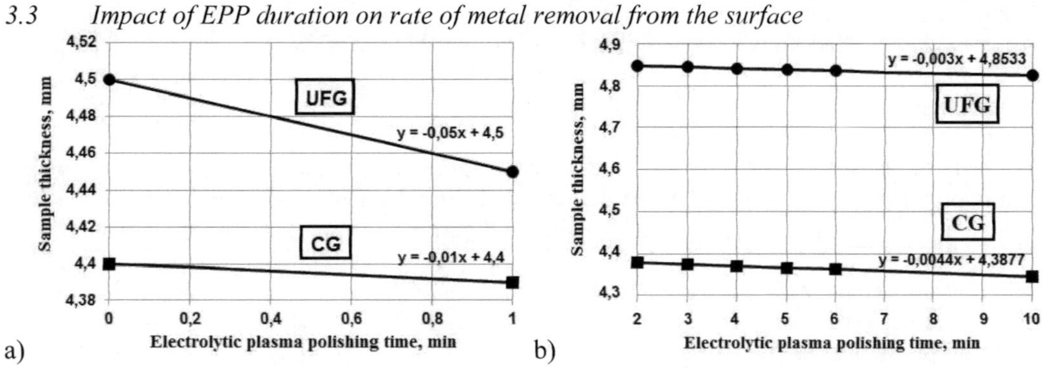

Figure 5. The thickness of CG and UFG samples during polishing after 1 minute (a); for 2 to 10 minutes (b).

It is established experimentally that during the first minute of polishing the thickness of samples reduces most intensively, when the metal removal rate from the surface of conventional and UFG alloy is 0.01 и 0.05 mm/min, respectively (Figure 5a). When the polishing duration increases to 10 minutes, the rate of metal removal decreases and gains some constant value of 0.03…0.04 mm/min (Figure 5b). Such behavior is typical also during chemical etching of surface of Ti and Ti alloys in solutions with fluorhydric acid, which is conditioned by formation of stable protective passivation film on the surface. This effect strengthens in the conditions of ultrafine-grained structure of material [11, 13].

4 Discussion

The experiment results obtained in this work testify to the fact that the previously developed electrolyte for CG Ti-6Al-4V [8] with fluoride salts is effective for EPP of UFG alloy surface, as formation of UFG structure in alloy does not change its chemical composition. The main difference in the EPP regimes applied to Ti alloy with UFG structure is in achievement of the polishing effect at lower voltage values, as compared to those applied to CG alloy (at 260 и 280 V, respectively). Taking into account that the shift voltage during EPP determines the energy of ions in the steam-gas shell, one may conclude that lower energy of ions is required for polishing of alloys with UFG structure. This allows assuming that lower energy consumption of EPP applied to Ti alloys with UFG structure is connected with increase in its electrochemical (corrosion) activity [11,13]. The mentioned effect of electrochemical activity enhancement can be explained by intensification of surface substructure heterogeneity and its thermodynamically instable state [13]. Moreover, transition of metal in the UFG state, for which high density of high-angle boundaries and defects is typical, is reflected in increase of electrons activity and enhancement of a diffusion coefficient in grain boundary area. Taking into account that the EPP occurs with formation and subsequent dissolution of oxide films, both described effects could impact the increase of the UFG structure alloy surface dissolution, which was observed in the experiment.

Special attention should be paid to the change in the chemical composition of the alloy surface after EPP in the electrolyte containing fluohydric acid. Fluoride ions in the surface can be assumed as a positive effect of EPP, for example, for medical items (implants) which remain in a human body for a long period. In particular, according to the authors of [14,15] change of the oxide layer properties during surface treatment by HF solutions has a positive impact on the surface biocompatibility. Formation of TiF groups contributes to formation of new bone tissue on the implant surface.

The main target of the electrolytic-plasma polishing is reduction of surface roughness in items and achievement of the set values of Ra. Such task is topical for different items applied in medicine and mechanical engineering. In this work the surface roughness in samples during EPP changes exponentially depending on the process duration (Figure 6). Such change in roughness is typical of many previously studied alloys. It is connected with distribution of material over the height of micro-roughnesses: there is less material at the top of micro-roughnesses than at their base. So at a constant dissolution rate of the material volume unit during EPP, primary change of the height of micro-roughnesses takes place with a higher rate [1].

In this work, it is shown that the UFG structure helps to accelerate the reduction of micro-roughness on the surface in the first 2 minutes of polishing, whereas the same Ra value in the CG alloy is achieved only after 4 minutes of polishing (Figure 4). A similar behavior was observed when measuring the metal removal rate, when in the first 1 minute of polishing of a CG and UFG surface it was different and amounted to 0.01 and 0.05 mm/min, respectively (Figure 5a). Perhaps this effect is associated with a more active process of metal dissolution on the crystalline defects, including grain boundaries. As is known, the total length of grain boundaries in the UFG alloy is much higher than in the CG. Similar behavior was observed in UFG pure titanium when a passivating film was formed in acid solutions. For example, in [11], the corrosion behavior of commercially pure titanium in UFG and CG states in solutions of hydrochloric and hydrofluoric acids was investigated. It was found that the increased corrosion resistance of UFG titanium is due to the faster formation of the passivation film. The proof of this effect was a lower density of corrosion currents, more positive corrosion potentials and low critical currents for UFG titanium. Further increase in the EPP duration already depends little on the material microstructure, as can be seen from Figure 4. Perhaps this is due to the consistent formation and entrainment of oxide films. It is obvious that the duration of the EPP process is set depending on the tasks set for a researcher, primarily related to achieving the target roughness parameters.

Thus, EPP of parts from Ti alloys with UFG structure with enhanced strength properties is a promising surface treatment technique, which allows achieving high surface quality which could not be achieved by other treatment techniques.

5 Conclusions

1. The electrolyte composition applied for EPP and its temperature ensure creation of conditions for steam-gas shell formation and, as a consequence, enhancement of surface polishing efficiency applied to Ti-6Al-4V with UFG structure. In the UFG alloy due to its enhanced electrochemical activity, micro-roughness sublimation, formation of oxide films and their further removal occur at lower voltages of 260 V as compared to the CG structure alloy, in which these processes take place at 280 V and higher.

2. It is shown that the UFG structure contributes to the acceleration of the reducing the surface micro-roughness in the first 2 minutes of the polishing process in comparison of CG alloy, which is due to the high density of grain boundaries and defects in the crystal structure. The total duration of the polishing process to achieve the specified roughness parameters (0.0040 ... 0.045 μm) constituted about 6 minutes for the CG and UFG states of the alloy.

Acknowledgements

The work was conducted with the support of the Russian Science Foundation under grant No. 16-19-10356 in USATU.

References

1. Gupta P, Tenhundfeld G, Daigle E O and Ryabkov D 2007 *Surf. and Coat.Tech.* **201** 8746
2. Yerokhin A L, Nie X, Leyland A, Matthews A and Dowey S J 1999 *Surf. and Coat.Tech.* **122** 73
3. Nestler K, Böttger-Hiller F, Adamitzki W, Glowa G, Zeidlera H and Schubert A 2016 *Proc. CIRP* **42** 503
4. Vaňa D, Podhorský Š, Hurajt M and Hanzen V 2013 *Int. Jour. of Modern Engin. Res.* **3** 788
5. Meletis E I, Nie X, Wang F L and Jiang J C 2002 *Surf. and Coat.Tech.* **150** 246
6. Zeidler H, Böttger-Hiller F, Edelmann J and Schubert A 2016 *Proc. CIRP* **49** 83
7. Smyslova M K, Tamindarov D R and Samarkina A B 2011*Avia.-kosmich. tech. and technolog.* **7** 25 (in Russian)
8. Patent RF 2006 *No* 2552203
9. Smyslova M, Samarkina A and Tamindarov D 2016, In: Proc. of 5th International Conference on Electrochemical and Plasma Electrolytic Modification of Metal Surfaces Kostroma, (Russia) May 16–20, 2016.
10. Semenova I P, Raab G I and Valiev R Z 2014 *Nanotechnologies in Russia* **9** 79
11. Balyanov A, Kutnyakova J, Amirkhanova N A, Stolyarov V V, Valiev R Z, Liao X Z, Zhao Y H, Jiang Y B, Xuc H F, Lowe T C and Zhu Y T 2004 *Scripta Mater.* **51** 225
12. Advances in Biomaterials Science and Biomedical Applications, (ed. by R. Pignatello), In Tech. 2013, 568p.
13. Sevidova E K and Simonova A A 2011 *Electr. treatment of mater.* **47** 70 (in Russian)
14. Ellingsen J E, Johansson C B, Wennerberg A and Holmen A 2004 *Int J Oral Max Impl* **19** 659
15. Davies J E 2003 *J. Dent. Educ.* **67** 932

Microstructural evaluation of an Al-Mg alloy after ECAP and post ECAP processing

P. Snopiński

Silesian University of Technology, Gliwice, Akademicka 2A st., Poland

E-mail: przemyslaw.snopinski@polsl.pl

Abstract. This work focuses on the effect of pre-deformation treatment and post-deformation aging on the microstructure, especially the precipitation kinetics of the Al-Mg aluminium alloy. The UFG material was produced by equal channel angular pressing which resulted in dramatic increase of mechanical strength. Important anomalies resulting from the severe plastic deformation appeared in the dilatometric curves, which must be influenced by the recrystallization and the precipitation processes. Furthermore, the acceleration effect of SPD processing on precipitation nucleation was confirmed.

1 Introduction

The Al-Mg alloys are suitable for marine applications due to their excellent properties such as low density, ease fabrication, mechanical strength and most notable resistance to corrosion. 5xxx series aluminium alloys are generally classified as a non-heat treatable. In contrast with other Al alloys, strength is achieved primarily by a solid solution strengthening, work or dispersion hardening. However, the Al-Mg alloys can actually form several precipitates and thus can be classified as heat treatable to some degree [1-4].

The precipitation behaviour of Al-Mg alloys has been a subject of intensive studies for a few decades. Scientists efforts were focused on the manner in which stable or metastable precipitates nucleate from a supersaturated solid solution. The first proposal of a possible precipitation sequence in Al-Mg alloys was suggested by Kelly et al.[5]:

$$SSS\alpha \rightarrow \beta' \rightarrow \beta \quad (1)$$

They suggested that two precipitate phases can form: 1) intermediate β'-phase and 2) the equilibrium β – phase (Al_3Mg_2). Both phases has a similar chemical composition but different crystal structure. Further work presented by Pollard [6] showed that previous precipitation sequence lack of formation of Guinier-Preston zones and thus, he suggested a modified sequence:

$$SSS\alpha \rightarrow \text{GP zones} \rightarrow \beta' \rightarrow \beta \quad (2)$$

He has explained his results in such a way that during ageing <50ºC, the formation of GP zones take place, which results in an increase of mechanical strength. Other scientists [7,8] reported the second type of GP zones can form β″ – phase. The culmination of researches discussed above by other scientists in this field [9-11] have yielded the following precipitation sequence that is commonly reported in the literature:

$$SSS\alpha \rightarrow \text{GP zones} \rightarrow \beta'' \rightarrow \beta' \rightarrow \beta \quad (3)$$

The increase in the aging temperature above 100 ºC led to a dissolution of GP zones and the β″ precipitates, thus allowing formation the intermediate (metastable and semi-coherent) β′ phase with a hexagonal structure. Such type of precipitates are formed by a nucleation and growth on the structural defects of the matrix [9-13].

Recently, there has been a growing interest in controlling the precipitation microstructures of aluminium alloys and thereby obtaining a desirable combination of strengthening from both grain refinement and precipitation hardening. One of the most promising methods for obtaining an increased strength as well as an increased ductility is a low-temperature ECAP processing and subsequent short heat treatment. Using this combination, strengthening processes such as the formation of very fine particles as well as softening recovery processes occurs simultaneously.

The understanding of the precipitation behaviour at the early stages of artificial ageing of SPD treated Al-Mg alloy appears to be important for clarifying the changes in the mechanical strength during ageing. The purpose of this paper is to investigate the microstructure of Al-Mg alloy subjected to ageing treatment after severe plastic deformation process.

2 Material and research methodology

The investigation has been carried out on commercial aluminium alloy—appointed by the standard EN 1706:2010 – EN AC 51100. The chemical composition of the alloys is given in Table 1.

Table 1. Chemical composition of the EN AC 51100 alloy

Element	Mg	Si	Fe	Cu	Al
Mass/%	2.86	0.07	0.07	0.01	Bal

The experiment described in this article was preceded by a selection of the most beneficial heat treatment conditions. The results were published in the previous articles [14,15]. The second stage of the experiment was to study the influence of the ECAP stain on the precipitation kinetics in an AlMg3 alloy. Cylindrical samples of 20 mm in diameter and 70 mm in length were ECAP processed with a constant ram speed of 20 mm/min using a die with an internal angle (Φ) of 120°. In order to decrease friction between ECAP sample and die molybdenum sulfide (MoS2) was used. Solution treated samples were then subjected to the ECAP process (4 passes, route Bc - the rotation of the sample was precisely checked between consecutive passes) and finally artificially aged at 160 °C.

The microstructure of the alloy was characterized by light optical microscope Axio Observer Image Analyser. The examinations of thin foils were made on the high-resolution transmission electron microscope JEM 3010UHR from JEOL, at the accelerating voltage of 300 kV. TEM specimens were prepared by cutting thin plates from the material. The specimens were ground down to foils with a maximum thickness of 80 μm before 3 mm diameter discs were punched from the specimens. The disks were further thinned by ion milling method with the Precision Ion Polishing System (PIPS™), used the ion milling device model 691 supplied by Gatan until one or more holes appeared. The ion milling was done with argon ions, accelerated by a voltage of 15 kV.

Static tensile test and microhardness were performed to investigate the precipitation kinetics of the artificially aged samples. Vickers microhardness (HV) was measured on cross-section plain designated as y-plane by imposing a load of 300 g for 15 s. Tensile tests were performed at room temperature using a universal tensile testing machine ZWICK Z/100.

The dilatometric analysis was performed using a TA Instruments Dilatometer under a protective atmosphere of pure argon. Thermal cycles consist of heating until 440 °C (10°C/min), followed by holding of 10 min at this temperature and a cooling with the same rate until room temperature.

IOP Conf. Series: Materials Science and Engineering **461** (2019) 012080 doi:10.1088/1757-899X/461/1/012080

3 Results

3.1 Structure

The microstructure of the solution treated sample prior ECAP consists of well-defined grains of different sizes as shown in Fig. 1a. The average grain size measured using the line interception method is ~ 330 μm.

As follows from the Fig. 1a during solution treatment, most of the Mg was brought into the α-Al solid solution. The ECAP process has changed the original coarse-grained microstructure of the AlMg3 alloy (Fig. 1b-c). Comparing to the initial microstructure, the ECAP processed samples grains are stretched, some areas of different crystallographic orientation – shear bands forms during ECAP shear deformation. It is evident that these shear bands nucleate on a grain boundary, inducing large offsets on it. Artificial ageing has a minor influence on the microstructure - no visible evidence of new grain nuclei (Fig. 1c).

a) b) c)

Figure 1. Structure of the EN AC 51100 alloy a) solution treated state, b) subjected to 4 ECAP passes, c) subjected to 4 ECAP passes and aged for 30 min

It is known from a nomenclature that the hexagonal β′-phase precipitates at 150 and 200 °C, while the direct formation of the equilibrium β-phase occurs at 250 and 300 °C. In our previous work, we found that β′-precipitates are responsible for precipitation strengthening of a solution treated Al-Mg alloy [9]. However, this investigation shows also that two types of precipitates exist in the microstructure of artificially aged sample (Figs. 2-3). The first one is the hexagonal (hp24) β′-precipitate (Fig. 2) responsible for the strengthening effect, while the second one is the β equilibrium precipitate with a spherical shape (Fig. 3). The presence of the equilibrium β-phase can be explained by dissolution of β′-precipitates during artificial ageing or independent appearance of β-phase precipitates, near existing strengthening precipitates.

a) b) c)

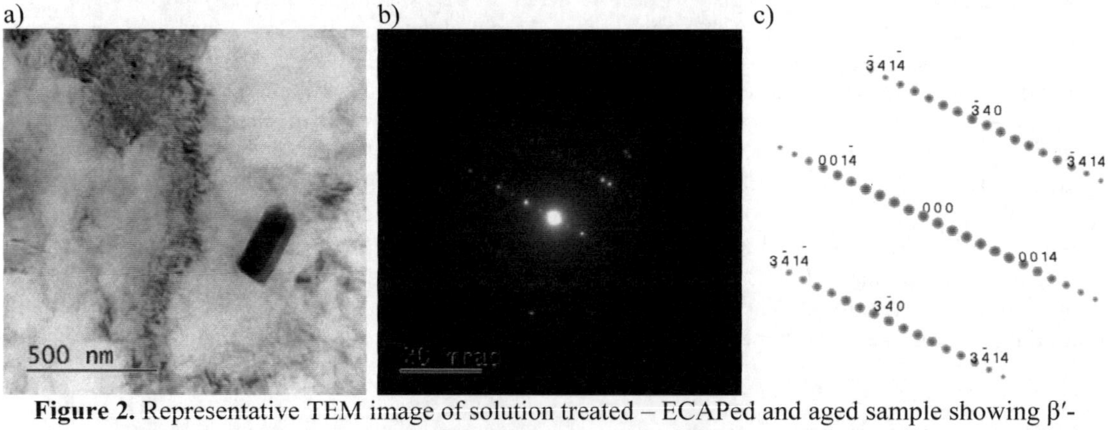

Figure 2. Representative TEM image of solution treated – ECAPed and aged sample showing β′-Al_3Mg_2 hexagonal preciptate, (b) diffraction pattern, (c) solved diffraction pattern

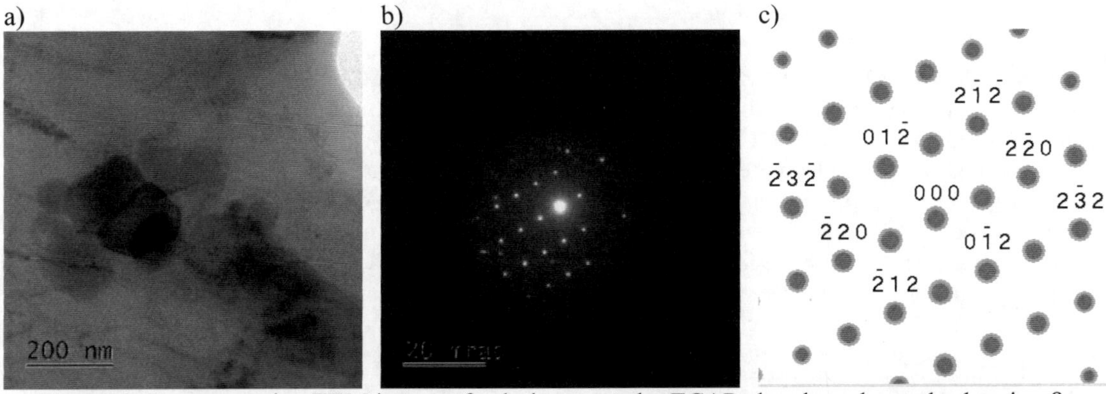

Figure 3. Representative TEM image of solution treated – ECAPed and aged sample showing β-Al_3Mg_2 fcc equilibrium precipitats, (b) diffraction pattern, (c) solved diffraction pattern

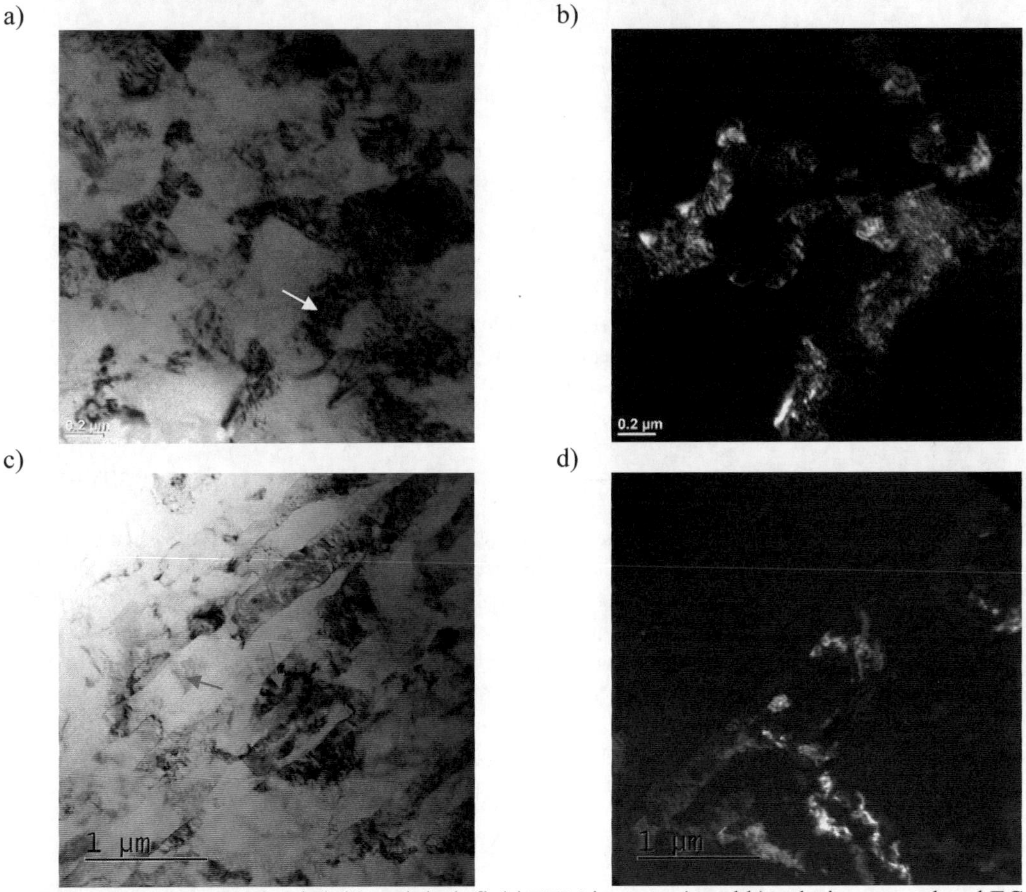

Figure 4. Representative bright and dark field TEM images a) and b) solution treated and ECAPed sample, c) and d) solution treated ECAPed and aged at 160 °C for 2h sample

TEM bright and dark field micrographs of the as-ECAP processed sample are shown in Figs. 4a-b. After four ECAP passes dislocation structure was formed (white arrow). Some equiaxial subgrains are also visible in the substructure. Artificial aging led to visible changes in the microstructure. Parallel bands with well-defined subgrains (red arrows) are also visible. There is also a noticeable dislocation density decrease (as a result of recovery and recrystallization processes).

3.2 Dilatometric measurement

Figure 5a-b shows, respectively, the variation of the ratio $\Delta L/L_0$ with a derivative of heating curve as a function of temperature. Some anomalies composed of a weak change in expansion in temperature ranges 150-190 °C and 330-350 °C are visible in (Fig. 5b), while the shape of the dilatometric curve of solution treated sample looks quite smooth (Fig. 5a). This first anomaly can be a result of the β′-phase precipitation. The precipitation process seems to be accelerated by the pre-deformation using ECAP procedure. The observation of the dilatometric curve of the ECAP processed sample reveals only some anomalies in the temperature range 390-400 °C, which can be a result of the recrystallization processes that taking place in a deformed sample.

a) b)

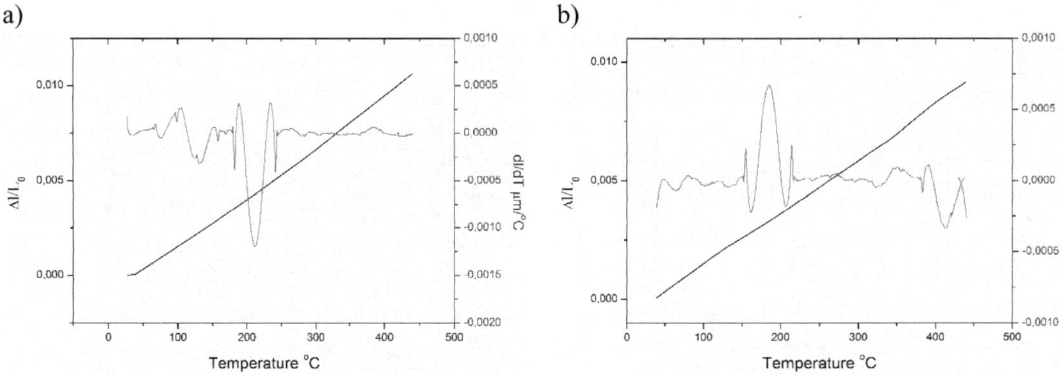

Figure 5. Dilatometric curves of an Al-Mg alloy a) solution treated and aged for 2h at 160 °C, b) solution treated and ECAPed,

3.3 Mechanical properties

Table 2 summarizes the results of the mechanical properties examinations. It is clearly visible that ECAP processing has an influence on the kinetics of precipitation process. The hardness and tensile strength increase slightly in the early stages of the post-ECAP ageing process, while further ageing led to mechanical properties decrease. The observed phenomenon – increase in the mechanical properties is due the precipitation of the hexagonal β′-phase. With an increase in ageing time the amount of equilibrium precipitates increases, thus the mechanical properties decrease. This is in good agreement with the TEM study in the previous paragraph where the presence of two type of precipitates in the sample after 2 hours of artificial ageing was observed.

Table 2. Summarized results of the mechanical properties examination

Aging time, [h]	0	0.5	1	2	4
Hardness, HV	112.3	115	112.6	111.1	111.2
Tensile strength, [MPa]	352	360	354	347	348

4. Conclusions

The whole results presented in this work reflect in particular the effect of the ageing time on the precipitation in Al-3%Mg alloy.

In this article, we found that during an artificial ageing process of the pre-deformed Al-Mg alloy two types of precipitates exists in the microstructure. The first one responsible for precipitation strengthening, that is semicoherent with the Al matrix β′-phase and the second equilibrium β phase precipitates.

The results of the mechanical properties tests confirmed a beneficial influence of the pre-deformation treatment on the precipitation kinetics of AlMg3 alloy. We found that after 30 min of artificial ageing the mechanical properties slightly increases, while further ageing has a negligible influence on the mechanical strength. The thermal expansion coefficients changes versus temperature showed the anomalies, that can be related to the formation of the phases during the artificial ageing process.

References

[1] Bensaada S, Bouziane M T, Mohammedi F 2011 Effect of the temperature on the mechanism of the precipitation in Al-8% mass.Mg alloy, Mater Lett. **65** 2829-2832,

[2] Nebti S, Hamana D and Cizeron G 1995 Calorimetric study of pre-precipitation and precipitation in Al-Mg alloy *Acta metall. Mater.* **43** 3583

[3] Hamana D, Bouchear M and Derafa A 1998 Effect of plastic deformation on the formation and dissolution of transistion phases in Al-12 wt.% Mg alloy *Mater Chem Phys* **57** p 99

[4] Tanski T, Snopinski P and Borek W 2017 Strength and structure of AlMg3 alloy after ECAP and post-ECAP processing *Mater Manuf Processes* **32** 1386

[5] Kelly A, Nicholson RB, Thomas G. 1936 Electron Microscopy and Strength of Crystals: John Wiley & Sons, Inc.,

[6] Pollard W 1965 Journal of the Institute of Metals **93** 339

[7] Starink M J and Zahra A M 1997 Low temperature of Al-Mg alloys *Phil. Mag.* **76** 701

[8] Starink M J and Zahra A M 1996 Precipitation kinetics of an Al-15%Mg alloys *Mater Sci Forum* **217** 795

[9] Tanski T, Snopinski P, Prusik K and Sroka M 2017 The effects of room temperature ECAP and subsequent aging on the structure and properties of the Al-3%Mg aluminium alloy *Mater Charact* **133** 185

[10] Hamana D, Baziz L and Bouchear M 2004 Kinetics and mechanism of formation of transformation of metastable β′-phase in Al-Mg alloys *Mater Chem Phys* **84** 112

[11] Jie J C, Wang H W, Zou C M, Wei Z J and Li T J 2014 Precipitation in Al-Mg solid solution prepared by solidification under high pressure *Mater Charact* **87** 19

[12] Slabanja M and Wahnstrom G 2005 Kinetic Monte Carlo study of Al-Mg precipitation *Acta Mater* **53** 3721

[13] Yukawa H, Murata Y, Morinaga M, Takahashi Y and Yoshida H 1995 Heterogenus distributions of magnesium atoms near the precipitate in Al-Mg based alloys *Acta metall. Mater.* **43** 681

[14] Krol M, Tanski T, Snopinski P and Tomiczek B 2017 Structure and properties of aluminium-magnesium casting alloys after heat treatment *J Therm Anal Calorim* **127** 299

[15] Snopiński P, Tański T and Sroka M 2017 The effect of heat treatment conditions on the structure evolution and mechanical properties of two binary Al-Mg aluminium alloys *Metalurgija* **56** 329

Evaluation of corrosion resistance of titanium/stainless steel diffusion joints bonded by interlayers made of pure metals

B Szwed[1] and M Konieczny[1]

[1]Kielce University of Technology, Kielce, Poland, EU

E-mail: bszwed@tu.kielce.pl

Abstract. The purpose of this study is to compare corrosion resistance of the joints between commercially pure titanium (Grade 2) and stainless steel (X5CrNi18-10) bonded by interlayers made of pure metals. Three different interlayers were used in the study, which were made of aluminium (Al 99.5), copper (Cu 99.99) and nickel (Ni 99.6). The diffusion bonding process has been employed to join those dissimilar materials. The microstructure of the joints were investigated using scanning electron microscopy equipped with an energy dispersive X-ray system (EDS). The phases of the reaction layers in the diffusion zone were identified by means of an X-ray diffraction (XRD).The structures of the joints varied importantly depending on the interlayer. The galvanic corrosion test was carried out in two solutions one in a 3,5% NaCl and second one in 1M Na_2SO_4 solution. The potentiodynamic polarization curves in a 3,5% NaCl solution show that the higher corrosion potential as well as a lower corrosion current were registered for the joints performed by nickel. The test carried out in 1M Na_2SO_4 solution show that that the lowest corrosion current was registered for the joints performed by copper and also all investigated samples have a tendency towards passivation.

1 Introduction

Titanium and its alloys have received great attention in nuclear and chemical industry owing to its high strength to weight ratio and excellent corrosion resistance [1, 2]. Because of it properties, titanium (Ti) is the most suitable material in reactor reprocessing plants for fabricating components like dissolver and evaporator, where high concentration of nitric acid and high temperatures are involved [3,4]. Because other components of reprocessing plants and dissolver are made of 304L stainless steel (SS), there is an obvious requirement to join Ti with SS, as well as economic reasons owing to the high cost of titanium [5]. Traditional fusion welding of dissimilar materials results in different problems like distortion of components, formation of stress concentration sites, development of chemical heterogeneities and a number of intermetallic phases that are formed in the weld pool. In addition, titanium and its alloys are chemically reactive, they are very difficult to weld, because they can easily pick up nitrogen and oxygen from the atmosphere [6, 7]. Hence, solid state diffusion bonding process is recommended for materials with extremely different physical and mechanical properties [8]. The use of appropriate intermediate materials can also inhibit diffusion of undesired elements [9]. Existing literature and previous attempts showed that to successfully joint titanium with stainless steel is necessary to use of appropriate intermediate materials. Aluminum can be considered as a useful interlayer due to the lowering of bonding parameters for solid state diffusion bonding and aluminum has certain erosion resistance and excellent plasticity [10, 11]. Pure nickel and nickel alloys can be used as a filler material between titanium and stainless steel due to satisfactory corrosion resistance for application at high temperature [12, 13]. Among these materials copper is the most useful metal because it does not form any intermetallic phases with iron (as does aluminum and nickel) [14, 15]. Eroglu et al. [16] reported that Cu-Ti base intermetallic phases have higher plasticity than the Fe-Ti base intermetallics. However when dissimilar materials are joined, particular attention should be paid to the potential corrosion issues arising from the galvanic couple. Galvanic corrosion is an electrochemical process resulting in accelerated and preferential corrosion of one of the metals. Dissimilar materials have different electrochemical potentials, this difference provides the thermodynamic driving force for the onset of galvanic corrosion [17-20]. Since investigated joints are composed of different materials and different structures, a clear estimation of the corrosion resistance of those joints are becoming an essential criterion in evaluating the performance of the joints. The present investigation reports the influence of different intermediate materials as a filler metal on the microstructure, tensile strength, impact energy and corrosion resistance of titanium-stainless steel joints bonded by Al, Cu and Ni interlayer.

2 Experiment procedure

The base materials used for a dissimilar joints were commercially pure titanium (Grade 2) and stainless steel (X5CrNi18-10), both received in the form of square rods having 10x10 mm width and 2000 mm length, and filler metal foils of 100 μm thickness. The nominal chemical composition at room temperature of these materials are given in table 1.

Table 1. Chemical compositions and mechanical properties of the base materials

Material	Titanium (Grade 2)	Stainless steel (X5CrNi 18-10)	Aluminum (Al 99.5)	Copper (Cu 99.99)	Nickel (Ni 99.6)
Chemical composition (wt. %)	Ti 99.654; Fe: 0.171; C: 0.024; N: 0.008; O: 0.142; H: 0.001	Fe: 71.495; C: 0.025; Mn: 1.460; Si: 0.390; P: 0.038; S: 0.012; Cr: 18.150; Ni: 8.050; Mo: 0.380	Al: 99.53; Fe: 0.21; Si: 0.16; Zn: 0.05; Cu: 0.03; Ti: 0.02	Cu: 99.99, approximately 0.001 of: Fe; Ni; Zn; Sn; Pb; Sb; As; S;	Ni: 99.57; Cu: 0.11; Co: 0.09; Si: 0.08; Mg: 0.07; Fe: 0.07; Al: 0.01

Specimens of 10 and 27.5 mm length were machined from the titanium and stainless steel rods. The square profile with 10x10 mm width were excised from the Al, Cu and Ni foils. The faces of the specimens were prepared by conventional grinding and polishing techniques and final polishing was made with 0.5 μm alumina suspension and colloidal silica with a grain size of 0.05 μm. All specimens were then cleaned in water and dried rapidly in air. The mating surfaces of the samples were kept in contact with steel clamp and inserted in a vacuum chamber. The bonding pressure of 2 MPa along the longitudinal direction was applied at room temperature. Diffusion bonding was carried out in a vacuum furnace Czylok PRC 77/1150. The bonding temperature for the samples with aluminum interlayer was 600 and 900 ^{0} C for specimens achieved using copper and also with nickel interlayer. The holding time for all samples were 60 minutes. Vacuum in the furnace was at the level of 10^{-3} Pa. The samples were cooled with the furnace. The specimens for metallographic examination were cut out longitudinally and their surfaces were prepared by conventional techniques, using sandpapers of 180 to 1200 grit, alumina suspension with a grain size of 0.5 μm and colloidal silica with a grain size of 0.05 μm. The polished surfaces of the brazed couples were also examined in a scanning electron microscope (SEM) JEOL JMS-5400 and JEOL JSM 7100F to obtain finer structural details in the diffusion zone. The composition of the reaction layers was determined in atomic percent using Oxford Instruments ISIS energy dispersive X-ray spectrometer (EDS) attached to the SEM. The results of the EDS analysis were compared with the binary and ternary phase diagrams of basic components. After the EDS analysis the X-ray diffraction (XRD) analysis conducted with a Seifert 3003T/T x-ray diffractometer, in order to confirm the validity of the assumed phases. The corrosion properties of the specimens were investigated using a CHI 1130A (CH Instruments Inc., Austin, Texas) electrochemical analyzer. The potentiodynamic polarization measurements were conducted at room temperature in a conventional three-electrode cell with the AgCl/Ag electrode as the reference electrode, the platinum wire as the counter electrode and the specimen as the working electrode. The specimens in the form of discs rotated with a speed of 12 revolutions/s. A potentiodynamic polarization measurements were performed in an applied potential range from -1.0 to 0.25 V with a scan rate of 3 mV/s. The tests were carried out in a 3.5 % NaCl and 0.5 M Na_2SO_4 solution acidified to pH 1. Prior to the tests, the specimens were immersed in a solutions for 1000 s to obtain a stable open circuit potential.

3 Results and discussion

The results of the investigations demonstrated significant diffusion changes and relatively wide diffusion zones on the boundaries with joined metals. The structures of the joints varied importantly depending on the interlayer as shows the micrographs and XRD analysis presented on the figure 1.

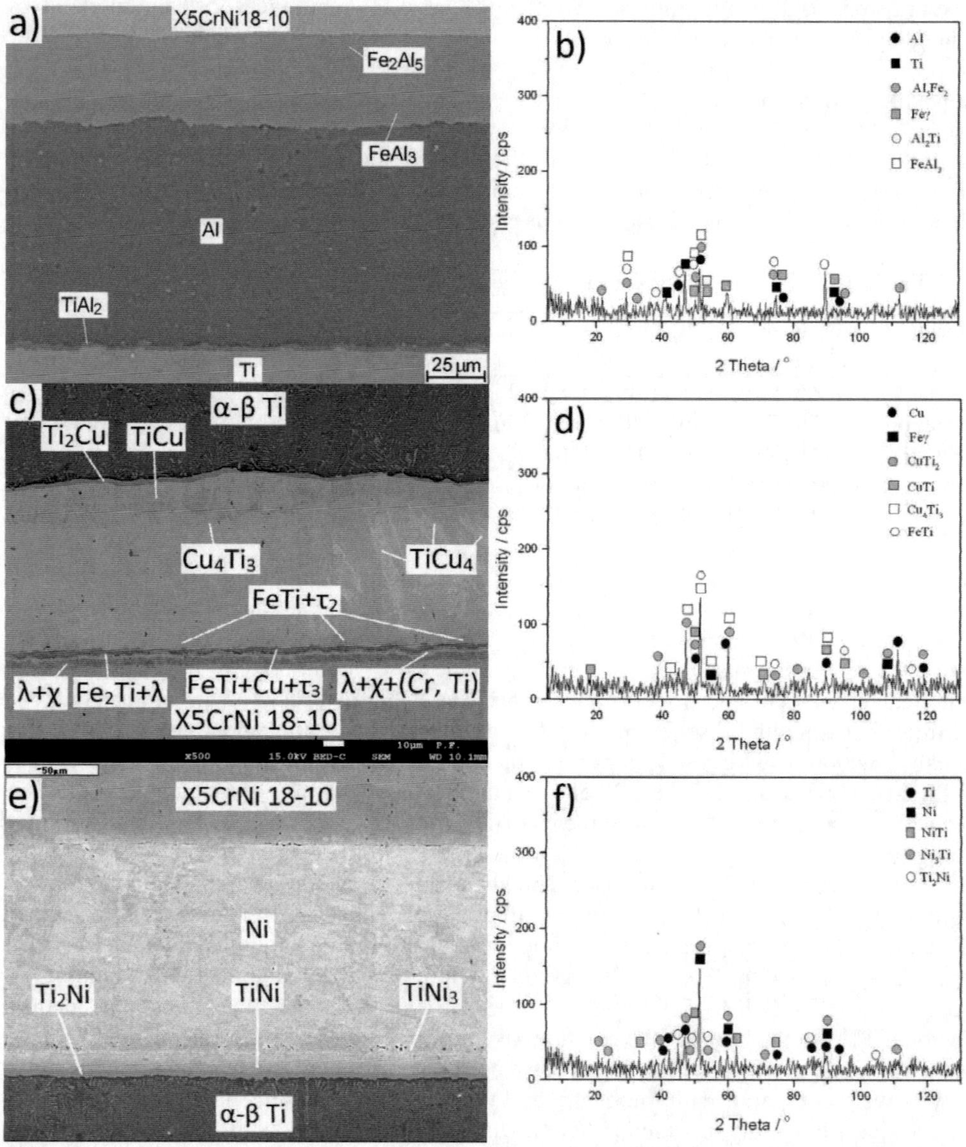

Figure 1. SEM images and XRD analysis of diffusion bonded joint by a), b) Al, c), d) Cu and e), f) Ni.

As the investigatione and the works [10,12,14] shows, that using aluminum as a filler metal results in formation intermetallic layer $TiAl_2$ at the titanium aluminum side of the diffusion joints. At the stainless steel aluminum interface were formed two layers of Fe_2Al_5 and $FeAl_3$ intermetallic phases. The intermetallic layers Ti_2Ni, TiNi, $TiNi_3$ were observed at the titanium side in the diffusion boned joints achieved by nickel filler metal. On the stainless steel nickel side, for the 850 °C brazing temperature, no reaction products were observed in the diffusion zone. The presence of Fe (10 at. %.), Cr (0.4 at. %.) and Mn (0.6 at. %.) in nickel indicates the substantial interdiffusion of those elements. The phases present in copper joints were intermetallics: $CuTi_2$, CuTi, Cu_4Ti_3, FeTi, and solid solutions based on intermetallic phases or substrate metals. All of the above phase layers have been confirmed in XRD analysis, as can be seen in figure 1 b), d) and f). The XRD analysis did not confirm the presence of the remaining phases from the Fe-Ti and Cu-Fe-Ti systems. Such a result could be caused by the fact that these phases occupied small volume fractions in comparison to the other recorded phases. The phases that were not detected during the XRD analysis had the following composition starting from the phase closest to the stainless steel: 64.46-65.88 at. % Fe, 27.86-29.72 at. % Cr, 2.73-2.75 at. % Ni, 2.48-2.72 at. % Ti, 0.61-0.79 at. % Cu and right behind her 56.68-57.30 at. % Fe, 31.13-32.89 at. % Cr, 6.9-8.14 at. % Ti, 2.67-2.85 at. % Ni, 0.58-0.74 at. % Cu. According to the chemical composition, it seems to be the $\lambda+\chi$. the difference in the color of the layers results from the different content of titanium and chromium. The next phase have a composition 54.4-54.71 at. % Fe, 28.79-33,31 at. % Ti, 8.74-12.74 at. % Cr with a small addition of nickel (3.01-3.39 at.%) And copper (0.22-0.65 at.%). Most likely it is a $Fe_2Ti+\lambda$. Another layer has been formed above the $Fe_2Ti+\lambda$ layer with the following composition 48.97-49,15 at. % Ti, 23.12-121.17 at. % Fe, 14.49-23.30 at. % Cu, 3.13-4.10 at. % Ni, 1.27-1.30% at. Cr, which may indicate a mixture of FeTi+Cu+τ3 phases. in the area between stainless steel and the copper interlayer, another layer was formed with the composition 39.58-47,29 at. % Ti, 34.86-46.73 at. % Cu, 10.45-14.68 at. % Fe, which additionally contained 1.88-2.07 at. % Ni and 1.17-1.29 at. % Cr. From the analysis of the Cu-Fe-Ti ternary phase diagram, this composition suggests a mixture of FeTi+τ2 phases. Due to high migration of copper in the temperature range of 850 to 1000 ^{0} C the diffusion of chemical species is easy through interlayer. Therefore titanium can migrate to the stainless steel side and iron can also migrate to the titanium side. Hence, the copper interlayer of 0.1 mm thickness cannot prevent the formation of brittle Fe-Ti base intermetallic phases, what can be achieved by using aluminum and nickel interlayers. Potentiodynamic polarization curves in NaCl solution for the bonded samples prepared by the Al, Cu and Ni fillers metals are shown in fig 2a.

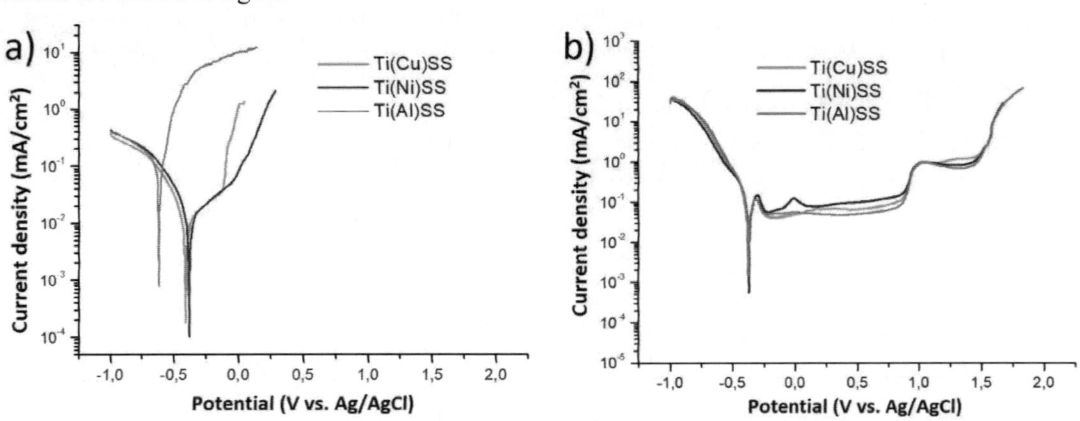

Figure 2. Polarization curves recorded in a) 3.5 % NaCl and b) 1M Na_2SO_4 (pH=1) solution for diffusion bonded joints.

The values of corrosion potential in NaCl solution were determined to be -0.618, -0.409, -0.383 V for the samples boned with the Al, Cu and Ni fillers, respectively in NaCl solution. The corrosion currents were estimated to be as follow: for joints with Al 0.10788 mA/cm^2, with Cu 0.01105 mA/cm^2 and with Ni 0.01057 mA/cm^2. The obtained polarization curves in Na_2SO_4 solution for the tested joints are very similar (Fig. 2b). Their corrosion potential is within the range of -0.371 to -0.377 V. The corrosion currents in Na_2SO_4 solution were estimated to be as follow: for joints with Al 0.14536 mA/cm^2, with Cu 0.13852 mA/cm^2 and with Ni 0.16254 mA/cm^2. The joints boned with copper as well with the nickel have shown very similar behaviour in terms of the corrosion potential, and dissolution current density. The highest corrosion potential as well as the lowest corrosion current was observed for the sample bonded by the copper interlayer in NaCl solution and for the joints examined in Na_2SO_4 solution such result was obtained for sample bonded by the nickel intermediate material, according to the literature studies [17-20] such a result indicates that those joints have the highest corrosion resistance among investigates diffusion boned joints. However, Lee et al. [17, 18] reported that Ag and alloy fillers have higher resistance to corrosion.

4 Conclusions

The use of appropriate intermediate material in diffusion bonded joint between titanium and stainless steel is critical factor to control the microstructure as also mechanical properties of the joint. As the investigation show, the phases present in joints were intermetallics and solid solutions based on intermetallic phases or substrate metals. The aluminum and nickel interlayer of 100 μm thickness effectively blocked the diffusion of titanium to stainless steel side, thus prevented from formation of Fe-Ti intermetallic phases on the boundaries of joined materials. Corrosion resistance tests showed that different results were obtained depending on the corrosive environment in which the samples were tested. However, joints boned with copper have shown very similar behavior to the nickel in terms of the corrosion potential, and dissolution current density. This shows that copper as well as nickel also can be successfully used in corrosive environment.

References

[1] Davis R 1990 ASM Handbook, Volume 02 - Properties and Selection: Nonferrous Alloys and Special-Purpose Materials (Portland: Asm International Handbook Committee) p 950

[2] Ghosh M and Chatterjee S 2003 Diffusion bonded transition joints of titanium to stainless steel with improved properties *Materials Science and Engineering A* **358** pp 152-158

[3] Möller F, Grden M, Thomy C and Vollertsen F 2011 Combined laser beam welding and brazing process for aluminum titanium hybrid structures *Physics Procedia* **12** pp 215-223

[4] Chularis A, Kolpacheva V and Tomashevskii M 1995 Electron structure and properties of intermetallic compounds in titanium metal dissimilar joints *Welding International* **9(10)** pp 812-814

[5] Kato H, Abe S and Tomizawa T 1997 Interfacial structures and mechanical properties of steel–Ni and steel–Ti diffusion bonds *Journal of Materials Science* **32(32)** pp 5225-5232

[6] Lee K, Park J, Lee J, Lee G, Kim W, Lim H, Rhee K, Lee B, Lee J K and Hong J 2011 Corrosion of Ti–STS dissimilar joints brazed by a Ag interlayer and Ag–Cu–(Pd) alloy fillers. *Journal of Nuclear Materials* **409** pp 183-187

[7] Yao W, Wu A, Zou G and Ren J 2008 Formation process of the bonding joint in Ti/Al diffusion bonding *Materials Science and Engineering A* **480** pp 456-463

[8] Dziadoń A, Mola R and Błaż L 2011 Formation of layered Mg/eutectic composite using diffusional processes at the Mg-Al interface *Archives of Metallurgy and Materials*, **56(3)** pp 677-684

[9] Konieczny M 2010 Mechanical properties and deformation behaviour of laminated titanium-intermetallic composites synthesised using Ti and Cu foils *Kovove Materialy-Metallic Materials* **48(1)** pp 47-53

[10] Szwed B and Konieczny M 2016 Influence of bonding pressure on the dissimilar joints of titanium and stainless steel with an aluminium interlayer *In METAL 2016: 25rd International Conference on Metallurgy and Materials. Brno Czech Republic May 25-27:*

TANGER pp 1552-1557

[11] Kundu S and Chatterjee S 2010 Interface microstructure and strength properties of diffusion bonded joints of titanium–Al interlayer–18Cr–8Ni stainless steel *Materials Science and Engineering A* **527(10)** pp 2714-2719

[12] Szwed B and Konieczny M 2016 Microstructure and Mechanical Properties of Joints of Titanium with Stainless Steel Performed Using Nickel Filler *Archives of Metallurgy and Materials* 61(2) pp 997-1001

[13] Kundu S and Chatterjee S 2008 Characterization of diffusion bonded joint between titanium and 304 stainless steel using a Ni interlayer *Materials Characterization* **59** pp 631-637

[14] Konieczny M and Szwed B 2016 Effect of cooling rate on titanium and stainless steel joints performed with copper interlayer *In METAL 2016: 25rd International Conference on Metallurgy and Materials. Brno Czech Republic May 25-27: TANGER* pp 1456-1461

[15] Elrefaey A and Tillmann W 2009 Solid state diffusion bonding of titanium to steel using a copper base alloy as interlayer *Journal of Materials Processing Technology* **209** pp 2746-2752

[16] Eroglu M, Khan I, and Orhan N 2002 Diffusion bonding between Ti-6Al-4V alloy and microduplex stainless steel with copper interlayer *Materials Science and Technology* **18** pp 68-72

[17] Lee K, Park J, Lee G and Rhee K 2013 Phase-dependent corrosion of titanium-to-stainless steel joints brazed by Ag–Cu eutectic alloy filler and Ag interlayer *Journal of Nuclear Materials* **439** pp 168-173

[18] Lee K, Park J, Lee G, Lee J, Kim W, Lim H, Rhee K and Hong J 2011 Corrosion of Ti–STS dissimilar joints brazed by a Ag interlayer and Ag–Cu–(Pd) alloy fillers *Journal of Nuclear Materials* **409** pp 183-187

[19] Elrefaey A; Wojarski L and Tillmann W 2012 Evaluation of Corrosion Performance of Titanium/Steel Joint Brazed by Cu-Based Filler Metal *Journal of Materials Engineering & Performance* **21** pp 707-713

[20] Nizamettin K, Behcet G and Fehim F 2007 Corrosion and mechanical-microstructural aspects of dissimilar joints of Ti–6Al–4V and Al plates *International Journal of Impact Engineering* **34** pp 1423-1432

Structural changes during the formation of diffusion bonded joints between titanium and stainless steel

B Szwed[1] and M Konieczny[1]

[1]Kielce University of Technology, Kielce, Poland, EU

E-mail: bszwed@tu.kielce.pl

Abstract. In the present investigation, transient liquid-phase bonding was performed between pure titanium (Grade 2) and stainless steel (X5CrNi 18-10) using nickel foil as an intermediate material. The process was carried out at the temperatures of: 950, 975 and 1000 ^{0} C for 60 minutes under the compressive stress of 2 MPa in a vacuum. The microstructure was investigated using light microscopy and scanning electron microscopy equipped EDS. The metallographic examination shows that the diffusion interfaces are free from cracks. However, Kirkendall voids were observed at the borders of materials. Especially large voids were formed in middle section of the joint prepared using the highest bonding temperature. The structure of the joints on the titanium side was composed of the eutectoid mixture $\alpha Ti+Ti_2Ni$ and layers of intermetallic phases Ti_2Ni, TiNi and $TiNi_3$. The nickel interlayer blocked the diffusion of titanium to stainless steel side at 950 °C bonding temperature and prevented from formation of the Fe-Ti intermetallic layers. The layer of a solid solution $\gamma Fe+Ni$ between nickel and stainless steel was observed at the 950 °C processing temperature. The mixture of $\lambda+FeTi+Ti$, FeTi+Ti, $\lambda+\alpha Fe$ and $\lambda+\chi+\alpha Fe$ phases was formed at 975 and 1000 °C at the nickel-stainless steel interlayer.

1 Introduction

In the recent years, there is a strong interest to produce joints between different dissimilar materials to satisfy require of modern industries. Titanium (Ti) and stainless steel (SS) have satisfactory specific strength, high melting point and excellent corrosion resistance. Because of that aerospace, chemical and nuclear industries strongly demand joints between titanium and stainless steel [1-4]. Many titanium joining methods, including the commonly employed joining processes, welding, brazing and soldering, all face the reactive nature of titanium. Joining by the conventional techniques also results in segregation of chemical species, stress concentration sites and formation of intermetallics at interfaces due to very low solubility of Ti and Fe [5-8]. Hence, it is of major importance to provide appropriate joining techniques in order to exploit the full potential of these hybrid structure. Diffusion brazing process also known as transient liquid phase bonding (TLP) is recommended for materials with extremely different physical and mechanical properties [9, 10]. TLP bonding is a process that joins materials using an interlayer. On heating, the interlayer melts and the interlayer element (or a constituent of an alloy interlayer) diffuses into the substrate materials, causing isothermal solidification. With this method a permanent connection is obtained below the melting point of base materials, furthermore the bond has a higher melting point than the bonding temperature [11, 12]. The most commonly used intermediate materials in this type of joining technique are: aluminum, copper, silver and their alloys [13-18]. Previous attempts [19] and literature data [20, 21] report that pure nickel and nickel alloys also may be considered as a useful filler material between titanium and

stainless steel due to the satisfactory corrosion resistance at high temperature applications comparing to copper that also is often used as an intermediate material.
The main aim of the study was to produce a permanent joints between pure titanium and stainless steel involving liquid phase form nickel interlayer. The present investigation reports the structural changes during the formation of the joints between those materials.

2 Experiment procedure

The base materials used in this work were pure titanium (Grade 2) and stainless steel (X5CrNi18-10), both received in the form of cylindrical rods having 8 mm diameter and 2000 mm length, and nickel foil of 100 μm thickness. Cylindrical specimens of 8 mm diameter and 10 mm length were machined from the titanium and stainless steel rods. The circular profiles discs with 8 mm diameter were excised from the nickel foil. The nominal chemical composition of these materials are given in table 1.

Table 1. Chemical compositions and mechanical properties of the base materials

Material	Titanium (Grade 2)	Stainless steel (X5CrNi 18-10)	Aluminum (Al 99.5)	Copper (Cu 99.99)	Nickel (Ni 99.6)
Chemical compositi on (wt. %)	Ti 99.654; Fe: 0.171; C: 0.024; N: 0.008; O: 0.142; H: 0.001	Fe: 71.495; C: 0.025; Mn: 1.460; Si: 0.390; P: 0.038; S: 0.012; Cr: 18.150; Ni: 8.050; Mo: 0.380	Al: 99.53; Fe: 0.21; Si: 0.16; Zn: 0.05; Cu: 0.03; Ti: 0.02	Cu: 99.99, approximately 0.001 of: Fe; Ni; Zn; Sn; Pb; Sb; As; S;	Ni: 99.57; Cu: 0.11; Co: 0.09; Si: 0.08; Mg: 0.07; Fe: 0.07; Al: 0.01

The faces of the cylinders were prepared by conventional grinding and polishing techniques and final polishing was made with 0.5 μm alumina suspension. To remove oxide layers from the base material, the samples were etched in acid solutions: titanium in an aqueous 5% solution of HF, stainless steel in an aqueous 10% solution of HCl, nickel in an aqueous 10% solution of HNO_3. The mating surfaces of the samples were kept in contact with steel clamp and inserted in a vacuum chamber. The external compressive stress of 2 MPa along the longitudinal direction was applied at room temperature. Joining process was carried out in a vacuum furnace Czylok PRC 77/1150. The lowest temperature at which the liquid phase appears was 942 ^{0} C according to the binary phase diagrams of base materials. In respect to this, the bonding temperature was established in the range from 950 to 1000 ^{0} C. The bonding process was divided into two stages. At first the temperature was initially raised to 900 ^{0} C and then the pressure was released to avoid an expulsion of liquid phases. In the second stage the temperature was increased to 950 ^{0} C, which was necessary for the initiation and rapid development of the structural processes at the interfaces of iron, nickel and titanium (and to appear liquid phase). The holding time for all samples was 60 minutes. Vacuum in the furnace was at the level of 10^{-3} Pa. The samples were furnace-cooled in vacuum till room temperature. The specimens for metallographic examination were cut out longitudinally and their surfaces were prepared by conventional techniques, using sandpapers of 180 to 1200 grit and alumina suspension with a grain size of 0.5 μm. The titanium side was etched in an aqueous solution of 95 ml H_2O and 5 ml HF. The stainless steel substrate was etched with a mixture of 90 ml C_2H_5OH, 10 ml HCl and 3 g $FeCl_3$. A solution consisting of 50 ml $C_2H_4O_2$ and 50 ml HNO_3 was used for etching the nickel interlayer. The samples were observed in a light microscope Nikon Eclipse MA200 to reveal the structural changes due to diffusion. The polished surfaces of the brazed couples were also examined in a scanning electron microscope (SEM) JEOL JMS-5400 to obtain finer structural details in the diffusion zone. The composition of the reaction layers was determined in atomic percent using Oxford Instruments ISIS energy dispersive X-ray spectrometer (EDS) attached to the SEM. The results of the EDS analysis were compared with the binary and ternary phase diagrams of basic components.

3 Results and discussion

To investigate structural changes during the formation of diffusion bonded joints between titanium and stainless steel with nickel interlayer observations were performed with an optical microscope. The microstructures of the TLP bonded joints are shown in fig. 1.

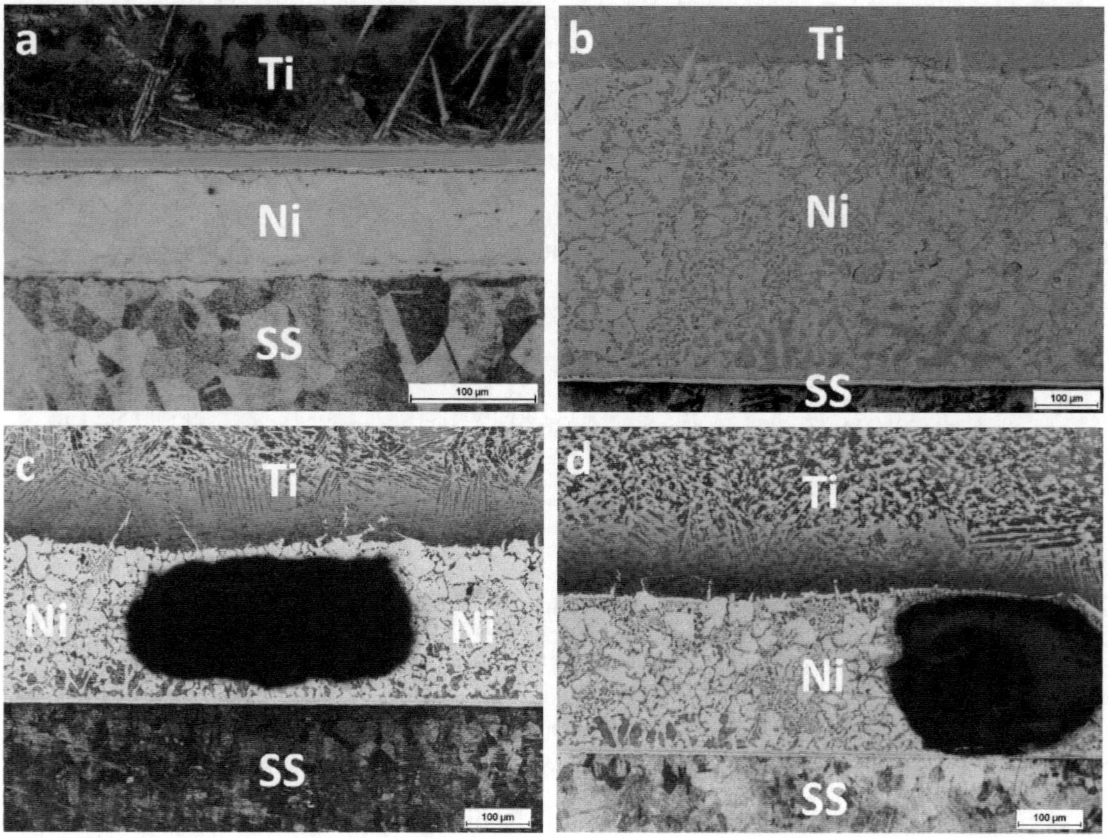

Figure 1. Light micrographs of the joints prepared at a) 950, b) 975, c) and d) 1000 °C for 60 minutes.

The results of the metallographic examination show that the diffusion interfaces are free from cracks and interface lines are clearly visible. However, Kirkendall voids were observed at the borders of materials. Especially large voids were formed in middle section and on the edges of the joint prepared using the highest bonding temperature (Fig. 1c, d). As can be seen on the figure 1a the liquid phase do not appears in samples prepared using 950 °C bonding temperature. It seems that the caused lies in the holding time. According to the results presented in the papers [11,12] the time is too short for liquid phase to emerge and should be extended, because it is just above of occurrence of the liquid phase in the Ti-Ni binary phase diagram. The titanium side of the joints is characterized by the α-β Ti structure. Because nickel is a β stabilizing element it lowers the eutectoid transformation temperature of Ti [1]. The occurrence of acicular α-β Ti results from the decomposition of β Ti during cooling with the vacuum furnace. Three distinct reaction layers were observed at the Ti-Ni interface. Those reaction products on the Ti-Ni side were present until the liquid phase appearance. The reactions with liquid transformed intermetallics to mushy stage, leading to this type of microstructure presented on the figure 1b. Samples prepared at 975 °C bonding temperature have identical microstructure as the joints made at 1000 °C. However, there were no large voids observed for bonds achieved at 975 °C during the microstructure investigation in the interlayer. The diffusion zone at the stainless steel-nickel interface is thinner than that at Ti-Ni side. The presence of distinct layer was revealed at the bonding temperature 950 °C. With the increase in the bonding temperature and appearance of liquid phase the layer on the border between stainless steel and nickel become smaller and more flat. The SEM investigation of the joints was performed in order to reveal more details in the reaction layers of the joint. Scanning electron microscopic images of the diffusion brazed joints are shown in fig. 2.

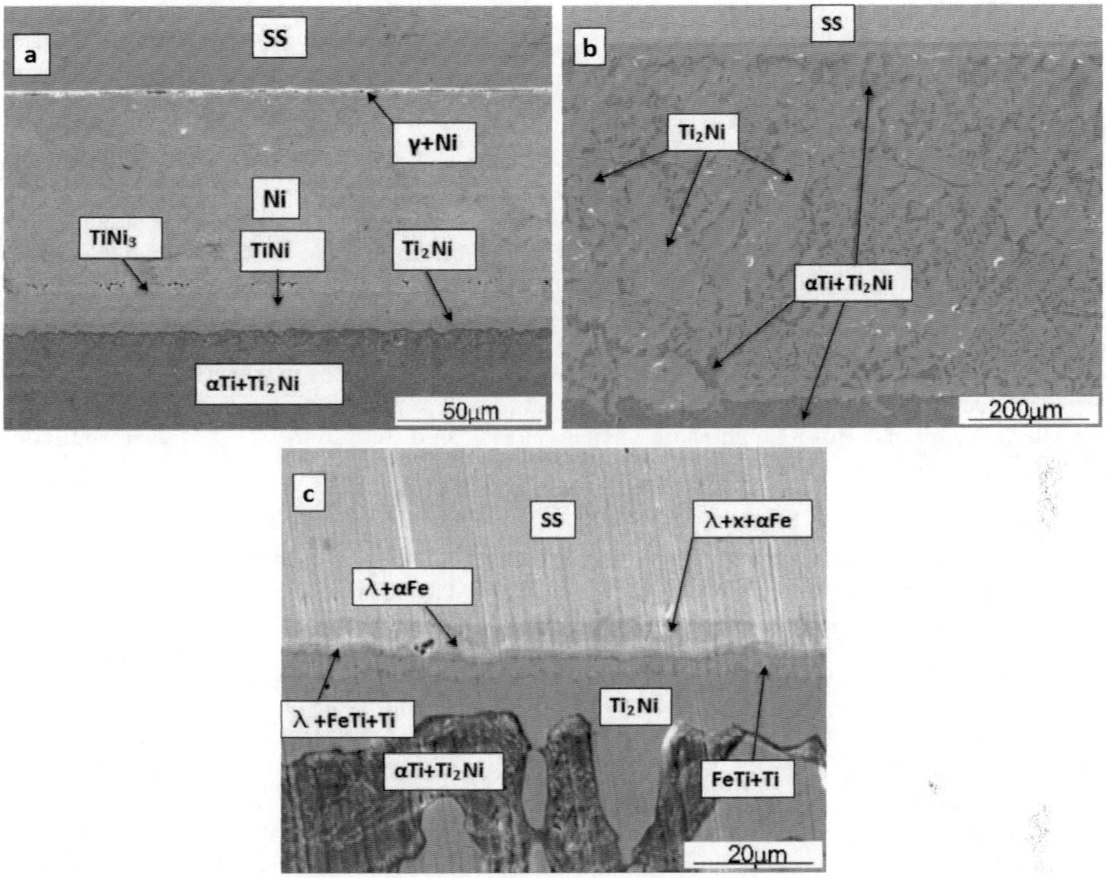

Figure 2. SEM images of the brazed joints processed at a) 950, b) and c) 975 °C for 60 minutes.

At the titanium-nickel interface three distinct reaction layers were observed for the sample achieved at 950°C processing temperature. The first reaction layer adjacent to the titanium side consisted of Ti (69.4-70.9 at. %) and Ni (bal.). According to the Ti-Ni binary phase diagram it is likely a Ti_2Ni intermetallic compound. The brightest layer on the nickel side consisted of Ni (73.1-75.9 at. %) and Ti (bal.). This composition corresponds to the $TiNi_3$ intermetallic phase. Between the two intermetallic phases, another reaction layer, consisted of Ti (50.5-51.3 at. %) and Ni (bal.), was revealed. This is likely the TiNi phase. For the joints prepared at 975 and 1000 °C brazing temperature in which the liquid phase occurred, a significant change was observed. The well-defined islands (shown in Fig. 2b) were composed of Ti (93.4 at. %) and Ni (5.6 at. %.) with small additions of Fe (1 at. %.). This microstructure is likely a mixture of $\alpha Ti+Ti_2Ni$. The bright area corresponds to the Ti_2Ni intermetallic phase with small amount of Fe (1.2 at. %.). The presence of a solid solution γFe+Ni between nickel and stainless steel was observed at the 950 °C processing temperature, until the appearance of the liquid phase. At the stainless steel interface four distinct reaction layers were observed for the 975 and 1000 °C bonding temperature (Fig. 2c). According to the literature the first layer adjacent to the steel was identified as a mixture of $\lambda+\chi+\alpha Fe$ phases [20, 21]. As reported Kundu and Chatterjee [21] the nickel concentration in this area is too low to form any nickel base intermetallic compound. The brightest layer has been recognized as λ+αFe phase combination. The fourth was identified as a combination of FeTi+Ti.

4 Conclusions

Diffusion bonding temperature is a critical factor to control the microstructure of the dissimilar joints, however, the holding time and the bonding pressure are equally important in the TLP bonding process. As the results show that in order to obtain a liquid phase at the bonding temperature 950 °C, the holding time should be extended to more than 60 minutes. Large voids observed in sample bonded at the highest temperature (1000 °C), probably occurred due to the fact that the cooling process was performed without external compressive stress. As the investigation also shown joining such a different materials from each other like titanium and stainless steel it is very difficult to avoid formations of intermetallic phases in the joint. The intermetallic layers Ti_2Ni, TiNi, $TiNi_3$ were identified at the titanium nickel side of the diffusion joint, however the TiNi, and $TiNi_3$ phases were not found at 1000 °C. There were only islands of αTi+Ti_2Ni in the Ti_2Ni matrix. At 950 °C bonding temperature a thin layer of a solid solution γFe+Ni appears between nickel and stainless steel. The mixture of λ+FeTi+Ti, FeTi+Ti, λ+αFe and λ+χ+αFe phases was formed at 1000 °C. This shows that the nickel interlayer of 100 μm thickness blocked the diffusion of titanium to stainless steel side at 950 °C. After the appearance of a liquid phase titanium atoms can diffuse to the stainless steel and form the FeTi intermetallic phase.

References

[1] Wilden J, Jahn S and Beck W 2008 Some examples of current diffusion bonding applications *Materialwissenschaft und Werkstofftechnik* **39(4-5)** pp 349-352

[2] Anas M, Atieh M, Tahir I and Khan A 2014 Application of Ni and Cu nanoparticles in transient liquid phase (TLP) bonding of Ti-6Al-4V and Mg-AZ31 alloys, *Journal of Materials Science* **49(22)** pp 7648-7658

[3] Davis R 1990 ASM Handbook, Volume 02 - Properties and Selection: Nonferrous Alloys and Special-Purpose Materials (Portland: Asm International Handbook Committee) p 950

[4] Lee K, Park J, Lee G, Lee J, Kim W, Lim H, Rhee K and Hong J 2011 Corrosion of Ti–STS dissimilar joints brazed by a Ag interlayer and Ag–Cu–(Pd) alloy fillers *Journal of Nuclear Materials* **409** pp 183-187

[5] Kamachi U, Rao M, Shanmugam K, Natarajan R and Baldev R 2003 Corrosion and microstructural aspects of dissimilar joints of titanium and type 304L stainless steel *Journal of Nuclear Materials* **321** pp 40-48

[6] Möller F, Grden M, Thomy C and Vollertsen F 2011 Combined laser beam welding and brazing process for aluminum titanium hybrid structures *Physics Procedia* **12** pp 215-223

[7] Fox G, Hahnlen R and Dapino M J 2013 Fusion welding of nickel–titanium and 304 stainless steel tubes: Part II: tungsten inert gas welding *Journal of Intelligent Material Systems and Structures* **24(8)** pp 962-972

[8] Lee S, Takahashi M, Kawahito Y and Katayama S 2015 Microstructural evolution and characteristics of weld fusion zone in high speed dissimilar welding of Ti and Al *International Journal of Precision Engineering and Manufacturing* **16(10)** pp 2121-2127

[9] Dziadoń A, Mola R and Błaż L 2011 Formation of layered Mg/eutectic composite using diffusional processes at the Mg-Al interface *Archives of Metallurgy and Materials*, **56(3)** pp 677-684

[10] Konieczny M 2010 Mechanical properties and deformation behaviour of laminated titanium-intermetallic composites synthesised using Ti and Cu foils *Kovove Materialy-Metallic Materials* **48(1)** pp 47-53

[11] Cook G O and Sorensen C D 2011 Overview of transient liquid phase and partial transient liquid phase bonding *Journal of Materials Science* **46** pp 5305-5323

[12] Norozi E, Atapour M and Shamanian M 2017 Effect of bonding time on the joint properties of

transient liquid phase bonding between Ti-6Al-4V and AISI 304 *Journal of Alloys and Compounds* **701** pp 335-341

[13] Szwed B and Konieczny M 2016 Influence of bonding pressure on the dissimilar joints of titanium and stainless steel with an aluminium interlayer *In METAL 2016: 25rd International Conference on Metallurgy and Materials. Brno Czech Republic May 25-27: TANGER* pp 1552-1557

[14] He P, Yue X and Zhang J H 2008 Hot pressing diffusion bonding of a titanium alloy to a stainless steel with an aluminum alloy interlayer *Materials Science and Engineering: A* **486(1-2)** pp 171-176

[15] Konieczny M and Szwed B 2016 Effect of cooling rate on titanium and stainless steel joints performed with copper interlayer *In METAL 2016: 25rd International Conference on Metallurgy and Materials. Brno Czech Republic May 25-27: TANGER* pp 1456-1461.

[16] Elrefaey A and Tillmann W 2009 Solid state diffusion bonding of titanium to steel using a copper base alloy as interlayer *Journal of Materials Processing Technology* **209** pp 2746-2752

[17] Atasoya E and Kahramanb N 2008 Diffusion bonding of commercially pure titanium to low carbon steel using a silver interlayer *Materials Characterization* **59** pp1481-1490

[18] Winiowski A 2013 Mechanical properties of joints of stainless steel and titanium brazed with silver filler metals *Kovove materialy-Metallic Materials* **51** pp 19-24

[19] Szwed B and Konieczny M 2016 Microstructure and Mechanical Properties of Joints of Titanium with Stainless Steel Performed Using Nickel Filler *Archives of Metallurgy and Materials* **61(2)** pp 997-1001

[20] Kundu S and Chatterjee S 2007 Structure and properties of diffusion bonded transition joints between commercially pure titanium and type 304 stainless steel using a nickel interlayer. *Journal of Materials Science* **42** pp 7906-7912

[21] Kundu S and Chatterjee S 2008 Characterization of diffusion bonded joint between titanium and 304 stainless steel using a Ni interlayer *Materials Characterization* **59** pp 631-637

IOP Conf. Series: Materials Science and Engineering **461** (2019) 012083 doi:10.1088/1757-899X/461/1/012083

Study of fatigue fractures of Al-Mn alloy

J Škoda[1], M Losertová[1], V Kubeš[1], K Konečná[1], P Hlinka[1] and P Czyz[2]

[1]VŠB - Technical university of Ostrava, Faculty of Metallurgy and Materials Engineering, Ostrava - Poruba, Czech Republic

[2] Hanon Systems Autopal s.r.o., Lužická 984/14, 741 01 Nový Jičín, Czech Republic

E-mail: jan.skoda@vsb.cz

Abstract. Aluminum alloys have become a standard application in automotive industry due to their strength to weight ratio and corrosion properties. Aluminum alloys based on Al-Mn (class 3XXX) that are applied as components in air conditioning modules in automotive as well as in construction and consumer industries are often submitted to vibrations which can eventually lead to fatigue cracks and in the end to complete failure of the components. The study of fatigue failures of cyclic loaded Al-Mn alloy was performed using fractographic analysis on scanning electron microscope JEOL JSM-649OLV. The loading was realized from 100,000 to 1,369,000 cycles at room temperature. The features of fatigue fracture surfaces were discussed comparing different tempers of the A3003 alloy, specifically in the temper 1/4 hard (H12) and full soft (O). The results for corresponding loading cycles were compared. Areas close to the crack initiation were characterized by radial marks corresponding to fatigue propagation. Fracture was propagated through the growing lines in form of fine striations with secondary cracks and microcracks between them. The striations were arranged to bands and corresponded to crack propagation at each loading cycle. For final fracture the typical quasi-cleavage morphology was observed.

1 Introduction

Aluminum alloys become a standard application in automotive industry due to their strength to weight ratio and corrosion properties. Beside their strength and corrosion characteristics, fatigue properties of the aluminum alloys are also important features for automotive use. Unlike steel, the fatigue behavior of aluminum alloys isn't still sufficiently explored and in some cases data on fatigue strength are missing completely.

Aluminum alloys based on Al-Mn (class 3XXX) that are applied as components in air conditioning modules in automotive as well as in construction and consumer industries, are often submitted to vibrations which can eventually lead to fatigue cracks and in the end to complete failure of the components. Because of this there is need for further investigation to complete the data about fatigue behavior of aluminum alloys, especially for Al-Mn based alloys as their applications include not only automotive, but also construction and consumer products.

Components in air conditioning and cooling modules are loaded with vibrations which can eventually cause fatigue cracks and in the end complete failure of the components. Failures of the component due to the vibrations can be then distinguished as high cycle fatigued that resulted from engine run vibrations and low cycle fatigued as consequence of the suspension vibrations due to vehicle ride.

The aim of the fractographic study was to evaluate the fracture surfaces after cyclic loading of the Al-Mn alloy (AA3003) in two tempers, namely H12 and O.

2 Experimental

Two sets of experimental specimens (table 1) in the tube shaped form of 19 mm in outer diameter and of 2 mm in wall thickness were submitted to fatigue tests. The first set was in H12 temper (work hardened to quarter-hard, not annealed after rolling) and the second one in O temper (fully annealed, soft) [1]. Microstructure and phase analysis was performed on selected specimens using Olympus GX51 optical microscope with DP1 digital camera and EDX microanalysis by the JEOL JSM-649OLV scanning electron microscope (SEM) equipped with the OXFORD INSTRUMENTS INCA x-act tool, respectively. Fatigue tests were realized on a vibrational electromagnetic device at frequencies of approximately 80 Hz and loading mode was controlled by acceleration adjustment. Fractographic study of failure surfaces after cyclic loading was carried out using SEM JEOL JSM-649OLV.

The specimens for the metallographic observation were cold embedded in non-conductive acrylic resin SpeciFix, manually grinded on abrasive SiC papers and polished on DP Nap cloth with Struers DiaDuo 3 μm diamond paste using Compact 1031 device. The phases were observed in etched condition after immersion in Keller's reagent for 3 seconds. The grain shape and size, grain boundaries and other microstructural characteristics were showed after electrolytic color etching in a solution composed of 5 ml 48% HBF_4 and 200 ml H_2O using Struers Lectropol – 5.

Table 1. Number of loading cycles for selected specimens.

Temper	Specimen	Number of fatigue cycles
H12	1	106,890
	2	487,080
	3	1,369,000
O	1	100,000
	2	302,000
	3	950,000

3 Results and discussion

The microstructure of the polished specimens of both sets was formed of an aluminum solid solution and primary phase particles that were defined as a ternary phase of Al-(Fe,Mn)-Si type [2] using EDX microanalysis. The intermetallic Al-(Fe,Mn)-Si particles can influence the recrystallization and precipitation behavior as well as the fracture toughness of the studied alloys and they are difficult to eliminate by homogenization at temperatures below 650 °C because of the risk of melting [3 – 6].

The color etched microstructures in the transverse cross section of H12 and O specimens are presented in figure 1. The work hardened temper (H12) showed equiaxed fine grained structure with recrystallized grains with average size of 20 μm (figure 1 a). The annealed microstructure (O temper) is characterized by coarse grains with average size of 200 μm (figure 1 b).

The comparison of fatigue behavior was performed with respect to both different tempers of the material (H12 and O, respectively) and the number of fatigue cycles. Fatigue crack initiation of both specimens set started at the locality with highest stress concentration, therefore at outer surface of the tubes. From here the crack propagated across the tube wall until the final fracture was accomplished. Macroscopically, the crack propagation is proved by radial steps with quasi-cleavage facets (figure 2a, 3a, 4a, 5a, 6a and 7a). Microscopically, the fatigue failure propagated through the growing lines in the form of fine striations between of which the secondary cracks and microcracks were observed (figure 2b, 3b, 4b, 5b, 6b and 7b). The morphology of final crack was characterized by mixed quasi-cleavage fracture, with ductile dimples inside that primary phase particles were observed. The occurrence of dimples is typical for ductile matrix.

Figure 1. Microstructure of the cross sections of Al-Mn tubes: a) fine grain size structure for H12 temper; b) coarse grain size structure for O temper.

For the H12 temper, as the number of the fatigue cycles increased the tendency to large steps development was remarked in initiation and propagation areas that could be related to fine grained microstructure and the strengthened temper. Unlike H12 fracture surfaces the fatigue areas of O specimens showed more rough relief that correspond to coarse grain size structure of soft temper. The occurrence of ductile striation confirmed the plastic deformation behaviour during fatigue loading. The secondary cracks or microcracks were generated after the magistral crack passing.
Although the striation represent certain response to each loading cycle and some differences in striation distances were remarked, it was not possible to determinate this characteristic as a proof related with the cycle number during fatigue loading. More detailed analysis of dislocation structures by means of transmission electron microscopy [7] would give more information to generalize the effect of number of loading cycles on the fracture feature.

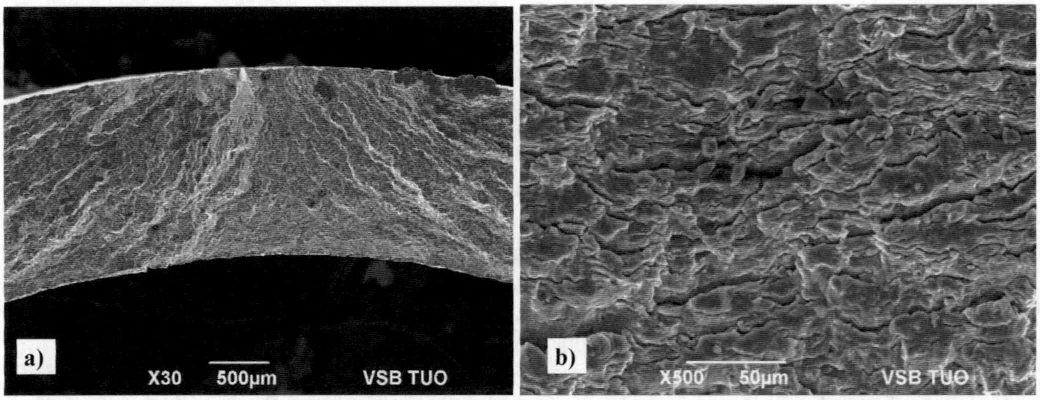

Figure 2. SEM fractography of specimen 1 fatigue tested in H12 temper, 106,890 cycles: a) view of crack initiation and b) detail of initiation area with steps, secondary microcracks and striations.

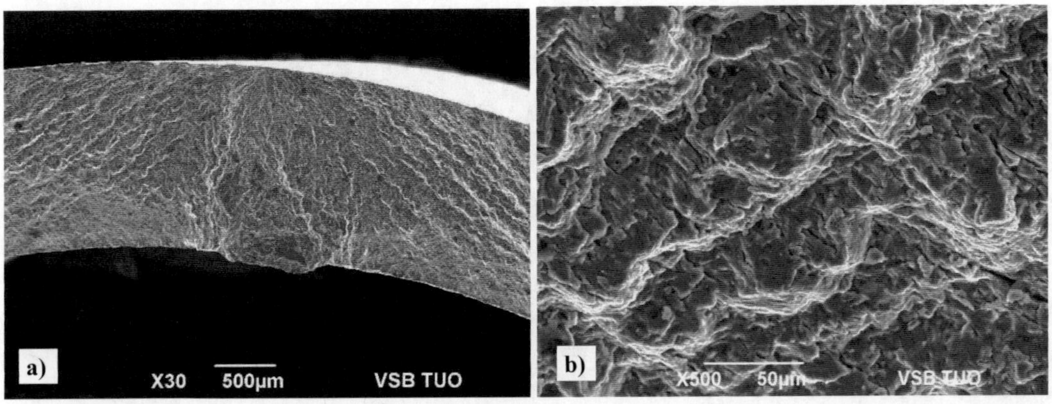

Figure 3. SEM fractography of specimen 2 fatigue tested in H12 temper, 487,080 cycles: a) view of crack initiation and b) detail of initiation area with steps, secondary microcracks and striations.

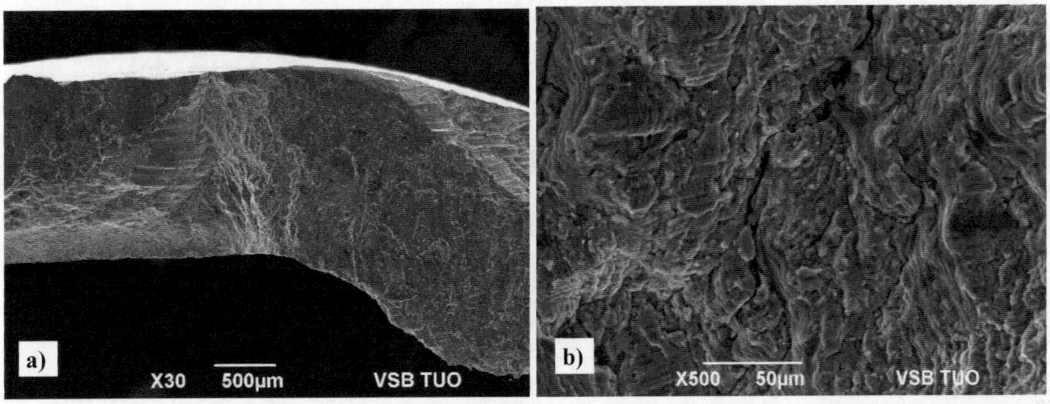

Figure 4. SEM fractography of specimen 3 fatigue tested in H12 temper, 1,369,000 cycles: a) view of crack initiation and b) detail of initiation area with steps, secondary microcracks and striations.

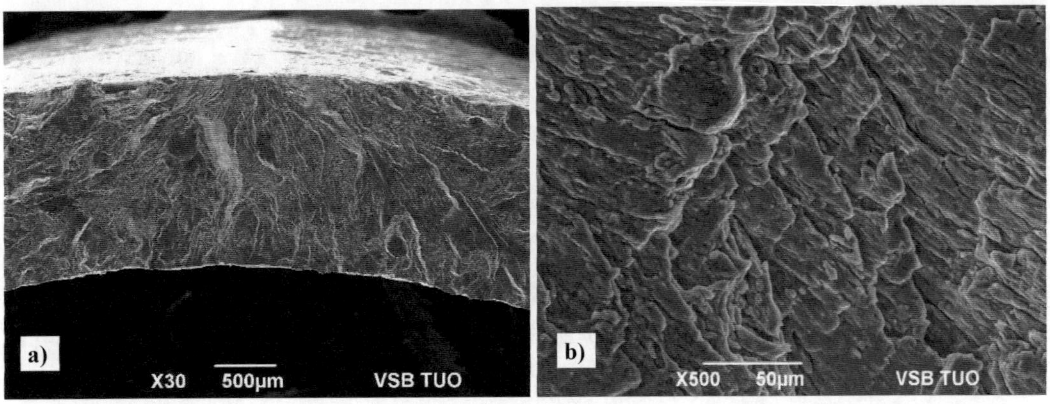

Figure 5. SEM fractography of specimen 1 fatigue tested in O temper, 100,000 cycles: a) view of crack initiation and b) detail of initiation area with steps, secondary microcracks and striations.

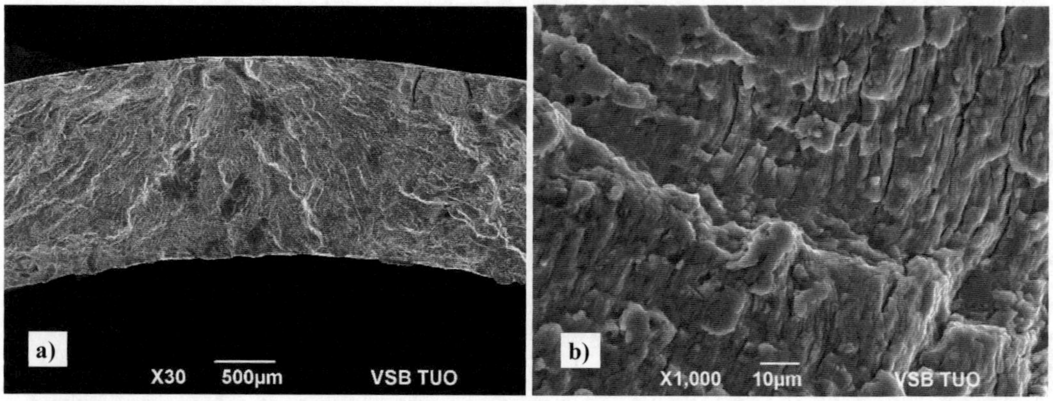

Figure 6. SEM fractography of specimen 2 fatigue tested in O temper, 302,000 cycles: a) view of crack initiation and b) detail of initiation area with steps, secondary microcracks and striations.

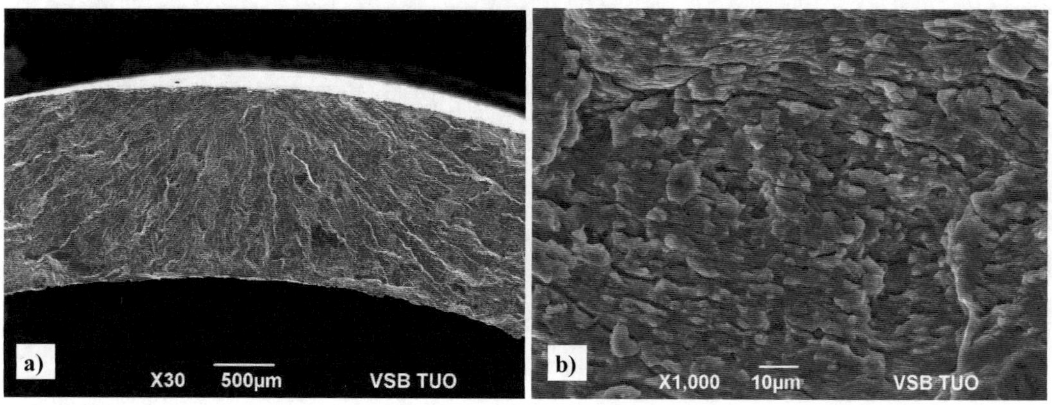

Figure 7. SEM fractography of specimen 3 fatigue tested in O temper, 950,000 cycles: a) view of crack initiation and b) detail of initiation area with steps, secondary microcracks and striations.

4 Conclusion

The fracture surfaces taken from the tubes of the Al-Mn based alloy were subjected to the fractographic analysis and investigated to find a correlation between the morphology of the fracture surface and the number of loading cycles completed. The macroscopic morphology of both temper states corresponded to grain size of microstructure.

The issue of studying the life and fatigue behavior of Al-Mn alloys is complex and cannot be generalized only on the base of literature data. To create more general theory and find some connections between the loading cycles and failure morphology of Al-Mn materials, it is necessary to receive and discuss various sets of results after low-cycle as well as high-cycle fatigue tests in different temper conditions. For better understanding of the fatigue propagation and striation during cyclic loading, it is necessary to complete the fracture analysis with the study of the dislocation structure using transmission electron microscopy.

Acknowledgements
This article has been elaborated in the framework of the project No. LO1203 "Regional Materials Science and Technology Centre - Feasibility Program" funded by Ministry of Education, Youth and Sports of the Czech Republic and co-financed by the European Social Fund and of the projects SGS SP2018/100 supported by Ministry of Education, Youth and Sports of the Czech Republic.

References

[1] Aluminium Alloy Data Sheet 3003. Atlassteels 2015 1 2 Online at: www.atlassteels.com.au

[2] Li Y J, Muggerud A M F, Olsen A and Furu T 2012 Precipitation of partially coherent α-Al(Mn,Fe)Si dispersoids and their strengthening effect in AA 3003 alloy *Acta Mater.* **60** 3 1004–1014

[3] Dehmas M, Aeby-Gautier E, Archambault P and Serriere M 2013 Interaction between eutectic intermetallic particles and dispersoids in the 3003 aluminum alloy *Metall. Mater. Trans.* A **44** 2 1059–1073

[4] Dehmas M, Weisbecker P, Geandier P, Aarchambault P and Aeby-Gautier E 2005 Experimental study of phase transformations in 3003 aluminium alloys during heating by in situ high energy X-ray synchrotron radiation *J. Alloys Compds.* **400** 116

[5] Li Y J and Arnberg L 2003 Evolution of eutectic intermetallic particles in DC-cast AA3003 alloy during heating and homogenization *Mat. Sci. Eng.* A **347** 1–2 130–135

[6] Huang H W and Ou B L 2009 Evolution of precipitation during different homogenization treatments in a 3003 aluminum alloy *Materials & Design* **30** 7 2685–2692

[7] Yaguchi H, Mitani H and Nagano K 2000 Fatigue-damage evaluation in aluminum heat-transfer tubes by measuring dislocation cell-wall thickness *Mater. Sci. Eng.* A **315** 1 6 189–194

Effect of Laser Shock Peening on Fatigue life of Austenitic stainless steels

Z Špirit[1,2], J Brajer[3], J Kaufman[3], M Chocholoušek[1], M Bohm[3], J Kott[1] and J Strejcius[1]

[1]Centrum vyzkumu Rez s.r.o., Morseova 1245/6, Plzen, Czech Republic, EU
[2]Department of Material Science and Technology, University of West Bohemia, Plzen, Czech Republic, EU
[3]Hilase, Institute of Physics of the ASCR, v.v.i. Czech Republic

E-mail: zbynek.spirit@cvrez.cz

Abstract. Laser shock peening (LSP) is an innovative surface treatment technique for strengthening metallic materials. The treatment improves fatigue performance and is achieved by introducing compressive residual stresses into the material surface layer. The compressive stresses are created over a significant surface area by means of high pressure shock waves, which are generated by confining the laser-induced plasma. In the present study, the LSP effect on fatigue life of austenitic stainless steels was investigated by changing the laser parameters. The tested material was austenitic stainless steel 08CH18N10T. The influence of LSP treatment on the fatigue life was examined through the three-point bend fatigue test.

1 Introduction

The LSP is one of innovative surface treatments whose principle was firstly formulated by the American scientist Whitem in 1963 in the USA. The greatest development of LSP technology began at the end of the 90's in the military and aerospace industries [1]. LSP basically achieves similar effects as conventional shot peening (SP) treatments existing in industry for over six decades. In traditional SP process, metallic, ceramic or glass balls acting as a miniscule ball-peen hammers make a small indentation on the metal surface on impact. The advantages over LSP are that it is cheap, it uses robust process equipment and it can be applied on large or small areas as required. The disadvantages of SP are that the process is semi-quantitative and depends on a metal strip or Almen type gauge, that is used to determine the SP intensity. This method cannot guarantee uniform and accurate SP intensity across the treated area. The next limitation of SP is that the depth of compressive residual stress usually does not exceed 0.25 mm in soft metals. Moreover, the surface roughness after SP is worse when compared to LSP [1, 2]. LSP process in comparison can produce a compressive residual stress more than 1 mm in depth, which is about five times deeper than the traditional SP process. SP may also damage the surface finish of metal components and can easily cause distortion of thin section, whereas the surface of a component treated with LSP loses little dimensional accuracy [3].
The principle of LSP is based on conversion of laser energy into a shock wave that results in material plastic deformation and redistribution of residual stresses. When the high energy laser pulses of short nanosecond durations focus on a water immersed metal material, the top surface (sacrificial layer) absorbs the laser energy and evaporates instantaneously through ablative interaction, as shown in figure 1. The resulting high-density vapor is immediately ionized (multiphoton ionization and

avalanche ionization) and forms a plasma plume which absorbs the rest of the laser pulse via the inverse Bremsstrahlung mechanism, the absorption of high-intensity radiation by electrons in the field of nuclei [4, 5]. High pressure of the expanding plasma generates a shock wave, which impinges on the surface of the metal material with an intensity of several GPa, far exceeding the yield strength of the metal. The shock wave propagates into the metal interior and loses energy as it creates a permanent strain on the metal material. The permanent strain remains after the shock wave passage. The strained region is subsequently constrained with surrounded metal material as a reaction to elastic strain, thus forming a compressive residual stress on the metal surface [5].
The LSP technology is used as a substitute for SP conventional treatments where higher standard of fatigue performance is required.

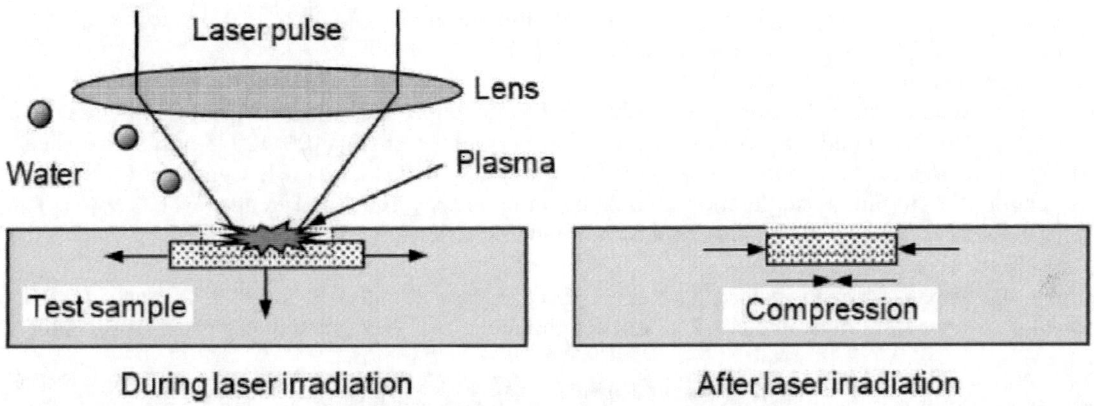

Figure 1. Principle of LSP process [5]

2 Material

Material for testing is austenitic stainless steel 08CH18N10T, which is used in nuclear industry. The material meets the required mechanical properties, microstructures and NDT inspections according to EN 10204/3.1. Microstructure after solution annealing is full austenitic with delta ferrite 3.5 wt. %, nitrides and carbonitrides. Chemical composition and basic mechanical properties are in table 1 and table 2, where YS is tensile yield strength, UTS ultimate tensile strength and HBW Brinell hardness.
Specimens for High Cycle Fatigue (HCF) three-point bending (3PB) test were manufactured from the material in accordance with the drawing design in figure 2.

Table 1. Chemical composition of 08CH18N10T

Material	Fe [wt.%]	C [wt.%]	Mn [wt.%]	Si [wt.%]	P [wt.%]	S [wt.%]	Cr [wt.%]	Ni [wt.%]	Cu [wt.%]	Mo [wt.%]	Ti [wt.%]
08CH18N10T	-	0.05	1.68	0.57	0.02	0.001	17.5	9.9	0.06	0.06	0.47

Table 2. Basic mechanical properties

Material	Temperature [°C]	YS [MPa]	UTS [MPa]	Elongation [%]	Reduction [%]	HBW (2.5/62.5)
08CH18N10T	30	248	565	54,5	76	140
	350	213	425	31	69	

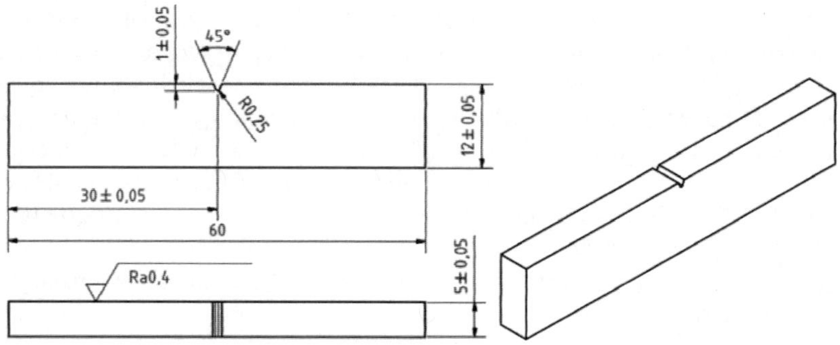

Figure 2. Specimen for HCF three-point bending test.

3 Experimental procedure

The aim of experiment is to compare the fatigue life on unpeened specimens with LSP treatment specimens. The specimens were manufactured from grinder sheet (100x120x5mm). The sheets surfaces were treated from both sides using L2 system – BIVOJ (HiLASE Centrum, CZ) with a wavelength of 1030 nm, pulse duration 10 ns, maximum energy 100 J and frequency 10 Hz [6]. The laser parameters of surface treatment are shown in table 3. After the LSP surface treatment (see figure 3), the sheets were cut to specimens.

Fatigue test were performed in load control on Electromagnetic pulsator for HCF testing RUMUL. Table 4 shows the testing parameters. Post-test fractographic analyses of fracture surfaces were made.

Figure 3. Sheets 6-9 after LSP treatment

Table 3. LSP parameters

Number of sheet	Energy in puls [J]	Beam diameter [mm]	Power density [GW/cm2]	Puls density [1/cm2]	Overlap in line [%]	Area overlap [%]	Tape	Number of passes
8	6.2	2.65	8.8	64	52.8	77.8	Vinyl	2
6	6.2	2.65	8.8	64	52.8	77.8	No	-
9	6.2	4.15	3.6	28.5	54.9	79.6	Vinyl	1
7	6.2	2.65	8.8	28.5	29.3	50	Vinyl	1

Table 4. HCF parameters

Load	Support span [mm]	Stress ratio	Gmax - Maximum stress in 3PB [MPa]	Frequency [Hz]	Test stop at delta frequency [Hz]	Test temperature [°C]
3PB	40	0,1	320	90-92	20	22

4 Results

The results of fatigue life of laser-peened and unpeened specimen are summarized in table 5. The average value of fatigue life cycles was detected, when the value of delta frequency decreased by 20 Hz. The level of maximum stress in 3PB load was 320 MPa for all testing specimens. All peened specimens achieved higher value of fatigue life cycles than unpeened specimens.

Table 5. Fatigue life of laser-peened and unpeened specimens.

Number of sheet	Material	The average fatigue life cycles	Standard deviation
10	08CH18N10T - unpeened (Gmax - 320 MPa)	148222	12016
8	08CH18N10T	485560	344386
6	08CH18N10T	338763	103763
9	08CH18N10T	246245	36980
7	08CH18N10T	433617	159238

Figure 4 shows achieved values of fatigue life cycles for each tested specimen. The horizontal lines show average value of fatigue life cycles of each sheet. The highest value of fatigue life cycles with highest standard deviation was measured on the samples from the sheet no. 8. The reason for such a high deviation is due to the fatigue life of specimen no. 8_3 at 1334622 received cycles.

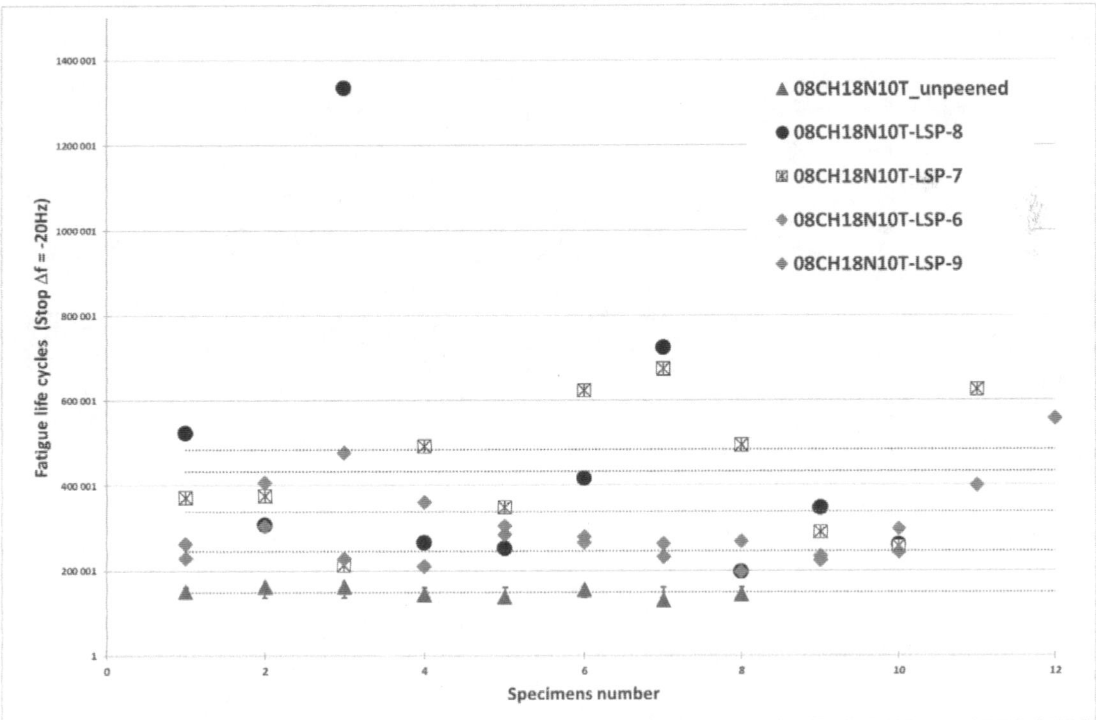

Figure 4. Fatigue life 08CH18N10T specimen at G_{max} = 320 MPa

Figure 5 shows crack propagation for selected specimens from each tested sheet. The number of cycles until the minimum of delta frequency corresponds to material crack resistance. The minimum of delta frequency was estimated as the start of crack initiation. Generally, the crack initiation threshold was higher for material with LSP treatment.

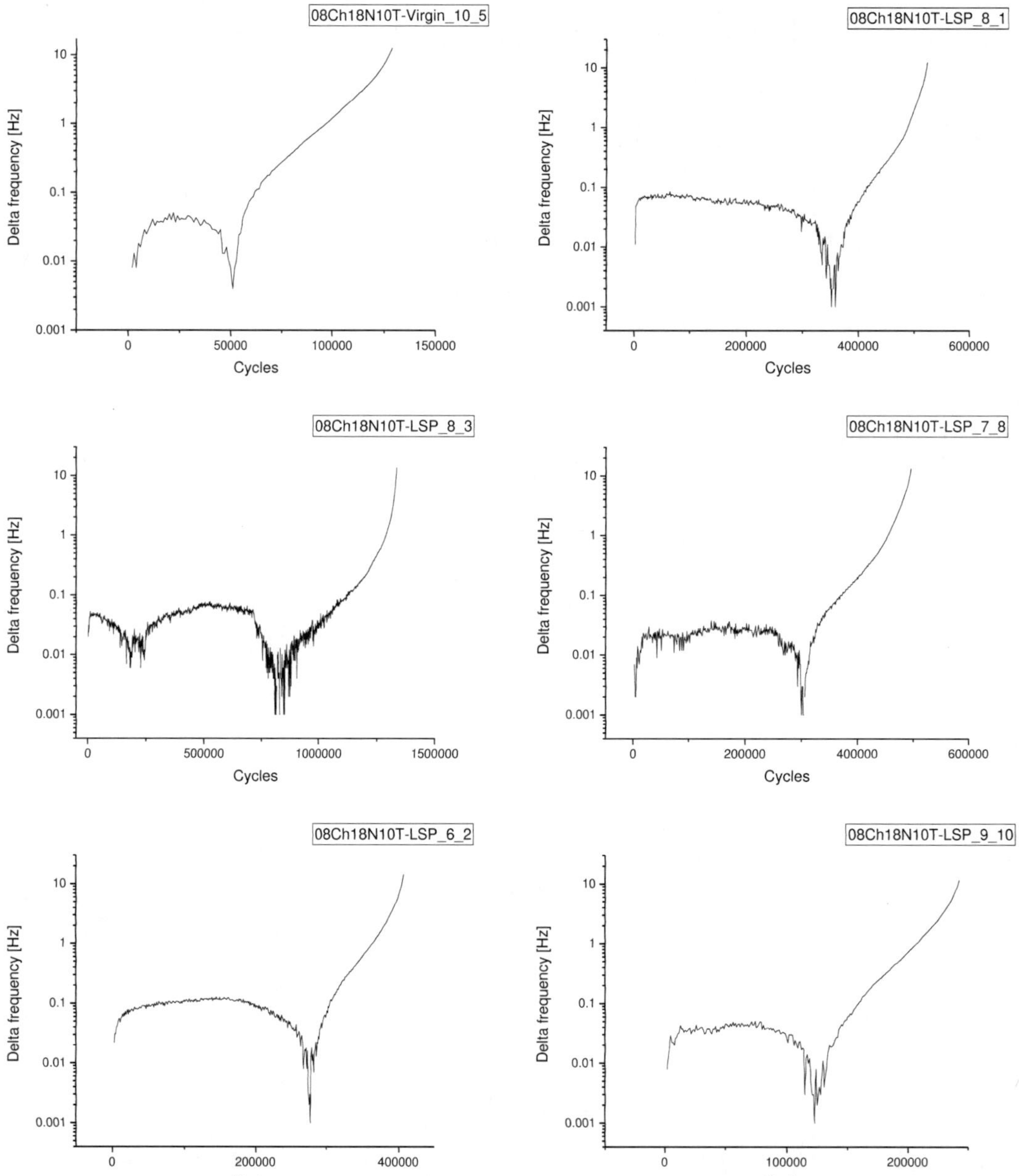

Figure 5. Crack propagation of specimens 10_5, 8_1, 8_3, 7_8, 6_2 and 9_10

Figure 6 shows fracture surface after HCF testing. The fracture surfaces of unpeened and peened specimens are transgranular and they show same cracking mechanism. The crack initiates from notch and grows perpendicularly to the notch. Final failure is ductile with dimples. Fracture surface of peened specimens shows rougher character with little distance between striation lines; it corresponds to a slower crack propagation.

Figure 6. Fracture surface after HCF testing specimen from sheets 10, 9, 8, 7 and 6.

5 Conclusions

The presented results indicate that the HCF performance on sheets from austenitic stainless steel 08CH18N10T was more improved by applying laser shock peening. The crack initiation resistance is higher than for unpeened specimens. Problems are with higher dispersion of the fatigue life cycles values. The dispersion could be caused by inhomogeneous pulse distribution during LSP surface treatment. All peened specimens had a higher fatigue life, the specimens from sheet no. 8 had on average by about threefold increase in fatigue life compared with unpeened.

Acknowledgements

The presented work was financially supported by the Ministry of Education, Youth and Sport Czech Republic - project LQ1603 Research for SUSEN. This work has been realized within the SUSEN Project (established in the framework of the European Regional Development Fund (ERDF) in project CZ.1.05/2.1.00/03.0108 and of the European Structural and Investment Funds (ESIF) in the project CZ.02.1.01/0.0/0.0/15_008/0000293).

References

[1] Champaigne J *History of Shot Peening Specifications*, Ninth International Conference on ShotPeening, Spring 2006

[2] Ding K and Ye L *Laser shock peening, Performance and process simulation*, Woodhead Publishing on behalf of The Institute of Materials, Minerals and Mining, Cambridge England, Spring 2006

[3] Claurer A.H, Gregory J.K,. Rack H.J, and Eylon D(eds) *Laser shock peening for fatigue resistance. Proceedings of Surface Performance of Titanium*, TMS, Warrendale, PA. Metal Society of AIME,217-30

[4] Brajer J *Application of laser shock peening*, CVUT.

[5] *Materials Reliability Program: Technical Basis for Primary Water Stress Corrosion Cracking Mitigation by Surface Stress Improvement* (MRP-267, Revision 2), EPRI Project, July 2016

[6] Pilar J, De Vido M, Divoky M, Mason P D (eds), *Characterization of Bivoj/DiPOLE 100: HiLASE 100-J/10-Hz diode pumped solid state laser*, [online] http://www.hilase.cz/?s=lsp [cit.2018-09-25]

Thermal stability and microstructure evolution of ultra-fine grained Al-Mg alloy

P. Snopiński, T. Tański

Silesian University of Technology, Akademicka 2A st., Poland

E-mail: przemyslaw.snopinski@polsl.pl

Abstract. The Al-Mg alloys are widely used in the automotive and marine industry due to their excellent properties such as corrosion resistance, good weldability and intermediate mechanical properties along with their low price. Magnesium has a high solubility in the aluminium ~17.35 wt. %. Its addition increases the work-hardening ability, thus an essential work hardening can be expected in those alloys. Refining the grain size of the aluminium-magnesium alloys to the ultra-fine (<1 μm) and even nanometric scale using severe plastic deformation processes become the most effective strategy to increase their mechanical strength, whereas the post plastic deformation treatments are strategies to increase ductility. Experiments were conducted to evaluate the mechanical behaviour of an AlMg5 alloy after ECAP processing and subjected to the post-ECAP annealing treatment. The initial grain size was ~270 μm and was reduced to ~20 μm by ECAP with post-ECAP annealing at 350 °C for 30 min. The annealing treatment led to some sort of bimodal grain mixture with the larger grains embedded in the original UFG structure that provide an excellent combination of strength and ductility.

1 Introduction

The aluminium-magnesium alloys are most widely used due to their excellent corrosion resistance, high ductility, low cost, low density and single phase structure (without precipitates). There are many attempts to improve the mechanical properties of Al-Mg alloys [1-5]. Grain refinement down to nanometric scale becomes the most effective strategy to increase their specific strength. Bulk nanostructured materials have motivated considerable attention due to their unusual mechanical and physical properties, e.g., very high strength in comparison with their coarse-grained alloys. Nanostructured materials are very promising for functional and structural applications, especially where its high strength to density ratio improves Energy efficiency. Such type of materials has also the possibility to significantly enhance the performance of nanomaterial-based devices. However, bulk nanostructured metallic materials have also some weakness - generally exhibit low uniform ductility, which hinders their industry applications [6-8].

Generally, the trade-off between high strength and ductility is a common material science engineering problem. These two very important properties are mutually exclusive. Therefore, since the last decades the engineers choose have been limited to strong or ductile materials, not both as it was required. Several research groups focused their efforts focused on the metallic nanomaterials with increased ductility [9-12]. There are some microstructural approaches that might improve ductility which includes: (i) bimodal grain-size distribution, (ii) post-deformation heat treatment (ageing) – generation of high fraction of nanoscale precipitates, (iii) introduction of large fraction of nanoscale twins, (iv) control of grain boundary character, but the applicability of this approaches depends on the composition of the alloy. Also, the bimodal grain size approach depends on the initial alloy state (severe plastic deformation processing). Despite varying degrees of success, most of the approaches have inherent processing and/or material limitations. For example, nanoscale precipitates can only be formed in precipitation hardenable aluminium alloys. Formation of a bimodal grain structure seems to be a most promising route to obtain simultaneously: excellent mechanical strength and increased ductility [13-16].

Since now, the significant experimental attempts have been devoted to developing bimodal grain structure in metallic non-ferrous alloys that exhibit simultaneously a high strength and increased ductility [7,11,12,15]. The challenge for producing desired microstructure remains, but with recent technological advancements (such as new SPD technologies), it is becoming increasingly more achievable to control and engineer the desired grain structure in metallic materials.

This paper shows the approach to develop a bimodal grain microstructure in a AlMg5 aluminium alloy and focuses on the microstructure-mechanical properties relationship of as fabricated bimodal material, and strengthening mechanisms.

2 Material and research methodology

The investigation has been carried out on commercial aluminium alloy—appointed by the standard EN 1706:2010 – EN AC 53100. The chemical composition of the alloys is given in Table 1.

Table 1. Chemical composition of the EN AC 51300 alloy (AlMg5)

Element	Mg	Si	Fe	Cu	Ti	Al
Mass/%	5.55	0.08	0.07	0.01	0.01	Bal

The work samples were subjected to the heat treatment prior plastic deformation. ECAP process was performed using a hydraulic press at room temperature via route Bc. Samples were pressed through an ECAP die with an abrupt internal angle of 120° up to two passes. The samples were processed through a die with a constant ram speed of 2 mm/s. To obtain a bimodal grain size distribution annealing was applied on the SPD processed samples. Microstructure observations of the AlMg5 alloy samples were carried out using light optical microscope Axio Observer A1 on the cross-section plane of the work specimens.

X-ray diffraction analysis were carried out on PANalytical X'Pert Pro diffraction system with cobalt anode ($K\alpha = 1.789$ Å) in order to determine the average dislocation density using a Williamson-Hall method.

The universal tensile testing machine ZWICK Z/100 was used to evaluate tensile properties. Tensile tests were performed at room temperature under a strain rate of 6.7×10^{-4} s^{-1}. The hardness was measured using Automatic Rockwell hardness tester ZWICK ZHR 4150 under a load of 60 kg. To obtain precise results, ten hardness measurements were taken on each sample.

3 Results

3.1 Structure

Optical microscopy images showing the microstructure of solution treated AlMg5 alloy are presented in Fig. 1a-b. As seen, the started material has an equiaxed coarse grain structure with an average size of about 270 μm. As follows from the Fig. 1a after solution heat treatment, most of the Mg was brought into the α-Al solid solution.

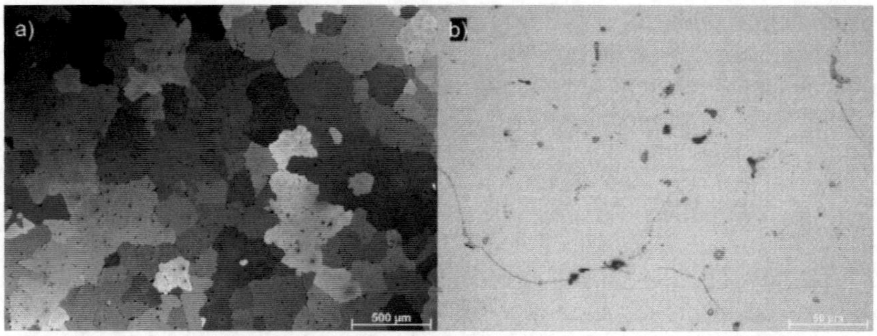

Figure 1. Structure of the EN AC 51300 alloy in the initial state a) polarized light image, b) bright field image

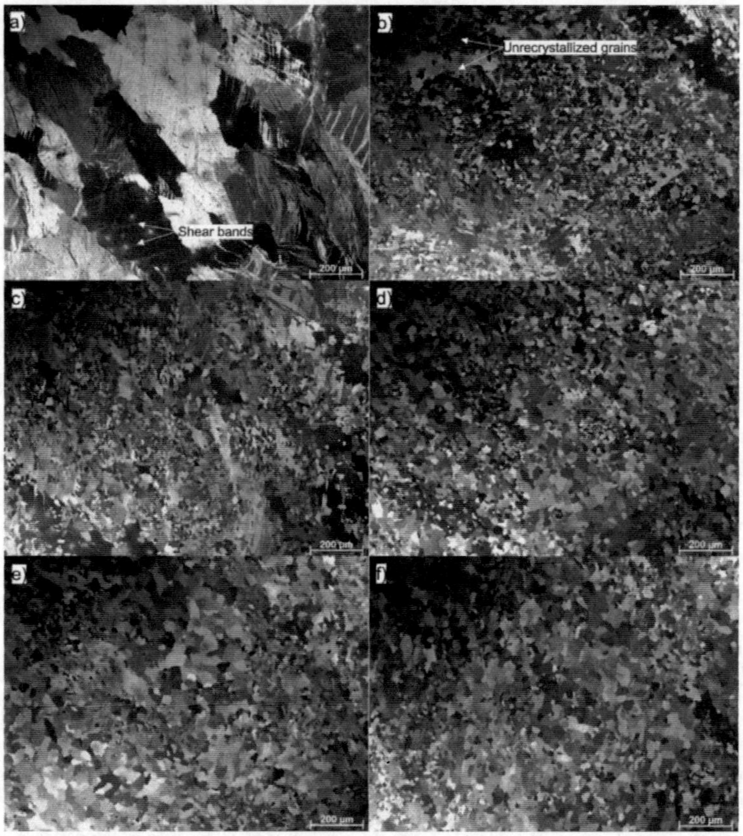

Figure 2. Structure of the EN AC 51300 alloy a) ECAP2x, b) ECAP2x + annealing 30 min, c) ECAP2x + annealing 1h, d) ECAP2x + annealing 2h, e) ECAP2x + annealing 4h, f) ECAP2x + annealing 8h

The ECAP process has changed the original coarse-grained microstructure of the AlMg5 alloy (Fig. 2a). The microstructure of the sample subjected to 2 ECAP passes consist of the elongated grains. The grains become fragmented through the mutual interactions of shear and microshear bands. In the structure of the post-ECAP annealed sample, a large number of fine-grains accompanied by a considerable fraction of micrometre-sized unrecrystallized grains (>400 μm) can be observed. This indicates that a bimodal grain structure was developed (Fig. 2b). After annealing at 350 °C for 1 h, the microstructure is still quite heterogeneous (Fig. 2c), which consists of fine-grained grains with larger grains >200 μm. This bimodal grain structure indicates straightforward that discontinuous

recrystallization occurred locally. Obviously, to facilitate full recrystallization to occur at this annealing temperature (350 ºC), annealing time should be prolonged. It is revealed that when the annealing time increases up to 2h, full recrystallization occurs and the average grain size is about ~23 µm (Fig. 2d). With further post-ECAP annealing, there is a slight increase in grain size to 28 µm ~30 µm after 4 and 8 h (Figs. 2e-f), while the microstructure become homogenous.
To investigate the grain structure and size of as deformed sample TEM was employed. In the Fig. 3 it can be seen that dislocation substructures and elongated sub-grains with a width of about 0.6 µm are formed after ECAP.

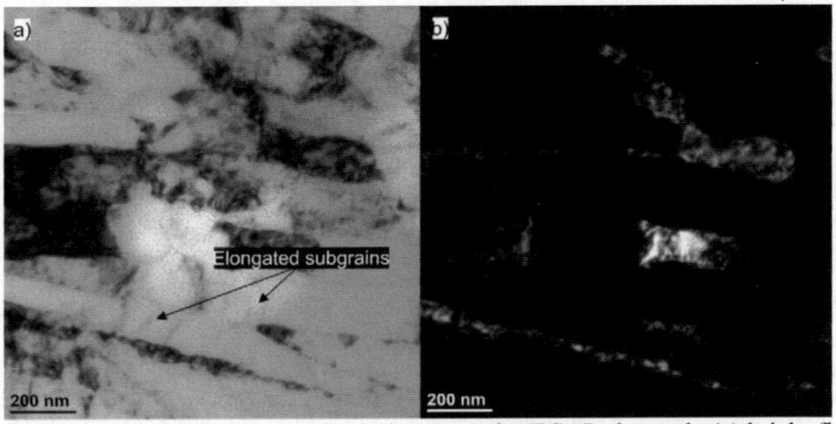

Figure 3. Representative TEM image of solution treated – ECAPed sample (a) bright field image, (b) dark field image

3.2 Mechanical properties

Tensile test curves of the as ECAPed and post-ECAP annealed samples are shown in Fig. 4. Corresponding mechanical properties obtained from Fig. 4 are summarized in Table 2. It is clearly seen that ECAP process leads to increase of mechanical strength (tensile/yield strength and hardness increase) and to significant drop of elongation to failure. Further processing result in a mechanical properties drop; whereas ductility increases significantly. Additionally, flow serrations during tensile deformation of as ECAPed and post-ECAP annealed samples are clearly seen on the tensile test curves. In the as ECAP processed sample A type of flow serrations are observed; whereas in the post-ECAP annealed samples, the combination of A, B and D types of flow serrations takes place.

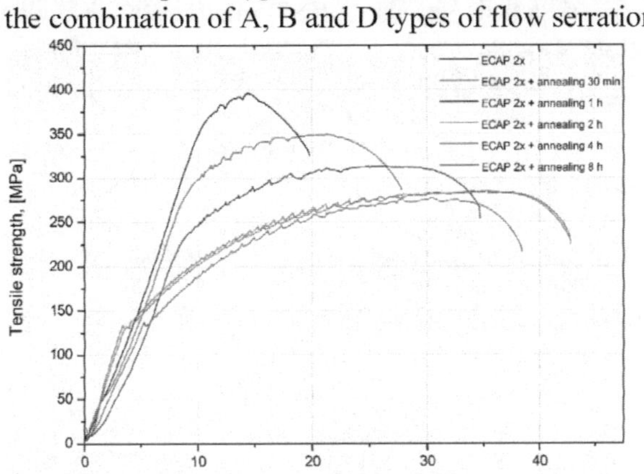

Figure 4. Representative tensile test curves of the AlMg5 alloy subjected to ECAP and post-ECAP annealing

Table 2. Summarized results of the mechanical properties examination

Annealing time, [h]	0	0.5	1	2	4	8
Hardness, HRF	93	78	76	70	72	70
Tensile elongation, %	7	14.7	24	30	33	32
Yield strength, [MPa]	335	275	230	124	124	124
Tensile strength, [MPa]	395	350	312	278	282	283

3.3 *Analysis of strengthening mechanism*

ECAP processing significatly affects the grain structure and the mechanical properties. The strengthening mechanisms are discussed in this section to investigate the relationship between the obtained microstructure and yield strength. Taking into account that different strengthening mechanisms act independently, thus having additive contributions, the overall YS of the AlMg5 alloy can be expressed as follows:

$$\sigma_{0.2}= \sigma_0+ \sigma_{ss}+ \sigma_{gb}+ \sigma_d \quad (1)$$

Where $\sigma_{0.2}$ is the yield strength, σ_0 is the flow stress of un-deformed pure Al (35 MPa), σ_{ss} is the solid solution strengthening, σ_{gb} is the grain boundary strengthening – due the grain refinement, σ_d is the dislocation strengthening.

Solid solution strengthening (σ_{ss}) is controlled by the magnesium solute content, which can be expressed as follows:

$$\sigma_{ss} = HC^n \quad (2)$$

where C is the concentration of an element in solid solution and n and H are constants equal to 1.19 and 13.3 Mpa/(wt.%). Substituting C=5.55 wt.% and the values of constants, the value finally estimates σ_{ss} = 102 MPa.

The most important and most visible from the microstructural point of view consequence of ECAP process is the grain refinement. The obtained refined microstructure shows improved YS which increase may be explained by Hall-Petch equation as below:

$$\sigma_{gb} = k_y d^{-0.5} \quad (3)$$

Where: k_y is the Hall-Petch coefficient – 0.13, D is an average grain size obtained by TEM measurement ~ 0.6 μm. The calculated contribution to the Yield strength reaches the value of σ_{gb}=165 MPa.

In the work-hardened materials dislocations mutually interacts and impede their motion. Therefore, increasing the dislocation density, the yield strength increases. The importance of dislocation strengthening may be expressed as follows:

$$\sigma_d = M\alpha Gb\rho^{1/2} \quad (4)$$

Where σ_d is the strength increment from dislocation strengthening, α is a constant 0.24, G is the shear modulus 24 GPa, b is the length of burgers vector of dislocations 0.286 nm for Aluminium, M is a Taylor factor (M=3 for untextured polycrystalline materials) and ρ is a dislocation density calculated using a Williamson-Hall method XRD $\rho = 0.205*10^{14} m^{-2}$. The overall contribution due the stored dislocations reaches value of σ_d=26 MPa.

Based on the calculations presented above the overall YS reaches value 328 MPa, whereas the measured YS value is ~335 MPa.

4. Conclusions

Microstructure and mechanical properties of EN AC 51300 alloy subjected to ECAP and post-ECAP annealing treatments was comparatively investigated. The main findings described in the article can be summarized as follows:

The elongated grains were gradually replaced by the new fine equiaxed grains accompanied by a considerable fraction of micrometre-sized unrecrystallized grains forming a bimodal grain distribution.
The micrometer-sized grains in the bimodal microstructure can effectively increase the ductility of the fine-grained AlMg5 alloy.
The ECAP process increases the aluminuim yield strength and hardness up to 335 MPa. The corresponding ultimate tensile strength_was increased from 398 MPa. However, the elongation was decreased to ~7% after 2 ECAP passes. The excellent strength of as ECAP processed AlMg5 alloy is due to the strain and grain size strengthening.
DSA process is observed during tensile deformation. The DSA process manifests itself in form of flow serrations on the tensile test curves

References

[1] Tanski T, Snopinski P and Borek W 2017 Strength and structure of AlMg3 alloy after ECAP and post-ECAP processing *Mater Manuf Processes* **32** 1386
[2] Tanski T, Snopinski P, Prusik K and Sroka M 2017 The effects of room temperature ECAP and subsequent aging on the structure and properties of the Al-3%Mg aluminium alloy *Mater Charact* **133** 185
[3] Krol M, Tanski T, Snopinski P and Tomiczek B 2017 Structure and properties of aluminium-magnesium casting alloys after heat treatment *J Therm Anal Calorim* **127** 299
[4] Snopiński P, Tański T and Sroka M 2017 The effect of heat treatment conditions on the structure evolution and mechanical properties of two binary Al-Mg aluminium alloys *Metalurgija* **56** 329
[5] G.E. Totten, D.S. MacKenzieHandbook of Aluminum: Physical Metallurgy and Processes Marcel Dekker, New York, USA (2003)
[6] Cepeda-Jiménez CM, García-Infanta JM, Pozuelo M, Ruano OA, Carreño F 2009 Impact toughness improvement of high-strength aluminium alloy by intrinsic and extrinsic fracture mechanisms via hot roll bonding *Scripta Mater* **61** 407
[7] Jayaganthan R, Brokmeier H-, Schwebke B, Panigrahi SK 2010 Microstructure and texture evolution in cryorolled al 7075 alloy *J Alloys Compd* **496** 183
[8] Dursun T, Soutis C 2014 Recent developments in advanced aircraft aluminium alloys *Mater Des* **56** 862
[9] Sabirov I, Murashkin MY, Valiev RZ 2013 Nanostructured aluminium alloys produced by severe plastic deformation: New horizons in development *Mater Sci Eng A* **560** 1
[10] Wang YM, Ma E 2004 Three strategies to achieve uniform tensile deformation in a nanostructured metal *Acta Mater* **52** 1699
[11] Zha M, Zhang H-, Yu Z-, Zhang X-, Meng X-, Wang H-, Jiang Q 2018 Bimodal microstructure – A feasible strategy for high-strength and ductile metallic materials *J Mater Sci Technol* **34** 257
[12] Zha M, Li Y, Mathiesen RH, Bjørge R, Roven HJ 2015 Microstructure evolution and mechanical behavior of a binary al-7Mg alloy processed by equal-channel angular pressing *Acta Mater* **84** 42
[13] Fan GJ, Choo H, Liaw PK, Lavernia EJ 2006 Plastic deformation and fracture of ultrafine-grained al-mg alloys with a bimodal grain size distribution *Acta Mater* **54** 1759
[14] Valdes-Tabernero MA, Sancho-Cadenas R, Sabirov I, Murashkin MY, Ovid'ko IA, Galvez F 2017 Effect of SPD processing on mechanical behavior and dynamic strain aging of an al-mg alloy in various deformation modes and wide strain rate range *Mater Sci Eng A* **696** 348
[15] Duan Y, Tang L, Xu G, Deng Y, Yin Z 2016 Microstructure and mechanical properties of 7005 aluminum alloy processed by room temperature ECAP and subsequent annealing *J Alloys Compd* **664** 518
[16] Tański T, Snopiński P, Hilser O 2017 Microstructure and mechanical properties of two binary al-mg alloys deformed using equal channel angular pressing *Mater Wis Werkstofftech* **48** 439

IOP Conf. Series: Materials Science and Engineering **461** (2019) 012086 doi:10.1088/1757-899X/461/1/012086

Damage criterion comparison at unidirectional and alternating torsion of samples

A A Bogatov[1], **S N Lezhnev**[2], **A O Tolkushkin**[1*]

[1]Ural Federal University, Russia, Yekaterinburg, Mira street 19

[2]Rudny industrial institute, Kazakhstan, Rudny, 50 years of October street 38

E-mail: a.o.tolkushkin@urfu.ru

Abstract. The paper presents the results of the analysis of computer simulation of unidirectional and alternating torsion of cylindrical samples in the software complex Deform 3D. The accumulated deformation degree and damage criterion were taken as compared parameters. Damage was calculated by two existing methods: Cockroft-Latham and Bogatov methods. The results of the analysis show that the proposed method for determining the damage criterion is more accurate when testing samples for alternating twisting than the most currently used method since it takes into account the number of loading stages and the plastic deformation amplitude.

1 Introduction

With the development of machine-building and metallurgical industries more high requirements for quality and performance are demanded from the metal products. One of the most important indicators of the quality of steel products is the ability to withstand high loads without destruction for long periods of time and often in adverse environmental conditions. Therefore, it is very important to accurately determine strain degree and tensile stresses value that the finished metal can withstand before the formation of macrofracture, which further leads to cracks, internal continuity failure of the metal and as a consequence its destruction. Damage criterion is an indicator to determine the critical strain degree and the magnitude of tensile stresses.

The aim of this work is to conduct the comparative analysis of damage criterion on the basis of computer simulation in the software complex Deform 3D in one of the mechanical tests types: torsion (unidirectional and alternating).

Currently in metal forming there is a variety of methods which allows to assess the damage degree of the materials. The most popular and frequently used method is the Cockroft -Latham damage criterion [1]. According to which the initiation of a macrofracture is occurring when the value of the plastic strain in the direction of the maximal principal tensile stress achieves critical value peculiar to each material and determined by the curves of hardening under tension or torsion of cylindrical samples:

$$\int_0^{\varepsilon_f} \sigma_{\max} d\varepsilon = C, \tag{1}$$

and the normalized Cockroft -Latham criterion in a dimensionless form:

$$\int_0^{\varepsilon_f}\left(\frac{\sigma_{\max}}{\sigma_s}\right)d\varepsilon = C, \quad (2)$$

where ε_f –maximum strain at failure,

$\sigma_{\max}$ –maximal principal tensile stress,

σ_s – strain resistance,

C – material constant.

However, the Cockroft -Latham criterion does not take into account the material loading history and the deformation behavior. Therefore, in this paper, to achieve set aim, it is proposed to use the method considered in the work [2, 3], which has an advantage over the Cockroft -Latham criterion, since it considers the material loading history and the deformation behavior, whether monotonic or alternating. It is assumed that metal has no damage before deformation. The moment of the microfracture appearance or sample separation during testing is associated with the achievement the value of damage ω = 1 (microfracture criterion). According to the proposed method, the influence of the deformation behavior of the samples on the plasticity is estimated using the indexes a and Λ_p, which are included in the metal damage accumulation model under monotonic (3) deformation.

$$\omega = \int_0^{\Lambda_i}\frac{a\Lambda^{a-1}}{\Lambda_p^a}d\Lambda . \quad (3)$$

where $\Lambda = \tan\gamma$ – shear strain degree,

Λ_i – plastic strain amplitude,

$a = a_0^{1+0,238\frac{\sigma}{T}}$ and $\Lambda_p = \Lambda_{p_0}\exp^{\left(\lambda\frac{\sigma}{T}\right)}$ – determining relations of the ductile failure mechanics known from the approximation of experimental data,

$\frac{\sigma}{T}$ – relative mean normal stress,

a_0, Λ_{p_0}, λ – coefficients, which depend on the steel or alloy grade, determined with the use of experimental data by the least-squares method.

The model of metal fracture under alternating plastic deformation is based on the idea that the intensity of metal damage accumulation at the beginning of each stage is zero and depends on the amplitude of alternating deformation Λ_i. During alternating deformation damage has the property of additivity, and for n deformation stages is determined by summation (4):

$$\omega = \sum_{i=0}^{n}\int_0^{\Lambda_i}\frac{a\Lambda^{a-1}}{\Lambda_p^a}d\Lambda , \quad (4)$$

It is assumed that the proposed method for determining the fracture of metals will allow to more accurately determine the value of the damage criterion for unidirectional or alternating torsion of cylindrical samples in comparison with existing methods.

2 Research methodology

Computer simulation of unidirectional and alternating torsion of cylindrical samples was carried out in the software complex Deform 3D. Steel 20 was selected as a material for the test samples (object type elastoplastic), for which the coefficients are equal to $a_0 = 2{,}8$, $\Lambda_{p_0} = 2{,}85$ and λ = -0,669. During the samples torsion the relative mean normal stress $\frac{\sigma}{T}$ and Lode coefficient μ_σ (used to estimate the stress state, characterizing the stress deviator and type of sample testing) are equal to zero, thus the indexes $a = a_0$, $\Lambda_p = \Lambda_{p_0}$ and λ coefficient is irrelevant. The samples were tested at constant values of thermomechanical parameters: σ/T = const; μ_σ = const; temperature Θ = const and shear-strain rate H = const.

During unidirectional or alternating torsion testing, it is more rational to use the samples with a neck in the central zone with a radius of *R*. In the work [4] it is shown that the optimal ratio of the neck radius *R* and the diameter of the sample in the smallest section *d* is 1,25. The use of such samples (Figure 1) allows not only to localize the strain, but also to more accurately determine the strain degree before the fracture occurs. For computer simulation, a solid model of the sample with the following dimensions was made: D = 10 mm, R = 10 mm, d = 8 mm, length of the working part of the sample l = 20 mm.

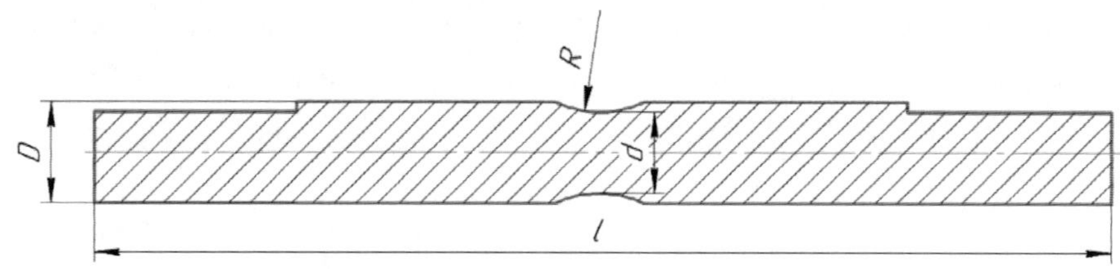

Figure 1. Torsion testing sample

For a more accurate determination of the indicators of the stress-strain state and the damage criterion on the solid-state model, a fine-mesh density window was created (Figure 2a). In both cases of unidirectional and alternating torsion of the samples, the total torsion angle was 180° (π rad). The amplitude of plastic deformation at alternating torsion was 45° (π/4 rad). The torsion speed was 10 RPM. To measure the angle γ the stripe on the sample was made (Figure 2b).

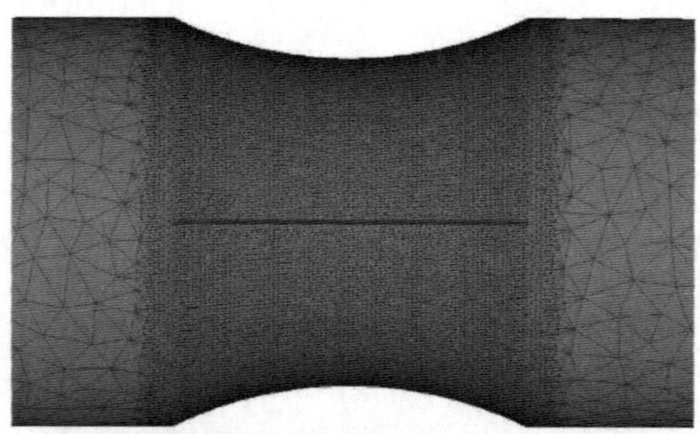

a)

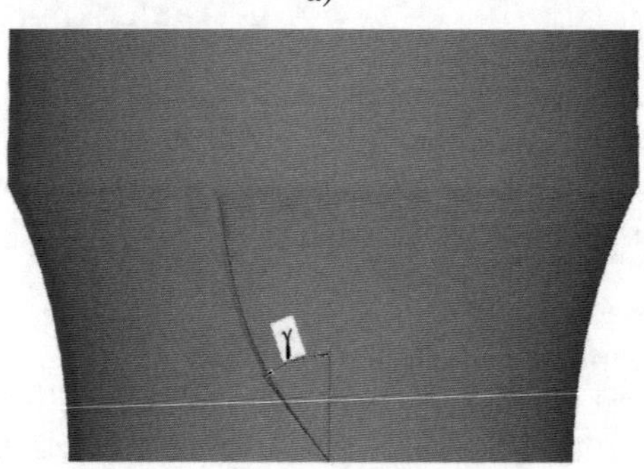

b)

Figure 2. Solid model of the sample: a – general view, b – γ angle measurement.

To determine the stress-strain state and the damage criterion in the sample cross section 21 points were plotted across the diameter in the central part (point 1 at the sample surface and point 21 at the sample axis) at a distance of 0.2 mm from each other (Figure 3). Point tracking allowed to record and import the values of the required parameters at any moment of time in a table form in order simplify the processing of the obtained after computer simulation data.

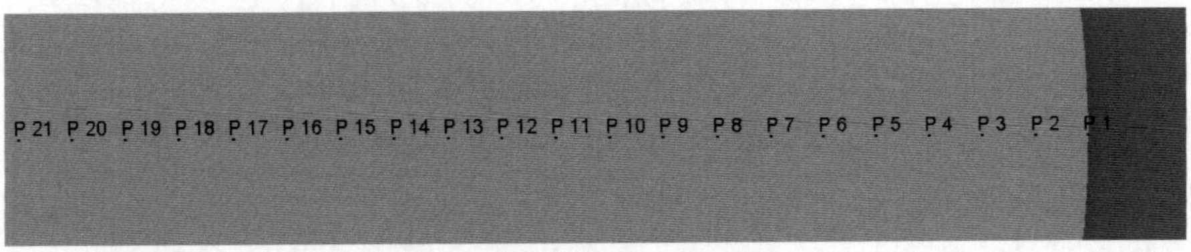

Figure 3. Point tracking

In order to simplify the calculations, the influence of temperature on the damage value and the accumulated strain degree was not taken into account.

3 Results and discussion

The values of all parameters are specified for the sample surface (tracking point 1).

The unidirectional torsion computer simulation results are presented in Table 1.

Table 1. The unidirectional torsion computer simulation results

Torsional angle	π/4	π/2	3π/4	π
γ	38,3	55,3	64,3	69,57
Λ_c	0,7898	1,4442	2,0778	2,6846
Λ_D	0,8081	1,4523	2,1129	2,7496
ω_c	0,0275	0,1491	0,4128	0,8459
$\omega_{C\text{-}L}$	0,2691	0,4962	0,7199	0,9163

where Λ_c – calculated shear strain degree depending on the γ angle;
Λ_D – shear strain degree according to Deform 3D;
ω_c – damage criterion, calculated by the method [2];
$\omega_{C\text{-}L}$ – Cockroft-Latham damage criterion.

As can be seen from Table 1, the difference between the values of calculated shear strain degree and shear strain according to the program data is not more than 2.5%, which makes it possible to judge that the convergence of theoretical calculations and computer simulation results is good.

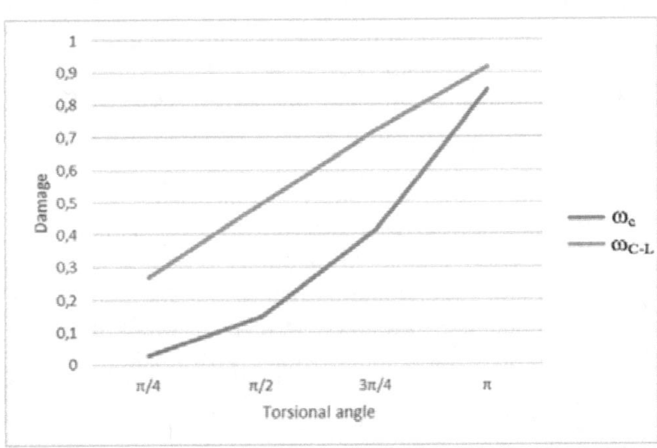

Figure 4. Damage criterion comparison during unidirectional torsion

Figure 4 shows a comparison graph of the damage values at unidirectional torsion. As can be seen from the figure, the damage curve according to the Cockroft -Latham criterion is linear and at the angle of twisting $\pi/4$ rad already has a higher value than by the calculated method. The convergence appears only at large torsional angles. From this it could be concluded that the proposed method allows more accurately determine the damage value at small torsional angles.

The alternating torsion computer simulation results are shown in Table 2.

Table 2. The alternating torsion computer simulation results

Torsional angle	$\pi/4$	$\pi/2$	$3\pi/4$	π
γ	38,27	33,4	30,1	29,07
Λ_c	0,7889	1,4483	2,0279	2,5839
Λ_D	0,8025	1,4755	2,1059	2,7163
ω_c	0,0274	0,0440	0,0556	0,0659
ω_{C-L}	0,2673	0,4920	0,7021	0,9059

During alternating torsion (Table 2), the difference between the values of calculated shear strain degree and shear strain according to the program data is not more than 5%, which also indicates a good convergence of theoretical calculations and computer simulation results.

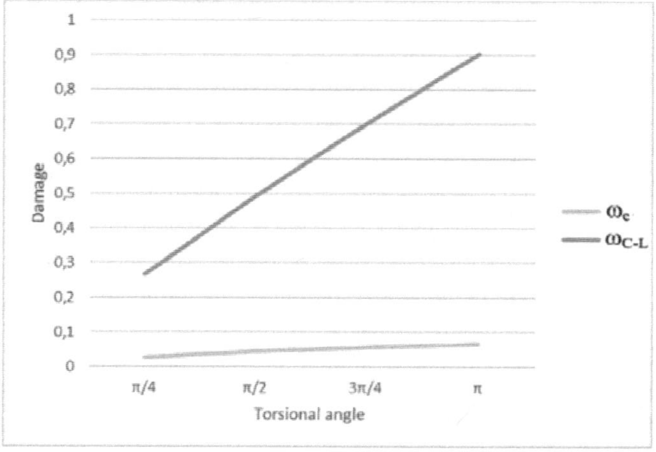

Figure 5. Damage criterion comparison during alternating torsion

Damage criterion comparison results at alternating torsion are presented in Figure 5. During calculating the damage at alternating torsion, it is assumed that the intensity of the accumulation of damage at the beginning of the *i-th* stage is zero and does not depend on the accumulated damage in the previous stages. Due to this, the sample could be subjected to alternating deformation by a greater number of deformation stages before the fracture than during monotonic deformation.

Conclusion

Computer simulation of the samples testing at unidirectional and alternating torsion was performed. The degree of shear deformation and damage criterion were chosen as the comparable parameters. The damage criterion was calculated using two existing methods. By comparative analysis results, it was determined, that the values of shear deformation degree according to computer simulation and theoretical calculation, in both cases of unidirectional and alternating torsion have good convergence. Comparison of damage calculation methods has shown that the use of the proposed in this paper method is more appropriate than the use of the existing method of Cockroft-Latham, because it considers the loading stages number and the plastic deformation amplitude and provides more accurate results at small torsional angles.

Acknowledgements
This article was prepared with the financial support of Competitiveness Enhancement Program – CEP 3.1.1.1-18

References

[1] Cockroft M and Latham D 1968 Ductility and Workability of Metals *Journal of the Institute of Metals* **96** pp 33-39
[2] Bogatov A A 2002 Mechanical properties and fracture models of metals (Ekaterinburg: Ural State Technical University – Ural Polytechnic Institute) p 329
[3] Bogatov A A 1997 On metal fracture under pressure working *Kuznechno-Shtampovochnoe Proizvodstvo* **7** pp 2-7
[4] Bogatov A A and Krinitsyn V A 1982 Method of the study of the material deformability (Ural Polytechnic Institute) C.A. 930064 (USSR)

The comparison of measured surface temperature of roll in stretch reducing mill with a simulation of process

J Turňa[1,2], R Ďurčík[1], P Buček[1], Ľ Parilák[1,3], P Maťaš[4],

[1] ŽP VVC s.r.o., Kolkáreň 35, 976 81, Podbrezová; Slovak Republic,
turna.jan@zelpo.sk

[2] Faculty of Materials Science and Technology in Trnava, Slovak University of Technology in Bratislava; Slovak Republic

[3] Faculty of Manufacturing Technologies of Technical University of Košice with a seat in Prešov; Slovak Republic

[4] Železiarne Podbrezová a.s., Kolkáreň 35, 976 81, Podbrezová; Slovak Republic

Abstract. This paper deals with the surface temperature of rolls during hot rolling of seamless steel tubes in stretch reducing mill in Zeleziarne Podbrezova. The main goal of our work was to measure the temperature of the rolls and compare it with the results coming from numerical simulation of hot rolling process. In general, the temperature of the tubes after stretch reducing should be higher than A_{r3} by 30 ÷ 50 °C. The surface temperature of selected roll (so called "B-roll") was measured by means of a contact thermocouple-based method. Boundary conditions for numerical simulation of hot rolling process was set up according to the actual process conditions during rolling.

1 Introductions

The rolling process in a stretch reducing mill on Figure 1 (SRW) marks the final hot forming operation in seamless steel tube production process.

In this process, a hollow tubular feedstock passes a series of 3-roll rolling stands with oval or circular rolling gap. The hollow undergoes gradual reduction of outer diameter and wall thickness, achieving final tube dimensions and final mechanical properties after exiting the mill and cooling on a cooling bed [1 ÷ 5]. The process itself does not use any internal tool; the rolls in the stands do all the job. In general, stretch reducing mills can be used to extend the production range for small- and medium-diameter hot rolled tubes [1 ÷ 7]. The main purpose is searching of the ideal value of HTC between roll-nozzle and feedstock roll.

Figure 1 Numerical model of a stretch reducing mill

2 Measurement methodology

When choosing the appropriate methodology of temperature measurement, two basic methods were at our hand: contact and contactless.

The contact method proved to be more suitable for measuring the surface temperature of the roll [8 ÷ 13]. Contactless measurement would be very difficult to use for several reasons (for instance the low emisivity of a rather glossy working surface of the roll, the presence of cooling water, steam etc.) Customized K-type thermocouple (TC) probe was used during our experiments. The standard deviation of thermocouple type K is ± 2.2 °C or 0.75 %. The measurement itself was carried out while the stretch reducing mill was put to a halt (cooling water and roll drives were turned off). The TC probe was put on the measurement spot of the working surface of the roll and the signal was recorded by a computer [11]. To ensure a reliable measurement, the following conditions need to be met [10–12]:

- Perpendicular orientation of the probe body with respect to the roll surface,
- The rectangular sensing ribbon should be aligned so that the longer side matches the roll circumference,
- The measurement spot should lie on the symmetry plane of the working surface of roll,
- The measurement should take at least 5 seconds while fulfilling all the conditions mentioned above.

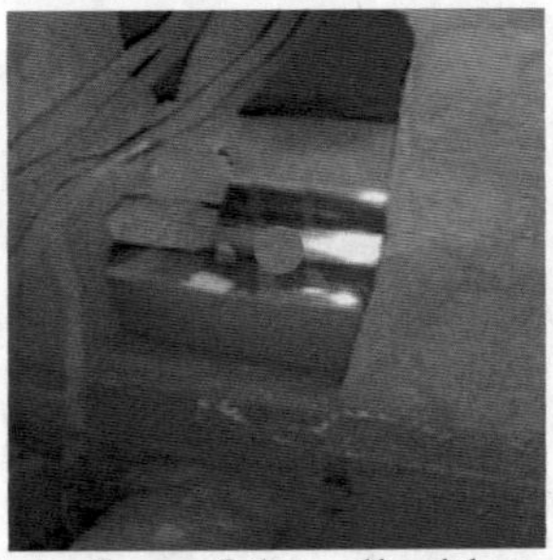

Figure 2. Detailed view on the working surface of the roll with red arrow pointing at the measurement spot when using contact temperature probe

2.1 *Result of a measured temperature*

The surface temperature of the roll was measured on the stand No. 3 and No. 5 under the specified conditions.

As can be seen in table 1 the surface temperature of the roll in stand No. 3 is on average temperature 99 °C after rolling campaign about 500 pieces of tubes. As can be seen in table 2 the surface temperature of the roll in stand No. 5 is on average temperature 97°C after rolling campaign about 500 pieces of tubes. The surface temperature of the roll after rolling the ideal number of pieces becomes stable and the temperature is in the range of ± 5 °C. On Figure 3 we see the illustrative measurement of surface temperature of rolls in the stretch reducing mill.

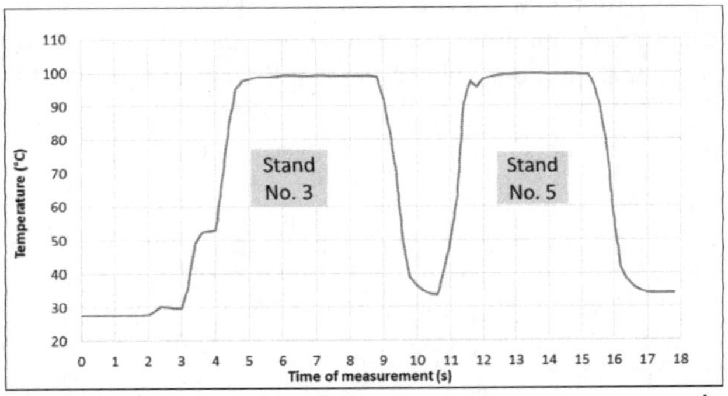

Figure 3. Illustrative record of surface temperature measurement in stands

Table 1. Values of measured temperature

Stand No.	T_{max} (°C)					
	Measurement No. 1	**Measurement No. 2**	**Measurement No. 3**	**Measurement No. 4**	**Measurement No. 5**	**Avg. temperature**
3	101	96	95	99	101	**99**
5	98	93	95	98	97	**97**

The rotating test bench is designed for investigation of cooling effects on the rotating cylindrical surfaces. The parameters of the bench are adapted to the main application which is design of cooling systems of rolling-mill working rolls [6–7]. The data obtained by the thermocouples during the roll cooling experiment gives information about temperature history at positions under the roll surface. Each data point is also connected to information about the angle of cylinder rotation and thus the data collected in "time" order can be converted into position order. The measured temperatures are used for calculation of heat transfer coefficient and surface temperature history [6–7].

3 Numerical simulation of SRW rolling process

3.1 Preparation of simulation

Simulation of SRW rolling process uses a 3D FEM model with given process and boundary conditions, considering significant heat transfer from the workpiece to the rolls and the environment (convection/HTC, radiation) and also spray water cooling of the roll surface (convection/HTC).

For given rolling simulation, process parameters for a given tube dimension were provided, such as the angular velocity of the rolls, initial temperature of the workpiece, etc. Various heat transfer coefficients (HTC) were used to monitor the heat transfer effects that were directly measured as the roll surface temperature using our sturdy TC contact probe.

3.2 Simulation run

A virtual rolling campaign with the series of twenty 16.4m long hollows was used to determine the roll surface temperature.

In this campaign, only stand No. 3 and stand No. 5 were considered. The actual length of the workpiece as well as the time of contact with the roll was recalculated for both rolling stands (as no full set of SRW stands was considered in simulations). The resulting contact time was 11 seconds, followed by 6 second time gap.

3.3 The temperature of roll through continual transition of tubes in stretch reducing mill

For roll in stand No. 3 the feedstock temperature was set to 925 °C, which corresponds to the temperature reading obtained from IMS temperature measurement system that senses temperature of input feedstock prior to entering the SRW mill (the actual temperature was approx. 940 °C; we considered some temperature drop due to 1st and 2nd rolling stand).

The temperature measurement point was selected at the center of the square of the working surface of the roll. On Figure 4 and Figure 5 we see the result from simulation for roll temperature in the stand No. 3.

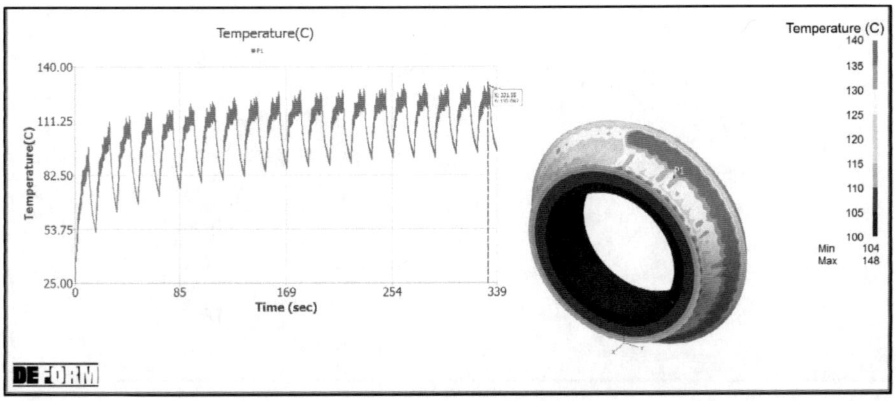

Figure 4. HTC roll-feedstock was measured to 13200 W/m^{2}K, maximal HTC roll-nozzle to be HTC 8100 W/m^{2}K at 100 °C

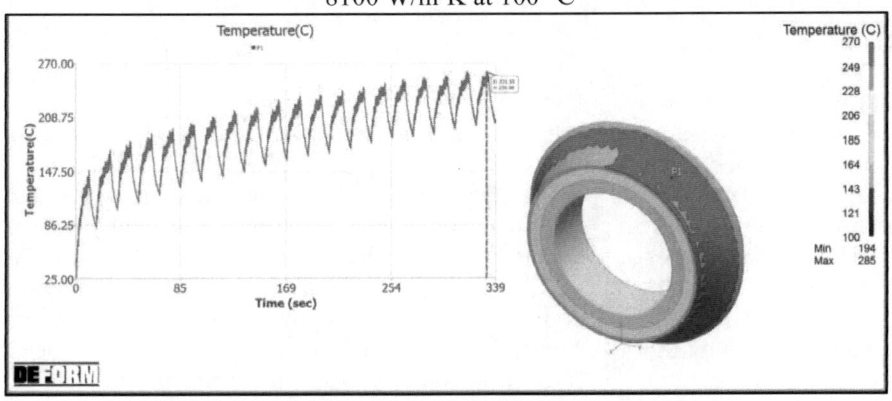

Figure 5. HTC roll-feedstock was raised to 22000 W/m^{2}K, maximal HTC roll-nozzle was measured to be 2400 W/m^{2}K considering that temperature 100 °C

For roll in stand No. 5 the feedstock temperature was set to 881 °C, which corresponds to the temperature reading prior to stand No. 5 using the pyrometer. On Figure 6 and Figure 7 we see the result from simulation for roll in the stand No. 5.

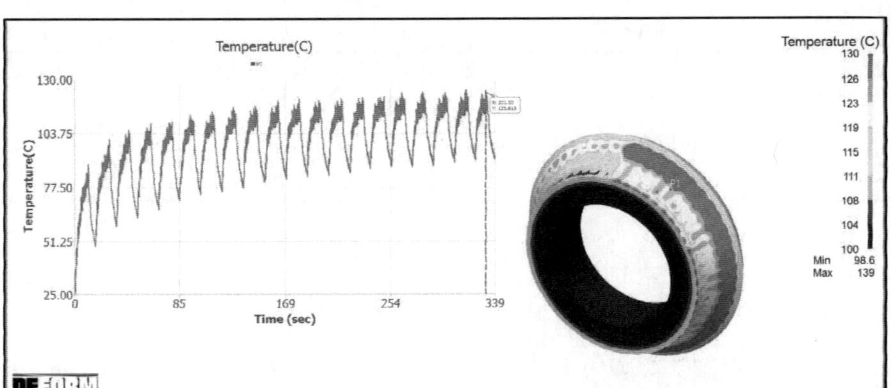

Figure 6. HTC roll-feedstock was measured to 12100 W/m^{2}K, maximal HTC between the roll and the nozzle to be HTC 8100 W/m^{2}K at 100 °C

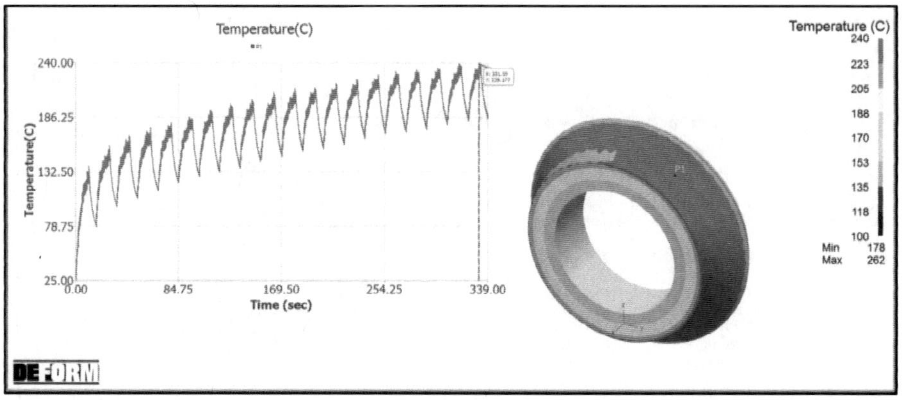

Figure 7. HTC roll-feedstock was raised to 20600 W/m^{2}K, maximal HTC roll-nozzle was measured to be 2400 W/m^{2}K at 100 °C

4 Results and Discussion

4.1 Result of a simulation

In the table 2 we see the resulting roll temperature (stand no.3 and stand no. 5) at the same point of continuous passage and permanent load.

We can say that the effect of HTC has a significant effect on the resulting temperature (average temperature from last peak, which see in table 2). With the same HTC roll-nozzle but with a different HTC roll-feedstock the difference in the resulting temperature on the stand No. 3 to 50 °C. Although, when changing the roll cooling intensity, the same HTC feedstock-roll (22000 W/m^{2}K) also has difference of up to 40 °C. As we can see from graphs above, the temperature itself is gradually settling (Figure 4, Figure 5, Figure 6 and Figure 7). This is due to the equilibrium state between the temperature received from the feedstock and the ambient temperature and reduced by the influence of the cooling of the roll by the medium.

Table 2. Average surface temperature on the roll

Surface temperature (°C)		**HTC feedstock ($T_{feedstock}$ – 925 °C) – roll in stand No. 3**	
		13200 W/m^{2}K	**22000 W/m^{2}K**
HTC roll–nozzle	**2400 W/m^{2}K**	133 °C	181 °C
	8100 W/m^{2}K	112 °C	141 °C
Surface temperature (°C)		**HTC feedstock ($T_{feedstock}$ – 881 °C) – roll in stand No. 5**	
		12100 W/m^{2}K	**20600 W/m^{2}K**
HTC roll–nozzle	**2400 W/m^{2}K**	123 °C	167 °C
	8100 W/m^{2}K	106 °C	132 °C

4.2 Discussion

When comparing the temperatures between simulation and the operating measurements, we can define boundary conditions to simulate the reduction process for a larger campaign.

The most suitable results were simulations using the HTC roll-nozzle 8100 W/m^{2}K and HTC feedstock-roll by HTC 13 200 W/m^{2}K. It is important to extend the rolling campaign to a similar number of pieces as it was actually rolled out using the same method. When comparing surface temperatures in simulations and experimental measurements, it is necessary to point to similar temperatures under the aforementioned boundary conditions. The temperature variation of the simulations to the measurements was caused by the transition of the feedstock over the remaining number of stands and reducing it. This effect caused a difference of 12 °C on average compared to the real measured values that were measured after the feedstock stretch reducing process. It is necessary to adjust the steady state from experimental measurement of HTC with real values of surface temperature of the rolls.

5 Conclusion

The results achieved were supported by results of modelling and simulation of heat transfer between the rolls and the tube [3].

The experiment confirmed the assumption that it is possible to obtain very similar values of the surface temperature of the roll as in the real measurement as well as in the simulation of the experimental rolling campaign. From this we propose to continue the experimental modeling of this process to a negligible difference of ± 5 °C.

Acknowledgement

This article was created in the framework of the research task of VÚ č. 1/2017 / ŽVVVC - "KOTAKON – Structural concept, research and development of boiler and construction grades".

References

[1] Počta B 1963 Steel tubes: Seamless tubes. In Czech. Praha. SNTL. 416 p.
[2] Ďurčík R et al 2017 Results of numerical simulation. (Internal report 20/2017/ŽPVVC, in Slovak). 2017. 46 p.
[3] Reškovič S et al 2010 Numerical design of a hot-stretch reducing process for welded tubes. *Materials and Technology.* vol. 44, no. 5, pp. 239-242
[4] The device IMS in stretch reducing mill. In Slovak. (SRW) PBTP-02.69-428/14
[5] Turňa J et al 2017 Measurement of roll temperature in stretch reducing mill. Hutnik: Wiadomości Hutnicze, vol. 84, no. 8, pp. 380-382.
[6] OMEGA – 8 Thermocouple Data Acquisition system [online]. [viewed 22.3.2018]. Available from: https://www.omegaeng.cz/pptst/tc-08.html
[7] Kotrbáček P et al 2015 High efficient and energy-efficient cooling systems for cooling profile rolls. In Czech. In Odborná konferencia ŽP VVC. Tále: HLP grafik s.r.o. 204.
[8] Horský J 2013 Cooling for profile rolls. In Czech. In Odborná seminár ŽP VVC 2013. Tále. HLP grafik s.r.o. 206 p.
[9] Pohanka M 2010 Technical Report – Design of Cooling Roll Solutions for Rolling Mill in ŽP a.s. in Czech. VUT Brno.
[10] Ďurčík R et al 2016 Results of Numerical Simulations of the KOTAKON. (Internal report 21/2016/ŽPVVC in Slovak). 2016. 46 p.
[11] Parilák Ľ et al 2016 Temperature measurement of rolled material in selected technology nodes in ŽP a.s. (Internal report 6/2016/ŽPVVC in Slovak). 2016. 35 p.
[12] Pohanka M et al 2016 Technical Report – Thermophysical properties of the working rolls of stretch reducing mill in ŽP a.s. in Czech. VUT Brno.
[13] Horský J et al 2016 Simulation of tube rolling and load of individual roll calibers. in Czech. VUT Brno.
[14] Aymone J, Bittencourt E, Creus J 2001 Simulation of 3D metal-forming using an arbitrary Lagrangian–Eulerian finite element method, *Journal of Materials Processing Technology* ISSN: 0924-0136
[15] Van den Boogaard A and Huétink J 2001 Numerical Simulation of Forming ProcessesUsing an Arbitrary Lagrangian Eulerian Finite Element Method, DIANA Computational Mechanics *Journal of Materials Processing Technology* ISSN: 0924-0136
[16] Huétink J, Vreede T, Van der Lugt J 1990 Progress in mixed Eulerian-Lagrangian finite element simulation of forming processes, *Numerical Methods in Engineering*

IOP Conf. Series: Materials Science and Engineering **461** (2019) 012088 doi:10.1088/1757-899X/461/1/012088

Refinement of a material model for springback after press forming

M. Urbánek, J. Mištera and M. Fedorko

COMTES FHT a.s., Prumyslova 995, Dobrany 334 41, Czech republic

miroslav.urbanek@comtesfht.cz

Abstract. The best tool for characterizing the behaviour of sheet metal during press forming is numerical modelling based on the finite element method (FEM). Using this technique, the response of the press-formed part can already be predicted during the process design stage. The main input data for the process simulation include the tool geometry and materials properties and process conditions, such as temperature, pressing velocity, lubrication and others. The resultant shape of the press-formed part is governed predominantly by the shape of the tools and the workpiece material. The input information on the workpiece material has the form of a material model whose accuracy dictates the accuracy of results. Where high-strength steels are used in the automotive industry, one should focus on springback which alters the shape of press-formed parts. Springback is an undesirable change to the part's shape caused by residual stresses being released after its removal from the tool. Tooling designers need to take springback into account by choosing an appropriate material model. Conventional hardening models, such as kinematic and isotropic hardening rules, are not adequate for modelling springback. Instead, more sophisticated models, such as the one introduced by Lemaitre and Chaboche are a better choice. In this paper, the options for measuring and FE modelling of the Bauschinger effect are discussed. The proposed idea was tested on a 1.5 mm sheet of DC01 deep-drawing steel.

1 Introduction

Sheet metal forming is a productive technology which has a widespread use in the automotive industry. Understanding the behaviour of metal sheet is a prerequisite for developing and fine-tuning a new manufacturing process. The shapes of forming tools are proposed by designers and production engineers who rely on FE modelling to verify a new concept. FE analysis depends fundamentally on its input data, such as process conditions, tool shape and materials properties. The knowledge of materials properties is embodied in material models which provide a mathematical description of the material's behaviour. The type of analysis determines how complex the material model needs to be. The characteristics required for sheet forming simulations include anisotropic properties, the work-hardening curve, the yield surface and the forming limit curve. If parting of materials is to be modelled, fracture criteria are needed as well. With temperature-dependent processes, the model must also account for quantities such as thermal conductivity and expansion and specific heat and, where relevant, temperature-dependent hardening curves and elasticity moduls. [1]

This paper is concerned with upgrading existing material models to account for springback in sheet metal. Springback is important at the moment the final press-formed part is removed from the die. The first scientist to explore springback was Johann Bauschinger, a German professor of applied mechanics, who discovered the dependence of plastic behaviour on deformation history. The effect is defined as the decrease of the elastic limit in a material under a load that acts in a particular direction, following previous plastic deformation in the opposite direction, [2].

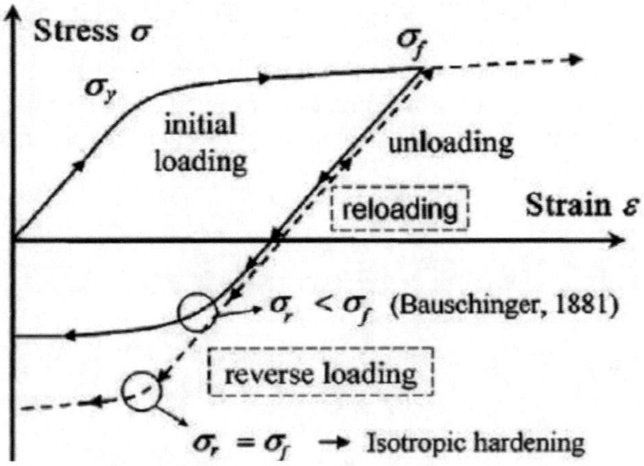

Figure 1. Schematic illustration of work hardening [3]

Figure 1 shows a stress-strain plot for a material to which an initial tensile load was applied. Under that load, the specimen experienced certain engineering strain. Stresses below the elastic limit have led to linear elastic strain that followed Hooke's law. Stressing beyond that limit caused plastic deformation. Once unloaded, the specimen recovered, following the line which is parallel to the initial stress-strain plot. Along this line, the elastic strain drops to zero again and some plastic deformation remains. In subsequent compression of the specimen, the initial response follows Hooke's law again but the load at the elastic limit is lower than in a specimen without this deformation history. These stresses meet the relationship $\sigma_r < \sigma_f$, where σ_r is the elastic limit of a material with the above-described deformation history. Detailed characterization of work hardening enables one to predict the post-forming elastic-plastic properties of a part. The most important manifestation of the Bauschinger effect is springback in deep drawing [4], [5].

2 Modelling the Bauschinger effect

The isotropic hardening rule is not suitable for modelling the Bauschinger effect. The reason is that in this model, the elastic-plastic behaviour of a material is altered equally in all directions (Figure 2a). The effect can be simulated using kinematic hardening (Figure 2b) but the approach is inadequate due to its unsuitable linear dependence of stress on strain beyond the plastic limit. For all these reasons, combined non-linear hardening models are recommended for reliable simulations of the Bauschinger effect, such as the model by Lemaitre and Chaboche.[6], [7]

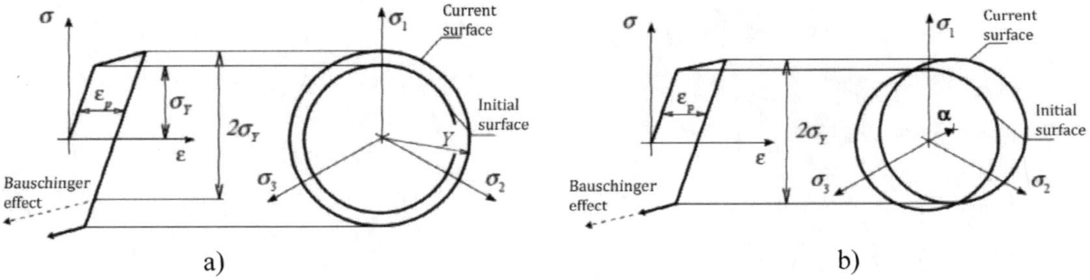

Figure 2. Schematic representation of hardening models [3]

3 Requirements on measurement

An instrument to be used for identifying material parameters related to the Bauschinger effect should be capable to impose alternating tensile and compressive loads beyond the yield point on flat test specimens in this particular case. The desired load vs. time diagram is shown in Figure 3 together with the resultant engineering stress curve. When a standard test bar is tested in tension and compression, problems arise during compressive loading. In the diagram, this problematic region is between points A and B where buckling occurs. Generally, buckling is associated with loss of stability. The load-carrying capacity is then determined by stability of the equilibrium condition instead of material strength.

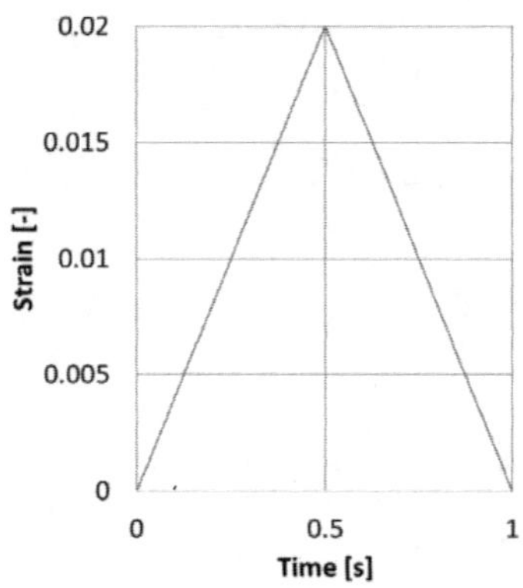

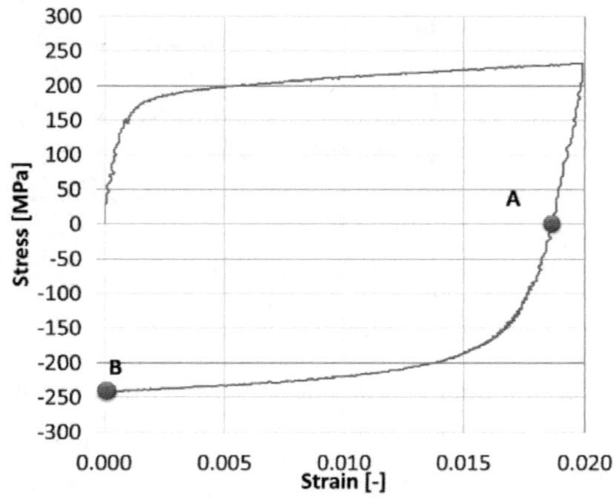

Figure 3. Load and stress plots

In order to provide stability throughout the test, a test fixture was employed. Its design enabled the parameters related to the Bauschinger effect to be measured, utilizing a DIC system as a control unit. The fixture consists of two plates put against the specimen faces which exert a desired contact pressure onto the specimen (Figure 4). Compression springs impose forces through holes in a plate which can be adjusted by screws. One of the plates has a stadium-shaped hole for measuring test strain by the DIC. The shape of the test specimen facilitates the desired stress distribution. The ends of the specimen are clamped in jaws which themselves can be attached to a test machine.

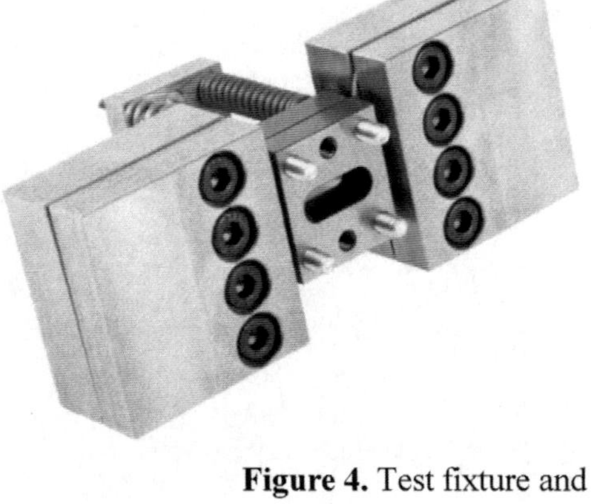

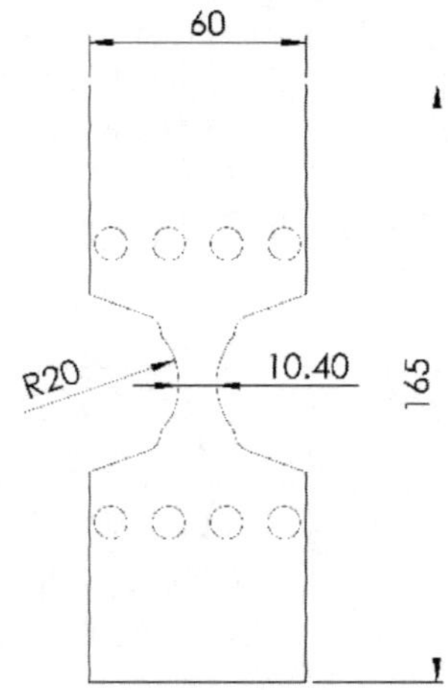

Figure 4. Test fixture and samples geometry

4 FE model

A FE model was constructed for verifying the behaviour of the test fixture and for checking its function and usability for measuring Bauschinger effect parameters. It was developed with the Abaqus/CAE 2017 software (Figure 5). Since major non-linearities occur during the test, the explicit algorithm was employed. Simulations of the test were performed on the gauge section of the specimen. The springs were modelled by 1D axial-type elements. Coefficient of friction f = 0.15 was considered for the contact between the plates and the specimen. The specimen material was defined using a modified Lemaitre and Chaboche model. Its parameters, adopted from literature on DC01 deep-drawing steel, are listed in Table 1. Other material properties for the simulation were taken from previous research [8] and the chemical composition of mild steel DC01 is shown in the Table 2.

Table 1. Model parameters

	Parameter	Value	Unit
Isotropic hardening law	R_∞	62.6	MPa
	b	25	-
Kinematic hardening law	C_1	7200	MPa
	γ_1	158	-

Table 2. Chemical composition

chemical element	C	Mn	P	S	N	Al
%	0.12	0.60	0.045	0.045	0.003	0.04

The simulation task had two time steps. The first one involved tension, the second one compression. The desired movement was defined through boundary conditions for the individual steps. The strain amplitude was 2 mm.

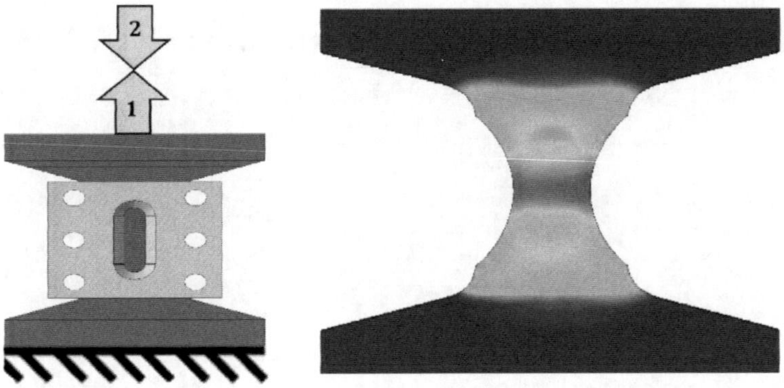

Figure 5. Boundary condition of FE model and strain field in the longitudinal direction

5 Results of measurement

To improve accuracy of the data from the literature, measurement was carried out using the test fixture on the basis of the FE model. Figure 6 shows the experimental stress-strain plot for the first loading cycle. The stress was obtained by dividing the reaction force by the cross-sectional area of the specimen. The reaction force was the difference between the load and the friction force. The friction force was determined by comparing the stress-strain curves in the first loading half-cycle with a standard tensile test curve. The friction force determined in this manner corresponds to the friction coefficient of 0.15. The experimental force-displacement plot for cyclic loading is shown in Figure 7 where the obvious similarity provides evidence of the repeatability of the test.

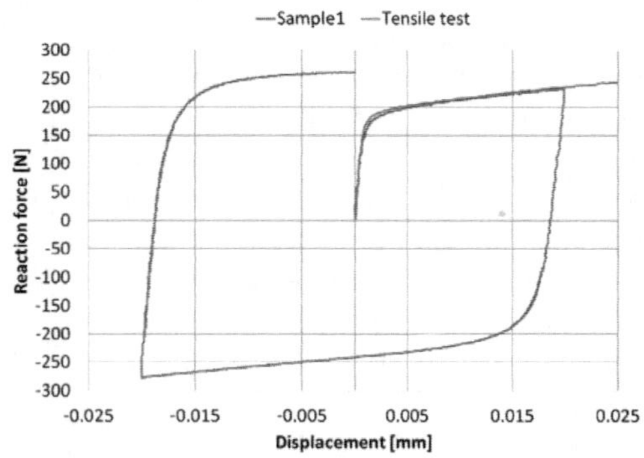

Figure 6. Comparison between a Bauschinger test and a tensile test

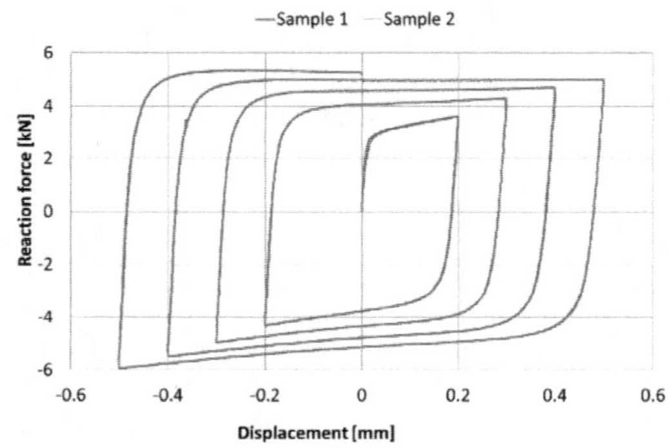

Figure 7. Repeatability of Bauschinger test

6 Results and discussion

For comparison of the measurement data and the computational hardening models, the graph in Figure 8 was constructed. It shows the experimental stress-strain curve and plots from FE simulations for three different material models according to Table 3. The basic isotropic hardening model fails to match the measured relationship and cannot be used for modelling the Bauschinger effect. The kinematic hardening curve captures this effect but fails to reflect other related phenomena. These basic models have proven inadequate for reliable characterization of the test. The behaviour of the modified Lemaitre and Chaboche model is in agreement with the measured data and is suitable for modelling. The discrepancies between this computational model and the experimental data are due to the values of parameters in Table 1. These are the initial values for an optimization exercise.

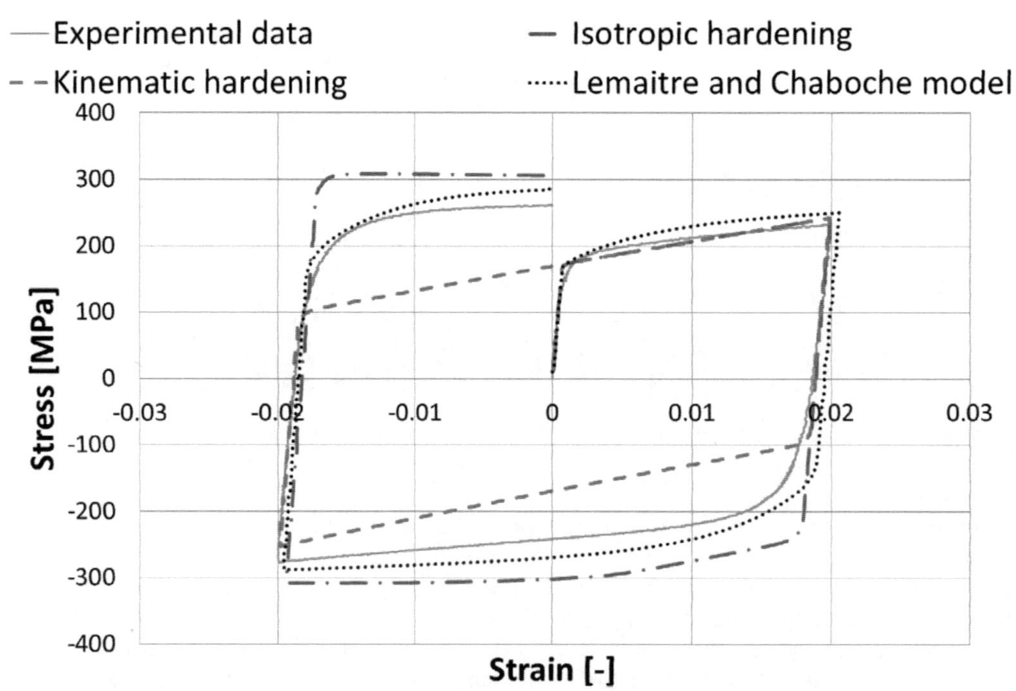

Figure 8. Comparison of hardening models and experimental data

Table 3. Material model parameters

Material model		Parameter	Value	Unit
Isotropic hardening		R_e	150	MPa
		$R_{p0.04}$	265	MPa
Kinematic hardening		Re	150	MPa
		$R_{p0.04}$	265	MPa
Modified Lemaitre and Chaboche	Isotropic	R_∞	62.6	MPa
		b	25	-
	Kinematic	C_1	7200	MPa
		γ_1	158	-

7 Conclusion

The Bauschinger effect in DC01 steel was measured using a newly-proposed test device. It comprised a special fixture with compression springs and the Mercury DIC system coupled with the test machine. The test control was provided by a virtual extensometer in the DIC system for a defined region of the specimen. The readings were then compared with simulations involving several hardening models. This comparison highlighted the inadequacy of standard hardening models for characterizing the effect in question. By contrast, the Lemaitre and Chaboche model yields a very good description of material behaviour under cyclic loading. The procedure was performed on DC01 deep-drawing steel of 1.5 mm thickness but its main potential is associated with future refinement for high-strength steels, such as DP1000, with thicknesses below 1 mm.

Acknowledgements

This paper was created by project Development of West-Bohemian Centre of Materials and Metallurgy No.: LO1412, financed by the MEYS of the Czech Republic.

References

[1] Holas J., 2010, Bauschinger effect and cyclic plasticity in structural alloys of magnesium, Brno, VUT in Brno, thesis

[2] Kim J. et al. ,2014, *International Journal of Plasticity*, Measurement of the Bauschinger behavior of sheet metals by three-point bending springback test with .... ,**Volume 59**, Shaanxi, Pages 84-107, ISSN 0749-6419.

[3] Mištera J., 2018, Advanced Sheet Metal Testing - Bauschinger Effect , WBU in Plzen, thesis

[4] Banabic D., 2010, Springer 2010, Sheet Metal Forming Processes, Berlin:. ISBN 978-3-540-88112-4.

[5] Arlat F. et al., 2012, *International Journal of Solids and Structures*, An application of homogeneous anisotropic hardening to springback prediction in pre-strained .... **Volume 49**, Issue 25, Gyeongbuk, Republic of Korea, p. 3562-72. ISSN 0020-7683.

[6] Merklein M. and Suttner S.,2014, Characterization of the Bauschinger effect and identification of the kinematic Chaboche model by .... Nürnberg: 1Institute of Manufacturing Technology LFT, FAU Erlangen-Nürnberg

[7] Yan J.,1998, Study of Bauschinger Effect in Various Spring Steels. Toronto: University of Toronto, ISBN 0-612-45601-3.

[8] Urbánek M, et al., 2018, *Numisheet 2018, Journal of Physics: Conference Series*, Sheet Thickness Reduction Influence on Fracture Strain Determination, **Volume 1063**, ISSN 1742-6596

Bulk nanostructured metals for advanced medical implants and devices

Ruslan Z. Valiev[1,2], Evgeny V. Parfenov[1], Georgy I. Raab[1], Irina P. Semenova[1], Ludek Dluhoš[3]

[1]Ufa State Aviation Technical University, 12 K. Marx str., Ufa 450008 Russia
[2]Saint Petersburg State University, 198504, Universitetskiy prospekt 28, Peterhof, St. Petersburg, Russia
[3]Timplant s.r.o., Sjednocení 77/1, Ostrava – Polanka, CZ 725 25, the Czech Republic

ruslan.valiev@ugatu.su

Abstract. Nanostructured metals (Ti and Ti alloys, stainless steels, Mg alloys) with enhanced static and fatigue strengths are promising materials for medical implants [1]. The use of severe plastic deformation (SPD) methods leads to significant strengthening of the metallic materials due to their nanostructuring when the formation of ultrafine grains is combined with the formation of nanostructural features – nano-phased precipitations, grain boundary segregations, nano-twins, etc. [2]. In the present article the recent developments from author and his colleagues on continuous SPD processing, i.e. equal channel angular pressing (ECAP)-Conform techniques, for producing nanostructured CP titanium are considered. The use of nanoTi rods with enhanced strength and fatigue life has enabled the fabrication of implants with improved design for dentistry and orthopedics. Furthermore, surface modification of nanoTi through chemical etching and bioactive coatings allows for considerable improvement of its biomedical properties. As a result of conducted studies, miniaturized dental implants and nanoTi plates with reduced thickness and enhanced osseointegration were manufactured and successfully tested in clinical trials.

1. Introduction

Presently, over 70% of implant devices are made of the metallic materials and many of metal and metals alloys that are used in the medical field include titanium and its alloys, stainless steels, Co-Cr alloys, Mg alloys and some other [3,4]. Research and developments of new types of implants with increased safety and efficiency are closely related to improving the mechanical properties of the materials, such as strength, plasticity and durability, no less than the functional ones – corrosion, toxicity level, osseointegration [4-6]. As is known, traditionally the properties of metals and alloys are improved through optimization of their chemical composition with application of plastic deformation and heat treatment.

In the last decade the nanostructuring of metals by SPD aimed at enhancing their properties has become also a developed area of modern materials science and engineering [2]. With regard to medical applications, the creation of nanostructures in metals and alloys by SPD processing can improve both mechanical and biomedical properties. This paper describes the results of our recent investigations relating to titanium and its alloys that are the most extensively used to fabricate medical implants and other items. The examples demonstrate that nanostructured metals with advanced properties offer opportunities to create a new generation of medical devices with improved design and functionality [7-9].

2. Experimental procedures

Rods of Grade 4 commercially pure (CP) titanium 12-mm in diameter, meeting all the specifications of the ASTM F67 standard for medical implants, were used in the study reported herein. The material impurity data was as following (in at.%) 0.197 C, 0.50 Fe, 1.035 O, 0.024 N, 0.094 H and the average grain size of the titanium rods in as-received condition equalled to ~25 μm. The rods were nanostructured via ECAP-C and drawing. The former processing technique is a comparatively recent modification of the standard ECAP method [10-12]. In the course of this process, a work-piece is forced through an ECAP die in a manner analogous to the Conform process, however in this case an upgraded ECAP design is applied for nanostructured materials to be produced with repeat-passes.

As-received Grade 4 Ti rods were subjected to ECAP-C in a die-set with a 120° intersection angle Φ through the B_c route. Subsequent drawing was carried out to a reduction ratio of 85%. The deformation temperature was equal to 200 °C. The processing details are presented in Refs. [11, 12]. The final processed rods had a length of 3 m with the diameter from 3 to 6 mm.

The microstructure was analysed by means of optical as well as transmission electron microscopy (TEM) in a JEOL JEM 2100 TEM with acceleration voltage 200 kV.

The cylinder-shaped samples (3 samples per each processing term as in table 1), 3 mm in diameter and of 15 mm gauge length were tensile-tested at room temperature in an INSTRON-type device with a primary strain rate of 10^{-3} s^{-1}. The tensile axis was parallel to the rod axis. Stress-controlled fatigue tests at ambient temperature were carried out to characterize the behaviour of nanostructured (NS) and conventional coarse grain (CG) CP titanium at a load ratio of R ($\sigma_{min}/\sigma_{max}$) = -1. Rotational bending testing with a frequency of 50 Hz was performed.

Surfaces of mechanically-polished CG and NS samples of Grade 4 Ti were acid-etched in 30% HNO_3 + 3% HF+H_2O for 20 minutes. Surface topography was examined with a LSM-5-Exciter laser scanning microscope (LSM). Surface profiles were analysed to determine the surface roughness parameter R_a and the size of etching dimples.

3. Results and discussions

3.1. Grain refinement and mechanical properties

The processing resulted in a large reduction in grain size, from the 25 μm equiaxed grain structure of the initial titanium rods to 150 nm after combined ECAP-C and drawing processing, as shown in figure 1. The selected area electron diffraction pattern, figure 1(c), further suggests that the ultrafine grains contained predominantly high-angle grain boundaries with increased grain-to-grain internal stresses [11, 12].

Figure 1. Microstructure of Grade 4 CP Ti: a) the initial coarse-grained rod; b, c) after ECAP+drawing (optical and TEM pictures)

IOP Conf. Series: Materials Science and Engineering **461** (2019) 012089 doi:10.1088/1757-899X/461/1/012089

The important thing was to produce homogeneous ultrafine-grained structure along the entire three-meter rod lengths to enable economical pilot production of implants and provide sufficient material for thorough testing of the mechanical and bio-medical properties of the nanostructured titanium. As is seen in table 1, the strength characteristics of nano Ti are very high and even considerably excel over the properties of two-phase Ti alloy. At the same time, as was demonstrated in recent studies [13], the nature of such high strength originates from several hardening mechanisms when dislocation and substructure hardening is introduced alongside with grain boundary hardening dealing with ultrafine grains and also the segregation of impurities at grain boundaries is possible.

Table 1. Mechanical properties of conventionally processed and nanostructured Grade 4 Ti produced by ECAP-C and drawing. Data on Ti-6Al-4V ELI alloy are presented to compare.

Processing/ treatment terms	UTS, MPa	YS, MPa	Elongation, %	Reduction of area, %	Fatigue strength at 10^7 cycles
Conventional Ti (as-received)	700±9	530±20	25±2	52	340
nano-Grade 4	1330±12	1267±23	11±1	48	620
Annealed Ti-6Al-4V ELI	940±10	840±18	16±1	45	530

3.2. Design of miniaturized implants

The simple rules can be used when it comes to redesigning the devices in terms of the influence of changing materials, such as replacing CG Ti by nanostructured CP Ti [9]. Fatigue performance must be retained with thorough account for the possibility of changing the device cross-sectional dimensions. The data on fatigue properties of nano Ti and initial Ti (see Table 1) [7] were used to calculate the design of dental implant with the diameter 2.4 and 2.0 mm (figure 2a) produced by the company "Timplant" s.r.o., Czech Republic [14,15] and nanoTi mini-plates (figure 2b) from the company «Conmet», Moscow (www.conmet.ru/e_main.html). The implants were tested to demonstrate exceptionally high properties [7].

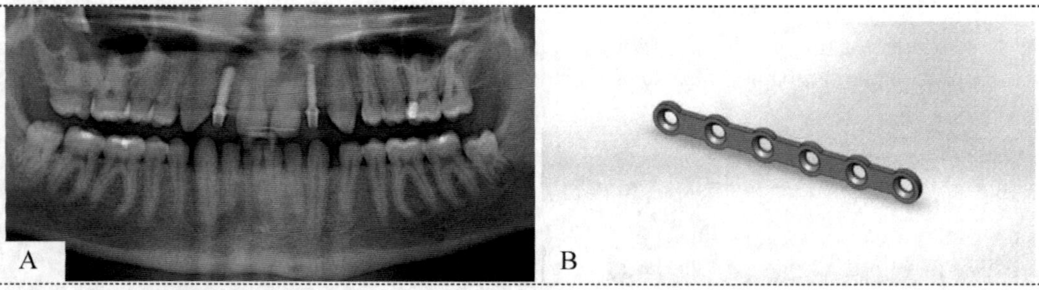

Figure 2. Nanoimplant of 2.0 mm diameter (www.timplant.cz/en/) from NS Grade 4 Ti in a panoramic X-ray radiograph after surgery (A) and the image of a miniplate with six holes made from NS Grade 4 Ti (B).

3.3. Surface modification of nano Ti implants

Surface properties are an important aspect of an implant design to ensure effective osseointegration. Pure Ti has very low bioactivity (i.e. bioinert material) and it does not bond directly to the human bone [6]. Extensive studies have shown that grain refinement down to the nanoscale in CP Ti can stimulate various bone-forming cell types to adhere and proliferate with increased efficiency [7,8,16-18]. Additional surface modification can further improve bioactivity of implants made from nano Ti. Two main approaches of surface modifications are studied in our works: chemical etching and deposition of bioactive coatings [7, 18-24].

The topography of etched surfaces is strongly determined by etching solution and etching time. Different solutions, such as acidic (H_2SO_4/H_2O_2) or basic (NH_4OH/H_2O_2) Piranha solutions can be used for etching of CP Ti [24] resulting in different surface topographies. Manipulation with etching time can substantially modify the surface topography and this effect is more pronounced in the nano Ti, as it has been very recently demonstrated in figure 3 [24].

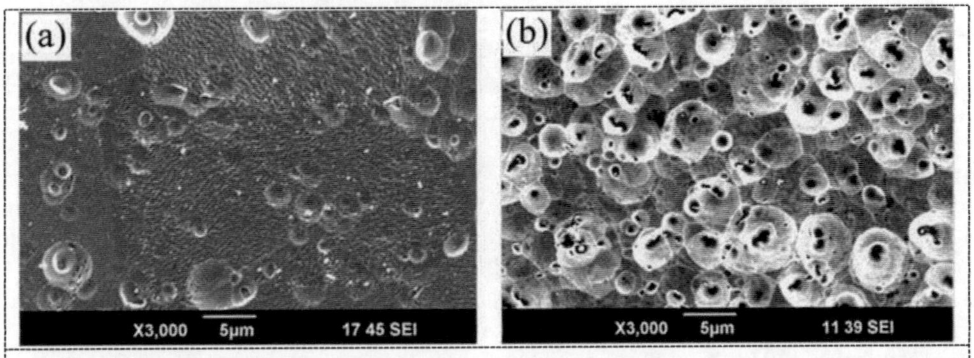

Figure 3. Surface of CG and NS samples of Grade 4 Ti after mechanical polishing and etching in the mixture of acids 30% HNO_3 + 3% HF+H_2O for 20 minutes: (a) CG Ti surface; (b) nano-Ti surface. SEM images.

Biocompatible coatings can also essentially facilitate integration of the nano Ti implants into a human bone [22, 23]. Presently, the research into synthesis of biocompatible coatings integrating the inorganic (Ca-, P- containing phases) and organic (biologically active and bioinert molecules) components on titanium implants appears to be topical state of the art [25, 26].

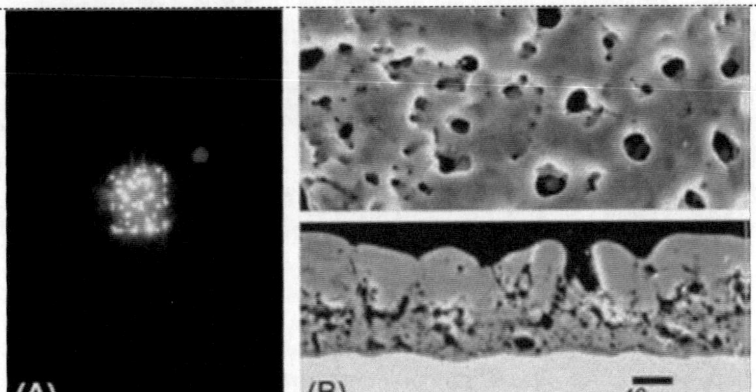

Figure 4. Photographs of microdischarges on the sample surface during PEO process (a); SEM image of PEO Ca-, P- containing coating on titanium, top view (above) and cross-section (below) (b) [28, 31].

During the last decade, significant attention is attracted to Ca-, P- containing coatings obtained by the method of plasma electrolytic oxidation (PEO) [20, 27, 28]. PEO process is an expansion of the traditional anodizing into the high voltages up to 600 V; these voltages promote microdischarges within the coating (figure 4); this results in its resoldifying and intensive growth [28, 29]. This method is applicable to both CG and UFG titanium [20, 30]. The coatings obtained by this method contain stable titania (rutile and anatase) tightly attached to the surface because of the process mechanism including electrochemical oxidation and numerous melting and crystallizing events at the microdischarge sites. This coating formation mechanism helps to develop coatings with regulated porosity, with the pore size from 0.1 to 10 μm [31]. This coating morphology provides gradual change of the elasticity modulus from the metallic implant to the bone; this enhances their biomechanical compatibility. High surface area of the PEO coating promotes osteoblast attachment on the implant surface. Applying varying pulse polarity during PEO and introduction of bioactive particles into the electrolyte helps to incorporate the anions and cations of the electrolyte into the coating; this provides Ca-, P- containing bioactive crystalline phases within the coating: hydroxyapatite, tricalciumphosphate, tetracalciumphosphate, perovskite [31]. Adhesion of the PEO coatings is higher than of the other coating types.

4. Summary and conclusions

Thus, nanostructured Grade 4 titanium appears to be the material with a very high potential for medical applications as its elements have no harmful effects on humans. It should be noted that nanostructured CP Ti also imparts good machinability. Additional research on nanostructured Ti and some of its alloys is foreseen to even further increase ultimate strength and yield strength as well as to develop a way to reduce the effective modulus of elasticity that could come close to the modulus of elasticity of jaw-bone [1, 7].

The recent results have shown also that ECAP-C with subsequent drawing provides new opportunities in the development of nanostructured Ti, providing the ability to enhance the strength uniformly in long-length rods that can be machined into dental implants and other devices for medical applications. Nanostructuring of CP Ti by SPD processing produces a material with mechanical properties superior to those of Ti-6Al-4V. The application of nano-Ti in maxillofacial surgery can be promising for producing miniature implant designs, in particular plates that will endure the same loads as the conventional items.

Chemical etching and deposition of bioactive coatings can be utilized for surface modification in nanostructured Ti. Nanostructured CP Ti with modified surfaces can show improved functional properties such as improved surface bioactivity and reduced Ti ion release into human body. This makes nanostructured Ti an attractive material for manufacturing implants. The first clinical studies have shown a significance of these research results [14, 32, 33].

A number of biomedical researches have been already made to date, which shows that further biological studies are still needed. This applies also to other nanostructured metals (β-Ti alloys, stainless steel, Co-Cr alloys, magnesium alloys etc.), which are now actively developing for biomedical applications [1].

Acknowledgments

The authors gratefully acknowledge the financial support through the RFBR project 16-02-00094 (RZV) and from Saint Petersburg State University in the framework of Call 3 project (id 26130576) (RZV).

References

[1] Lowe T C, Valiev R Z 2014 Frontiers of bulk nanostructured metals in biomedical applications, Chapter 1 in *Advanced biomaterials and biodevices,* ed A Tiwari and A N Nordin (Wiley-Scrivener Publ., USA) pp. 3-52

[2] Valiev R Z, Zhilyaev A P, Langdon T G 2014 *Bulk Nanostructured Materials: Fundamentals and Applications* (John Wiley & Sons, Inc., Hoboken, USA)

[3] Geetha M, Singh A K, Asokamai R, Gogia A K 2009 *Prog. Mater. Sci.* **54** 397
[4] Chen Q, Thouas A 2015 *Mater. Sci. Eng. R* **87** 1
[5] Niinomi M, *Met. Mat. Trans A* 2001 **32A** 477
[6] Brunette D M, Tengvall P, Textor M, Thomsen P 2001 *Titanium in Medicine* (Springer, Verlag, Berlin) p. 1019
[7] Valiev R Z, Sabirov I, Zemtsova E G, Parfenov E V, Dluhoš L, Lowe T C 2018: Nanostructured pure Ti for development of miniturized biomedical implants, Chapter 2.9 in *Titanium in medical and dental applications* (1st Edition, ed. by F. Froes, M. Qian, Woodhead Publishing).
[8] Estrin Y, Lapovok R, Medvedev A E, Kasper C, Ivanova E, Lowe T C 2018 Mechanical performance and cell response of pure titanium with ultrafine-grained structure produced by severe plastic deformation, Chapter 4.4 in *Titanium in medical and dental applications* (1st Edition, ed. by F. Froes, M. Qian, Woodhead Publishing) p 419
[9] Semenova I P, Klevtsov G V, Klevtsova N A, Dyakonov G S, Matchin A A, Valiev R Z 2016 *Adv. Eng. Mater.* **18** (7) 1216
[10] Raab G I, Valiev R Z, Lowe T C, Zhu Y T 2004 *Mater. Sci. Eng. A* **382** 30
[11] Gunderov D V, Polyakov A V, Semenova I P, Raab G I, Churakova A A, Gimaltdinova E I, Sabirov I, Segurado J, Sitdikov V D, Alexandrov I V, Enikeev N A, Valiev R Z 2013 *Mater. Sci. Eng. A* **562** 128
[12] Polyakov A V, Dluhos L, Dyakonov G S, Raab G I, Valiev R Z 2015 *Adv. Eng. Mater.* **17** 1865
[13] Dyakonov G S, Mironov S, Semenova I P, Valiev R Z, Semiatin S L 2017 *Mater Sci Eng A* **701** 289
[14] www.timplant.cz/en/
[15] Mishnaevsky Jr. L, Levashov E, Valiev R Z, Segurado J, Sabirov I, Enikeev N, Prokoshkin S, Solov'yov A V, Korotitskiy A, Gutmanas E, Gotman I, Rabkin E, Psakh'e S, Dluhoš L, Seefeldt M, Smolin A 2014 *Mater. Sci. Eng. R* **81** 1
[16] Park J W, Kim Y J, Park C H, Lee D H, Ko Y G, Jang J H, Lee C S 2009 *Acta Biomater.* **5** 3272
[17] Lai M, Cai K, Hu Y, Yang X, Liu Q 2012 *Colloids Surf. B Biointerfaces* **97** 211
[18] Nie F L, Zheng Y F, Wei S C, Wang D S, Yu Z T, Salimgareeva G K, Polyakov A V, Valiev R Z 2013 *J. Biomed. Mater. Res.* Part A **101A** 1694
[19] Polyakov A V, Dyakonov G S, Semenova I P, Raab G I, Dluhos L, Valiev R Z 2015 *Adv. Biomater. Dev. Med.* **2** 63
[20] Yao Z Q, Ivanisenko Y, Diemant T, Caron A, Chuvilin A, Jiang J Z, Valiev R Z, Qi M, Fecht H J 2010 *Acta Biomater.* **6** 2816
[21] Zemtsova E G, Arbenin A Y, Valiev R Z, Orekhov E V, Semenov V G, Smirnov V M 2016 *Surf. Mater.* **9(12)** 1010
[22] Nazarov D V, Zemtsova E G, Valiev R Z, Smirnov V M 2015 *Materials* **8(12)** 8366
[23] Shin M H, Baek S M, Polyakov A V, Semenova I P, Valiev R Z, Hwang W B, Hahn S K, Kim H S 2018 *Sci. Rep.* **8** 9907
[24] Nazarov D V, Zemtsova E G, Solokhin A Yu, Valiev R Z, Smirnov V M 2017 *Nanomater.* **7** 15
[25] Dorozhkin S V 2015 *Mater. Sci. Eng. C* **55** 272
[26] Wu S, Liu X, Yeung K W K et al 2014 *Mater. Sci. Eng. R* **80** 1
[27] Yerokhin A L, Nie X, Leyland A et al 1999 *Surf. Coat. Tech.* **122** 73
[28] Parfenov E V, Yerokhin A, Matthews A 2009 *Surf. Coat. Tech.* **203** 2896
[29] Parfenov E V, Yerokhin A, Nevyantseva R R et al 2015 *Surf. Coat. Tech.* **269** 2
[30] Sharkeev Y P, Legostaeva E V, Eroshenko Y A et al 2009 *Composite Interfaces* **16** 535
[31] Yerokhin A, Parfenov E V, Matthews A 2016 *Surf. Coat. Tech.* **301** (2016) 54.
[32] Nazarov D V, Zemtsova E G, Solokhin A Y, Valiev R Z, Smirnov V M 2017 *Nano* **7** 15
[33] Fernandes D J, Elias C N, Valiev R Z 2015 *Mater. Res.* **18** 1163

IOP Conf. Series: Materials Science and Engineering **461** (2019) 012090 doi:10.1088/1757-899X/461/1/012090

Low Temperature Superplasticity of High-Strength Ultrafine-Grained Al 7050 alloy

E.V. Bobruk[1,2], M.Yu. Murashkin[1,2], I.V. Lomakin[2], V.U. Kazykhanov[1], R.Z. Valiev[1,2]

[1]Ufa State Aviation Technical University, 12 K. Marx str., Ufa 450008 Russia
[2]Saint Petersburg State University, 198504, Universitetskiy prospekt 28, Peterhof, St. Petersburg, Russia

ruslan.valiev@ugatu.su

Abstract. Ultrafine-grained (UFG) Al 7050 alloy of the 7xxx series demonstrates very high strength at room temperature with UTS=845 MPa and hardness Hv=255, which considerably exceeds the value after conventional heat treatment T73, i.e. Hv=165. It is also revealed that with increasing temperature the alloy exhibits rather high ductility with elongation to failure δ=170% at 120 °C and strain rate of 5×10^{-4} s^{-1} and typical superplastic behavior with δ=480% and δ=690% at temperatures 170 °C and 200 °C correspondingly. Such values of superplasticity (SP) temperature are considerably lower than that typical of this alloy with micron-sized grains (T_{SP}=450-500 °C). Such reduction in superplasticity temperature made it possible to preserve ultrafine grains in the alloy structure and provide its high-strength state after SP straining.

1. Introduction

Over the last 20 years, a breakthrough has been achieved in producing Al-based alloys using severe plastic deformation (SPD) techniques [1]. Along with high mechanical strength and even «superstrength», Al alloys with such a microstructure reveal a range of remarkable performance properties such as enhanced fatigue strength, fracture toughness, electric conductivity, etc. [1-3].

Al-based alloys with the UFG structure can also demonstrate superplastic behavior at temperatures much lower and at strain rates higher (more than 10^{-2} s^{-1}) than those appropriate for the coarse-grained materials [2, 4-5]. The general principles of SP, as demonstrated by the UFG materials, have been summarized in a number of reviews and books [1, 2, 4, 5]. The reports note that low-temperature and high-rate SP is very attractive for commercial superplastic forming, as it enables reduced power consumption, increases the service life of a shape-forming tool and reduces the timeframe for production. Besides, lower SP temperature induces a reduction in the microstructural degradation in the UFG materials; as a result, the produced components retain their high level of mechanical and performance characteristics. Recent studies on the model alloys of the Al–Zn and Mg–Li systems showed that the formation of the UFG structure with grain boundaries containing layers formed by segregations of alloying elements may ensure the signs of SP in these alloys, even at room temperature (RT) [6-7]. Good ductility values are observed in the commercial UFG alloy 1570 of the Al–Mg–Sc system (up to 300% at 200°C), [8] in which the ultimate tensile strength at RT is 950MPa [9].

High-strength aluminum alloys of the system Al-Zn-Mg-Cu (7xxx series) are widely known and at present, there is an avalanche of experimental data on mechanical and corrosion behavior. Despite the fact that a large amount of literature deals with the aging of coarse-grained 7xxxx series Al alloys, there is a lack of systematic research on the UFG states and their influence on mechanical and functional properties, including the manifestation of SP at low temperatures and high strain rates.

2. Experimental procedures

In this work, commercial aluminum alloy 7050 of the 7xxx series was used as the material under investigation. The chemical composition of the alloy is Al-6.4Zn-1.7Mg-2.3Cu-0.12Zr-0.10Fe-0.05Si-0.02Ti (wt. %).

After quenching, the disk-shaped samples 20 mm in diameter and 1.4 mm in thickness were subjected to high pressure torsion under a pressure of 6 GPa at RT for maximum refining of the grain structure. As was experimentally stated, the hardness of the UFG alloy discs with the chosen sizes stabilized on the spot located at a half radius when n ~10.

Thus, all the studies on the structure and mechanical properties of the UFG alloy were performed with precision on the half radius of the samples.

A number of initial samples were subjected to conventional treatment T73 that is recommended for hardening of the 7xxx series alloys in [10-12]. The treatment included quenching and two-stage aging at a temperature of 110°C during 6 h and then 160°C during 16 h.

The thermal stability of the UFG alloy was also investigated, the alloy blanks were annealed at 120, 170 and 200 °C for 10 hours, the hardness was measured after 1, 2, 3, 5, 8 and 10 hours of exposure.

Hardness was determined at room temperature according to the Vickers method (Hv) using a Buehler Omnimet Micromet-5101; 10measurements were taken in each area, at a load of 0.5N and a 15 s holding time under load.

The tensile tests were carried out on the test machine Instron. The tests were computer controlled, the temperature, strain and stress were registered. The mechanical tests of the alloy samples with a gauge part of 2.0×1.0×0.8 mm were conducted at RT, 120, 170 and 200 °C in the range of strain rate of $10^{-1}...10^{-4}$ s^{-1} to measure the total elongation to failure.

The microstructures were examined with a JEOL JEM 2100 transmission electron microscopy (TEM), operating with an accelerating voltage of 200 kV. The samples intended for foils were first cut using electrical discharge machining, then mechanically thinned to a thickness of 0.15 mm, and finally electrically polished in the solution of 80 % methanol and20% nitric acid on the TenuPol-5 unit at a polishing temperature from –25 °C.

3. Results and discussions

3.1. Grain refinement and thermal stability

Figure 1(a,b) displays TEM images of the homogeneous microstructure after HPT at RT obtained via electron microscopy. The initial coarse-crystalline structure transforms into the UFG structure with an average grain size of 140±15 nm. The circular distribution of spots on the diffraction pattern indicates high-angle misorientations of the grain boundaries formed during HPT (figure 1a). Recent studies into the fine structure showed that the formation of UFG microstructure during HPT at RT in the 7xxx series alloys is accompanied by the formation of nano-sized particles of the $\mu(Zn_2Mg)$ phase as well as the segregation of Zn and Mg atoms mostly at grain boundaries. As was shown in [11], the presence of grain boundary segregations in the UFG Al alloys may favorably influence the strength and high-strain-rate superplasticity.

After HPT the hardness of the UFG alloy 7050 constituted Hv 255 and this exceeds the alloy hardness after conventional heat treatment T73 (Hv 165) by ~ 55 %.

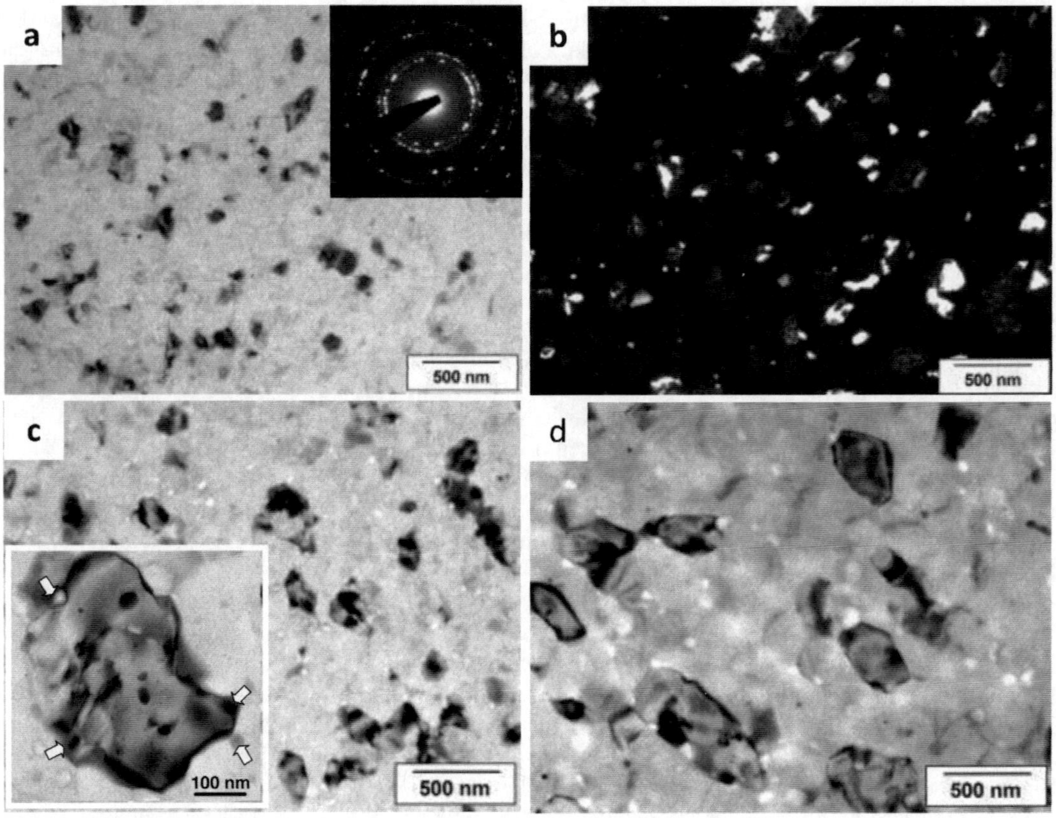

Figure 1. Microstructure of UFG Al 7050 alloy after 10 rotations of HPT at RT: (a) bright-field and diffraction pattern; (b) dark-field TEM analysis; and further annealing at 170 and 200 °C correspondingly (c. d) bright-field

No noticeable changes were registered in the UFG structure after annealing for 10 hours at 120°C. The average grain size was 160±10 nm. Long-term annealing at 170 and 200 °C led to a more significant increase of ultrafine grains (figure 1c,d) and their average grain size increased to 230±20 nm and 460±15 nm correspondingly. In addition to grain growth, annealing resulted in coarsening of the particles of $\mu(Zn_2Mg)$ phase located at GB (figure 1c). Also, annealing led to coarsening of particles in ultrafine grains.

Evolution of the UFG structure well correlates with the nature of the change in the alloy hardness (figure 2). As a result of annealing at 120°C, when microstructural changes were minimal, the alloy hardness decreased to 230 Hv. As the heat treatment progresses at elevated temperatures, the hardness of UFG alloy monotonically decreases and reaches the level of hardness after treatment T73 after 8 hours of annealing at 170 °C and at 200 °C during 2 hours.

Figure 2 demonstrates thermal stability of ultrafine-grained Al alloy after annealing at 120, 170 and 200 °C during 10 hours. It was established that the UFG aluminum alloy retains its thermal stability up to temperatures of 170 °C for 10 hours. At a temperature of 200 °C, noticeable softening begins after 4 hours of annealing; its hardness reduces to 150 Hv.

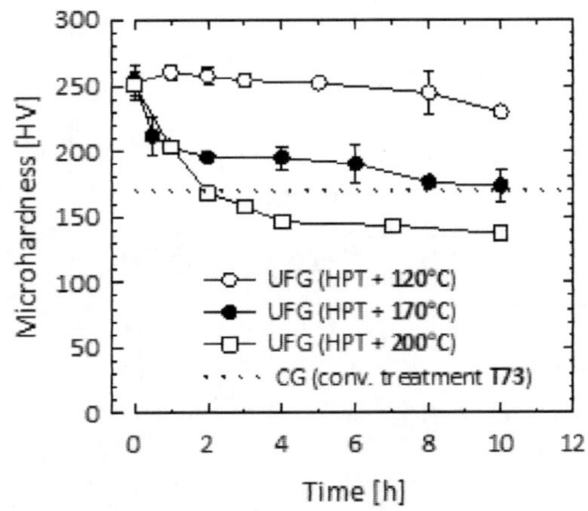

Figure 2. Thermal stability of ultrafine-grained Al alloy after annealing at 120, 170 and 200 °C during 10 hours.

3.2. Mechanical properties

The formation of UFG state with an average grain size of 150±6 nm and a shape form of 1.2 resulted in the growth of ultimate tensile strength (UTS) to 845±5 MPa, of yield strength (YS) to 818±9 MPa at RT. After conventional treatment T73 the following values were registered in the 7050 alloy: UTS =580±5 MPa, YS = 516±10 MPa.

To reveal the ability of high-strength material to SP in the lower temperature zone, the tensile tests were performed with strain rates of 10^{-1}, 10^{-2}, 5×10^{-3}, 10^{-3} 5×10^{-4} and 10^{-4} s^{-1}at 120, 170 and 200 °C. The temperatures of the thermal stability from 120 to 200 °C were chosen according to the obtained results of microhardness change after annealing (figure 2).

Figure 3demonstrates the typical stress-strain curves taken after mechanical testing (at 120, 170 and 200 °C) of the UFG alloy processed by HPT at RT. As the strain rate reduces, the ductility of the alloy increases. The maximum values of ductility 170 % are achieved at a strain rate of 5×10^{-4} s^{-1}. At a testing temperature of 120 °C the ultimate tensile strength retains high in the whole range of rates. The hardness after testing at 120 °C is 230 Hv for strain rate 10^{-2} s^{-1} and 200 Hv for rate 5×10^{-4} s^{-1}.

The growth of the testing temperature to 170 °C results in the achievement of unusually high strain rate ductility - 480 %, the strength retains at a rather high level, as shown in figure 3. When the strain rate reduces, the ductility decreases, which is evidently connected with the recovery processes taking place during the long-term static testing. However, at 170 °C in the whole rate range the high ductility values are achieved. The hardness after testing at 170 °C is 195 Hv.

The further increase of the temperature of tensile test to 200 °C leads to the growth of ductility to 690 %, the strength values reduce considerably, as shown in figure 3. The hardness after testing at 200 °C decreases to the hardness values after heat treatment T73 and is 165 Hv.

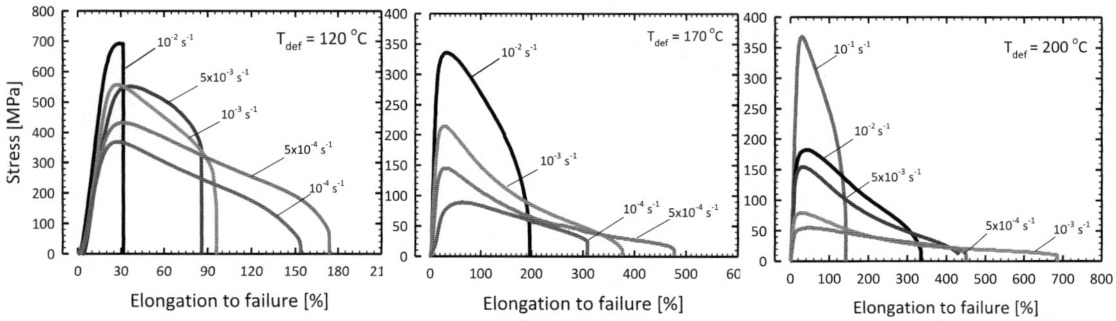

Figure 3. Engineering tensile curves of UFG Al 7050 alloy. Tensile testing at: 120, 170 and 200 °C

Figures 4 demonstrates the curves of flow stress and elongation vs strain rate that are plotted against the results of static testing executed at 120, 170 and 200 °C. When the deformation temperature increases, the stress reduces, which is typical of superplastic materials [2,4,13].

The strain rate decrease during testing at 120 °C in the whole range of rates results in the reduction of the flow stress and the increase of the ductility, as shown on figure 3.

The testing temperatures of 170-200 °C are marked with the reduction in the flow stress value in the rate range from 10^{-2} to $5x10^{-4}$ s^{-1}.

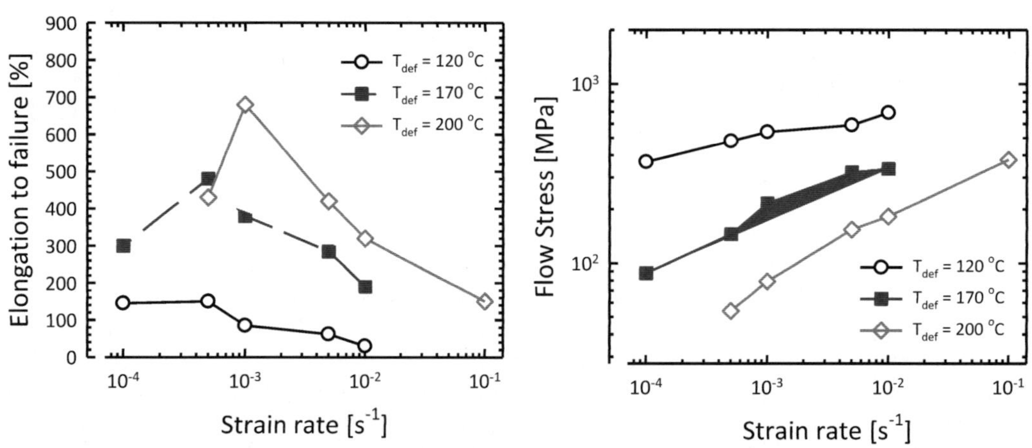

Figure 4. Dependence of elongation and flow stress on strain rate based on the results of static tests at 120, 170, and 200 °C of UFG Al 7050 alloy

The maximum elongations are achieved at 170 °C and a strain rate of $5x10^{-4}$ s^{-1} and at 200 °C and strain rate of 10^{-3} s^{-1}. It should be noted that at 170 °C there are no clear maximums of ductility. However, currently, in a number of production-oriented industries the task has been set to refocus manifestations of superplasticity towards the region of low temperatures and high strain rates [14,15,16]. Therefore, according to the results of these studies, it is possible to determine the strain rate range where the material exhibits high values of plasticity (elongation near 300 %) and retains high values of strength (Hv near 190-200): strain rate from 10^{-3} to 10^{-4} s^{-1} at a temperature of 170 °C.

4. Summary and conclusions

1.The formation of UFG state with an average grain size of 140±15 nm resulted in the growth of microhardness in the 7050 alloy to Hv255±12, in UTS to 845±5 MPa, in YS to 818±9 MPa at RT. After conventional treatment T73 the values are Hv=160, UTS=580±5 MPa, YS=516±10 MPa.
2.The ultrafine-grained aluminum alloy has been established to retain its thermal stability up to temperatures of 170 °C for 10 hours. At a temperature of 200 °C, noticeable softening begins after 4 hours of annealing, its hardness reduces till 150Hv.
3.Tensile tests are performed at 120, 170 and 200 °C in the strain rate range from 10^{-2} to 10^{-4} s^{-1} to determine the high values of ductility and strain rate sensitivity. The maximum elongations of 170 % are achieved at 120 °C and a strain rate of 5×10^{-4} s^{-1}, 480 % - at 170 °C and a strain rate of 5×10^{-4} s^{-1}, at 200 °C and a strain rate of 10^{-3} s^{-1} the elongations achieved 690 %. The ultrafine-grained alloy retains an increased level of hardness after mechanical tests in the temperature range 120-200 °C.

Acknowledgments
The authors gratefully acknowledge the financial support through the Ministry of Science and Higher Education of the Russian Federation under grant agreement No. 14.586.21.0061 (unique project identifier RFMEFI58618X0061).

References

[1] Sabirov I, Murashkin M Yu, Valiev R Z 2013 *Mater. Sci. Eng.* A **560** 1
[2] Valiev R Z, Zhilyaev A P, Langdon T G 2014 *Bulk Nanostructured Materials: Fundamentals and Applications* (John Wiley & Sons, Inc., Hoboken, USA)
[3] Sabirov I, Enikeev N A, Murashkin M Yu, Valiev R Z 2015*Bulk Nanostructured Materials with Multifunctional Properties*(Springer Cham Heidelberg New York Dordrecht London)
[4] Kawasaki M, Langdon T G 2007 *J. Mater. Sci.* **42** 1782
[5] Kawasaki M, Langdon T G 2016 *J. Mater. Sci.* **51** 19
[6] Edalati K, Masuda T, Arita M, Furui M, Sauvage X, Horita Z, Valiev R Z 2017 *Sci. Rep.* **7**: 2662
[7] Edalati K, Horita Z, Valiev R Z 2018 *Sci. Rep.* **8:**6740
[8] Mishra RS, Valiev RZ, McFadden SX, Islamgaliev RK, Mukherjee AK 2001*Phil. Mag. A*. **81**37
[9] Valiev R Z, Enikeev N A, Murashkin M Yu, Kazykhanov V U, Sauvage X 2010*Scr. Mater.* **63** 949
[10] García–Infanta J M, Zhilyaev A P, Sharafutdinov A, Ruano O A, Carreno F 2009*J. All. Comp.* **473** 163
[11] Lin Y C, Jiang Y Q, Zhang J L, Chen X M2018*Adv. Eng. Mater.* **20(1)** 1
[12] Hidalgo–Manrique P, Orozco–Caballero A, Cepeda–Jiménez C M, Ruano O A, Carreño F2016 *J. Mater. Sci. Tech.* **32** 774
[13] Kumar P, Kawasaki M, Langdon T G 2016*J. Mater. Sci.* **51** 718
[14] Kaibyshev O A 1992 *Superplasticity of alloys, intermetallides and ceramics*(Berlin, Springer-Verlag)
[15] Barnes AJ 1999*Mater. Sci. Forum* **357-359** 3
[16] Sanders DG 2001*Mater. Sci. Forum* **357–359** 17

IOP Conf. Series: Materials Science and Engineering **461** (2019) 012091 doi:10.1088/1757-899X/461/1/012091

Evaluation of dielectric properties of 3D printed objects based on printing resolution

P Veselý[1], T Tichý[1], O Šefl[2], E Horynová[1]

[1] Dept. of Electrotechnology, Czech Technical University, Technická 2, 166 27 Praha, Czech Republic
[2] Dept. of Electrical Power Engineering, Czech Technical University, Technická 2, 166 27 Praha, Czech Republic

E-mail: veselp13@fel.cvut.cz, tichyto4@fel.cvut.cz, seflondr@fel.cvut.cz, eva.horynova@fel.cvut.cz

Abstract. Fused Deposition Modelling (FDM) is one of the most common methods of 3D printing used in many fields of industry, especially in development departments. Since plastics are fundamental materials for electronics industry, the aim of this work is to examine dielectric properties of objects printed from such materials. This work's contribution is the evaluation of the dependency of the printed objects properties on printing quality and the use of 3D printed plastic components in electronics.
For the experiment, three commonly used materials in FDM were chosen – PLA, ABS and PET-G. The materials were pure without any additional admixtures and the relevant test samples were printed with different printing resolution (height of one layer). The following properties were examined – permittivity, dissipation factor and dielectric strength. The results showed that permittivity slightly decreased with increasing height of one layer. Dissipation factor varied significantly in the measured range and there was no apparent dependency on the printing resolution. Rather, it was an indicator of the printing quality. Dielectric strength also slightly decreased with the decreasing resolution; however this parameter was governed primarily by the employed material. Generally, an improvement of the dielectric properties of these materials is required due to a relatively small dielectric strength, for example by adding admixtures to the base material or better printing quality.

1 Introduction

Classic methods of manufacturing (mould pressing, cutting etc.) are time-consuming and expensive, so other methods are required for a quick prototyping. With an expansion of Additive Manufacturing (AM) and 3D printing technologies, the time of development of electronic devices rapidly decreased. [1]

Fused Deposition Modeling (FDM) is one of the methods of 3D printing that uses mainly thermoplastic materials. Its principle is simple and, in general, it can print any material that melts and then solidifies again. Obviously, there are requirements on its viscosity during liquid state to achieve good adhesion to the printing substrate and cohesion of layers. Standard FDM uses material in the form of a string supplied on a spool (filament). Commonly used materials are thermoplastic polymers like acrylonitrile butadiene styrene (ABS), polyethylene terephthalate glycol-modified (PET-G), poly-lactic acid (PLA), polyamide (nylon), co-polyester (CPE), polycarbonate (PC) etc. [2]

These materials are dielectric and some of them find use in electronics industry as insulators, e.g. wire insulation, boxes and construction parts. Among the other materials' electrical properties, dielectric strength presents a very important parameter that determines their viability in power electronics, where high voltages typically occur. Other relevant dielectric parameters are permittivity and dissipation factor. Both these factors determine the behavior of the material under electric field and the losses caused by alternating current flow.

FDM does not create ideally homogenous solid components; therefore there are differences in properties of final objects in comparison with standard methods of manufacturing, e.g. moulding. The differences are caused by air-gaps present within the 3D object (between lines in X and Y axis and layers in Z axis) or defective cohesion of lines and layers etc. Some of the influencing parameters are the printing resolution (height of one layer), Z axis offset and the temperature of the extrusion head. [3]

According to [4], objects printed by FDM from materials ABS and PLA have acceptable dielectric properties and dielectric strength for lower voltage level applications. The contribution of this study is to examine the mentioned dielectric properties of materials designated for FDM, respectively the properties of printed objects from these materials in dependency on printing quality. In this experiment, the quality is defined by the printing resolution.

The comparison of the measurement results with the known values of chosen parameters and the results from similar work [4] should lead to better understanding of how to use these materials and FDM in electronics industry, especially for low-frequency and power applications.

2 The experiment description

For the experiment, three commonly used materials mentioned above were chosen – ABS, PLA and PET-G. Filaments (1.75 mm) were delivered directly from the producers of 3D printer filaments; they should not contain any additional admixtures.

Table 1. Measured samples

Sample identification	Amount	Material	Printing resolution (μm)	Printing temperature (°C)	Manufacturer of filament
PLA50	10	PLA	50	220	Prusa Polymers, Czech Republic
PLA100	10	PLA	100	220	
PLA150	10	PLA	150	220	
PLA200	10	PLA	200	220	
ABS50	10	ABS	50	255	
ABS100	10	ABS	100	255	
ABS150	10	ABS	150	255	
ABS200	10	ABS	200	255	
PET50	10	PET-G	50	235	
PET100	10	PET-G	100	235	
PET150	10	PET-G	150	235	
PET200	10	PET-G	200	235	

The samples were designed as circles with a diameter of 5 cm and width of 0.6 mm. The width was chosen as the least common multiple of the four chosen widths of one layer – 50, 100, 150 and 200 μm. Altogether, ten samples of each combination (material and printing resolution) were prepared. The simplified sample identification can be seen in Table 1.

Prusa i3 MK3 that utilized the RepRap concept was chosen as the 3D printing device, as it represents one of the most common printers. [5] The digital model of the sample was converted to printer instructions by open-software Slic3r with the following settings for all materials – 100 %

filling density with fill angle 90°, one line of perimeter and default speed of printing recommended by printer producer. The printing temperatures were slightly changed from the recommended values according to the observation during the printing of the first few samples, as can be seen in notation in Table 1. The printer was carefully calibrated in Z axis to get the best possible results.

2.1 *Permittivity and dissipation factor measurement*

For low frequency and low permittivity measurements, the parallel plates method (or capacitor method) is optimal. [6] Therefore, a special electrode system was used. It consisted of two parallel circle electrodes with the same diameter as the sample and a spring system for fixation of the sample. The system was connected to LCR meter APPA 703 by 4-wires probe. After putting the sample between the electrodes, capacitance and dissipation factor were measured with measuring frequency 100 Hz, 1 kHz and 100 kHz. From capacitance, permittivity can be calculated as:

$$\varepsilon_r = \frac{C \cdot w}{\varepsilon_0 \cdot (\pi \cdot \frac{d^2}{4})} \quad (1)$$

where C *(F)* is the measured value of the capacitance, w *(m)* is the width of the sample, $\varepsilon_0 = 8{,}85 \cdot 10^{-12}$ F.m^{-1} is the permittivity of vacuum and d *(m)* is the diameter of the sample.

2.2 *Dielectric strength measurement*

The measurement of the dielectric strength was carried out on an electrode system connected to a 50-Hz AC voltage source. The system consisted of two cylindrical electrodes with a diameter of 6 mm, between which the samples were inserted. Afterwards, the source was connected to the upper electrode and the applied voltage was being increased linearly (2 kV/sec) until a breakdown occurred. The value of breakdown voltage of the sample was equal to the value of RMS voltage measured at the moment of breakdown. A calibrated voltmeter METEX M-3850 with a high voltage probe Agilent 34136A was used for voltage measurement. All of the experiments were realized inside a container filled with natural insulating oil. The oil had a similar permittivity as the samples, which helped to eliminate the possible surface discharges. The samples were immersed in the oil only during the test (~1 min), which prevented any significant oil ingress into the material.

3 Results

The results from the permittivity measurement are illustrated in the following figures. Figure 1 represents the dependence of permittivity of each material on the measuring frequency for printing resolution 150 μm. Such a resolution was chosen as it had the lowest failure occurrence during the printing. It is apparent that the value of permittivity slightly decreases with increasing frequency for all three examined materials. In Figure 2, the dependency of permittivity on printing resolution is depicted for measuring frequency 100 Hz. Except for ABS, the permittivity of the examined samples was larger for smaller height of one layer.

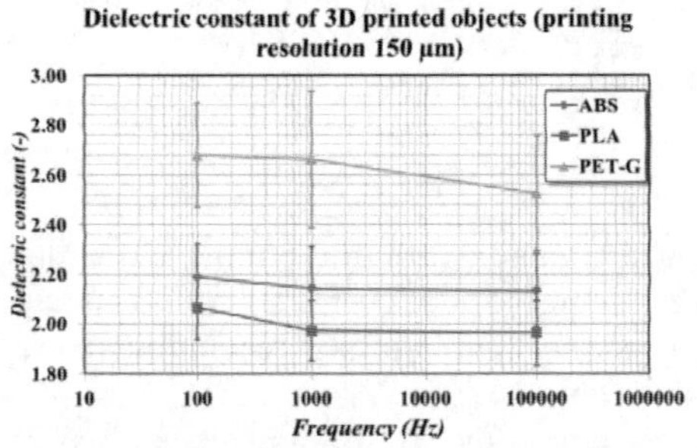

Figure 1. Permittivity of 3D printed objects in dependency on frequency

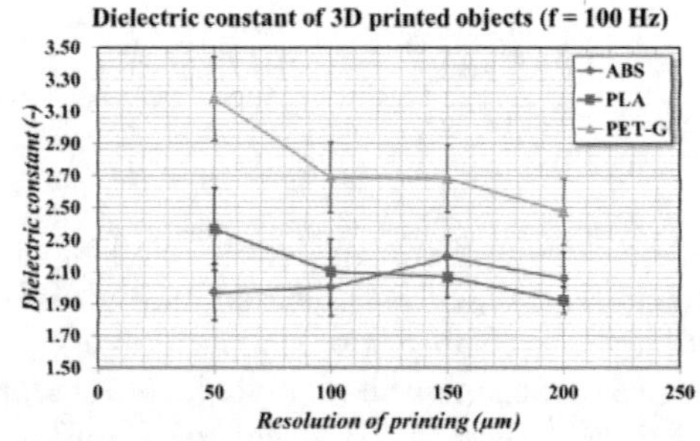

Figure 2. Permittivity of 3D printed objects in dependency on printing resolution

The results from dissipation factor measurement are shown in Figure 3 and Figure 4. No apparent dependency on printing resolution was found; nevertheless dissipation factor slightly decreased with frequency. The largest difference occurred between 100 Hz and 1 kHz for the PET50 samples; however, these samples also had the highest standard deviation. In general, high value of deviation was calculated for most of the dissipation factor measurements, therefore the data were not statistically credible.

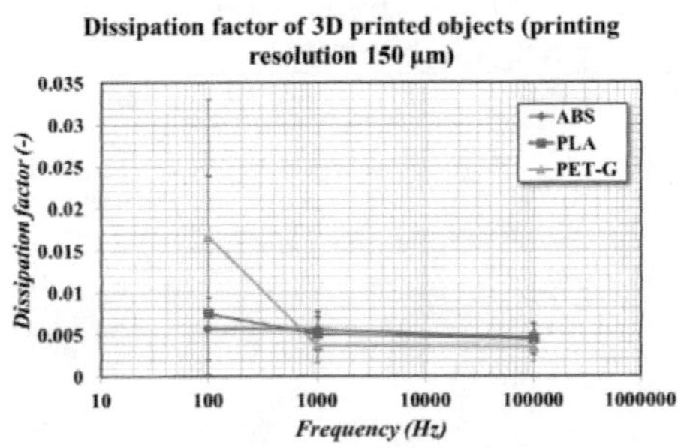

Figure 3. Dissipation factor of 3D printed objects for printing resolution 150 μm

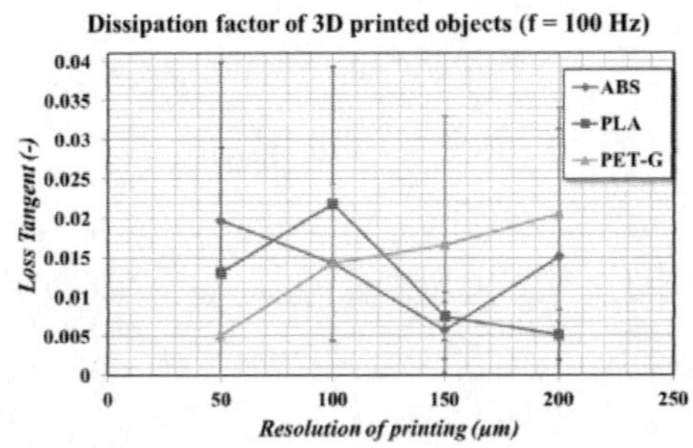

Figure 4. Dissipation factor of 3D printed for measuring frequency 100 Hz

The data of the permittivity measurement were statistically tested by Student's t-test with significance level $\alpha = 5$ %; p values for differences between the highest and lowest printing quality for permittivity measurement with the measuring frequency 100 Hz (p_{DC}) can be found in the summary in Table 2. For the dissipation factor measurement, p values are not listed in the table.

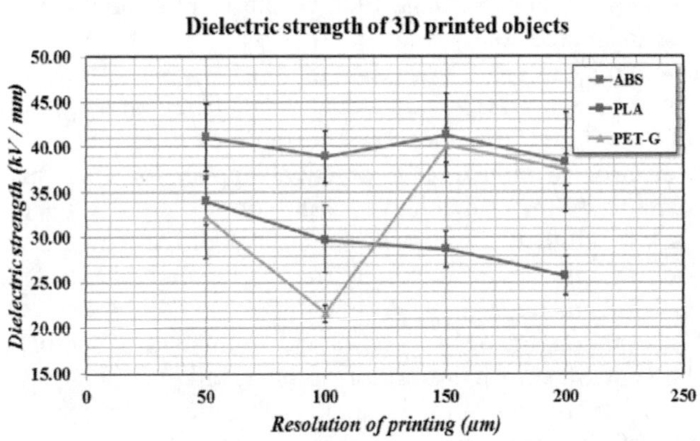

Figure 5. Dielectric strength of 3D printed objects

The results of dielectric strength measurement can be seen in Figure 5, which represents the dependence of dielectric strength on the printing resolution. The two-parameter Weibull distribution was applied on the data (according to the standard IEC 62539:2007). This distribution is commonly used for describing a process of electric degradation, but in the case of the linearly increasing voltage with time, it is possible to interchange time axis with field strength axis. Scale (α) and shape (β) parameters of the Weibull distribution can be seen in Table 2. The estimated scale parameter was regarded as the dielectric strength. In the case of dielectric breakdown tests, the value of the shape parameter should be much larger than 1 (at least 10), as that means the results are distributed in a narrow band around the mean value.

ABS had the highest value of dielectric strength, but it also showed no apparent differences between the resolutions. For PLA samples, the value of dielectric strength was lower in comparison with ABS and the dielectric strength had decreasing character with increasing resolution. For PET-G, the dielectric strength was significantly worse for higher printing resolutions.

Table 2. Results of dielectric parameters measurements for materials suitable for 3D printing (measuring frequency 100 Hz for permittivity and dissipation factor, 50 Hz for dielectric strength)

Sample	Average permittivity (-)	Statistical significance p_{DC} (-)	Average dissipation factor (-)	Weibull shape parameter ≡ Dielectric strength (kV/mm)	Weibull scale parameter (-)
PLA50	2.36 ± 0.26	0.00006	0.013 ± 0.016	34.1	12.9
PLA100	2.10 ± 0.20		0.022 ± 0.018	29.8	16.3
PLA150	2.07 ± 0.13		0.008 ± 0.003	28.7	10.2
PLA200	1.92 ± 0.08		0.005 ± 0.003	25.8	7.8
ABS50	1.97 ± 0.18	0.28	0.020 ± 0.020	41.1	7.9
ABS100	2.00 ± 0.18		0.014 ± 0.010	38.9	4.5
ABS150	2.19 ± 0.13		0.006 ± 0.004	41.3	8.3
ABS200	2.06 ± 0.16		0.015 ± 0.016	38.3	6.7
PET50	3.18 ± 0.26	0.000001	0.005 ± 0.009	32.3	4.1
PET100	2.69 ± 0.22		0.014 ± 0.010	21.6	13.4
PET150	2.68 ± 0.21		0.017 ± 0.016	40.2	11.8
PET200	2.47 ± 0.20		0.021 ± 0.014	37.5	12.1

4 Conclusions and discussion

In the experiment, the dielectric properties of 3D printed objects made of three commonly used materials in FDM were evaluated. For PLA and PET-G, the results showed that permittivity slightly decreased with the increasing height of one layer. This may have been caused by larger air gaps between the lines and layers of the object. For ABS, the data were not statistically significant due to the bad printing quality of surface layers, so no evident dependency on height of one layer was found out. The highest value of permittivity was measured on samples from PET-G, but it is still relatively small value for use in electronics industry for practical application like bushing condensers. An improvement of their permittivity may be achieved with some additives, for example the corundum powder (Al_2O_3). It is a task for further research, because the additives could also negatively influence the process of 3D printing (melting point, viscosity etc.). The slight dependence of permittivity on frequency indicates the polar character of the materials.

Measurement of dissipation factor was burdened with a considerable error. Therefore, no concrete result can be concluded. According to the observation during measurement, dissipation factor was rather an indicator of poor printing quality. The bigger dissipation factor, the worse the object was with rough surface or with more air gaps.

In general, the lower values of permittivity, compared to other measurements [7], [8], and relatively high deviations, especially for high printing resolution with height of one layer 50 μm, may indicate poor reliability and repeatability of manufacture by cheap printers using the RepRap concept. Therefore, the dependency of the dielectric properties of 3D printed objects on the printing resolution should be confirmed by further research using industrial printers.

In the case of dielectric strength measurement, ABS was evaluated as the most suitable material as insulation for electronic devices from all the examined materials. On the other hand, printing problems with ABS related to its properties during the liquid state cause unstable quality and no reproducibility. Therefore, no significant dependency of dielectric strength on the printing resolution was observed. The results of dielectric strength measurement for PET-G showed an unexpected but statistically significant difference between the printing resolutions. The dielectric strength of objects printed in high resolution (50 and 100 μm) was up to 2 times worse compared to low resolution (150 and 200 μm). It can be explained by poor printability of PET-G for high resolution due to the lower speed of the filament movement through the extruder, different heating of the material and thus different properties in the liquid state that might cause worse adhesion to the substrate or cohesion of layers. Only the dielectric strength of PLA decreased with lower resolution, which was expected - the thickness of air gaps increased with lower resolution. Thus, the observed dependency proved decent printing repeatability of PLA.

In comparison with [4], similar results of dielectric strength were achieved for PLA and ABS. Both materials exhibit the dielectric strength above 30 kV/mm; however the dielectric strength of ABS is higher than PLA.

Based on acquired results, all examined materials are acceptable as insulation materials for lower voltage level devices. The use of higher printing resolution is recommended for increasing the dielectric strength. However, the improvement of the materials, respectively the printed objects, is required due to relatively small dielectric strength in comparison with commonly used insulation materials like transformer paper or XLPE. It can be achieved by better printing quality or admixtures.

Measurement error could be also caused by degradation by the environment. Thermoplastic materials are susceptible to moistening and photodegradation. Climatic tests of the objects and the influence of environment on the electrical and other properties might be also the subject of further research.

5 Acknowledgement

The authors gratefully acknowledge financial support from Grant Agency of the Czech Technical University in Prague, grant No. SGS15/196/OHK3/3T/13.

References

[1] W. Gao *et al.*, 'The status, challenges, and future of additive manufacturing in engineering', *Comput.-Aided Des.*, vol. 69, pp. 65–89, Dec. 2015.

[2] T. D. Ngo, A. Kashani, G. Imbalzano, K. T. Q. Nguyen, and D. Hui, 'Additive manufacturing (3D printing): A review of materials, methods, applications and challenges', *Compos. Part B Eng.*, vol. 143, pp. 172–196, Jun. 2018.

[3] O. A. Mohamed, S. H. Masood, and J. L. Bhowmik, 'Optimization of fused deposition modeling process parameters: a review of current research and future prospects', *Adv. Manuf.*, vol. 3, no. 1, pp. 42–53, Mar. 2015.

[4] X.-R. Li *et al.*, 'Analysis of Morphology and Electrical Insulation of 3D Printing Parts', presented at the 2018 IEEE International Conference on High Voltage Engineering and Application, Athens, Greece, 2018.

[5] E. Sells, S. Bailard, Z. Smith, A. Bowyer, and V. Olliver, 'RepRap: The Replicating Rapid Prototyper: Maximizing Customizability by Breeding the Means of Production', in *Handbook of Research in Mass Customization and Personalization*, 0 vols, World Scientific Publishing Company, 2009, pp. 568–580.

[6] O. V. Tereshchenko, F. J. K. Buesink, and F. B. J. Leferink, 'An overview of the techniques for measuring the dielectric properties of materials', in *2011 XXXth URSI General Assembly and Scientific Symposium*, 2011, pp. 1–4.

[7] E. Huber, M. Mirzaee, J. Bjorgaard, M. Hoyack, S. Noghanian, and I. Chang, 'Dielectric property measurement of PLA', in *2016 IEEE International Conference on Electro Information Technology (EIT)*, 2016, pp. 0788–0792.

[8] G. Boussatour, P.- Cresson, B. Genestie, N. Joly, and T. Lasri, 'Characterization of biodegradable and biosourced polylactic acid (PLA) substrate in a wide frequency range (0,5–26 GHz)', in *2017 IEEE MTT-S International Microwave Workshop Series on Advanced Materials and Processes for RF and THz Applications (IMWS-AMP)*, 2017, pp. 1–3.

IOP Conf. Series: Materials Science and Engineering **461** (2019) 012092 doi:10.1088/1757-899X/461/1/012092

Comparison of WC and (TiW)C in nickel-based alloy coatings prepared by laser cladding

M Vostřák, Š Houdková, M Hruška

The University of West Bohemia, New Technology Research Centre, Univerzitní 8, 306 14 Plzeň, CZ

E-mail: mvostrak@ntc.zcu.cz

Abstract. For wear demanding applications, the coatings, consisting of WC in the nickel-based alloy are often used due to their excellent properties. Nevertheless, there are still limits of these coatings, mainly deterioration of tungsten carbide due to heat load during laser cladding. The suppress the limits, a new type of coating material, consisting of the $(TiW)C_{1-x}$ in nickel-based alloy, was designed. The goal of this study is to compare the microstructure and mechanical properties of new $(TiW)C_{1-x}$ based coating with a usual WC-based coating to confirm its improved behaviour.

1 Introduction

Laser cladding is a technology capable of producing compact functional coatings from a variety of materials. For example, it is widely used for the production of a wear resistant layer using the metal matrix composites (MMCs). These materials excel in their high wear resistance due to the tough matrix and hard reinforcing particles [1].

Among MMCs, the combination of Ni-based alloy and tungsten carbide is widely used. The role of laser cladding processing parameters as well as the role of powder composition on the final appearance of tungsten carbides in the coating and thus on its final functional properties as is hardness was analyzed in various studies [2–6]. However, tungsten carbide also has some disadvantages which can limit its use. Tungsten carbide suffers from extensive degradation during laser cladding process and has inferior thermodynamic stability compared to many other carbides [7]. New types of carbides and MMCs coatings are being developed to overcome the limitations associated with tungsten carbide based coatings [7].

The aim of this study is to compare microstructure and mechanical properties of the new experimental $(TiW)C_{1-x}$ based coating with previously tested commercially available WC-based coating. Both coatings were deposited by 5kW solid state disc laser Trumpf TruDisk.

2 Experimental procedure

2.1 Feedstock powders

For the WC based coating, the commercially available feedstock powder from the Oerlikon Metco company designated as MetcoClad 52052 was used. The powder is composed of 60 wt% of WC spherical particles blended with 40 wt% of gas atomized NiCrBSi (8.0Cr3.5Si1.6B0.3C Ni bal.) particles. The particle size distribution lies between 45-106 μm. The WC was manufactured in a

IOP Conf. Series: Materials Science and Engineering **461** (2019) 012092 doi:10.1088/1757-899X/461/1/012092

unique way, resulting in a non-acicular shape with higher hardness than conventional fused and crushed WC [8].

The new experimental powder from the Oerlikon Metco company is composed of 60 wt% of angularly shaped $(TiW)C_{1-x}$ carbides blended with the same Ni self-fluxing alloy as in the previous case. The density of $(TiW)C_{1-x}$ carbide is 10.1 g/cm^3, lower than the density of WC which is 15,63 g/cm^3 [7]. This means, that the volume percentage of carbides in the new power is higher than in MetcoClad 52052.

2.2 Laser cladding process and parameters

Laser cladding was executed using the solid state laser Trumpf TruDisk 8002 with Precitec coaxial 4-way cladding head YC52.

From the experimental powder, two variants of the coatings were cladded using different heat input. The laser cladding process parameters were chosen as follows: powder feed rate F = 540 mg/s, process speed S = 8.3 mm/s and laser power for variant **1** P = 2400 W and variant **2** P = 1400 W.

For the comparison, it was chosen the tungsten carbide based coating cladded in the previous study. The processing parameters for this coating were as follows: F = 475 mg/s, S = 8.3 mm/s, P = 3000 W. The overlapping ratio was 50% of the single bead width.

The coatings were deposited onto plates of carbon steel (EN10083 2: C45) with dimensions 100 × 100 × 20 mm to ensure sufficient heat dissipation. Substrates were pre-heated to 350 °C prior to deposition to avoid cracking of the coating due to the high thermal gradient. After the deposition, the samples were left to cool down at room temperature.

2.3 Characterization

The microstructure was evaluated on the coatings′ cross-sections (ground and polished by automatic Leco grinding and polishing equipment) using digital optical microscope HIROX KH 7700. The thickness of the coating was measured as the minimal distance from the interface of the coating and the substrate to the top of the coating. It was averaged from 5 measurements. The microhardness HV1 was evaluated on the coatings' cross-sections. For each coating, the matrix of indents' locations (6 x 6) was designed to cover the area of the two overlapping tracks.

3 Results and discussion

3.1 Structure of the coatings

The thickness of the $(TiW)C_{1-x}$ based coating is similar for both variants, namely 2450 μm for the variant **1** with high heat input and 2588 μm for the variant **2**. The structure of the $(TiW)C_{1-x}$ coatings on the cross sections longitudinal and traverse to the direction of cladding is shown in figure 1.

The structure of both variants of the coating differs significantly. In the coating cladded with high heat input marked as variant **1**, high amount of undissolved carbide particles is near the coating surface and near the interface with the substrate. In the middle of the coating, the majority of carbide particles is dissolved in the matrix and solidified again. The reason for different carbide dissolution in the coating is the speed of solidification. The surface and interface area are cooled more effectively than the middle of the coating and thus solidified earlier. The detailed microstructure in these three areas is shown in figure 2.

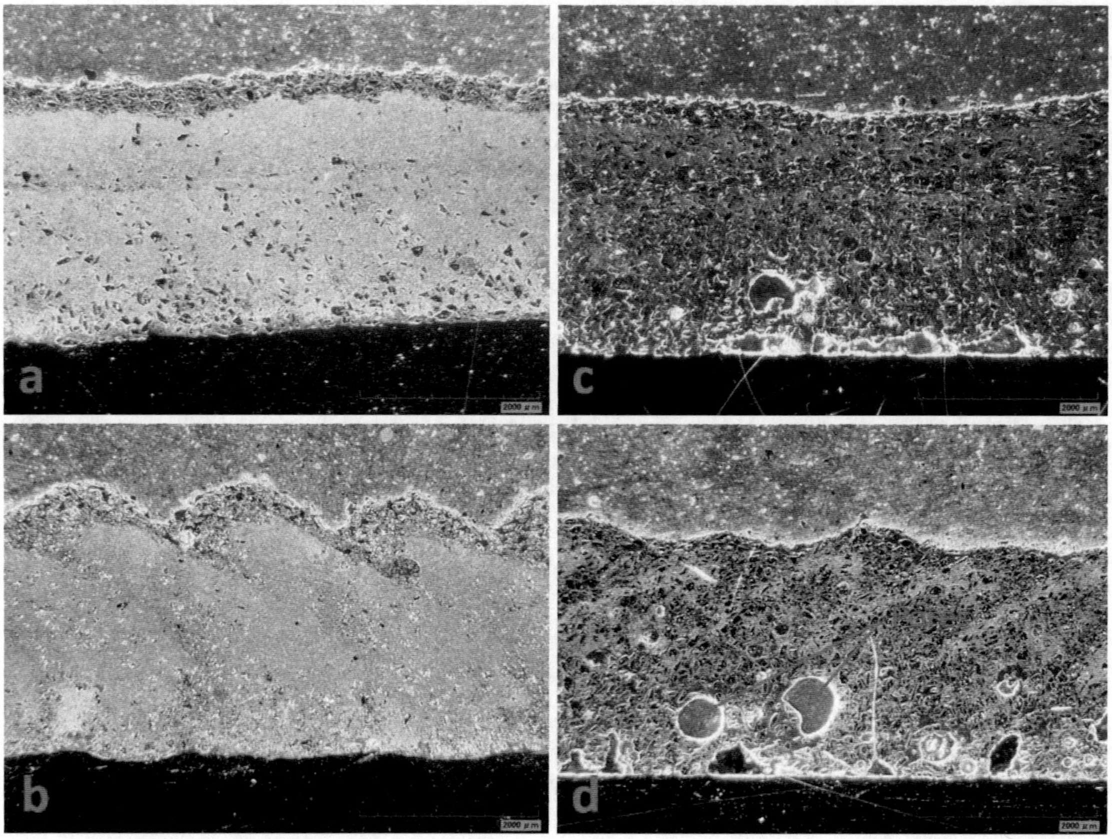

Figure 1. The cross-section of the $(TiW)C_{1-x}$ based coatings, variant 1 with high heat input: a) longitudinal and b) transverse; variant 2 with low heat input: c) longitudinal and d) transverse.

On the contrary, in the variant **2** of the coating, cladded with lower heat input, the majority of the carbides is preserved undissolved through the whole coating. The reason is that thanks to lower heat input, the temperature is not as high as in variant **1** and the whole coating solidifies more rapidly. The detailed microstructure of the coating is again shown in figure 2. However, the variant **2** of the coating contain many imperfections – cracks and big pores up to size 1 mm appearing primarily on the interface with the substrate. The interface of the coating with the substrate, in this case, is insufficiently melted and it is not as homogenous as in variant **1**. This is due to low heat input, the energy is insufficient to effectively to melt higher volume of the substrate and to heat the melt pool to a higher temperature. Due to rapid solidification, the gas bounded on powder surface does not have sufficient time to escape the melt pool and creates pores. The imperfections in the coating can be clearly visible on the cross-section on figure 1 c) and d) and on figure 2 f).

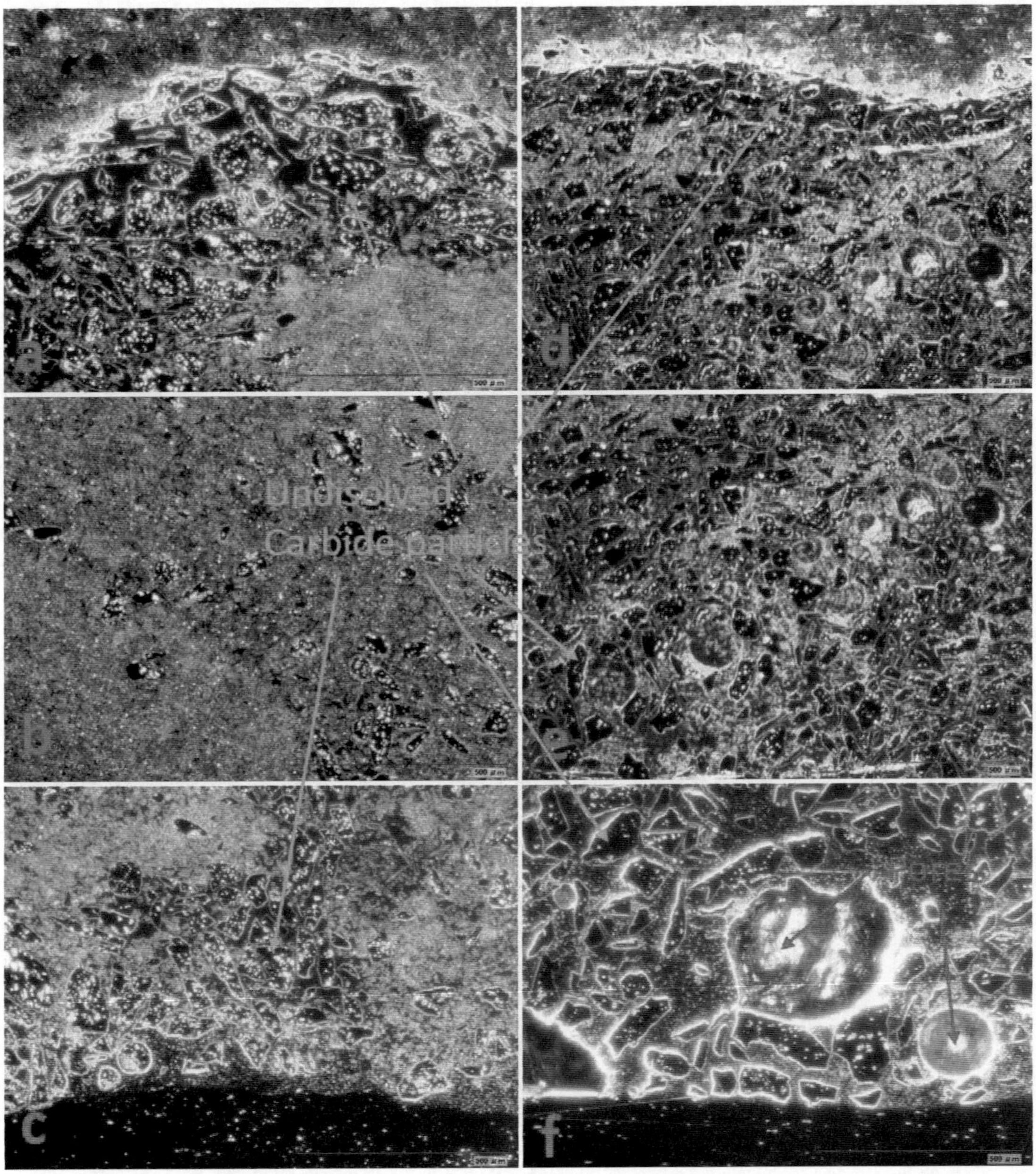

Figure 2. Detailed microstructure of the $(TiW)C_{1-x}$ based coatings, variant 1: a) top, b) middle, c) interface with the substrate; and variant 2: d) top, e) middle, f) interface with the substrate.

For the comparison, the microstructure of the WC based coating is shown in figure 3. Similarly to $(TiW)C_{1-x}$ coating variant **1,** more undissolved carbide particles remain near the coating surface and interface with the substrate. However, the majority of the carbides are dissolved in the matrix and the amount of undissolved WC particles is significantly lower than in above-presented coatings. The reason is that tungsten carbide has inferior thermodynamic stability. The second reason is that this coating contains lover volume percentage of carbide particles that the new experimental one. The carbides with matrix are mixed in the same weight ratio but the tungsten carbide has a higher density. The dissolution of carbide would significantly affect the final coating mechanical properties.

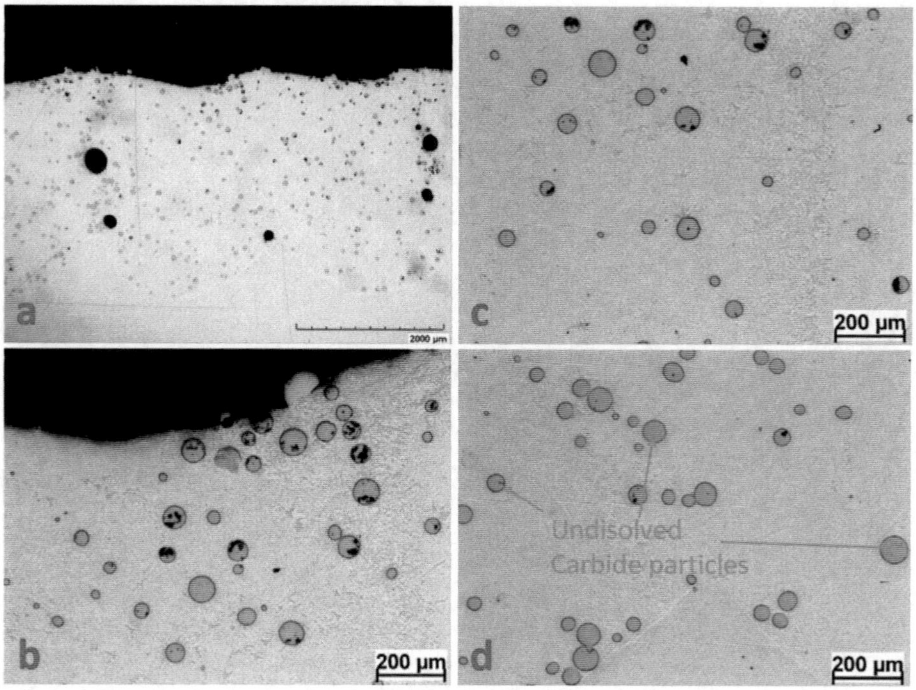

Figure 3. The microstructure of WC based coating: a) overall transverse, and details b) top, c) middle and d) interface with the substrate.

3.2 Hardness

The dependence of the hardness HV1 on the distance from the interface with the substrate for all three coating is presented in figure 4.

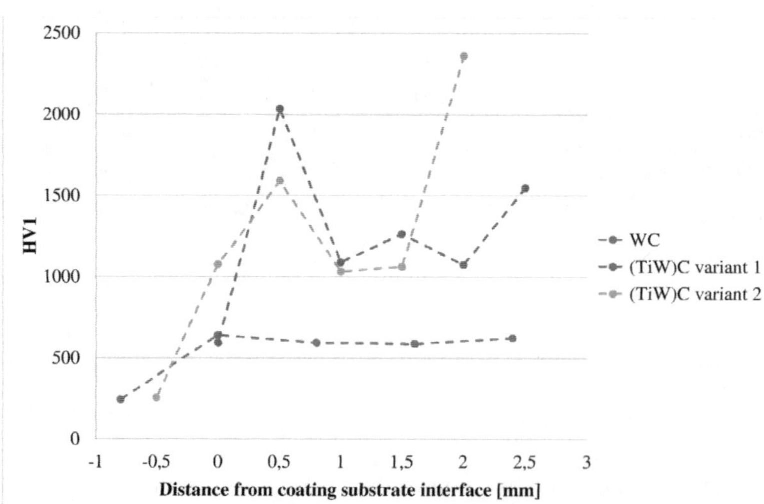

Figure 4. Microhardness depth profiles of tested coatings.

The hardness of the new experimental coating is significantly higher than the WC based coating. The average hardness HV1 of $(TiW)C_{1-x}$ based coatings is in the range 1070 – 2370 and for the WC based coating only 600 – 640. The main reason is the dissolution of the carbide particles. The hardness profile reflects the amount of undissolved particles. The role of carbide dissolution on the hardness of WC based coating is discussed in a previous study [2]. The second reason is the new carbide itself has a higher hardness than the tungsten carbide. The average microhardness HV0.05 of WC is only 2200 and of $(TiW)C_{1-x}$ is 2818 [7].

4 Conclusion

In this study, the coatings from a new experimental powder containing $(TiW)C_{1-x}$ in Ni alloy were created by laser cladding technology. The new coatings were compared to the coating cladded from a commercially available powder containing WC in Ni alloy.

In the new experimental $(TiW)C_{1-x}$ based coating, a high amount of carbide particles remain undissolved comparing to the WC based coating. The dissolution also differs for the two variants of the $(TiW)C_{1-x}$ coating. With high heat input, the undissolved particles are mainly near the surface and interface and in the middle, more carbides are dissolved in the matrix. In the variant cladded with low heat input, the carbide particles remain undissolved through the whole coating.

The hardness HV1 of the new coating is in the range 1070 – 2370 and is much higher than the hardness of WC based coating. The hardness depth profile reflects the dissolution of the carbides.

The new coating is very promising and more detailed analysis of mechanical properties such as wear resistance is underway.

Acknowledgments

The research outcome was developed within the CENTEM project, reg. no. CZ.1.05/2.1.00/03.0088, co-funded by the ERDF as part of the Ministry of Education, Youth and Sports OP RDI programme and, in the follow-up sustainability stage, supported through CENTEM PLUS (LO1402) by financial resources from the Ministry of Education, Youth and Sports under the "National Sustainability Programme I" and project SGS-2016-005.

References

[1] Nurminen J, Näkki J and Vuoristo P 2009 Microstructure and properties of hard and wear resistant MMC coatings deposited by laser cladding Int. J. Refract. Met. Hard Mater. 27 472–8

[2] Vostřák M, Houdková Š, Bystrianský M and Česánek Z 2018 The influence of process parameters on structure and abrasive wear resistance of laser clad WC-NiCrBSi coatings Mater. Res. Express 5 096522

[3] Luo X, Li J and Li G J 2015 Effect of NiCrBSi content on microstructural evolution, cracking susceptibility and wear behaviors of laser cladding WC/Ni-NiCrBSi composite coatings J. Alloys Compd. 626 102–11

[4] Fernández M R, García A, Cuetos J M, González R, Noriega A and Cadenas M 2015 Effect of actual WC content on the reciprocating wear of a laser cladding NiCrBSi alloy reinforced with WC Wear 324–325 80–9

[5] Deschuyteneer D, Petit F, Gonon M and Cambier F 2015 Processing and characterization of laser clad NiCrBSi/WC composite coatings - Influence of microstructure on hardness and wear Surf. Coatings Technol. 283 162–71

[6] Amado J M, Tobar M J, Yáñez A, Amigó V and Candel J J 2011 Crack Free Tungsten Carbide Reinforced Ni (Cr) Layers obtained by Laser Cladding Phys. Procedia 12 338–44

[7] Fiala P, Hepp R and Zikin A 2017 Alloyed Carbides Beyond WC as a New Material Platform for Solving Challenges in Hardfacing ITSC 2017 Proceedings of the Conference in Düsseldorf / Germany on June 7 – 9, 2017 (Düsseldorf)

[8] Oerlikon Metco 2014 Material Product Data Sheet Tungsten Carbide – Nickel Alloy Powder Blend for Laser Cladding vol 40

The effect of yttrium addition on the chemical composition of the non-metallic phase in liquid steel

S Gerasin[1], D Kalisz[1], P L Żak[1] , K. Kuglin[1] and J Iwanciw[2]

[1] AGH University of Science and Technology, Faculty of Foundry Engineering, 23 Reymonta Street, 30-059 Kraków, Poland

[2] retired

E-mail: dak@agh.edu.pl

Abstract. The introduction of yttrium as an alloy additive to steel is connected with the formation of non-metallic inclusions resulting in the losses of this element. The presence of yttrium is expected to minimize the precipitation processes caused by sulphur and oxygen in steel. Therefore the order in which the additives are introduced during refining processes decides about yttrium final content in steel in bounded and dissolved form. For defining the amount of the generated non-metallic phase and kinetics of deoxidation and desulphurization processes authors used a non-commercial computer program based on the worked out calculation models. The process was analyzed in view of the order of introducing additives: Al, Y and Ca to preliminarily deoxidized steel. It was revealed that yttrium dosing is connected with the formation of oxide and sulphide phases. The introduction of yttrium prior to aluminium brings about huge losses, mainly as far as oxides formation is concerned. In the case of higher participation of sulphur, this effect is even stronger. When yttrium is introduced after aluminium and calcium, yttrium is used for the formation of its sulphide, and in this way the manganese sulphide formation is reduced. The simulations showed that yttrium should be introduced at the end of the refining process after adding aluminium and calcium. In this case the yttrium losses for the formation of its sulphide and oxide and least.

1 Introduction

The modification of steel composition with alloy additives can be realized after taking into account potential melting losses of added metal in the form of liquid or solid dross, mainly oxides. In particular cases attention should be also paid to the formation of sulfides and nitrides. Such losses are observed mainly during preliminary deoxidation of steel during flow-off, though they may also take place at any stage of secondary processing, e.g. flow-off, or at any stage of secondary processing of steel, during argon blowing and refining. Processes leading to the formation of dross are in line with the tendency of the system to establish the state of thermodynamic equilibrium in the system where the total free energy reaches minimum at a given temperature, pressure and mass of particular components. As a consequence the four-phase system, consisting of liquid steel, liquid slag, solid or liquid non-metallic precipitates forming a dispersed and gaseous phases should be accounted for. In the near-real conditions observed in the metallurgical processes no such a state of equilibrium is generally established. More realistic and important image of alloy addition processes can be obtained by limiting investigations only to the analysis of equilibrium between liquid steel and non-metallic dispersed phase.

IOP Conf. Series: Materials Science and Engineering **461** (2019) 012093 doi:10.1088/1757-899X/461/1/012093

The deoxidant [R] and oxygen [O] are dissolved in liquid steel with oxide. The product of the reaction can [1-5]:
- remain in the form of small particles suspended in liquid steel, because small particles flow off very slowly,
- form a particle of liquid oxide solution with other oxides. A solution of some oxides has a definitely lower liquidus temperature,
- form a compact or loose aggregate with other non-metallic particles as a result of operation of capillary forces.

The analysis of deoxidation, refining and introduction of alloy additives to steel requires taking into consideration two basic elements: thermodynamic characteristic of phases appearing in the process, which allows for determining the states of equilibrium, and kinetic description of elementary processes. On this basis particular elementary processes can be qualitatively evaluated in at a given moment of process, and the final effect can be determined [1-3, 6, 7].

Yttrium is highly active in steel and shows a considerable chemical affinity, especially to oxygen and sulphur. In reaction with dissolved elements forms solid crystalline precipitates: oxides and also sulfides, oxysulfides, nitrides and carbonitrides [8]. For this reason attention should be paid to the loss of elements reacting with oxygen and sulphur dissolved in metal bath.

Hence the aim of this paper is mainly to work out a method simulating the evolution of non-metallic precipitates content in the deoxidation process and during introduction of an additive, e.g. yttrium to steel. In this way the yttrium dross as an alloy component of steel can be predicted.

2 Description of the computer simulation program for deoxidation reactions

The WYK_STAL computer simulation program worked out by J. Iwanciw and others [9-11] was used for real-time simulating and aiding secondary metallurgical processes. This program calculates the successive stages of a system consisting of liquid steel, non-metallurgical precipitates, liquid slag and gas.

The starting point for the simulations is liquid steel of definite weight, chemical composition and temperature. When deoxidizing steel the whole mass of iron alloys of a given composition can be added directly during flow-off, or in portions; in the boundary case the dosing can be represented as continuous. The processes taking place among introduced steel components at that time are very dynamic as the system is subject to strong mixing energy of the streaming liquid steel. Chemical reactions take place, and in a relatively simple model they can be reduced to the oxidation of steel components, mainly the non-metallic ones. Slag and non-metallic precipitates have a similar chemical character, i.e. they are oxides solutions. Mass proportions of slag and precipitates can be assessed on the basis of oxygen concentration in steel during the flow-off. The total mass of slag is a sum of slag-generating additives (lime), furnace slag which entered the ladle during the flow-off and precipitates which few out from steel to slag.

The WYK_STAL program allows to calculate pure elements or multi component additives which are to be admixed to the liquid metal bath, process simulation, accounting for the calculated additives and checking out the composition after the simulation. This program is equipped with additional modules and such options as deoxidation, desulphurization, formation of nitrides or verification of simulation results.

Hence this program can be used for simulating, secondary refining technologies and establishing the composition of steels which have never been produced in the steel mills before. This program makes use only of own databases and user process data in the form of text files. A specific process can be defined on the basis of the following files, which have to be worked out and used:
(a) process parameters, (b) required composition of metal, (c) initial composition of metal,
(d) initial composition of slag, (e) additives files, (f) configuration files.

The program calculates the successive states of a system made of liquid steel, liquid slag, non-metallic inclusions and alloy additives.

For simulating the behaviour of alloy additives when introducing them to steel, authors developed the files for computer program to make it work for the yttrium oxides, sulphides and nitrides formation concurrently with the reactions relating to the basic oxides formation. Apart from this an elementary thermodynamic parameters base was created for the formation of particular compounds, including the changes of standard free energy of formation, and parameters of oxygen and sulphur interaction with the components of the solution. Moreover the database for yttrium oxides and sulphides as well as the parameters of yttrium interaction with steel components was updated. The base of parameters for oxides and nitrides formation base on the literature data [1, 9-13].

3 Calculation results

The process of non-metallic phase formation as a result of introduction of deoxidizing additives was analyzed. Authors analyzed the basic system Fe–O–S-Al-Y, where 30 kg aluminum was added in the fifth minute, and 76 kg Yttrium in the twentieth minute of the process.

The program allows for choosing a model for thermodynamic calculations:

- Model a: activity of the formed compound $a = 1$;
- Model b: computatuions for partition coefficient L_S.

For simulating the deoxidation and desulphurization process, the following data were assumed:

- process duration: 30 minutes;
- weight of metal in ladle: 140 000 kg;
- weight of slag: 100 kg;
- temperature of metal at the beginning of the process: 1670 °C;
- pressure of gaseous phase 1 atm;
- initial maximum oxygen bounded in inclusions: 0.0001%;
- initial total oxygen content in steel: 0.01%;
- initial slag composition: CaO – 45%; Al_2O_3–2%; MgO–9%; MnO–5%; SiO_2–12%; FeO–27%;
- change of additives introduced to slag: 100 kg CaO was added in the first minute, 100 kg SiO_2 was added in the second minute.

The steel composition chosen for the calculations is shown in the Table 1.

Table 1. Steel composition used during calculations

Element	C	Mn	Si	P	N	S	Fe
wt. %	0.054	0.05	0.23	0.007	0.005	0.01	balance

The results of calculations for model a are presented in figures 1 – 3.

At the initial stage of the refining process the non-metallic inclusions consist of FeO and MnO, the formation of which is hindered by the addition of aluminum and then yttrium. The introduction of yttrium at the final stage results in the formation of oxide and sulfide inclusions, where the participation of oxygen in the system is much lower. This low oxygen content in the system is caused by the preliminary deoxidation with the use of aluminum. The slag composition was observed to change after adding elements. At the final stage of refining small amounts of yttrium sulfide and oxide were present, being a consequence of precipitation and absorption of the outflowing non-metallic inclusions by slag.

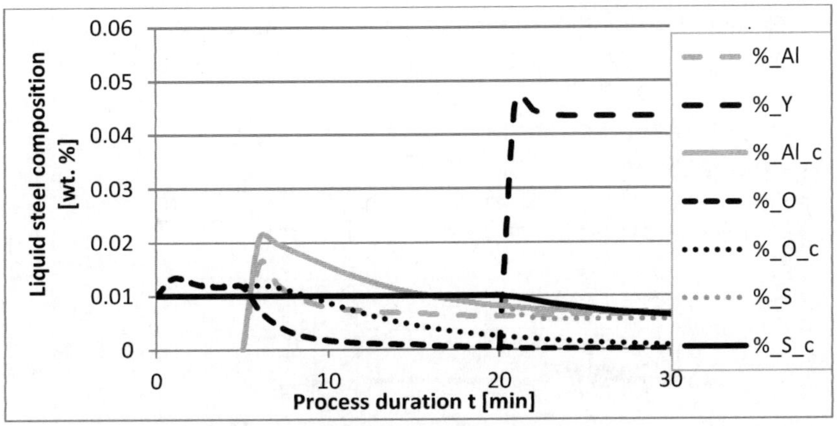

Figure 1. Change of element content in liquid steel vs process duration

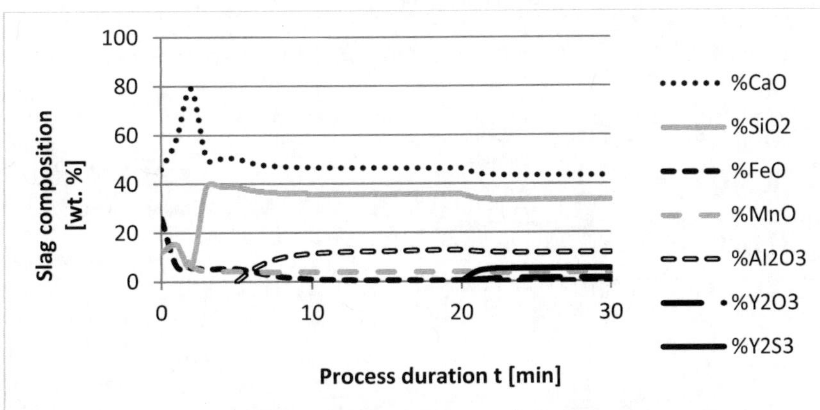

Figure 2. Change of chemical composition of slag vs process duration

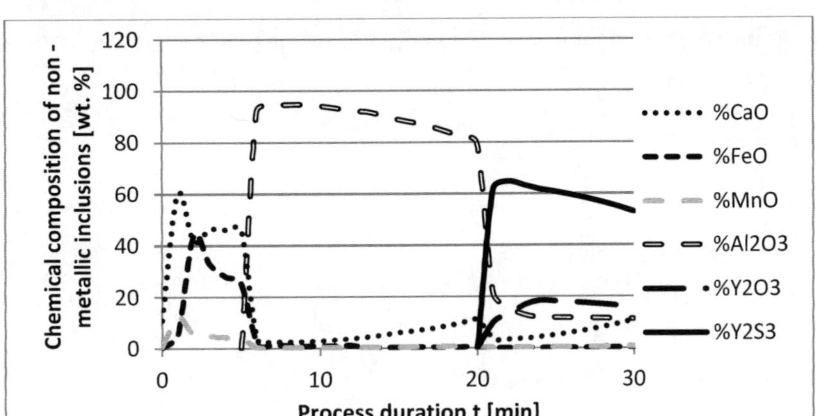

Figure 3. Change of chemical composition of non-metalic inclusions vs process duration

Analogous calculations were performed for model b, accounting for partition coefficient L_S. The results of calculations are presented in fig. 4-6.

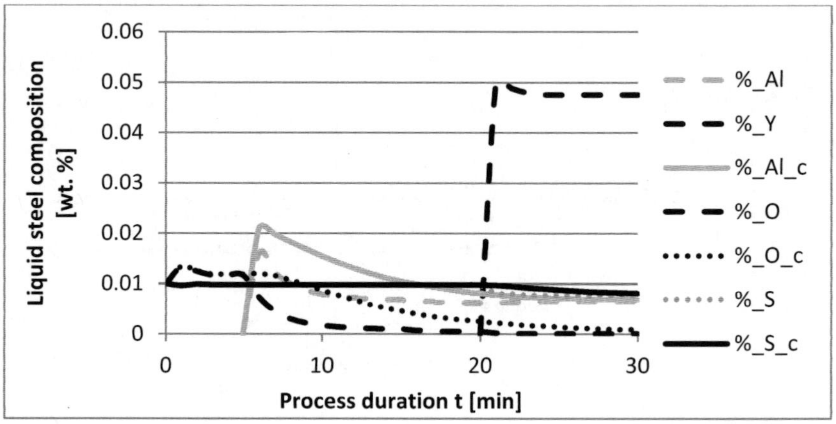

Figure 4. Change of element content in liquid steel vs process duration

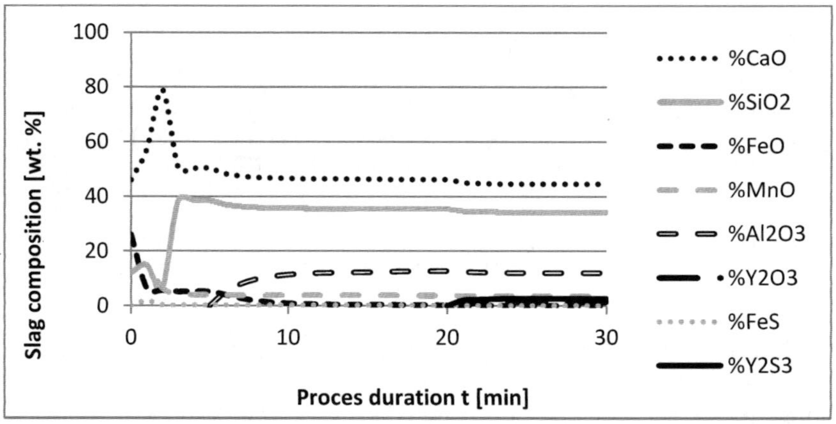

Figure 5. Change of chemical composition of slag vs process duration

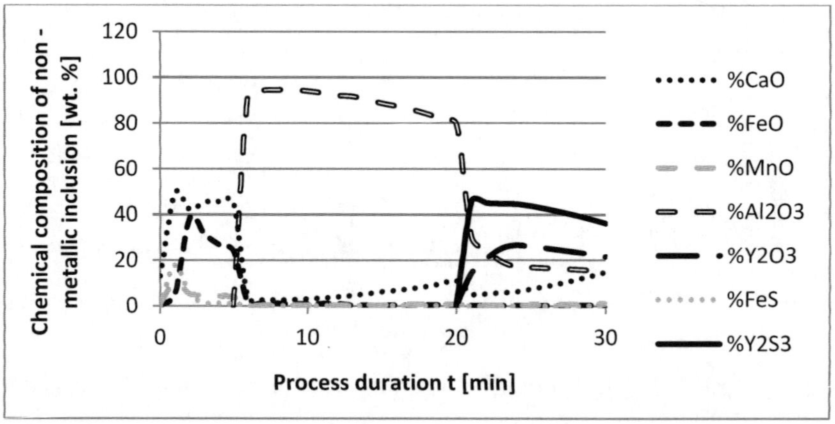

Figure 6. Change of chemical composition of non-metallic inclusions vs process duration

A very similar course of plots representing the chemical composition of elements dissolved in liquid steel in model a was observed in this case. FeS was identified in the slag. The presence of parameter L_S in the calculations had an influence on the chemical composition of non-metallic inclusions containing yttrium. The participation of Y_2S_3/Y_2O_3 in inclusions was lower as compared to the results of calculations performed for model a. Apart from this, the sulfur content in liquid steel was higher.

4 Conclusion

Yttrium in liquid steel plays the role of a complex deoxidizer and desulphurizer. Its addition also affects the composition of steel, slag and the formed non-metallic inclusions.

Computer calculations of inclusions formation in thermodynamic equilibrium conditions allow for determining the chemical composition of the formed non-metallic phase in a given deoxidation situation. On this basis the amount of oxygen dissolved in metal bath and chemical composition of refined liquid steel can be also established.

The results of calculations showed that the products of chemical reactions taking place during preliminary deoxidation mainly consisted of oxides, though the participation of specific phases depended on the yttrium addition.

Acknowledgements
The authors acknowledgement the AGH University of Science and Technology for its financial support within grant no. 11.11.170.318.14.

References

[1] Kalisz D 2013 Termodynamiczna charakterystyka powstawania fazy niemetalicznej w ciekłej stali [in Polish] (Krakow: Wyd. Nauk. Akapit) chapter 1 pp 15-50
[2] Jowsa J 2008 Inżynieria procesów kadziowych w metalurgii stali [in Polish] (Częstochowa: Wyd. Politechnika Częstochowska)
[3] Lis T 2009 Metalurgia stali o wysokiej czystości [in Polish] (Gliwice: Wyd. Politechnika Śląska)
[4] Kalisz D and Żak P L 2016 PSG method for simulating agglomeration of Al_2O_3 inclusion in liquid steel *Acta Physica Polonica A* vol **130** pp 157-159
[5] Kalisz D and Żak P L 2015 Modeling solute segregation and formation of non-metallic inclusion during solidification of a titanium – containing steel *Kovove Mater.* vol **53**(1) pp 35-41
[6] Kalicka Z 1998 Rola układów dyspersyjnych w opisie rafinacji ciekłej stali (Kraków: Uczelniane Wydawnictwo Naukowo-Dydaktyczne)
[7] Liu H L, Liu Ch J, Jiang M F 2012 Effect of rare earths on impact toughness of low carbon steel *Materials and Design* vol **33** pp. 306-312
[8] Longmei W, Ting D and Kexiang Y 1987 Study of thermodynamics and phase equilibria in order to predict the behavior of yttrium in iron and steel *Inorganica Chimica Acta* vol **140** pp 189-191
[9] Iwanciw J 1985 Proc. Conf. CIBEH on Informatics in Metallurgical Industry (Kozubnik) Conditions and possibilities of mathematical models application to control of steelmaking
[10] Iwanciw J 2002 Simulator of steelmaking processes for work in real time, Komitet Metalurgii PAN vol **1**, (Krakow: Wyd. Nauk. Akapit) p 186
[11] Iwanciw J 2002 Software manual for WYK_STAL simulation program
[12] Du T and Wang L 1985 Thermodynamics of Fe-Y-S, Fe-Y-O and Fe-Y-S- O metallic solutions *J. Less Common Met.* **110** pp 179-185
[13] Gerasin S 2017 Doctoral Dissertation: Evolution of chemical composition and model of growth non-metallic inclusions in liquid steel with yttrium content (Krakow: AGH-UST)

Thermal expansion evolution of metastable β Ti-15Mo alloy during linear heating

P. Zháňal[1,2], P. Beran[3], T. Hansen[4], J. Šmilauerová[1], J. Stráský[1], M. Janeček[1], P. Harcuba[1]

[1]Department of Physics of Materials, Charles University, Ke Karlovu 5, 12116 Prague, Czech Republic
[2]Research Center Rez., Ltd., Hlavní 130, 25068 Husinec-Řež, Czech Republic
[3]Department of Neutron Physics, Nuclear physics institute CAS, Hlavní 130, 25068 Husinec-Řež, Czech Republic
[4]Institut Laue-Langevin, BP 156, 38042 Grenoble Cedex 9, France

E-mail: pavel.zh@karlov.mff.cuni.cz

Abstract. Thermal expansion measurement was employed for characterization of phase transformations occurring in metastable β titanium alloy (Ti-15Mo) during linear heating with three different heating rates. Effects of the transformations on materials dimensions is discussed and compared with changes of lattice parameters of present phases obtained from in-situ powder neutron diffraction. The shrinkage of the material with increasing temperature at about 300 °C was attributed to the decrease of lattice parameter of β phase, which occurs due to diffusion of alloying elements out of ω phase. Complete dissolution of ω phase was detected at 560 °C regardless of the heating rate.

1 Introduction

Titanium alloys are used in a wide range of components for aerospace and chemical industry and as alternative structural materials for automotive sector due to their high specific strength and excellent fatigue and corrosion resistance [1,2]. Moreover, due to biocompatibility and partial or total superelasticity, β-Ti alloys are also considered for biomedical applications [3,4]. The only disadvantage, which hinders wider use of titanium, is its comparatively high cost, which is caused by high processing costs.

Titanium is a polymorphic material. The high-temperature phase has a body-centered cubic structure (bcc), so called β phase. The ambient temperature phase has a hexagonal close-packed structure (hcp) and is referred to as α phase. In recent years, the main interest has been focused on development and investigation of the so-called metastable β alloys. These alloys contain sufficient amount of β stabilizing elements to retain the high temperature β phase at room temperature after quenching. β phase is in a metastable state and ultimately equilibrium α + β composition can be achieved by annealing treatment [1,4,5]. Several other structural phases can be found in metastable β Ti alloys. In the investigated Ti-15Mo, hexagonal (non-close packed) ω phase forms by a diffusionless shuffle transformation first described in [6]. The phase formed during quenching is referred to as athermal ω (ω_{ath}). Athermal ω particles have similar chemical composition and are coherent with β matrix [7]. ω particles are prolate spheroids and their main semi-axes have a few nanometers [8,9]. By heating up to certain temperature (around 110 °C in Ti-15Mo), β stabilizing elements (Mo in our case) are rejected out of ω particles. From the temperature at which the diffusion starts to play a role, ω phase is referred to as isothermal ω (ω_{iso}). ω_{iso} particles grow during annealing/heating and become more chemically stabilized i.e. they differ from surrounding β matrix in composition [10]. ω particles serve as preferential nucleation sites for α phase particles precipitation during annealing at suitable temperatures (above about 400 °C). Mechanical properties of metastable β titanium alloys are mainly determined by the distribution and the size of metastable ω particles or stable α precipitates in β matrix. The formation of the precipitates, their morphology, volume fraction, the size and the distribution are influenced by phase transformations occurring during heat treatment. Therefore, a thorough understanding of these processes is necessary to optimize the mechanical properties of the material through thermomechanical treatment.

This paper presents a study of phase transformations occurring in the metastable β titanium alloy Ti-15Mo (15 wt % (8.1 at. %) of molybdenum). Ti-15Mo alloy was chosen for this experimental study, because it is designed for biomedical use [3], and it represents a simple binary system that has been already widely investigated [11].

2 Materials and methods

The metastable β-Ti alloy (Ti-15Mo) used in this study contains 15 wt % (8.1 at %) of molybdenum and negligible amounts of impurity elements (less than 0.2 of O, 0.1 of C and Fe, 0.05 of Ni and 0.015 of H in wt % [12]). The material was solution treated (ST) at 900 °C for 4 hours in a quartz tube filled with high-purity Ar and quenched in water. This condition corresponds to the ST initial state for the investigation.

In-situ thermal expansion measurement (dilatometry) and in-situ powder neutron diffraction were employed to investigate phase transformations in Ti-15Mo alloy.

2.1 Dilatometry

Evolution of thermal expansion of Ti-15Mo was measured in-situ during linear heating with the heating rates of 1, 5 and 20 °C/min up to 850 °C utilizing the Linseis L75 PT vertical dilatometer. Simultaneous measurement of standard sapphire specimen was utilized to determine the thermal expansion of the apparatus, which was utilized to correct thermal expansion of investigated samples. Dilatometry is a technique for characterizing dimensional changes of a material caused by physical or chemical processes. From the dilatometric curve it is possible to analyze temperature and kinetics of phase transformations or to calculate the coefficient of thermal expansion.

2.2 Neutron diffraction

Neutron diffraction experiments were performed at the Institut Laue-Langevin (ILL), Grenoble, France, at the instrument D20 with an extremely high neutron flux utilizing neutrons with the wavelength of 1.54 Å. The data were obtained within [13] proposal. The detector at D20 covers a scattering range of 153.6 °, and allows to obtain the diffraction pattern in the order of seconds as a function of temperature, pressure or other parameters. The acquisition time of a single spectra was 30 s and the heating rate of 5 °C/min was utilized.

3 Results and discussions

3.1 *The influence of heating rate on phase transitions measured by in-situ dilatometry*

The evolution of thermal expansion of Ti-15Mo was measured in-situ during linear heating with the heating rates of 1, 5 and 20 °C/min up to 850 °C. The results of all measurements of thermal expansion are shown in figure 1. All curves deviate from linearity for all studied heating rates. The deviation from the linear trend is shifted to higher temperatures for faster heating rates. Between 400 - 450 °C the material starts to expand faster, up to about 560 °C, where the curves return to linearity up to 850 °C. The point of return of the curves to the linearity at 560 °C corresponds to a complete dissolution of ω particles [8].

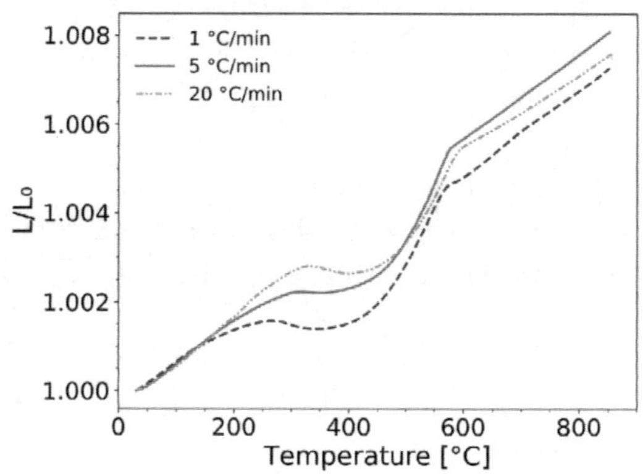

Figure 1. Thermal expansion of Ti-15Mo during heating with different heating rates.

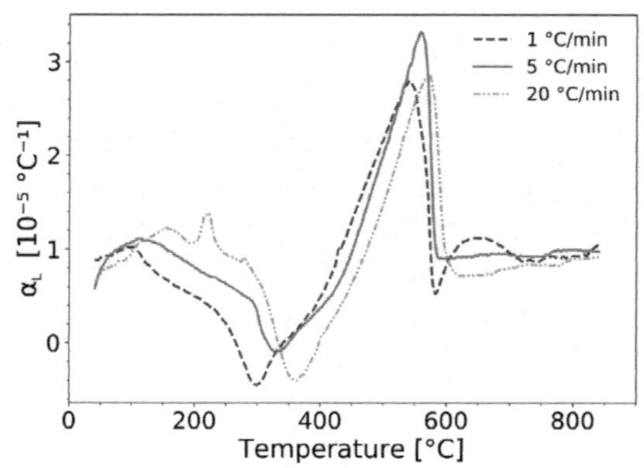

Figure 2. Thermal expansion coefficient of Ti-15Mo during heating with different heating rates.

Additional information can be obtained from the derivative of thermal expansion, which represents the linear expansion coefficient (α_L) of the material. The definition of the linear expansion coefficient is given by the equation:

$$a_L = \frac{1}{L_0}\frac{dL}{dT} \quad (1)$$

The evolution of α_L for all three heating rates is displayed in figure 2. Unfortunately, the heating rate of 20 °C/min is already beyond the performance ability of the apparatus, which resulted in influencing the results at lower temperatures. The furnace was unable to heat the specimen at low temperatures fast enough. The correct heating rate was reached at about 250 °C, which is manifested by a peak at this temperature, which does not appear in the other curves. Otherwise all curves show similar trends. Interestingly, the values of global minima of α_L observed in temperature range 300-400 °C are negative, which corresponds to shrinking of the material with increasing temperature for all heating rates. This behaviour can be attributed to the decrease of lattice parameters of both β and ω phases, which was also observed during isothermal ageing of LCB alloy at 300 °C [14]. At higher temperatures α_L increases and the dissolution of ω particles at 560 °C causes an abrupt decrease of α_L. Between 560 and 750 °C there is an apparent bump at α_L evolution measured with the heating rate 1 °C/min. This bump is probably connected to α phase formation and its subsequent dissolution towards β-transus as the slowest heating rate provides the longest time for α phase to precipitate. The last observable change of α_L at all curves, which probably corresponds to complete dissolution of α phase (β-transus), occurs at about 760 °C. At temperatures above 800 °C the values of α_L are nearly equal for all heating rates indicating that no additional transformations occur in the material.

3.2 *Evolution of lattice parameters determined from powder neutron diffraction*

Powder neutron diffraction of Ti-15Mo was measured in-situ during linear heating with the heating rate of 5 °C/min up to 850 °C. The evolution of the lattice parameter of β phase (bcc) is shown in figure 3 and the development of lattice parameters of ω phase (hexagonal elementary cell) is compared in figure 4. The scatter of the values of lattice parameters of ω phase at low temperatures (bellow 300 °C) is caused by extremely small dimensions of ω particles, which hinders fitting of diffraction patterns. ω particles start to grow at about 300 °C and the values of their lattice parameters are determined more easily. As a result the scatter of the values of lattice parameters is negligible for higher temperatures. The absence of values of lattice parameters of ω phase above 560 °C (see figure 4) is due to complete dissolution of this phase at 560 °C.

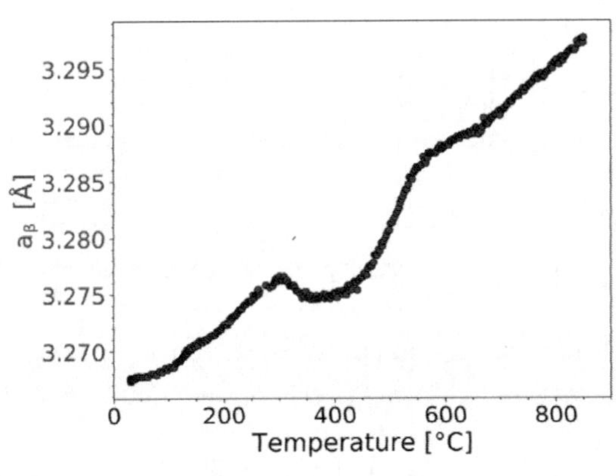

Figure 3. The evolution of a_β during heating with heating rate of 5 °C/min.

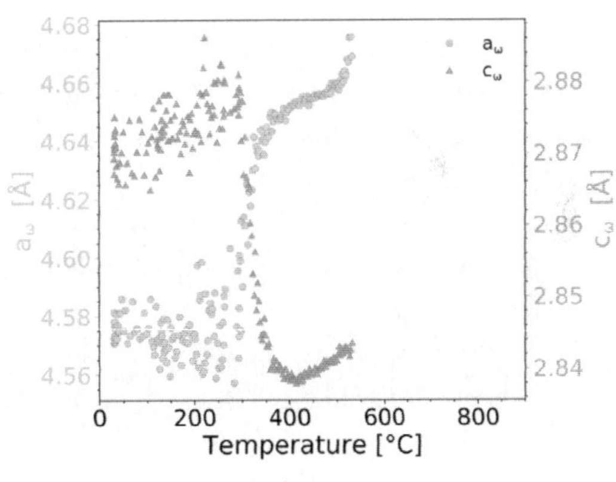

Figure 4. The evolution of a_ω and c_ω during heating with heating rate of 5 °C/min.

The comparison of evolution of volume of elementary cells of β and ω phases is shown in figure 5 as blue dots and green triangles, respectively. The increase of a_ω above 300 °C prevails over the decrease of c_ω and the volume of the ω cell increases. The volume of the elementary cell of β phase obviously follows the trend of its lattice parameter (cf. Figures 3 and 5).

The evolution of the lattice parameter of β phase has a dominant effect on thermal expansion of the material. This can be easily understood from figure 6, where the comparison of relative values of thermal expansion (red curve) and the evolution of relative lattice parameter of β phase (blue dots) is presented. The lattice parameter closely follows the trend of dilatometric curve, even the values are very similar.

The evolution of lattice parameters of α phase (observed between 580 and 720 °C), was omitted because no distinct changes of samples dimensions were detected in this particular temperature range.

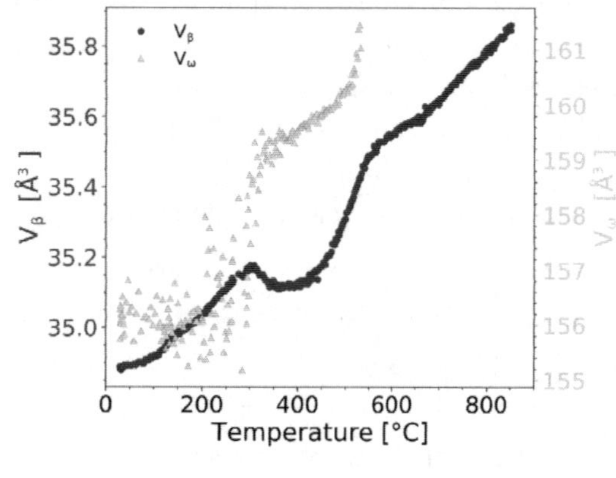

Figure 5. The comparison of evolution of volume of elementary cell of β and ω phases during heating with heating rate of 5 °C/min.

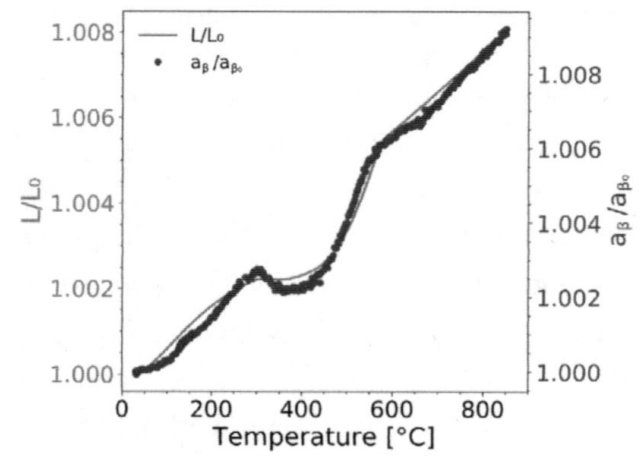

Figure 6. The comparison of thermal expansion and evolution of relative lattice parameter of β phase during heating with heating rate of 5 °C/min.

4 Conclusions

The main results of the present work can be summarized as follows:

- Dilatometry proved to be a suitable method for the detection of phase transformations occurring in titanium alloys.
- Around 300, 330 and 360 °C for the heating rates of 1, 5 and 20 °C/min, respectively, the material exhibits a decrease in length with increasing temperature for all measured heating rates.
- The dissolution of ω phase, which causes an abrupt decrease of α_L, was detected at the same temperature as in electrical resistance experiments (560 °C) presented in our previous work [8].
- The evolution of lattice parameters of β and ω phases during heating with the heating rate of 5 °C/min was determined from neutron powder diffraction experiment.
- The diffusion of alloying elements out of ω particles greatly affects lattice parameters of both β and ω phases.
- The thermal expansion of Ti-15Mo during heating is primarily affected by evolution of lattice parameter of β phase.

Acknowledgements

This work was financially supported by the Czech Science Foundation under the Project 16-12598S and within the program of Large Infrastructures for Research, Experimental Development and Innovation (project No. LM2015050) and project LTT17019 financed by the Ministry of Education, Youth and Sports, Czech Republic.

Pavel Zháňal would like to express gratitude for financial support by the Ministry of Education, Youth and Sport Czech Republic - project LQ1603 Research for SUSEN and to the SUSEN Project (established in the framework of the European Regional Development Fund (ERDF) in project CZ.1.05/2.1.00/03.0108 and of the European Structural and Investment Funds (ESIF) in the project CZ.02.1.01/0.0/0.0/15_008/0000293).

References

[1] Leyens C and Peters M 2003 *Titanium and titanium alloys: fundamentals and applications* (Weinheim: Chichester: Wiley-VCH ; John Wiley)

[2] Duerig T W and Williams J C 1984 Overview: Microstructure and Properies of Beta-Titanium *Beta-Titanium Alloys in the 1980's* ed R R Boyer and H W Rosenberg (Atlanta: The Metallurgical Society of AIME) pp 19–69

[3] Martins Júnior J R S, Nogueira R A, Araújo R O de, Donato T A G, Arana-Chavez V E, Claro A P R A, Moraes J C S, Buzalaf M A R and Grandini C R 2011 Preparation and characterization of Ti-15Mo alloy used as biomaterial *Materials Research* **14** 107–12

[4] Banerjee S and Mukhopadhyay P 2007 *Phase transformations: examples from titanium and zirconium alloys* (Amsterdam ; Oxford: Elsevier/Pergamon)

[5] Donachie M J 1988 *Titanium: A Technical Guide* (AMS International)

[6] De Fontaine D 1970 Mechanical instabilities in the b.c.c. lattice and the beta to omega phase transformation *Acta Metallurgica* **18** 275–9

[7] Nag S, Devaraj A, Srinivasan R, Williams R E A, Gupta N, Viswanathan G B, Tiley J S, Banerjee S, Srinivasan S G, Fraser H L and Banerjee R 2011 Novel Mixed-Mode Phase Transition Involving a Composition-Dependent Displacive Component *Physical Review Letters* **106**

[8] Zháňal P, Harcuba P, Hájek M, Smola B, Stráský J, Šmilauerová J, Veselý J and Janeček M 2018 Evolution of ω phase during heating of metastable β titanium alloy Ti–15Mo *Journal of Materials Science* **53** 837–45

[9] Prima F, Vermaut P, Texier G, Ansel D and Gloriant T 2006 Evidence of α-nanophase heterogeneous nucleation from ω particles in a β-metastable Ti-based alloy by high-resolution electron microscopy *Scripta Materialia* **54** 645–8

[10] Devaraj A, Nag S, Srinivasan R, Williams R E A, Banerjee S, Banerjee R and Fraser H L 2012 Experimental evidence of concurrent compositional and structural instabilities leading to ω precipitation in titanium–molybdenum alloys *Acta Materialia* **60** 596–609

[11] Nag S, Banerjee R and Fraser H L 2005 Microstructural evolution and strengthening mechanisms in Ti–Nb–Zr–Ta, Ti–Mo–Zr–Fe and Ti–15Mo biocompatible alloys *Materials Science and Engineering: C* **25** 357–62

[12] Disegi J 2009 *Implant Materials - Wrought Titanium – 15% Molybdenum* (Synthes)

[13] Zháňal P, Beran P and Hansen T 2018 Study of Phase Transformations in Ti-15Mo Alloy Leading to Inverse Resistivity Evolution *Institut Laue-Langevin (ILL)*

[14] Šmilauerová J, Harcuba P, Pospíšil J, Matěj Z and Holý V 2013 Growth of ω inclusions in Ti alloys: An X-ray diffraction study *Acta Materialia* **61** 6635–45

IOP Conf. Series: Materials Science and Engineering **461** (2019) 012095 doi:10.1088/1757-899X/461/1/012095

Determining of the machine compliance using instrumented indentation test and finite element method

D Zíta[1], P Hanus[1,2], E Schmidová[1] and J Zajíc[1]

[1]University of Pardubice, Studentská 95, 53210 Pardubice, Czech Republic

[2]petr.hanus@upce.cz

Abstract. At present time the importance of indentation measurements once again increases. As modern and perspective measurement method it allows to measure various mechanical properties while only needing micro volumes of materials. While standard hardness measurement methods examine only the plastic response of sampled material, Depth Sensing Indentation (DSI) testing is founded on real-time recording of applied force and resulting travel of indenter. This allows to evaluate both plastic and elastic response of examined material during the indentation process which in turn allows for calculation of various static material properties in very small volumes of samples. The big advantage of this approach is saving of time (especially on sample preparation), however the data analysis – and thus results – could be influenced by certain factors. Those worth noting are surface quality of the sample, wear of the indenter head, its shape and material, and also the compliance of testing machine's frame. If we consider the frame of testing machine having non-zero compliance, inevitably there will occur not only displacement of indenter but also displacement of frame itself, especially noticeable while using high loads. This negatively influences the results of such measurement. This work's aim is to calculate the compliance of test machine's frame using Finite Element Method model and 1 mm in diameter steel ball indenter. The result of this work is addition of virtual spring to the FEM model and fine tuning of its stiffness. This should allow us to minimize the influence of frame's compliance on the results of future measurements.

1 Introduction

Indentation tests are one of the most common techniques for the local mechanical characterization of materials. Different numerical models have been developed to predict uniaxial mechanical properties using different indenter geometries – mainly spherical, cone, pyramidal and flat-ended cylinder. Elastic-plastic models are established to simulate the response of load-displacement curve during the loading-unloading process, detail discussions reflect connected problems [1, 2, 3]. The indentation of an elastic-plastic material is a problem difficult to be solved analytically, because of the unknown shape and extension of the plastic field [4, 5].

Different effects are the part of current research in the contact mechanics nowadays, to provide precise estimation of local mechanical parameters by indentation test. Effect of edge compliance on the contact mechanics of rigid spherical indenters in adhesive contact has been already investigated as a function of applied load [6]. Due to the considered adhesive interaction, qualitative and quantitative changes were observed in the stress and deformation fields. The effect of the edge compliance was also detailed in measurable quantities like indentation depth and contact stiffness. An inverse procedure, starting from experimentally measured pairs of load and depth of penetration, followed by an extensive finite element modelling of the indentation process gave an accurate reproduction of elastic–plastic properties [7, 8]. Some current results indicate the significant decrease of indentation resistance compared to real tensile tests [9]. The same problem, i.e. calibration issues in real indenter geometry and frame compliance in spherical nanoindentation was investigated using two different methodologies; fitting the Hertz equation to force–depth data for indentation and profiling the residual impression for indentation depths, which led to a large discrepancy in effective radii [10]. Research of composites is mainly focused on the influence of the tip defect on the determination of the mechanical properties using Berkovich, Vickers, Knoop and spherical indenters [11].

Finite element analysis presents the approach which requires suitable experimental validation. Considering the tests particular conditions, indentation process needs to be defined according the testing conditions. Relatively close agreement between experiment and simulation can be achieved considering the strain incremental extension. Comparatively less success is attained in matching numerical and experimental results presumably because of the effects of anisotropy. Stain rate dependence and/or heterogeneities at the microstructural level not captured by the constitutive theory used in the simulations. The sensitivity of numerical results to indenter imperfection and indenter elasticity has been subjected to current research. The effect of friction and specimen creep during indentation on load-displacement predictions is also assessed [12].

Surface quality conditions, temperature as well as a local material heterogeneity can influence the accuracy of obtained material parameters. Presented work was devoted to problem of loading system compliance, methodology of numerical assessment was suggested, and final experimental validation was performed.

The basic foundations for this article are measurements curves obtained from fully-instrumented universal hardness measurement machine Zwick ZHU/Z2.5Z. These measurements were conducted using 1 mm in diameter spherical indenter pushed into bainite steel, using maximum force of 300 N. The output of the measurement is force-depth loading curve. This dependency is however influenced by compliance of the frame of the test machine. To acquire the compliance of the frame we use Finite Elements Methods model and using this we calculate expected real behavior of used samples. In next step the FEM model is adjusted by addition of spring element simulating the compliance of the machine's frame. The final goal is to fine tune parameters of this added spring to obtain as similar as possible values in real-life measurement and in FEM model.

2 Model

All of the performed FEM calculations were done using commercially available solver ABAQUS 6.17. Two different samples were simulated, the first one representing linear-elastic material, the second one being elastic-plastic material based on real tensile test results of bainite steel. The first one is used to validate the model by comparing the results of numerical calculation and analytic solution based on Hertz theory. Both models use ideally rigid spherical indenter with 1 mm diameter and axisymmetric cylindric sample with height of 4.4 mm and radius also 4.4 mm.

Boundary conditions are very similar to all models. Zero displacement in the direction perpendicular to the axis of symmetry is used for nodes on this axis (AB), zero displacement in direction perpendicular to baseline is prescribed for nodes on this baseline (BD) – figure 1. Loading is applied by forced displacement of the indenter. For model with added spring representing compliance of frame of test machine nodes on the baseline are connected to one node of spring using stiff kinematic coupling. The other node has zero displacement and zero rotations in all directions. This represents spring with one degree of freedom – displacement in direction of the indenter displacement.

Between the surface of indenter and upper surface of sample the contact "surface to surface" is prescribed. With elastic sample the coefficient of friction is f=0, in other cases it is f=0.1.

The mesh of the sample is made up of three- and four-nodes 2D axisymmetric elements (CAX). Smallest dimension of elements (in area of contact between sample and indenter) is 3 μm. The total number of elements is 28 800.

The results of this simulation are shown on figure 2 – Von Mises stress with maximum load and after loading.

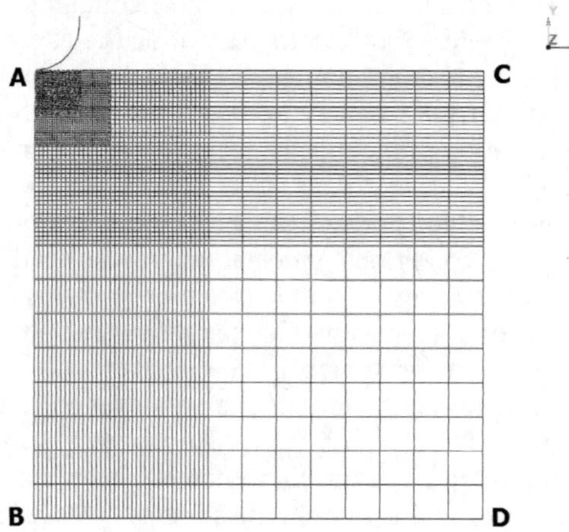

Figure 1. FEM mesh

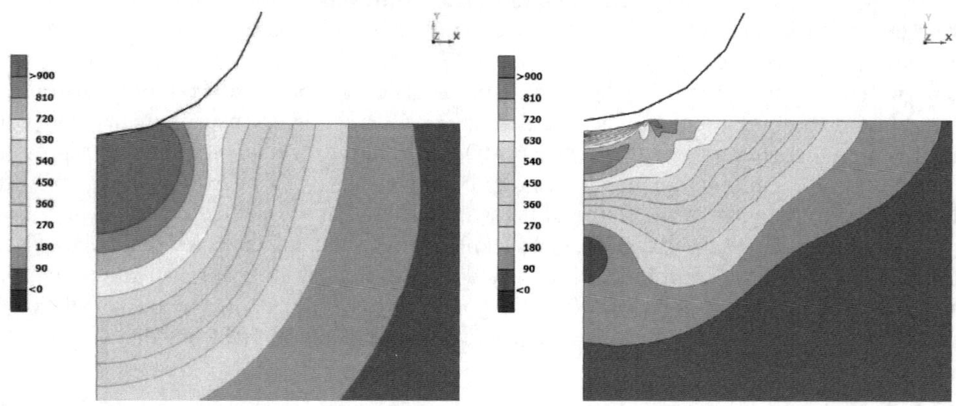

Figure 2. Von Mises stress with maximum load (left) and after loading (right)

2.1 *Linear elastic sample*

To validate the model the results of numerical calculations were compared with analytical solution of equations (1), (2) and (3) based on Hertz theory [13]. Material of the sample is in this case linear-elastic with Young's modulus $E = 210$ MPa.

$$\sigma_x = -p_{max} \cdot \left[\left(1 - \left|\frac{z}{a}\right| \cdot \tan^{-1}\left|\frac{a}{z}\right|\right) \cdot (1+\nu) - \frac{1}{2\cdot\left(1+\frac{z^2}{a^2}\right)} \right] \tag{1}$$

$$\sigma_z = \frac{-p_{max}}{1+\frac{z^2}{a^2}} \tag{2}$$

$$\sigma_{Tresca} = \sigma_x - \sigma_z \tag{3}$$

where z is depth below surface, a is radius of contact area, p_{max} is maximum pressure on the center of contact surface and σ_{Tresca} is the equivalent Tresca's stress. On figure 3 the good match between numerical and analytical solutions for depth of indentation 10 μm is clearly visible.

IOP Conf. Series: Materials Science and Engineering **461** (2019) 012095 doi:10.1088/1757-899X/461/1/012095

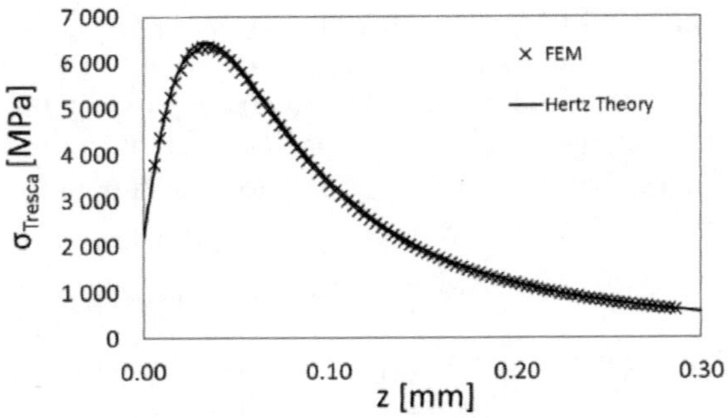

Figure 3. Comparison of analytical and FEM solution

2.2 Bainite steel, elastic–plastic sample

Elastic-plastic sample has linear behavior up to yield strength with Young's modulus $E = 210$ GPa. After this point the stress-strain dependency is taken directly from results of tensile strength test – as can be seen on figure 4. Stiffness characteristic of the spring representing compliance of the frame is presented on figure 5.

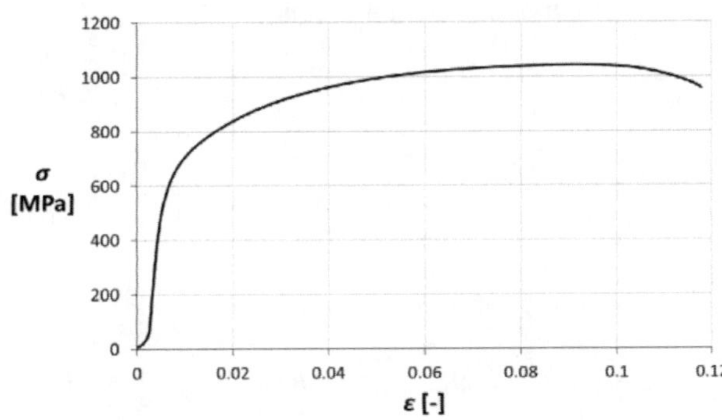

Figure 4. Tensile test curve (True Stress - Strain)

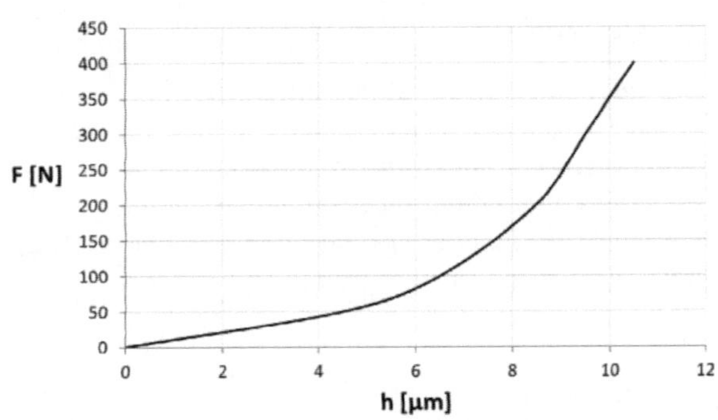

Figure 5. Characteristic of the spring

3 Results

Figure 6 shows comparison between FEM analyses with completely rigid frame and with added spring representing real frame's compliance and also with real test record. It can be clearly seen that there is a significant difference between real test and model using completely rigid frame. On the other hand, the model with added spring (representing frame's compliance C_f – figure 7) shows very good match with real test. This is caused by deformation happening not only to sample but also to test machine's frame and to the indenter.

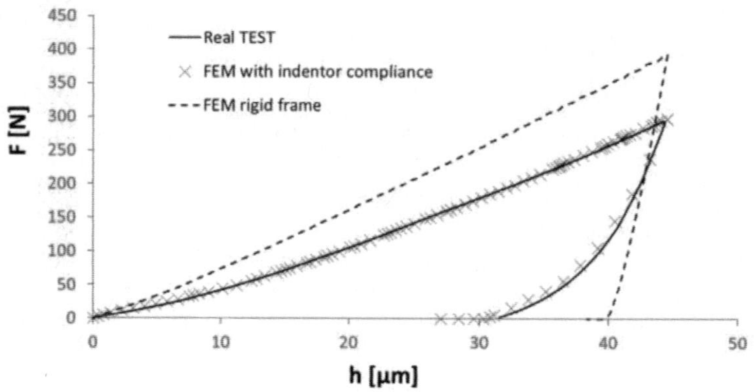

Figure 6. Indentation curves

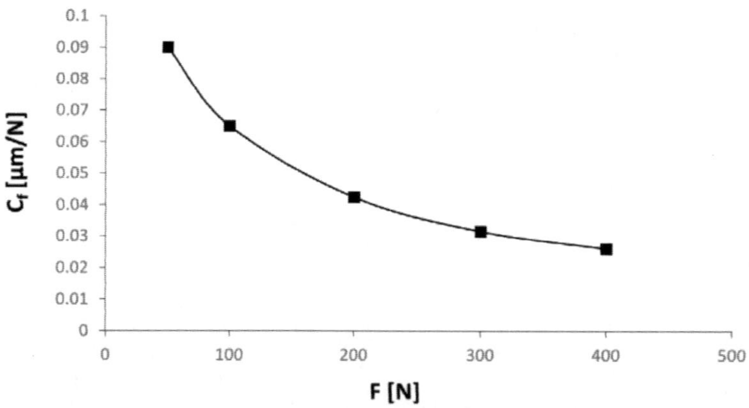

Figure 7. Indenter frame compliance C_f

The curve on figure 6 was constructed using iteration methods – by minimizing of difference between measured indentation curve and indentation curve obtained from FEM analysis with added spring. Effects of surface roughness and indenter wear are both included in compliance of frame. We expect negligible effect of surface roughness as the samples were finely grinded (roughness approximately *Ra* = 0.8). For the purpose of FEM simulation, we used ideal shape of indenter.

4 Summary

It is necessary to mention that the computed compliance of the frame also contains compliance of the indenter and also the effect of surface roughness and free travel of parts of test machine. The next planned step is to evaluate the compliance with real-life experiment which will show the impact of said parameters.

Acknowledgements
Made with support of project no. FV30149 "3D printing of numerically optimized steel parts using DMLS technology".

References

[1] Chen R et al. 2015 Numerical simulation of spherical indentation method to identify metal material mroperties *Adv. Mat. Res.* 1119 pp 779-82

[2] Lee H, Lee J H and Pharr G M 2005 A numerical approach to spherical indentation techniques for material property evaluation *J Mech. and Phys. of Sol.* 53 p 2037-69

[3] Kang J J, Becker A A, Wen W and Sun W 2018 Extracting elastic-plastic properties from experimental loading-unloading indentation curves using different optimization techniques *Int. J. Mech. Sc.* 144 pp 102-9

[4] Riccardi B and Montanari R 2004 Indentation of metals by a flat-ended cylindrical punch *Mat. Sc. and Eng.* A 381 pp 281-91

[5] Puchnin M, Anisimov E, Cejp J, Kunka I and Vicens S 2014 Advantages of express-methods in investigation of mechanical and physical properties of aluminum alloys *Man. Tech.* 14 pp 234-38.

[6] Stan G 2018 The effect of edge compliance on the adhesive contact between a spherical indenter and a quarter space *Int. J. Sol. and Struct.* pp 1-11

[7] Beghini M, Bertini L and Fontanari V 2006 Evaluation of the stress–strain curve of metallic materials by spherical indentation *Int. J. Sol. and Struct.* 43 pp 2441-59

[8] Pham T H, Phan Q M and Kim S E 2018 Identification of the plastic properties of structural steel using sherical indentation *Mat. Sc. and Eng.* A 711 pp 44-61

[9] Wu J, Wang M, Hui Y, Zhang Z and Fan H 2018 Identification of anisotropic plasticity properties of materials using spherical indentation imprint mapping *Mat. Sc. and Eng.* A 723 pp 269-78

[10] Kang S, Kim Y, Lee Y, Kim J and Kwon D 2012 Determining effective radius and frame compliance in spherical nanoindentation *Mat. Sc. and Eng.* A 538 pp 58-62

[11] Chicot D, de Baets P, Staia M H, Puchi-Cabrera E S, Louis G, Perez Delgado Y and Vleugels J 2013 Influence of tip defect and indenter shape on the mechanical properties determination by indentation of a TiB2–60%B4C ceramic composite *Int. J. of Ref. Met. and H. Mat.* 38 pp 102-10

[12] Syngellakis S et al. 2018 Finite element simulation of spherical indentation experiments *Int. J. Comp. Meth. and Exp. Meas.* 6 (4) pp 749-63

[13] Johnson K L 1985 Contact mechanics, (Cambridge University Press, Cambridge)

Extrusion of Al-Si alloy compacts under conventional conditions and in the semi-solid state

K Zyguła[1], O Lypchanskyi[1], M Wojtaszek[1], T Śleboda[1],

[1]AGH University of Science and Technology, Faculty of Metals Engineering and Industrial Computer Science, Av. Mickiewicza 30, 30-059 Cracow, PL

kzygula@agh.edu.pl

Abstract. In this research work the results of hot extrusion process of aluminium alloy compacts conducted at the temperature of conventional hot extrusion as well as at the temperature range corresponding to the semi-solid state were presented. Initial material was prepared by uniaxial pressing of Al17Si5Fe3Cu1.1Mg0.6Zr alloy powder. The extrusion process was carried out under isothermal conditions, and the temperature stability was strictly controlled. The changes of extrusion force in relation to punch displacement were recorded during hot deformation to demonstrate the effects associated with the appearance of the liquid phase in the volume of the processed material. After extrusion the samples were cooled in water as well as at room temperature. The influence of temperature of extrusion and the resulting fraction of the liquid phase in the processed material on its microstructure and chosen mechanical properties were determined. Mechanical properties of the examined alloy after deformation were carried out by measuring the hardness and in uniaxial compression tests. Additionally, one of the compacts was remelted and cast, what made possible comparison of the microstructure and chosen properties of the obtained products. In the result of the investigations it has been proved, that extrusion in semi-sold state results in reduction of forces necessary to complete deformation process and leads to obtaining the material with good mechanical properties and fine-grained, favourable microstructure.

1. Introduction

Aluminium alloys with addition of silicon known as 3xxx or 4xxx series alloys are commonly used in casting of car engine cylinder head [1]. Moreover, extruded aluminium alloys found its application in automotive industry as car frame, resulting in its weight reduction. Due to the combination of low density and excellent corrosion resistance such alloys are widely used in aerospace industry [2,3].

Semi-solid processing, also know as thixoforming, is a technology which combines traditional hot forming with casting process. Most important assumption of semi-solid method is deformation between solidus and liquidus temperature of the processed material. Aluminium alloys are widely used in semi-solid processing. Components obtained by such method have good mechanical properties and advantageous microstructure. Moreover, thixoformed products can be successfully T6 heat treated to improve their properties [4,5,6]. Hot forming of metals above the solidus line leads to change in the deformation mechanism. Deformation occurs along the grain boundaries, where the liquid phase is located. Because of decrease of loads during deformation in the semi-solid range, the process is economically attractive, and gives possibilities to obtain near-net shape products [7,8].

2. Experimental research

2.1. Aim and scope of study

The aim of the research was to evaluate the microstructure and chosen mechanical properties of compacts made of Al17Si5Fe3Cu1.1Mg0.6Zr (Al17Si) alloy powder, extruded in isothermal conditions in both the temperature of conventional hot forming processes and in the semi-solid state. The investigations involved production of compacts by hot pressing process and then extrusion in various temperatures. To determine the temperature of transition into semi-solid state, differential scanning calorimetry and thermographic tests were carried out. The microstructure of the investigated materials after deformation was analysed by light microscopy. To study the influence of the temperature of extrusion on the mechanical properties of the examined material, hardness measurements and uniaxial compression tests were conducted.

2.2. Material for the investigations

The initial material used in hot pressing process was Al17Si alloy powder with chemical composition given in table 1. A size of powder particles did not exceed 60 µm. Compaction process was carried out on hydraulic press. The temperature of hot pressing was 500 °C. Alloy powder was held at this temperature for 10 minutes and then compacted under the pressure of 150 MPa for 3 minutes. Additionally, for comparison of the microstructure and chosen properties, one of the compacts was remelted and cast. The microstructures of the compact and cast material were shown in figure 1. During slow cooling of cast Al17Si alloy, a growth of precipitation of primary Si blocks, Si eutectic needle and other phase identified as thick δ Al_4FeSi_2, $Al_5Cu_2Mg_8Si_6$ or Al_2Cu spread over Al matrix was noted [9]. Al17Si alloy compact had homogenous and fine-grained microstructure. Spherical participations were uniformly distributed in Al alloy matrix.

Table.1 Chemical composition of Al17Si alloy.

Si	Fe	Cu	Mg	Zr	Al
17.0	5.0	3.0	1.1	0.6	Balanced

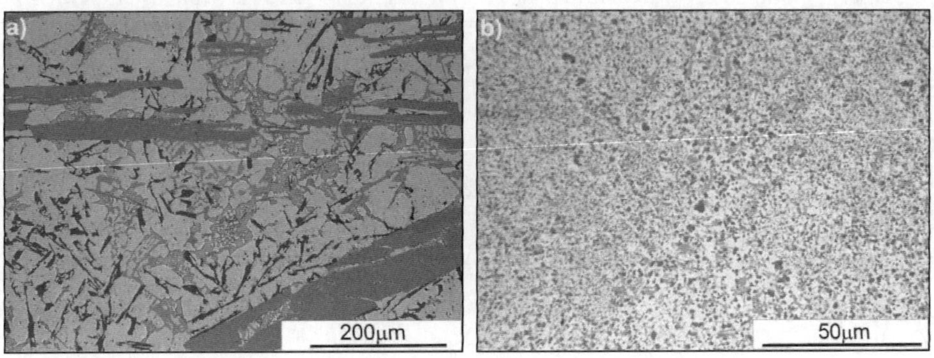

Figure 1. Microstructures of a) cast Al17Si alloy and b) Al17Si alloy compact. Cross-section, unetched.

2.3. Designation of the solid to semi-solid transition temperature

To find the temperature of the first appearance of the liquid state in Al17Si alloy, the thermographic and differential scanning calorimetry (DSC) methods were used. In the thermographic method, the compact with the attached thermocouple was melted in the furnace. The temperature changes in function of time were recorded. Heating curve was presented in figure 2a. During DSC analysis, heat flow changes in a function of the temperature was recorded. To analyse the heat flow curve Origin 8 Pro software was used.

In the result of the investigations, two temperatures of the transition into semi-solid state were obtained. In the first method, the effect related to the beginning of the occurrence of liquid phase was noticed at 505 °C, while in the case of DSC method at about 510 °C. The difference between these two results is thought to be due to the size of the sample used for the tests. Since the DSC method is considered to be more accurate, the temperature of 510 °C was chosen as the solid to semi-solid state transition temperature. The results were consistent with those presented in other research works [9, 10].

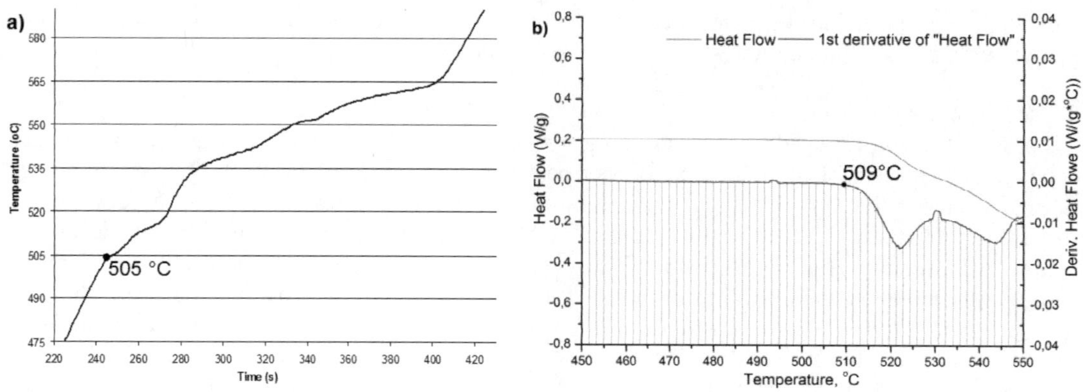

Figure 2. Results of determining the solid to semi-solid transition temperature by a) thermographic method, b) DSC method.

2.4. Extrusion process

The obtained compacts were hot formed in direct extrusion process in isothermal conditions on hydraulic press with hot extrusion system consisting of set of tools and temperature control system. The process was conducted at conventional hot forming temperature as well as in semi-solid state. Extrusion parameters were applied according to table 2. Material obtained after deformation had shape of rod with diameter of 18 mm.

Table 2. The parameters of hot extrusion of Al17Si compacts.

Temperature	Punch velocity	Extrusion coefficient	Die angle	Cooling
490 °C, 510 °C, 520 °C, 530 °C	0.15 mm/s	3.8	45°	Water / room temperature

The changes of extrusion forces in relation to punch displacement during deformation process were recorded. The results were presented in figure 3. Based on the obtained relationships, the value of force corresponding to the stabilization of extrusion process was determined. In the result of increasing temperature from 490 °C - which was assumed as conventional hot forming temperature - to the temperatures where liquid state occurred, the extrusion stabilization forces decreased gradually from 96 kN to 30 kN. These results indicated, that during the material transition from solid to semi-solid state there was a partial change in the deformation mechanism to slip along grain boundaries.

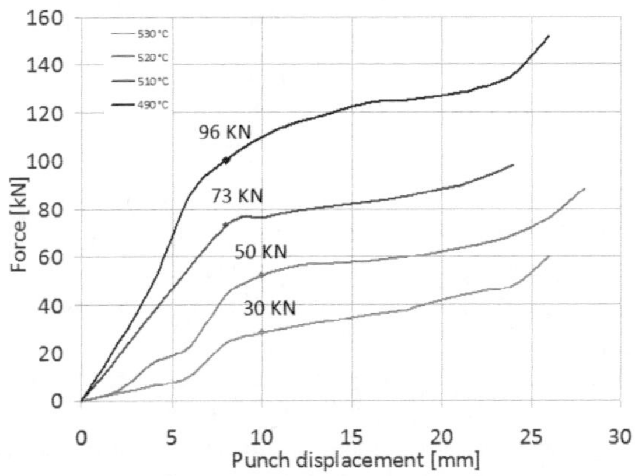

Figure 3. The effect of temperature on the changes in extrusion force in relation to the punch displacement, recorded during hot extrusion of Al17Si alloy compacts.

2.5. Metallographic investigations

The observations of the microstructure of Al17Si alloy compact before and after hot extrusion was conducted using optical microscope LEICA DM4000M. Chosen microstructures were shown in figure 4. The extruded material had very similar microstructure regardless of the extrusion temperature or type of cooling. A fine-grained microstructure was observed in both cases, the dark primary Si particles as well as Si eutectic and other complex phases were uniformly distributed in the α-aluminium matrix. In the case of longitudinal cross-sections material grains did not elongate in the direction of the material flow. Slight grain growth was noticed in the materials deformed at higher temperature or in those cooled at lower cooling rates.

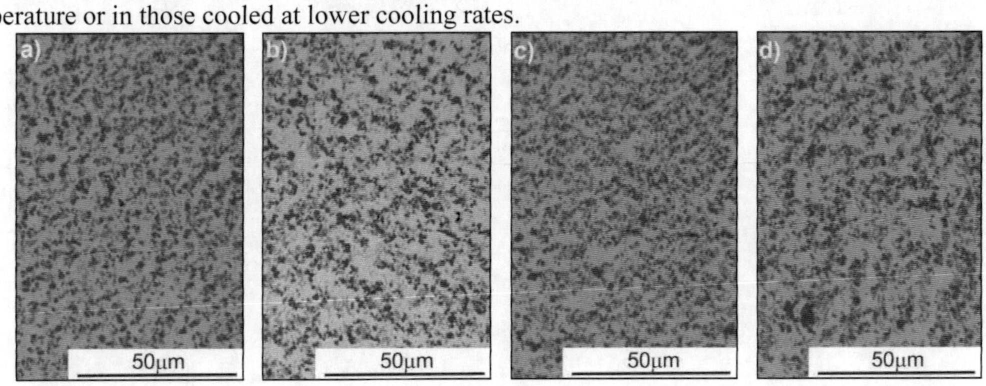

Figure 4. Microstructures of Al17Si alloy after hot extrusion at a) 490°C, b) 520°C cooled in room temperature and c) 510 °C, d) 530 °C cooled in water. Cross-section, unetched.

2.6. Mechanical properties

2.6.1. Hardness measurements

Hardness measurements were conducted by the Vickers method on the compacted and cast Al17Si alloy as well as on the samples after hot extrusion. Hardness was measured on cross-section in the central area of the sample. Table 3 presents average HV_1 values with standard deviation.

Comparing cast Al17Si alloy and Al17Si alloy powder compacts, material obtained from alloy powder had significantly higher (by about 65 HV_1) hardness than cast material. As the result of rapid cooling in water directly after hot extrusion process, hardness of the samples increased by about 10-20 HV_1. In the case of cooling at room temperature, hardness decreased in relation to compact by about

20 HV_1. This effect is related to the growth of Al α-phase grains during slow cooling. Major differences in materials hardness in relation to the changes in extrusion temperature were not observed.

2.6.2. Uniaxial compression tests

The strength of the cast and PM material after hot extrusion was determined in the uniaxial compression tests, which were carried out at room temperature on Instron 1196 machine. Samples with dimensions 12 mm in height and 10 mm in diameter were machined from the investigated materials. Compressive strength of the tested materials are presented in table 3. Figure 5a shows recorded stress- strain curves.

Basing on the results of the tests it was found, that the compressive strength values are comparable for materials produced at conventional hot forming temperature and in the semi-solid state. No significant differences in the nature of stress – strain curves of the extruded materials were observed. A similar material behaviour was noticed in the case of the samples extruded in semi-solid state. Samples deformed at the temperature of conventional hot forming showed better ductility in compression tests. Cast Al17Si alloy had significantly lower mechanical properties than alloy compacts. Figure 5b shows the comparison of chosen mechanical properties.

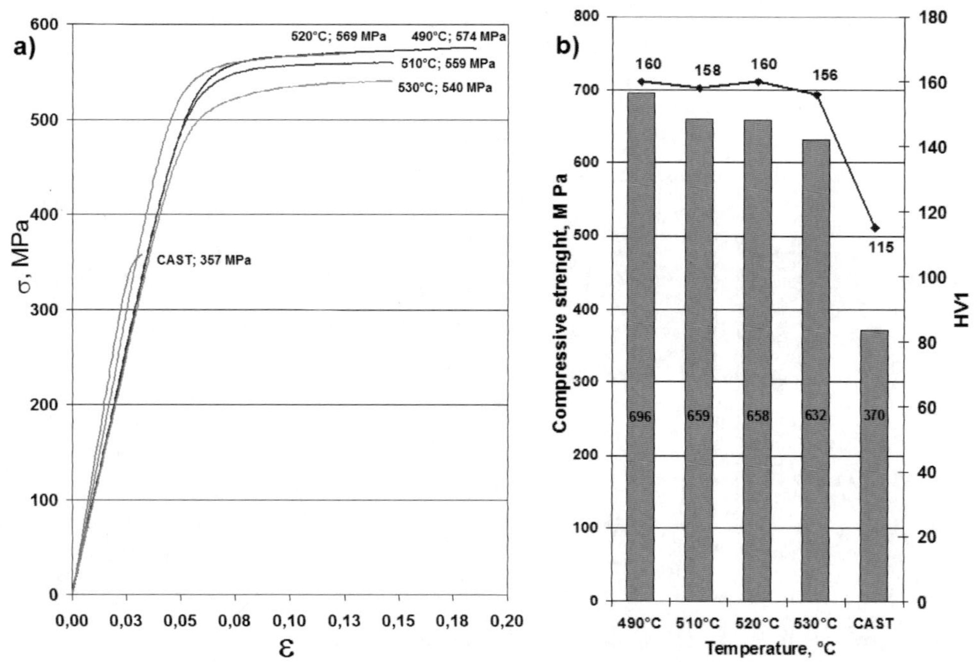

Figure 5. Mechanical properties of the investigated materials: a) stress-strain curves for cast and extruded Al17Si alloy compacts and b) influence of temperature of extrusion of Al17Si alloy on chosen mechanical properties

Table 3. Chosen mechanical properties of cast and PM Al17Si alloy before and after hot extrusion.

	Compact	**490 °C**		**510 °C**		**520 °C**		**530 °C**		**Cast**
HV_1	180 ±5	160 ±2	192 ±2	158 ±2	191 ±6	160 ±2	200 ±2	156 ±1	190 ±5	115 ±9
Compressive strength, MPa	-	696 ±22		659 ±14		658 ±27		632 ±15		370 ±6

3. Conclusions

Basing on the results of the investigations the following conclusion can be drawn:

- In the result of DSC and thermographic method it was found, that liquid phase appears in the volume of material at the temperature of 505-510 °C.
- Gradual increase of the extrusion temperature under isothermal conditions led to significant decrease in the level of the forces necessary to carry out deformation process. The nature of force changes indicates the presence of a liquid phase in the material volume at the temperatures above 510 °C.
- Metallographic analysis showed significant differences in the state of the cast material microstructure and materials made by powder metallurgy technology. The use of the fine powder fraction as the starting material resulted in homogeneous microstructure of the material with a significantly smaller grain size. However, significant changes in the microstructure due to increase of the extrusion temperature were not observed.
- The hardness of the samples extruded in the temperature range assumed during the tests was comparable. Differences in hardness resulted from the method of samples cooling and it was approximately 30 HV_1 higher for water-cooled materials, compared to samples cooled at the room temperature.
- The compressive strength of the extruded samples decreased slightly as the process temperature increased. The average compressive strength values for the samples obtained by the powder metallurgy method were significantly higher than for those obtained by casting.
- Isothermal extrusion of powder compacts, carried out under conventional hot forming conditions as well as in the semi-solid state, led to obtaining pore-free products with favourable microstructure and attractive mechanical properties.

References

[1] Grieb M B, Christ H J and Plege B 2010 *Procedia Eng.* **2** pp 1767–1776
[2] Miller W S, Zhuang L, Bottema J, Wittebrood A J, De Smet P, Haszler A and Vieregge A 2000 *Mater. Sci. Eng., A* **280** pp 37–49
[3] Fridlyander I N 2001 *Int. J. Fract.* **43** pp 6-10
[4] Pola A, Tocci M and Kapranos P 2018 *Metals* **8** 181
[5] Hassas-Irani S B, Zarei-Hanzaki A, Bazaz B and Roostaei A A 2013 *Mater. Des.* **46** pp 579–587
[6] Bolouri A, Yong P J and Chung G K 2014 *Int. J. Adv. Manuf. Technol.* **70** pp 2139–2149
[7] Akbarzadeh E, Picas J A and Baile M T 2015 *Mater. Des.* **88** pp 683–692
[8] Güven E A 2016 *Trans. Indian Inst. Met.* **69** pp 935–940
[9] Timmermans G, Froyen L and Humbeeck J V 2000 *J. Matter Sci.* **35** pp 3289–3299
[10] Samuel F H 1998 *J. Matter Sci.* **33** pp 2283–2297

AUTHOR INDEX

AUTHOR INDEX

AUTHOR INDEX